26
Manuals
171

WAR OFFICE

# TEXTBOOK

OF

# SMALL ARMS

### 1929

LONDON:
PUBLISHED BY HIS MAJESTY'S STATIONERY OFFICE
To be purchased directly from H.M. STATIONERY OFFICE at the following addresses:
Adastral House, Kingsway, London, W.C.2; 120, George Street, Edinburgh;
York Street, Manchester; 1, St. Andrew's Crescent, Cardiff;
15, Donegall Square West, Belfast;
or through any Bookseller.

1929.

*Price* 5s. 0d. *net.*

57—480

## EDITOR'S PREFACE

The Textbook of Small Arms, which appears to have been first printed in 1863, was a thin volume specially intended for Officers under instruction at the School of Musketry at Hythe. Reprints or revised editions, called for by changes in armament, appeared in 1868, 1877 and 1880. The next edition, that of 1884, was almost entirely re-written; it bore the title of "Treatise on Military Small Arms and Ammunition . . . a Textbook for the Army," by Lieut.-Colonel H. Bond, R.A., Assistant Superintendent Royal Small Arms Factories. This work contained abbreviated ballistic tables and showed the method of using them. A fresh edition appeared in 1888.

The adoption of the magazine rifle and a smokeless propellant made a new book necessary, and the name Textbook was again used in 1894 for an enlarged work. The next edition, that of 1904, was an almost completely new volume, double the size of the last; for this Captain W. B. Wallace, Inspector of Small Arms, was responsible; it contains some remarks on automatic rifles. A fresh version of the same book was issued in 1909, the ballistic tables being separately bound.

For the present edition the name Textbook has been retained; it has, however, been almost completely re-written. The extended use and the rapid development of all kinds of Small Arms which have taken place during and since the Great War, have made it necessary for its scope to be greatly widened in the endeavour to provide a general survey of all matters concerning Small Arms. It would be impossible to deal exhaustively with each particular topic within the limits of a single volume, nor could any single writer cover authoritatively the whole range of differing subjects which have now to be included. It has been thought well to allow considerable latitude in method of treatment in the different sections of the book. Principles of design and methods of calculation have been treated suggestively rather than dogmatically, and endeavour has been made to bring out the interest of the various subjects. The object has been to provide an introduction to technical matters for the practical soldier rather than a complete work of reference for the technical expert only.

In dealing with rifles and pistols, and with the great subject, new to the Textbook, of machine guns, the system adopted has been that of describing the different types of mechanism employed, and of indicating the variants within those types, rather than of attempting to describe exhaustively each separate variety of weapon. The Chapters on ammunition contain, among other things, much information on manufacture, inspection, and storage. Grenades are for the first time included in the Textbook.

The gradual obsolescence of the *arme blanche* has been thought to justify the historical treatment of the sword by an admitted authority on that subject.

iv

Although general laws always hold good, rifle ballistics require somewhat different treatment from those of large ordnance. It seemed therefore superfluous merely to repeat what already appears on this subject in the Textbook of Ballistics and Gunnery. Instead of doing this, an attempt has been made to show the great interest of the subject, and to indicate some other lines on which it may be treated, it being understood that the factors to be taken into account are many, and that the results of calculation will be correct in proportion as they approximate to actual results, but, apart from this, have no inherent infallibility.

The Editor desires to acknowledge his great indebtedness to the various helpers, official and unofficial, who have contributed to the work.

## CORRIGENDA.

Outside cover—*For* " FOR " *read* " OF."

Page 405 (last column), under Norway.

*For* " 1910 " *read* " 1894 ".

Page 22, under "*Krag-Jorgensen Rifle*"

*For* " Norway, 1910 " *read* " Norway, 1894 ".

# TEXTBOOK OF SMALL ARMS

## CONTENTS

Part I—Small Arms
" II—Small Arm Ammunition
" III—Ballistics of Small Arms
" IV—Appendices and Ballistic Tables

## PART I

| Chapt. | Sect. | Small Arms | Page |
|---|---|---|---|
| I. | | **The Rifle** | |
| | 1 | The history of the rifle | 1 |
| | 2 | A review of the systems of modern rifles and a comparison of their principal components | 10 |
| | 3 | The British service rifle | 41 |
| | 4 | Outline of manufacture and inspection of the British service rifle | 49 |
| | 5 | Considerations affecting the accuracy of the rifle | 50 |
| | 6 | Sights | 58 |
| | 7 | Rests for rifle shooting | 65 |
| | 8 | Definitions pertaining to the rifle and rifle shooting | 67 |
| II. | | **The Sword, Lance and Bayonet** | |
| | 1 | The history of the sword, lance and bayonet | 69 |
| | 2 | Outline of the manufacture and inspection of the British service bayonet | 84 |
| III. | | **Revolvers and Self-Loading Pistols** | |
| | 1 | The revolver | 86 |
| | 2 | Self-loading pistols | 95 |
| IV. | | **Grenades** | |
| | 1 | The history of grenades | 105 |
| | 2 | Considerations affecting the design of grenades | 116 |
| | 3 | Fuzes, igniters and bursters | 118 |
| | 4 | Means of projection | 124 |
| | 5 | British and foreign grenades | 131 |

| Chapt. | Sect. | | Page |
|---|---|---|---|
| V. | | MACHINE GUNS AND LIGHT MACHINE GUNS | |
| | 1 | The history of machine guns | 150 |
| | 2 | The characteristics of an efficient automatic weapon | 154 |
| | 3 | Features of design found in modern automatic weapons | 159 |
| | 4 | Tests and adjustments pertaining to automatic weapons | 193 |
| | 5 | Mountings | 198 |
| | 6 | Blank firing attachments | 202 |

# PART II

## Small Arm Ammunition

| Chapt. | Sect. | | Page |
|---|---|---|---|
| I. | — | HISTORY OF THE DEVELOPMENT OF THE SMALL ARM CARTRIDGE | 205 |
| II. | — | NOTES ON THE DESIGN OF THE MODERN MILITARY CARTRIDGE | 208 |
| III. | — | EXPLOSIVES | |
| | 1 | General remarks | 215 |
| | 2 | Historical outline | 216 |
| | 3 | General characteristics of small arm propellents | 218 |
| | 4 | Manufacture of small arm propellents | 221 |
| IV. | — | SMALL ARM CARTRIDGE MANUFACTURE | |
| | 1 | General considerations | 223 |
| | 2 | General factory organization | 224 |
| | 3 | Case manufacture | 225 |
| | 4 | Bullet manufacture | 230 |
| | 5 | Cap manufacture and filling | 233 |
| | 6 | Loading | 235 |
| V. | — | PROOF OF SMALL ARM AMMUNITION WITH SPECIAL REFERENCE TO ·303-INCH MK. VII | |
| | 1 | Selection of proof | 237 |
| | 2 | Accuracy trials | 238 |
| | 3 | Velocity and pressure trials and the use of standard ammunition | 239 |
| | 4 | Freedom from defects and functioning trials | 242 |
| VI. | — | INSPECTION OF ·303-INCH MK. VII AMMUNITION | |
| | 1 | Necessity for complete inspection | 247 |
| | 2 | Break down and chemical analysis | 248 |
| | 3 | Weighing | 249 |
| | 4 | Gauging | 250 |
| | 5 | Visual examination | 253 |
| | 6 | Packing | 253 |
| | 7 | Assessment of quantities | 255 |
| | 8 | Inspection of packing accessories | 255 |
| | 9 | Storage | 256 |

| Chapt. | Sect. | | Page |
|---|---|---|---|
| VII. | — | Miscellaneous Military Ammunition | |
| | 1 | The ·303-inch Mk. VI cartridge | 257 |
| | 2 | The ·303-inch armour-piercing cartridge | 257 |
| | 3 | Tracer ammunition | 258 |
| | 4 | Special ammunition for the Royal Air Force | 259 |
| | 5 | Rifle grenade cartridges | 261 |
| | 6 | The ·303-inch blank cartridge | 262 |
| | 7 | Drill ammunition | 263 |
| | 8 | Pistol ammunition | 263 |
| | 9 | Miniature rifle ammunition | 265 |

PART III

Ballistics of Small Arms

| Chapt. | Sect. | | Page |
|---|---|---|---|
| I. | — | Interior Ballistics (Descriptive) | 267 |
| II. | — | Exterior Ballistics (Descriptive) | 275 |
| | 1 | The sub-division of the trajectory into three parts | 276 |
| | 2 | The parabolic or unresisted trajectory, and the recognition from earliest times of the effect of a resisting medium | 277 |
| | 3 | Early experiments from Newton's time up to the introduction of Rifled Ordnance, with remarks on the velocity of sound in air | 279 |
| | 4 | Bashforth's work with his Chronograph | 280 |
| | 5 | Ballistic Tables and the value of "C" | 285 |
| | 6 | The wind problem | 285 |
| | 7 | The principle of the rigidity of the trajectory | 287 |
| | 8 | The twist of rifling, drift, yaw and the rotation of the earth | 288 |
| III. | — | Interior Ballistics (Numerical) | 292 |
| | 1 | The principle of mechanical similitude | 294 |
| | 2 | The practical method used by Housman | 295 |
| | 3 | Rational or formal methods of calculation following Charbonnier and Hezlet's methods | 297 |
| | 4 | The energy contained in the powder and the amount carried away by the bullet | 300 |
| | 5 | Mr. F. W. Jones' articles in Arms and Explosives | 303 |
| | 6 | The Monomial method | 305 |
| IV. | — | Exterior Ballistics (Numerical) | 308 |
| | 1 | Methods not requiring ballistic tables | 309 |
| | 2 | The Ballistic Tables and their use | 317 |
| | 3 | Given the values of C, R and V, to construct a full range table | 323 |
| | 4 | Detailed numerical example of all the trajectories of one rifle, for elevations from 0° to 90° | 326 |
| | 5 | Practical methods of determining the value of the Ballistic Coefficient | 329 |
| | 6 | Long range fire analysis in two arcs | 332 |
| | 7 | The Ballistics of low power weapons and revolvers | 336 |
| V. | — | The "Le Boulénge" Chronograph | 337 |
| VI. | — | Instruments for Measuring the Pressure in the Rifle and how to use them | 341 |
| VII. | — | Ballistic Pendulum and Theory of Recoil | 347 |
| VIII. | — | Probability of Fire | 352 |
| IX. | — | The Strength of Guns | 359 |
| X. | — | Wounding Effects of Bullets | 362 |
| XI. | — | Definitions and Units | 366 |

## PART IV

### APPENDICES

| | | PAGE |
|---|---|---|
| I. | Range table of ·303 Mk. VII from Hythe firings | 371 |
| II. | Abstract of results of Ordnance Committee calculations | 372 |
| III. | Ordnance Committee calculations plotted and pricked off | 373 |
| IV. | Table II used for V = 2600 f.s. C = 0·284 | 374 |
| V. | Use of S. & T. tables of Table I to determine R. and L. when V, C and $v$ are given by formulæ | 374 |
| VI. | Table I | 375 |
| | Table Iᴀ | 381 |
| | ,,  II | 387 |
| | ,,  III | 393 |
| | ,,  IV | 395 |
| | ,,  V | 398 |
| VII. | Index to Formulæ, Part III, Chap. IV. | 401 |
| VIII. | Details of the rifles of various Powers | 403 |
| IX. | ,, ,, machine guns and Light machine guns of various Powers | 408 |
| X. | ,, ,, small arm ammunition of various Powers | 418 |

# TEXTBOOK OF SMALL ARMS

## PART I

### CHAPTER I—SECTION 1

### THE HISTORY OF THE RIFLE

I

The following is an outline of the development of firearms.

The composition of gunpowder is first given by Roger Bacon, who recorded it in the year 1248; it is almost certain that he was its discoverer. The belief that it was known to the Chinese and other Eastern nations at an earlier date seems not to be justified. They used incendiary mixtures, but not explosive ones, and had not differentiated saltpetre from other salts. This was only accomplished a few years before Bacon's discovery.

It was not he, however, who applied gunpowder to the propulsion of missiles; this was a later invention, and is usually attributed to Bernard Schwarz, a German monk, who lived in the fourteenth century. There is no trace of firearms before 1300. The tradition that the idea of a gun arose from the explosion of gunpowder in a mortar, the lid of which was violently blown off, may very well be true. There is evidence that the earliest guns were pot-shaped vessels, and that cannon were first evolved from this form. The earliest missiles were arrows padded out to fit the bore.

The hand-gun (as the artilleryman still calls it) was derived from the cannon and does not appear till late in the fourteenth century. At first it consisted of a short iron tube, prolonged behind the breech into a rod which was used to manipulate it and was tucked under the arm when the piece was fired; the charge was ignited by applying a match to a touch-hole on the upper side of the piece.

The uncertainty of ignition of the earliest hand firearms was so great, and the ingredients of early gunpowder so impure, that their efficiency can only have been of a low order. But their force was at least sufficient to penetrate armour, and their noise demoralized the mounts of the men-at-arms. It may be observed that there were influential believers in the superiority of the bow even so late as the reign of Elizabeth.

To the primitive hand-gun a wooden stock had been added by 1400; it was straight, and rested on the top of the shoulder; the shorter curved stock, to rest against the breast, was a later development. The wooden stock seems to have led to the touch-hole being placed at the side, and furnished with a pan to hold the priming, and a pan cover; the match was now held in a cock or split lever attached to the stock and hinged so as to lower its end into the pan when actuated by the movement of a simple system of levers controlled by the movement of the "tricker." The match lock so arranged was known as the "harquebus." Similar weapons are still in use in many parts of the world.

The drawbacks of carrying, and of keeping dry, lengths of slow match hung to the girdle, with a shorter piece having two smouldering ends held in the fingers, were obvious enough; but the essential improvement of producing the fire only at the moment at which the piece was to be fired took many years to assume a practical form. The wheel lock, too costly and too complicated for the ordinary soldier, was produced in Germany early in the sixteenth century, and was prominently in use for pistols and for sporting weapons, until after 150 years it was definitely superseded by the flint lock. In the wheel lock, the cock was armed with pyrites, which, when lowered, it pressed against the serrated edge of a steel wheel projecting through the bottom of the pan. The action of pulling the trigger released a spring, the wheel revolved, and sparks were struck which ignited the priming. The simpler snap-hance lock, developed during the seventeenth century into the flint lock, dates from the same period. It also had a cock carrying pyrites; when the cock fell, this struck against a piece of steel, throwing sparks into the pan. The first use of flint dates from about 1600; the flint lock approached its perfection when the motion of the cock in falling not only struck sparks against the hammer, but also opened the pan which had up to that instant protected the priming. A good flint would fire 30 shots with some certainty before it was worn down.

The flint lock came into the British Army about the time of the Restoration; all Marlborough's wars were fought with it, and it changed little until after Waterloo, when Forsyth's invention of ignition by the percussion of fulminate (which dates from 1805) gradually brought about its extinction. The percussion system was not adopted for the British Army till 1836. The percussion cap has proved itself perfectly adaptable to the change from muzzle loader to breech-loader and magazine rifle, and from black to smokeless powder.

## II

The hand-gun had been in systematic use for more than a hundred years before rifling was invented by some unknown genius. It is clear that the invention dates from the early days of the sixteenth century, and from the school of armourers in Nuremburg; but authentic details of its origin are lacking. The knowledge that spinning a projectile adds to its steadiness in flight belongs to primitive times. The difficulty in the case of firearms lay in the application of the principle. Spiral grooving would spin a ball if the ball were gripped by it firmly from breech to muzzle; this required that the ball should fit tightly when rammed down from the muzzle in loading. But powder was foul, and, in early days, made of very impure ingredients; the crust left on the bore made it a slow and laborious process to force the ball down after once firing. This difficulty was perennial, and though rifling came into use on the Continent for sporting weapons, for pistols, and for target purposes, for which accuracy was of more importance than rapidity of fire, the invention was 200 years old before any material success attended it for military use. The endeavour had many times been made. The mountaineers of the Continent used their hunting rifles in the field with success on many occasions, but the rifle was essentially a specialist's arm. Christian IV. of Denmark had a number of wheel-lock rifles made for his troops soon after 1600, but it does not appear that his experiment met with any success. Marshall Puységur in the seventeenth century recommended the arming of two men in each company of the French infantry with rifles, and some of their cavalry regiments were armed with the rifle before 1680.

Rifling does not seem to have been known in this country till about 1580, when the invention was probably some 60 years old. The first scientific investigator of the subject was Benjamin Robins (1707-1751) who proved that the very wild flight of musket balls was due to their receiving a rotary motion from their last contact with the bore on leaving the muzzle, and also that a rifle ball in its flight kept foremost that part of the ball which was uppermost in the barrel after loading. His prophecy in his "New Principles of Gunnery," 1742, is too remarkable not to be quoted. "Whatever State," he says, "shall thoroughly comprehend the nature and advantages of rifled barrel pieces, and having facilitated and completed their construction, shall introduce into their armies their general use, with a dexterity in the management of them, will by this means acquire a superiority which will almost equal anything that has been done at any time by the particular excellence of any one kind of arms."

Not long after this utterance British troops began to use rifles in the field. In the backwoods of America the settlers used their hunting rifles with so much effect that the only effective rejoinder was to pit rifle against rifle; for this purpose Jaegers were recruited on the Continent. With these as a foundation, the Royal American Regiment, afterwards the 60th Rifles, was raised in 1756. Colonel Bouquet, who commanded it, obtained 16 rifles for use in the Battalion in 1758; the use of this arm was afterwards extended, and the officers carried rifles; that carried by Wolfe (who died in 1759) is still preserved in the Museum of the R.U.S.I.; it is of American make. In 1794, one battalion of the regiment was armed throughout with the rifle, and its 5th battalion was raised expressly as a rifle battalion in 1798, and was dressed in green. The rifles used by this regiment appear to have been made on the Continent, but no doubt some American rifles were used.

A remarkable interlude in the eighteenth century was afforded by Lt.-Col. Patrick Ferguson of the 71st Highlanders, a man who suffered the usual experience of those who are too far ahead of their time. He invented, and had made in this country, a breech-loading rifle fit for service purposes, with which he gave successful demonstrations in 1776 at Woolwich. In this arm the breech was opened and closed by a vertical plug threaded with a steep screw, so that it only needed a single turn of the lever forming the trigger guard to open or close it. He used this rifle himself in the American war, and armed some of his men with it, but his force met with disaster and he himself was killed at King's Mountain in 1780, and no further use was made of the first breech-loader and the first rifle of English manufacture used in war by British troops.

In 1800 the Rifle Brigade was raised and armed with the Baker rifle, the production of the Whitechapel gunmaker of that name. It was shorter and lighter than the musket, having a 30-inch barrel. This rifle was afterwards issued to light infantry companies and other units, and was the only rifle in use in the British Service up to 1838. In that year the Brunswick rifle superseded it, a rifle with ignition by percussion and having a belt cast on the ball, which a deep notch across the muzzle guided into the two deep grooves of the rifling. This rifle was not a success; it also suffered much from difficulties caused by the fouling. Particulars of these rifles are given on page 5, together with those of the British Service rifles which preceded the introduction of magazine arms. An attempt to arm troops with a better rifle than the Brunswick was made by the issue to the 1st Batt. Rifle Brigade during the Kaffir War (1846-52) of Lancaster rifles firing a flat-based conical bullet cast with two wings to fit the grooving, and some useful work was done with this rifle. But it was not free from the defects of its predecessors.

## III

Many endeavours had been made to increase the rapidity of loading the rifle by devising a projectile which would pass easily down the bore and yet, when fired, expand into the grooving. Delvigne and Thouvenin in France led the way to real improvement (1830-50), and the first solution was found by Minié, who in 1849 produced a cylindro-conoidal hollow-based bullet which fired satisfactorily from a rifle with an ordinary breech. This bullet had a deep taper hollow in the base, with a hemispherical iron cup fitted to it; this was forced into the hollow by the explosion, so expanding the bullet.

Minié's principle met with immediate success. It was adopted in this country in 1851. The "English Minié" rifle, however, was not found to be very satisfactory, and was never generally issued, though it was used by some of the troops in the Crimea. As a makeshift arrangement a number of 1842 pattern muskets were rifled and issued to the Royal Marines. These were known as Sea Service rifles.

In 1852 a committee was formed to decide on a new rifled arm for the British forces, and after trying many rifles of private manufacture recommended the Enfield rifle of 0·577 calibre, fitted with a leaf backsight which could be folded down either forward or backward, and in 1853 the Pritchett bullet (suggested to Pritchett by Mr. Metford) which had a hollow base, without plug or cup to assist expansion, was adopted for it. Other alterations were afterwards made, and in its final form (1859) the bullet, which had a plug fitted to its hollow base, and was heavily lubricated, was of only ·55-inch diameter, so that it had to expand 0·027-inch into the bore of 0·577-inch, and, further than this, to expand into the grooving.

The Enfield rifle was first issued in 1855, and was used in the Crimea; a shortened form of it, with a 5-grooved barrel 29 inches long, was made for the Navy in 1858, and was issued to the rifle battalions.

The use of a bullet of soft lead, expanding into a barrel much larger than itself, led to difficulties, and these were accentuated by the want of uniformity in the manufacture of the barrels. In 1854, therefore, the Commander-in-Chief, Lord Hardinge, asked Mr. Joseph Whitworth, who was a pioneer of accurate measurements and workmanship but not an expert in small arms, to experiment on the subject. His recommendation, after many trials, was a reduction of bore from ·577-inch to ·450-inch, with the use of a longer bullet of the same weight as that of the Enfield and the use of a spiral having one turn in 20 inches. This was a very sound change, more so than the proposal to make both the bullet and the grooving of hexagonal form. But the Whitworth rifle, both in flatness of trajectory and in accuracy, was greatly in advance of the Enfield, as shown by a trial in 1857, which gave the following results :—

| Distance, Yards | Angle of Elevation | | Mean Deviation | |
| --- | --- | --- | --- | --- |
| | Whitworth | Enfield | Whitworth | Enfield |
| | | | Feet | Feet |
| 500 | 1° 15′ | 1° 32′ | 0·37 | 2·24 |
| 800 | 2° 22′ | 2° 45′ | 1·00 | 4·20 |
| 1100 | 3° 8′ | 4° 12′ | 2·62 | 8·00 |
| 1400 | 5° 0′ | — | 4·62 | — |
| 1800 | 6° 40′ | — | 11·62 | — |

The Whitworth rifle was never adopted for the Service, though a number were made at Enfield and the Rifle Brigade was for a time armed with them. It was some years before the reduced calibre was adopted. But one very important result followed from Whitworth's improved methods. From 1860 all parts of the rifles made for the Army were so dimensioned as to be interchangeable. Interchangeability of parts had been introduced in America by Hall in his flint lock breech-loading rifle, known as U.S. model 1819, a remarkable arm of which several thousands were made, but had not hitherto been attempted in England.

Polygonal grooving and mechanically-fitting bullets, such as Whitworth's, were soon proved to be unnecessary. Mr. Metford, C.E., in 1865 finally solved the difficulty caused by the fouling, attaining great success with a rifle of ·450 bore, having five very shallow grooves, firing a bullet of lead hardened with antimony and wrapped in a sheath of paper. With proper wadding the bore was swept out by every shot, and fouling did not accumulate.

Full advantage, however, was not taken of this method in the Martini-Henry, adopted in 1871, after most exhaustive trials by a committee appointed in 1866. The grooving of this rifle was in fact heptagonal, with a ridge in the angle between each two sides: it fired a bullet of cylindrical section, but the rifling was too deep to be kept clear of fouling. Yet, speaking generally, it did good work for the Army for many years.

In 1883 a committee was appointed to consider the production of an improved Martini-Henry rifle, and in 1886 they recommended a reduction of the bore to ·402 with shallow grooving, for which the Martini action was retained. A number of these rifles were made, but they were never issued except for experimental purposes.

In 1886 reports of experiments made on the Continent with reduced calibres led to a reopening of the question of calibre, and in 1887 the Committee made experiments with a rifle and ammunition invented by Colonel Rubin, which had a calibre of only ·3-inch firing a bullet of lead jacketed with copper. In June, 1887, the Committee recommended the small calibre for adoption, and in January, 1888, they recommended a magazine rifle of ·303 calibre having the Lee breech action and Metford's shallow segmental rifling. Smokeless powder was under experiment, but not yet available for adoption, and the charge was 70 grains of black powder compressed into pellet form and giving a velocity of 1850 feet a second to a bullet of 215 grains. Cordite was introduced in 1892 and raised the velocity to 2,000 ft.-secs. Its erosive effect on the bore made it desirable in 1895 to revert to the more usual form of grooving with an almost square edge, which was found to be more durable. The necessity to avoid angles in which powder fouling might accumulate no longer existed, since the new propellent gave no fouling.

PARTICULARS OF RIFLES IN THE BRITISH ARMY 1800–1888

| Name | Baker | Brunswick | Minié Rifled Musket | Enfield | Snider | Martini-Henry |
|---|---|---|---|---|---|---|
| Year | 1800 | 1838 | 1851 | 1853 | 1867 | 1871 |
| Breech system | M.L. | M.L. | M.L. | M.L. | B.L. | B.L. |
| Ignition | Flint | Percussion | Percussion | Percussion | Central fire | Central fire |
| Length | 3' 9½" | 3' 10" | 4' 7" | 4' 7" | 4' 7" | 4' 1½" |
| Weight (without bayonet) | 9·5 lbs. | 9 lbs. | 10 lb. 8¾ oz. | 8 lb. 14½ oz. | | 9 lbs. |
| Barrel:— | | | | | | |
| Length (inches) | 30 | 30 | 39 | 39 | | 33¼ |
| Calibre (inches) | ·615* | ·704* | ·702* | ·577 | | ·450 |
| Rifling:— | | | | | | |
| Grooves | 7 | 2 | 4 | 3 | | 7 |
| Spiral:— | | | | | | |
| 1 turn in inches | 120 | 30 | 72 | 78 | | 22 |
| 1 turn in calibres | 195 | 43 | 103 | 135 | | 49 |
| Bullet:— | | | | | | |
| Form | sphere | sphere | Cylindro-conoidal hollow base with iron cup. | Cylindro-conoidal hollow point. Hollow base with solid plug. | | Cylindro-conoidal with paper patch. Slight hollow in base. |
| Weight (grains) | about 350 | about 530 | 680 | 530 | | 480 |
| Material | soft lead | soft lead | soft lead | hardened lead | | hardened lead |
| Approximate muzzle velocity (ft. per sec.) | 1,200 | 1,200 | 1,200 | 1,200 | | 1,350 |
| Sighted to (yards) | 200 | 300 | 1,000 | 1,000 | | 1,000 |

* Dimensions and calibres of British firearms prior to about 1855, when Whitworth introduced accurate measurements and good machinery, must be taken as being only approximate, since all arms were hand-made, and those produced in quantity for the Army did not approach in accuracy of dimensions the sporting arms made for private individuals. This applies especially to smooth-bore weapons.

## IV

The convenience of inserting the charge at the breech instead of at the muzzle was very early recognized. A very early method applied to cannon was that of loading the charge into a detachable chamber wedged into place; and examples of breech-loading firearms are found as early as the first half of the sixteenth century. Robins in 1742 alludes to rifles charged at the breech through an opening at the side filled by a screw. Colonel Ferguson's rifle, in which one turn of the lever embodying the trigger guard opened the breech by lowering a screw plug in rear of it, has already been mentioned. Many different systems of breech-loading were manufactured and many more proposed up to the middle of the nineteenth century, but the prime difficulty with them was the impossibility of checking the leakage of gas from the chamber in the absence of any obturating contrivance. This difficulty persisted even after the method of inserting a made-up cartridge, which contained its own fulminate for ignition, had become practical. Dreyse's needle gun, invented in 1838 and adopted in 1842 by Prussia, was on this principle. A needle-shaped striker, piercing the cartridge at the base, and striking a disc of fulminating material within it, ignited the charge. It was true that the needles rusted and might break, and that there was a leakage of gas at the breech, but the gain of rapidity in loading outweighed all such drawbacks, and the "Zundnadelgewehr" was used with conspicuous success in the wars of 1848, 1866 and 1870 by Prussian and German troops. This rifle had a breech action on the bolt principle.

The revolving principle was inapplicable to arms requiring a powerful cartridge. Lefaucheux's method of hinging the barrels and clamping them down against a false breech, though admirable for shot guns, was unsuited to military arms. Many forms of breech-loader were tried, with hinged or falling blocks, and with various arrangements, such as that of Westley Richards, whereby a cap had to be placed on the nipple, the flash of which penetrated the paper of the cartridge, or (as in the Sharps) the paper at the base of the cartridge was cut off by the sharpened edge of a rising block. To check the leakage of gas, a wad at the base of the cartridge, a disc of copper and other such devices were tried. But the breech-loader was not a real success before the production of the metallic cartridge containing its own ignition.

In 1864 the importance of arming British troops with a breech-loader was confirmed by a special committee. The Ordnance Committee, after investigating some fifty different actions, selected Mr. Jacob Snider's horizontally hinged breech block. This was held in place by a projecting stud (Mark II) or a spring catch (Mark III) and had a hook extractor to withdraw the cartridge by engaging in the rim. This system owed its success to the adoption of Colonel Boxer's improved cartridge, which had walls of thin coiled brass, as well as paper, an iron disc for base, and a central fire cap. The muzzle-loading Enfields were converted to this system and, when the cartridge had been perfected, gave better accuracy as breech-loaders than in their original form.

The conversion of the Enfield rifle to breech-loading was expressly a temporary expedient, and in October, 1866, a special committee was appointed to produce a breech-loading arm. After examining the merits of a very large number of arms and kinds of ammunition, and carrying out many necessary trials, in which the question of the breech action was wisely dealt with separately from that of the barrel and rifling, they recommended the action of Mr. Martini, as modified at Enfield, a hammerless action which could neither be put to half cock nor bolted, so that it was fully cocked unless the striker was right forward. It fired a central fire cartridge of coiled brass with an iron base, made in a bottle shape to reduce its length, and having a wad of wax behind the bullet.

The Martini was one among many actions in which a block, either hinged at the rear end or moved downwards by a lever, was made use of. The bolt action, in which the block is moved horizontally away from the breech, and locked or unlocked by a partial turn of a lever, as in the ordinary door-bolt, was not looked upon with favour for the central fire cartridge though it had been used for the Prussian needle gun. It was found to be sometimes liable to accidental discharge on the forcible closing of the breech. This drawback was in the course of time removed by the addition of a device whereby the point of the striker is withdrawn behind the face of the bolt except when the action is safely locked, and with this modification bolt actions are now in universal use for military weapons.

A fresh committee assembled in 1883 to consider the production of an improved Martini-Henry rifle, and three years later recommended a reduction of the calibre to ·402, retention of the Martini action with a lengthened lever to give more powerful ejection, and the use of a cartridge case of solid drawn brass. A considerable number of these rifles were made for trial, but they were never issued to the troops as an arm. The Enfield-Martini, as it was called, marks a stage in the evolution of the modern rifle.

Weapons containing a supply of cartridges in a tube or box in the weapon itself had appeared at a comparatively early stage in the history of the breech-loader. Such were the Henry and Spencer rifles of 1860, the Winchester of 1867, and the Vetterli adopted by Switzerland in 1867. The French Kropatschek (1878) and the German Mauser (1884) were developed as magazine rifles, but they were all of calibres larger than ·400 inch. Austria in 1886 adopted a Mannlicher rifle of ·433 bore, fitted with a straight pull bolt and having Lee's box magazine, afterwards so extensively used by other countries. Following the lead of Colonel Rubin in Switzerland, Germany adopted a magazine rifle with a reduced calibre (·311-inch). This was in 1888. The reduction of calibre allowed the cartridges to be so much smaller and lighter that the difficulty of ammunition supply, that bugbear of magazine systems, was no longer formidable.

The committee which had recommended the Enfield Martini was directed to investigate the question of the adoption of a magazine rifle, and in 1887 they decided that the best of the systems before them was Lee's magazine combined with his bolt action. This was therefore ready to be adapted to the new ammunition of reduced calibre, and the first trials of the combination were made in 1888 and proved satisfactory. It speaks well for the thoroughness of the work of this committee that the Lee action and magazine, with modifications only of detail from their early form, should have proved themselves thoroughly efficient and reliable in the Great War thirty years later.

The ·303 rifle has undergone a number of alterations since its first adoption, and its various revisions have produced a number of different Marks of the rifle. The magazine of the Lee-Metford Mark I held only 8 cartridges; that of Mark II held 10, in two columns. In 1895 a new form of safety catch was adopted, and in the same year the rifling was altered. With smokeless powder, the difficulties caused by powder fouling had vanished, and with them the chief *raison d'être* of the segmental grooving of Metford pattern, which was found to be too easily obliterated by the erosion due to cordite. Recourse was therefore had to the earlier pattern of shallow grooving with square edges which had been, but for the powder fouling, successful in earlier days; the lands were made of equal width with the grooves, and their depth increased from ·0045 to ·0065. The modified pattern was known as Lee-Enfield Mark I. The next alteration was to omit the cleaning rod. In 1894 a Lee-Metford carbine was introduced, and in 1896 a Lee-Enfield carbine; these had barrels $9\frac{1}{2}$ inches shorter than the rifles, and their foresights were protected by wings.

Following the experience of the South African War, there was a demand for a shortened rifle suitable alike for infantry and for mounted troops. It was evident, too, that the system on which the magazine of the rifle had been designed to operate needed modification. The cartridges had to be inserted into it singly from the pouch, and the supply in the magazine, put out of action by the cut-off, was to be kept in reserve for moments of intensive fire. This system, when opposed to the fire of troops continuously using the magazine, and refilling it with five cartridges at a time from a clip or charger, gave little advantage over the breech-loader without the magazine. The committee appointed in January, 1900, produced the rifle, approved in December, 1902, known as the "rifle, short, magazine, Lee-Enfield, Mark I." This was shorter by 5 inches than the Lee-Enfield, was stocked to the muzzle, and had sights of an improved pattern, the foresight being protected by wings. Charger guides were fitted, one of them on the bolt head, and a 10-round magazine, the cartridges being issued in chargers holding 5; the cut-off was retained, but fire was normally through the magazine. A Mark II rifle was a similar rifle made by conversion from Lee-Metford Mark II* and Lee-Enfield rifles. A Mark III rifle (January, 1907) embodied charger guides on a bridge spanning the body, foresight of blade form, and U-shaped backsight. For the Territorial Army M.L.M. Mark II and M.L.E. Marks I, I* and II* rifles were converted to charger loading and fitted with the improved sights.

A new pattern of rifle was under trial in 1913–14, having a front-locking bolt action of modified Mauser pattern, and a calibre of ·276 inch. The cartridge was semi-rimless and bottle-shaped, the bullet, of 165 grains weight, received from the charge a velocity of 2,800 f.s. The barrel was much thicker and stronger than that of the S.M.L.E. and its length was 26 inches; it had a light nosecap and a naked projecting muzzle to which the bayonet was attached. In place of the ordinary backsight, a light aperture sight was fitted ; this was placed on the bridge over the action, being housed between projecting flanges. This rifle met the demand for a weapon ballistically more efficient than the S.M.L.E., but the flash and the violence of the blast issuing from the muzzle were unduly great. Between the mouth of the chamber and the front end of a double-column magazine, front-locking actions necessarily interpose a space in which the locking recesses lie, and across which the cartridge has to be thrust before it begins to enter the chamber ; this adds to the length of the movement made by the hand in opening and closing the breech. The rifle, however, did well in its trials for rapidity and accuracy of fire. But it was found that, with the large charge and high pressure used, the heating of the barrel was very great, and that if a cartridge loaded with a cordite charge remained for more than quite a short time in the chamber, dangerous pressures were produced. The best way to deal with this difficulty was under consideration in the summer of 1914. During the war a large number of rifles were made to this pattern, but with the modifications to the magazine, chamber and barrel necessary to enable them to take the ·303 Mark VII cartridge. These were known as 1914 pattern rifles. They proved to be admirable for sniping, being extremely accurate, but except for this purpose were not issued to troops fighting abroad. With many of these rifles, there was some difficulty in adjusting the magazines to feed the Mark VII cartridge correctly into the chamber, owing to its rim, which tended to upset its alignment during its forward travel.

A Mark VI S.M.L.E., recently approved, embodies the modifications shown by war experience to be desirable. It has a barrel and body of increased strength and stiffness, with a projecting muzzle and virtually no nosecap, it has various improvements in the detail of the action and safety catch, and it is fitted with an aperture sight. The bayonet is of a new pattern, cruciform in section, lighter and shorter than those hitherto used, and designed only for thrusting.

V

The development of military firearms has made astonishing progress since the period of the harquebus. Then, as until the middle of the nineteenth century, each item of the charge had to be inserted separately into the piece ; first the powder, then the wadding, then the ball. Further, the fine powder for the priming had to be placed in the pan, and the smouldering slow match, carried continually lighted, had to be fitted into the cock to precisely the right length. The match and the pan required constant attention, to ensure that burnt ash did not impede the fire of the one, nor loose grains of spilt powder remain where they might prematurely ignite the other. Sighting hardly existed ; fighting was at close quarters, and efficiency required no more than a decent levelling of the piece. The 32 motions of loading and firing demanded the utmost coolness and precision from those who had to carry them out in the face of the enemy. The rate of fire was proportionately slow.

The flint lock, which struck the fire only when it was needed, began the process of simplification. But the flint itself required adjustment and renewal at intervals and the inconvenience of priming remained. The invention of the cartridge further reduced the number of movements. To bite off the end of the paper receptacle, use some of its contents for priming the pan, pour the rest down the muzzle, and then to put in the ball and ram it down in one motion with all that remained of the cartridge, was a great saving of time. The introduction of the iron ramrod by Frederick William of Prussia gave an advance in rapidity ; it came into use in the British Army in 1752. When the percussion cap came in, not only was the trouble caused by priming eliminated, but also most of the uncertainty of fire due to the exposure of the pan to wet and wind. But the cap had still to be carried and placed on the nipple.

Breech-loading and the central fire cartridge gave the soldier a case ready charged, in which almost all the operations of loading had been already carried out in the factory. Still, with the Snider, the cartridge had to be laid in the breech and pushed home with the thumb, and the hammer raised to full cock after closing the breech, before the rifle could be fired; after firing, the hammer had to be half-cocked, the breech opened, the cartridge extracted from the chamber and brought into the trough of the action and then ejected by turning the rifle over sideways, before another round could be loaded. The Martini-Henry definitely ejected the empty case by the movement of the lever which opened the breech; the same motion cocked the action, but the fresh cartridge had still to be pushed forward into the chamber by the thumb, a process sometimes requiring effort. Finally, the bolt-action rifle conveyed the cartridge into the chamber in the process of closing the breech and cocking the action, without a separate motion, and the magazine mechanism actually placed it where it could be so conveyed, leaving no motion special to loading to be done while the magazine contained a cartridge. It remains only for automatic mechanism to open and close the breech, and the manipulation of the trigger is left as the sole movement demanded from the soldier. Even this is avoided in the fully automatic weapon, which actually releases the sear for itself. To direct the aim and to control the bursts of fire are then the only remaining functions of the firer, and the correct operation of the mechanism and the maintenance of the ammunition supply are his only anxieties.

Increased speed of fire has accompanied the simplification of loading processes. With the long bow, a rate of fire of 12 shots a minute was possible enough; and the much slower rate of the harquebus was one of the serious difficulties which delayed the coming of firearms. The rifle again had to justify its still slower rate of fire; with the flint-lock musket, in which the ball was small enough to be dropped down the bore (so that if the muzzle were lowered it was liable to roll out), the speed of fire was at most 5 shots a minute; with the rifle of the same period only one shot could be fired in $2\frac{1}{2}$ minutes. This rate was, however, greatly improved upon with later muzzle loaders. 12 shots a minute and more could be fired with the breech-loader, and this rate is much exceeded with the magazine rifle, with which so far as actual mechanical manipulation is concerned, the rate of fire may be anything up to 60 shots a minute, and with which smokeless powder enables every shot to be effectively aimed in firing in the field at a rate of 20 rounds or more to the minute. Without smokeless powder, indeed, the increased speed of fire due to the magazine would have been of small advantage. With black powder the smoke of the previous shot frequently enough obscured the target for the individual firer; it almost always did so in collective fire. It also advertised conspicuously the firer's position.

The development of fire power has had far-reaching effects on formations and on tactics. The maintenance of continuous fire on the frontage occupied by a body of men has been a primary requirement from the earliest times of the use of firearms by formed troops; when achieved it was effective against infantry of any arm. To obtain continuous fire with the harquebus, it was necessary to form troops in from 8 to 12 ranks, which fired successively, the rank which had just fired retiring to clear the front for the next, or the rear rank moving up and forming in front of it. In this way the necessary time for reloading and priming was obtained for each rank. The gradual reduction in the number of ranks, as the weapon became handier and the loading arrangements more rapid, is very noticeable. We see them diminished to 6 and to 4, and during the eighteenth century to 3; the development of the line of 2 ranks only was taught at Shorncliffe under Sir John Moore, and the two-rank formation was an improvement which gave Wellington a tactical advantage in the Peninsular War. With this formation, defence against cavalry was provided when necessary by forming fours and closing on the centre of the company; the front rank knelt and the enemy were confronted with a hedge of fixed bayonets. It was long before fire power was so far developed as to make it unnecessary for the infantry soldier to carry the long pike-like bayonet, but since the introduction of magazine fire either a single rank or a rank of men in open order can sweep their front effectively with continuous fire, a fire far more effective both for accuracy, range, and power, than that of previous generations.

The perfection of fire effect of small arms seems to have been reached when a few machine guns, posted at intervals, give to a wide front absolute protection against the advance of infantry or mounted troops.

## CHAPTER I—SECTION 2

## A REVIEW OF THE SYSTEMS OF MODERN RIFLES AND A COMPARISON OF THEIR PRINCIPAL COMPONENTS

All modern military rifles are breech-loaders and are fitted with magazines that will contain five or more cartridges. The breech actions, without exception, are operated on the "bolt" system. In this system a hollow cylinder of steel, with a solid end and fitted with a handle, is locked into position behind the cartridge in the breech-chamber and, together with the metal of the cartridge case, closes the breech against all escape of the gases of explosion in a backwards direction. The pressure developed by the burning powder is in the neighbourhood of 18 tons to the square inch, and a considerable margin of safety has to be allowed. The pressure is resisted, taken up, and transferred to the action-body by one or more "lugs" which work in slots in the action-body or against resistance pieces. The lugs are part of the bolt and are formed when the cylinder is machined to shape in manufacture. Theoretically, from a mechanical point of view, the best designs are those in which they are disposed symmetrically so that the pressure is taken up evenly and are placed as near the bolt-head as possible in order that the bolt itself shall not be subject to stresses which are bound to be taken unevenly by a long column of metal free to move, within narrow limits, in any direction. The bolt is hollow in order that it may contain the firing-pin, or striker, and the main-spring which actuates it; the front end, or bolt-head, being pierced to allow the striker point to reach the cap in the base of the cartridge. The bolt is also fitted with a hook, or claw, to engage with the rim or groove on the base of the cartridge and extract it from the chamber after firing. A safety device, in the nature of a lock to prevent accidental discharge of the weapon, is usually fitted to the rear of the bolt. The action body, in which the bolt works, is screwed on to the barrel. The interior is grooved to allow the lugs to pass when the bolt is worked and is recessed so that the lugs will be given a substantial bearing surface when in the firing position. The right side of the body is usually cut away to allow the fired case to be ejected to that side.

Barrel form and figure of rifling in all modern rifles are very similar. Differences, in outside contour of the barrel, from a straight taper from reinforce to muzzle, are sometimes in the nature of steps with the object of breaking up the barrel vibrations on firing. In length barrels vary between 24 and 32 inches, in calibre of bore between $\cdot 256$-inch and $\cdot 315$-inch, and in weight between 2 lbs. 3 ozs. and 3 lbs. 3 ozs. In all but two cases the figure of rifling is that known as concentric or "Enfield" with four grooves (five in the case of the British Short Magazine Lee-Enfield). The two exceptions are of segmental or "Metford" figure. Of these the Danish Krag has six grooves and the Japanese rifle four grooves. The twist of the rifling is to the right, or clockwise, except in the case of the British and French rifles. The Italian rifle has a progressive twist.

Trigger actions are, in all cases, on the "double-pull" system. In this system the trigger mechanism is so arranged that, after a long, light pull the motion is stopped, the firing pin being released by a further short and heavier pull.

Though the breech actions are all on the bolt system much ingenuity has been exercised in designing them to function with speed, regularity and safety. There are many and important differences between the various types, and rifles fall, naturally, into classes according to type. The main division is between "straight-pull" actions in which the bolt is drawn straight back to extract the fired case and pushed straight forward again to feed in a fresh cartridge and lock the breech, and the ordinary "turning-bolt" actions, in which the breech is first unlocked by turning the bolt upwards and then the cartridge extracted by drawing the bolt backwards. A fresh cartridge is fed in and the breech locked again by a reverse movement. Of straight-pull rifles the only ones now in use are the Austrian Mannlicher and the Swiss Schmidt-Rubin. The Ross, once the arm of the Canadian forces, is now obsolete.

Turning-bolt actions may be divided into three classes: Mauser and Mauser types, Mannlichers, and other types. In the last category are the Lee-Enfield, the Lebel, the Nagant and the Krag-Jorgensen.

The Mauser action is that most generally in use, more than twenty powers being armed with this type. Fig. 1, illustrating a typical Mauser action, has been reproduced from the descriptive handbook of "The Mauser Magazine Rifle (Model 1904) and its Ammunition" (also see Plate I).

*Mauser Rifles.*

| | | | | |
|---|---|---|---|---|
| Argentine | .. Pattern 1891 | .. Cal. 7·65 mm. | .. ·301 inch. |
| Belgium | .. 1889 | .. 7·65 ,, | .. ·301 ,, |
| Bolivia .. | .. 1891 | .. 7·65 ,, | .. ·301 ,, |
| Brazil .. | .. 1904 | .. 7·00 ,, | .. ·276 ,, |
| Chile .. | .. 1904 | .. 7·00 ,, | .. ·276 ,, |
| China .. | .. 1893 | .. 7·00 ,, | .. ·276 ,, |
| Columbia | .. 1891 | .. 7·65 ,, | .. ·301 ,, |
| Eucador | .. 1891 | .. 7·65 ,, | .. ·301 ,, |
| Germany | .. 1898 | .. 7·90 ,, | .. ·311 ,, |
| Luxemburg | .. 1896 | .. 6·50 ,, | .. ·256 ,, |
| Mexico .. | .. 1902 | .. 7·00 ,, | .. ·276 ,, |
| Peru .. | .. 1891 | .. 7·65 ,, | .. ·301 ,, |
| Portugal | .. 1904 | .. 6·50 ,, | .. ·256 ,, |
| Persia .. | .. Various old patterns. | | |
| Spain .. | .. 1896 | .. 7·00 mm. | .. ·276 inch. |
| Sweden .. | .. 1906 | .. 6·50 ,, | .. ·256 ,, |
| Turkey .. | .. { 1905 And other patterns. | .. 7·65 ,, | .. ·301 ,, |
| Uruguay | .. 1895 | .. 7·00 mm. | .. ·276 inch. |
| Yugo Slavia .. | .. 1899 | .. 7·00 ,, | .. ·276 ,, |

*Mauser Type Rifles.*

| | | | | |
|---|---|---|---|---|
| Italy .. | .. Adopted 1891 | .. 6·50 mm. | .. ·256 inch.* |
| Japan .. | .. { "Year 38th" Pattern (1907) | 6·50 ,, | .. ·256 ,, |
| U.S.A. .. | .. Adopted 1903 | .. 7·62 ,, | .. ·300 ,, |

The Mauser bolt is of strong and simple construction, in one piece, without a separate, movable bolthead. The locking lugs are opposite one another at the front end ($a, a$) (Fig. 1). On the German bolt (pattern 1898) and on the pattern 1904 bolt, is an extra lug ($a^1$) engaging in a recess in the cylindrical part at the rear end of the body. This acts as an additional safeguard in case the front lugs should break. The lever, or handle, is straight and stands out at right angles from the rear end of the right side of the bolt. The end is shaped to a round knob.

At the back end of the bolt is cut a cam-shaped recess ($d$) which receives the stud of the cocking piece and on turning up the bolt-lever slightly withdraws the striker. On the opposite side is a small notch ($d^2$) for the tooth of the safety-bolt. The small rib ($a^2$) is special to the German and 1904 pattern rifles. It acts as a guide in withdrawing the bolt. When the bolt is closed it lies underneath and supports the extractor. The face of the bolt is recessed to take the base of the rimless cartridge.

The striker has a short point and a collar ($e$) against which the mainspring bears. The rear end has three interrupted grooves for connection with the cocking-piece. The rear portion of the pin is flattened on two opposite sides to prevent it from turning in

* The Italian so-called Mannlicher-Carcano has the Mannlicher clip system of loading; but the bolt action is of the Mauser type, adapted by M. Carcano of the Turin Small Arms Factory.

the bolt-plug. The pattern 1889 (Belgian) rifle has a striker differing somewhat from later designs. The rear end is threaded, and a rib formed on the pin takes the place of the flats in preventing the striker from turning.

The mainspring is of ·06-in. diameter coiled wire with 28 to 31 coils in the different patterns. Uncompressed it is about 5 inches long.

The cocking-piece (D) is provided with interrupted lugs so that it can be connected to the striker pin by giving it a quarter-turn. An incorrect connection of these parts is impossible, as the rear groove on the striker and the corresponding bearing in the cocking-piece are broader than the two front ones. In the pattern 1889 a female thread takes the place of the interrupted lugs. On the underside of the cocking-piece is a projection ($f$) which travels in a groove, cut for it in the tang of the body. This projection engages with the sear nose when the bolt is pushed forward, so that the bolt travels forward whilst the striker remains stationary, thus compressing the spring fully for firing. The front top surface of the stud ($g$) is chamfered to correspond with the sloping surface ($d$) at the end of the bolt. On turning up the bolt the cam surface ($d$) forces back the stud ($g$), gives a preliminary compression to the spring, and withdraws the point of the striker from the fired cup.

The bolt-plug (E) screws loosely into the rear end of the bolt by means of a buttress (reinforce) thread, and closes the opening through which the striker is inserted in assembling. It serves for a seat for the mainspring. It takes no part in the turning movement of the bolt, the cocking-stud ($f$) working in the slot ($f^1$). Peculiar to the German pattern and the pattern 1904 is the strengthening of the front part of the bolt-plug in semi-circular shape. This flange acts also as a shield to protect the user against accidental back-blast. In the German and pattern 1904 rifles also there is on the left side of the plug a pin catch ($p$) which, in a certain position, engages with the bolt and keeps bolt-plug and bolt securely locked against accidental unscrewing during the opening and closing movements. In other patterns the front of the stud ($g$) on the cocking-piece rests in a small slot on the rear of the bolt and prevents the bolt-plug from turning when the safety bolt is not holding back the cocking-piece. The top of the bolt-plug is drilled with a cylindrical hole for the bolt of the safety catch (F) which is shown in position on the assembled firing pin in Figs. G and $G^1$.

The safety bolt consists of a thumbpiece and a spindle which works in the hole in the bolt-plug. When the bolt is closed and the thumbpiece is turned vertical, as shown in the figures, the flange ($h$) on the safety bolt is brought in front of the top of the cocking-piece and forces it back a little, withdrawing the stud ($f$) from contact with the sear. On turning the thumb piece over to the right the cocking-piece is still locked as above, but at the end of the stem, where it is not cut away, enters into the slot ($d^2$) on the end of the bolt and locks the bolt and bolt-plug together, preventing the former from being revolved. The safety bolt is retained in the two safety positions by the top of the cocking-piece bearing on shallow depressions on the flange of the safety bolt. In pattern 1889 (Belgian) the safety bolt is retained in position by a small pin fitted in the under side of the thumbpiece, actuated by a spiral spring and bearing in a groove cut across the top of the bolt-plug. This practice is followed in the U.S.A. bolt.

The extractor is a strong, broad, long spring (H) terminating in a claw, the front of which is chamfered to allow it to ride over the base of the cartridge, when the hook engages in the groove at the base of the cartridge and holds it rigidly against the bolt head. The extractor is provided with an undercut groove which engages on the dovetail projections on the spring band ($j$) which revolves in a groove on the bolt. A projection ($k$) works in a groove at the bolt head and through this medium the pull of the bolt is transmitted to the extractor. The right-hand lug lies under the extractor in a recess when the bolt is drawn back. In the pattern 1889 the extractor is a short spring let into the side of the bolt head between the two locking lugs. It revolves with the bolt, which is an objectionable feature owing to the fact that the rear end of the barrel has to be weakened by being cut away to afford clearance for the end of the extractor, and also on account of the friction introduced between the extractor and the base of the fired cartridge.

The body is screwed to the breech end of the barrel. On either side of the boltway a groove is cut in which work the bolt-lugs and the extractor. The groove on the right side of the body is partly cut away to allow the cartridge to be ejected to the right. At the front end of the body, connected with the grooves, are the recesses for the locking lugs. The rear part of the body, through which the bolt moves, forms a complete cylinder, prolonged behind into a tang which is recessed for the cocking-stud to move over it. In front of the cylindrical portion are vertical grooves to hold the charger in loading. The tang is bored and threaded for the rear action screw which helps to hold action, stock and trigger guard together. In the tang is an opening through which projects the sear-tooth. In some early patterns another slot is cut in the body, just behind the magazine way, for a tooth on the front end of the sear which engages in a slot in the rear of the underside of the bolt when the trigger is pressed. This is to prevent the accidental discharge of the arm when the bolt is not fully home. In pattern 1889 and in pattern 1904 a cross groove is cut, just behind the magazine way, for the additional safety lug on the bolt. In the Spanish pattern the left side of the body is cut away to give clearance for the thumb in pushing the cartridges out of the charger into the magazine. Some Turkish patterns are fitted with a cut-off similar to that of the Lee-Enfield, Mark III. It should be noted that during the Great War the Turks were provided with Mausers of different patterns from their standard patterns 1890 and 1903.

The bottom of the action-body is cut away for the magazine. In pattern 1889 and later patterns the magazine and trigger guard are in one piece and secured to the stock and action body by means of screws which are clearly shown in the section (J) which figures the whole bolt mechanism. The five cartridges lie in two columns and are pressed up against the overhanging edges of the magazine by the pressure of a ribbon spring secured to the bottom plate and the magazine platform. The platform is provided with a raised rib near its left side which raises the column of cartridges resting on it to the level of the other column and enables the cartridges in both columns in turn to be caught by the advancing bolt and fed into the chamber. The bottom plate is held in position by a small stud to the rear of the magazine opening. The stud can be depressed by the point of a bullet, when the plate can be slid backwards and taken out, together with the spring and platform. As a war measure the German rifle was provided with a deep curved box magazine to hold twenty cartridges. It could be attached in place of the bottom plate of the standard magazine. It was provided with a simple cut-off in shape of a pin which could be inserted from side to side over the cartridge column. In the pattern 1889 (Belgian), the magazine is a detachable box of sheet steel, holding five cartridges in single column. The platform is pressed upwards by a system of levers and flat springs.

All Mauser rifles are charger loaded, the chargers consisting of strips of thin sheet steel with the edges turned over to engage in the cartridge grooves. The cartridges are held with sufficient firmness in the charger by a spring of wavy ribbon steel secured by small tangs to the inside of the charger. The cartridges are swept into the magazine by one motion of the thumb, the charger being thrown out by the first forward motion of the bolt.

The trigger mechanism is on the "double pull" system. The sear ($k_1$) projects into the groove cut in the tang and is part of a bar ($m$) pivoted near its front end to the body. It is actuated by a spiral spring ($l$) let into it, one end of which bears against the body. The bar portion of the sear has a vertical slot cut in it, in which the trigger ($n$) is pivoted. On pressing the trigger the forward of the two projections on the upper surface of the arm presses against the body and the sear is depressed, a sliding movement being felt. The leverage is then transferred to the rearmost projection, a slight check is felt and on further pressing the trigger the sear parts smartly from the cocking-piece stud and the striker flies forward to fire the cartridge. There is no half-cock, nor can the striker be placed at full-cock without opening and reclosing the bolt.

An arrangement is provided for stopping the backward travel of the bolt to prevent it being accidentally withdrawn or jerked out. This consists of a lever hinged on the left side of the body. On it is a tooth which projects, through a slot, into the groove for the left lug of the bolt. The lever is pivoted on a pin and is kept pressed against the body by a small flat spring let into it. The bolt is stopped in its backward travel by the left lug coming in contact with the tooth. The retaining-bolt can be swung away from the body so that the bolt can be withdrawn when necessary for cleaning, inspection, etc.

The ejector is pivoted on the same pin as the retaining-bolt and forms part of it. It is a triangular shaped flat piece of steel actuated by a spring inside the retaining-bolt lever. A slot is cut for it in the left locking-lug and in the face of the bolt-head so that when the bolt is drawn back the ejector springs into this slot. The base of the cartridge then strikes against the ejector; but as the extractor continues to draw back the right side of the cartridge the latter is swung round and thrown out of the action to the right. There are some small differences in retaining-bolt and ejector in the different patterns of Mauser rifles but they are differences of form and not of principle.

The *United States Magazine Rifle*, Model 1903, has an action of Mauser type, but differing from those previously described in the following particulars (Fig. 2, reproduced from the U.S.A. Official Handbook, also see Plate II).

The striker is in two parts. The striker rod (B) has a recess and collar formed on its forward end. The striker proper ($B^1$) is bored out in the rear and has a groove cut into which the collar of the pin fits. A sleeve ($B^2$) fits over the front end of the pin and the rear end of the striker covering the joint and preventing accidental separation. The striker rod is liable to break at the recess if not properly heat-treated. The striker has annular grooves cut on its surface to retain lubricating oil. The mainspring bears against the end of the striker sleeve. It is of ·05-inch wire with 39 coils and is set to an uncompressed length of 4·25 inches.

The cocking-piece (D) ends in a knob which serves to cock the action without raising the lever, in event of a miss-fire. The parts of the cocking-piece, the stud ($g$) and the nose ($f$), are different in shape from those described previously in true Mauser-made actions; but they work in a similar way.

The bolt-plug (E) is fitted with a small spring catch on the left which prevents the bolt-plug from unscrewing when the breech is open. This corresponds to the pin ($p$) on the German and 1904 Mauser bolt-plugs (see Fig. 1).

The safety lock consists of a thumbpiece with the words "ready" stamped on the right side and "safe" on the left, and a spindle, to which it is firmly riveted, which works in a hole on the top of the bolt-plug. There is a small spring stud on the underside of the forward end of the thumbpiece which works in a groove in the bolt-plug over which it projects, the groove being notched in three positions, top, right and left. The front end of the stem is cut away in two places. When the thumbpiece is turned over to the right, showing the word "safe," the part of the stem or spindle not cut away turns into a recess in the rear of the bolt, preventing the bolt lever being raised. The bottom portion of the thumbpiece has two circumferential ribs which are cut away in one place. These ribs turn into recesses cut in the cocking-piece and prevent the latter going forward except in the position "ready." If the thumbpiece is turned to "ready" the rifle can be fired and the bolt opened for loading. If the thumbpiece is placed upright the breech can be opened but the rifle cannot be fired. In the "safe" position the rifle cannot be fired nor can the breech be opened.

The cut-off and retaining bolt are ingeniously combined in one spindle and thumbpiece. The spindle lies longitudinally in a recess in the left side of the body. A spring stud, similar to that on the safety-bolt, limits its action. On opposite sides of the thumbpiece are the words "on" and "off," the "on" side being polished to distinguish it at a glance from the other. The spindle is cut away longitudinally in two places—in one place for its whole length and in the other for a limited amount of its length. When the thumbpiece is put down (Off) that part of the spindle which is not cut prevents the bolt from being withdrawn sufficiently far to enable the cartridges to rise from the magazine. When the thumbpiece is put right up (On) that portion of the stem which is partly cut away projects into the boltway and allows the bolt to come back far enough to clear the cartridges in the magazine. When the thumbpiece is in a horizontal position that part of the spindle completely cut away coincides with the bolt-way and the bolt can be completely removed from the action-body.

The ejector is fitted to the left side of the body immediately in front of the cut-off. It is similar in shape to the usual Mauser ejector, but has no spring. It works through the slot cut in the left lug of the bolt.

The *Italian Mannlicher Carcano* pattern 1891, (Fig. 3, reprinted from T.B.S.A., 1909, also see Plate III), has an action of Mauser type adapted by M. Carcano, of the Turin Small Arms Factory. The Mannlicher clip system of loading, similar to that employed in the Dutch and Roumanian rifles, is employed to fill the magazine. The action differs from the typical Mauser in the following particulars:—The bolt-lever, which is not at the extreme end of the bolt, projects at right angles about one-third of the length of the bolt from the rear. The rear of the body does not form a complete cylinder, but is cut away at the top to allow the bolt-lever to pass. Under the front end of the body is a transverse rib, following Mannlicher practice, which fits into a groove in the stock, and serves to transfer the shock of recoil to the stock.

The bolt-plug (E) fits into the rear end of the bolt, and a groove ($a$) inside the rear end of the latter admits of the stud ($c$) entering the slot ($b$). The mainspring bears on the front of the plug and the striker passes through the hole in the centre of it. The bolt-plug acts as a safety arrangement in the following manner. When the plug is in its normal position the stud ($c$) is resting in the recess at the extreme end of the slot ($e$), and when the cocking-piece and striker are held back by the sear the mainspring tends to drive them forward. If now the bolt-plug is pressed forward and turned round, by means of the finger-piece, so that the stud ($c$) rests in the recess ($d$), the back of the bolt-plug bears against the front of the cocking-piece, and the latter has no tendency to fly forward. If the trigger is now pulled cocking-piece and striker remain at rest. If the stud ($c$) is turned into the end of the slot ($b$) and drawn out through the groove ($a$), the bolt-plug, cocking-piece and striker and mainspring are withdrawn from the bolt.

The extractor is a medium length, narrow spring inserted into its groove from the front of the bolt. The extractor revolves with the bolt, which is an objectionable feature. The extractor lies under the right-hand lug, and has a shoulder on it which springs up under the lug and takes the pull during extraction. At the shoulder is a small groove to take the end of a screw-driver or punch in depressing the shoulder when removing the extractor from its groove.

The retaining bolt is on the right side of the body. Its upper end passes through a slot into the right-hand groove for the bolt-lug. It is slotted to connect with an arm on the right side of the trigger. To withdraw the bolt the trigger must be pulled when the retaining bolt is lowered and the bolt can pass over it.

The magazine holds six cartridges, and is similar to that of the Dutch Mannlicher.

The ejector rests in a bearing in the front of the sear and is pressed upwards by the sear spring. Its top end passes through a slot in the body, enters a groove on the underside of the bolt when the bolt is withdrawn, and strikes against the left hand, lower part of the base of the cartridge, throwing it out to the right.

The *Japanese Year 38th Pattern Rifle* is of Mauser pattern, and differs from typical Mausers in the following particulars:—

The striker and cocking-piece are made in one. The striker is of large diameter for two-thirds of its length, and is bored to a depth of 4·2 inches from the rear to take the mainspring. A cocking-toe is formed on its rear end. In the interior at the rear end are cut two longitudinal guide grooves, disposed at right angles to one another, in which a stud on the locking-bolt moves.

On the rear of the bolt the usual chamfered recesses are cut, in order to withdraw the striker from the fired cap on the first raising of the lever of the bolt. As there is no bolt-plug these recesses are cut somewhat deeper than in typical Mauser practice.

The locking-bolt is in the form of a cylindrical cap on the end of the bolt, with a stem which fits into the striker and bears against the mainspring. It can be placed in the safety position only when the action is cocked. To lock the action the locking-bolt is pressed forward and then turned to the right, the movement to the right being limited by the travel of a stud, on the under surface of the locking-bolt, in a groove in the body.

By pressing forward the locking-bolt the mainspring is compressed, and by turning the locking-bolt a small stud on the rear part of the stem, which had engaged with a guide grooves in the striker, turns the striker, and the cocking-toe is cleared from the sear. A ledge on the forward end of the locking-bolt engages with a stud on the bolt, which is thereby locked. The locking-bolt is held by the weight of the mainspring with a stud on the rim of the locking-bolt forced into a recess in the groove at the rear of the body. When the locking-bolt is disengaged the weight of the mainspring is transferred from the point of the stem of the locking-bolt to a circumferential rib on the bolt which engages with a rib on the inside of the cylinder of the locking-bolt.

The knob of the bolt lever is egg-shaped.

A bolt cover is provided. It is semi-circular in cross-section and is retained in guide grooves on either side the body, in which it moves. It encases the whole of the top of the body when the breech is closed. The bolt lever passes through a hole in the cover and draws the latter backwards and forwards.

The magazine and trigger-guard are not in one piece. The box is made of sheet steel and holds five cartridges in two columns. In other respects it is similar to typical Mauser magazines.

*Mainspring Compression.*—In all Mauser actions on the first motion of the lever to open the action the cocking-piece and bolt-plug are prevented from turning with the bolt owing to the cocking-stud projecting into the groove cut in the tang of the body. The tooth on the cocking-stud is forced back by the cam-shaped recess on the rear of the bolt, slightly withdrawing the striker and partially compressing the mainspring.

In the Belgian, Spanish and Turkish, and Japanese patterns, on locking the bolt when there is about an inch of travel before final closing, the cocking-stud engages with the sear so that the cocking-piece and striker are held back whilst the bolt and bolt-plug are pushed forward, thus completing the compression of the mainspring.

In the 1904 and German patterns of the Mauser, and in the U.S.A. and Italian actions, the cocking-stud does not engage with the sear until the bottom of the bolt-lever meets the cylindrical portion of the body. The mainspring takes its final compression when the bolt lever is turned down. This compression is materially assisted and made easy by the lugs on the bolt travelling along the cam-shaped entrances to their recesses and being thereby forced forward.

*Primary Extraction.*—In all Mauser actions primary extraction of the fired case is attained by the leverage exercised by the lower end of the lever, at the point of junction with the bolt, moving against an inclined plane cut on the rear face of the cylindrical part of the body. In the U.S.A. action the inclined plane is a spur or continuation of the lever, formed round the bolt, which works against a cam surface on the rear of the body.

In the Italian action the rear end of the body is not cylindrical but is slotted to allow the bolt-lever to pass. The necessary leverage to move the fired case is given by the left-hand lug on the bolt working against a cam-surface at the front end of the left-hand groove in the body.

This completes the descriptions of Mauser actions. The barrels, sights, stocks, etc., of the rifles above described will be found classified and described at the end of this chapter.

*Mannlicher Rifles* (turning-bolt action). (Fig. 4 and Plates IV A and B).

|  | Pattern. | m.m. | in. |
|---|---|---|---|
| Greece | 1903 | 6·5 | ·256 |
| Holland | 1895 | 6·5 | ·256 |
| Roumania | 1893 | 6·5 | ·256 |

In the Mannlicher bolt the locking lugs are close to the head on opposite sides of the bolt; but disposed a little further back than those of the Mauser to allow of the fitting of a separate bolt-head which does not rotate with the bolt when it is actuated. The advantages of a separate bolt-head are many and considerably outweigh the advantages claimed for the solid bolt-head, the most important of which are strength and the impossibility of the bolt-head being lost. In order to get the mainspring and striker into the bolt the front or rear end must have an opening as large in diameter as that of the mainspring. These openings are closed either by a bolt-head or a bolt-plug. A bolt-head is lighter and less complicated than a bolt-plug and the extractor can be attached to it and can thus be prevented from rotating with the bolt and, at the same time, be of light and simple construction. The disadvantage of having the extractor rotate with the bolt has already been dealt with. Another advantage of the separate bolt-head is that if the end of the bolt be damaged by defective cartridges or in any other way, a repair can be more cheaply effected by exchanging the bolt-head than by renewing the entire bolt cylinder.

The disadvantage of a separate bolt-head is that the lugs have to be disposed further back on the bolt to allow for the head and are thus further removed from the barrel and cartridge, and the action body has to be lengthened to allow of proper working of the action. Also, if the bolt-head becomes loose, through wear, the shooting of the rifle is impaired.

The Mannlicher bolt is simple in construction and, like the Mauser, can be stripped and assembled without the use of tools. The position of the two locking lugs has already been indicated and reference to figures (A) and (B), Fig. 4, will obviate further description. (A) represents the Dutch and (B) the Roumanian bolt. The Greek bolt resembles the Roumanian closely except for differences that will be detailed as they are come to. The Dutch bolt has a strengthening rib ($a$) in line with the lever. The lever projects at right angles to the bolt about one-third of the length of the bolt from the rear, and terminates in a knob. The knob of the Greek bolt is hollowed out for lightness. It is larger than in either of the other patterns.

There is a cam recess ($b$) at the end of the bolt similar to that described in the Mauser bolts and serving a similar purpose. The cocking-piece stud ($c$) enters it on firing, and the leverage given by the inclined plane on turning up the lever in the first motion of reloading forces back the cocking-piece, slightly withdraws the striker, and gives the first compression of the mainspring. At the end of its travel along the inclined plane the tooth of the cocking-piece rests in a shallow groove ($d$) in which it is held by the weight of the mainspring and prevented from turning when the bolt is withdrawn. On the Greek bolt there is also a projecting cocking-piece stop stud in front of the cam which prevents the cocking-piece being overturned to the right when in the cocked position for insertion in the rifle.

As the striker and mainspring are inserted from the front there is no bolt-stop. A segmental recess ($d^1$) on the rear end of the bolt receives the end of the safety bolt. It is larger in the Roumanian and Greek than in the Dutch pattern. A small gas escape is provided behind the left-hand lug. On the Roumanian and Greek bolts there is a groove ($e$) to provide a clearance for the ejector fitted to those types. There is a hole in the rear of the bolt through which the end of the striker works.

The bolt-heads, $A^1$ and $B^1$, for the above bolts are provided with tenons which fit into the front ends of the bolts. The studs ($f$) work in circular grooves on the inside of the bolts. The bolt-heads can only be withdrawn when the studs are turned, so that they are opposite clearances provided for them. The tenons are slotted at ($g$) for the flats on the front of the strikers. The extractor (C) fits into a groove in the side of the bolt-head.

The mainspring consists of 26 coils of ·056 wire. The rear end bears against a seating in the bolt cylinder.

The striker, D, passes through the mainspring, bolt, and cocking-piece, and screws into the nut, E. In the Greek pattern the attachment is by interrupted lugs. The front end of the mainspring bears against the collar ($h$). The flat ($j$) works in the slot ($g$) in the bolt-head and prevents the cocking-piece from turning when the stud is clear of the groove in the tang, the bolt being drawn to the rear.

The cocking-piece (F) fits on to the striker and butts up against the shoulder ($k$) on it. It is prevented from turning by the end of a screw ($l$) bearing on the flat on the striker. The stud ($m$) travels in the groove in the tang of the body and engages with the bent of the sear. The rib ($n$) is bored cut from the rear for the stem of the safety bolt.

In the Greek pattern the underside of the top rib of the cocking-piece is grooved on the right side to clear the stop-stud on the bolt, and the striker hole is bored with interrupted rings for attachment to the striker.

The stem of the safety bolt (H) fits in its seating in the rib of the cocking-piece. It is operated by means of a finger-piece. A small spiral spring fits over the stem and keeps the safety bolt pressed to the rear. When the striker is cocked, and the finger-piece is pointing to the left, the stem of the safety bolt is able to move forward over the bolt, as the half that is cut away (o) is underneath, next to the bolt. When the finger-piece is turned over to the right, the small cam (p) on the end of the safety bolt engages the front of the recess ($d^1$) in the bolt and forces safety bolt and cocking-piece back, disengaging the stud on the latter from the sear bent. The safety bolt and cocking-piece cannot now move forward when the trigger is pressed. In the Roumanian and Greek pattern the finger-piece can be turned over to the right, and the bolt locked, when the cocking-piece is in the fired position. The end of the stem of the safety bolt then engages in the recess (q), a clearance (r) being provided in the stem for the division between the recess. In the Dutch pattern the bolt cannot be locked in the fired position.

The extractor (C), fits in a groove in the bolt-head. The shoulder (s) fits against a corresponding shoulder in the bolt-head, and takes the strain during extraction. The front terminates in the usual claw.

In the Dutch and Greek pattern an undercut stud on the ejector (t) slides backwards and forwards in an undercut groove in the bolt-head, its travel being limited by a screw working in a groove. The back of the ejector strikes the retaining bolt when the bolt and bolt-head are drawn to the rear. The ejector slides forward, and the front of the stud strikes the left rear of the cartridge, ejecting it to the right. The front edge of the stud is bevelled off to enable the base of the top cartridge as it rises out of the magazine to push back the ejector.

In the Roumanian action the ejector is pivoted in a slot on the underside of the body immediately behind the magazine way. It is similar in shape to the ejectors already described for typical Mauser actions. The tail of the ejector works in the groove (e) in the bolt. When the bolt is drawn back so far that the tail is at the front end of the groove it is depressed and the small tooth on the front being raised into the bolt-way and into the slot (v) in the bolt-head, strikes the base of the cartridge and tilts it out of the rifle, to the right. The bolt-head is prevented from turning by the ejector or, in the Roumanian pattern, by a stud which takes its place, working in the left-hand groove of the body.

The retaining bolt (J) is pivoted on a vertical pin on the left side of the body near its rear end. A tooth projects into the left groove in the body and arrests the backward movement of the left lug. In the Dutch rifle it is the ejector that strikes the tooth. A small spiral spring keeps the tooth up to its work. To withdraw the bolt the thumb-piece on the rear end of the retaining bolt is pressed inwards against the body, when the tooth is removed from the boltway and the bolt is free to move backwards out of the body.

The body screws on to the barrel in the usual manner. On either side of the boltway is a longitudinal groove for the lugs on either side of the front end of the bolt. At the front end of these grooves are cam-shaped grooves which lead to recesses above and below the boltway. On firing, the lugs resist the backward pressure of the cartridge case by taking a bearing against the rear face of the recesses. The groove on the right is partly cut away to facilitate loading the magazine and to permit of ejection of the fired cartridge. Behind the recesses the bottom of the body is cut away for the magazine. Behind the magazine way the body does not form a complete cylinder, but is cut away at the top to permit of the passage of the bolt-lever. The rear part of the body forms a tang in which a groove is cut for the stud of the cocking-piece. Near the front end of this groove is an opening through which the bent of the sear projects. At the front end of the magazine-way is a projection, the rear face of which is grooved to form a guide for the bullets of the cartridges as they rise in the magazine, it also serves to keep the trigger guard at the correct distance from the body. Underneath the front part of the body is a boss which fits in a recess in the stock and transfers the shock of recoil from the barrel and body to the stock. It also takes the front guard screw. In the Roumanian rifle there is a narrow slot just in rear of the magazine-way for the ejector to work in.

In the Dutch and Roumanian rifles the magazine works on the clip system. Each clip, which is of sheet steel, with deep sides, contains five cartridges. The clip is placed in the magazine together with the cartridges, which are forced up out of the clip by the action of the magazine spring against a lever termed an "elevator." When all the cartridges have been pushed up out of the clip the clip falls out through the bottom of the magazine, in which a hole is left for the purpose. The advantages of this system are that the loading is easier than when the cartridges have to be swept out of a charger by the action of the thumb, and the clips and cartridges can be easily removed from the magazine when necessary. The disadvantages are that the clips are more bulky and heavier than chargers and are more liable to be bent and distorted. The magazine must be made deeper than in the best designs of charger-loading systems, to hold the same number of cartridges, as they lie in the magazine box in single instead of in double column. With a charger extra pressure can be applied by the thumb to overcome extra friction caused by rust or dirt between the charger and the cartridges; but in the case of the clip the strength of the magazine spring cannot be temporarily increased to overcome any abnormal resistance. With charger-loaded magazines a partially emptied magazine can be filled up with separate cartridges, if desired. With clip-loading the magazine cannot be charged until the empty clip has fallen out of the bottom of the box. In addition, the large hole in the bottom of the box is undesirable in that it is liable to let dirt into the magazine, particularly when firing over a parapet or when firing prone.

In the Dutch and Roumanian rifle the lower part of the box is formed in one piece with the trigger guard. In the Roumanian pattern a small platform is pivoted on to the end of the elevator and actuated by a spiral spring. In the Dutch pattern there is no platform, the end of the elevator being rounded off to work against the cartridges, which are not so well supported as by the platform.

The Schœnauer magazine of the Greek rifle is charger-loaded and is designed on an ingenious rotary system, which, however, presents no advantages over the more simple charger-loaded box magazines with rising platform and is more liable to become clogged with dirt that cannot easily be cleared from the mechanism.

A rectangular box forming the bottom part of the magazine is fitted into the bottom of the body and held in position by a bottom fixing plate, pivoted in the centre at the bottom of the box; each end of this plate, when revolved, enters radial locking grooves, cut in the front and rear of the downward extensions of the bottom of the body, which forms the top part of the magazine. The magazine bottom-fixing plate is retained in position by a spring which enters a recess in the plate.

In the rectangular box, forming the bottom part of the magazine, a rotary platform is fitted. The platform is bored to receive front and rear axis studs attached to a spiral spring. The axis studs are grooved to receive, and prevented from turning by, two retaining pins, fitted in the axis hole of the platform. The platform is provided externally with five grooves to receive the cartridges.

As the cartridges are charged, the magazine platform is rotated and the spiral spring coiled, its tension being thereby increased (a certain amount of tension is on the spring when assembled with studs to the platform). By the expansion of the spring the platform is rotated, and the cartridges forced up to a cartridge stop projecting into the boltway. The thumb-piece of the cartridge stop projects through a slot cut in the top on the right side of the body, where it is pivoted. When charging the magazine, the cartridge stop is depressed by the cartridges. If it is desired to remove the cartridges from the magazine, the cartridge stop is depressed by pressure on the thumb-piece, when the cartridges are ejected out of the magazine.

The trigger mechanism of all Mannlicher rifles is on the "double-pull" system. Its construction can be clearly seen from the figure.

The mainspring is given an initial compression and the striker withdrawn from the fired cap by the first motion of turning up the bolt, when the cocking-piece is forced back by the action of the cam-shaped recess on the rear of the body cylinder working against a tooth on the cocking-piece, which enters a groove at the end of the recess on the completion of its travel. Final compression is given by turning down the lever when the bolt has been returned. During this turning movement the cocking-piece stud engages on the sear bent, and the lugs on the bolt travel along the cam-shaped grooves leading from the grooves in the bolt-way to their seatings when in the cocked position. The reverse action, when unlocking the bolt, gives primary extraction of the cartridge.

This completes the description of Mannlicher turning-bolt actions. The barrels, sights, stocks, etc., of the above rifles will be found classified and described at the end of this section. The Straight-pull Mannlicher will be described later in this chapter, together with other actions of its class.

*Other Turning-Bolt Actions.*

The other turning-bolt actions, which do not fall into either of the classes dealt with are the Lebel (French), the Krag-Jorgensen (Denmark and Norway), the Nagant (Russia), and the Lee (Great Britain). The Lee action is fully described in Section 3. The Lebel, Nagant, and Lee bolts have separate bolt-heads. The bolt-heads of the Lebel and Nagant carry the locking-lugs and rotate with the bolt. This system would appear to have no advantages, except for repair purposes, over that in which there is no separate bolt-head, as it is desirable that the bolt, especially the front part, should be as solid and rigid as possible. In the Lee action the bolt-head does not rotate with the bolt, but the lugs are disposed at the rear of the bolt cylinder. This is not a desirable arrangement as, on firing, the greater part of the body and bolt are thrown into a state of tension and compression respectively. This strain, acting on the unsymmetrical central part of the body, causes lateral vibrations, which have to be compensated for by displacing the foresight laterally. The unsymmetrical incidence of the strain also has a disturbing influence on the accuracy of the rifle. The British action has, however, advantages which may compensate for the fact that, mechanically, its design is unsound. There is no deep cylindrical portion in front of the action body as in the continental forms, in which dirt can accumulate. The form of the action and the shape and disposition of the bolt-lever allow of extremely rapid fire. It is possible to deliver roughly directed fire at the rate of 60 rounds a minute, a rate which is often attained and sometimes passed in routine tests on the range of the Small Arms Ammunition Inspection Department. With no other action is it possible to attain anything like this rate of fire. Competition shooting has proved that, with good ammunition, and when adjusted carefully, the British rifle is capable of a degree of accuracy which compares favourably with that of any Continental Service weapon.

The Krag-Jorgensen has no separate bolt-head but there is one lug only which engages in a recess below the entrance to the chamber. This unsymmetrical arrangement is objectionable because it must give rise to considerable vibrations in a vertical plane.

*Lebel Rifle* (Fig. 5. and Plate V).

    France.    Pattern 1886.    8 mm.    ·315 inch.

This was the first small bore rifle to be adopted by any nation and with it smokeless powder was first used. A later model of this rifle introduced in 1907 is similar to the 1886 model, except as regards the magazine.

The bolt (A) is a strong cylinder, bored out from the front for the mainspring. It has a straight lever terminating in a knob. On the opposite side to the lever is a groove (10) into which the nose of the sear projects. On the left side, with the lever in the raised position, is the groove (11) for the ejector (34). On the same side is cut, at the rear end of the cylinder, a cam-shaped recess (12) for the similarly shaped projecting tooth (28) on the cocking-piece. When the bolt-lever is raised the end of the tooth (28) rests in the recess (13). The projecting rib (14) extends beyond the face of the bolt, and serves to connect the bolt-head and bolt by means of the recess (15) fitting over the stud (17) on the bolt-head. A screw passes through the hole (18) in the bolt and enters the hole (19) in the bolt-head tenon, which in turn enters the mainspring channel of the bolt-head; the latter is therefore forced to turn with the bolt. This rib also acts as a guide to the bolt, the small rib (16) travelling along the left top edge of the body.

On the left of the bolt-head B is the ejector groove (11A). On the right, the extractor (20) fits into an undercut groove, the rear end being splayed out slightly to resist the pull during extraction. The two lugs (21) enter the recesses (3) in the body on the right and left of the boltway, and prevent the bolt being forced to the rear on firing. The face of the bolt-head is cupped out to a depth equal to the thickness of the cartridge rim; the extractor, therefore, will not rise over the rim of the cartridge until the cartridge is almost home in the chamber. The bolt-head is bored out for the striker, the hole in rear being oval to receive the part (24) on the striker. The groove (10) on the bolt for the sear nose is continued for a short distance along the bolt-head at (10a).

The mainspring consists of 19 coils of wire, 0·05 inch thick, set to a length of 3·9 inches.

The striker has a shoulder (22) for the mainspring to bear against, and two slots (23), near the rear end, fit into the T-shaped recess in the striker knob (26). The part that enters into the bolt-head is thinned down in front of the shoulder at (24). The point is further reduced in two steps at (25).

The cocking-piece D has a projection (27) in front, working in the opening between the sides of the body. Underneath the projection is a tooth (28) which is shaped to fit into the cam recess (12) at the back of the bolt cylinder. The bottom front corner (29) of the cocking-piece is the full cock bent. The notch (30) is a second full cock bent. The notch (31) affords clearance for the bent of the sear, when the cocking-piece is forward in the fired position. The top is hollowed out and roughened to form a comb for drawing back, or letting down the striker, but the mainspring is too strong for this to be done with safety. The striker passes through the cocking-piece, its knob fitting into the recess (33) in the rear end of the latter and being locked with a half turn.

The extractor (20) is a short flat spring terminating in the usual claw, which projects over the face of the bolt-head; it is dovetailed into its groove in the bolt-head. The breech end of the barrel is bevelled off for a quarter of its circumference to afford clearance for the claw of the extractor.

The ejector is a small pin (34) screwed into the body on the left side, it projects into the bolt-way and works in the slot (11) and (11a) in the bolt and bolt-head.

The body has vertical sides which slope inwards at the top. The body is prolonged in front at (2a), and forms a cylindrical reinforce for the barrel. The rear end forms a long tang which is prolonged forwards at (40) between the sides of the body. The bolt slides in a longitudinal groove, cut in the upper part of the body, and the lugs (21) on the bolt-head turn into recesses (3) at the front end of the bolt-way. The entrances to these recesses are rounded off to allow of the bolt coming back during primary extraction. At the front end, below the projection (2a), is an opening (4) for the magazine tube. The bottom of the body is closed by a plate (5) to which is attached the sear and the cartridge elevating mechanism, which are situated below the bolt-way. The projection (6) on the front end of the plate fits into a recess in the body, and a screw passes through the body and through a screw-hole (7) in the rear end of the plate. The body is cut away on the right side to allow of the cartridge cases being ejected to the right. The rear part of the bolt-way is open on top to allow of the passage of the rib (14), on the bolt, and also of the bolt-lever, which turns down in front of the shoulder (8). Two recesses (9) and (9a) are cut in the bottom of the right side of the body for the handle of the carrier axis-pin lever (49).

The trigger mechanism is on the double pull system and is very similar to that of the Mauser and Mannlicher actions, except that the sear arm is short so that the sear is disposed rather in front of the trigger than behind it, and the spring is V-shaped instead of being spiral. The trigger is little curved and is awkward in shape.

The magazine of the 1886 pattern is tubular and consists of a longitudinal hole bored in the fore-end, lined towards the rear with a short steel tube. It holds eight cartridges which are pressed towards the rear by a steel spring consisting of 78 coils of wire ·03-inch thick. The spring terminates in a steel plunger which is prevented from coming out of the tube by a small shoulder in the latter. The cartridges are raised to the level of the chamber by a carrier scoop actuated by the bolt striking against a projecting lever. The magazine is loaded by depressing the scoop with the bullet of the cartridge and passing it into the magazine. On the right side of the bottom of the action body a chequered button, connected with the carrier axis-pin, projects. When this is pushed to the front the depressor lever is moved so that it cannot be struck by the bolt. The magazine can then be retained full and the rifle used as a single-loader. The magazine of the later 1907–15 pattern is of the more usual fixed vertical box type, holding 5 rounds.

Besides being slow to load, a tubular magazine is objectionable because the balance of the rifle shifts as the column of cartridges is pushed to the rear and the cartridges loaded and fired. The necessary raising platform, also, is complicated and very liable to be jammed by dirt.

On first raising the bolt the mainspring is compressed nearly to its full extent by the cocking-piece being forced backwards by the cam recess on the rear of the bolt cylinder. Final compression is given on closing the bolt by the lugs working along the curved entrances to their recesses. Primary extraction is given by the front end of the projection on the bolt being forced back by a curved face on the body.

This completes the description of the Lebel action. The barrel, stock, sights, etc., will be described at the end of this section.

*Krag-Jorgensen Rifle* (Fig. 6 and Plate VI).

| Denmark | Pattern 1889 | 8 mm. | ·315 inch. |
|---------|--------------|-------|------------|
| Norway  | 1910         | 6·5 mm. | ·256 inch. |

The bolt (1) is of simple construction with no separate bolt-head. A single locking lug (2) is situated at the head of the bolt. On the right side, when the bolt is closed, is a solid rib (3) which bears against a shoulder in the body, and assists in taking the shock of recoil. The lever terminating in a knob, is at the rear end of the bolt cylinder, and is set back a little. In the Norwegian pattern it is turned down but not to the same degree as in the British rifle. The usual cam-shaped recess is cut in the back of the bolt. A gas escape hole (4) is bored in the right side of the bolt cylinder, close to the lug.

A flange (5) runs partly round the rear end of the bolt and serves to retain the bolt plug. The bolt can be stripped without the aid of tools.

The bolt is bored out from the rear for the mainspring, which is of wire ·053-inch in diameter, with 28 coils set to a length of 4·5 inches.

The striker (6) is in two pieces, the point forming the front is secured to it by a sort of knuckle joint. This is as described in the U.S.A. rifle except that there is no sleeve. The mainspring bears against the rear end of the front part of the striker.

The rear portion of the striker screws into the cocking-piece (7) from which the cocking-stud (8) projects downwards, travelling in a groove cut for it in the tang of the body. The front of the stud is shaped to fit into the cam recess of the bolt. The stud is provided with a half bent, as well as a full bent, and the cocking-piece has a roughened thumbpiece. In the Norwegian action there is no half bent. The cocking-piece ends in a knob.

The striker passes through the plug which fits into the rear portion of the bolt, and against which the mainspring bears. The sleeve is bored out for the cocking-piece, and has a slot in its underside for the cocking-stud to travel in. On the top it extends over the bolt, travelling in the slot cut in the body cylinder; a groove is cut in this portion of the sleeve, in which a corresponding flange on the rear of the bolt works, and prevents the pressure of the mainspring from forcing the sleeve out to the rear.

The extractor (9), which is a long flat spring terminating in the usual claw, is pivoted by a screw to the bolt sleeve, and fits over the bolt rib, when the bolt lever is raised. It has a secondary spring (10) mortised into it, which fits under a projection in the body and prevents the extractor from rising during primary extraction. The extractor claw projects over the face of the bolt-head recess. When the bolt is closed the end of the extractor fits into a recess in the body, but when the bolt-lever is raised the projection on the front of the rib of the bolt bears against a projection on the extractor, guiding the latter while the bolt is being drawn back. A slot in the cylindrical portion of the body assists in performing the last-mentioned function. The extractor of the Norwegian rifle is wider than that of the Danish.

There is no retaining bolt, but the lug on the bolt coming up against the resisting shoulder stops the backward movement of the bolt. To remove the bolt it must be drawn back, the extractor lifted so as to clear the top of the body, and the bolt lever turned to the left—this brings the lug opposite the slot in the body cylinder, when the bolt can be withdrawn.

The ejector (11) is a spring, dovetailed into the bottom of the bolt-way. Its front end projects upwards and enters a groove in the head of the bolt, when the latter is drawn back. On the base of the cartridge striking the ejector, the cartridge is rotated upwards and to the right, and is ejected from the action.

There is no locking or safety bolt in the Danish rifle. The striker can be placed at half and full cock without opening the action. The Norwegian rifle is fitted with a safety bolt of the Mauser pattern.

The body has no features calling for special mention, except that the recess (12) for the locking lug is at the bottom of the front end of the body and is readily accessible for cleaning purposes.

The magazine (13) is a horizontal box under the body, closed by a door on the right side. The spring, lever, and platform which press the cartridges out into the bolt-way are attached to the door. When the door is open the platform is held back against it, allowing for the free introduction of the five cartridges it will hold. The magazine can be charged with the bolt either open or closed and can be replenished at any time by one or more cartridges. This type of magazine is very slow compared with those which are charger or clip loaded. A charger is provided for this magazine but it is a clumsy arrangement provided with a handle. The handle has to be raised from the side of the charger box and the rifle tilted to the left before the cartridges can be spilled into the magazine. A cut off is provided by which the cartridges can be prevented from rising into the bolt-way.

The trigger mechanism is on the double pull system. The sear is actuated by a flat spring.

The first compression of the mainspring is given in the manner described for Mauser rifles and the final compression by the sliding of the lug along the cam-shaped entrance to its recess.

Primary extraction is given by the bolt lever being forced back by working along the curved surface at the back of the body.

This completes the description of the Krag-Jorgensen action and magazine. The barrel, sights, stock, etc., are described at the end of this section.

*The Nagant Rifle* (three line) (Fig. 7 and Plate VII).

    Russia        Pattern 1900        7·62 mm.        ·30 inch.

The bolt of this rifle bears some resemblance to that of the French Lebel but is of far more complicated construction. There is a separate bolt-head (1) which revolves with the bolt (2) and carries two lugs. A connecting bar lies underneath the bolt, holds the bolt-head to the bolt, acts as a guide to the cocking-piece and helps to retain the bolt in the body. When the bolt is closed the lugs are horizontal instead of being one above the other as in Mauser and Mannlicher actions. The striker (3) is in one piece and is actuated by a main spring of 28 coils of ·05 inch diameter wire set to a length of 4 inches. The cocking-piece calls for no special comment except that it can be pulled back by hand and revolved to the left when it fits into a recess on the rear end of the bolt which prevents the rifle from being fired or the bolt from being opened.

The extractor is small and fits into a groove in the bolt-head.

The magazine is of box form and is made in one piece with the trigger guard. It is fitted with an interrupter (4) to prevent the possibility of double loading. The interrupter is a plate which works in a slot on the left of the body in a manner somewhat similar to the cut-off of the British rifle. On loading the magazine the cartridges are forced down from the charger and press the interrupter outwards and are thus able to pass it. In loading and firing the cartridge next below the top one is held down by a tooth on the plate until the top cartridge has been forced into the chamber and the bolt lever turned down. No other rifle is fitted with this device. The ejector is a projection on the edge of the interrupter.

The trigger mechanism gives a single pull. The trigger is provided with a tooth (5) at the top which projects into the bolt-way and acts as a retaining arrangement. The trigger must be drawn back to depress this tooth and allow the front end of the connecting bar to pass over it, when the bolt can be withdrawn from the body.

Withdrawal of the striker from the fired cap is arranged for and first compression of the mainspring is given, in the usual manner, by the cocking-piece nose working in a cam recess on the bolt end. Final compression of the main spring is given by the final turning down of the bolt and the action of the lugs on the sloped entrances to their seatings.

Primary extraction is effected by the front of a rib on the bolt working along an inclined plane in the body.

The barrel, sights, stock, etc., of the Nagant rifle are dealt with at the end of this section.

*Straight-Pull Action Rifles—*

| | | | | |
|---|---|---|---|---|
| Austria | Mannlicher | Pattern 1895 | 8 mm. | ·315 inch. |
| Switzerland | Schmidt-Rubin | Pattern 1909 | 7·5 mm. | ·295 inch. |

Compared with turning-bolt actions, straight-pull actions are complicated both in design and functioning. There are more moving parts bearing against one another and consequently more friction. In turning bolts, primary extraction is obtained by the direct leverage of the bolt handle being transferred to cam surfaces on the bolt and body.

With a straight-pull bolt this additional leverage is more difficult to obtain, but to a limited extent it can be introduced during the unlocking of the action at the beginning of the backward movement of the bolt handle. In the Schmidt-Rubin, Austrian Mannlicher, and Ross, the locking of the action is effected by rotating lugs. The necessary turning movement of that portion of the bolt carrying the lugs is obtained during the unlocking of the action, at the beginning of the backward movement of the bolt handle, through the medium of helical grooves. A certain amount of additional leverage for primary extraction is obtained by cutting the lugs and their seatings in the body on a screw pitch. This, however, is not so effective as the direct cam action obtainable with a turning bolt. In theory, straight-pull bolts can be operated a trifle more quickly and are more easily worked without removing the rifle from the shoulder, but it is doubtful if there is any real advantage. Certainly no straight-pull rifle can be worked at the speed of the Lee-Enfield. Straight-pull rifles are said to be less likely to jam when exposed to sand and mud as there is no turning movement to draw deleterious material down between the bolt and the left side of the body.

*The Straight-Pull Mannlicher* (Fig. 8—from the Austrian Official Handbook, also see Plate VIII).

In this action the bolt is a hollow cylinder, reinforced at the rear end, where is the lever, terminating in the usual knob. The lever projects at right angles to the bolt. On either side of the bolt are ribs (*a, a*), which work in grooves in the body and prevent the bolt from turning. On the underneath of the front of the bolt are two feathers (*b*), which act as a retaining arrangement by coming in contact with two horns on the trigger when the bolt is drawn backwards. The trigger must be pushed forward to lower the horns and allow the feathers to pass, when the bolt may be withdrawn.

The underneath of the rear reinforced portion of the bolt is cut away for the stud of the cocking-piece. The recess for the cocking-piece is separated from that for the tail of the bolt-head by a collar (*d*) secured by a screw, the point of which projects into the firing pin hole, and, bearing against a flat (*e*) on the firing pin (B), prevents the latter from turning. The safety bolt is pivoted in the left side of the reinforce. Inside the middle portion of the bolt are two helical feathers (*f*), which work in corresponding grooves (*g, g*) in the tail of the bolt-head, and rotate it in opening and closing the bolt. A groove is cut on the inside of the right rib for the extractor.

The bolt-head, C, consists of a head which projects beyond the face of the bolt cylinder, and the tail which enters the cylinder. The bolt-head has cam-shaped locking lugs on either side, which enter the recesses of the body by way of the cam-shaped grooves and, support the bolt-head in the firing position. A groove is cut in the head for the ejector to work in. The rear end of the tail has two external helical grooves, already mentioned, in which work the feathers in the inside of the bolt cylinder. The helical grooves have each a small groove (*h*) leading out of them in the direction of the length of the bolt, one to the front and one to the rear. The groove to the front is on the top of the tail, that to the rear is on the right side when the bolt is opened.

The bolt-head contains the mainspring of ·04 wire, coiled to a length of 4·9 inches, and the striker. The rear end of the bolt-head tail is closed by a screw plug, D, against which the mainspring bears, the striker passing out through the plug. The other end of the mainspring bears against a collar on the striker. The extractor is a long flat spring, which lies in the right rib of the bolt cylinder. The portion which projects fits over the right locking lug and terminates in a broad claw. The other extremity, has a small nib on its underside, which engages in the two longitudinal grooves in the tail of the bolt-head above mentioned. When the action is closed, the nib is engaged in the longitudinal groove on the top of the bolt-head tail; when the bolt is drawn back, the nib rises out of this groove, and when the bolt-head has turned a quarter of the circle from right to left falls into the other groove. The right lug is then embraced by the head of the extractor and the extractor is drawn back by the bolt. The claw engages the top cartridge in the magazine immediately it is pushed forward by the bolt in advancing, and holds it in all the backward and forward motions of the bolt, thus rendering double loading impossible.

The cocking-piece screws on to the end of the striker and works in the rear end of the bolt cylinder. In its left side is a groove in which the locking bolt engages, when it is employed to lock the action with the spring eased. At full cock the locking bolt, when used as a safety bolt, is interposed between the front face of the cocking-piece and the rear face of the bolt cylinder; the tooth of the bolt is cam-shaped, and when pushed into position forces back the cocking-stud from engagement with the sear.

The body calls for no special comment, except that the tang groove for the cocking-piece stud to work in is undercut on either side to take the feathers (*b, b*) on either side the front end of the bolt. Underneath the front part of the body is a downwards projection which transfers the shock of recoil from the barrel and body to the stock. A similar, but larger projection forms the prolongation of the front of the magazine.

The sear consists of two components, the body and the bent; they are both pivoted on the same pin, which passes through the action body. The bent fits in a slot in the body.

The ejector is pivoted to the front of the sear body. Its bottom-end is pressed forward by a small spiral spring, the other end of which presses the sear and sear bent backwards. The upper end of the ejector is slightly depressed by the bolt; the spring is partly compressed, and tends to keep the sear and sear-bent up to their work.

The trigger is in the form of a bell crank lever, the long arm projecting downwards through the trigger guard, the short arm terminating in a hook which engages the rear of the sear. At the angle is the crosspiece with the two horns which project into the boltway and prevent the withdrawal of the bolt, as already described. The trigger is not pivoted to the body in any way, but is supported in its groove by the sear.

When the trigger is pressed, the bent of the sear is depressed, releasing the cocking-stud and allowing the striker to fly forward; at the same time the front of the sear body is raised into the boltway behind the safety projection on the back end of the underside of the bolt, preventing any backward motion of the latter while the arm is being fired. Further, as the bolt is pushed forward, this safety projection sliding over the projecting front portion of the sear, prevents the latter rising, and consequently prevents the bent of the sear from being lowered, by pressing the trigger, until the bolt is completely closed.

The magazine and guard are in one piece. The former is clip loaded with five cartridges, and is similar in most particulars to that already described for the RoumanianMannlicher.

The bolt is actuated by a straight backwards-and-forwards pull and push. When the lever is pulled to the rear, the bolt cylinder cannot revolve owing to the ribs on it workin ؛ in the grooves of the body, and the feathers, in the undercut grooves in the tang. The bolt-head, on the other hand, cannot come to the rear until the locking lugs have been disengaged from their recesses in the body, and this is effected by the turning motion given to the tail of the bolt-head by the helical feathers in the inside of the bolt cylinder, working in the bolt-head tail. Primary extraction is given by the cam shape of the ends of the grooves in the body in which the locking lugs work. The first motion of the bolt to the rear partly compresses the mainspring. As soon as the locking lugs are disengaged from the body they are in prolongation of the ribs in the bolt cylinder, and the whole bolt can then be drawn to the rear.

This completes the description of the Austrian Mannlicher action. Barrel, sights, stock, etc., will be referred to in their proper place at the end of this section

*The Schmidt-Rubin* (Fig. 9 and Plate IX).

In this action the body departs from the shape which, with minor differences, has been seen in all rifles hitherto discussed. Behind the magazine way it forms a complete cylinder for a length of $4\frac{1}{2}$ inches and is thus of considerable length and clumsy. On the right side, and forming part of the body, is a smaller cylinder opening into the larger one. The bottom of this small cylinder is slotted for the tooth of the retaining bolt. The main cylinder is slotted for the sear and grooved for the striker stud to travel. This is in place of the grooved tang found in most other rifles. The main cylinder is also grooved longitudinally on each side for the bolt lugs to work in. Those grooves lead into the lug seatings which are cut on a screw pitch. The rear part of the body forms a short tang, which is bored for one of the three action screws.

The bolt which is made up of the bolt cylinder C, the locking sleeve D, and the bolt cap E, is operated by means of the action rod F. The bolt cylinder is bored out from the rear for the striker and mainspring, and is slotted on the right side to admit the tooth $(a)$ of the action rod. A circular flange affords a bearing for the front end of the locking sleeve. In front of this flange the bolt cylinder is grooved on both sides to allow it to pass through the turned-in sides of the magazine.

On the left is a deeper groove for the ejector. On top a little to the right is a flat for the extractor. The front end of the bolt cylinder is permanently screwed in. The face of the cylinder is recessed for the head of the cartridge, but the rim of the recess is not cut away at the bottom to allow the heads of the cartridges to rise up at once under the claw of the extractor when loading, therefore double loading is possible. The bolt cap E screws on to the rear end of the bolt cylinder. It is bored out for the striker, and has a shoulder inside, against which the rear end of the mainspring bears. At the rear end is a broad flange, with a clearance for the action rod. It is slotted out for the striker stud. At right angles to this slot is the safety slot for the striker stud. On the right side is a rib with an undercut groove for the stud $(d)$ on the action rod. The front end forms a bearing for the rear end of the locking sleeve. The locking sleeve fits loosely behind the flange on the bolt cylinder, the diameter of the latter being slightly reduced from $(e)$ to $(e^1)$ to lessen friction. At the front end are the two lugs, which resist the backward pressure of the cartridge on firing. The top lug is slightly in advance of the other, and their front and rear faces are cut with a screw pitch. A helical slot for the stud $(f)$ on the action rod runs from $(g)$ to $(h)$. A recess in this slot affords a seating for the stud $(f)$ when the action rod is drawn back. The action rod works in the cylinder in the body. On the right, at the rear end, is a lever for operating the rod, to the end of which lever, on either side, vulcanite knobs are fixed by screws. Underneath is a groove for the tooth of the retaining bolt. Within this groove are two projections ($i$, $k$) against which the tooth of the retaining bolt bears, in the open and closed positions, and holds the action rod steady. On the left a dovetail stud ($d$) fits in the undercut groove on the bolt cap. The rib ($l$) works in the slot between the cylinder in the body, in which the bolt works, and the cylinder for the action rod. The stud works in the helical groove ($g$) ($h$) in the locking sleeve, which it causes to revolve. The stud ($a$) enters the bolt cylinder through the slot ($m$), and lies in front of the head ($n$) on the striker.

The striker G is divided into two parts, the front part has a short point at the front end, and a head ($n$) at the other. The rear part has a button at one end, which fits into a suitable recess in the head, and underneath carries the striker stud, which works in the slot (14) in the bolt cap. The striker can be drawn back by means of the ring on the rear end.

The mainspring consists of 15 coils of flattened steel wire $\cdot 06 \times \cdot 04$ set to a length of 2·3 inches.

The extractor, H, is a flat spring terminating in the usual claw, projecting beyond the face of the bolt. It is provided with a stud with circular stem and oval head. Near each end are two small projections. To fix the extractor, place the stud in the oval hole in the bolt cylinder, with the extractor at right angles to the latter, then turn the extractor parallel with the bolt, and the head of the stud will lock under undercut grooves, and the projections will spring into seatings cut for them. A shallow recess enables the end to be raised when removing the extractor.

The ejector is a pin with a broad head and flat point, which passes through the left side of the body. It is secured by a keeper screw.

The retaining bolt, I, is pivoted underneath the small cylinder on a projection, which fits into the slot through which the axis pin passes. At the front end is a spiral spring which fits in a hole and bears against the bottom of the cylinder. On the rear end is a tooth, which is pressed upwards by the spring into the groove in the action rod. When the latter is fully drawn back this tooth locks into the front end of the groove. To withdraw the bolt push it forward slightly, and depress the tooth by pressing on the thumbpiece.

The trigger mechanism is on the double pull system and differs but little from those already discussed under turning bolt Mannlicher actions. The sear bar is pivoted at its forward end and the sear is kept up to its work by a spiral spring disposed vertically. The sear bar is rather longer than is usual. The trigger is of the usual shape and is provided with two humps for taking the two pulls. It is pivoted to the rear end of the sear bar.

The magazine is made of sheet steel and holds six rimless cartridges. The bottom is strengthened by a steel strap brazed on. In the bottom are two openings for the escape of dust, which, however, is also liable to find its way in through these openings. The magazine box is inserted through an opening in the trigger guard and is held in position by a spring catch on the right side of the box engaging on the trigger guard plate. The platform is shaped to raise one of the two columns of three cartridges higher than the other and present the cartridges in the two columns alternately at the top of the magazine. The platform can be slipped out of the box to the rear and is actuated by a zig-zag wire spring with three coils. The magazine can be filled either by the insertion of single cartridges or from a charger. The charger is made of papier mâché strengthened and protected at the bottom by a tinned iron strip, two tongues on either side of which are turned over to retain the cartridges.

The action of the mechanism is as follows :—In the first motion of drawing the action rod to the rear the tooth ($f$) moves in the straight path of the slot ($g$) of the locking sleeve ; the tooth ($a$) bearing against the head of the striker begins to draw it back, and the projection ($i$), in the groove of the action rod, depresses the tooth ($f$), which is constrained to move in a straight line, on account of the rib (1) moving in the slot between the two cylinders, bears against the curved part of the slot on the locking sleeve, and rotates the sleeve, turning the lugs from in front of their seatings into the diagonal grooves in the body. On account of the lugs and their seatings being cut on a screw pitch, the entire bolt is withdrawn about 1/16th inch, this forms the primary extraction. The tooth (*f*) has now got to the back end (*h*) of the slot in the locking sleeve, and has drawn the striker fully back. The whole bolt is now free to come back; as it does so, the lugs (19 and 19a), on the bolt sleeve, pass along the diagonal grooves and further rotate the locking sleeve, until the recess at the end of the locking sleeve groove slips in front of the stud (*f*). The latter is pressed forward into this recess by the head on the striker, actuated by the mainspring. On continuing the backward motion, the projection in the action rod groove depresses the retaining bolt tooth, and the latter then strikes the end of the groove, and arrests the backward motion. The dovetailed stud (*d*) on the action rod working in the groove on the bolt cap, serves to keep the action rod parallel with the bolt.

On pushing forward the handle of the action rod the retaining tooth in the action rod groove is depressed by the projection, and the whole bolt mechanism moves forward. When the lugs pass down the diagonal grooves the locking sleeve is rotated, so that the recess (*o*) no longer retains the stud. When the flange on the bolt cap arrives within 1/16th inch of the cylinder on the body, the sear bent engages the striker stud and the lugs on the bolt cylinder have arrived at the entrance to their seatings. On pushing home the action rod the stud (*f*) passes along the groove in this locking sleeve and revolves the sleeve, placing the lugs in their recesses, and fully closing the bolt. As the bolt is finally closed the extractor claw springs into the groove round the head of the cartridge, which has been pushed forward out of the magazine by the bolt, and the retaining bolt tooth rises behind the projection in the action rod groove, and prevents the action rod from slipping back.

By drawing back the striker, and revolving it so that the striker stud points to the left, the latter enters the safety slot in the bolt cap. The point of the striker cannot then reach the cap of the cartridge and the bolt cannot be drawn back, as the rear face of the striker stud is engaged by the shoulder in the groove in the bolt cap.

This completes the description of the Schmidt-Rubin action. Barrel, sights, stock, etc., are referred to at the end of the section.

*The Ross* (Fig. 10 and Plate X).

The Ross was, previously to the early days of the Great War, the arm of the Canadian forces, but is now obsolete. There have been various marks, some with solid lugs and some with lugs interrupted as in the breech mechanism of big guns, and with differences in sights and in other particulars.

The essentials of the action are, a bolt (1) and handle machined solid from one piece, and a turning portion (2), carrying the lugs (3) on the front end and containing the mainspring and striker. The lugs, striker and cocking-piece, which is secured by a screw thread to the end of the striker, are similar in design to those of the Mauser bolt. The Ross bolt is provided with ribs which work in longitudinal grooves in the body and prevent the bolt from turning. On the inside is a screw thread (4) which engages with a similar screw thread on the turning portion and gives it the necessary turning movement to lock and unlock the lugs. The lugs work in cam-shaped recesses in the fore-end of the action-body in the usual way.

The screw-thread on the turning portion is not cut right round it but is on two spiral ribs.

The extractor, which has the usual claw end at the front, is provided with a long tail on which is a lug which engages with the rear edge of the bolt head and takes the pull during extraction. A cam surface works against this lug and causes the claw to grip the case more firmly during extraction.

The trigger (5) is a bent lever pivoted to a frame (6) which also carries the sear (7). When the trigger is pressed the lower end of the sear is forced forward and the nose (8) therefore depressed from engagement with the bent. At the same time the stop (9), which is pivoted on the sear, is raised and engages behind the two lugs (10) on the underside of the bolt—thus preventing the bolt being withdrawn at the same moment as the trigger is pressed and serving as additional safety lugs. A double pull in the trigger action is arranged for by two ribs on the upper end of the trigger in the same manner as in the S.M.L.E. Mark III.

The magazine consists of a box enclosed in the stock, and retained by the trigger guard. It is provided with means for charger-loading, and is of interesting design. The platform (11) is supported by the spring (12) which works on a telescopic piston or plunger (13) pivoted at (14) and (15). Its movement is controlled by two arms, one of which (16) (that on the far side) is shown in the sketch. The separate axes on which these arms pivot both in the body of the magazine and in the platform are so arranged that the platform remains parallel to the axis of the rifle throughout the greater part of its movement.

In an earlier model of this rifle no provision was made for charger-loading, but, by means of a lever, the platform of the magazine was connected to a finger-piece conveniently placed on the right side of the stock to be operated by the fingers of the left hand in the firing position. By depressing the finger-piece the magazine platform was lowered and cartridges could then be dropped into the magazine with the right hand.

*Barrels.*

The length of a rifle is governed by the length of the barrel with which it is fitted. Barrels as short as 20 inches can be made to shoot well; but a certain amount of length is necessary to give a reasonable handle for the bayonet. The majority of countries use rifles of about 4·25 feet, with barrels measuring about 31 inches. Great Britain and the United States, however, are armed with short rifles just over 3·5 feet long with barrels shortened to correspond. The Swiss Schmidt–Rubin, which was a short rifle in the 1900 pattern, has been re-designed with a barrel of 30·75 inches. A well-designed short rifle has the merit of being handier for snap-shooting than a long rifle. It also obviates the necessity of providing a carbine for the mounted services. One arm for all services is a great convenience in details of storing, issue, and upkeep.

The weight of the barrel, within limits, has little effect on the weight of the complete rifle. The S.M.L.E. and the United States rifles are by no means the lightest military rifles. The weight is governed, to some extent, by the consideration that a light rifle gives an unpleasantly heavy recoil with cartridges of the usual military specification. Most modern rifles weigh, complete, without bayonet, between 8·5 and 9 lbs., yet there is more than a pound difference between the lightest and the heaviest barrels.

The following table shows the barrel length and weight, and the relation between these dimensions, of a few foreign rifles for comparison with our own Service rifle :—

| Rifle. | Bore. | Weight. | Barrel. | | |
|---|---|---|---|---|---|
| | | | Length (in.) | Weight (lbs.) | Lbs. per in. |
| S.M.L.E. | ·303 | lb. oz.<br>8  10 | 25·19 | 2·156 | ·083 |
| U.S.A. | ·300 | 8   8 | 23·79 | 2·906 | ·122 |
| Roumanian | ·256 | 8  13 | 28·56 | 3·093 | ·108 |
| Nagant | ·315 | 8  15 | 31·50 | 3·093 | ·098 |
| Japan | ·256 | 8  12 | 31·30 | 2·969 | ·095 |
| Chile Mauser | ·276 | 8  13 | 29·20 | 2·719 | ·093 |
| Lebel | ·315 | 9   3 | 31·50 | 3·187 | ·115 |

It will be seen from the above table that the S.M.L.E. has the lightest barrel in actual weight and also the lightest barrel relative to its length.

When firing, the barrel commences to vibrate before the bullet has left the muzzle. This vibration exercises a very disturbing effect on the accuracy of the shooting, on account of the difficulty of ensuring that the fore-end shall influence the vibrations to the same extent for every shot. With a heavy barrel the vibrations are not so severe and the influence of the fore-end is less important.

On account of the heavy pressures set up on firing, but which fall off very rapidly, barrels are made thickest at the breech-end. This thickened portion is called the "reinforce." The effect of this thickening of the metal on the external contour of the barrel is exaggerated by the size of the chamber necessary to accommodate the case. The exterior of the barrel from just in front of the reinforce to the muzzle is usually slightly tapered. In the German, Spanish and Turkish Mausers the barrel is turned down externally in steps, with the idea of breaking up the vibrations on firing. Barrels are browned or blued to protect them from rust on the outside and to give them a dull non-reflecting surface. In some cases the outer contour is slightly increased at appropriate points to afford a seating for the sights.

The degree of twist given to the rifling would seem to be governed a good deal by the individual preferences of the designers. It is essential that it should be enough to give the bullet sufficient spin to keep it end on in its flight; but usually there is an ample margin beyond the minimum necessary to give stability. In most rifles the twist is about 1 turn in 31 calibres. The most rapid is the French with 1 turn in 30 calibres and the slowest the Krag-Jorgensen, with 1 turn in 37·5 calibres. The Italian Carcano rifle is peculiar in that it has a progressive twist.

Full particulars of barrel length, twist of rifling, etc., will be found in Appendix VIII at the end of this book.

*Stocks, Handguards and Furniture.*

The stock is the wooden part of the rifle, in which the body and barrel are embedded for protection and convenience in handling. The wood most often used is walnut. It must be thoroughly seasoned and dry when manufactured, or much trouble will be experienced from warping.

The stocks of military rifles are usually made in one piece, but those of the British and French rifles are in two pieces. A one-piece stock has many advantages. The butt cannot become loose and shaky; it is easier to take off the rifle; the body of the rifle can be made smaller, as it is not necessary to provide a bearing for the butt; the body, stock and trigger guard can be securely joined together by two screws; and, lastly, no stock-bolt is required. On the other hand, a two-piece stock is economical, both in service and during manufacture, for if either part becomes unserviceable it can be replaced.

The fore-end is attached to the barrel by means of two bands. Where handguards are provided these are also secured by the band or bands appropriate to their length. The front of the fore-end is usually protected by a nose-cap, which may be a development of the upper band. The bar on which the bayonet is secured is usually on the upper band, the ring of the bayonet guard going over the barrel. In the S.M.L.E. Mark III and German and 1904 Mausers this bar is on the nose-cap, and the bayonet does not touch the barrel, and so has less effect on elevation and accuracy of fire when fixed.

The butt is bent down below a prolongation of the axis of the barrel, so that when the rifle is at the shoulder in the firing position the head has not to be bent inconveniently low to align the sights for the shortest range. A rather straight butt is convenient for aiming at extreme ranges. In Mauser rifles the butts are straight, thus making for economy in timber with a one-piece stock, and no additional long-range sights are provided.

The stocking up of the rifle is important for reasons already given. It is necessary, particularly with light barrels, to make provision to ensure that the vibrations of the barrel when firing shall be influenced by the stock in exactly the same way for every shot. In the Danish Krag and the Belgian Mauser the barrel is surrounded by a tube which touches at the breech and muzzle ends only. This is not the best expedient, as it increases the expenses of manufacture and repair. The casing is liable to be injured by blows, and as the sights are fixed to it they are liable to be thrown out of alignment. The barrel is more difficult to strip and rust is liable to set up under the casing. In modern rifles, such as the S.M.L.E., the Schmidt-Rubin and the U.S.A., the fitting of stock and handguards is carefully designed, so that the barrel and woodwork shall not touch in unnecessary places. The Schmidt-Rubin has a barrel-sleeve of bronze under the upper band which holds the barrel. In this rifle the barrel touches wood in no place after the reinforce.

The following table gives particulars of stocks and furniture :—

| Rifle. | Stock. | Grip. | Handguard. | Buttplate. | Nose-cap. | Cleaning Rod. |
|---|---|---|---|---|---|---|
| Mausers | 1 piece | Half pistol in German and '04 pattern | To lower band | Steel | Yes | Full length in Turkish, Belgian and Italian. Rest half length. |
| U.S.A. | 1 piece | Straight | To upper band | Steel. Butt trap for cleaning gear | No | No. |
| Mannlichers— | | | | | | |
| Straight pull | 1 piece | Pistol | To backsight | Steel | No | No. |
| Turning bolt | 1 piece | Straight. Pistol in Greek | To lower band | Steel | Semi in Dutch | Half length. |
| Krag | 1 piece | Pistol in Norwegian | No | Steel | No | No. |
| Lebel | 2 piece | Straight | No | Steel | Yes | No. |
| Nagant | 1 piece | Straight | Between bands | Steel | Small | Yes. |
| Rubin | 1 piece | Straight | To upper band | Steel | No | No. |
| S.M.L.E. | 2 piece | Pistol | To nose-cap | Iron or brass. Trap for cleaning gear | Yes | No. |

*Sights.*

The foresight usually consists of a barleycorn or blade dovetailed into a foresight block at right angles to the axis of the barrel. By adjusting the foresight to one side or the other the rifle can be made to shoot correctly for direction. In the Schmidt-Rubin carbine this lateral adjustment is made by sliding the foot of the foresight blade along a diagonal slot in the block. This method would appear to have no advantages over that usually followed, and the length of the sight base would be altered for each adjustment. Many nations use foresight blades or barleycorns of different heights, so as to be able to get the lowest elevation on the backsight correct, as minute variations in stock and fit of the bolt may alter the jump of the rifle. When the lowest graduation of the backsight is correct, the others should also be correct, as they are a definite height above the lowest. These heights are carefully worked out by repeated trial when the sighting of the arm is being settled. Any alteration in the ballistics of the cartridge used means that all rifles must have their sighting altered to correspond. An advantage of adjustable foresights over fixed ones is that they can be hardened and tempered so as to be less liable to deformation from an accidental blow. In most carbines and in some rifles, notably the S.M.L.E. and the U.S.A. rifle, the foresight is protected by wings or by a hood which forms part of the nose-cap or the upper band. The block for the blade or barleycorn is usually made part of a ring which fits over the barrel, and is either soldered and screwed or pinned to the barrel.

Backsights for military rifles usually consist of a leaf pivoted at its front or rear end to a bed, which is brazed or screwed and soldered to the barrel or formed on a tube as for the foresight. The necessary elevation to the sights can be given in several ways. In all cases a sliding piece is carried on the leaf. In some sights the leaf is pierced, or formed by two pillars, the sliding piece being cut with a V or U notch through which aim can be taken. In such cases there is usually another notch cut in a cap on the end of the leaf which can be used with the leaf down for short ranges, the leaf being raised into a perpendicular position for longer ranges. The backsight of the Charger Loading Lee-Enfield rifle was a good example of this form, except that, for convenience in fixing a wind gauge, the short-range notch was cut on a bar formed at right angles to the slide. The aperture backsight is similarly mounted in the short, magazine, Lee-Enfield Mark VI. In other cases, as in the S.M.L.E. Mark III, the leaf is solid, and is raised or lowered by the action of the sliding piece on a ramp formed on the backsight bed. Sights so made are very strong, and little liable to deformation. Provision is made for a positive locking of the sight at each range by the provision of notches on the side of the leaf in which a spring-actuated pin engages. The pin can be freed when it is required to move the sight, by means of a thumbpiece on the side of the slide. The backsight of the German Mauser is of this form. Another form of sight is that used on the Schmidt-Rubin, and some others in which the notches are cut in the sides of a projection above the bed, which take the place of the ramp. The sight can be set to any graduation by means of a lever connected with the leaf. The V is cut in the end of the leaf, which is turned up at an angle of about 45°.

A wind gauge is provided on the U.S.A. rifle and on the S.M.L.E. rifles up to the original Mark III. There is no wind gauge on the Mark III*, on the later Mark III, nor on the Mark VI. In the sight of the U.S.A. rifle there is an automatic correction for drift. This is not made on any other sight.

Peep-hole or orthoptic backsights have many advantages over those of the open or V-pattern. A sight of this pattern has long been used for the long range sight of the British rifle, but so far the United States has been the only nation to provide a peep backsight for ordinary use. There are alternative U notches on this rifle. In the S.M.L.E. Mark VI a peep-sight has been provided of sound construction and placed where such a sight should be—as near the eye as possible. The advantages of a peep-hole sight are that the eye has to focus two things only, the foresight and the object aimed at. A great deal of the difficulty of shooting is occasioned by the impossibility some men find in getting both backsight and foresight reasonably clear. With a peep even of comparatively large dimensions most men find little difficulty in seeing both object and foresight without enough blur to cause a bad aim. It is difficult, without deliberately trying, to aim with incorrect elevation, as the eye automatically centres the foresight in the hole. The elimination of optical difficulties decreases the time taken in aiming. The disadvantage of a peep-sight, when the aperture is small, is that, in a bad light, indistinct objects are rendered even more indistinct by the fact that the aperture cuts out a certain amount of light. With the provision of an aperture of reasonable dimensions, say $0 \cdot 1$ inch, this objection should disappear without detracting from any of the advantages of this form of sight.

On a few patterns of rifle a long-range sight is provided for use at extreme ranges. In the straight pull Mannlicher and the Krag-Jorgensen, this takes the form of a supplementary V on the side of the backsight slide, the foresight being a knob fixed to the side of one of the bands. The British long-range sight is described in the next section. It is not fitted to the Mark III*, the later Mark III and the Mark VI rifles.

*Carbines.*

Most countries which use long rifles provide a carbine for use by mounted troops. These differ little in action from the rifles of the countries concerned, but the barrels are necessarily shorter and bands, swivels and stocks are of a form convenient for mounted use. The following nations provide carbines for mounted troops :—France (old pattern 3-clip magazine; new pattern 5-clip magazine—of box form), Austria, Spain, Holland, Roumania, Italy (bayonet permanently attached and can be folded back against fore-end), Switzerland.

38

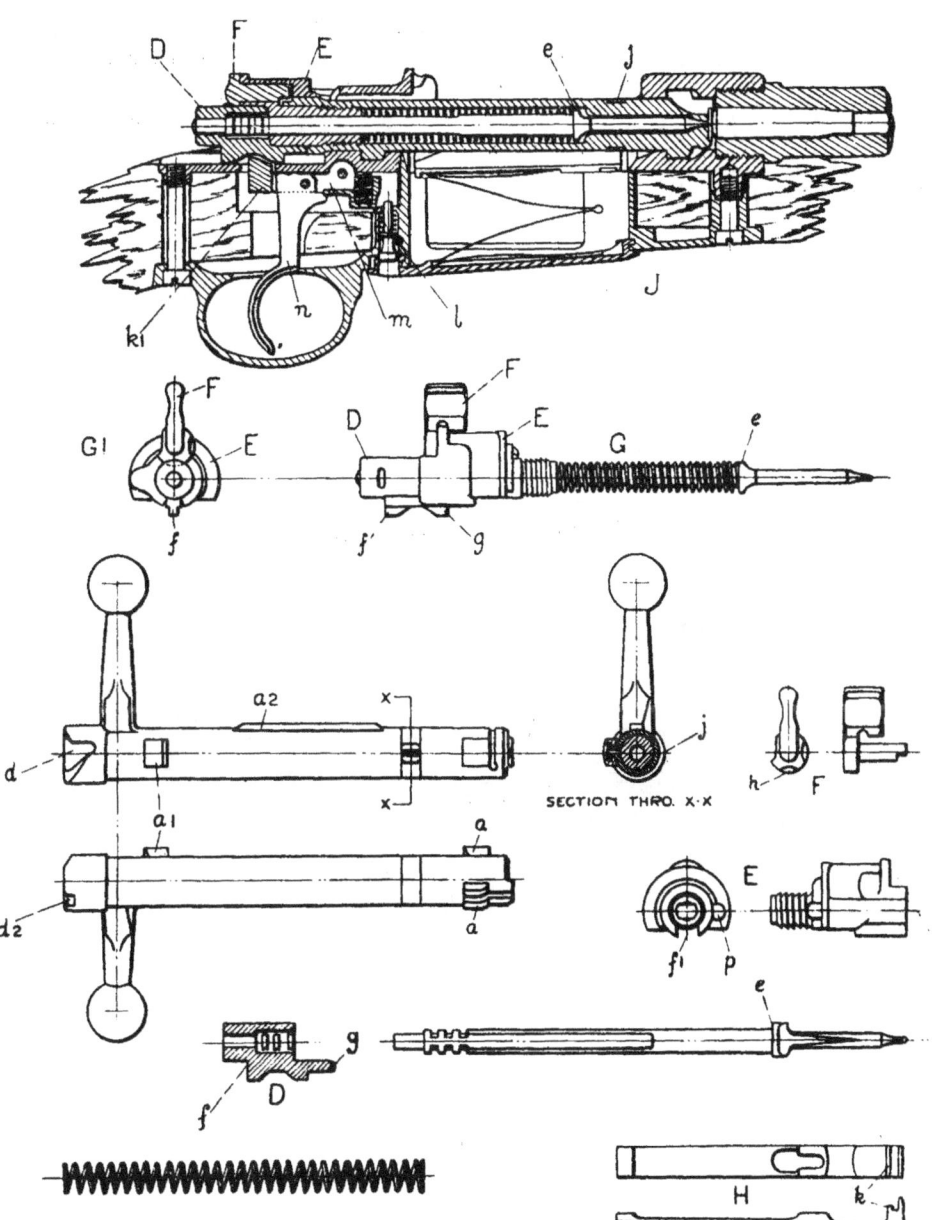

Part I, Chap. I, Fig. 1.

Mauser (Model 1904).

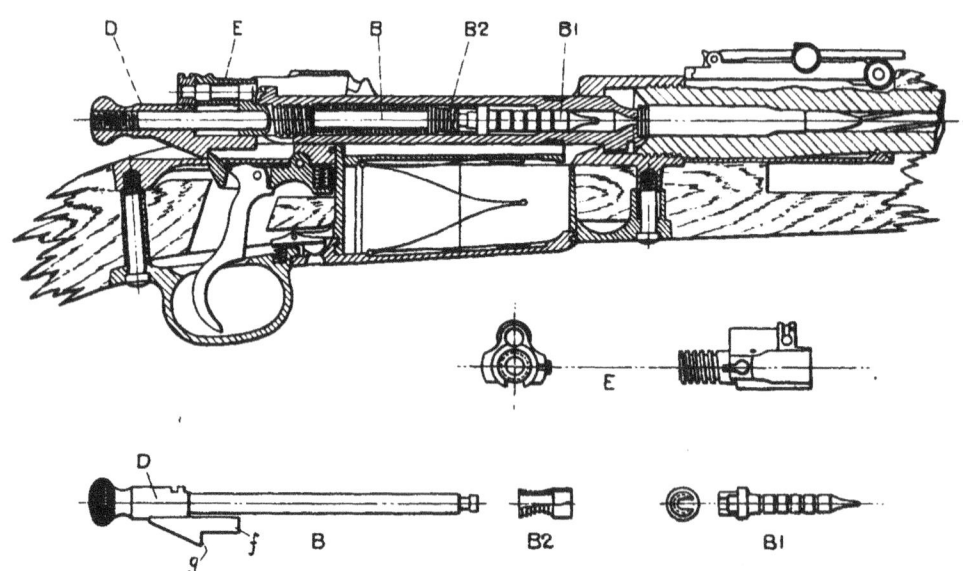

Part I, Chap. I, Fig. 2.
United States Rifle (1903).

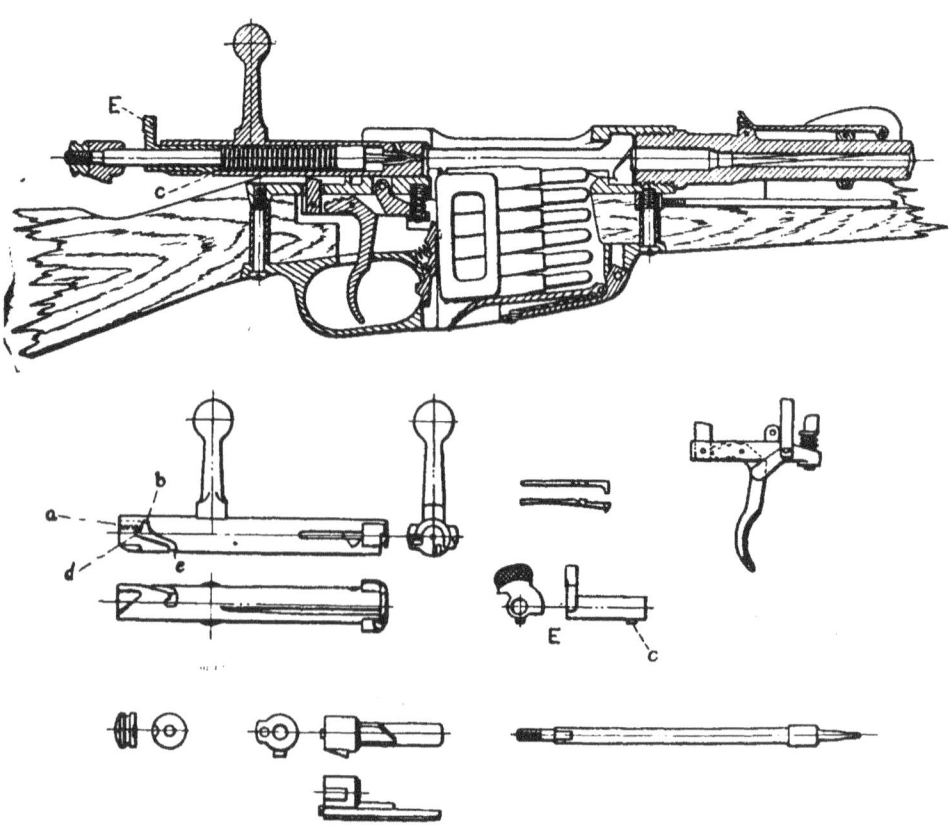

Part I, Chap. I, Fig. 3.
Mannlicher Carcano.

35

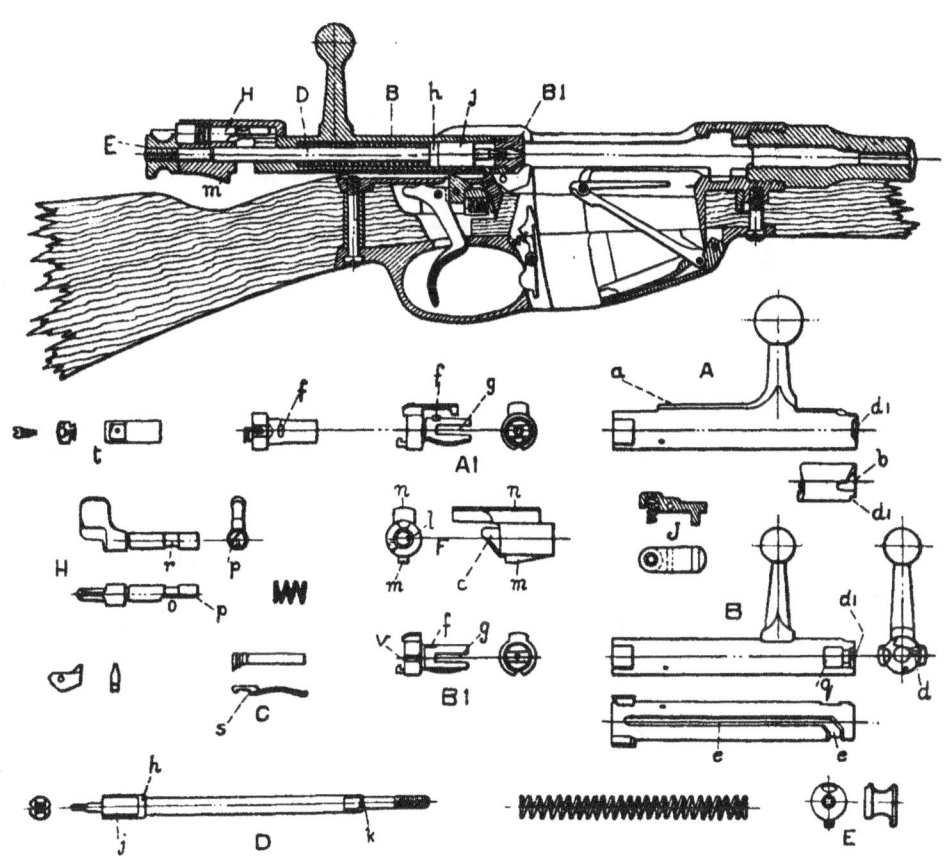

Part I, Chap. I, Fig. 4.
Mannlicher (turning bolt action).

36

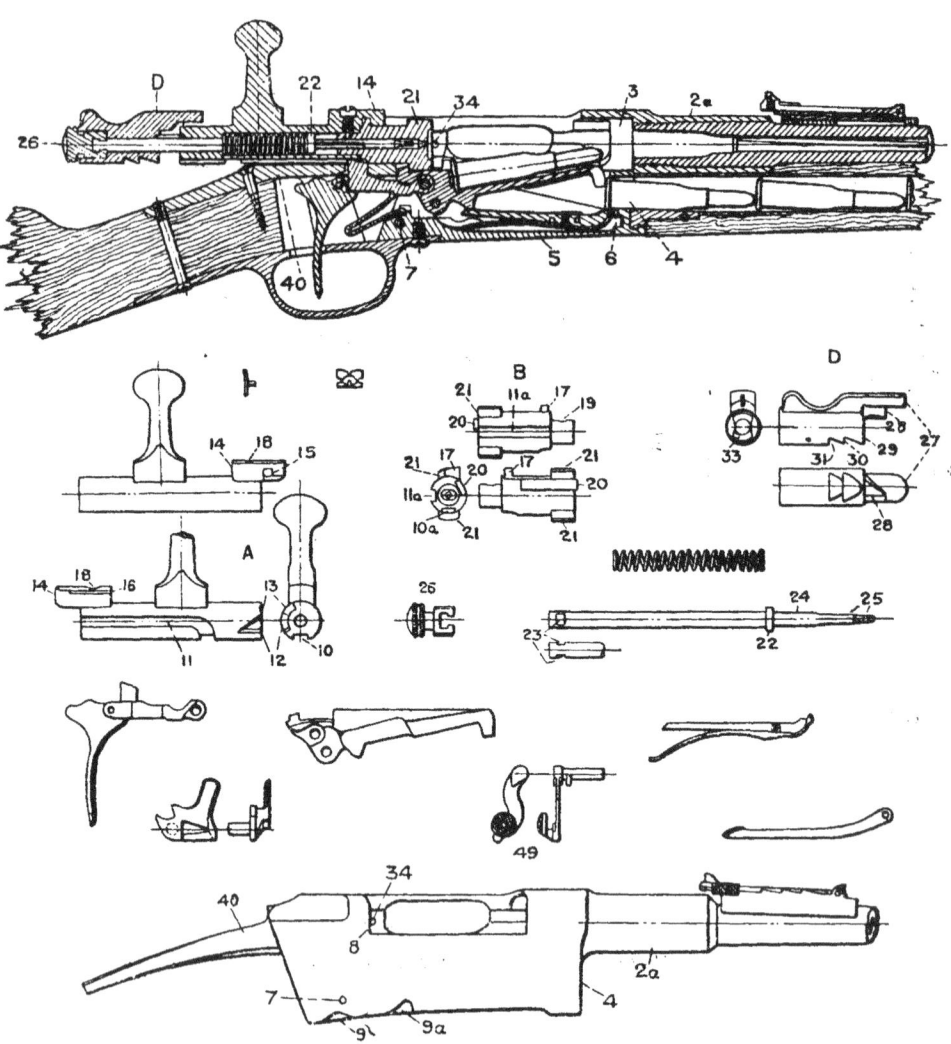

Part I, Chap. I, Fig. 5.
Lebel.

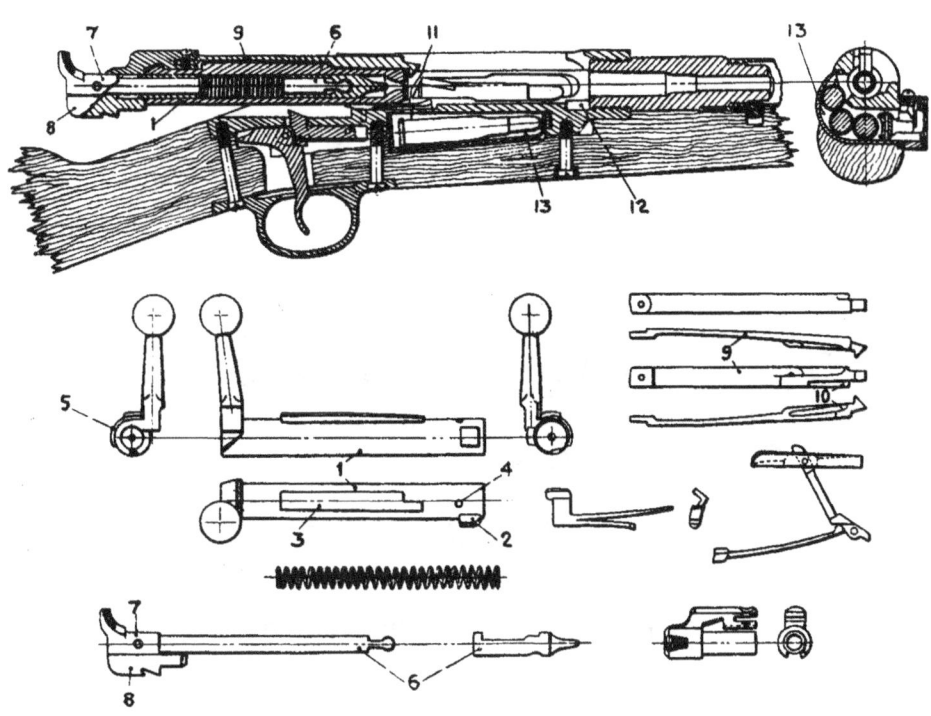

Part I, Chap. I, Fig. 6.
Krag-Jorgensen.

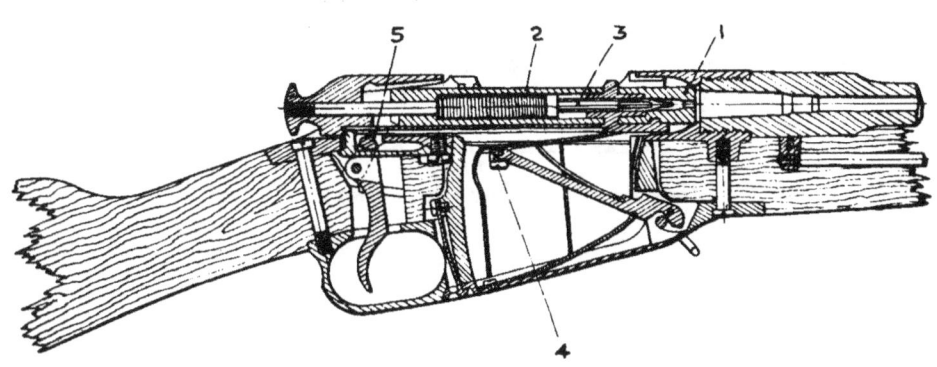

Part I, Chap. I, Fig. 7.
Nagant.

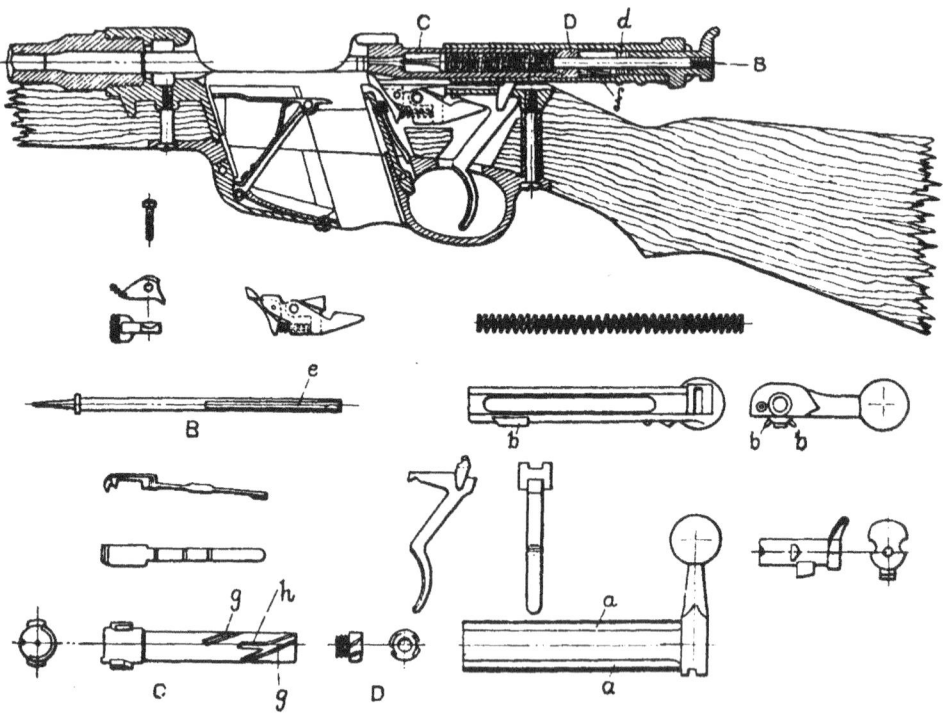

Part I, Chap. I, Fig. 8.

Mannlicher (straight pull action).

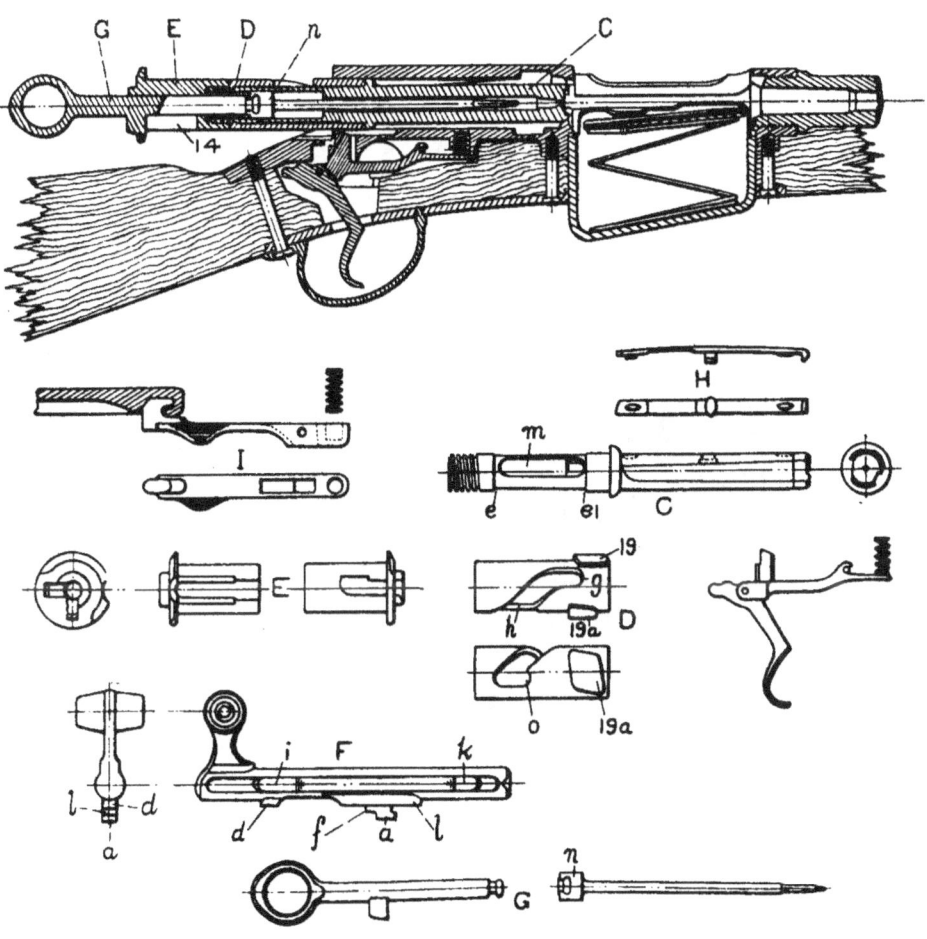

Part I, Chap. I, Fig. 9.

Schmidt-Rubin.

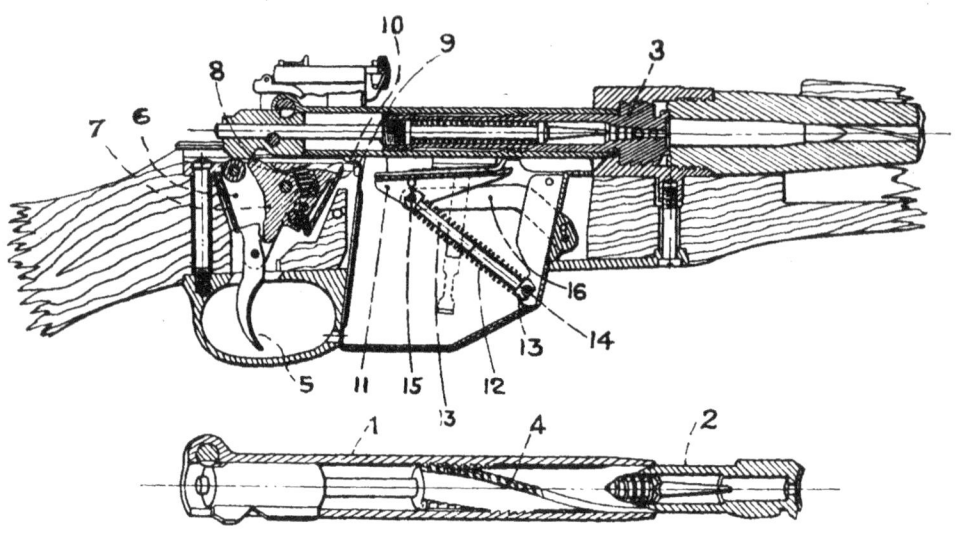

Part I, Chap. I, Fig. 10.

Ross.

[*To face page* 40.

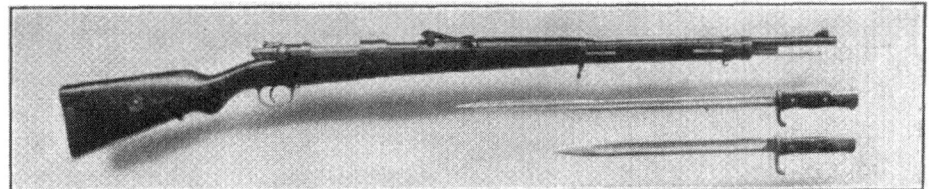

Part I, Chap. I, Plate I.—Mauser, 1898 (Germany).

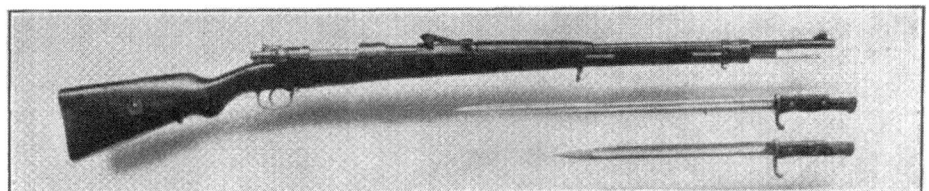

Part I, Chap. I, Plate II.—Springfield, Model 1903 (U.S.A.).

Part I, Chap. I, Plate III.—Mannlicher-Carcano (Italy).

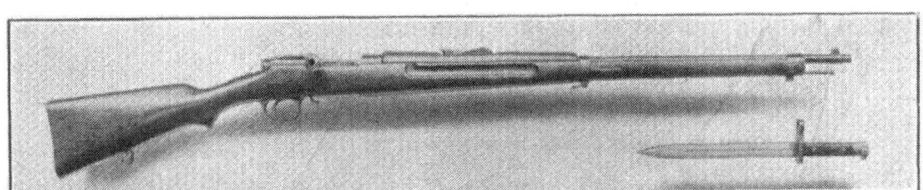

Part I, Chap. I, Plate IVa.—Mannlicher-Schonauer (Greece).

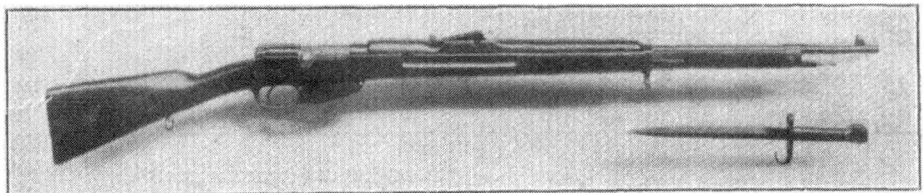

Part I, Chap. I, Plate IVb.—Mannlicher (Holland).

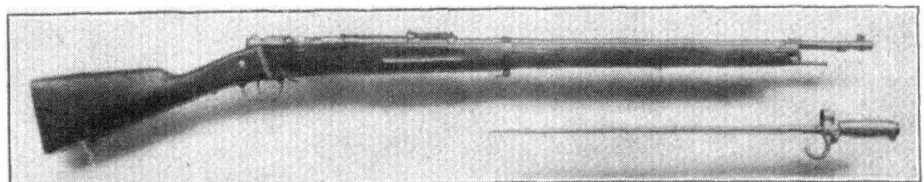

Part I, Chap. I, Plate V.—Lebel, 1886 (France).

[*To face page* 41.

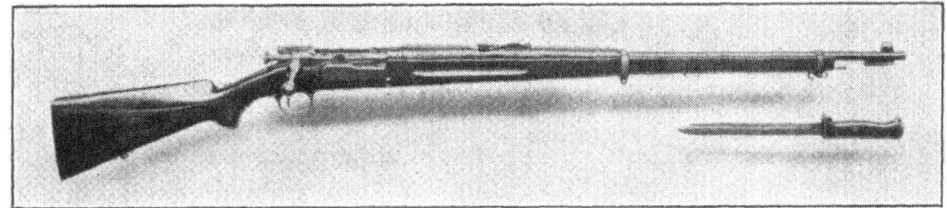

Part I, Chap. I, Plate VI.—Krag-Jorgensen (Norway).

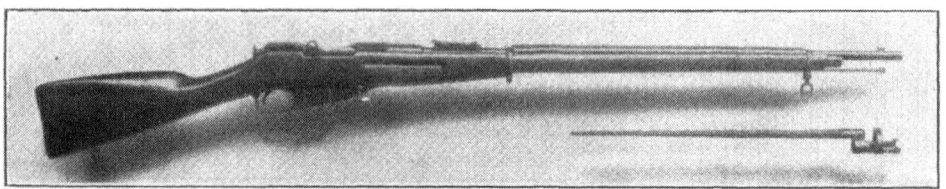

Part I, Chap. I, Plate VII.—Nagant (Russia).

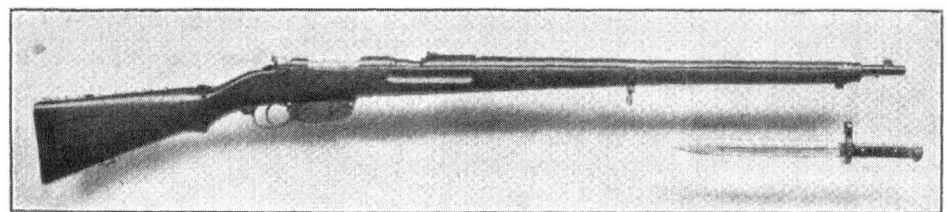

Part I, Chap. I, Plate VIII.—Mannlicher, Straight-pull (Austria).

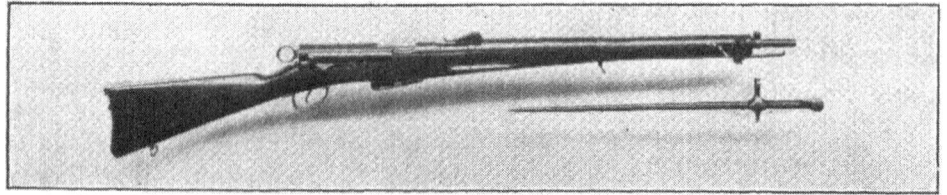

Part I, Chap. I, Plate IX.—Schmidt-Rubin (Switzerland).

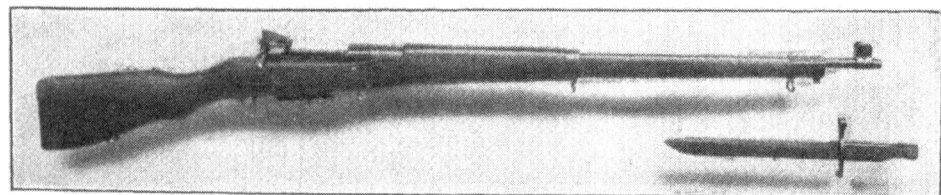

Part I, Chap. I, Plate X.—Ross.

## CHAPTER I—SECTION 3

## THE BRITISH SERVICE RIFLE

The short, magazine, Lee-Enfield rifle was approved on 23rd December, 1902, to take the place of the magazine Lee-Metford and magazine Lee-Enfield (familiarly known as the " long rifle "), various marks of which had been the service weapon since 1888. Of the rifle, S.M.L.E., there are six marks, Marks III, III*, and IV now being in use.

| | | |
|---|---|---|
| Mark I | .. | Approved 1902. New rifle. |
| Mark II | .. | A conversion from long rifle. Similar to Mark I. |
| Mark III | .. | Approved 1907. Improvement on Mark I. |
| Mark IV | .. | Conversion from old marks. Similar to Mark III. |
| Mark III* | .. | Approved 1918. No cut off. No long-range sights. |
| Mark V | .. | Provisionally approved 1922. Improvement on Mark III. Peephole backsight. (Superseded before being produced in quantity.) |
| Mark VI | .. | Improvement on Mark V. Not yet issued. |

All the above marks have the Lee bolt action, similar to that approved in 1888 for the Lee-Metford magazine rifle, Mark I. Details of the Mark III and Mark III* are as follows. (*See* Fig. 11 and Plate XI.)

The bolt (1) is simple in construction. It is cylindrical and has a bent lever at the rear end, terminating in a round knob for convenience in handling. The shape and position of the lever are very convenient for rapid manipulation. On the right side of the bolt cylinder is formed a solid rib (2), which works in a slot in the rear of the body and acts as a guide when the bolt is worked backwards and forwards. On the opposite side to the rib, and having its rear face level with the rear face of the rib, is a solid lug (3). The lug and the rear face of the rib engage against appropriate bearings when the bolt is locked and support the bolt on firing. The disadvantages and advantages of having the lugs so far removed from the face of the bolt-head have been discussed in the previous chapter. The rear faces of the rib and lug are cut on a screw pitch corresponding to the slope of their seatings on the body. On turning down the bolt this gives the necessary leverage to force the cartridge home. The lug working in its screw-pitched recess in the body provides the leverage for primary extraction on turning up the bolt. The bolt-cylinder is bored out to take the striker (4) and mainspring. The rear part of the boring is constricted to the diameter of the striker, to form a seating for the rear of the mainspring and to act as a guide for the striker. Underneath the rear end of the bolt-cylinder is a recess (5) formed by a long groove on the right and a short groove on the left, connected together in front by a cam-shaped face and separated by a stud (6). On raising the bolt lever the cocking-piece stud (7), which is resting in the right-hand or long groove, is forced backwards by the cam-shaped face until it rests in the left-hand or short groove, thus partially compressing the mainspring and withdrawing the point of the striker from the face of the bolt-head. This is necessary to prevent the striker point from firing the cap of the next cartridge as it is fed upwards from the magazine and pushed forward into the chamber.

The bolt-head (8) has a tenon which screws into the bolt-cylinder. On it is a solid projection which has a hook (9) on its right side which engages with a rib on the right side of the body and prevents the bolt-head from turning with the bolt-cylinder. On the side of the bolt-head is a hole to allow for the escape of gas in event of a blow-back or burst case. The projection on the bolt-head is slotted to take the extractor, which is a short steel bar with the usual claw at the end. It is pivoted on a screw at the rear of the slot and is kept up to its work by a small V-spring let into the slot above it.

The striker is in one piece and has a collar (10) against which the front end of the mainspring bears. The front face of the collar seats against the rear end of the bolt-head tenon in the " fired " position and thus limits the protrusion of the striker. Originally there was, in front of the collar, a small stud which fitted into a recess cut for it in the rear of the bolt-head tenon. This stud was later discarded, but both the modified and unmodified types of striker are still in use in the service. The end of the striker is screw-threaded for attachment to the cocking-piece (11).

The mainspring (12) is of coiled steel wire set to a length of $3\frac{1}{2}$ inches.

The cocking-piece has a long tongue projecting to the front and lying against the under side of the bolt when it is assembled. The front end of the tongue forms the full bent (13) and a groove cut across it the half bent (14). A stud (7) on the upper side of the tongue works in the two grooves already mentioned in the rear of the bolt-cylinder. On the left side the tongue is recessed in two places for the locking-bolt to engage in. The rear of the cocking-piece was originally formed with a circular projection, cut away on the left-hand side and roughened to serve as a grip for finger and thumb when cocking without operating the bolt. The Mark III* rifle was fitted, as an alternative to facilitate manufacture, with a cocking-piece which had its rear end formed as a flat piece, grooved on each side to give a grip for finger and thumb. Both these types of cocking-piece are now interchangeable in the Marks III and III*.

The bolt may be easily stripped for cleaning and examination. The striker keeper screw (15), which retains the striker in position in the hole drilled and tapped for it in the rear portion of the cocking-piece, having been removed, the bolt-head is unscrewed. Those strikers which are fitted with a stud on the front face of the collar can be unscrewed by means of the bolt-head, but to avoid the possibility of damage to the bolt-head tenon this method should not be employed, and the special tools with which armourers are provided should be used.

The body is cut away on the right side for the greater part of its length to allow the projection on the bolt-head to work backwards and forwards. The hook on the projection of the bolt-head engages in a slot or rib, which is cut away at the rear end just sufficiently to allow the hook on the bolt-head to be free. There is a small retaining catch forming a continuation of the rib at this point. When the bolt is drawn back as far as possible the hook can be forced up over this catch and the bolt removed from the body. The retaining catch is a small spring secured by the rear axis screw on the right side of the body. The rear of the body does not form a complete cylinder but is slotted out at the top to afford passage for the rib on the bolt and at the bottom for the bolt lug and cocking-pieces. The right-hand side of the rear of the body forms the right resistance shoulder for the rib on the bolt, and opposite this on the left side is the slot into which fits the bolt-lug, the rear face of the slot forming the resistance shoulder on this side of the body. It is cut on a screw pitch to assist in forcing the cartridge home in the chamber and in the final compression of the mainspring when the bolt is turned down into the locked position for firing. The recess for the lug is at the rear end of the body, opposite the resistance shoulder, the entrance to it is cut on an incline in the usual way to give the leverage necessary for primary extraction.

In front of the resistance shoulder, at a sufficient distance to give clearance for the bolt-head projection when the bolt is being removed, a charger-guide, in the form of a bridge (16) is riveted to the left and right sides of the body. Immediately in front of the charger-guide, the left side of the body is cut away in a semi-circle to allow the thumb to sweep the cartridges out of the charger into the magazine. The front end of the body is a complete cylinder into which the barrel screws. As no space has to be provided for the lug recesses the bolt-head enters a little distance only into this cylinder, which is not much recessed and is readily cleaned. Beneath the barrel chamber the action body is sloped off (17) to provide a way or guide for the cartridges entering the chamber from the magazine. The right side of the body is slotted to take the cut-off. There is no tang as in Continental actions. The place of the tang is taken by a socket (18) which is part of the body and projects downwards. Into it the butt fits and in the centre of it is a hollow, threaded boss (19) for the stock-bolt (2). The rear end of the body, including the upper surface of the socket, is grooved to allow passage for the lug and cocking-piece tongue. The usual opening beneath the body is provided for the magazine.

At the rear end, on the left of the body, two holes are drilled for the locking-bolt (21) and locking-bolt safety-catch (22). The locking-bolt is a stem, fitted with a roughened thumbpiece by which it may be actuated. The stem fits into a hole in the body leading into the groove for the cocking-piece tongue. The end of the stem is cut away so that when the thumbpiece is in the forward position the cut-away portion is level with the floor of the groove for the cocking-piece tongue and the cocking-piece can pass over it. When the thumbpiece is drawn to the rear the solid portion of the stem rises in the groove and engages in one of two recesses cut in the cocking-piece tongue, according to whether the latter is in the cocked or fired position. The bolt can thus be locked fired, or cocked. When the stem of the locking-bolt engages in the front, or cocked position recess, it draws back the cocking-piece slightly, removing the bent from contact with the nose of the sear. On the stem of the locking-bolt, close to the thumbpiece, is cut a steep-pitched thread (23). On this thread works the arm of the safety-catch. On the end of this arm is a short stem which fits in the hole entering the bolt-way of the body. When the thumbpiece of the locking-bolt is in the forward position this stem is within its hole and clear of the bolt in the bolt-way. When the thumbpiece is drawn to the rear, the threads on the locking-bolt stem and safety-catch arm push the stem forward so that its end enters the short groove on the end of the bolt and prevents the latter from being rotated and drawn back. By the combined action of locking-bolt and safety-catch both cocking-piece and bolt are positively locked against any possibility of accidental opening or discharge.

The ejector is a small screw which projects slightly into the bolt-way on the left side. On drawing back the bolt the back edge of the rim of the cartridge case catches against the end of this screw and is thrown out of the rifle to the right. In practice this action only takes place in the case of a bulletted round, when the case is held on the bolt face until the bullet is clear of the breech. An empty case, being shorter and therefore being clear of the barrel sooner, is normally thrown out to the right by the action of its rim frictioning against the sloping portion of the groove hollowed out in the left side of the body immediately behind the breech.

The trigger mechanism is on the double-pull system already described in the account of Continental rifles, but differs from them in most particulars, save for the provision of two ribs (24) on the upper part of the trigger. The sear (25) is a two-armed, bell-cranked lever, pivoted to the projection beneath the body on the same screw which holds the retaining catch. It is pressed to the rear and upwards by a U-shaped spring (26) which also serves to keep the magazine-catch up to its work. The long, upper, arm passes through a hole in the body into the groove for the cocking-piece tongue, and engages with the full-bent on the latter when the bolt is pushed forwards. The short arm projects downwards. The trigger is pivoted on a pin which passes through the trigger guard (27). The two ribs are on the front surface of the upper part of the trigger. On pressing the trigger the lower of the two ribs engages with the short arm of the sear and causes the latter to revolve on its axis until the end of the long arm has come close to the edge of the bent. The pull during this movement is light as the rib is close to the trigger pivot and great leverage is obtained. The fulcrum is then transferred to the upper of the two ribs which, being further from the pivot, affords less leverage, and a stronger pull is therefore necessary to make the sear move the small remaining distance which releases the cocking-piece and allows it to fly forward. The motion imparted to the sear by the motion of the trigger acting through its upper rib is, however, more rapid, and the sear is thus drawn smartly off the bent.

The action of the bolt mechanism has already been indicated in the description of the parts. The complete sequence is as follows :—

On raising the bolt-lever the cocking-piece is prevented from turning with the bolt by the fact that the tongue is engaged in the groove in the body. The bolt-head is prevented from turning by its hook engaging with the rib on the right side of the body. As the bolt is turned the cam-face at the end of the two grooves on the rear of the bolt forces back the stud on the upper side of the tongue of the cocking-piece. This draws the end of the striker clear of the end of the bolt-head and partially compresses the mainspring. As the turning movement continues the sloping face of the lug working against the sloping face of the recess in the body causes the whole bolt to move to the rear, effecting primary extraction. When the bolt has been turned as far as it will go the rib touches the left side of the body and is opposite the gap in the rear of the body. The lug is now in the groove for the cocking-piece. The bolt can be drawn back until the projection on the bolt-head strikes against the resisting shoulder. This acts as a retaining arrangement. The stud on the cocking-piece tongue has now fallen into the recess in the front end of the short groove on the bolt and the cocking-piece cannot revolve and is retained in position for entering its groove in the body. On pushing forward the bolt the full bent of the cocking-piece engages the end of the sear and the mainspring is further compressed. As the bolt is driven forward the stud between the long and short grooves on the bolt passes the stud on the now stationary cocking-piece. On turning down the bolt, the bolt is forced forward by the action of the sloping faces on the rear of the lug, and the rib working against their bearings on the body. In the complete action of closing the bolt the free forward travel is 3 inches, the travel after the sear has engaged with the bent when the pressure of the mainspring has to be overcome, is $\frac{1}{2}$-inch, and the final forward movement, on turning down the lever, $\frac{1}{8}$-inch. When the action is cocked the stud on the cocking-piece tongue lies in the long groove in the body and the cocking-piece and striker are free to fly forward when the sear is released from the bent by pressing the trigger. Should the trigger be pulled when the bolt is not completely closed the stud on the cocking-piece tongue strikes against the stud between the two grooves on the bolt and either causes the bolt to close, automatically, before the striker point reaches the cap of the cartridge, or else the two studs meet full face and the striker is prevented from flying forward. If the action is then closed by hand the sear falls into the half-bent and the action is locked owing to the two studs lying side by side, preventing the rotation of the bolt. It is possible to cock the action fully by drawing back the cocking-piece.

The magazine (28) is a detachable sheet-steel box, strengthened by two flutings on either side. It contains ten cartridges in two columns which are fed up as required by the action of a zig-zag ribbon steel spring (29). The platform (30) is so formed that the left side is higher than the right. The left-hand column of cartridges is thus presented to the bolt first and then the right-hand column, cartridges being pushed forward alternately from each column until the magazine is empty. The sides of the rear end of the box are extended slightly upwards and turned in to retain the cartridges. In the No. 1B (stamped with the figure 4) magazine there are small inturning projections made by turning over the top of the sides of the box in front. These serve to keep the platform in position when the magazine is empty. In earlier marks of magazine a stop clip is pivoted on the right side of the box, in front, which helps to keep the platform in position when the magazine is empty and keeps the bullet of the upper cartridge of the right-hand column in position when the magazine is full. This clip can be drawn down to the front when the magazine has been detached from the body, and the platform and spring can be withdrawn for examination and cleaning. The spring is secured to the No. 3 platform by a tongue of metal turned over on the right-hand side and by two rivets. In earlier marks it is secured by two tongues of metal on each side. Downward turned tongues of metal on the front and on the left-hand side at the rear of the platform serve as positioning guides. A small turned down tongue on the right side at the rear serves the same purpose. At the back of the box is a rib in which is cut a tooth (31) to engage in the magazine retaining spring catch. In the No. 1A and 1B (stamped 3 and 4) magazines there is also a small auxiliary spring which bears against the front of the trigger guard. Into the front of the box is hooked and secured the magazine platform auxiliary spring (32) which serves to keep the front end of the platform at a proper angle when the magazine is full and also protects the front of the box from being dented by the points of the bullets.

The cut-off is pivoted to a vertical screw in the projection on the right side of the body. It works in a slot parallel to and below the rib on the body for the bolt-head hook. It is provided with a cylindrical thumbpiece, bored out for lightness and ribbed on top for the thumb to grip. It is spring-tempered and set to press upwards, a small projecting flat on the upper surface acting as a catch against the side of the body and holding the cut-off open or shut. In the shut position the cut-off holds down the cartridges in the magazine out of the path of the bolt and acts as a platform for single loading. The hole in the rear of the cut-off is for convenience in manufacture only. In the Mark III* rifle there is no cut-off. In none of the marks is any provision made to indicate that the magazine is empty, as is provided in the United States rifle and in some Continental arms.

The trigger guard is attached to the body by a screw (33) passing up through a collar (34) let into the fore-end in front of the magazine, and by a small transverse screw (35) passing through ears on the bottom of the socket of the body.

The barrel, which screws into the body in the usual manner, is strongly reinforced at the breech end, which is formed into a flat on its upper surface known as the Nock's or "Knox" form (36), from an old-time gunmaker named Nock, who first devised this method of ensuring the correct breeching up of barrel to body necessary to bring the sights vertical. It is 25·1 inches long overall and weighs 2 lbs. 2½ ozs. This is the lightest barrel used in any service arm.

The rifling is of the Enfield figure with five grooves of ·0065 inch mean depth. The width of the lands between the grooves is ·0936 inch. The twist of the rifling is one turn in 10 inches left hand. The left-handed twist was originally adopted in order to compensate for the drift due to the rotation of the earth in the Northern hemisphere. It also has the effect of twisting the butt of the rifle away from the firer's cheek instead of against it.

The foresight (37) is of the "blade" pattern and consists of a plate dove-tailed into the foresight block (38) at right angles to the axis of the barrel. It is capable of lateral adjustment. The foresight block is formed with a band which fits the barrel and is kept in position by a key and cross-pin. It is set approximately ·015 inch left to counteract the lateral throw of the rifle due to vibrations set up on firing. The backsight (39) is attached to a bed (40) which encircles the barrel, to which it is fixed by a cross-pin in the middle. It is also supported by the sight spring screw. The sides of the bed are raised to form a ramp (41). The leaf (42) is a solid piece of steel pivoted to the bed in front and kept in position by a spring fitted into the bed. It can be turned over on to the hand guard and rebounds into correct position when it is brought past the vertical. It is graduated from 200 to 2,000 yards. On the top left side of the leaf are lines representing every 25 yards. On the top right side the lines represent every 100 yards. The odd figures from 300 to 1,900 yards are omitted. A slide (43) fitted with a spring catch and a fine adjustment worm-wheel (44) enables the sight to be set at any elevation. The right side of the leaf is cut with screw-thread notches, and in these the fine adjustment worm-wheel engages. By pressing a catch on the left side of the slide the fine adjustment is released, and the slide may be moved quickly along the leaf by the action of the thumb only. The periphery of the worm-wheel is divided by 10 thumb-nail notches, the distance between each notch representing 5 yards in range, *i.e.*, 5 notches equal 25 yards, or one division on the left side of the leaf. One complete revolution of the fine adjustment worm-wheel moves the slide 50 yards.

A wind gauge was originally fitted on the rear end of the leaf, but has been discarded. It was held in position by the wind-gauge screw. The scale was marked in divisions representing 6 inches deflection on the target at 100 yards. Each quarter-turn of the wind-gauge screw represented 1 inch of deflection for every 100 yards of range at which the sight was set. At each quarter-turn a friction spring engaged in a nick inside the head of the screw, checking its rotation. A U-shaped notch was cut in the top edge of the slide, and the face was roughed to prevent the reflection of light.

In the Mark III* and in the later Mark III rifle there is no wind gauge. Its place is taken by a cap which is attached to the leaf by means of a screw. It is provided with a U-notch and roughened on the face. There are two patterns of backsight cap which differ slightly in form.

Long-range sights were provided in the earlier Mark III, giving elevations from 1,700 to 2,800 yards. The backsight consisted of an aperture attached to the left side of the body. It was carried on a bar terminating at the upper end in a cup-shaped button through which a peep-hole was bored. It was pivoted on the stem of the locking bolt and kept in position by a spring. The foresight, known as the dial sight, was attached to the left side of the fore-end, and consisted of a dial on which the ranges were marked, a pointer, and a bead which acted as a foresight.

No long-range sights are fitted to the Mark III* rifle or to the later Mark III.

The stock is in two pieces. The fore-end (45) is held to the barrel by a nose cap (46) and outer band (47), which are fitted with swivels. The swivel of the nose-cap is a piling swivel, *i.e.*, cut away in the centre. A swivel is also fitted on the butt. Naval service swivels are made slightly larger than for land service. The barrel being comparatively light, accuracy is liable to be detrimentally affected by a badly fitting fore-end. In the assembled rifle there are three important metal-on-wood bearing points where even bearings must be ensured. They are as follows :—

(1) The thrust of recoil is received by the stock, through the medium of the sear lugs on the body, on the resistance shoulders formed a little in front of its rear end. It is essential that the thrust should be taken up evenly on both sides.

(2) The barrel must be held firmly down on the fore-end at the reinforce. This is effected by the fore-trigger-guard screw, which is fitted with a collar which limits the amount of " crush " which can be obtained on the wood by tightening. It must be noted that in the case of a shrunk fore-end there is a danger of the screw being screwed up tightly against the collar without pulling the barrel down tightly on the fore-end. Careful fitting is therefore necessary. It is also important to remember that, since the trigger is mounted on the trigger guard, a loose fore trigger-guard screw may affect the " pull-off " of the rifle by allowing the front end of the trigger guard to drop, and thus slightly affecting the relative positions of the trigger and the tail of the sear.

(3) The barrel is caused to bear lightly on the woodwork $\frac{1}{2}$ inch in rear of the inner band (48) by means of a spring acting through the medium of the latter.

Between (2) and (3) the woodwork is hollowed out so as to be clear of the barrel, and from the inner band forwards the barrel is held away from the fore-end by the fore-end spring stud, the hole in the nose-cap being slightly oval in form to give the necessary clearance.

The foregoing is a brief description of the service stocking. For match shooting under N.R.A. conditions, with private rifles, certain other methods have been evolved. These other methods have for their object the stiffening of the barrel and the damping of the vibrations set up on firing in order to attain a relatively high standard of accuracy for the special conditions of target shooting. They consist, briefly, in adopting some means of packing the barrel between the fore-end and fore-hand guard. The service stocking was evolved with the object of ensuring a consistently satisfactory standard of accuracy under service conditions.

A backsight protector formed with two upstanding ears, roughened on top so as not to reflect the light, is let into the fore-end and secured by a vertical screw and nut.

The nose-cap completely encircles the barrel at the muzzle and is provided with a tang which projects backwards under the fore-end and carries the piling swivel at its rear end. Immediately in front of the piling swivel is a sword bar for the attachment of the pommel of the bayonet and in front below the muzzle is a boss on which the ring of the bayonet cross-piece fits. The sword bayonet is thus fixed underneath the rifle to the nose-cap only and does not touch the barrel. The nose-cap is provided with high wings roughened on top, which protect the foresight. It is pierced on either side beneath the wings for lightness.

The hand-guard (49) extends the full length of the barrel and is divided into two parts by a saw cut opposite the backsight bed. This is for convenience in fitting and removing. The rear portion fits over the barrel and is held in position by means of a spring riveted on to it. The front end of the hand-guard is strengthened by a sheet steel cap which fits under a recess in the nose-cap. A groove is cut in the correct position for the jointed outer band, a slot being formed to give clearance to the hinge. The hand-guard does not touch the barrel as the groove is of greater diameter than the barrel. An inner band is carried permanently on the barrel; it is grooved out so as to touch the barrel in two places, and is fixed in the groove of the fore-end in rear of the lower band by means of a screw, the head of which bears against a strong spiral spring.

The butt is attached to the socket on the action body by means of a stock bolt (20) that is inserted through a hole drilled longitudinally from the butt end. It is made in three ordinary lengths, long and short butts being marked by the letter L or S stamped on the wood on the top of the butt. For special use during the Great War, butts shorter than the ordinary short butt were made. These were termed "bantams," and stamped with the letter "B."

The butt plate is of brass, forgeable alloy or malleable iron. It is finished polished or zinc electro-plated according to the materials employed in manufacture. It is fitted with a trap for the insertion of oil bottle and pull-through. A marking disc is fixed with a screw to the right side. The "grip" is of a special "semi-pistol" shape.

The Rifle, Short, Magazine, Lee-Enfield Mark V was an improvement on the Mark III, but although a certain number were produced in 1925 none were issued, and it was later discarded in favour of the more fully developed Mark VI. The Mark V differed from the Mark III in several particulars, the chief of which were (1) the adoption of an aperture backsight located on a specially-designed bed on the body behind the bridge charger guide; (2) the making of the hand guard in one piece completely covering the barrel.

The Mark VI is the outcome of experiment, since the Great War, but as yet has not been produced in quantity. At the present juncture a detailed description cannot be given, but the essential features in which it differs from the Mark III are as follows:—

1. The aperture backsight of the Mark V has been retained in a modified form.
2. The nose-cap is of very much lighter design than that of the Mark III.
3. The method of stocking has been simplified.
4. The barrel is considerably heavier.
5. The bayonet is much shorter and lighter than that used with the Mark III, and fits directly on to the muzzle of the barrel, which projects a short distance in front of the fore-end.
6. As in the Mark V, the hand guard is in one piece.
7. Without the bayonet the rifle is about 8 ozs. lighter than the Mark III. The weight of the bayonet is about 1 lb. less than that of the bayonet of the Mark III.

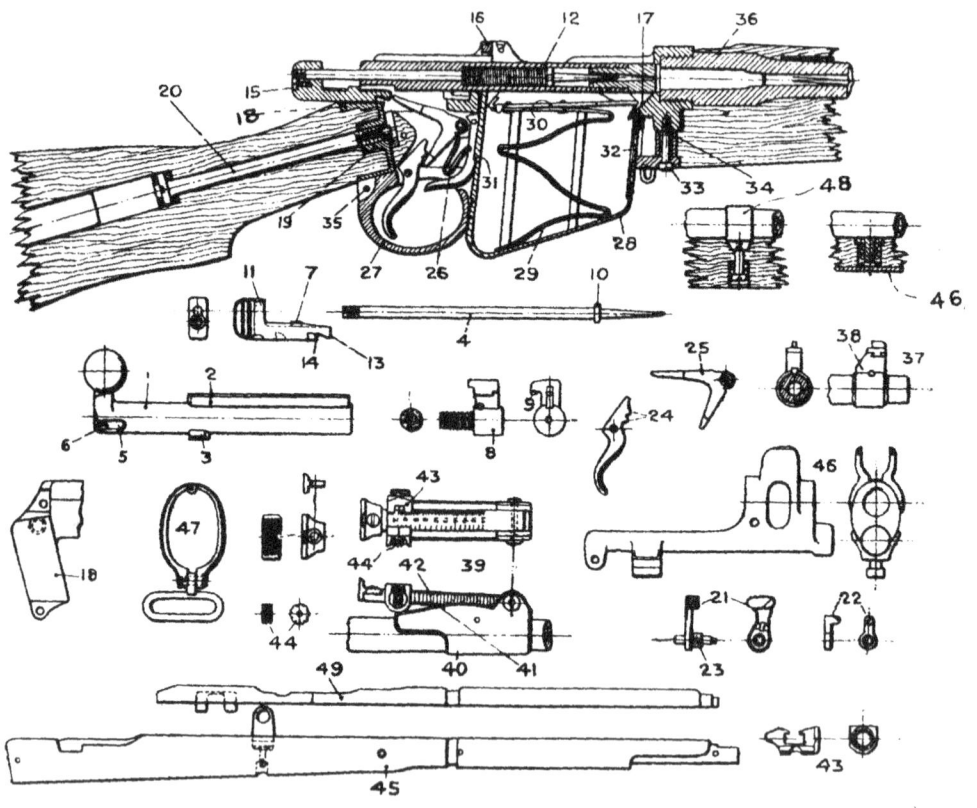

Part I, Chap. I, Fig. 11.

Short, Magazine, Lee-Enfield Mark III.

[To face page 48.

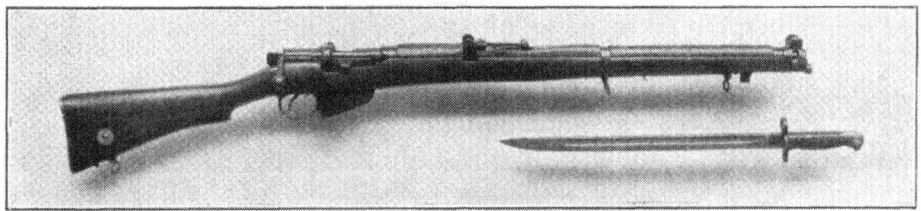

Part I, Chap. I, Plate XI.—S.M.L.E., Mark III (Great Britain).

## CHAPTER I—SECTION 4

## OUTLINE OF MANUFACTURE AND INSPECTION OF THE SERVICE RIFLE

In the Rifle, Short, Magazine, Lee-Enfield, there are, in round numbers, one hundred and thirty parts. The manufacture of each of these parts is governed by details laid down in specifications and drawings that are prepared after the preliminary business of deciding upon and approving the design of the rifle. The specification gives exact particulars of the materials to be employed in the making of each part; the stages of manufacture at which each part will be viewed and inspected; details of markings, etc.; and the conditions governing the acceptance of the finished weapon. The drawings show the exact shape and dimensions of each part with the amount of "toleration" allowed, and details of assembling.

The rifles are made either by The Royal Small Arms Factory or by private firms of manufacturers. The object of the manufacturer is to turn out a serviceable weapon, well up to the specification, at as low a cost as possible. In order to ensure the necessary qualities of accuracy of shooting, safety and wearing qualities of the various components and interchangeability of parts, an independent inspection is given to the rifle by a special department charged with that duty.

The materials employed are steel, iron, brass and wood.

Steel is a variety of iron containing carbon to the extent of from $0 \cdot 1$ to $2 \cdot 0$ per cent. Other substances are present in varying proportions, but the carbon is the most important ingredient, and it is to the presence of carbon that steel owes its most valuable property, that, by certain heat treatments and methods of cooling, it can be brought to any desired state of hardness or softness within very wide limits. Steel is classified as "mild" up to about $\cdot 25$ per cent. carbon, "medium carbon" up to about $\cdot 7$ per cent. carbon, and "high carbon" above about $\cdot 7$ per cent. carbon. As the property of "hardening" increases, up to a certain point, with the percentage of carbon contained, a high carbon steel is employed for such important components as the bolt, cocking-piece and sear, which have to withstand considerable pressures or have wearing surfaces. When the component is required for work of a springy nature, such as cut-off, retaining spring for bolt-head, or sear-spring, a special spring steel is employed, as this quality of spring cannot be imparted to a mild steel. In all about sixteen different varieties of steel are employed in the manufacture of different parts of the rifle. Of these the highest carbon content is $1 \cdot 2$ in a steel used for certain springs, and the lowest $0 \cdot 2$; an extra mild steel is, in some cases, used as an alternative to wrought iron for the bolt-head.

The specification for the steel from which the barrels are made calls for a composition as follows:—

| | |
|---|---|
| Carbon | From $0 \cdot 5$ to $0 \cdot 6$ per cent. |
| Silicon | Not above $0 \cdot 25$ per cent. |
| Manganese | From $0 \cdot 5$ to $0 \cdot 7$ per cent. |
| Phosphorus | Not above $0 \cdot 03$ per cent. |
| Sulphur | Not above $0 \cdot 03$ per cent. |

To these and most of the other steels employed a very rigid tensile strength test is applied. In the case of barrel steel the yield point must not occur at under 26 tons to the square inch and the test piece must show an extension of not less than 15 per cent. before fracture. In laying down the specification for the various kinds of steel two points have to be considered. One is the function of the components in the finished rifle, and the other is the work to be done on it in the manufacturing processes, and, conjointly with this, the cost of material and the wear of tools. Where possible, a cheap and easily worked mild steel is used.

The iron and steel is obtained from the manufacturers in various forms, depending on the subsequent operations. For example, the action-bodies are made from long, rectangular bars, the triggers from bars whose cross-section resembles that of the finished component, and the barrels from moulds, or round bars about a foot long. Magazine cases are stamped out of steel sheet and subsequently bent into shape and brazed, the smaller pins are made from steel wire, as also is the mainspring, whilst the springs for the magazine rear hand-guard, etc., are made from ribbon-steel.

The principal manufacturing processes, once the material has been delivered at the works, are forging and machining. Forging is used to bring the metal to a shape roughly approximating to that of the finished component and is also useful in improving the homogeneity of the steel. Were it not for this latter consideration much work might be saved by machining straight from the rough steel in very many cases where the component is not of eccentric form.

Drop forging is generally employed. In this the bar, heated to a cherry red, is brought roughly to shape between two dies specially prepared for each component, one being held in the anvil and the other in the hammer. The hammer is actuated by steam or electricity. Several operations are generally necessary to bring the forging to its finished form. The preliminary stages in the manufacture of the barrel are either forging or rolling or both. In either case special tools are used to elongate the mould gradually to approximately the correct shape and length. The "Knox form" is obtained by drop-forging.

The machining of rifle components is very representative of the best class of modern repetition work and is performed on simple and capstan lathes, and on milling, drilling, profiling, slotting and boring machines. Most components have to undergo several operations, some of them a very large number; the finished action-body, for instance, being the result of about one hundred and fifty separate operations.

Drilling consists of cutting a hole by means of a drill, the point of which is shaped to form the cutting edge or edges.

Boring consists of enlarging a circular hole by means of a boring bar, the sides of which form cutting edges, or from the sides of which separate cutters project.

In milling, a hardened steel wheel with a series of cutting edges on its rim or on one or both its sides, is revolved, and the work is automatically fed forward or rotated as the cutting proceeds.

In shaping, the work is held stationary whilst the cut is being effected by the tool. The tool is given a reciprocating motion but the cut is only made during the forward stroke.

Slotting is somewhat similar to shaping, but the tool is given a vertical motion, and the process is used for enlarging a circular hole that has been drilled previously, in any direction by traversing the work.

It is obvious that there must be a relative motion between the tool and the work, and, further, that one or both of them must be moved to give the requisite feed to bring or keep the tool in contact with a fresh surface or portion of the work. Much ingenuity has been expended on the design of machines to perform these motions automatically or semi-automatically, enabling a large number of components to be made with a minimum of labour. It should be understood that the cut, or amount removed, is necessarily small, and that it is because of this, as well as for reasons connected with the shape of the component, that the large number of operations previously mentioned are necessary to complete some portions of the rifle. It is one of the chief concerns of the factory staff to arrange the sequence and number of these operations so as to keep down expenses to a minimum.

A machine once adjusted to perform a certain operation is, of course, not capable of repeating the same indefinitely and reproducing in large numbers components in any particular stage, all absolutely identical. The wear of cutters and fixings proceeds slowly but surely, and it becomes necessary to fix manufacturing limits within which the serviceability of the article is not appreciably affected. These limits, which are generally within one or two-thousandths of an inch, vary with the importance of the component and the stage at which it has arrived.

In order to facilitate the work of cutting the component to shape it is usual, when possible, to reduce the hardness of the material temporarily. This is effected by annealing, which consists in heating the metal and allowing it to cool slowly.

No object would be served by detailing the operations of machining the various parts, and there is, indeed, no space available in this work for such a list of mechanical details. It may be mentioned, though, that the amount of metal removed in the process of converting the forging into the finished component is often greater in weight than the component when finished. To revert again to the action-body. The forging for this weighs in the neighbourhood of 5 lbs. The finished body weighs 1⅛ lbs. only.

The machined components are finished in various ways depending on the use to which they are put in the completed arm and the position they occupy. Most of the metal work which shows on the exterior of the weapon is either blued, blacked or browned to preserve it as far as possible from rusting, and to prevent the reflection of light. These processes are often a part of the hardening or tempering processes which must follow on machining to restore either the hardness or temper lost in the working of the steel. Hardening is effected by heating the component to an appropriate temperature and "quenching" it by suddenly dipping it in oil or some other liquid. This leaves the metal so brittle that small components can be snapped betweeen the fingers and a further tempering operation is necessary. This consists of again heating the metal to a less heat than before and allowing it to cool. The amount of temper is governed by the heat to which the article is raised and the rate of cooling. In case-hardening a very hard surface is given to such parts as are made of mild steel or iron, and do not contain sufficient carbon to enable them to be hardened as just described. The conditions under which such components are usually used are such that a hard steel would be liable to fracture. In case-hardening the components are packed in a box containing bone dust or some other organic source of carbon and are subjected to a lengthy heating at a high temperature. They are then quenched.

"Oil-blacking" and "blazing-off" are effected by heating the component, dipping it in oil, and burning off the oil that remains after draining. This leaves a carbonaceous deposit on the surface. Browning is a chemical process, the surface being oxidized by acids. Blueing is effected by the oxidation of the surface by heat.

The boring and rifling of barrels was formerly regarded as a "mystery," knowledge of which was confined to a few only. It is now a mechanical operation differing only in degree from the making of other components. Naturally, as the barrel is the most important component, considerable care is bestowed on it; but as, after certain limits of accuracy in manufacture have been passed, there is no means of knowing whether the barrel will be an exceedingly straight shooting one, or whether it will only pass the ordinary tests satisfactorily, there is nothing to be gained by an over-elaboration of detail.

When it comes from the forge the barrel is rough-turned to exterior shape and drilled. The drilling may be done "straight through" from end to end in one operation, or the drilling may be started from either end and meet in the middle. In drilling and boring the bits and bars slide horizontally along the bed of the machine whilst the barrel is caused to revolve. A lubricant is pumped into the hole as the work proceeds, and washes out the "swarf." The bore must be straight and concentric with the exterior. The straightness is tested by light and shade effects caused by the multiple reflections of the muzzle down the bore. The slightest want of concentricity in these repeated circles is an indication that some correction must be applied. This correction may be applied by means of blows struck on the exterior of the barrel with a copper or brass hammer, but modern manufacturers favour the use of a special barrel straightening machine. To test the concentricity of the bore with the exterior the barrel is spun upon plugs entering the bore.

The grooves of the rifling in the bore are cut one at a time. A very light cut is made, only about one-thousandth of an inch being removed at a time, so that it requires from five to seven cuts to form each groove. The cutting tool is pulled through the barrel and is, at the same time, given the necessary spiral movement. Much ingenuity has been expended in making the operation of rifling as nearly automatic as possible.

The rifled barrel is then screwed at the end to form an attachment to the body. In this operation care must be taken that the screw thread is in proper relation to the Knox form, so that the barrel and sights come upright in breeching up. The cartridge chamber is next rough bored, and the sights, back and front, are fitted.

The barrel and action are assembled preliminary to the proof, which is the firing of a cartridge giving a pressure 25 per cent. above that of the ordinary service cartridge. As well as testing the strength of barrel and action, the proof charge has the effect of bringing the lugs of the bolt into close and intimate relationship with the groove and resistance shoulder of the body. An uneven bearing at these points may affect the shooting of the rifle and for this reason, in addition to the fact that cartridge head space may be affected by a change of bolt, care should always be taken to ensure that a rifle has its own bolt. The proof charge also sets up the chamber which is the reason why it is left slightly smaller than the finished size before the proof charge is fired. After proof, the barrel, action-body bolt and bolt-head are examined to see that they are sound, and they are then marked with the proof mark. The chamber is next brought to finished dimensions and the barrel is browned. The barrel and body, which have been numbered, together with the bolt, are breeched up again, the barrel having previously been lapped or polished with a lead plug carrying fine emery and oil. The lead plug is made by pouring molten lead down the barrel on to a plug of tow on the end of the lapping tool. In lapping the bore is given a fine polish. Care must be taken that the bore is not unduly enlarged. The stock and furniture, magazine, etc., are next assembled to the completed barrel and action, and the rifle is finished.

The stock—butt and fore-end—and the handguard, are of walnut. The wood, selected as to straightness of grain, is delivered square cut and roughly to form. In normal circumstances it is stored for a period of three years for seasoning. The rough blocks are turned roughly to the contour of the finished component in copying lathes or moulding machines. A wheel and a revolving cutter are arranged so that their movements conform exactly. The former runs over a revolving iron dummy having the shape of the finished stock, and the latter cuts a similarly revolving block of wood to the same shape as the dummy. The finish turning and final " marking off " operations are done on similar machines. The wood is finally hand smoothed by being held against revolving brushes loaded with powdered glass, or against a rapidly-revolving canvas band dressed with powdered glass. The finished woodwork is soaked in linseed oil for half an hour. Profiling, drilling and slotting all have their place in the manufacture of the wooden components, just as they do in those of metal, and all work must be within narrow limits of accuracy in order that the various parts may be assembled without much trouble in the fitting.

Throughout the work, and in every stage of it, gauging forms a most important part of the operations of manufacture. The machine hands are provided with gauges to check the work as it comes from the machines, manufacturers' viewers and foremen keep a constant check on the work by means of other gauges, and, finally, the viewers of the Army Inspection Department gauge all the parts. A special set of reference or master gauges in the hands of the Inspection Department is used, with the greatest reservation, for the special purpose of checking the working gauges and preventing the latter from being used when wear or other causes have rendered them unreliable. The gauges employed number, all told, about 1,250.

The stages at which Government inspection by the Inspection Department shall take place is rigidly laid down in the specification. The barrel, for instance, must be submitted for view in nine stages of manufacture. These are: finished bored, rifled, screwed, first chambered, sights fitted after proof, browned and finished chambered, breeched up, and finished sighting. The bolt and body are each inspected twice. The woodwork is inspected in the finished stage only. In addition, the complete rifle is inspected, and a certain number are picked for 600 yards' shooting tests.

The points of importance in the finished viewing and final inspection of the rifle are :—

(a) To see that all components are complete, and are marked with a view mark, and number and view mark where applicable.

(b) To see that the various parts fit and bear correctly. In this connection the fit of the barrel, fore-end and nose-cap, or " stocking up," is of special importance. The object is to ensure that the fore-end does not exert a variable influence on the vibration of the barrel. Any difference in this particular must affect the accuracy of the weapon, and is therefore carefully checked.

(c) The working of the action is examined, and the magazine, charger-guide and extraction tested with the aid of dummy cartridges and fired cases.

(d) The weight of the pull-off and certain springs is tested.

(e) The general finish is examined, and the sights are tested. The working of the slide and any other movable portions of the backsight are tested.

All rifles are fired at 100 feet for accuracy, and to test the setting of the sights. Trial shots are first fired, and, if necessary, the foresight blade is adjusted laterally, within limits, and it may be replaced by a higher or lower one, if necessary. Six heights of foresight blade are provided, varying by increments of 0·015-inch from 0·955 inch to 1·03 inch above the axis of the barrel. Five rounds are then fired, and four out of the five shots must come within a rectangle 1 inch broad and 1½ inches high. Failing this the rifle is rejected.

Ten per cent. of the rifles are then fired at 600 yards, any which gave doubtful results at the shorter range being included. In this test nine out of a group of ten shots must lie within a two-feet circle.

The shooting tests are done from a mechanical rest, designed to approximate as closely as possible to the conditions in which a rifle is ordinarily fired, as regards points of support, recoil, etc. Means are provided for laying the rifle by means of hand wheels, and, with the aid of a special telescope laid on the blade of the foresight and bearing on the upper surface of the backsight cap, a true, full regulation sight is obtained. After laying, and before firing, the telescope is removed.

After passing all tests satisfactorily the blade of the foresight is fixed in position by the front edge of the foresight block being punched into a recess cut in the lower portion of the blade.

The rifles are once more given a general inspection, and are then packed and passed into store.

## CHAPTER I—SECTION 5

### PRACTICAL CONSIDERATIONS AFFECTING THE ACCURACY OF THE RIFLE

It is only within recent times that the movement of the rifle on firing and its effect on the line of projection of the bullet, have been well understood. Early artillerists could not trace any fall of the ball from a cannon in the early part of its flight, partly because the fall was in fact small, and no doubt partly also because, when the gun had been carefully aligned on the mark, it was assumed that the projectile quitted the muzzle in that line ; this encouraged, if it did not originate, the theory of a " point blank " range, over which no fall occurred.

The true explanation, that the whole weapon moved before the missile had quitted it, was long in coming ; its formulation is due to Metford, who was puzzled to find that the sighting was incorrect when the sighting line was made parallel to the axis of the barrel.

A service rifle fired from the shoulder recoils about 0·1 inch before the bullet quits the muzzle, the point of resistance to its movement being below the centre of gravity, which has then already begun to rise. The barrel is not, as is often supposed, a rigid body incapable of bending ; it is very sensitive to stress. Such pressure as can be put on it by a finger will bend it perceptibly. Like any other metallic rod or tube, a rifle barrel vibrates when struck, and, owing to its proportions it more closely resembles a rod than a tube in the manner in which it vibrates. It is with its vibrations as a rod rigidly fixed at one end that we are more immediately concerned. In such a case there are two types of transverse vibration which can be set up (1) a fundamental vibration, in which the whole length of the barrel vibrates as a single unit, there being only one node or point at which the barrel is still, the point at which it is fixed (the breech), (2) a series of overtones in which the barrel is divided longitudinally into a number of vibrating sections, each terminating in a node at the end nearer the breech. These two types of vibration can, and usually do, co-exist. The frequency of the fundamental vibration depends on the proportions of the barrel, and that of each particular overtone is always in a fixed ratio to that of the fundamental.

The shock of the explosion naturally sets up these vibrations in a violent form, and they are affected in a greater or less degree by irregularities in the shape of the barrel, by the external attachments, and by the manner in which it is stocked up. The compounded effects of the fundamental vibration and the overtones, at the moment when the bullet quits the muzzle, on the inclination of the last few inches of the barrel in relation to the axis of the bore before firing, constitute the main contributory factor in the "jump" of the rifle, which will be dealt with later.

It is essential that the rifle when in use should be periodically overhauled if its accuracy is to be maintained. The fore end, if it warps at all, is liable to put pressure on the barrel at one or more points, and may alter or limit its vibrations, and that not consistently. The bands holding the barrel and stock together will have a similar effect if they bear tightly on the barrel. Supposing that the screw by which the barrel is held in the action be not perfectly true, it will be subjected to unsymmetrical stresses when fired. If the lugs on each side of the bolt, or the shoulders in the body on which they bear, be not perfectly symmetrical so as to take their bearing simultaneously on firing, the small lateral movement which takes place before they can share the work equally, will be enough to give a lateral movement to the barrel and will affect the flight of the bullet in a lateral direction. The use of a bolt belonging to another rifle may thus affect the sighting. A rifle clamped in a rigid rest will not shoot to the same sighting as it will when held in the hands, and the effect of resting it (for instance) near the muzzle, or upon the trigger guard will as a rule be to vary the sighting.

Variations in the vibration may also arise from other causes. As the bullet passes up the barrel, the gases behind it are exerting pressure in all directions upon the chamber and upon the bore; the whole bore is, in fact, expanded in a minute degree in rear of the bullet. The bullet fits the bore tightly, and so far as possible seals it, creating heavy friction during its passage. Any variation from one shot to another in the amount of this friction will affect not only the velocity, but also the amount of movement which has been imparted to the rifle before the bullet has reached the muzzle.

The actual and relative effect of these influences will, of course, vary largely under different conditions. A heavy barrel is less affected than a light one by equal stresses. A heavy charge will set up more vibratory and general motion than a light one in the same barrel. When the velocity is low, on the other hand, the motion will have more time to develop before the projectile emerges, and consequently, the whole disturbance may be equally great, though less violent and rapid.

The basis of the sighting scale is obviously the zero position of the sights, when the axis of the barrel receives no tilt to allow for any fall of the bullet, but discharges it parallel to the line of aim. This parallelism is established by adjusting the backsight where the adjustment admits) until a point on the target is struck which corresponds with the prolongation of the axis of the barrel, the centre of the group of shots fired being only so much below the point aimed at as is due to (a) the height of the tip of the foresight above the axis of the barrel, and (b) the fall by gravity in the distance traversed by the bullet.

(a) Is ascertained by direct measurement of the weapon. It will be found to be about 1 inch in the S.M.L.E. and about 2 inches in rifles fitted with telescopic sights.

(b) Is ascertained from the formula for fall by gravity, $h = \tfrac{1}{2}gt^2$, where $h$ is the amount of the fall in feet for the distance, and $t$ the time of flight for the distance in seconds; $g$ being the ordinary value of the acceleration due to gravity, 32·2 f/s. per second.

It is convenient to fire at a small mark at 12½, 25 or 50 yards, at which the fall is but small. Thus, if at 25 yards a cartridge, such as Mark VII be fired, the muzzle velocity of which is 2,440 ft.-secs., this figure may without material error be taken as the average velocity over this short distance. The time of flight will then be, for 75 feet, 75/2440 = 0·03075 seconds, which, squared and multiplied by $\tfrac{1}{2}g$ (= 16·1) gives $h$ as 0·0152 feet = 0·183 inches.

Then, assuming that the tip of the foresight as 1 inch above the centre of the bore, when the adjustment of the sights is such that the bullets strike 1·183 inches below the point of aim, the parallelism between the line of sight and the prolongation of the axis of the bore is complete, and the zero or basis of the sighting scale has been arrived at.

It is not always possible to lower the backsight as far as the zero point, but this does not prevent the zero from being ascertained. Thus, in the case supposed above, if the rifle be fired at 25 yards with the sight set at the sighting known to be correct for 200 yards, the bullet will strike above the proper zero point by 6·5 minutes, the actual elevation for 200 yards, minus half-a-minute (or 1 inch at 200 yards) allowance for the height of the foresight. The value of a minute of angle is at 100 yards 1·047 inches, and at 25 yards 0·262 inch, so that the strike should be at $0·262 \times 6 = 1·57$ inches above the proper zero point. In this way, by firing at 25 yards, and using the range table, the proper adjustment of the sights for any longer range can be ascertained.

It will have been understood from what has been already said, that the prolongation of the axis of the bore as above mentioned is not the same line as appears when the rifle is held in a vice or a rest and the line of the sights, when correctly set to zero, compared with the direction of the centre of the bore viewed centrally from the breech when the bolt has been removed. This is because of the vibrations and movements already referred to which operate to disturb the rifle before the bullet has made its exit. The amount of the discrepancy between what has been called the " constructive " zero, *i.e.*, the mechanical parallelism of the sights with the bore, and the actual zero as found by firing, can be ascertained by careful inspection. Setting the sights to their actual zero point (or to some known higher elevation in reference to it), the rifle, with the bolt removed, is supported in a vice or on a rest. The eye then looks through the bore (which for accurate observation should be filled at the breech and muzzle with metal plugs, each having a small hole in its centre) and the line of the bore is accurately directed upon a mark on (say) a sheet of cardboard at 25 or 50 yards. Without moving the rifle, the aim is then examined, and the point at which it is directed is marked on the target. The angle by which this point differs from the line of prolongation of the bore (corrected for height of foresight), can then be measured; this is the amount of the "jump." It may be either positive (upwards) or negative (downwards) in direction. In the S.M.L.E. with Mark VII ammunition, it will be found to be between 4 and 5 mins. negative. With the same rifle and Mark VI ammunition, it is about 7 mins., and positive. An obvious instance of positive jump may be seen in many revolvers, which have their foresights noticeably higher than their backsights.

The amount of jump varies somewhat as between different weapons of similar type fired with similar ammunition. It will also be found to vary in the same weapon when the amount of the charge is varied or when bullets of different weights are used. Where the rifle is unsymmetrical laterally, as when it has a heavy projecting bolt head, some small amount of lateral jump will be found. A weight, such as that of a bayonet, attached to the barrel at the muzzle, is sure to have a considerable effect on the jump.

Compensation is a term which is applied to a particular effect of jump. If there are fired from a rifle loads otherwise similar, but graduated so as to give a series of different velocities increasing or decreasing by small amounts, it is usually found that the shots are thrown successively higher or successively lower. The degree to which the change extends is of course limited, and when the top or bottom of the particular phase of variation has been reached the movement is reversed, after passing through a phase in which the changes are but small.

Taking then a given rifle, and firing charges differing only in that they give successive increments or decrements of (say) 20 ft. secs. of velocity, we shall find one of the three following effects :—

    i. A tendency for the shots to strike higher and higher.
    ii. A tendency for them to strike lower and lower.
    iii. A more or less neutral phase, transitional between i and ii, in which the variation is comparatively small.

Now there are inevitable variations of velocity between successive shots with precisely similar loads, however carefully the ammunition may have been prepared. It is obvious that other things being equal the effect of + or − variations of velocity from the standard will be that the individual shots will have trajectories slightly less or slightly more curved than the average trajectory, and that the cone of fire will accordingly be somewhat elongated

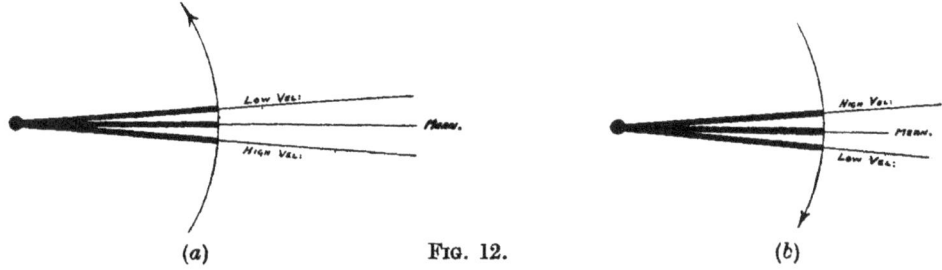

(a)    Fig. 12.    (b)

vertically. If, however, it should happen that the variation in jump due to the variation in velocity is of such a nature, with the particular rifle and charge used, that when the bullets having low velocity emerge from the muzzle they are delivered at a slightly higher angle vertically than shots of normal velocity (see Fig. 12 (a)), and similarly if the bullets having higher velocity are directed slightly lower, the changes thus occurring in the angle of departure will tend to bring together the trajectories of the different bullets, and they will in their flight intersect the normal trajectory at some distance in front of the muzzle.

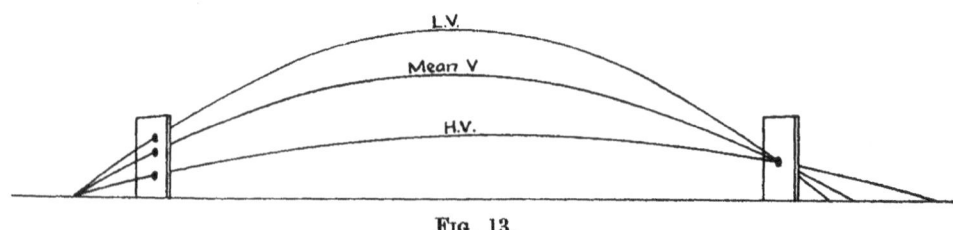

Fig. 13.

The effect of this will be to give, in the region of that distance, a much closer vertical concentration to the group of shots than would otherwise be possible with cartridges varying so much in velocity. In the case supposed, the irregularities of velocity are to a great extent compensated by the variations in jump; hence the term "compensation."

A conspicuous instance of this compensation occurred with the S.M.L.E. rifle at the Bisley Meeting of 1920, when it was found that ammunition (Mark VII) issued made quite erratic shooting in the vertical sense at 200 yards, and was incapable of shooting into the bullseye, while at 600 the groups were much better in proportion to the distance, and at 900 and 1000 yards it was as accurate as ammunition giving much more regular velocities. The range at which compensation is most effective with the S.M.L.E. and Mark VII ammunition is about 900–1000 yards; with the longer Lee-Enfield and Mark VI it was about 1500 yards.

It will be appreciated that such a result is only obtained when the proportions of the rifle and the constituents of the cartridge co-operate in the direction required. They may equally well have a neutral or an unfavourable influence. If the shots of higher velocity are directed above the normal and those of lower velocity below it, the inaccuracies due to the varying velocities will be exaggerated at all distances from the muzzle onwards (*see* Fig. 12 (*b*)). If on the other hand the bullets emerge with the vibrations at or near a neutral phase (as in iii above) the variations of velocity will be neither corrected nor emphasized.

It is evident that the amount of the compensation, and therefore the distance at which it is most effective, will be greatest if with the charge giving standard velocity the muzzle is in the mean position between the extremes of the amplitude of its vibrations, and is therefore in most rapid motion. To diminish the range at which compensation occurs, it would therefore be necessary that the bullet of mean velocity should make its exit at a part of the curve in which the muzzle is approaching the end of a vibration and is in a phase of less rapid translation. It is not, however, usually found possible to adapt the vibrations perfectly from this point of view, since precedence has to be given to the wider considerations of the weight and proportions of the barrel and of the arm generally.

The condition of the bore is of great importance to its accuracy of shooting. It gradually deteriorates, suffering from wear as well as usually from avoidable damage.

Wear necessarily occurs from the following causes :—

i. The friction of the bullet.

ii. Erosion or the washing away of the surface of the bore by the propellent gases.

iii. Use of the pull-through, patches, etc., in cleaning.

iv. Fouling, which may be classified as follows :—

 *a*. Internal : the forcing into the pores of the metal of the products of combustion of the propellent and the cap. This trouble shows itself in " sweating " or the appearance of a dark deposit on the surface of the bore, which promptly sets up rust ; this is prevented by the early use of boiling water after firing and by daily cleaning for some days.

 *b*. Metallic (also known as nickeling) : the deposit of particles of the bullet envelope, chiefly on the lands. This occurs much more readily when the surface of the bore has lost its polish and become roughened. Metallic fouling from bullets with cupro-nickel envelopes consists of particles of those two metals, and can be dissolved by suitable chemical treatment ; it can also be dragged out by using the pull-through, or polished out by lapping. The two last methods are not without effect on the steel. This fouling affects the mean point of impact and is prejudicial to accuracy ; when excessive, it may lead to a bulge. It appears chiefly towards the muzzle, and may accumulate in lumps visible to the eye.

There is still much obscurity as to the factors governing the deposit of metallic fouling. It is certainly more readily deposited by some batches of bullets than by others. It appears to become greater with an increase of temperature, whether due to hot weather, rate of fire, or temperature of propellent. The effect seems to be analogous to that of a bearing which " seizes," the surfaces in contact having become heated after lubrication has failed. The surfaces of the bullet and the bore both become extremely hot from friction, and it is improbable that the beeswax in the cannelure of the bullet can have any lubricative effect in the upper part of the barrel.

Preventible wear and damage arise in many cases from :—

i. Excessive use of the gauze in cleaning.

ii. " Cord wear," or the wearing of a groove by frequent sidewise pressure with the pull-through or cleaning rod at either breech or muzzle. It becomes noticeable at the breech from the fired case showing signs of being expanded into the groove ; at the muzzle it leads to marked inaccuracy in shooting.

iii. Scratches or cuts caused by the use of improper implements or tools for cleaning or removing obstructions.

iv. Pitting of the surface, due to rust or bad steel.

Barrel injuries, mostly accidental, are as follows :—

i. Bends : a bent barrel can in some cases be straightened.

ii. Bulges : these are local expansions of the bore, and occur in any part of the barrel, but mostly towards the muzzle. They are presumably due to the bullet meeting some slight obstruction and to the consequent development of a local pressure which stretches the steel beyond its elastic limit.

iii. Puckering : an inward bulge or dent.

iv. Enlarged chamber. This form of bulge is very rarely found ; it may be due to bad steel or to excessive pressure. A similar effect may be produced by excessive use of the chamber stick.

v. Obstructed bore. This is usually due to a broken pull-through, etc., or to a bullet or part of one remaining in the bore. Special tools are issued for removing obstructions. These are: Rods, clearing, Arms, with a plain plug and screw bit ; a spoon bit is still found, but it is obsolescent. The obstruction should be jammed down with the plain plug, and the screw bit then used.

The optical examination of the barrel to ascertain its condition is known as viewing. The procedure is as follows. The barrel is held up to the light in such a manner that a slight shadow is thrown (as from a window bar or some dark edge) into its interior. In a straight barrel the edge of the shadow will show as a portion of a perfect cone, and this does not alter its shape if the barrel is revolved or reversed end for end. Should such faults as bulges, puckers, bends, etc., be present, the lines of the cone are distorted. At the same time other faults, such as cuts, scratches, pitting, fouling, wear, etc., should be looked for.

It should be understood that if a rifle is to be in condition to give fine shooting results it will need special attention in many details. Not only must the sights be kept perfectly clean, unrubbed and undamaged, but similar attention must be given to the woodwork, so that the stock may not be warped, nor liable to warp under the influence of rain or sun. The wood should be as waterproof as possible, for which purpose linseed oil or some preparation which will fill the pores of the wood, as cannot be done by the thinner oils used for cleaning or lubricating, is necessary. The muzzle of the barrel must not be pinched in the nose-cap so as to check its freedom to vibrate. Screws should be watched, as they tend to work loose with firing ; it is, however, a mistake to strain them by exerting too much force by tightening them. The condition of the pull-off is of importance ; it tends to vary with use, and may easily become heavy enough to make a perfect release of the trigger difficult. It should lift a weight of about half a pound more than the minimum weight allowed, which for the S.M.L.E. is 5 lbs.

## CHAPTER I—SECTION 6

### SIGHTS

In discussing the various types of sights described below it is desirable to bear in mind one of the properties of the human eye, since it is only by so doing that the principles underlying the various designs can be fully appreciated. The human eye, like the photographic camera, which is exactly similar to it in principle, is unable to focus simultaneously objects at different distances from it. From this it follows that in aligning two or more objects, such as the open sights on a rifle with the target, only one of them can, at any given moment, appear perfectly defined. In tracing the design of rifle sights from the simpler to the more complicated it will be found, therefore, that the attention of inventors has been directed, firstly, towards bringing all objects which have to be seen into as nearly as possible the same focus, and, secondly, towards magnifying the image of the target so as to make it more readily visible and increase the accuracy of aim.

*The Open Sight.*

This is the type used on nearly all military rifles. As is well known, it consists of a foresight near the muzzle of the rifle, and a backsight, the essential part of which is as a rule a notch of some sort. The foresight usually has some bold upstanding shape, which can be easily seen against the target, such as the broad blade on the S.M.L.E. The shapes of backsight and foresight that have been tried with the object of securing their more accurate alignment with one another have been innumerable. No shape seems to have any special advantage, and certainly none can give anything approaching perfect alignment owing to the impossibility of focussing more than one of the three objects, one of which, the backsight, is quite near the eye.

The position of the backsight is important. If it is too near the eye errors of alignment from lack of definition will be increased. On the other hand, the effect on the aim of a given error of alignment between foresight and backsight is increased in inverse proportion to the sight radius, *i.e.*, the distance between the foresight and the backsight, so that this must not be unduly decreased by moving the backsight too far from the eye. The compromise which has been generally agreed upon is to have the backsight about 15 inches from the eye. Possibly a rather greater distance might be an advantage.

The backsight notch, or other device, is usually cut in the top edge of a slide mounted to move up and down on a folding leaf, though other designs, such as that on the S.M.L.E., are sometimes used. Scales are usually cut on the leaf and are graduated in terms of range according to the cartridge which it is intended shall be used. Arrangements for moving the backsight laterally in order to allow for variations of the wind, etc., are not usually found on open backsights.

*The Aperture Sight.*

As the name of the sight implies, the backsight in this case consists essentially of a small circular aperture through which aim is taken. The backsight is best situated just in front of the eye; as near it, in fact, as is consistent with the firer's safety from being hit when the rifle recoils. When aiming, no attempt is made to focus the edges of the aperture, but, since it is circular, it is quite easy to put the foresight and target sufficiently near the centre for the error involved to be negligible. Since the nearest object which needs to be focussed is the foresight, it follows that the lack of definition is much less than is the case when using open sights. This lack of definition is further reduced by what is known as the orthoptic effect of the rear aperture. This acts by cutting out the more peripheral of the rays which would otherwise pass into the eye, so that the blurring of any object which is out of focus is reduced. The principle is exactly similar to that of stopping down the lens of a camera in order to obtain what is known as depth of focus, *i.e.*, reasonably good simultaneous definition of objects at different distances. The orthoptic effect of the aperture is similar to that of the pupil of the eye, and is of greater relative benefit to a man who normally has a large pupil and little pigment in his eye. The optical advantages of aperture sights are particularly marked in the case of elderly men, who may find shooting with open sights very difficult. Age results in very little diminution in the accuracy of the aim when aperture sights are used.

The size of the aperture is important; too small an aperture unduly reduces the light and makes the target difficult to see; too large an aperture reduces the orthoptic effect, and also makes centring of the foresight and target more difficult. No one size can be the best for all individuals or under all circumstances, so that some arrangement for changing the size of the aperture is theoretically desirable. An average size used for target shooting is $0 \cdot 06$-inch, while for shooting in the field, $0 \cdot 08$-inch at least, is needed.

It will be evident that the aperture sight permits of much more accurate aim being taken than does the open sight. It has, however, one disadvantage, which is that it is more difficult to find the target quickly when aiming; though if the exact position of the target is known (as in firing a number of shots at the same mark) it is possible to aim quicker than with the open sight. This defect of the aperture sight is due to the blocking out of part of the field of view by the metal plate or disc in which the aperture is cut. The nearer the aperture is to the eye, the larger is the field of view, and, therefore, the less noticable is this defect.

The original Lyman backsight, which is largely used on sporting rifles, especially in America, is designed to obviate this defect. The aperture is surrounded only by a very thin ring of metal, and is mounted on the top of a stem without surrounding structure. When aiming, all that is seen is a hazy ring just sufficient to enable the foresight and target to be centred, but too faint to interfere in any way with the view. Although this sight achieves its object to a large extent, it does so only by a partial sacrifice of the orthoptic effect. Also, in the case of the average soldier, there might be some risk that in moments of excitement he would forget to aim through the aperture at all.

In bolt action rifles the aperture backsight is normally fixed over the rear end of the body. Attempts to bring it nearer to the eye have been made by fixing it to the cocking-piece or to the rear end of the bolt. The cocking-piece position is generally ruled out because of the slowing of the lock time which results. If the sight is fixed to the rear end of the bolt proper, there is difficulty in making the sight large enough to admit of the range of adjustment necessary for military purposes; also special arrangements have to be made to ensure the bolt resuming exactly the same position each time it is closed.

Modern aperture backsights are generally made adjustable vertically by means of a screw with a milled head. A similar lateral adjustment is in most cases also provided. There are usually vernier scales, graduated approximately to minutes of angle, to read off the amount of movement. Another scale, graduated in terms of range, is generally provided in addition. Sometimes a clicking arrangement is also fitted, corresponding with the graduations on the sight, so that small adjustments can be made quickly without stopping to read the scale.

Aperture sights are fitted to the U.S.A. Springfield, the 1914 Enfield rifle and the new Mark VI. They were also fitted to some models of the Ross rifles. Aperture sights have been in use for some centuries, having frequently been fitted to cross-bows at least as early as the sixteenth century.

*Sights involving various Optical Principles.*

Various devices have been invented with the object of providing, by means of lenses or mirrors, simultaneous focussing of both the foresight or aiming point and the target without magnification of the latter. They have all, however, certain practical defects, and, as they are none of them used on rifles at present, they will not be described here.

*The Galilean Telescopic Sight.*

The Galilean telescope embodies the principle on which most low-power field glasses are constructed. This sight magnifies the image of the target, but does not give perfect definition of both target and aiming point. A convex lens, which must be of greater focal length than the sight radius, is attached near the muzzle of the rifle. The aiming point is attached to or ground into the front surface of this lens. The lens is generally mounted in a tube in order to protect it from the rain. The effect of this convex lens is to make the rays of light from the target converge. The backsight consists essentially of a very small aperture, which usually has a concave lens just behind it. The effect of this concave lens is to render the rays from the target again approximately parallel. It is illustrated in Fig. 14.

In order to enable the firer to see the target clearly the concave rear lens should have a focal length equal to the difference between the focal length of the front lens and the sight radius. On the other hand, the aiming point would be seen most clearly with a convex rear lens of focal length equal to the sight radius. It will be evident, therefore, that there must be a distinct amount of blurring of either the aiming mark or the target. The amount of blurring that actually occurs is, however, greatly reduced by the small size of rear aperture which can advantageously be used. A diameter of ·04 inch is an average size. The actual rear lens used in practice depends on whether it is desired to give the better definition to the target or to the aiming point. Usually the target is selected in order that it may not readily become invisible in a bad light, and then a concave rear lens, of slightly longer focal length than that theoretically required to define the target, is used. If it is decided to define the aiming point it may be that no lens at all will be required in the rear sight. Perfect definition of the target cannot be obtained if the aiming point is to be reasonably visible and *vice-versa*.

The field of view of this sight is very limited. It may be calculated by finding the angle subtended at the backsight by the front lens, and dividing this by the magnification.

Accidental rotation of the front lens must be carefully guarded against as it generally introduces a serious error if it occurs. Rotation of the rear lens, however, if it is situated behind the aperture, has no effect on the line of sight.

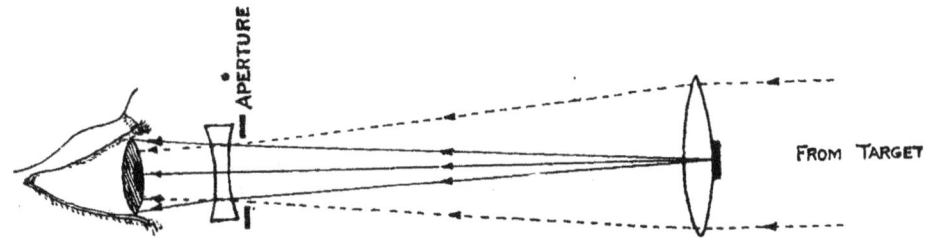

Fig. 14.—The Galilean Telescopic Sight.

The continuous lines represent the rays from the aiming mark, the dotted lines those from the target.

The length of sight radius is important with this sight, because, for any given magnification, the lack of definition is inversely proportional to the sight radius. Thus much more blurring occurs with this sight when used on the S.M.L.E. which provides a 30-inch sight radius, than is the case with the match rifle fired from the back position, when the sight radius is about 45 inches.

The construction of the backsight is exactly the same as that of the aperture sight described above, except that provision has to be made for accommodating the rear lens (which may be very minute), and for protecting it from rain. The sight is, therefore, comparatively easy to fix rigidly—a marked advantage as compared with the terrestrial telescopic sight described below.

The Galilean telescopic sight is a comparatively recent introduction. It was invented by the late Dr. Common, who, used it without a rear lens. Mr. Maurice Blood perfected its construction for match rifle shooting, and it is now universally employed at Bisley for that purpose. It was used in various forms on snipers' rifles during the Great War, but its imperfect definition, small field of view, and fragility were found to be drawbacks.

*The Terrestrial Telescopic Sight.*

The principle of this sight is that of the ordinary terrestrial telescope. A convex front lens or objective, as it is called, of shorter focal length than the sight radius is used. The parallel rays from the target pass through this, and form an inverted real image at its focus. Here is situated the aiming point which consists of a pointer, of fine crossed lines (known as "cross wires"), or of some such device. The inverted image of the target is viewed by means of a system of four lenses, comprising the erecting lenses and eyepiece. Sometimes in place of the erecting lenses a prism is employed after the fashion of the ordinary prism binoculars, though this method is not generally so satisfactory in the case of telescopic sights. Since the aiming point and the inverted real image of the target lie in the same focal plane, it follows that they both appear perfectly defined when viewed by the firer. This is the only sight in use at present with which all the objects which have to be seen whilst aiming are perfectly defined. It is illustrated in Fig. 15.

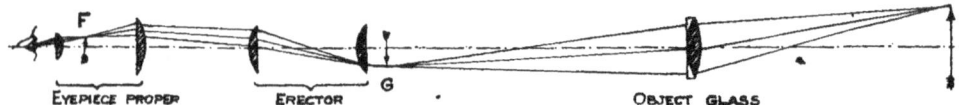

Fig. 15.—The Terrestrial Telescopic Sight.
The "cross wires" are situated at G.

Almost any desired amount of magnification can be obtained with this sight. The field of view is usually much greater than is the case with the Galilean for a similar magnification. The illumination with a good sight of low power is such that accurate shooting can be continued in a worse light than with any other form of sight. The one possible exception is when the object glass cannot be properly shaded, in shooting towards the sun from a position in bright sunlight at an object in deep shade. In this exceptional case open sights are better.

In order to keep the telescope from contact with the face, on recoil, the focus for the eye is situated at some little distance from the rear lens. This gives another advantage to this sight since it affords a certain amount of lateral relief for the eye, i.e., when the eye is situated at the correct distance from the eyepiece, it may be moved from side to side, or up and down within a small circle without affecting the aim. The size of this circle varies with the construction of the sight; it is usually about $\frac{1}{8}$-inch to $\frac{1}{4}$-inch in diameter. This feature obviously increases the speed with which aim can be taken.

Accidental rotation of the whole telescope in its mountings, or of the objective, or of the cross wires usually produces serious error for optical reasons. It should be clearly understood that canting the rifle when aiming can cause no error of this kind, but only the ordinary error of cant because by the act of aiming, the telescope must be rotated with the line of sight as the axis of rotation.

Telescopic sights are constructed so as to be completely enclosed in a tube, which is usually about 8 to 12 inches in length. Telescopes of the full length of the barrel were used at one time, but their use was abandoned for sporting and Service purposes as they were not found to be sufficiently strong. The telescope tube is mounted on a base rigidly attached to the action of the rifle. A dovetail arrangement is generally introduced in order to permit the sight to be readily attached and detached. Adjustment for elevation is usually provided by moving the cross wires up or down by means of an elevating screw, which has a scale inscribed on a drum head on top. Lateral adjustment, if provided for, is sometimes carried out by means of a pivoting arrangement of the base. Frequently, however, terrestrial telescopic sights have no provision for lateral adjustment, owing to the great difficulty of preventing the sight working loose when it is fitted. The reason for this is the much greater weight of telescopic sights as compared with the separate backsight or foresight of any of the other sights described; the result is that, if even a minute amount of play occurs in the mountings, a violent hammering action is set up by the recoil, which is apt to upset the adjustments, and may even break up the telescope itself. One method of obviating this difficulty is that adopted in the case of the Winchester telescopic sight used on the Springfield rifle. This consists in avoiding the shock of the discharge by simply allowing the telescope to slide forward in its mountings when the rifle is fired. This sight is capable of very high accuracy, as is proved by the results of target shooting in America, but it requires a great deal of care in order to get the best results; also, as it has to be reset by hand after each shot, it is quite useless for rapid fire. Another method that has lately been tried is to clamp the lateral adjustment rigidly by means of a locking screw after it has been set; this is the principle of the Noske mounting which has lately been introduced with apparent success in America. The 1918 model Service telescopic sight has no lateral adjustment proper, but the lateral zero may be shifted to suit any particular rifle by unclamping a prism set base up in front of the objective and rotating it slightly. A complete turn of this prism through 360° causes the cross wires to describe a circle 15 feet in diameter at 500 yards. The normal position of the cross wires is at VI o'clock on this circle. Hence each turn through one degree gives about a quarter of a minute of deflection; thus a slight alteration in the vertical zero is introduced with every change of the lateral zero.

It will be apparent from the above that the great defect of the terrestrial telescopic sight is the difficulty of keeping it rigid on its mountings. This is particularly important because of the shortness of the sight base, which is usually about 6 inches only, so that the effect of any lack of rigidity is greatly magnified by comparison with that occurring with other types of sights. The telescopes themselves are generally so much better made than was formerly the case that they do not very frequently give trouble, if properly mounted. Another objection which has been taken to this sight is that it does not in any way follow the curves of a barrel which has become hot as the result of firing, and has been bent by a warping fore-end. How much importance there may be in this is a matter for determination by experiment with the particular rifle used; the simplest remedy is to stock up the rifle so that the fore end does not touch the barrel.

There can be no doubt that this sight is incomparably the best when great accuracy is needed at objects which are difficult to see, but it has to be treated as a rather delicate instrument reserved for a special purpose. It is usually carried in a leather case slung over the shoulder, and is only attached to the rifle when required for use.

Telescopic sights are believed to have been used as early as 1775 by the Americans during the American War of Independence. They are mentioned by Colonel Beaufoy in "Scloppetaria," published in London in 1808.

*The Line of Sight.*

In practically all rifles the sights are fixed above the barrel, and, therefore, the line of sight is a certain distance above the commencement of the bullet's trajectory. This has an appreciable effect upon the distance of the bullet from the line of sight during its flight and, consequently, has some influence upon the chance of hitting a target of given size at a distance other than that for which the sights are set.

This can best be demonstrated by means of an actual example. Suppose that it is desired to set the sights of a rifle so as just to keep the trajectory within a 6-inch target when aim is taken at the middle of it. Take, for example, a modern high-velocity cartridge with angles of elevation and drops in inches as follows:—

|       |           |          |
|-------|-----------|----------|
| 50 yards | 1·0 min. | 0·5 in. |
| 100 ,, | 2·2 ,, | 2·3 ,, |
| 150 ,, | 3·4 ,, | 5·3 ,, |
| 200 ,, | 4·6 ,, | 9·6 ,, |
| 250 ,, | 5·9 ,, | 15·4 ,, |
| 300 ,, | 7·2 ,, | 22·6 ,, |

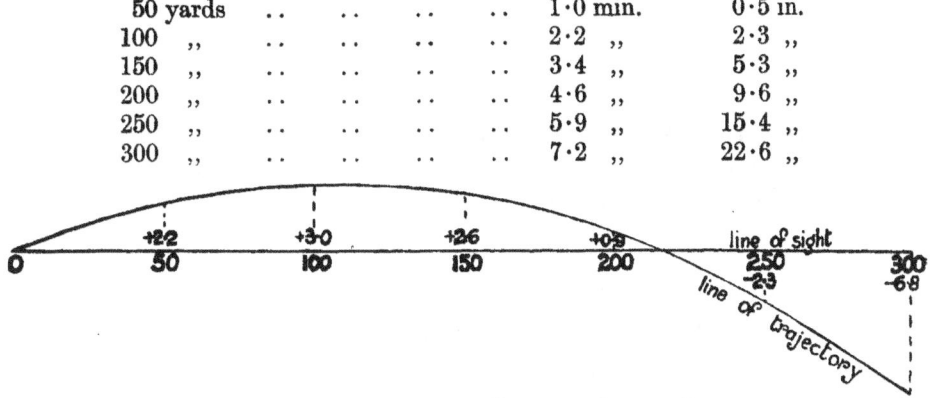

Fig. 16.—6-inch Danger Zone, when Line of Sight and Line of Trajectory intersect at the Muzzle

Range in yards.   Heights in inches.   (Exaggerated.)

Then the trajectory, when the line of sight intersects the muzzle of the rifle, and the sights are set for about 215 yards, is as shown in Fig. 16.

If, however, the line of sight be supposed to pass 3 inches above the muzzle of the rifle (as might be arranged by means of a telescopic sight), then the trajectory is as follows in Fig. 17, the sights being set for about 255 yards.

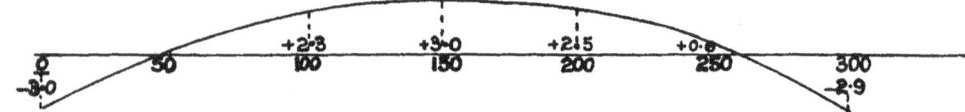

Fig. 17.—6-inch Danger Zone, when Line of Sight passes 3 inches above Line of Trajectory at the Muzzle.

Range in yards.    Heights in inches.    (Exaggerated.)

It will be seen that in the first case the 6-inch target would be hit at any range up to 260 yards, whilst in the latter the range within which a hit would occur is 300 yards, thus representing a clear gain of 40 yards for the raised line of sight.

In connection with such problems, it must never be forgotten that, on account of the combined error of man and rifle, a large proportion of advantage has to be deducted from the size of target that can be allowed for the purpose of calculating the trajectory error.

*The Effect of Cant.*

When a rifle is canted laterally during the act of aiming, the line of sight must clearly be the axis around which rotation takes place. The result is that the point at which the barrel is directed above the target (and consequently the point of impact of the bullet,

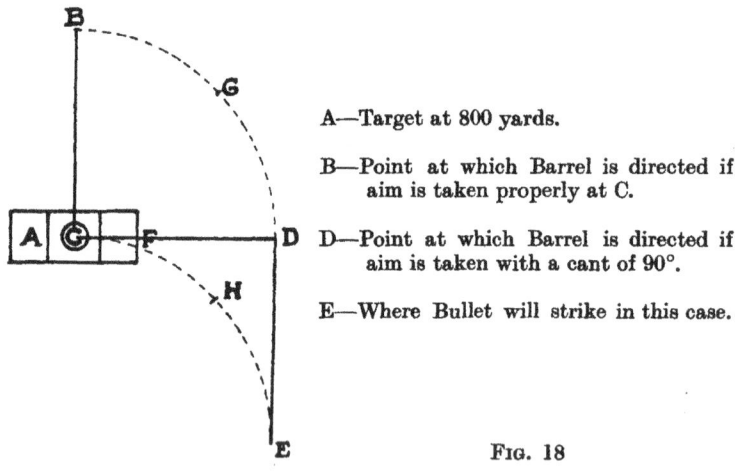

A—Target at 800 yards.

B—Point at which Barrel is directed if aim is taken properly at C.

D—Point at which Barrel is directed if aim is taken with a cant of 90°.

E—Where Bullet will strike in this case.

Fig. 18

which is a constant amount below this point) is moved a variable distance laterally. There is also, as is obvious from Fig. 18 (from Ommundsen and Robinson, page 272), a slight vertical error, but this is quite negligible for any amount of cant that occurs in practice.

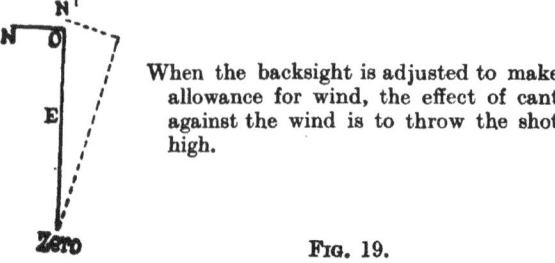

When the backsight is adjusted to make allowance for wind, the effect of cant against the wind is to throw the shot high.

Fig. 19.

There is only one condition under which canting the rifle whilst aiming can produce a vertical error; this is when there is a considerable lateral allowance for wind on the sights, as in Fig. 19. In this case the vertical error may be either up or down, according as the cant is with or against the wind. The occurrence of vertical error in this way was described by A. G. Foulkes in "The Theory and Practice of Target Shooting," London, 1895.

The calculation of the exact error produced by canting is a matter of trigonometry resulting in a simple formula.

If the elevation on the backsight is E degrees and the angle of cant is B degrees, and deflection produced by cant is C minutes, then

$$\tan C = \sin B \tan E.$$

Hence
$$C = \frac{3600}{3437} BE,$$

or nearly enough
$$C = BE.$$

The working rule is therefore, "Multiply the elevation in degrees by the cant in degrees to get the number of minutes deflection."

## CHAPTER I—SECTION 7

### RESTS FOR RIFLE SHOOTING

The object of employing a rest when shooting is either to minimize or else to stabilize the amount of error due to aim. With the very best of rests it is possible to eliminate this source of error to a very great extent, but not entirely. In the final analysis the human being who operates the machine can always make a mistake, even though it be a small one.

On the Continent, but especially in America, rest shooting has been standardized as a form of sport. Now, however, the American War Office have concluded that Dr. F. W. Mann's form of mechanical rest is the only one worth considering, when the greatest possible accuracy is wanted.

This Mann rest is the 6-point rest with the best-conditioned set of six points yet devised. Before describing it a very brief description may be given of other forms of rest, but it must always be remembered that every rest has its own peculiarities, and that the sighting to give a central bull's-eye with one rest may give a very wide shot with another form of rest, although the groups of each may be as tight. This is owing to the fact that every rifle barrel vibrates when fired and that every particular control of the barrel interferes to some extent with these vibrations, and therefore with the line of departure.

The elementary forms of artificial rest are such as control the natural pulsations of the human body, and include such as utilize the support of a tree-trunk or a sandbag. These are all liable to cause trouble if the gun itself makes contact with the support, because the contact is not always in the same place. Good forms of such rests are:—

(a) A sandbag under the wrist for any position.
(b) An alpenstock or pole in the fingers of the left hand, leaving the thumb and forefinger to hold the gun, especially standing.
(c) When sitting, a strap to hold the knees from sagging apart, thus affording firm support for the elbows on the inside of the thighs.
(d) Table rests with muzzle support; these, however, require care in use to keep the contact identical from shot to shot.

For general use a wrist rest for the left wrist in a perfectly comfortable position is probably capable of as good work as any. It has the additional advantage of shooting to the same sighting as free holding.

Slide rests and similar mechanical contrivances requiring a change in the stocking up of the rifle are not altogether satisfactory. The most usual example is the Whitworth rest, in which the rifle is stripped of its wooden fore-end and the barrel is nutted or bedded down on to a heavy mass of metal free to recoil in a solid V-shape gutter. After each shot the gun and bed complete are run up again to the original position in the gutter, ready for the next shot. Such a rest is often called a fixed rest.

There are two serious objections to the Whitworth rest :—

(1) A highly skilled artisan is required to bed down the barrels and to watch them continually.

(2) The resulting shooting is not the shooting either of the cartridge in a perfect barrel or of the cartridge in its own stocked-up action.

There remains, then, the true geometric method known as the six-point rest, of which Dr. Mann's pattern is the most perfect. It was described in scientific language and for the first time in Thomson and Tait's "Treatise on Natural Philosophy," (Cambridge, 1879) Vol. I, pp. 149 *et seq*. In a footnote it is stated that this method was taught to Lord Kelvin by Professor Willis thirty years before. It is believed that the Enfield rest, designed by Colonel Watkin in 1903, is the first public example of this rest, although Dr. Mann made one (and independently) some years before for his private use.

The principle is of the utmost simplicity. If six definite separate points control the position of a rifle, it can be replaced on the rest time after time in exactly the same position. It is, of course, essential, in order to attain this, that the points should rigidly maintain their relation to one another and to the platform to which they are fixed, while the platform itself must not move with relation to the surface on which it rests.

The reason for requiring these six definite points is that every body has six degrees of freedom, three of translation and three of rotation.

In translation it can move—

(1) to and fro,
(2) up and down,
(3) right or left.

In rotation it can rotate—

(1) for elevation or depression,
(2) for traverse right or left,
(3) clockwise or counter-clockwise.

In Plates XII and XIII the most elementary skeleton of a six-point rest is shown. It is merely a box nailed down on to a firm table and provided with two V-cuts and a knob of wood. The six points of contact are clearly numbered, and a rifle is shown in position. Provided the rifle is "dumped" into the two Vs, and is then run up till the trigger guard touches No. 5, and is then slightly rotated clockwise till the side of the trigger guard touches No. 6, the rifle must be in one and only one exact position. If it is then loaded and fired, taking reasonable care to keep the six contacts made, it will throw the best group of which the gun is capable.

Plates XIV and XV show an elaboration of the system enabling the position of the group to be varied by elevating and traversing the rest. The actual dimensions of the Vs and of the trigger lump were arrived at by trial and error, using every military rifle available. The Mannlicher, with its fixed box magazine, will not fit properly into it, but all the others drop in and out without difficulty and shoot well from it.

Plates XVI and XVII show the Watkin or Enfield rest. In Plate XVI the man's left forefinger is on point 5, the run-up point for which contact is made in S.M.L.E. by the butt swivel.

In Plate XVII the rifle is being fired in the correct way with the least possible interference. The elevating and traversing wheels are clearly visible. The six points are as follows :—Two on the front support, one below and one on right, the left-hand one being only a spring. At back there are similarly one on right top and one at bottom for the toe of the butt to rest on. The fifth contact is on the right side near the bottom of the butt and the sixth is at the swivel. The big spring arm at the back is only for recoil-taking.

[*To face page* 66.

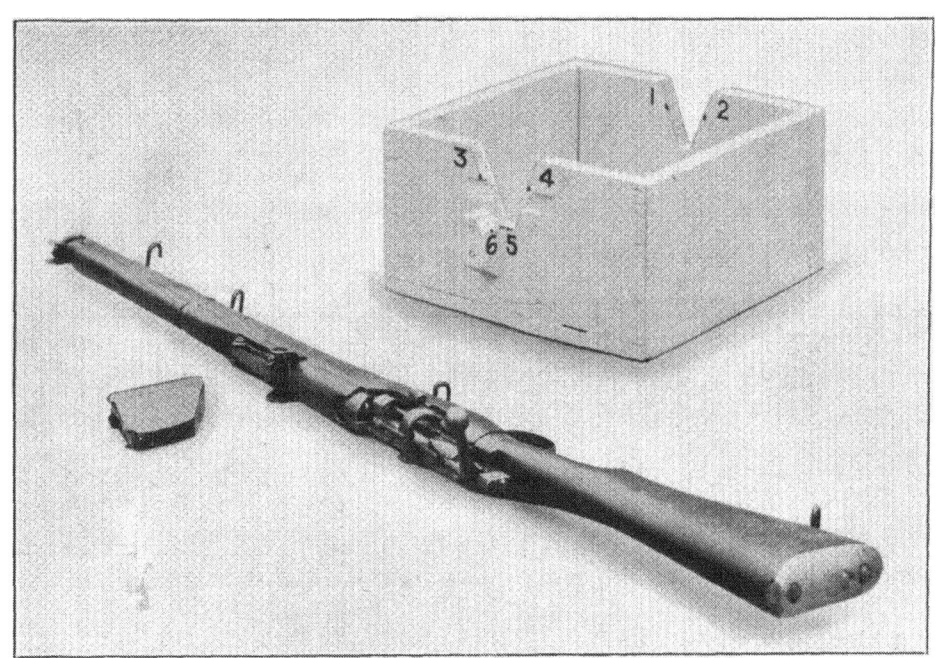

Part I, Chap. I, Plate XII.

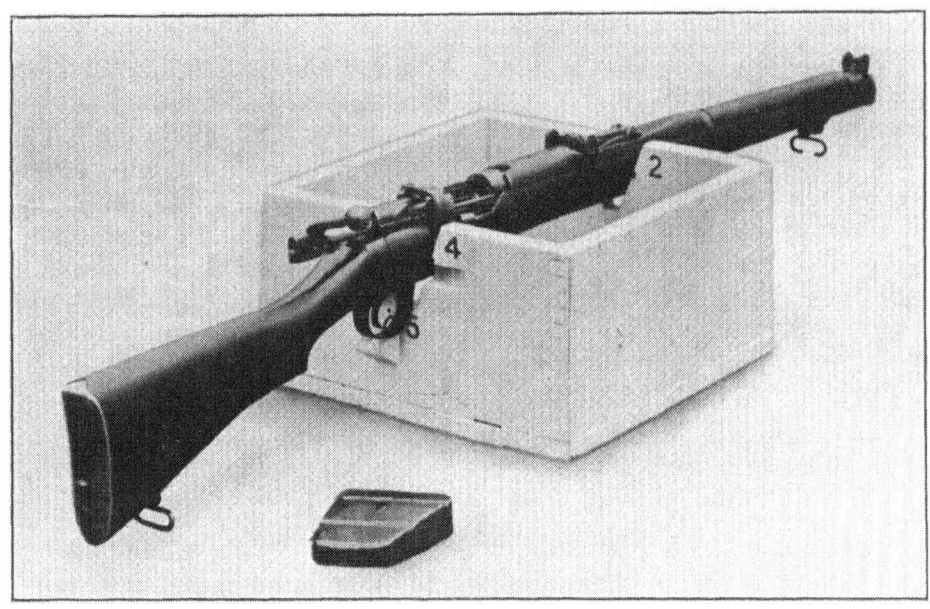

Part I, Chap. I, Plate XIII.

Part I, Chap. I, Plate XIV.

PART I, CHAP. I, PLATE XV.

[*To face page* 67.

Part I, Chap. I, Plate XVI.

Part I, Chap. I, Plate XVII.

and the spring latch on the top is to hold the rifle in place. With skilful and thoughtful handling this rest is capable of wonderful shooting. To make it fool-proof, it should be kept scrupulously free from grit and each contact should be provided with an electric contact to ring when made. The recoil pad is the most fruitful source of trouble as its pressure may lift the butt slightly or give it a jerk on the first motion of recoil.

Dr. Mann solved the problem of conditioning six contacts in the best way, if regard be only taken of the accuracy of the cartridge. The Watkin rest tests the combination of stocked up rifle and cartridge together, but neither separately. For many purposes it is essential to test the cartridge alone. For these purposes Dr. Mann used a barrel weighing with its adjuncts some 15 lbs. By preference as much of the weight as possible was placed in the barrel, say 10 to 12 lbs., by using a very thick tube or by casting lead into a tube surrounding a thin barrel. The barrel was mounted in two concentric rings of solid metal, one near the breech and one near the muzzle. Any convenient action and bolt was fitted to the barrel and the whole was mounted on a heavy V gutter capable of being elevated and traversed as in Plates XIV and XV. The gutter carried a front stop up to which the front ring was run and also a lateral stop for revolution to ensure that the trigger guard or the like was always revolved to the same o'clock. Hardly any care was required to release the trigger without moving the gun, and in the result the actual shooting of the cartridge was registered with hardly any error of mounting or of laying. The shooting of such a rest has to be seen to be appreciated.

Taking actual experimental figures shot out at Woolwich as a guide, some real notion of the relative merits of various methods of testing accuracy may be obtained. For convenience they may be tabulated as below:—

Method   I.—Mann rest.
         II.—Whitworth rest in first-class condition.
        III.—Watkin's Enfield six-point rest (*a*) with S.M.L.E. specially stocked up by an expert; (*b*) with selected S.M.L.E. as issued.
        IV.—Expert marksman with wrist rest and aperture sight, using rifles as in III (*a*) and III (*b*).

The results at 500 yards in easy weather for a ten shot group of best quality ·303, Mark VII ammunition, measured by group diameters in inches, are:—

| Method. | Best Group. | Average Group. |
|---|---|---|
| I | 6 | 10 |
| II | 7 | 12 |
| III and IV (*a*) | 7 | 12 |
| III and IV (*b*) | 11 | 18 |

To reduce these group diameters to figure of merit it suffices to divide by $3\frac{1}{2}$ so that a 7-inch group has a figure of merit of 2·0 inches.

## CHAPTER I—SECTION 8

### DEFINITIONS PERTAINING TO THE RIFLE AND RIFLE SHOOTING

*Barrel*, a steel tube closed at one end by the *bolt*. The closed end, the *breech*, has a *chamber* cut in its interior surface to take the cartridge. The whole interior surface of the tube, from a little forward of the chamber to the *muzzle*, is cut with spiral grooves known as *rifling*.

The *bolt* is a device for closing the breech of the barrel. In modern rifles it is hollow and contains the firing pin and spring. It is fitted with means for withdrawing the fired cartridge from the chamber. It is held against the backward thrust of the explosion by lugs which form part of the material from which it is made and which bear against surfaces on the *body*.

*Body.*—The housing for the bolt. It is attached to the breech of the barrel by a screwed portion.

*Primary extraction* is the first backwards motion or loosening of the cartridge in the chamber, effected while disengaging the lugs of the bolt from their seating.

*Action.*—The combination of body and bolt, together with the trigger and other mechanism for functioning the rifle.

The *bore* is the interior of the barrel from the front end of the chamber to the muzzle. Sometimes used to denote the *calibre*.

*Calibre* is the diameter of the bore measured across the *lands*. In British arms the calibre is indicated in inches. In Continental arms the calibre is indicated in millimetres.

The *lands* are the portions of the bore left between the grooves of the rifling.

The *lead* (pronounced *leed*) is the conical progression of the bore from the chamber to the rifling.

*Head space* is the distance between the front face of the bolt when closed home and the rear face of the breech.

*Free travel* is the distance the bullet has to pass over when it leaves the cartridge case before it engages with the rifling.

*Stock*, the woodwork of the rifle. In the British Service rifle consists of butt, fore-end and handguard.

*Stocking-up*, the process of fitting the woodwork to the assembled barrel and action.

*Charger* is a holder which contains a number, usually 5, cartridges for the magazine of a rifle. On loading the cartridges are swept out of the charger into the magazine, the charger falling away.

A *clip* is a holder which contains a number of cartridges, usually 5, for the magazine of a rifle. On loading, the clip and cartridges are inserted into the magazine and the clip drops out at the bottom when the magazine is empty.

*Units of Measurement.*—Scientific measurements are usually made in accordance with the metric system, and on the Continent all ballistic measurements are made and recorded in this system. Ballistics being a practical science, measurements in this country are made in the units commonly in use. They are :—

*Distance.*

The *yard* for range.
The *foot* for height of trajectory and, in certain cases, for range.
*Inches* for the calibre of the rifle.

*Mass.*

*Tons* and *hundredweights* (*cwts*) for weights of ordnance.
*Pounds* and *ounces* for weights of rifles and projectiles.
*Grains* for weight of charges.

NOTE.—In ballistic calculations the weight of the projectile is always taken in pounds avoirdupois (7,000 grains—1 lb.), but in tables of weights of rifle projectiles the weight is usually given in grains. For instance, the weight of the projectile of the Mark VII cartridge is 174 grains and is always so stated.

*Angles.*—Angles of projection, etc., are measured in degrees and minutes. The circle is divided into 360°. A right angle thus contains 90°. Each degree is divided into 60 minutes (') and each minute can be further subdivided into 60 seconds ("). To avoid confusion the minute is usually referred to as a " minute of angle."

A minute of angle is subtended by 1·047 inch for each 100 yards of range. For rough working it is sufficient to regard a minute of angle as being subtended by 1 inch for each 100 yards of range.

## CHAPTER II—SECTION 1

## THE HISTORY OF THE SWORD, LANCE AND BAYONET

*The Origin of the Sword and Early History.*

In early ages primitive man lived chiefly by hunting, and the comparative weakness of his physical armament against the teeth, claws and horns of the animals that he pursued was compensated for by the possession of prehensile hands and the wit to use them. He could cast stones and wield a weapon, at first a branch or bone, and it was the development of what was at first instinct into intelligent application that started him upon the long sequence of invention and experiment that has led to the highly complicated engines of war of the present day.

His first non-missile weapons were of course instruments of concussion and in the development of his armoury the sword came comparatively late.

The club or hammer gave rise to the flint axe, and at the same time he learnt to thrust with the spear.

The knife or dagger preceded the sword, at first probably in the form of a pointed stick with which he skewered or cut up his meat, and so long as he was dependent on wood or flint, the use of a sword or knife as an efficient cutting implement was delayed. Without doubt a kind of wooden sword was introduced and specimens of such a weapon have been found in all continents and are in use among some savage tribes to this day, but the sword could not have been of great use to man until he had learnt to work in metal.

The sword of the Bronze Age was leaf-shaped, two-edged, pointed and with the blade and hilt cast in one piece. It was generally about 24 inches long.

A large number of specimens have been found and are to be seen in most museums of importance; they are often remarkably well-preserved and exhibit great beauty of form.

With comparatively slight variations the bronze leaf-shaped sword survived into historic times and only disappeared with the discovery of forged iron, which provided a much keener and stronger blade. The leaf-shape may be traced in the Greek Xiphos, of which the short Roman sword was but a subsequent variation. The sword as a long, heavy, trenchant weapon is first heard of with the Barbarian invasions.

*The parts of the Sword.*

It was with the introduction of the iron or steel blade that the sword resolved itself into the component parts which it still retains, namely the blade and the hilt. The blade is secured to the hilt by a narrow extension known as the " tang." The balance of this weapon was helped by the addition of the " pommel " and the hand protected by a cross-guard called the " quillons." The tang was enclosed in wood, horn, bone or other substance to provide a satisfactory " grip." It is the subsequent transformation and development of these parts to suit the weapon for use under varying conditions and for varying purposes changing with the evolution of society and, above all, with the progress and vicissitudes of tactics, that form the history of the sword.

*The Blade.*

The blade can be either curved or straight. The object of the curve is to increase the cutting power by reducing resistance, but it is clear that the curve must always be at the expense of the thrust; the curve almost destroys the use of the sword as a defensive weapon. The curved blade has always been the favourite of the Eastern races. It can be traced in the " Akinakès " of Greek historic times, and examples in the hands of the Dacians can be seen represented on the Trajan column (A.D. 101–106). But only in the form of the falchion is it to be found in Western Europe in the Middle Ages, when the straight blade was practically universal. There was a fine example depicted on the wall of the Painted Chamber at Westminster (c. 1225) and an actual specimen attributed to the early years of the 14th century is preserved in the Norwich Museum. (Laking, European Armour, Vol. I.)

It was not until the 15th century that the curved weapon appeared with any frequency in European armies. The encroachments of the Turks (with whom the curved sword was ever a favourite weapon) in Eastern Europe may have had much to do with this fact, and the elegant and rich decoration of oriental specimens probably encouraged its importation through Venice to more western lands. Certainly MS. illuminations of the 15th century depict men of rank carrying obviously oriental specimens. At the end of the 15th century the Italian wars familiarized European armies with the curved sabre-like weapon of the *Stradiots*, who were Albanian light-horse employed as mercenaries.

The curved sword as a heavy cavalry sabre was more widely adopted after the Turkish wars in Hungary in the 17th century, and soon every European army had its regiments of Hussars. But apart from the tendency always to copy the arms and uniform of the last successful army on the Continent, no matter whose it was, the cavalry tradition from the time of Marlborough to comparatively recent times was "shock tactics." Hence the survival for so long of the curved sabre.

The section of the blade depends upon whether it is intended to use it for cut or thrust. The best cut is from a wedge shape, hence the back edged blade; the best thrust is from a blade of triangular and square section, gradually brought to a point. Necessarily, for war, the sword blade is fashioned to deliver and to meet both cut and thrust. The grooving of the blade is, of course, a question of weight and balance.

*The Hilt.*

The modifications of the hilt were confined to slight changes in the form of the pommel or quillons and the length of the grip, as subsequent illustrations will show, until the development of the defensive use of the sword and the art of fence. The gradual disuse of armour in the 16th century, leaving the hand unprotected, naturally turned man's mind to the invention of a defensive hilt to replace the steel gauntlet. The responsibility of guarding the knuckles and wrist was therefore left to the hilt itself, and accordingly all kinds of bars, rings and twisted guards to protect the hand against all varieties of points and cuts appear, which in their fullest form are seen in the basket hilt of the so-called Scottish "claymore" and in the swept hilts of 16th century rapiers. But, be it remembered, the complicated hilt was far more the result of the increased use of the sword in private life than of its use in war.

*The Lance.*

Before going on to consider the evolution of the sword in Western Europe, from classical times and throughout subsequent centuries, it is convenient to refer here to the early history of the lance. The "lance" is the name given to the "spear" when used by the mounted man, and the spear itself was in early years but a longer form of the javelin. It consisted of two parts, the wooden haft and the head. Early specimens of heads may be seen in most museums and do not vary in great degree. The spear or lance of classical times was a light weapon often provided with a thong for attachment to the wrist, and in the hands of cavalry was used with an overhand thrust. The most serious factor in the development of the lance was the introduction of the stirrup during the Barbarian Invasions which accompanied and succeeded the decline and fall of the Roman Empire. This was of great moment, for it enabled the rider to balance himself differently and so couch his lance and charge home with it with the full force of his onset instead of thrusting with the strength of his arm alone. At the same time, be it noted, he was able to stand in his stirrups and swing a heavier sword.

In old MSS. we note the lance heads often represented with short bars or lugs. It is probable that the lance was in much earlier use than the sword and, in fact, throughout history the most effective weapon for offence both with mounted men and infantry (when it was usually called a pike) has been the lance. The rifle with fixed bayonet is, of course, the modern form and lineal descendant of the pike. The importance of the lance is shown by its representation in MSS., coins and seals as the arms carried by kings and princes. In the Bayeux tapestry William the Conqueror is portrayed generally with lance or mace and only once with the sword.

The mediæval use of the lance for shock tactics in the hands of heavy cavalry will be referred to later. It became heavier as time went on, and the leaf-shaped head was reduced soon to a pointed ferrule mounted on the head of a shaft which swelled towards the hand where it narrowed for the grip and then was increased in diameter for the sake of balance at the butt. A guard to the hand called a "vamplate" was often added, which was of smaller size and flatter shape than that used in tournaments. Such was the war lance of the 15th century which in every respect was lighter than the lance of the joust. Philippe de Commines mentions that at the battle of Fornovo (1495) the Italian cavalry used "bourdenasses," hollow lances, light and easily handled and gaily painted, but of no use in serious fighting. When the heavily armed men-at-arms ceased to figure in European warfare the lance suffered a temporary eclipse, but it was reintroduced from the East in a lighter form, and as such has remained to the present day. The tendency of all European armies since 1870 has been to increase the lance strength of cavalry, the idea being that the use of cavalry, as distinguished from its use as mounted infantry, is by shock, and for shock the lance is far more effective than the sword. The experience of the Great War has not been favourable to the lance, and its use by British cavalry has now been definitely abandoned.

*The Evolution of the Sword.*

We will now consider the sword in the centuries which follow the fall of the Roman Empire, and trace its various forms, called for by the changing tactics of European warfare.

Until comparatively modern times, the sword always tended to be an individual weapon, at first worn only by chiefs and men of rank. It is true as we learn from Polybius (c. 204–122 B.C.) that the Roman Army, which was the first drilled force as we understand the term, adopted it as part of their establishment for all ranks, but the Roman army was exceptional, and when we again meet with such a drilled force, with such a discipline, fire arms and bayonets are the effective weapons of infantry.

The Romans too, as a highly civilized power, strove to shorten their campaigns and to force an issue as quickly as possible, hence the Roman Legion always had in view the hand-to-hand fight which would give victory and be immediately followed by the conclusion of a peace. The barbarian enemy could afford to remain at war, which may be said to have been their main occupation.

The Roman sword (Fig. 1) was short, having rather a broad blade with two parallel edges and a point formed by an obtuse angle. The hilt was cruciform, and the length of the whole weapon seldom more than 28 inches. At a later period, as shown by the Trajan column, it was lengthened, and a weapon with one edge, called the "spatha," was introduced.

The Roman legionary cast his *pilum* or javelin and then closed with the enemy with his sword. The Romans relied on the thrust rather than on the cut, as the wound by the point was delivered more quickly and was far more deadly, and the result of Cæsar's command to his cavalry to thrust at the faces of Pompey's men at Pharsalia is a well-known instance of the Roman appreciation of the method of attack.

In no period of the history of war was the sword so effectively used. After the Roman Legion disappeared in the 5th century, and as long as the ascendancy of the horseman endured, the lance and bow replace the sword and *pilum* as the principal weapons.

The use of the sword as a secondary and a personal weapon (*i.e.*, not part of the shock tactics of the mass of an army, but as a weapon of the individual to be resorted to after engagement at close quarters), was destined to last throughout the middle ages.

In the latter days of the Roman Empire the tactics and arms of her enemies had improved ; long experience of Roman tactics had made them more efficient, and at the same time the national composition and the quality of the Roman legions had deteriorated. The time arrived when it was the turn of the Romans to learn from their enemies and to adopt tactics better suited to an army now adulterated with such a large proportion of foreigners. Among many changes, the most notable was the increased proportion of cavalry to infantry. At the battle of Adrianople (377 A.D.) they succumbed to the cavalry of the Goths. These were armed with lance and sword and Sir Charles Oman calls the Gothic cavalryman " the lineal ancestor of all the knights of the Middle Ages, the inauguration of that ascendancy of the horseman which was to endure for a thousand years." (*Art of War*, p. 14.)

*The Early Middle Ages.*

As has been said above, the introduction of the stirrup enabled the mounted man to use longer and heavier weapons. But in England during the Early Middle Ages, before the Norman Conquest, the people of this country fought on foot. Changes in England always took place later than on the Continent, owing to her insular position and absence of contact with other nations. Their weapons differed from those of their neighbours the Franks, who used a heavy sword known as the *scramasax* and the famous *francisque*, a variety of axe.

*The Sword in pre-Conquest England.*

The Anglo-Saxon used sword and spear. There are several specimens of the latter in the London Museum. The heads were generally of the leaf pattern, and in some MSS. they are depicted with short bars or lugs issuing from the neck immediately below the head, as we remarked in the drawings of the Bayeux tapestry. The Saxon sword (Fig. 2) was a short double-edged and pointed weapon known as the *seax*, of which we illustrate an example not unlike a Cingalese dagger.

The Vikings, who began to appear in the 8th century, fought in battle as foot soldiers and only used horses which they commandeered on landing for covering the ground in their rapid forays. They used sword and axe. The sword commonly described as the " Viking sword " is probably a form used by them only in later times, that is to say, in the 10th and 11th centuries. The earlier form was short and leaf-shaped, an enriched hilt of the later long-bladed type is in the Ashmolean Museum, Oxford, and exemplifies the typical quillons and spreading pommel.

The Dane was only beaten by the cavalry which the organization of feudalism had developed on the Continent. But in England both sides continued to fight on foot, and the Danes introduced their long-handled axe into this country where it became a national weapon. It was an Anglo-Danish force fighting on foot with sword and axe which was defeated by the Norman army arranged in three lines respectively of archers, spear-men and cavalry with lance and sword. Only after the battle of Hastings were cavalry introduced into England and the Anglo-Norman armies henceforth had their due proportion of heavy mounted men.

*The Sword at Hastings*, 1066 A.D.

The sword at the time of the battle of Hastings (1066 A.D.) had a broad blade of a lozenge section with two cutting edges which at the extremity drew to a point (Fig. 3). The quillons were inclined to be thick and rather short, and the pommel was very distinctive, being tri-lobed or oval. The Bayeux tapestry, worked about 1125, well illustrates the arms of the period, and Harold is shown with a true lobated pommelled sword.

*The Sword in the Byzantine Army of the 9th Century.*

The wars of the Crusades followed, but it is worth while to look back for a moment at the Byzantine armies of the previous period which offer a most interesting study of the first highly-developed military organization since the disappearance of the Roman legions.

Undoubtedly they were better disciplined and better led than their contemporaries in Western Europe. The Byzantine generals combined the use of cavalry and infantry and still used some of the latter to fight in the manner of the old Roman legion. They used the bow, lance and sword, and the last was a somewhat larger variety of the Roman form.

*The Saracens.*

The Byzantine armies taught the Saracens much, and when the forces of Western Europe commenced their campaigns in Asia Minor they found the rapidly moving Saracen armies, with their skilful use of mounted archers, more than a match for them. Only when the Western cavalry could charge home, use their crossbows and crush the enemy by sheer force did they succeed (*e.g.* battle of Arsuf, 1191 A.D.) and then their lances and straight swords took a heavy toll of life. It should be noted that in this battle Richard used a front line of infantry kneeling, armed with the lance to receive Saracenic cavalry. This precedent was followed throughout the ensuing centuries.

*The 12th and 13th Centuries.*

From the time of the Norman Conquest until the end of the 12th century the heavy feudal cavalry was the main factor in all the battles of Western Europe. The idea of a cavalry shock at a particular moment in a battle, and for a particular purpose at a definite point (distinct from the pursuit of a beaten enemy), charging with drawn sword was not destined to appear until comparatively modern times. The feudal warrior thought that for all purposes, and at all times, the soldier should be mounted. The sword remained the ideal weapon of the individual warrior to be resorted to when his lance was broken in the charge and both armies were locked together and at handstrokes.* No doubt the mace or the axe was often used, but the knight died fighting with sword in hand. How personal a weapon the sword was can be gathered from the numerous legends of famous swords often bearing the traditions of miraculous powers by saints or heroes. Weyland's sword from the Norse *Sagas*, the *Excalibur* of Arthur, and the *Joyeuse* of Charlemagne come readily to mind.

The Battle of Bouvines (1214) is the typical battle of the 13th century in the West and is an example of hand-to-hand individual fighting which decided the battle after the initial charge with lances. For this reason it has been cited as a victory of the sword.

The Battle of Benevento (1226) is also often quoted as an example of a victory by the sword but probably only because the tale has come down to us that the French received the command to "give point" with their swords when in close combat with the Germans, who only used their great swords to cut.

The sword of the 13th century (Fig. 4) had a short grip, longer quillons than many of the earlier examples, and a pommel in the shape of a ball or of the distinctive brazil nut pattern.

As before, the blades were two-edged and straight, though slightly tapering towards the end and terminating in a rounded angle.

*The 14th Century.*

At the end of the 13th century the feudal force of mailed cavalry was at the height and also at the end of its power. The rise of the long bow in England was soon to prove fatal

---

* *Cf.* The illustration of two mounted men fighting in the Psalter of St. Albans, preserved in Hildesheim (*Tapisserie de Bayeux*, by Commandant Lefebre des Noëttes, page 14, published at Caen, 1912).

to the French cavalry, and at Courtrai in 1302 the latter suffered their first defeat at the hands of a force of plebeian infantry, with front rank kneeling with the pike, the rear rank standing with the pike at the carry and behind them the bowmen bending their bows without interruption.

The English archer revolutionized warfare in Western Europe.

Our King Edward III and his leaders appreciated the tactical importance of a defensive battle by the skilful disposition of the archers and dismounted men-at-arms on a carefully chosen position. Crecy (1346) became the tactical precedent. This type of fighting exactly suited, then as now, the national character. Archers had already proved their worth at Falkirk in 1298, and at Bannockburn the English light horse had died by hundreds before the lines of the Scottish pikemen; the system of meeting an offensive by dismounted defence on rising ground with well-protected flanks had already been used at Dupplin (1332), but at Crecy the defeat of the most famous chivalry of Europe was spectacular. The same tactics were again overwhelmingly successful at Poitiers (1356), and at Agincourt (1415) the French cross-bow men were put out of action by English bowmen before they had become engaged and the French cavalry failed to account for more than 100 men. These battles taught the French that a battle could not now be won by charging cavalry.

It must be noted that the English campaigns in France in the 14th and 15th centuries were the most important events in the military history of the time. Spain had not yet come to the fore and elsewhere the art of war was almost non-existent. In the disturbed politics of Central Europe the heavy cavalry still remained the essential factor, but they were overwhelmed at Sempach by the Swiss mountaineers and never stayed to try conclusions with Zizka's armed wagons, which without exaggeration were the forerunners of the tank. While in Italy, professional armies had arisen under *Condottieri*, who, if we are to believe Macchiavelli, made so successful a science of treachery and strategy that they could conduct campaigns without bloodshed.

The decline in the importance of cavalry in the pitched battles of the 14th century did not mean the disuse of the lance, for this as we have seen was cut short and used by the dismounted men-at-arms to form a defensive front. When so used, it came to be called the "Pike." When the enemy's attack had burst upon the line and the two forces were locked together (in cases where the archers had not broken down the attack before reaching hand-strokes) the dismounted man-at-arms drew his sword or axe, and as at Auray (1364) fought out the conclusion hand to hand and foot to foot. The archer meanwhile threw down his bow and fell upon the flanks with a short sword, or more often a *maul* or axe. The introduction of plate armour, which was rapidly progressing at this date, was concurrent and probably connected with the use of heavier hand weapons.

The sword of the man-at-arms of the 14th century was longer and heavier than that of the 13th. He also made use of the "estoc," or thrusting sword, and frequently carried both at his saddle. A typical example of the period shows a longer, more tapering grip, with slender but strong quillons, usually straight, but often drooping slightly towards the point or turned down at the extremities. We illustrate an example (Fig. 5) which shows its severe but elegant proportions. All have the usual wheel pommel.

*The 15th Century.*

With the arrival of the 15th century the grip became even longer and more tapering, and the pommel, though still often of the wheel form, which lent itself to highly ornamental enrichment, particularly in the insertion of the owner's coat-of-arms in enamels, was often pear-shaped. The straight slender quillons gave great dignity to the weapon, and a variety known as the "bastard" or "hand-and-a-half" sword, enabled the wielder by virtue of its additional length of grip to use both hands on the hilt in the downward stroke. This form of sword is well shown in the effigy of Lord Berdolf in Dennington Church, Suffolk. (Figs. 6, 7 and 7a).

The true single-handed sword was often used with the first finger or the two first fingers over the quillons and we find either one or two rings just below the quillons and forming part of the shoulder of the blade added to protect these fingers. These are the first additional guards to the simple cross-guard. Towards the end of the century a knuckle guard is added and gradually the hilt becomes more and more complicated.

Specimens of the curved scimitar of the 15th century are rare, but they are constantly seen portrayed in Missals.

The French King Charles VII (who reigned 1422 to 1461), in his efforts to retrieve the disasters of the English war and reconstitute his kingdom as an united and centralized nation, organized his army on a regular footing.

By the *Ordonnance sur la Gendarmerie* (1439) he established a force of cavalry wearing body armour, and armed with a heavy lance and sword, and in 1448 he created the "*francs archiers*" to supp'y an arm in which France had always been fatally weak. These were efficiently equipped by regulation and carried a short sword as also did the billmen who were becoming more and more a feature of the large armies of the second half of the 15th century, and were the forerunners of the Swiss halberdiers and pikemen of later days. The Burgundian wars demanded a larger army and continuous recruiting, so that it was necessary to raise as many mercenaries as possible.

The use of mercenaries increased in every country as warfare became professionalized. The mediæval archer needed no more training than the skill he acquired on the village green, and the mounted man learnt his work riding in the chase or in the tournament, but the musketeers and bombardiers were necessarily professionals and the more systematized tactics of the day required drilled troops in organized numbers under experienced commanders.

*The Italian Wars,* 1494–1529.

At the end of the 15th and during the first-half of the 16th centuries, Italy was the battle ground of Europe, and the great armies which took the field required the brain of the strategist as well as the eye of the tactician. At the same time, several new arms made their appearance. The man-at-arms clad in his heavy armour disappears (he was too easy a prey to firearms) and the more mobile demi-lancers or light armed cavalry (the successors of the *hobilars*), and the *Stradiots*, light Albanian horse, armed with the sabre, are the new cavalry. Scanderbeg, the great Albanian hero, wielded the famous sabre of enormous size which is still preserved at Vienna.

At the commencement of the 16th century, the Swiss mercenary infantry came on to the scene with halberds, pikes and short swords of the *landsknecht* type. The *landsknechts* were the German heavy infantry of the time, and some corps carried enormous two-handed swords, of which the blades often had waved edges (Fig. 8a). Troops so armed appear in engravings and pictures of the first half of the 16th century, but cannot have been very effective, owing to the extended formation necessary for the manipulation of their weapons. They received double pay and always formed the colour guard. The short *landsknecht* sword with a guard usually of two rings was purely an infantry weapon (Fig. 8).

Long after firearms had assumed great importance in war, the spear or lance, in the form of the pike, remained the staple infantry weapon and the mainstay of the battle. It was a light straight pole of ash, 16 to 18 feet long, with a small lozenge-shaped spearhead without barbs. Borne by armoured men in close order, its length allowed the pike heads of four ranks to be presented to attacking cavalry. Even one or two ranks of "shot," themselves incapable of withstanding a charge, could obtain shelter close to the front rank of kneeling pikemen, beneath their projecting weapons. With the advent of firearms, body armour became obsolescent, but so complete was the resistance offered to cavalry by pikemen that the pistol was adopted for mounted troops as a weapon giving a better probability of breaking their solid ranks than the sword, which, however, was retained for the ensuing hand-to-hand fight.

It was the pike as a weapon of offence and defence which provided the basis for tactics, which were modified as the fire power of guns and small arms increased, until the work of developing a composite weapon combining the pike with the firearm, was completed about the year 1700.

Before proceeding to review the history of the sword in the armies of the 17th century, a word must be said on its use as a private weapon, for in this use it had undergone rapid developments.

*The Rapier of the 16th Century.*

During the 14th and 15th centuries the usual sword for civilian wear was the basilard, a broad-bladed weapon with a rounded point and short-curved quillons. It appeared in its most ornate form in Italy as the *cinquedea*. But in the 16th century the rapier was introduced. Its development was coincident with the progress of duelling and the art of fence, which was now studied scientifically for the first time. Various schools developed various types of rapier hilts. To the quillons were added the additional defences of knuckle-guard and thumb-ring which had already been foreshadowed in the second-half of the 15th century, but the 16th century was the great period of the development of the hilt. Most rapiers are of the type known colloquially as " swept " hilted, which gave the maximum protection to the hand (Fig. 9). The protective twists of the hilt defences to the hand even began to encroach on the blade. The cup-hilted rapier (Fig. 10) is another variety and often attained very great beauty of design. In a sense the cup was in the nature of a miniature buckler.

The rapier lent itself to ornamentation of a most elaborate character. The hilts were often of steel, chiselled with consummate artistic skill, and the forging and tempering of blades became an art which made Toledo, in particular, famous for their manufacture. Germany, too, produced fine sword blades, among which may be mentioned those of Solingen. The Italian swordsmith Andrea Ferrara, of Belluno (1530–1583), has acquired an almost legendary fame, and his name is found on countless numbers of swords, mostly forged long after his death, and in places remote from where his family still carried on the business. A great many Scottish basket-hilted broadswords of the 17th century, popularly known as " claymores," bear his name.

From the middle to the end of the 16th century, cavalry were also armed with the *estoc* (Fig. 11), or purely thrusting sword, of which a common form was especially used by the Saxon soldiery.

The Scottish " claidheam-mor " was a two-handed sword of the end of the 16th century and forms an interesting class of its own.

The basket hilt first appeared on the " Schiavona " (Fig. 12), introduced through Venice in the 15th century by the Slav immigrants, on the other side of the Adriatic. In a modified form this type of hilt became the cavalry broadsword of the 17th century— a stiff, straight sword, often with a back-edge. The heavy, straight-bladed " tuck " of the cavalry of the time of the Civil War in England was usually furnished with a hilt of the basket type (Figs. 13 and 14).

*The 17th Century.*

The wars in the Low countries in the latter years of the 17th century produced large numbers of writers on tactics and weapons. The steady improvement of firearms was, however, having its effect. Conservative writers continued to advocate the wearing of armour on the limbs, and the use of the lance and bow, but the test of war sealed its fate, and the cavalry in half-armour could make, in the tactics of the time, greater use of the pistol and broadsword, while the infantry became pikemen and musketeers.

The Thirty Years' War (1618–1648) in Germany had an immense influence on European methods of warfare, as we read in the literature of the rival schools of Dutch and Swedish tacticians. Gustavus Adolphus, King of Sweden, was a great military reformer, and he carried out his rapid movements by using light field pieces and light cavalry who wore buff coats, helmets and cuirasses, and charged with pistol and broadsword. Gustavus Adolphus's tactics set the precedent for the armies in our Civil War in England, and the cavalry of both Cavaliers and Roundheads, similarly organized and equipped, played a most important part in them. They charged in double line, firing on contact, then, wheeling round on either flank, recharged with their swords held at arm's length to give point.

The 17th century was essentially the period when the sword in the hands of the cavalry was used most effectively. The battle tactics did not conceive victory by cavalry alone, but by using cavalry for shock at suitable moments in a battle, in which infantry played the chief part. The infantry throughout most of the century were a mixed force of pikemen and musketeers, the former in a continually decreasing proportion, but it was not until the introduction of the flint-lock, as a service weapon, at the end of the century that the infantry became the main strength of an army as a force using the musket as its distinguishing weapon. The invention of the bayonet meant that the musketeer combined the services of the pikeman with his own.

In becoming an ordinary service arm of the trooper, as distinguished from the personal weapon of the knight and gentleman, the sword lost much of its beauty. The knightly sword of the middle ages was remarkable for its elegance of form and severe proportions, the rapier of the 16th century for its craftsmanship and richness of decoration, but the troopers' sword of the 17th century was neither elegant nor rich, though it made history when Cromwell's cavalry thundered into battle at Dunbar (1650) and swept all before them.

At the end of the 17th century the straight cavalry sword gave way to the heavy curved cavalry sabre which came from Hungary with the Hussars. The " hanger " was another sword of this date. It was the name of the short sword, always in vogue, and especially used on board ship. The rapier remained as a private weapon. In Italy the cup-hilted rapier remained in favour throughout the first half of the century, but the plain hilt with a single bar from the pommel to near the quillons which were formed by two wings in front of the pas d'ane (this had now become two slender rings for forefinger and thumb) was introduced from France and became the gentleman's weapon of the second half of the 17th century. The blade was usually triangular in section and of the form known as " Colichemarde " which is a French corruption of Königsmark. The small sword of the 18th century, which still exists as the court sword of to-day, was a smaller form of this weapon.

With the Restoration (1660) the present British regular army came into existence, and the form of the regulation sword varied with the dress regulations issued from the Horse Guards by the Board of General Officers, and later by the War Office when that Department came into existence. The manufacture of swords by the thousand according to sealed pattern did not prevent the use of a great variety of forms, from the trooper's great sabre to the bandsman's flat curved sword and the pseudo-Roman cruciform sword of the Commissariat corps of the mid-19th century. To go into all the types in use in the 18th and 19th centuries would far exceed the limits of this introduction.

*The 18th and 19th Centuries.*

We illustrate (Fig. 15) the sword of the cavalry of the time of William and Mary, an English weapon made at Shotley Mill in Yorkshire, the fine sword of the French cavalry at Waterloo (Fig. 16) (to be compared to the cavalry sword of to-day), and our own cavalry sword of the same time (Fig. 17), a curved sabre which still survives in the full dress sword of our generals. Finally, we illustrate some of the regulation swords used in 1910; the Highland regiment staff-serjeant sword (Fig. 18), the staff-serjeant's sword, dismounted service pattern 1897 (Fig. 19); the cavalry sword, 1899 pattern (Fig. 20), and the cavalry sword, 1908 pattern, Mark I (Fig. 21).

Throughout the ages the sword has exercised a hold on popular imagination that no other weapon has ever equalled. In speech and literature from the psalmist and the prophets down to the political orator of the present day, it has been used as a synonym of war itself. It has been, and is used, to confer the honour of knighthood and to symbolize civic dignity. But it is because it was for ages the weapon with which the issue was contested, man to man, in hot blood that it has appealed to the minds of generations. It has neither the brutality of the weapons of contusion, the mace and the axe, nor can it kill men from afar and in cold blood like arrow, bolt or bullet, but its keen blade, requiring skill to forge and dexterity to use, made it the weapon of the gentleman, and, in company with the lance, the arm of that force which from of old has represented the best blood of the nation.

*The Bayonet.*

We have referred to the use by infantry of the lance or spear which was afterwards called the " pike." The idea of converting a firearm into a pike by fitting a blade into its muzzle must often have occurred to the inventive, and this arrangement appears to have been introduced in the 16th century for sporting purposes. It was difficult to use such a weapon with effect in the soldier's hand-to-hand fighting, for the match lock musket was heavy and unhandy, and the left hand was encumbered with a length of match with both ends smouldering as well as with the rest. The coming of the fusil or flint lock opened the way for the bayonet, which was first applied to it about the year 1647, if Marshal de Puységur is not mistaken in speaking of its use by troops under his command at that time.

The name first appears as meaning a knife or dagger to hang at the girdle, apparently unconnected with fire arms. Its origin is uncertain. Another form of the word is " bagonet," which is found before 1700 and survived, in America at least, till after the middle of the 19th century.

The bayonet seems at first to have generally taken the form of a two-edged and pointed dagger about a foot long, with a plug handle tapered so as to fit into the muzzle of the piece; quillons were soon added. In our own army it was used by the Tangier Regiments in 1663, and was issued to the Regiment of Dragoons in 1672, though from them it was afterwards withdrawn. After 1678 the British bayonet was made with a single edge, a guard being added, no doubt to make it convenient for use as a short sword. The 7th Fusiliers, when raised in 1685, had the bayonet, and the Foot Guards were armed with it in 1686. There is no mention of it in the official drill book of 1684, but Speed in 1688 gives the words of command " Draw your dagger," " Screw it into the muzzle of your firelock," etc., and a similar drill appears in the " General Exercise of the Prince of Orange, 1689."

The plug bayonet was for use primarily when the soldier had exhausted his ammunition, since it blocked the bore of the piece. Endeavours were soon made to avoid this limitation; de Puységur tells of a bayonet fixed to the barrel by means of rings as early as 1678. At Killiecrankie (1689) the drawbacks of the plug bayonet were conspicuous. General Mackay's infantry were overwhelmed by the Highlanders' charge in the interval between ceasing fire and fixing their bayonets. Mackay promptly devised a method of fastening the bayonet by rings. The ring bayonet, however, was never generally adopted.

The socket bayonet soon appeared, but did not fully meet requirements till it had been perfected by the addition of a curved neck joining the blade to the socket, so that the former stood well clear of the line of the bore. De Puységur states that in the war of 1688 the French proposed to suppress pikes and matchlocks, and to use only the firelock with the bayonet " à douille," but that the want of uniformity in the size of the musket barrels made the scheme ineffective, so that it was abandoned; that subsequently other nations having taken to socket bayonets, the French had to return to them; and that by 1701 pikes were no longer in use. It seems clear that the socket bayonet was not in use in the British Army till after 1700.

In or about 1706 was published " The Duke of Marlborough's new Exercise of Firelocks and Bayonets," based upon the Pike Exercise, from which such commands as Charge, Advance, Trail, and Slope are survivals.

The factor of length in the bayonet arm was at first of great importance in resisting cavalry, and the blade of the bayonet was accordingly soon lengthened. In view of the primary importance of the thrust, it ultimately assumed the triangular reed-shaped section which was the usual form until the advent of the magazine rifle 35 years ago, when a blade of proportions not very different from those of the original plug bayonet again came into use. This had become feasible because close order formations to resist cavalry were no longer needed; the rapidity of fire of the breechloader, and still more of the magazine rifle, being a sufficient defence. A substitute for the pike is thus no longer needed, and the use of the bayonet is in the main for hand-to-hand fighting by the individual, for which purpose handiness is of much greater importance than length.

Attempts have been made at various times to devise a bayonet which should be an integral part of the fire arm. Thus we find 18th century blunderbusses fitted with a hinged spike to fold under the barrel, which springs into position when a catch is released; pistols, too, have been made with a dagger-like spike similarly arranged. A long steel spike lying in a socket in the fore end, of the infantry arm, and sliding forward so as to project in front of the muzzle, is a variant long ago produced. Such devices have not so far met with favour. The weight of a bayonet strong enough for service use is substantial, and it has been found to be carried with less effort by the soldier on his waist belt than as an inseparable part of his firearm; it has also been thought an advantage that he should not be entirely disarmed if he should be separated from his rifle. It would apparently require a considerable alteration in the weight of the rifle or the bayonet or both to make any such system practicable.

The idea of arming troops with a sword which could be affixed to the muzzle and serve as a bayonet is an old one, but was never generally adopted. The Rifle Brigade were armed with the sword bayonet in 1800, and its use has continued in Rifle Regiments and certain other troops until recent times. In modern conditions its additional weight is a great drawback. It is very rarely that a sword could be of use to the infantryman, whose bayonet, being of dagger form, can in the last resort be used as a hand weapon.

Among the bayonets used with the Martini-Henry rifle were the bayonet, long, C (Fig. 22), and the sword bayonet, M.H. rifle, converted (Fig. 23). The latter was originally used with the muzzle-loading rifle of 1860 (converted to Snider in 1867) and was converted for use with the M.H. rifle. With the L.M.L.E. rifle the sword bayonet 1888, Mark Ic (Fig. 24), was used. The present service bayonet in use with the S.M.L.E., Mark III, is the sword bayonet 1907, Mark Ic (Fig. 25).

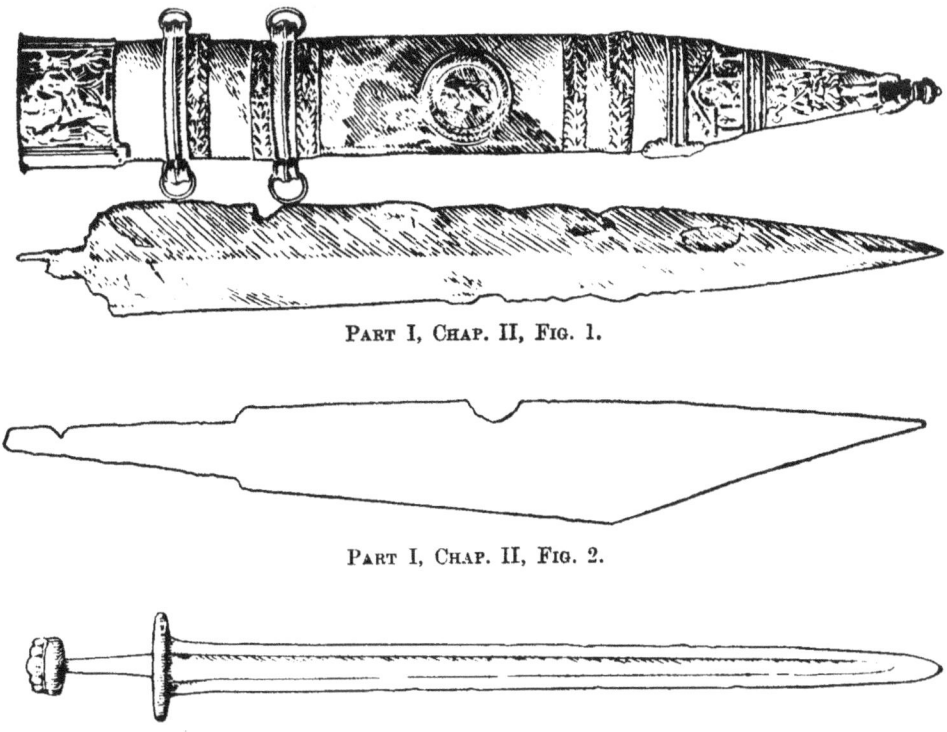

Part I, Chap. II, Fig. 1.

Part I, Chap. II, Fig. 2.

Part I, Chap. II, Fig. 3.

80

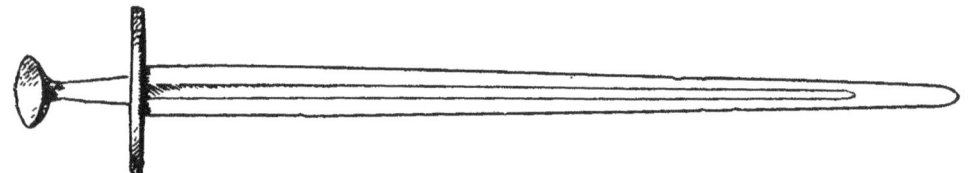

Part I, Chap. II, Fig. 4.

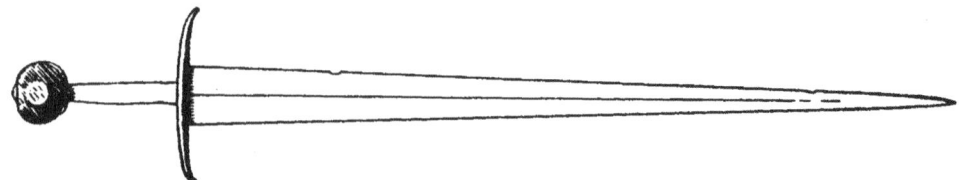

Part I, Chap. II, Fig. 5.

Part I, Chap. II, Fig. 6

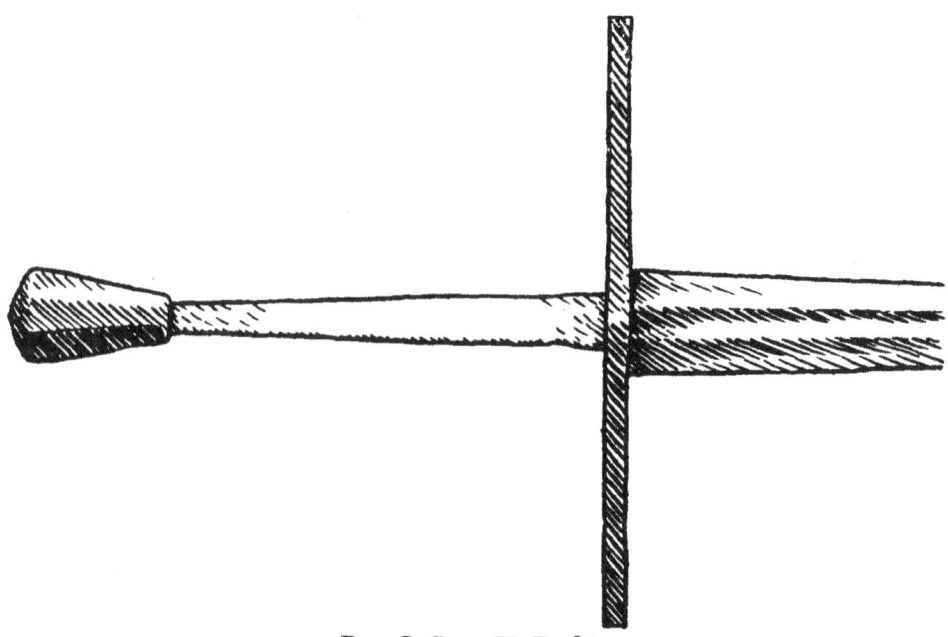

Part I, Chap. II, Fig. 7.

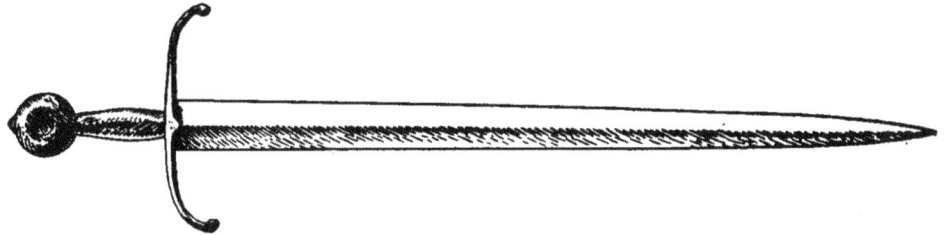

Part I, Chap. II, Fig. 7a.

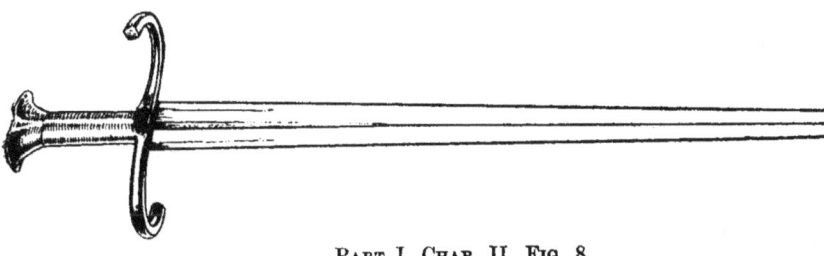

Part I, Chap. II, Fig. 8.

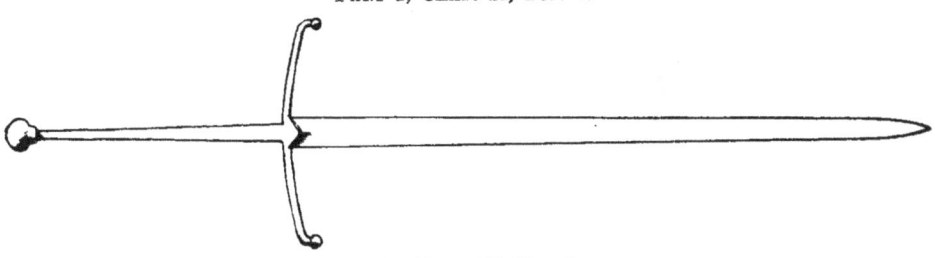

Part I, Chap. II, Fig. 8a.

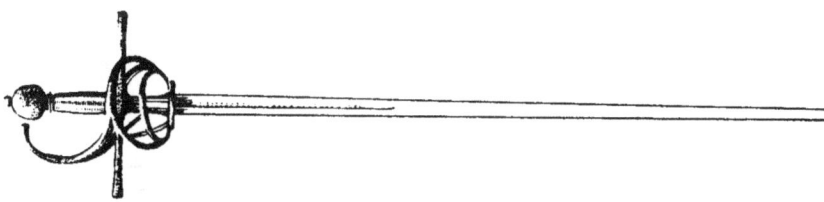

Part I, Chap. II, Fig. 9.

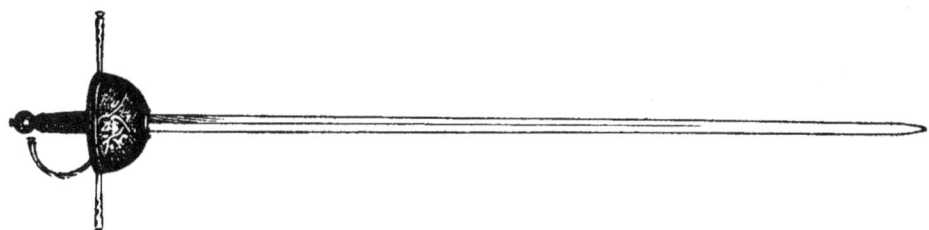

Part I, Chap. II, Fig. 10.

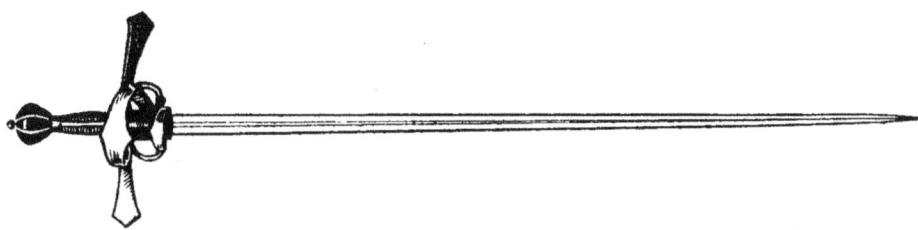

Part I, Chap. II, Fig. 11.

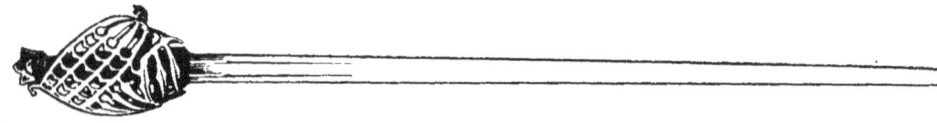

Part I, Chap. II, Fig. 12.

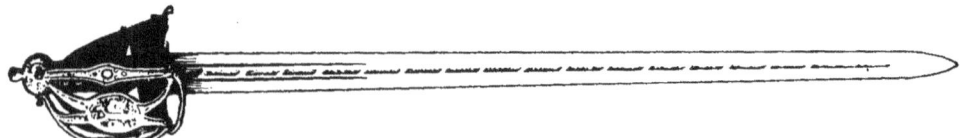

Part I, Chap. II, Fig. 13.

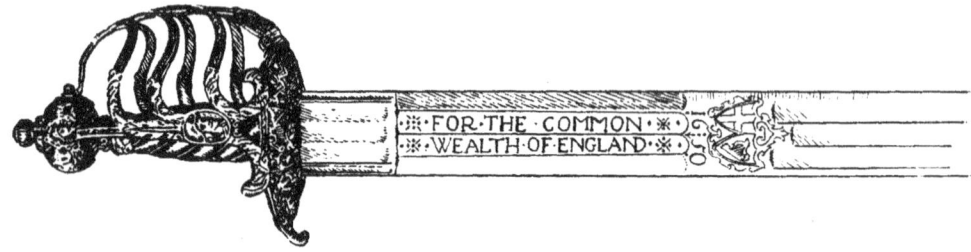

Part I, Chap. II, Fig. 14.

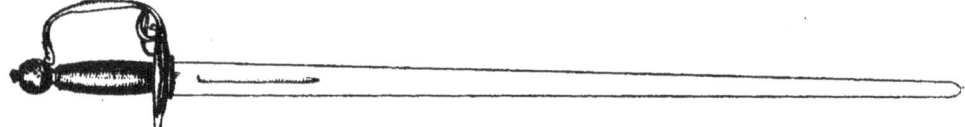

Part I, Chap. II, Fig. 15.

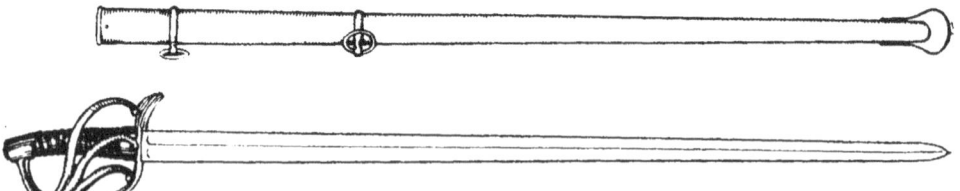

Part I, Chap. II, Fig. 16.

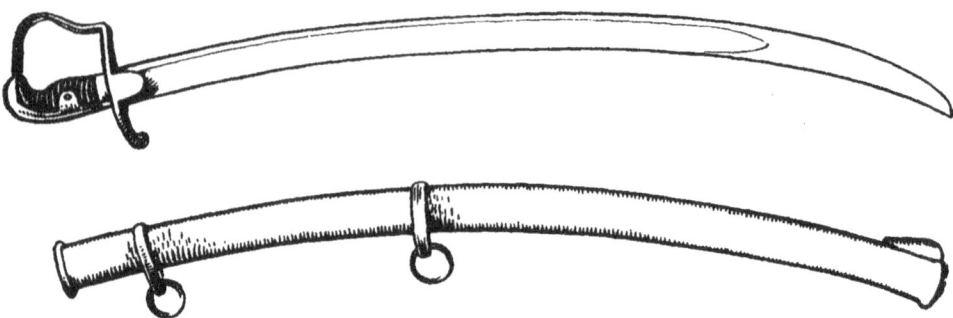

Part I, Chap. II, Fig. 17.

83

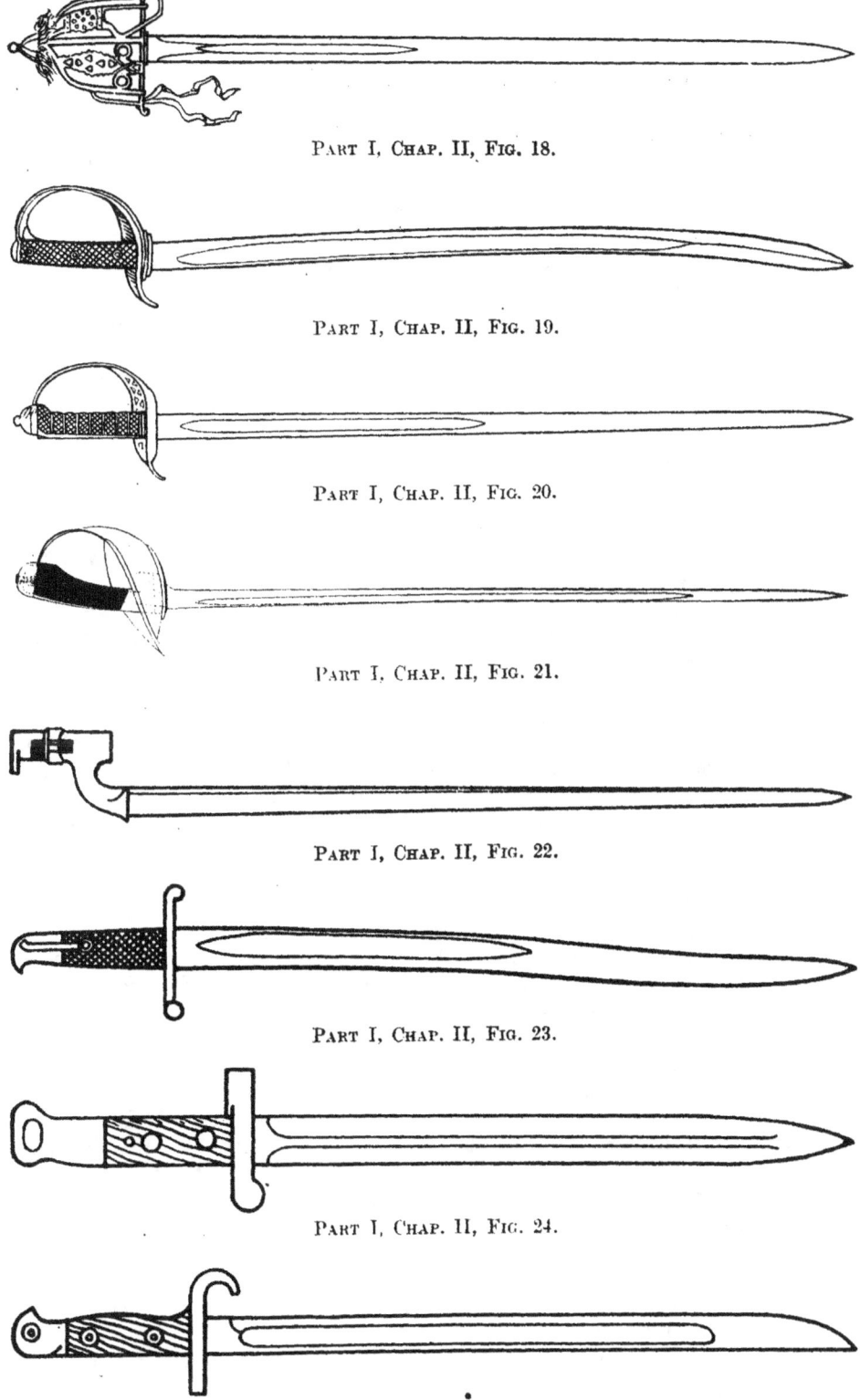

Part I, Chap. II, Fig. 18.

Part I, Chap. II, Fig. 19.

Part I, Chap. II, Fig. 20.

Part I, Chap. II, Fig. 21.

Part I, Chap. II, Fig. 22.

Part I, Chap. II, Fig. 23.

Part I, Chap. II, Fig. 24.

Part I, Chap. II, Fig. 25.

## CHAPTER II—SECTION 2

### OUTLINE OF MANUFACTURE AND INSPECTION OF THE SERVICE BAYONET

The sword bayonet, pattern 1907, Mark I/C/ for rifles, short, M.L.E., is just under 22 inches long, and weighs, complete, between 16 and 18 ounces.

The blade and tang are manufactured in one solid piece from the best cast steel. The steel employed must be from Swedish or other high-class approved ores only, and the following analysis is recommended, though the only actual requirements are that the phosphorus, sulphur, copper and other impurities are within the stated limits.

| | |
|---|---|
| Carbon | 0·90 per cent. to 1·20 per cent. |
| Silicon | Not above 0·20 per cent. |
| Manganese | 0·15 per cent. to 0·45 per cent. |
| Phosphorus | Not above 0·20 per cent. |
| Sulphur | Not above 0·20 per cent. |
| Copper and other impurities | Only traces. |

The principal manufacturing processes in the making of the blade and tang are milling, grinding and polishing. In milling a hardened steel wheel, with appropriate cutting edges on its rim, or on one or both its sides, is revolved and the work is automatically fed forward or rotated as the cutting progresses. Milling machines, in some cases, have the wheels arranged in batteries of two or more so that the work on several blades is performed in one operation. In grinding and polishing the work is held against a revolving wheel supplied with a suitable abrasive.

The flat bars of cast steel are first sawn to length. Then the sides of the tang are milled and the rough bars brought to the correct overall length. The backs and points of the blades are next roughed to shape. A pair of gang mills of six wheels each is used to mill the backs and mill the blades to width. A straddle mill working on both sides reduces the roughed-out bar to approximately the correct wedge shape, and it is then flat milled on both sides.

The next step is milling the tang back and front. The back is milled flat and the front to the curved figure necessary to give a comfortable hand grip. Other milling machines, flat and rotary, put on the correct figure to take the cross-piece where blade and tang join. The blade is then reduced to thickness and hand ground on an emery disc. Considerable skill is necessary in this operation, which is one of the few actual hand processes in the whole course of manufacture.

The fullers, or recesses in the sides of the blades, are next milled and ground. The whole blade is then polished. One or two small processes, such as drilling the tang for the grip screws, are completed and the whole blade and tang is finished so far as actual machining is concerned. The blades are now hardened by being heated and plunged into oil or water, or through oil into water, according to the steel used, and are subsequently tempered to stand the severe test which they have to go through before they are passed into service.

The cross-piece and pommel are made of the best wrought iron or mild steel. The cross-piece is worked to shape in several milling machines. The pommel is milled and slotted, the bolt hole drilled and drifted to size and the front, back, and end profiled to shape. The profiling machines are particularly interesting. The cutting tool is guided by a wheel which presses on a dummy of the correct shape which is caused to revolve at the same rate as the work.

The grips are made of walnut wood and are shaped in ordinary woodworking machines. The bolt and nut and the necessary screws for holding the grips in place are of mild steel. The bolt spring is of the best No. 20 gauge steel wire.

Assembling is all hand work. The pommel is brazed to the tang and the cross-piece to the shoulder of the blade. The brazing has to be well done to stand the tests of the

Inspection Department. The grips are oiled on the undersides before being fitted into place, and are fixed with two screws and nuts.

Before being assembled the pommel and cross-piece are browned and the bolt is blacked or browned. Browning is a chemical process in which the component is dipped in an acid mixture which gives the steel a dark, dull appearance. Oil blacking is effected by heating the component, then dipping it in oil and burning off the oil that adheres to the surface. This leaves on the surface a black carbonaceous deposit. Subsequent to these operations the components are "bobbed" or polished.

The blade and mountings are viewed in the shops in all stages of manufacture, gauges conforming to the dimensions laid down being used. In most cases the manufacturer submits the bayonet to the same tests as those used by the Inspection Department.

Before being accepted and passed into service the bayonets must be inspected and proved by the Chief Inspector of Small Arms or by an officer deputed by him for this duty. Inspection is carried out at various stages of manufacture. If one-fourth of any delivery is found inferior to the sealed pattern, or contrary to the terms of the specification governing manufacture, the whole consignment is liable to rejection.

The blades are first submitted to view by the Inspection Department in the hardened and tempered stage, in batches of not more than 200, with tangs finished and ready for the assembling of cross-piece and pommel. One blade in each batch is submitted to an overtest, rendering it unserviceable, to ascertain that hardening and tempering are correct. This test consists of reducing the length of the blade $2\frac{1}{4}$ inches by bending in a powerful machine. The blade must not break or take a permanent set of more than $\frac{1}{16}$ of an inch.

If the trial blade stands this severe test the remainder are tested in an ingenious machine that strikes them with a severe blow. With the back and edge the blow is struck on an oak block; with the flats it is delivered on the iron table of the machine. The machine consists of an iron arm, to which the blade is fixed, which is held back in a vertical position, by means of a pawl, against a spring with a tension of from 17 to 19 lbs. The iron arm is $14\frac{1}{4}$ inches long, and weighs about $14\frac{1}{2}$ lbs. The tension of the spring ceases when the arm has reached a position about $22\frac{1}{2}°$ from the horizontal, the arm and the blade completing the blow by the momentum they have acquired. When the flats are struck the blow is not so severe, the arm starting from a position $45°$ from the horizontal, and not from the vertical.

The blades are also tested in the bending machine used to test the first blade to destruction. In this case a weight of 110 lbs. is used, and the blades must straighten out after having been bent $\frac{3}{4}$ inch in both directions. They are further sprung round a curved block on both flats, and must not take a permanent set. The greatest depth of curve in this test is 1·75 inches. The blades are trough gauged, and must not be more than 0·02 inch narrow in any part. They are then gauged and examined for grinding, and a percentage is weighed. The blades are marked with an X on the right flat to denote the convex side for subsequent bendings, and are further marked with the examiner's mark on the left side of the X. The weight of the blade must be between $10\frac{3}{4}$ and 12 ounces.

All the small parts of the bayonet, the grips, bolt, spring, screws and nuts, are inspected before assembling. The grips are gauged for size and shape and position of screw-holes, and for depth of recess for nut and screw head. The springs are kept under compression for a period of not less than 14 hours, after which they are examined and gauged. They must not have taken any appreciable set. The screws and nuts are gauged for diameter, length and thread. The grips are marked with the examiner's mark on the outside.

The bayonets are next inspected rough-mounted, assembled with bolt. The set of the pommel and cross-piece, which have been brazed on, are gauged in a receiver. To test the soundness of the brazing, the blade is struck on the edge, by hand, on an oak block. Weak brazing can be instantly detected by this method, the weakly-brazed part shifting from position. After striking the bayonet is examined and gauged for position of grip screw holes in the tang, size and figure of the bolt, bolt recess, distance of bolt from the top end of the neck of the mortise in the pommel, size of sword bar mortise and thickness of the cross-piece. They are also gauged with a sword bar mandril. In some cases they are sprung round the curved block on the left flat, and may be trough gauged if considered necessary. They are marked with the examiner's mark on the right flat, on the right side of the X.

The final inspection is in the browned and finished stage, with the blade polished and the edge ground to an angle of 36°, commencing about 2 inches from the cross-piece and extending to the point, the ground edge being left 0·01 inch in thickness. The blade is examined for flaws and for the finish of the polishing. It is sprung round the test block on the left flat, gauged in the trough, any blade 0·02 inch narrow in any part being rejected. It is gauged for thickness, a toleration of 0·01 in thickness being allowed throughout. The thickness of the fullering is gauged and must not be thicker at the point end than at the shoulder. All the details of the assembling and fixing of the small parts and furniture of the hilt are checked in this examination, note being taken that there are no sharp edges on the cross-piece or pommel. The completed bayonets are weighed and the examiner slightly rivets over the screwed end of the bolt. All bayonets that have passed the tests and examinations are then marked with the proof mark and with the examiner's mark under the proof mark on the right flat. The left flat is marked with the " crown G.R. " over, and the date mark under, the pattern mark (1907).

The bayonets are then greased and packed and passed into store.

## CHAPTER III—SECTION 1

### THE REVOLVER

#### General Considerations

The revolver is essentially a weapon for quick use at close quarters. Although capable of a relatively high degree of accuracy, it should be looked upon more as a defensive weapon than as an arm of precision, and its capacity of delivering a knock-down blow within the limits of its normal short fighting range is a matter of greater importance than precision at long ranges. Similarly, the conditions of its use are such that only fixed sights are employed, and it is most important that the weapon should have a correct set or alignment and balance in the hand. In practice it is used instinctively, that is to say, aligned and discharged as a shotgun is used upon moving game rather than consciously " sighted," as is the rifle.

In a well-designed arm the relation of the curve or slope of the butt to the barrel is such that when the arm is extended naturally the long axis of the barrel is naturally " on " the target parallel with the line of sight, without any material flexure of the wrist joint. Pistols in which the stock is set almost at a right angle to the barrel such as some types of self-loading pistol, do not permit this natural alignment, and point low, so that a conscious correction of a natural pointing alignment has to be made to bring the foresight into view. In a moment of excitement this correction is probably forgotten, and the shots go low. With a well-designed curved or sloping grip the shooter, even if he is excited, will naturally tend to deliver his shots along a plane parallel to the line of sight, and will place his shots in the neighbourhood of the point he is looking at even without conscious aim. A pistol, therefore, should " point " naturally, and the function of cocking by trigger pressure in the revolver should be carried out with a minimum disturbance of grip and alignment.

*History of the Revolver.*

Guns and pistols with multiple revolving barrels or several chambers and a fixed barrel date back to the 16th century. The word " pistol " is actually derived from " Pistoia " in Etruria, where hand firearms (for use on horseback) were first manufactured in 1540. This term is now generally used in connection with any single-handed firearm. In nearly all these early arms only one wheel or flintlock is employed, and the barrels had to be rotated by hand and the lock re-cocked for each discharge. The invention of the percussion cap made the development of revolving firearm mechanisms simpler, as it eliminated the trouble of priming charges and the independent motions of the pancover.

The first percussion revolvers were of the type known as the " pepperbox," in which several barrels rotated round a central spindle. The nipples were set at right angles to the bore and the caps were struck by a hammer falling vertically upon them but intercepting the line of sight. Many variations of this type of revolver were produced between 1825 and 1835, when Colonel Samuel Colt patented his well-known type of revolver.

The first Colt revolvers differed from the " pepperbox " types in that a revolving set of cylinders and a single fixed barrel were used. This was no novelty but was a return to the older flintlock revolvers of the Collier type, but Colt introduced central fire nipples and a hammer below the line of sight. His revolver was also far lighter than the cumbersome " pepperboxes " and it was easily sighted on the target. The " pepperboxes " had the customary ratchet arrangement for rotating the cylinder, but this was in their case actuated by the pull on the trigger, $i.e.$, " double-action " and there was no means of cocking other than by this " double-action " pull. The Colt, however, was cocked by the thumb in the ordinary way and the movement of the hammer rotated the cylinder which was further locked in position by a stop which was independent of the ratchet gear.

Thus, owing in part to the mechanical refinement of the independent cylinder stop which registered the chamber in alignment with the barrel, but far more to the fact that, when cocked, Colt's single-action revolver was no harder to aim and fire than any ordinary pistol, the Colt became deservedly well known for its accuracy and reliability.

The British and Continental makers continued to make revolvers dependent on a long double-action trigger and incapable of being independently cocked by the hammer, until 1851, when the English maker, Adams, produced a solid framed revolver which was a true " double-action " in that the cylinder could be rotated and the arm discharged either by trigger pressure or by cocking the hammer. During the Crimean war both Adams and Colt revolvers were carried by the officers, and in 1853 Colt established a factory in London which operated for some two years. In 1862 Colt and Adams revolvers were in use in the Navy, a number of Adams being converted to breech loading in 1868.

The development of the various breech-loading cartridges modified pistol design. Revolvers were made for rim fire cartridges in America, while the Continental makers turned out pin fire and even " needle-gun " types of revolver. There was no great difference between the standard revolver of this 1870 period and the previous single-barrelled muzzle-loading type, which although commonly termed muzzle-loading, was actually loaded from the front of the cylinder and not through the barrel. A firing pin was added to the hammer and the cylinders were bored straight through to take rim or central fire cartridges, and the ram-rod loading lever was converted into or replaced by an ejecting rod device attached to the barrel. Loading was accomplished through an opening or gate in the right side of the cylinder shield and both loading and ejecting were slow. Slowness of reloading had been the bane of the revolver as a cavalry weapon in its muzzle-loading days and remained so during this period.

In America, the " spiritual home " of the revolver, during the Civil War, Mosby's Guerillas, a light cavalry force, were revolver-armed and carried two revolvers in saddle holsters as well as two similar belt revolvers for each man. But four revolvers never equalled the true fire value of a repeating carbine; shortness of range, inaccuracy and slow reloading handicapped them. Cartridge loading had modified the difficulty, but it was not until self-extracting mechanism were brought out that the reloading difficulty could be considered to have been much diminished.

The earliest self-extracting revolvers were of British and Continental origin, and the Galand, Thomas, and Tranter revolvers were typical of the period of 1875.

The first self-extracting revolver officially issued for the Army in this country was the Enfield, a weapon of ·442-inch calibre, rim fire; this was issued later in a ·450-inch calibre for a central fire cartridge, but was very unsatisfactory and was subsequently withdrawn. The majority of the self-extracting revolvers of this period were shaky and complicated, and the attempt to provide self-extraction was only successful at the cost of rigid alignment of barrel cylinder and standing breech.

In general the Service revolvers of most Powers were not very widely different at this period. They were for the most part simple double-action six-chambered solid-framed revolvers of ·450-inch or 11 mm. calibre. Nearly all suffered from the same defects, such as loss of gas between chamber face and barrel, indifferent alignment of cylinder and barrel, and all were of more than doubtful accuracy. They were, however, reliable within the limits of their performance and inventors were busy trying to improve them.

In the meantime certain changes had occurred in the gun trade. The arms industry, in so far as revolvers were concerned, tended to settle into the hands of big firms provided with well-equipped manufacturing plant, and the smaller makers disappeared or were absorbed. By 1890 the revolver manufacture had practically concentrated in the hands of the Colt Firearms Manufacturing Co. and Smith & Wesson of America, Webleys of Birmingham, and the few big semi-official manufacturing concerns of the Continent, such as the Manufacture d'Armes de St. Etienne, the Fabrique National of Belgium and the Austrian Steyr works.

In 1890 there were only two good models of self-extracting Service revolver. These were the Webley and the Smith and Wesson, and both were of the familiar hinged-barrel "break action" type. In both the ejection was functioned by a central extractor rod sliding in the cylinder axis and operated by a cam lever in the knuckle joint of the frame and barrel. The main difference was in the fastening of the strap or projection of the barrel which locks to the standing breech. In the Smith and Wesson this was a simple T-lever attached to the strap; in the Webley it was a stirrup fastening or barrel catch attached to the standing breech. Of the two weapons the Smith and Wesson was rather more delicate, and the Webley was adopted as the official Service revolver in August, 1890. This was the original Webley, Mark I, a short 4-inch barrelled pistol of ·441 calibre firing the ·450/·476 cartridge. Since then the list of marks has gone through Mark I, II, III, IV, V and VI, and the earlier three marks are obsolete.

The Army regulations lay down that an officer may use any make of pistol provided that it takes the standard Service cordite pistol cartridge of ·455 calibre. In practice this means that Webleys, Colts, Smith and Wesson's and Webley-Wilkinson's are all in use. During the war ·455 calibre arms of many makes were officially issued to all ranks, but later the Royal Small Arms Factory at Enfield made its own, a model of the Mark VI Webley.

Great Britain is now the only Great Power whose official hand weapon is still the revolver, the other nations having successively adopted self-loading pistols of various calibres, but the experiences of the European War amply justified the British hesitancy in adopting self-loaders. The American Expeditionary Force took the field with both revolvers and self-loaders. Practical experience in the field showed that the revolver was in every way the more reliable weapon though both these arms used similar cartridges. The subject will be more fully discussed in the succeeding chapter on self-loading pistols. The Webley and Webley-Wilkinson revolvers are the same in principle, that is to say, they are double action, break down, self-extracting pistols with a safety stirrup fastening which holds the barrel strap into engagement with the standing breech. If this is not fully closed it intercepts the fall of the hammer and the arm cannot be discharged. The older Webleys differ in the detail of the cylinder-dismounting mechanism, in the shape of their butts, and in barrel length, but there was no material change of lock mechanism or general dimensions of components until the introduction of the Webley-Scott, Mark VI, which was designed to standardize all components.

The Colt's and Smith and Wesson's now in use are double action, solid framed, extractor-fitted pistols of the "swing-out" cylinder type, Smith and Wesson having abandoned their breakdown action in 1902 in favour of the swing-out model. Neither of these arms is a self-extractor like the Webley in which the cartridges are automatically ejected when the action is broken, but both need separate pressure applied to the ejector rod.

In accuracy there is nothing to choose between the performance of the three makes, but the big grip fitted to the Colt New Service model makes it a difficult weapon to use for quick, accurate, double-action shooting. The Smith and Wesson errs in the opposite direction, and though its double-action mechanism is the quickest and smoothest of the three, the sharp-shouldered frame and the quick taper of the grips render recoil painfully noticeable, so that there is a tendency for the grip of the hand to move between shots.

Rapidity of fire is an essential in pistol shooting, and though the double action of a revolver may well be ignored when it is considered as a target arm, it is of the highest importance when the pistol is considered from the active service point of view as a weapon.

Both the Webley, Colt, and the Smith and Wesson embody safety locks in their mechanism, so that an accidental fall in which the cocking-piece of the hammer comes into sudden contact with a stone or similar hard object will not result in the striker being driven against a cartridge, thus discharging the weapon. The same mechanism functions in similar cases when the arm meets with accident while at full cock.

The solid frame principle of the Colt and Smith and Wesson revolvers is theoretically superior to the Webley breakdown design. It gives increased rigidity and should help to maintain the cylinder in close contact and precise alignment with the barrel. This, however, is not borne out in practice, and the break-down type is just as accurate and durable as the solid frame and is capable of firing equally heavy charges.

The most common fault of revolvers is the wear of the cylinder ratchet or lifting pawl, so that the cylinder is not rotated a full sixth of a revolution every time that the trigger is pulled, and the cylinder fails to align perfectly with the barrel, while the independent cylinder stop fails to engage truly with the notches provided in the cylinder circumference. This is due as a rule far less to wear than to careless handling of the arms, as by closing the action with the hammer cocked and kindred errors of treatment. The effect of this lack of alignment is that when fired a shaving is taken off one edge of the bullet, a greater escape of gas takes place at the junction of the barrel and chamber, and the precision of the arm is impaired, although it still remains a serviceable arm for use within the limit of its normal active service usage at close ranges.

The Service revolver is designed to stand a great deal of knocking about, and has ample strength to resist not only normal treatment, but even violence and neglect, without marked deterioration of its shooting qualities at ranges up to 50 yards. Owing to the short sight base and the general limitations of all one-handed weapons, a personal factor customarily limits their effective use to something less than a quarter of their mechanically effective fighting range, which, if this personal factor were absent would, in the case of the ·455 Service revolver, be at least 300 yards. The extreme range with a 35° angle of projection is no less than 1,500 yards, but, despite these ballistic possibilities, it should be clearly understood that the pistol can never be a successful competitor against the carbine. As soon as the limitations of the one-handed arm are reached, it develops into the hybrid, magazine, shoulder-stocked, low-velocity automatic gun, which, however, for convenience, we group under the heading of automatic rifles, such as the Thompson "Sub-machine Gun" or the Bergman "Musquete."

The value of the calibre of self-loading pistols and revolvers has been much obscured by theory, but practice of recent years has amply proved that small calibre plus high velocity, although developing many foot-pounds of energy, yet lacks stopping or shocking value. There have been many attempts to substitute a high-velocity cartridge of ·38 calibre, as a Service equivalent to the traditional ·455. In practice it has been found that the small calibre sometimes fails to stop its man and that the large-diameter leaden plug of the ·455, moving even 300 or 400 feet a second slower than the high-velocity, small-calibre projectile, is yet far more effective. Recent experiment has, however, developed a new experimental ·38 cartridge whose efficiency is, so far as ballistic tests can ascertain, not less than that of the ·455.

A hit with a ·455 anywhere literally knocks an adversary over. This quality of efficiency depends to some extent on the massive soft-lead bullet and the relatively low velocity rather than on any inherent magic in the calibre, for the ·455 or ·45 self-loading pistol firing a lighter nickel-covered bullet at a higher velocity cannot be depended on to produce equal shock effect. The efficiency of the ·455 revolver cartridge is due to combination of the large calibre with the soft material, the mass, and the relatively low velocity of the projectile. These combine in such a way that the adversary experiences in his body the maximum development of shocking as distinct from penetrative effect. This is just what is wanted in an active service revolver.

A drawback of the revolver is its relatively slow process of loading when compared to the swift exchange of magazines accomplishing this function in automatic pistols. The objection is theoretical rather than practical, for, although no quick-loading device sold by accessory dealers has ever been of any real use, the United States army has issued an excellent half-moon clip holding three rimless cartridges, so that two clips loaded the cylinder. This device was initiated in order that the Standard U.S.A. ·450 self-loader pistol cartridge could be used in all revolvers and self-loading pistols standard to their forces. There is no reason why a similar clip holding a rimless cartridge with a lead rather than nickel covered bullet, should not be adapted to all British Service pistols, if one were considered necessary.

The Webley pistol Mark IV is still used, but the Mark VI, made either by Webley's or at the R.S.A.F. Enfield, is now the standard Service revolver, and differs only from its predecessors in having a 6-inch barrel, a differently shaped grip, and a removable foresight blade.

*Webley Revolver, Mark VI, ·455-inch bore : 6 shot.* (Fig. 1 and Plate I.)

The barrel (1) is of ·455 bore rifled with seven grooves ; the foresight block forms part of the barrel, but the blade is separately inserted in a longitudinal curved slot in the block and retained by a set screw passing through both blade and block. The barrel is extended to the rear in the form of a strap. Beneath the barrel is a projection forming the joint ; in this projection the cylinder axis (2) is fixed parallel to the barrel. The holster guide (3), prevents the cylinder catching when putting the revolver in the holster.

The body (4) contains the action, forms the stock, and supports the bases of the cartridges. The front end is slotted out and forms a joint for the barrel, the latter being pivoted on an axis pin (5) retained in position by a screw. The sides of the body forming the stock are closed by vulcanite plates (6) held together by a screw (7) with countersunk cup and nut. The opening in the underside of the body is closed by the trigger guard (8) and attached to it by two screws. A hardened-steel shield (4a) is fitted to the body for the cartridge heads to bear against ; it has a slot for the pawl (27) to work through.

The barrel catch (4b) is pivoted to the body by a screw (4c) ; its upper end is pressed forward by a V-shaped spring on the right side of the body. The upper end passes over the barrel strap and holds it securely down. The lower end forms a thumb-piece, a downward pressure on which releases the barrel strap. The bottom end of the latter is rounded off, so that it can snap into position on the barrel being jerked upwards. The backsight V is cut in the top of the barrel catch. If the latter is not in position the hammer nose cannot reach the cap of the cartridge.

The cylinder (9) contains six chambers and fits on to the cylinder axis (2). The rear is recessed out for the extractor (10). A projection on the front of the cylinder contains a groove (11) for the lip (12) of the cylinder cam (13), which prevents the cylinder coming off its axis when the revolver is open.

The extractor (10) is a plate with six recesses round its edge, so that it fits under the rims of all the cartridges in the chambers. On its face are the ratchet teeth by which the cylinder is revolved, on its reverse a steady pin entering a recess in the cylinder. The extractor is mounted on a stem with flats on two sides to prevent it from turning in the cylinder. On the stem is the spiral extractor spring (14), which bears against the cylinder in rear and in front against the nut (15) screwed on the extractor stem. The extractor is operated by the extractor lever (16), which has an oval hole in its centre. The extractor lever is mounted on the barrel axis pin (5) and works in a vertical slot in the barrel joint, an arm (17) reaches up through a slot and bears against the end of the nut on the extractor stem. In a narrow slot in the rear of the extractor lever is the extractor lever auxiliary (18), which is pressed to the rear by a small spiral spring; this presses the extractor lever to the front as far as the oval slot will allow.

The action of the extractor is as follows :—When the barrel has been depressed a short distance the tooth (19) on the extractor lever catches against the bottom of the body and arrests the motion of the extractor lever. The arm (17) of the extractor lever in its turn arrests the motion of the extractor stem; therefore, as the barrel and cylinder are revolved forwards and downwards the extractor is forced out of the cylinder and ejects the cartridges but as the barrel approaches its lowest position the corner (20) of the barrel joint passing over the tooth (19) presses it to the rear, and when it has forced the tooth (19) into the groove in the body the extractor lever is free to rotate. The extractor spring then re-asserts itself, driving the extractor home and causing the extractor lever to rotate on the axis pin.

The hammer (21) is pivoted on the screw (21A), its nose passes through an opening in the body so as to be able to strike the cap of the cartridge in the uppermost chamber. The hammer catch (22) is pivoted on a screw (23); it is pressed forward by a small spiral spring carried in a recess in the hammer. The bent (24) in the hammer is used when the weapon is used single-action, *i.e.*, when cocked by hand before pressing the trigger.

The trigger (25) is pivoted on the screw (25A); at the rear end is the trigger nose (26). Alongside the trigger nose is pivoted the pawl (27). On top of the trigger is the cylinder stop (28) which enters into the six elongated grooves on the outside of the cylinder, and prevents the pawl from rotating the latter too far. The trigger catch (29) is a separate component pivoted on the screw (25A) in a recess in the trigger; it is controlled by a small flat spring screwed on to the top of the trigger. As the trigger is drawn back the trigger catch descends slightly and bears against the front part of its opening in the body, and as the trigger continues to revolve, the end of the trigger catch spring slips into a notch near the axis of the trigger catch, and causes its front end to start up and enter into one of the small external recesses in the cylinder, which is then entirely prevented from moving until the pawl again acts.

The mainspring (30) has two branches, the upper branch operates the hammer by means of the swivel (31). The lower branch rests upon the mainspring auxiliary (32) which bears upon the pawl just in front of its pivot and so keeps it pressed to the front; it also keeps the trigger pressed downwards.

On cocking the hammer by hand the projection in which the bent (24) is cut, catches under the trigger nose (26) and raises the rear end of the trigger; this causes the pawl to rotate the cylinder by means of the ratchet on the extractor until the cylinder stop (28) arrests the motion; at the same time the trigger catch (29) enters its seating in the cylinder holding it securely. The trigger nose then falls into the bent (24), the hammer being fully cocked and the mainspring compressed, as its lower arm is raised by the mainspring auxiliary. On pressing the trigger the hammer falls and fires the cartridge; the head of the hammer then rests against the top of the body. On releasing the trigger the mainspring auxiliary is lowered and the hammer head falls back a short distance from the body. When in this position the hammer, should it receive an accidental blow, is prevented from being forced forward, by the hammer catch meeting the nose of the trigger.

On pressing the trigger in order to fire without previously cocking the hammer, the trigger nose (26) bears against the bottom of the hammer catch (22) and causes the hammer to rotate. The pawl, cylinder stop, and trigger catch act as before. When the cylinder is revolved so that the next chamber is opposite the barrel, the end of the hammer catch slips off the trigger nose and the hammer falls as before. In both cases, when the pressure of the finger on the trigger is released, the trigger nose (26) presses the hammer catch (22) inwards so as to get past it.

*Swing-out Types. Colt. ·455.*

The Colt New Service is shown in Fig. 2 and Plate II. It is a substantial arm weighing $2\frac{1}{2}$ lbs., and is made in two barrel lengths, $4\frac{1}{2}$ inches and $5\frac{1}{2}$ inches. On cocking, either by the thumb or by the trigger pressure, a movement of the rebound lever depresses the cylinder bolt freeing the cylinder. A pawl attached to the trigger bears on the ratchet and revolves the cylinder in the usual manner. At the same time a projecting pin on the right side of the trigger engages a small bell crank lever, which is pivoted on the hammer axis pin. This is connected to and retracts a small intercepting slide moving in the body of the right side of the frame, which terminates in a block situated between the frame and the breast. This block is an automatic safety device ensuring that unless the trigger is pulled the face of the hammer cannot come far enough forward for the firing pin to reach the cap of the cartridge. In the same way if the latch pin, which secures the closing of the swing-out system, is not fully home and the cylinder locked, its projection back into the lock chamber intercepts the path of the positive lock member and the weapon cannot be cocked or fired.

This safety mechanism is somewhat delicate, but the block is visible between the hammer and the breech as a member moving with the trigger action; should no movement of the block be apparent the arm should be dismantled and the parts adjusted, as although it may appear to function it will not fire.

*The Smith and Wesson ·455 (Plate III).*

This revolver is outwardly very similar to the Colt, but is lighter and rather more delicate in construction. The cylinder revolves counter-clock wise. The cylinder latch lock presses forward instead of pulling back and is operated by a spiral wire in place of a flat spring. The detail refinement of manufacture is very marked and the general finish of a high standard. The arm is, however, not so generally popular as other types, for owing to the relatively small butt it is mainly preferred by men with small hands.

*The French Modéle d'Ordonnance, 1892, 8 mm. (·315 inch).*

The French service revolver (Plate IV) is a six-chambered weapon with an ejection system similar to the Colt but with the cylinder swinging out to the right instead of the left. This makes it far less convenient to load. It fires a long central fire high velocity rimmed cartridge with a copper covered bullet and is capable of good range and fair precision. The calibre is, however, too small. A feature of interest is its ready demountability. The side plate of the action is hinged and the removal of one screw permits the whole to be opened like a book and the action displayed for dismounting or cleaning. It was, however, not much used on service, French officers preferring the more portable self-loading pocket pistols which were of equal value.

A certain number of single action solid frame, non-ejecting ordnance type revolvers of ·450 calibre were used by details of German and Austrian troops during the war, but these were old weapons still in service. The Italians also used a constabulary type ·380 calibre revolver as a service issue. The Japanese cavalry are equipped with a Smith and Wesson type revolver manufactured in Japan. The Russian revolver, the Nagant of ·3 inch calibre, was in appearance very similar to the French modèle d'Ordonnance, but incorporated an ingenious system meant to prevent gas escape between cylinder and barrel. The chambers were recessed at the forward end to admit the coned face of the breech end of the barrel. When in alignment a breech block wedge lifted by the trigger mechanism pushed the cylinder forward. The cartridge, whose case extended beyond the nose of the bullet, entered the cone of the barrel and made a gas-tight joint. The arm was complicated and delicate, and it is doubtful if any real advantage was gained. The weapon now appears to have been largely superseded by Mauser self-loading pistols.

The future trend of revolver design is indicated by the ·38 inch Experimental models recently tested at Woolwich. These were largely the outcome of the demand for a reliable police revolver of less bulk and weight than essentially military weapons, yet possessing adequate stopping power. A new type of ·38 cartridge with a 200 grain bullet was found to give results superior to the commercial ·38-inch types of ammunition.

This revolver (which is still in the experimental stage) is of the familiar break down type with the stirrup fastening catch. The lock work is of new design, as is the cylinder stop mechanism, and a detachable side plate admits of ready access to the mechanism. In addition to the improved details of mechanism the whole pistol has been designed from a practical point of view. The dimensions, shape of grip, details of knurling on the butt strap, and general balance in the hand have all been worked out by practical experiment with a special "try-pistol." The result is a weapon with a low sited barrel which should give excellent "instinctive aim" and which embodies most of the good points of many makes and models. The reduction in weight of a model with a five-inch barrel is approximately three-quarters of a pound as compared with the Service weapon. The ammunition is also lighter, affording a saving of some twelve ounces for every fifty rounds.

In the American army in 1918 the suitability of the pistol for trench warfare was considered so pronounced that every effort was made to equip every man in the field with either a revolver or a self-loader. The same general tendency toward an increase in pistol armament was noticeable, although to a lesser extent, in the armies of other combatants; and there is no doubt that the closing stages of the Great War showed that the self-loading pistol and revolver were weapons whose potentialities had been overlooked or underestimated.

BALLISTIC DATA OF REVOLVERS AND REVOLVER CARTRIDGES.

| Cartridge. | Calibre of revolver. | Weight of bullet in grains. | M.V. f/s. |
|---|---|---|---|
| British Service, Mark II, with Mark VI revolver | ·455 | 265 | 600 |
| | (Ballistic coefficient = 0·2) | | |
| Colt, U.S. | ·45 | 255 | 770 |
| Winchester (with smokeless powder and nickel covered bullet).. | ·44 | 200 | 918 |
| S.W. Special | ·44 | 246 | 750 |
| | ·38 | 158 | 850 |
| Colt, new Police | ·38 | 150 | 580 |
| S.W. | ·38 | 145 | 630 |
| S.W. | ·32 | 85 | 630 |
| Long Rim Fire (with 6-inch barrel) | ·22 | 35 | 770 |

DATA FOR ·455 WEBLEY REVOLVER, SERVICE AND TARGET MODELS.

| | Service, Mark IV. | Service, Mark VI. | Target, Model. |
|---|---|---|---|
| Length of barrel | 4 inches | 6 inches | 7½ inches |
| Muzzle velocity in foot-seconds | 570 | 600 | 640 |
| Striking energy at 100 yards in ft./lbs. | 125 | 142 | 166 |
| Extreme range at 35° elevation (yards) | — | 1,300 | 1,550 |
| Sight base, in inches | 5 | 7 | 8·5 |
| Drop in feet of bullet below line of sight at— | | | |
| 20 yards | 0·16 | 0·12 | 0·11 |
| 50 yards | 1·0 | 0·78 | 0·7 |
| 100 yards | 4·0 | 3·2 | 2·8 |
| 200 yards | 16·0 | 13·0 | 11·2 |

N.B.—± 30 f.s. in muzzle velocity must be allowed for variations in ammunition.

*Reduced calibre devices.*

Webley and Morris tubes, adapters for the ·22-inch rim fire cartridge, and kindred devices have been tried for practice purposes. They are of little value as they involve

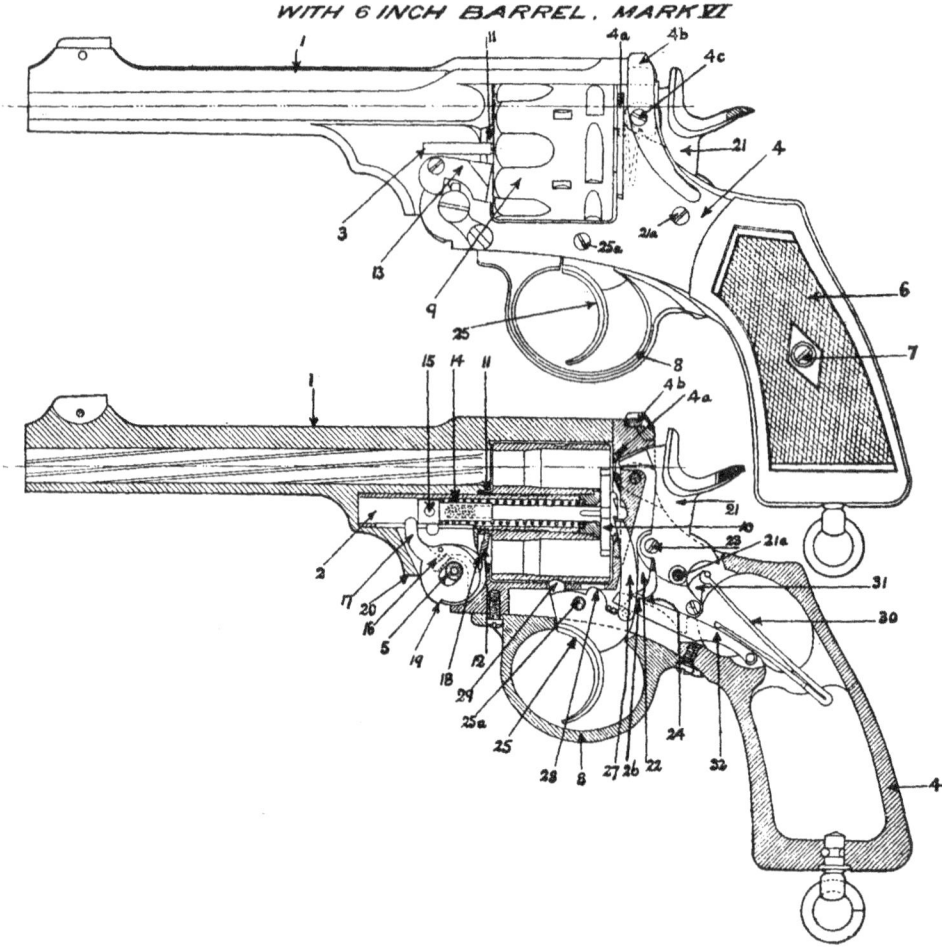

Part I, Chap. III, Fig. 1.

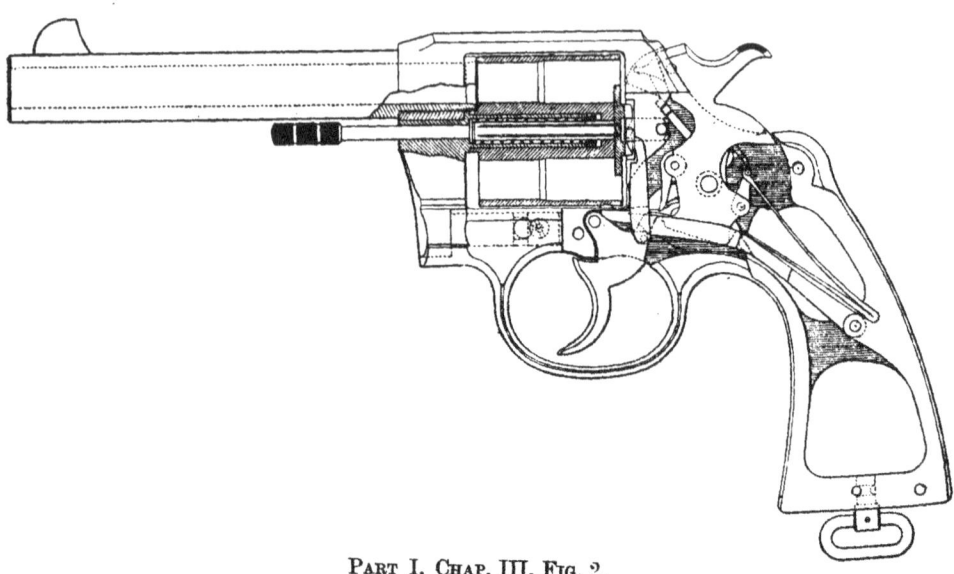

Part I, Chap. III, Fig. 2.
Colt ·455, with Side Plate, Latch and Stocks Removed;

[*To face page* 94.

·455-INCH WEBLEY, MARK VI.

PART I, CHAP. III, PLATE I.

COLT ·455.

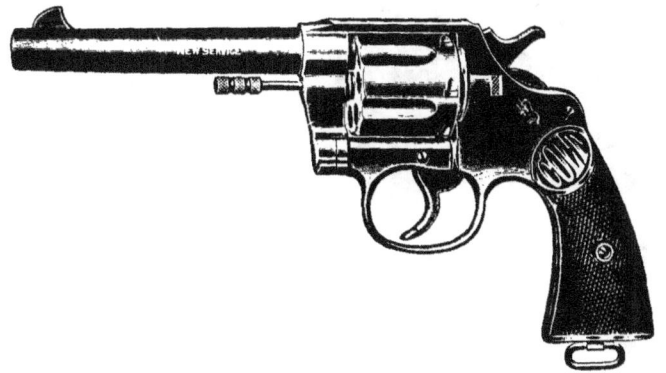

PART I, CHAP. III, PLATE II.

[*To face page* 95.

·455-INCH SMITH AND WESSON.

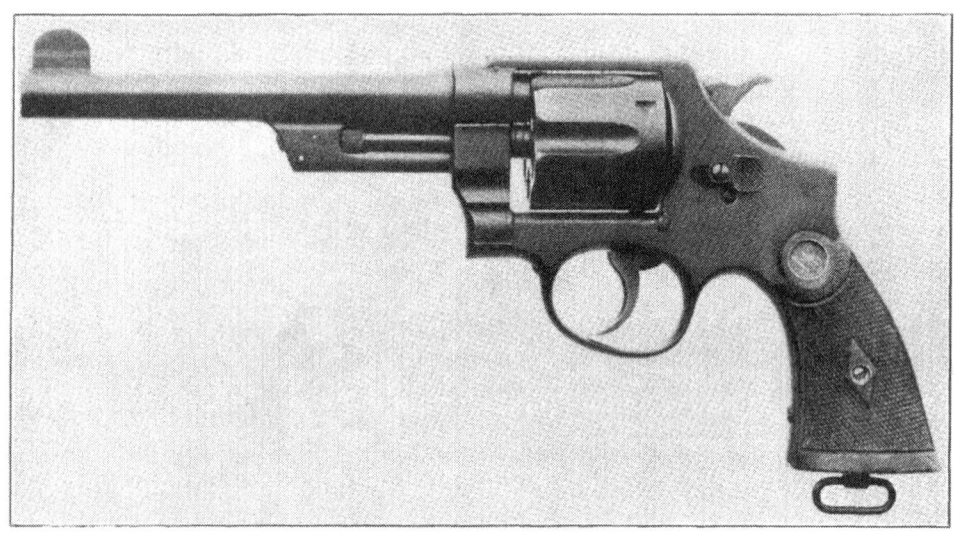

PART I, CHAP. III, PLATE III.

8 MM. FRENCH MODÈLE D'ORDONNANCE.

PART I, CHAP. III, PLATE IV.

the use of sight adapters and also convert the weapon for the time being into a single loader.

A special model of the Service revolver calibred and adapted for the ·22 rim fire is probably the best of these devices. The conditions of use of a miniature cartridge in revolver shooting are so different from the use of the standard ammunition, that it is doubtful whether practice with the ·22 is of real value.

Adapters on the tube principle have been applied to self-loading pistols as well. The magazine is loaded with steel cartridges bored to receive a percussion cap and a single pellet. The recoil is necessarily insufficient to operate the action and the devices are not satisfactory either as regards accuracy or functioning.

## CHAPTER III—SECTION 2

### SELF-LOADING PISTOLS

The most important development of modern small arms since the evolution of the metallic cartridge case and the perfection of magazine, breech-loading systems has been the application of the waste energy of the discharge to reload, recock, and discharge the arm. Such a weapon is termed "automatic," but in so far as pistols are concerned, the term "automatic" normally refers to actions which are self-loading and self-cocking, but need individual trigger pressure to discharge each round. The disturbance of aim caused by recoil is such that a fully automatic pistol would be useless, in that only the first shot would strike the target, while the others would fly high in the air; special mechanism is therefore arranged so that the automatic functions are strictly limited to ejection, recocking and reloading, while independent trigger pressure for each shot is required. Such an arm is more truly called a "self-loader," for recently new arms such as the Thompson sub-machine gun and Bergman muskete have been evolved, which are really fully automatic pistols or semi-carbines firing pistol ammunition. These represent the pistol at its highest mechanical development of purely automatic functioning, when the arm itself not only ejects, reloads and recocks, but fires its successive rounds so long as the trigger is held back by the firer.

In terminology it has therefore been thought better to class the pistols popularly known as "automatics" as "self-loaders" and to reserve the term "automatic" as a designation for the new arms which are not true machine guns or rifles and in which not only the function of loading, but that of discharge, is automatically carried out by mechanism so long as the trigger is pressed.

*History.*

The self-loading pistol has an obscure history, as it appears to have been thought of long ago. It appears to have been an English invention, for as early as March 2nd, 1663–64, Sir Robert Moray, F.R.S., reported to the Royal Society: "that there was come to Prince Rupert a rare mechanician who pretended . . . to make a pistol shooting as fast as it could be presented and yet to be stopped at pleasure; and wherein the motion of the fire and bullet within was made to charge the piece with powder and bullet to prime it and to bend the cock " (" Ref. History of the Royal Society," by Thos. Birch, D.D., Vol. 1).

This wording is sufficiently precise to indicate a complete comprehension of the automatic principle, but it was not until the evolution of the fulminate primer combined in the metallic cartridge case that automatic or self-loading weapons could hope to be effective. Two hundred years later the idea was taken up again. The earlier inventions did not mature beyond the patent specification and experimental stage and were mainly gas-operated—that is to say, powder gas under pressure was trapped from the barrel and utilized to function the arm by means of a piston and lever mechanism. The earliest specimen of a partly automatic pistol appears to be a revolver invented in 1863 by Orbea, of Eibar. In the specimen preserved in the Museo de Armeria, at Eibar, it is a simple solid-frame revolver of ·450 calibre possessing an attachment to the barrel so arranged that a piston actuated by the powder gas ejected the empty case of the previous round from the chamber. The gas-operating system has now been abandoned for pistols and recoil systems are universal.

In 1893 appeared the Borchardt, the predecessor of the modern German service pistol, the Luger Parabellum. As a weapon it was not a success, for it was both delicate and complicated, but, nevertheless, it embodied nearly all the distinguishing characteristics of succeeding self-loaders. It took small-calibre jacketed ammunition with rimless cartridges, had an eight-shot magazine in a hollow handle, was equipped with a detachable shoulder stock, was elaborately sighted, and used the toggle-joint mechanism still embodied in the Parabellum.

In quick succession came the Mannlicher and Bergman self-loading pistols (1894), the Charola-Anitua (1897), the Simplex (1897) and others, but it was not until the Mauser self-loading pistol carbine combination (1898) made its appearance that the self-loading principle could be deemed to have made a definite place for itself other than as an interesting novelty. The Mauser sprang into fashion with the Boer War, and from that time the eventual development of the self-loading pistol as a military weapon was assured.

Most of these early pistols were of the locked-breech type and used powerful small-calibre, high-velocity cartridges. The simple " blow-back " system was first popularized by the Browning pistol, invented by an American but manufactured in Belgium (1898). This pistol achieved instant success, and millions were sold throughout the world.

In 1900 the Webley Fosbery self-loading revolver was introduced. This was a hybrid arm utilizing the energy of recoil to re-cock and to rotate the cylinder of a revolver. It was not successful, for it was necessary to fire it with a rigid arm in order to ensure sufficient resistance to the backward traverse of the barrel and action in the slotted frame. It was also awkwardly high above the hand, and consequently not easy to shoot straight with. It embodied, however, excellent principles in that the use of the revolver principle enabled ordinary service ammunition to be used and eliminated that disadvantage of most self-loading weapons, the unreliable grip magazine or charger.

In 1903, Browning's patents were taken up by the Colt Firearms Company, who produced an improved pocket pistol of first-class manufacture. This design of pistol has now been in use for twenty years, and no important modifications of the original design have been made, and it is still almost as good as any other on the market.

The self-loading principle is essentially a 20th-century development. The germ was there in the last decade of the 19th century, but it has taken the first two decades of the new century to try out the commercial development of the new system. The greatest war of modern times has taken place, and the new arm has been exhaustively tried in the field, apart from its extensive use in peace-time on both sides of the Atlantic. It follows, therefore, that a writer on hand weapons of to-day has an indirect responsibility thrust upon his shoulders which is greater than weighing the differences between military rifles. The pistol, revolver, or self-loader, is a weapon of opportunity and also of self-defence in emergency; therefore, certainty and reliability must be the only genuine criterion. To this end it may be emphasized that six certain shots out of a reliable revolver are worth twice their number of problematical shots from even the best of self-loaders. The officer who has to depend on his weapon, be it service arm or pocket pistol, will be well advised to choose a revolver in place of a self-loader, despite the alluring conveniences of the latter.

Self-loaders are magazine or clip-fed and hold, as a rule, eight or nine rounds in comparison with the six shots of a revolver of equivalent calibre. They are flatter, more compact, and more convenient, but less trustworthy.

If we compare the relative advantages and disadvantages of self-loading pistols and revolvers of service calibre ·455 we find the following :—

(1) *Number of Rounds.*—Self-loader is superior.

(2) *Speed of Fire.*—The self-loader possesses only a theoretical advantage, for the disturbance of aim due to recoil is such that a service revolver can be fired equally quickly if a practical modicum of precision—say, hitting a man-sized target at 12 yards—is included in the conditions of test.

(3) *Reloading.*—Self-loaders are superior if spare magazines are supplied. Magazines are, however, one of the weakest points about self-loaders and any slight deformation of the lips or feed produces jams. The clip-loading pistols have not proved markedly effective in action, and the virtues of the robust clip-loading revolver outweigh both magazine and clip-loading types of self-loader.

(4) *Ballistic efficiency.*—On figures, the self-loader is undoubtedly superior. There is no gas leak between chamber and barrel, and consequently no loss of velocity. On the other hand, according to accepted practice, nickel covered bullets have to be used in order to secure this high velocity. A coated bullet is required in order that it may be tightly crimped in the cartridge case and further because the abrupt feed from magazine to chamber tends to deform a soft lead bullet. In practice, the leaden revolver projectile of equal mass has greater effect and shocking power despite its ballistic shortcomings.

(5) *Reliability.*—The revolver is infinitely more reliable. This arises from many causes. There are fewer and simpler working parts. The action is not dependent on the cartridge. Where a missfire occurs, a self-loader until cleared by double handed action (a matter of some seconds) is entirely useless. Those few seconds may be fatal. A revolver under kindred circumstances can be fired again in half-a-second, a fresh chamber being rotated and aligned with hammer and barrel by trigger pressure alone. It should be remembered that even under the best possible conditions of manufacture, no test has yet been devised which eliminates the risk of a batch of powderless pistol cartridges being issued. The weight of propellent is so small that errors cannot be mechanically detected. Mud, dust, wet, extremely low temperatures or desert sand, all find the self-loading pistol far more vulnerable than the revolver. There are more delicate friction surfaces, more crevices and cavities where " service conditions " can impair mechanism.

The years of experience behind revolvers have produced an arm which will stand infinite neglect and yet be serviceable. The self-loader still needs care, continual lubrication and attention not always compatible with a hard campaign. Then there is the question of " moral "—one can never be quite certain about even the best self-loaders. With a good revolver there is no doubt.

(6) *Recoil.*—In earlier days, it was held that the spring absorbing mechanism of a self-loader lessened recoil in comparison with revolvers. This is doubtful, for the damping effect of springs and friction only absorbs a small quantity of energy and in the end the recoil, however delayed, has to be absorbed by the muscles of the firer. It is probable that with a self-loader an almost identical amount of mechanical energy is transmitted as recoil and that more shock is experienced than with a revolver of equal weight firing an identical charge. This is because in a self-loading pistol the impulse of recoil is first transmitted to the recoil slide and moving portions alone. These represent only a proportion of the weight of the whole arm, but this weight is free to move and takes at first the whole energy of recoil, delivering it as a hammer blow against the frame and remainder of the pistol, when the moving parts reach the limit of their travel on the ejection stroke. In practice, springs, friction and various deterrent devices smooth out part of this shock, but whereas in a revolver the total weight of the arm is immediately available as a recoil

absorber, in the self-loader the hand of the firer receives a succession of recoil effects. This blow normally takes effect with the moving weight above and rather behind the natural grip, so that the disturbance of aim is greater. The return journey of the slide again introduces a further shock in the opposite direction and more effort is required to re-align the weapon on the target than is necessary with the revolver. Actually the practised shot will be able to fire a ·455 revolver more rapidly than a ·455 self-loader.

(7) *Dimensions.*—The self-loader is more compact than the revolver and can be made rather lighter. These advantages apply rather to pocket types than to the service models, and there are compensating disadvantages in that the big self-loader is not so readily drawn from its holster and that the balance is often such that the butt does not present itself readily to the hand. A pocket self-loader is, however, better to carry in certain countries where thin clothes are worn, as it does not bulk as largely as a revolver of equal calibre. Reliability of functioning diminishes in proportion to the reduction of calibre.

(8) *Safety.*—The revolver with its double action and its immediate exposure of all chambers to examination is safer than the automatic, which may have a cartridge overlooked in the chamber when the magazine is withdrawn and which may easily be set aside cocked. A multiplicity of safety devices are incorporated with modern self-loaders, but these make for complexity at the expense of robustness. Some weapons are provided with a device ensuring that if the magazine is withdrawn a cartridge overlooked in the chamber cannot be discharged. Others, in addition to this, have a safety bolt or latch on the action, a half-cock to an external hammer and a "grip safety" inhibiting discharge unless the arm is gripped.

These safeties have obvious reasons for existence, but it is probable that the multiplication of safety devices loses more lives than it saves, and that good training in the use of an arm should reduce the number of safeties needing independent release to one, which is all that is required on an ordinary hammerless sporting gun. This might well be a "grip safety" provided that it is so designed as to be grit, mud and sand proof.

An automatic internal safety insuring that an overlooked cartridge in the chamber cannot be fired when the magazine is withdrawn is also valuable, but it is too much to expect a man to make sure that the safety latch is down, and that the hammer is not at half-cock, as well as relying on the grip safety.

(9) *Calibre.*—Theory and practice show a wide diversity, because a surplus of foot pounds of energy are of no use at all unless that energy is converted into shock effect on or in the body of an adversary. The ·455 revolver and the ·455 self-loader are of equal calibre. At ten yards the momentum of the bullet of the self-loader is far greater, but in practice the revolver bullet is superior in shock effect on the human body. In the same way the German Luger service self-loader of 9 mm. (nominally called ·380", actually ·347") is actually, owing to its high penetration, less effective than self-loading pocket pistols of the same calibre firing a shorter cartridge of less velocity. Experience has certainly shown that the relative advantage, as a military weapon of defence, of a self-loading pistol compared with a revolver of equal calibre, is more apparent in theory than in practice.

For service use the ·45 or ·455 self-loader is the only reliable calibre, but in a well-designed pistol it should be possible to return to a suitable hard lead projectile and to abandon the present cupro-nickel jacketted bullet in favour of one with greater capacity for deformation on tissue. It should be noted that the larger the calibre the less likelihood there is of jams, for the larger cartridge has a greater margin of power.

In pocket arms only the ·380 short self-loader need be considered, for pistols of this calibre are now made but little larger than the ·32 self-loader and vest pocket ·25 self-loader, both of which calibres are too small to rely upon for any effect at all unless a vital spot is hit.

### Conclusion.

No self-loading pistol is yet as reliable as a revolver, nor does it hit as effective a blow. If a pocket arm is to be carried with only a problematical likelihood of its being needed, a self-loader is more convenient and less bulky. Should there be any real likelihood of a pistol being required and two shots needed, a reliable revolver of pocket size should be carried instead. It will be found that this opinion is concurred in by the arm experts of British, American and German technical journals.

There are many makes and calibres of self-loading pistols manufactured in Great Britain, the United States of America, Germany, France, Austria, Belgium, Spain, Italy and Japan. These arms can be divided into three broad categories :—

(1) Those in which the breech is mechanically locked to the barrel during the passage of the bullet up the barrel, *i.e.*, having locked breech actions.

(2) Those in which a temporary or deterrent locking device secures the breech to the barrel for a short period, *i.e.*, semi-locked actions.

(3) Those in which there is no lock but the mass of the breech and recoiling parts, aided by a spring, are sufficient to hold the action closed until the bullet has left the barrel, *i.e.*, blow-back actions.

The latter principle is the most widely used for small calibre weapons, for in small calibre weapons firing a moderate charge the weight of the breech-block, held up against the breech by means of the powerful recoil spring, is quite enough to prevent the breech opening before the bullet has left the barrel.

Theoretically, a charge of powder fired in a " pistol " whose weight is exactly the same as that of the bullet, and which is equally free to move, shoots the bullet as far in one direction as the " pistol " is thrown by the recoil in the opposite direction. In other words, the momentum of the pistol is equal to the momentum of the bullet, and if their masses are equal then their velocities will also be equal but in opposite directions.

In actual practice the inertia of the bullet is very many times less than that of the recoiling parts and the resistance of the recoil spring. When the cartridge is fired the light bullet, possessing less resistance than any other part of the weapon, is therefore, driven out by the gas pressure before the same force has had time to overcome the inertia of the breech and spring. This " hesitation " on the part of the recoiling parts is only momentary, lasting an infinitesimal fraction of a second, but it is sufficient for the bullet to have left the barrel before the breech begins to open. If the cartridge is a very long or powerful one, or if the barrel is very long, some arrangement has to be made that will hold the barrel and breech block together until the bullet has left the barrel, otherwise the base of the cartridge would be unsupported while the gas pressure was still binding the wall of the case to the wall of the chamber. This would result in a separated case, a portion remaining in the chamber, and the weapon would jam.

The ·450 and ·38 Colts and Webleys, the Mauser ·301 and Luger-Parabellum pistols all have some such breech-locking device. The Savage ·38 has a semi-lock or delaying action, and smaller weapons of the pocket type nearly all depend upon heavy breech blocks, and are on the " blow-back " principle.

The locked-breech pistols usually provide for either (*a*) the breech and barrel recoiling together in the same plane for a short distance (Parabellum, Mauser and big Bayard), or (*b*) the barrel and breech bolt, being connected by a sleeve or slide, recoiling a certain limited distance together until the former is automatically raised or lowered from engagement with the sleeve (Webley and Colt).

The Savage has a rotary principle of its own. A similar system is more fully employed in the 1916 Steyr.

In the Colt locked breech pistols method (*b*) is employed, and the barrel is caused to drop from engagement with locking recesses in the sleeve by the action of a link (or links) pivoting on a fixed axis in the body. This system has proved very efficient in practice.

In the Webley locked-breech models the squared sides of the barrel have inclined planes milled upon them, which secure the barrel to the frame. As the action closes, the pressure of the bolt lifts this barrel, which is free to move upon its planes in the slots, and raises it into engagement with the breech-block slide. In practice this type of weapon is far too easily put out of action by dirt or fouling and is not reliable.

In the Luger-Parabellum the barrel and breech-block recoil after firing as one piece, until the projections of the toggle-joint in the breech bolt are lifted by the curved incline on the frame, which bends the " knee " of the toggle-joint, causing its front arm to pull back the breech block and its rear or fixed arm to compress the recoil spring, which is engaged with a hook lever hanging from its centre. The firing spring and striker are contained in the breech-block and are compressed by the first upward retracting movement of the front arm of the toggle. The return movement is accomplished by the compressed spiral recoil spring in the butt, which pulls down the toggle, straightening out the knee and locking the bolt and barrel together again. The arm is fairly reliable, but the same mechanism would be unduly cumbrous in a pistol of heavier calibre.

*Action of the Colt Self-Loader.* (Fig. 3 and Plate V.)

Following the discharge of the first cartridge, the recoil drives back the slide. During the first part of its travel the barrel, which is pivoted on a link, is carried back with the slide, locked to it by lugs on the top of the barrel which engage with recesses in the slide.

The extent to which the barrel can move backward is limited by the radius of the link arm, which, moving on its axis, pulls the barrel downwards, disconnecting it from the slide. The slide, to which is attached the cartridge extractor, continues to move, cocking the hammer by pressing it back and rolling over it. The sear then engages with the bent of the hammer and retains it cocked. The slide toward the end of its travel thrusts the base of the empty cartridge against an ejector block on the frame, which jerks the empty case clear of the extractor claw and out of the action. Finally, the slide travel is stopped by the recoil-spring housing butting against the frame.

The backward movement has also compressed the recoil spring and disconnected the trigger mechanism, so that no involuntary pull or jerk on the trigger can release the hammer while the slide is back.

The next cartridge impelled by the magazine spring now rises between the recoil slide and the open mouth of the chamber. The recoil spring then reasserts itself and forces the slide forward again so that it pushes the cartridge from the magazine into the open chamber mouth and, coming in contact with a projection on the upper face of the barrel, pushes the barrel forward so that the link arm is pulled up and the top surface of the barrel once again locks into the notches on the inside of the recoil slide. At the finish of the forward movement a recess in the under-surface of the slide comes over a small lever in the frame, which is able to spring up into the space. This lever is the disconnector, which when pressed down disconnects the trigger mechanism from the scear and the hammer at all positions of the slide except when the breech is fully closed. Thus the pistol cannot be fired unless the breech is safely closed.

When the last cartridge in the magazine has been fired, the magazine platform lifts a stop which holds the action open and the slide at the end of its backward stroke. The insertion of a new full magazine releases this stop, and the slide flies forward, loading the first of the cartridges from the fresh magazine into the chamber.

On the left-hand side of the pistol is a thumb safety catch. A half-cock notch is provided on the hammer, and there is also a grip safety on the butt which ensures that unless the pistol is gripped it cannot be discharged.

The foresight is fixed, but the backsight is capable of lateral adjustment.

The Colt is probably the most reliable of large-calibre self-loading pistols. The butt is correctly designed and slopes properly to the hand, the material is reliable and the workmanship accurate, parts are interchangeable, and it will stand active service conditions fairly well.

*The Action of the ·455 Webley-Scott.* (Fig. 4 and Plate VI.)

When a cartridge is fired the breech slide is driven backward. On its inside lateral surfaces are diagonal-inclined grooves in which diagonal-inclined planes on the barrel engage. The rearward movement of the slide forces the barrel back, but these diagonal slots translate the backward movement into a downward one until the barrel is freed from its engagement with the slide. The barrel stops at the end of its short diagonal movement, but the slide continues to recoil, cocking the hammer, extracting and ejecting the empty case and disconnecting trigger and hammer action, as is usual in self-loading pistols. On its return it reloads. The recoil spring in the Webley-Scott is not of the usual spiral concentric type, parallel with the barrel, but is a flat spring housed beneath one of the ebonite grips of the butt and acted on by a lever which projects through the frame and engages with the slide. This system has the disadvantage that any injury to the fragile ebonite grip exposes the mechanism and renders the pistol unuseable. The angle of the butt is not good and the functioning of the weapon not very reliable. The same faults of design occur throughout the series of pocket models and medium calibres. In practice the weapon has not proved suitable for active service conditions on land, though it was retained in the Navy during the Great War.

*Action of Luger-Borchardt-Parabellum.* (Plate VII.)

This pistol employs a hinged breech block acting on the "toggle-joint" principle. A toggle joint may be compared to the human leg. So long as the leg is held straight and rigid it will resist very heavy pressures applied to the sole of the foot. If, on the other hand, a force is applied to bend the knee joint then the leg can no longer resist and yields to the pressure by flexing at the knee.

In the Parabellum the barrel and breech recoil together as a straight unit for a short distance, until curved lugs on the frame "bend the knee" of the toggle joint upward and allow the breech bolt to move backward against a spring acting on the shorter arm of the fixed lever of the toggle joint. During the first half inch of travel a short coil spring driving the firing pin, situated inside the breech bolt, is compressed. There is no external hammer and the trigger mechanism acts by lateral pressure on a sear, whose tail projects through the side of the frame above the trigger guard.

The pistol is well designed, but has several serious weaknesses. The enclosed hammer-firing pin mechanism is frail and the side plate, covering the sear mechanism, admits sand or mud and is easily put out of action under active service conditions.

Several models of the weapon are made differing in length of barrel, in sighting, and in small details of safety catch and trigger. The long-barreled type 7·5-inch barrel is ambitiously sighted to 800 metres, at which range bullets would still penetrate a French steel helmet. The ordinary magazine holds eight shots, and a drum magazine holding 32 shots is also used with the weapon. When the drum is used a detachable stock is fitted.

The slope of the butt and general balance of the weapon are markedly good, but the small calibre 9 mm., actually ·347-inch, limits its efficiency. It is accurate and has well-designed sights, which are practical as well as theoretically efficient.

*Pistols of Foreign Powers.*

During the Great War the supply of service type pistols used by the various combatants was unequal to the demand. As a result various arms of different calibre and manufacture were issued. The German service pistol was the Luger-Parabellum, but in the field the Mauser 9-mm. (Plate VIII), Steyr Mannlicher 9-mm. (special) and Mauser ·32-inch as well as small quantities of Walther, Dreyse and Frommer pistols of ·32-inch calibre were used.

The French ordnance revolver was largely superseded by commercial types of ·32 self-loading pistols of indifferent Spanish workmanship, F. N. Brownings and other similar arms. Under the conditions of the Treaty of Versailles the German and Austrian factories are forbidden to manufacture or sell pistols or revolvers of the military 9-mm. calibre. They have since produced a number of very well-designed simple and efficient self-loading pistols of ·32-inch and 7·63 mm. In many cases the dimensions of these new pistols are such that they will accommodate a 9-mm. barrel and magazine without material alteration of the manufacturing plant.

Profiting by experience of large scale manufacture during the war the Spanish arms factories have been turning their attention to the production of military models. The Belgian-made Browning F.N. firing the long 9-mm. "Bayard" cartridge is now being supplemented in the Spanish services by an "Astra" pistol taking the same cartridge.

In the Italian forces self-loading pistols of a blow-back type, firing a cartridge of the same calibre and size as the Luger-Parabellum but with a weaker propellent, are in use.

The U.S. Army is equipped with the Colt ·45 self-loader. This pistol is the same as the Colt self-loader used in the Canadian forces, but the latter is adapted for the ·455 self-loader cartridge as used in the Webley ·455 self-loader. This cartridge is not interchangeable with the ·45 U.S. cartridge. The performance of the latter is the better.

The Japanese forces use the Nambu pistol made in Japan. There are two calibres, the 8-mm. 8-round model with holster stock and a 7-mm. 7-round officers' model. The bullet is not cupro-nickel covered but of hardened lead.

Belgium, Holland and some South American States are equipped with the military model Browning F.N. firing the long 9-mm. cartridge.

Other South American States are equipped with the ·38 Colt self-loader.

DATA FOR SELF-LOADING PISTOLS.

| Make of Weapon. | Nominal Calibre and Designation of Cartridge. | Number of Shots in Magazine. | Weight of Bullet in Grains. | Nominal M.V. in ft./sec. | Weight of Empty Pistol in ozs. | |
|---|---|---|---|---|---|---|
| Colt | ·45 U.S. Government | 7 | 200 | 850 | 39 | NOTE. |
| Colt | ·45 | 7 | 230 | 800 | 39 | |
| Colt | ·455 S.L. | 7 | 224 | 750 | 39 | There is often confusion |
| Webley | ·455 S.L. | 7 | 224 | 710 | 36 | between the cartridges for |
| Colt | ·38 Auto. | 8 | 130 | 1,175 | 34 | 9 mm. or ·38 pistols and |
| Luger-Parabellum | ·9 mm. Parabellum | 7 | 125 | 1,040 | 31 | the "·380." The ·380 auto. is the same as the |
| Bayard | ·9 mm. Bayard | 8 | 120 | 1,134 | 33 | 9 mm. short Browning used |
| Browning | ·9 mm. Browning long | 7 | 110 | 950 | 32 | in Continental pistols, and is |
| Webley | ·38 Auto. | 8 | 128 | 1,000 | 31 | for pocket pistols. Military |
| Steyr | ·9 mm. Steyr | 8 | 116 | 1,150 | 18½ | models take the 9 mm. long |
| Webley | ·9 mm. long | 8 | 110 | 1,005 | 32 | or military Browning or |
| Colt | ·380 Auto. | 7 | 130 | 860 | 22 | special cartridges called after |
| Savage | ·380 Auto. | 10 | 130 | 860 | 19 | the make of pistol, i.e., |
| Mauser | ·63 mm. Mauser | 10 | 86 | 1,350 | 40 | Bayard, Steyr, Parabellum, |
| Mauser | ·32 Auto. | 8 | 74 | 940 | 20 | etc. The ·38 auto. is a |
| Colt | ·32 Auto. | 8 | 74 | 938 | 23 | special long American cart- |
| Webley M.P. | ·32 Auto. | 8 | 74 | 900 | 18 | ridge used in the Colt and |
| Browning | ·32 Auto. | 7 | 74 | 950 | 20 | Webley semi-military type |
| Average | ·25 Auto. | 7 | 50 | 700 | 13 | pistols. |

The ballistics of self-loading pistols vary according to the quality and make of ammunition used, for bullet weights and loads are varied by the different manufacturers. The average muzzle energy developed by self-loading pistols, according to calibre, is approximately as follows :—

|  | Ft.-lbs. |
|---|---|
| ·25 | 60 |
| ·32 | 150 |
| ·380 | 160 |
| ·9 mm. | 300 |
| ·38 | 350 |
| ·45 | 360 |
| ·455 | 300 |

During the closing stages of the Great War, and later, automatic gun-pistols or pistol carbines were developed. The first step in this direction was the provision of a 32-shot magazine or drum-feed to fit the long-barrelled Luger-Parabellum, issued to N.C.Os. of German machine-gun units. These arms were ordinary Lugers fitted with 9 or 11½-inch barrels, a tangent rear sight and a detachable wooden stock. In practice, detachable stocks fitted to pistols are of little value. The weapons are too light for accurate shooting and the combination is badly balanced and badly proportioned. In the case of the Parabellum the rearward movement of the barrel introduced another disturbing element. It was, however, a weapon of importance for trench warfare and from it was evolved a new type of machine-pistol using the same cartridge and the same type of drum magazine, but built on the lines of an automatic carbine. It was known as a " Bergman Muskete."

In a subsequent American development, the Thompson sub-machine gun, a similar type of arm fed from a helical drum magazine was introduced. The promoters claim for this weapon that a new principle of breech action dependent on the adhesive action of plane surfaces under pressure, is employed. Removal of the inclined planes acting as a lock converts it to a plain blow-back action similar to the Bergman machine pistol, and it functions as well without as with the locking gear. Several similar types of weapon with spiral rotating or other delaying, rather than locking, actions, have been developed on the continent.

Colt ·455 Self-Loader.

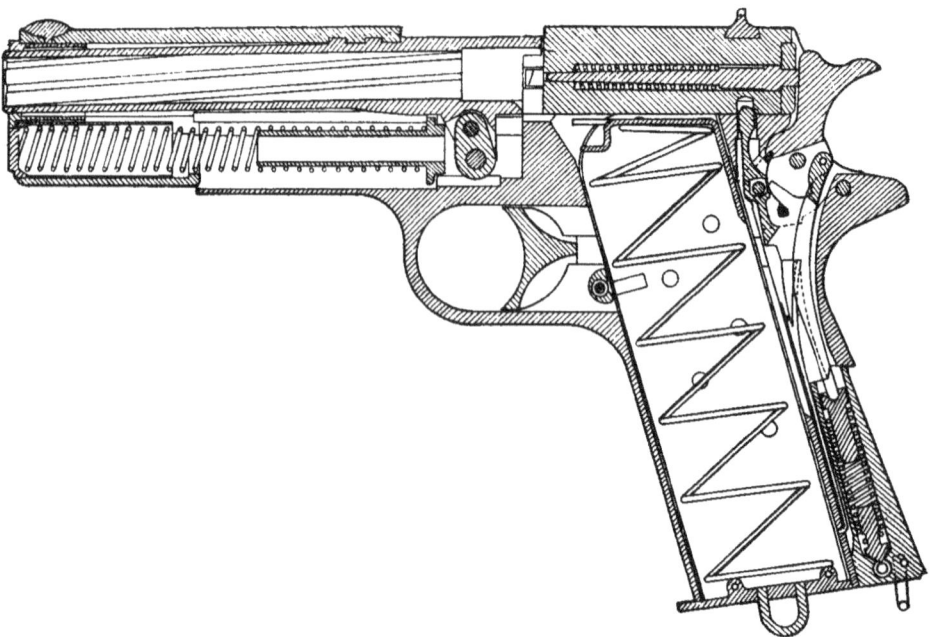

Part I, Chap. III, Fig. 3.

·455-inch Webley and Scott Self-Loader.

Part I, Chap. III, Fig. 4.

[*To face page* 104.

Part I, Chap. III, Plate V.
Colt ·455 Self-Loader.

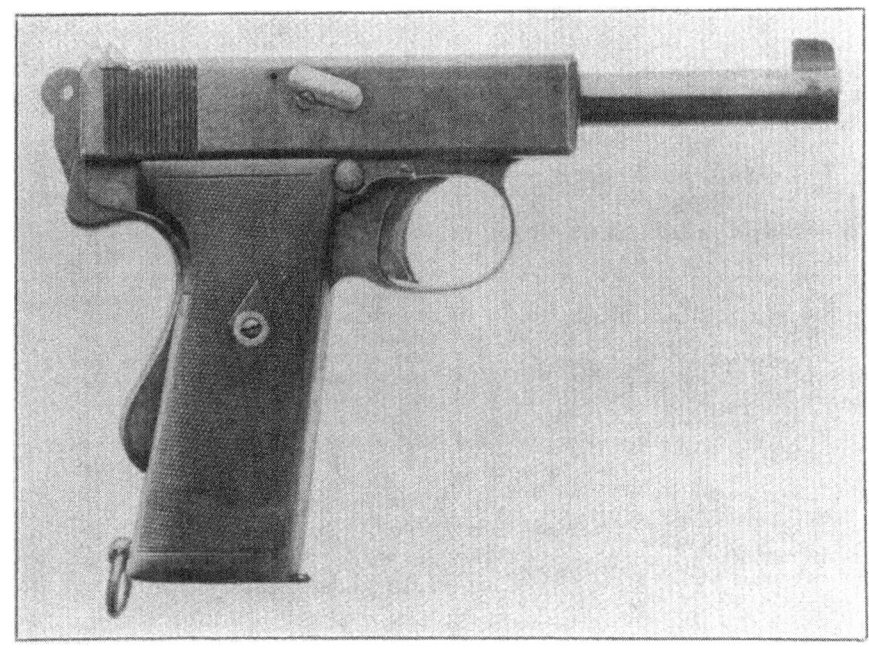

Part I, Chap. III, Plate VI.
·455 Webley and Scott Self-Loader.

[*To face page* 105.

Part I, Chap. III, Plate VII.
Luger-Borchardt-Parabellum.

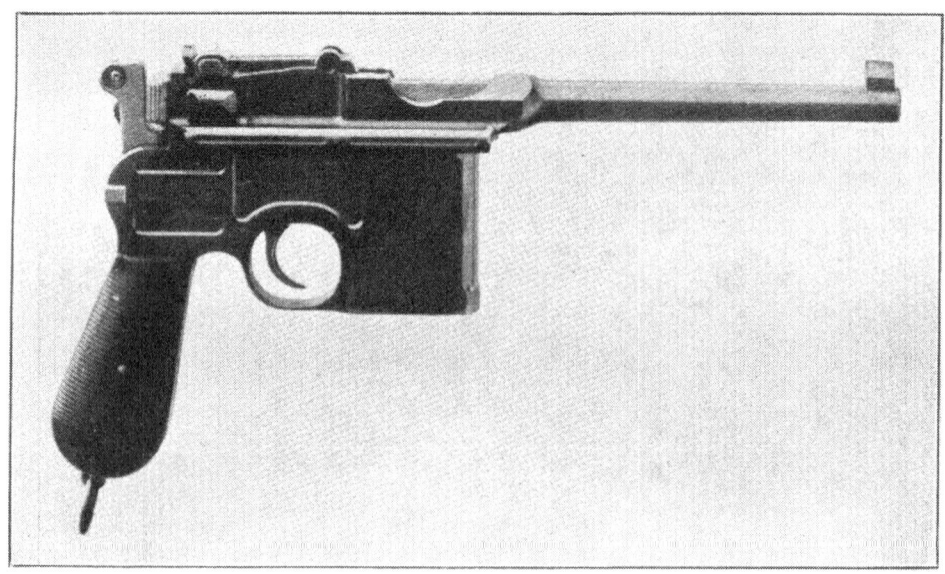

Part I, Chap. III, Plate VIII.
9 mm. Mauser Self-Loader.

## CHAPTER IV—SECTION 1

### HISTORY OF GRENADES

The grenade as a weapon of war dates from the 15th century; it may possibly have been invented, or discovered by accident, even earlier than this, but its history is somewhat obscure. The earliest grenades of which there is any record were made of baked earth; later types had cases of wood or brass and were sometimes filled with bullets of iron, lead, or even glass.

Baptista della Valle, writing at the commencement of the 16th century, describes the preparation of grenades to be thrown by hand. These grenades measured about 3¼ inches in diameter and were similar in shape to a pomegranate, from which fruit the weapon took its name.

It is also recorded by Du Bellay that grenades were made at Arles in 1536.

Hand grenades were first used in Italy in the defence of fortifications; they were employed in 1562 at the siege of Rouen and at the siege of Famagusta in 1571. There are further references to their use in the assault on Wachtendoek in 1606, in the defence of Regensburg in 1634 and in the defence of Vienna in 1683. Vauban used no less than 20,000 in the siege of Namur.

The grenade of the 17th century generally consisted of a hollow cast-iron sphere containing powder and provided with a fuze; the whole weighing about 2½ lbs.

When the fuze had been ignited by means of a slow-match the grenade was whirled round to get the fuze well alight and was then hurled at the enemy. It often happened that grenades burst prematurely, shattering the hand of the thrower, who would also often be struck by fragments should he not have thrown the grenade sufficiently far. With the primitive material then available, skill and courage were absolute essentials for the grenade thrower, and only the bravest and most determined men were suitable for the purpose. Beginning with 4 per company of French Infantry in 1667, the employment of "Grenadiers" spread until they were formed into special companies, battalions, and even regiments. Other countries copied the system and the grenadiers became, on account of their being picked and specially trained men, the choice troops of their respective armies. Cavalry regiments were also provided with grenades and were known as Horse Grenadiers.

The military works of the time show that the grenades were sometimes made use of in pitched battles but met with little success, and as fire arms improved, gradually fell into disuse. By 1760 they had completely disappeared from battlefields.

As to their employment in the Napoleonic wars (wars of movement as opposed to the " siege " operations of the two centuries preceding) we know nothing authentic, as they are scarcely ever mentioned in the accounts.

We next hear of grenades in the siege of Sevastopol, where both French and Russians made use of them. For lack of metal shells the besieged used glass bottles filled with powder and provided with a fuze inserted in the neck.

In the American Civil War, a bomb not unlike a hand grenade and provided with a percussion fuze, played a part in the close combat; in this connection it is interesting to note that a grenade having a primitive percussion fuze appeared in the 16th century.

In the wars of 1864, 1866 and 1870—again chiefly wars of movement—grenades do not seem to have been used at all.

They were, however, used by our troops in the Sudan in 1884–5; these grenades consisted of cast-iron spheres of about the size of a tennis ball, were filled with powder and provided with the usual simple fuze.

The Russo-Japanese war marks the real renaissance of the grenade.

At the siege of Port Arthur both sides found the need of a weapon of this description and showed considerable ingenuity in improvising them from a variety of materials. The Russians used old iron cases or mountain artillery shells; later the brass cases of Q.F.

Artillery cut down to a height of 4 inches, filled with dynamite or gun-cotton, and fitted with a 15-seconds length of Bickford's Safety Fuze. The first Japanese grenades were made from old preserve tins or bamboo tubes filled with a pound of gun-cotton and fitted with a similar fuze and detonator to that used by the Russians. When first introduced the fuzes were often cut too long and the grenades were thrown back by the enemy. Fuzes were lighted by means of matches, slow-match, or a cigarette, but later a rough percussion igniter was improvised by means of a rifle cartridge, which acted as a primer, and steel wire. The Japanese also invented a grenade consisting of a prism of picric acid between two cakes of gun-cotton; the whole was wrapped in paper bound round with string and weighed about 1 lb. In due course grenades were regularly made by both sides.

At Mukden the Japanese used a metal cylinder fitted with a wooden handle and a percussion fuze; later patterns had a guide tape fixed to the handle to ensure the grenade falling on its nose. The Russian patterns were generally similar.

Eventually the Marten Hale grenade was adopted by the Japanese. This grenade consisted of a brass tube $1\frac{3}{4}$ inches in diameter by 6 inches long and contained one-third of a lb. of tonite. The top of the tube was closed by a wooden plug, to which was attached the slinging cord, 18 inches in length. This cord was unravelled at one end to act as a tail. The lower end of the tube contained a detonator, cap, and direct impact striker mechanism. About 1 inch above the base was a steel collar formed with 24 segments. The complete grenade weighed 1·37 lbs. As a result of the Russo-Japanese war most nations experimented with grenades and produced various designs. Marten Hale produced his rodded rifle grenade.

The French in Morocco used improvised grenades consisting of two melinite cartridges tied together and fitted with a piece of palm branch as a tail; they had a time fuse which was lighted in the ordinary way, and in addition a percussion fuse consisting of a nail with a roughened point fixed in the mouth of a detonator. As a result of experience with this grenade they produced a new type. This grenade consisted of a spherical cast-iron shell of $3\frac{1}{16}$ths inches diameter containing 0·24 lbs. of the powder used for gun ammunition; the complete grenade weighed 2·6 lbs. Ignition was caused by means of a friction tube and a five-seconds length of fuze mixture. The thrower of this grenade was provided with a wrist strap carrying a hook, to which the eye of the friction wire of the grenade was attached preparatory to throwing. On releasing the grenade in the act of throwing, the friction wire was jerked out of its tube and ignited the fuze.

Another grenade produced about this time was the Aasen. The hand grenade consisted of a metal body fitted with a short wooden handle and a fabric tail somewhat like an inverted parachute. It had a percussion fuze. It was chiefly remarkable for its safety device, which consisted of a cord 10 metres long; this cord was rolled up inside the handle; its inner end was attached to the safety release, while the other was provided with a loop to be held by the thrower. On throwing the grenade the cord became unwound, and on coming to the end of its tether, jerked out the safety pin, freed the striker and fell free. The grenade then exploded on impact.

In 1913, Germany adopted a rodded rifle grenade somewhat similar to the Hale. This grenade was provided with a powder pellet safety device (a variation of that used in some of their artillery fuzes). With this arrangement the movement of the striker was obstructed by a powder pellet which was lighted by an auxiliary striker and cap on discharging the grenade. The pellet having burned away, the main striker was free to move and cause the explosion of the grenade on impact.

About the same time the No. 1 Mark I hand grenade (Fig. 1) was adopted in this country. This grenade had a tubular brass body surrounded by a segmented cast-iron ring; this ring was intended to break up on the detonation of the charge and also, by its weight, ensured that the grenade fell on its nose. To the base of the grenade was fixed a cane handle 16 inches long and near the end of this was attached a silk braid tail 1 yard in length. The head of the grenade was provided with a detonator chamber and was covered by a brass cap which carried a steel striker riveted to its centre; this cap was rotatable to one of three positions and constituted one of the safety devices. The second safety device was provided by a safety pin. The cap having been set to the " fire " position and the safety pin removed, the grenade was thrown by means of the handle. The tail caused the grenade to fall nose first, and on impact the cap was driven on to the body of the grenade; the striker needle thus entered the detonator and fired the charge. The latter consisted of 4 oz. 2 dr. of lyddite with a primer of compressed picric acid.

The outbreak of war in 1914 found Germany well prepared relatively to the Allies as regards grenades. The latter were forced to improvise, and a large number of patterns were made. Of British grenades alone, some 25 patterns and a still larger number of Marks of explosive grenades were produced between 1914 and 1918. It was very soon found that the handle of the No. 1 Mark I was too long for use in a trench and the Mark II with an $8\frac{1}{4}$ inch handle was introduced. Of this large number of grenades only those of the greatest interest from the design point of view can be described in detail.

The No. 3 was a rifle grenade of the Hale type (Fig. 2). As a safety device the striker was held by two retaining bolts which, in turn, were kept in place by the screwed hub of a wind-vane; this wind-vane rotated while the grenade was in flight, and, being mounted on a screw-thread, moved back, allowing the retaining bolts to fall out and free the striker. The mechanism then functioned on impact. A 10-inch rifle rod was fitted and gave a range of about 200 yards.

The No. 20 grenade (Fig. 3) was similar to the No. 3, but had, instead of the wind-vane safety device, a releasing socket which set back on discharge and allowed the retaining bolts to fall clear. It was found that the hand grenades in use at this period had certain disadvantages. The wooden handles made them awkward to carry and difficult to use from a narrow trench. This latter fact, combined with the use of a percussion fuse, had been the cause of a large number of accidents owing to grenades being struck against the back of a trench when throwing.

The No. 5, or Mills pattern, grenade (Fig. 4) avoided these troubles, as it had no handle, and was fitted with a time fuze, which, moreover, was not ignited until the grenade had left the hand. This proved to be one of the most satisfactory types of grenade produced during the war by any nation.

Modified for use as a rifle grenade, it became the No. 23, and again modified for use with a cup discharger, it became the No. 36. In both forms it could be used as either a hand or rifle grenade.

Grenades Nos. 6 to 9 and 12 to 16, were improvised and are chiefly interesting as examples of what can be done when materials are scarce. Their issue was a temporary measure only.

The No. 12 was the so-called " hair-brush " grenade, and weighed 3 lb.

The No. 19 (Fig. 5) was of the direct impact percussion type; it had a pear-shaped, segmented cast-iron body and a short wooden handle with webbing tails.

The No. 22 (Fig. 6) was a cheap and simple form of rodded rifle grenade. The body was of cast-iron with shallow segmenting grooves. The igniter consisted of a S.A. cartridge case with a No. 8 Mark VII detonator fitted inside it. The striker mechanism was of the direct impact type, consisting of a sheet steel cap with a striker point riveted to its centre. The cap was held in place on the body of the grenade by four spring arms. On impact, the cap was forced on to the body and the striker point driven into the cap of the igniter. No safety device was provided, except that due to the grip of the spring arms on the body of the grenade, and many prematures occurred owing to the cap setting back on discharge. In later issues the caps were made of lighter material, but the grenade was eventually withdrawn owing to this failing. A 15-inch rod fitted with a copper gas check was used and gave a range of 350 yards, but the gas-check was later abandoned as it was found to give no advantage in practice.

The No. 34 (Fig. 7), the so-called " egg " grenade, was an example of the small and light grenade intended to give a greater range than the heavier types. The body was of cast-iron and more or less oval in form. It was fitted with an Adams striker mechanism and took the igniter set of the No. 5 grenade, modified as regards the cap holder. Before throwing, the fuze was ignited by driving the exposed end of the striker against the boot or other hard body and thus firing the cap; the fuze then burned for 5 seconds (in late Marks, 7 seconds). Later patterns of this grenade were reduced in weight to 11 ozs.

In addition to the purely explosive grenades already mentioned, other types, intended for special purposes, were developed in the course of the war. Of these the smoke grenade was the most important. Grenade No. 27 was an example of this type. Intended for use as either a hand or rifle grenade, it consisted of a tinned sheet steel canister filled with white phosphorus and fitted with an Adams striker mechanism. The burster consisted of a cap, a length of safety fuze and a detonator. The detonator served to split open the canister and scatter the contents; the rapid burning of the white phosphorus on exposure to the air then produced a dense cloud of white smoke.

A later pattern of this grenade was produced under the No. 37; it was designed for use with the cup discharger or as a hand grenade, and was of stronger construction than the No. 27. It weighed $1\frac{1}{2}$ lbs. as compared with the $1\frac{1}{4}$ lbs. of its predecessor. Another important type of grenade was that designed for signalling purposes. These grenades consisted of a tubular body into which were packed a paper parachute and the particular signalling device required, *i.e.*, one or more smoke or light-producing candles, and a very small charge of gunpowder. The grenades were fired from a rifle in the usual way. A fuze, ignited at the moment of discharge, was timed so that the grenade should reach the highest point in its flight before the flame was communicated to the gunpowder charge in the grenade. The firing of the charge caused the contents of the grenade to be ejected from the tubular body and at the same time ignited the signal candle or candles. On the parachute opening, the signal candle hung suspended from it and drifted slowly to the ground.

A type of grenade worthy of mention is the anti-tank grenade. These were intended to be fired at a low angle by means of a rifle rod, and it was necessary to obtain a direct hit if they were to be effective. The difficulty of doing this and the limitations as regards weight made it of doubtful value. It is interesting, however, as showing the uses to which grenades have been put.

### PART I, CHAP. IV, FIG. 1.
### HAND-GRENADE (MARK I).

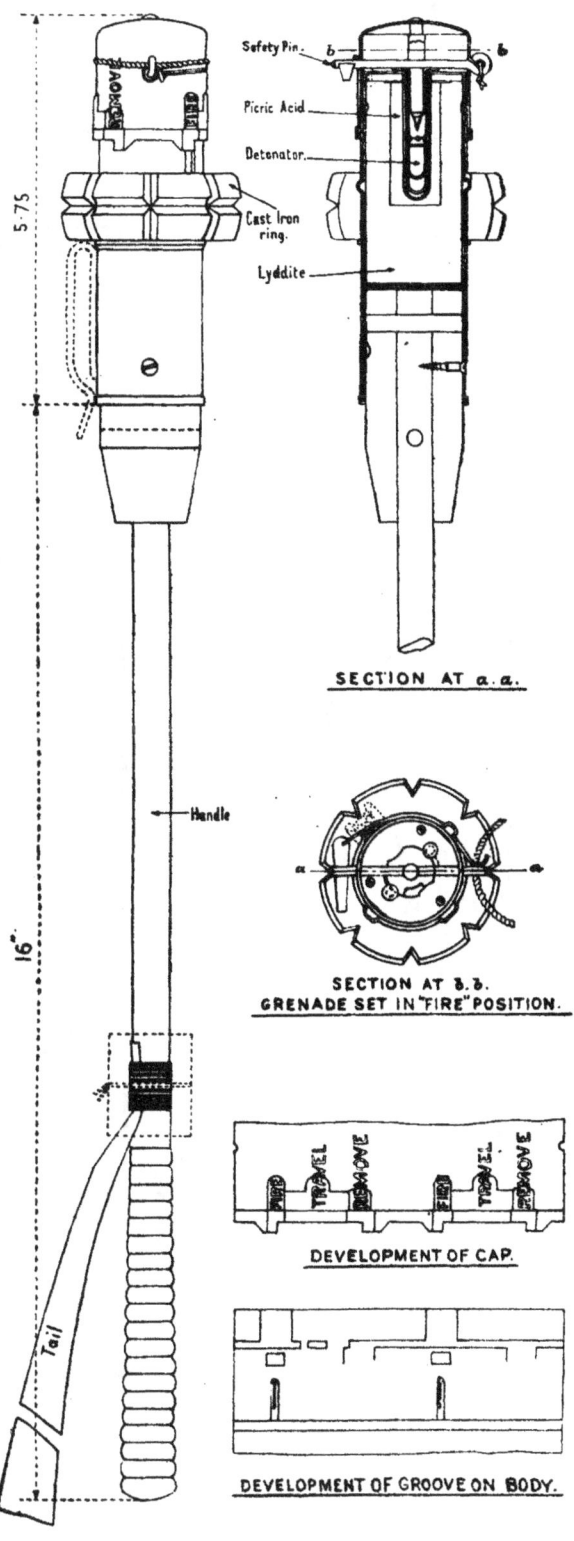

PART I, CHAP. IV, FIG. 2.
GRENADE, ·303-INCH RIFLE, NO. 3, MARK II|c|.
Scale ¼.

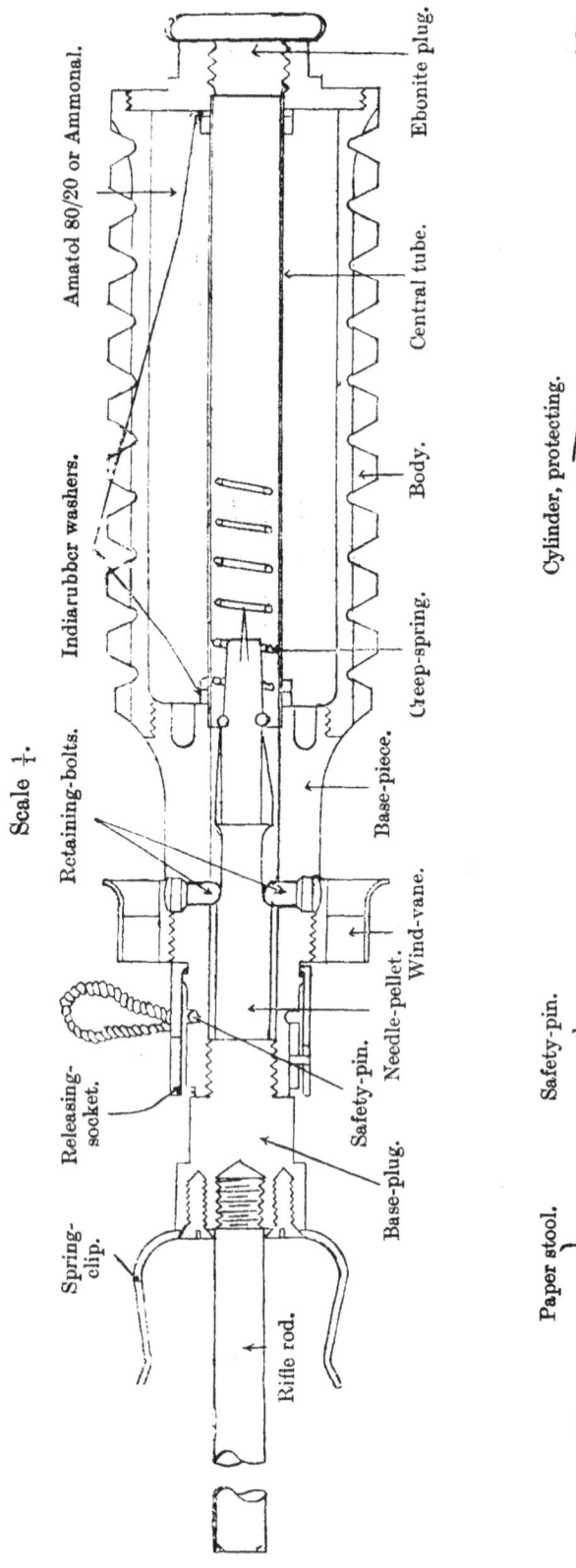
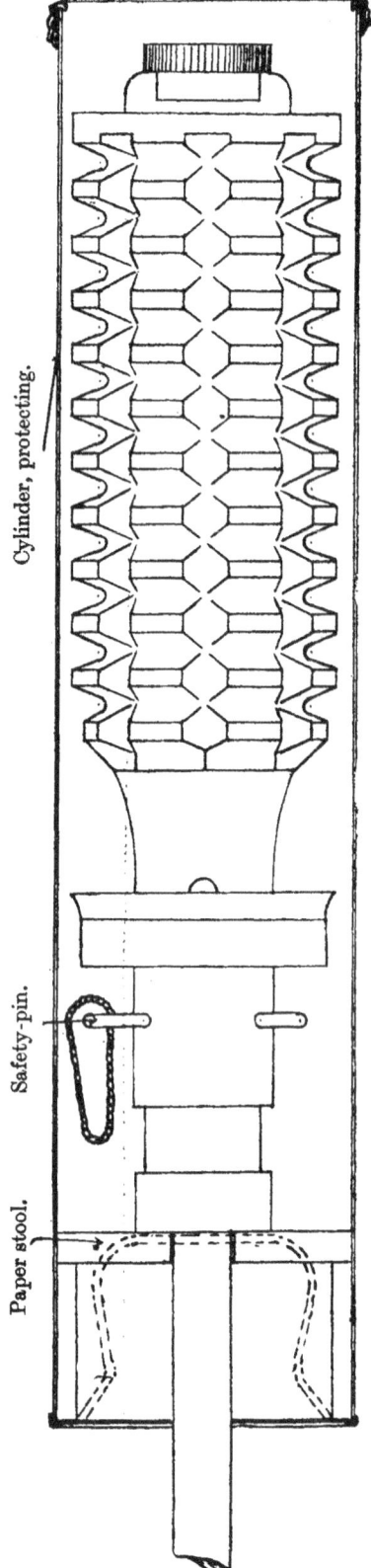

Part I, Chap. IV, Fig. 3.

Grenade, ·303-inch Rifle, No. 20, Mark I |L|.

Scale ¼.

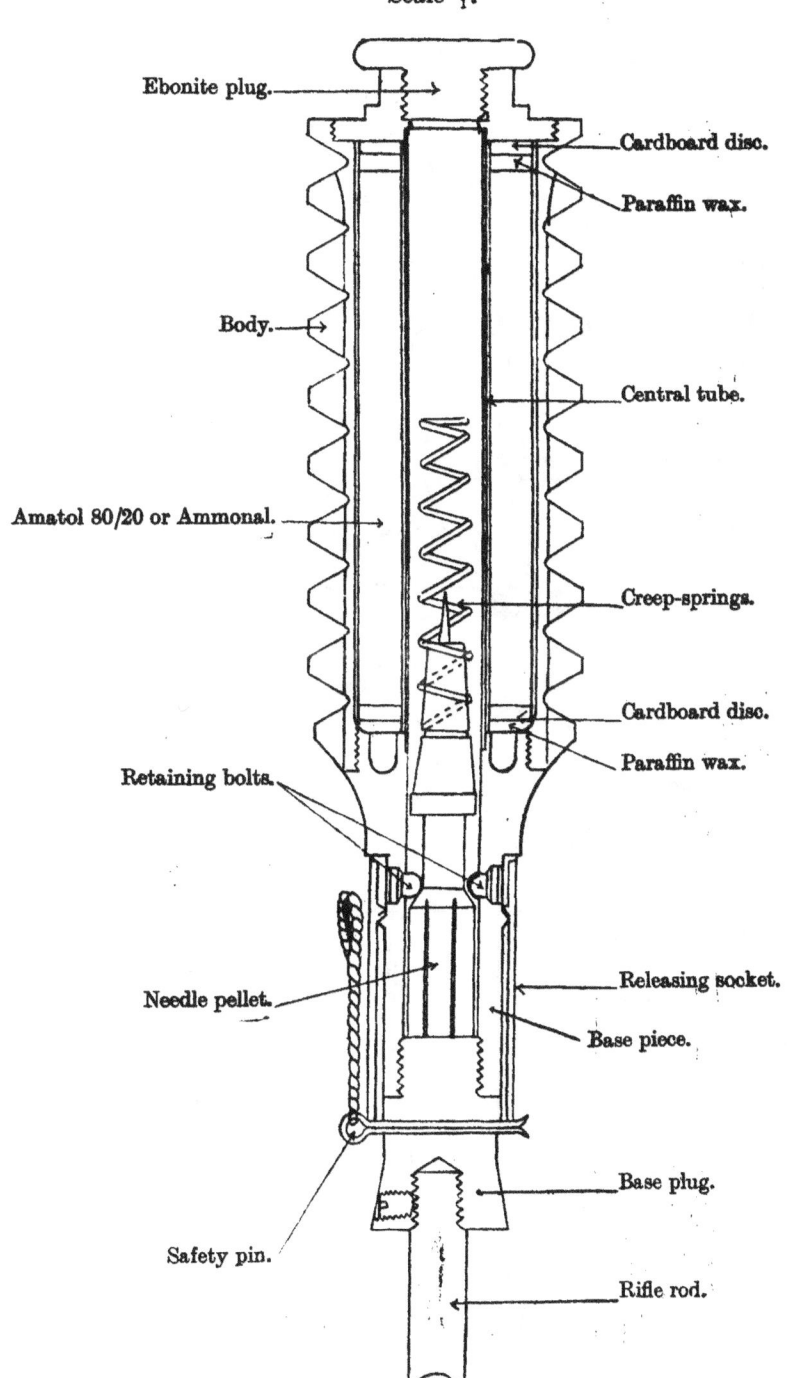

Part I, Chap. IV,—Fig. 4.

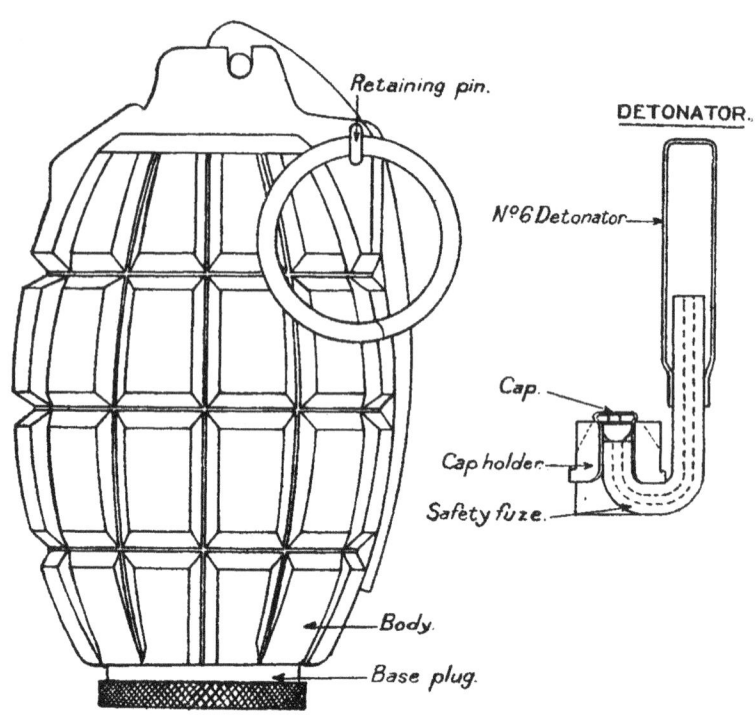

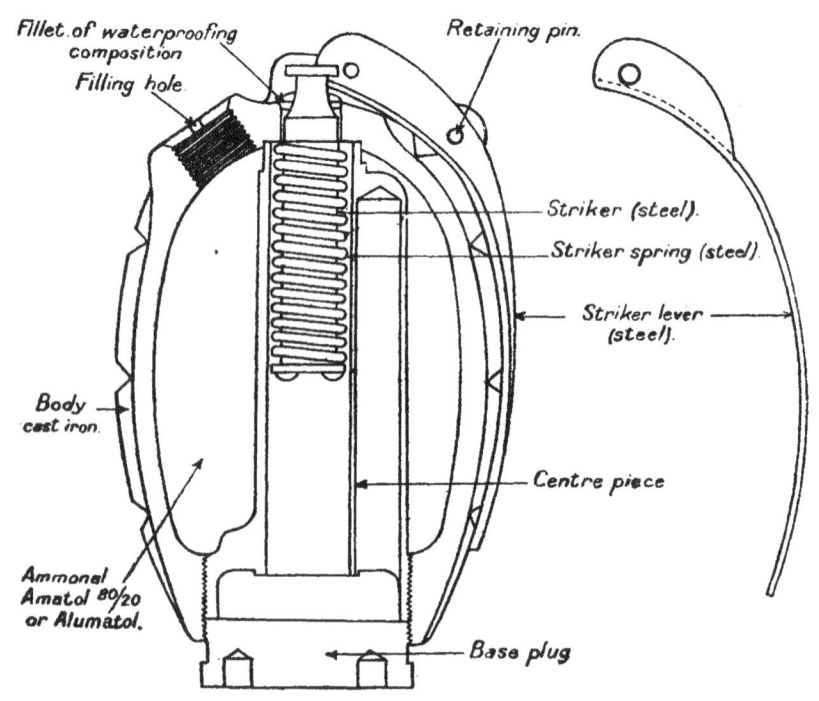

Part I, Chap. IV, Fig. 5.

Grenade, Hand, No. 19, Mark 1 | L |.

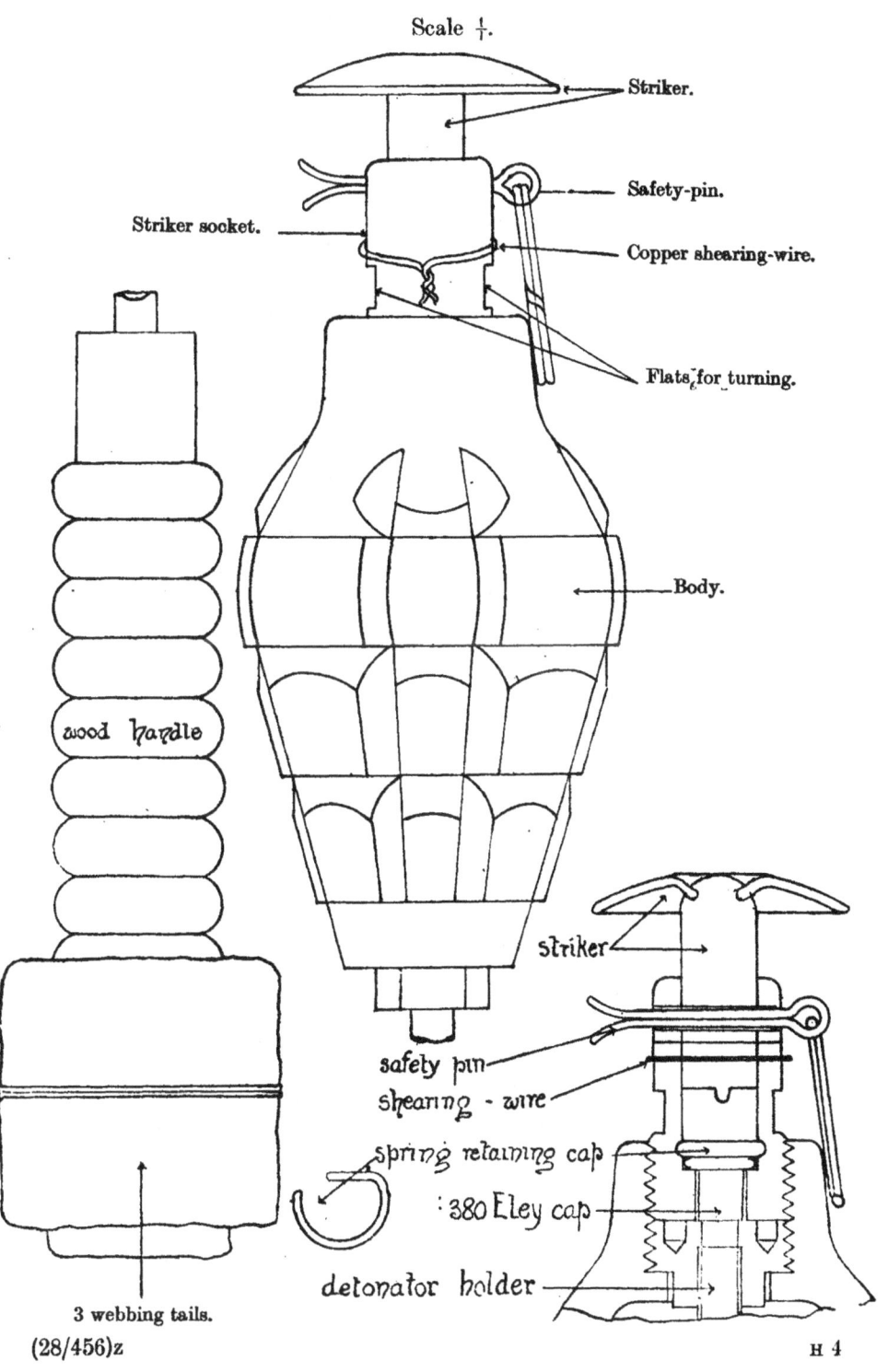

PART I, CHAP. IV, FIG. 6.

GRENADE, ·303-INCH RIFLE, NO. 22, MARK II | L | .

(With detonator fitted.)

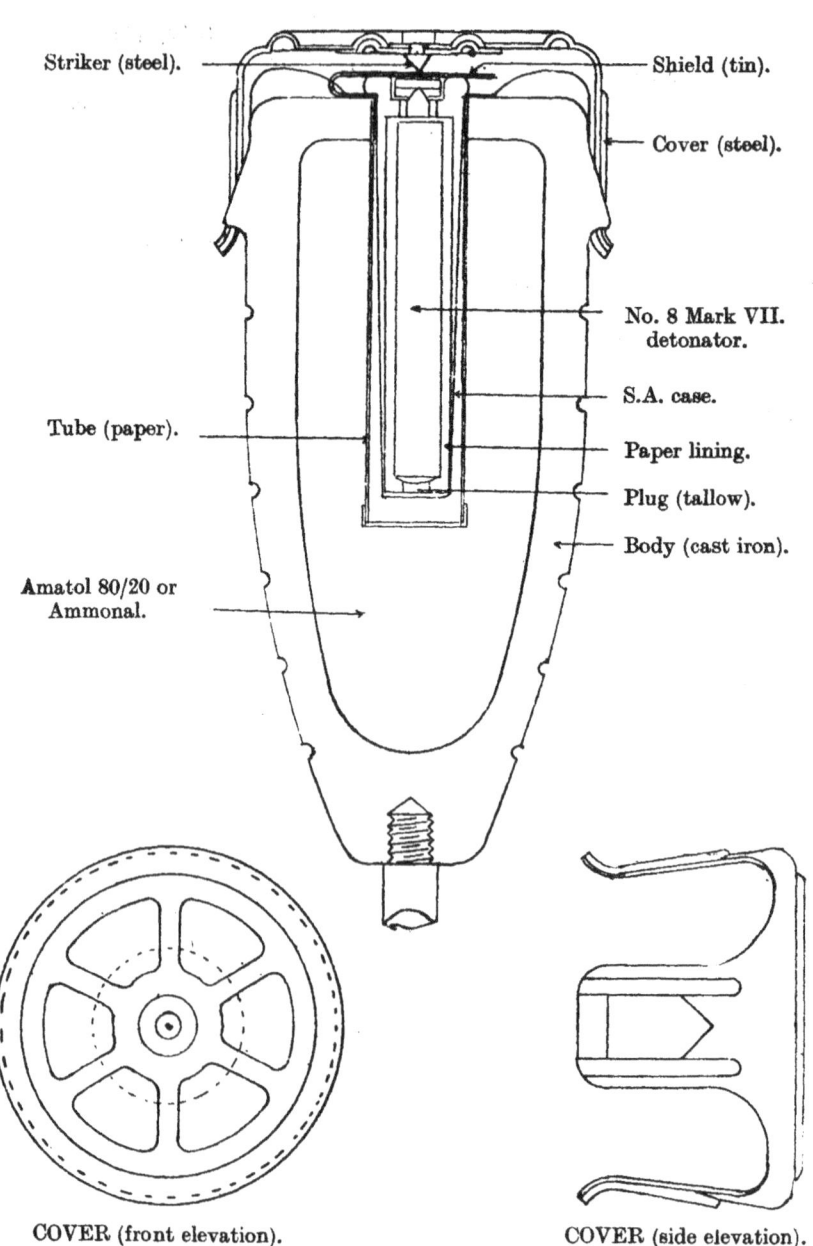

COVER (front elevation).   COVER (side elevation).

Part I, Chap. IV, Fig. 7.

Grenade, Hand, No. 34 | L |.

(With detonator fitted.)

Scale $\frac{1}{1}$.

MARK I.

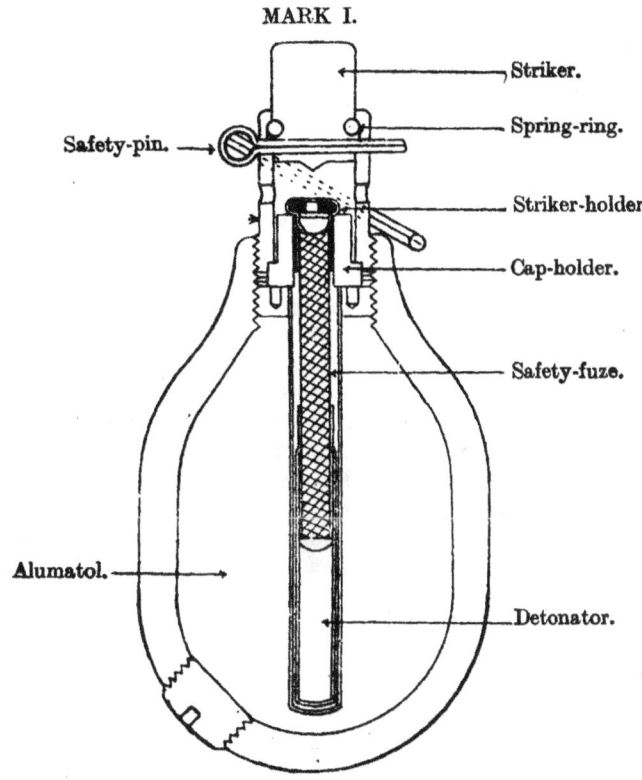

MARK II.

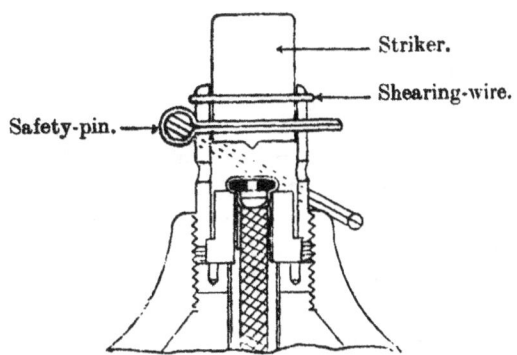

## CHAPTER IV—SECTION 2

### CONSIDERATIONS AFFECTING DESIGN

The design of a grenade requires the consideration of a number of factors which, in some cases, interact upon one another. Three factors of this kind are range, weight and danger zone.

In order to give flexibility in the use of the weapon it is generally desirable that the range should be the maximum obtainable with accuracy.

In the case of grenades fired from a rifle, however, there is a limitation in the strength of the rifle barrel and stock. Given that these parts are already stressed to their safe limit, further increase of range can only be obtained by reducing the weight of the grenade. Reduction of weight would also enable a large number to be carried without increase of load and is, therefore, desirable from the point of view of transport.

Against this, however, is the fact that with reduction in weight comes reduction in the size of the danger zone—other things being equal—which may be a disadvantage. The lighter grenade will probably also be affected by the wind to a greater extent and accuracy may suffer. If the same grenade is to be used equally as a hand or rifle grenade the matter is still further complicated. It may be desirable that a hand grenade should have a small danger zone in order that the thrower or his comrades should not run the risk of being hit by fragments of their own grenade.

The French, making no attempt at compromise, adopted three distinct types of grenade. The first, styled an offensive grenade, has a small danger zone depending largely upon the blast effect of the explosion, and may be safely used by a thrower in the open. The second type is a defensive grenade giving considerable splinter effect with a relatively large danger zone, and is only used by troops under cover. These two are purely hand grenades. The third type is a rifle grenade only, and has a danger zone equal to that of the defensive grenade; in this case, however, the longer range affords a measure of security to the firer.

The German stick hand grenade is likewise a weapon with a comparatively local effect, whereas their rifle grenades have a larger danger zone.

On the other hand, considerations of manufacture and supply favour a single type of grenade for all purposes.

Closely connected with the size of danger zone is the question of fragmentation.

A grenade which breaks up into a large number of small pieces will give a close "pattern"; the chances of obtaining a hit are greater, but the killing power of the fragments will be less owing to their small weight and rapid loss of velocity. On the other hand, a grenade which breaks up into a relatively small number of large fragments will have a more open pattern and the chances of a hit will be less, while the killing power of the fragments will be greater and the danger zone larger. As a hand grenade the first type would be preferable. It is of little use to increase the size of fragments beyond a certain size as the "pattern" becomes so open that the chances of obtaining a hit are very small. Probably the best compromise is arrived at when the fragments are of about $\frac{1}{2}$-inch size. The size of fragments can be controlled and depends chiefly upon three things—the nature of the explosive, the material of which the body of the grenade is made, and the thickness and design of the walls.

It must be remembered, however, that (unless the grenade is particularly designed to avoid it) there will be certain parts, such as base plug, filling screw, and parts of the striker mechanism which will not be broken up by the explosion, and, consequently, will have greater range and killing power than the normal fragments; this must be considered in designing hand grenades, and also as regards the fixing of danger areas in peace training with any grenade.

When a grenade is intended for use as either a hand or rifle grenade, the question of fuze design is important. If a time fuze is to be employed it must allow a time of burning at least equal to the maximum time of flight at any range at which the grenade is to be fired, or air bursts will occur. On the other hand, a fuze of this length may possibly be too great when the grenade is used for throwing, and may enable the enemy to pick it up and throw it back.

To overcome this difficulty requires either an adjustable fuze, such as is used in artillery shells, or a percussion fuze.

The percussion fuze possesses the great advantage that it is independent of the time of flight, and the grenade on striking the ground explodes at once. There is no difficulty in designing an effective and safe fuze of this kind for rifle grenades, but the problem is one of considerable difficulty where the hand grenade, or the combined hand and rifle grenade, is concerned.

The requirements of mechanical design and tactics respectively are so interwoven that it is difficult, if not impossible, to draw a dividing line between them.

Tactical needs, possibly vaguely expressed, call forth a new weapon, and in its design mechanical considerations will probably be given greatest weight. It will, however, subsequently be found in use that certain of the mechanical features embodied are inconvenient from the service point of view, and modification ensues. Although this process of evolution must necessarily follow it can still be said at the outset that there are certain fundamental requirements which the design should satisfy. In the case of the grenade they may be summarized as follows :—

1. It must be effective as a weapon, first and foremost.
2. It must be certain in its action, *i.e.*, reliable.
3. It must be proof against prematures at all times.
4. It must be absolutely safe when properly handled.
5. It must be safe if dropped accidentally.
6. It must be, as nearly as possible, foolproof.
7. It should be easy and cheap to manufacture.

In addition to these preliminary requirements there are the Service requirements, which take into account the purposes for which the weapon is required in the field, and the conditions under which it will be stored and used in peace and war.

The following requirements of a modern hand grenade embody the experience gained in the Great War 1914-1918 :—

### A.—H.E.

1. *Weight.*—Under $1\frac{1}{4}$ lb.
2. *Safe.*—They must not be liable to detonate if dropped, *e.g.*, if the thrower is hit in the act of throwing.
3. *Arming.*—They must " arm " between 15 and 20 feet off the throwers' hand. If by imparting extra spin they can be made to arm quicker, so much the better.
4. *Action.*—They must detonate on percussion.
5. *Reliability.*—They must give a minimum of " blinds." Waterproofing is a factor in this.
6. *Safety of " Blinds."*—They should not be so sensitive that a blind will explode on the slightest touch. If possible they should stand a drop of 1 foot without exploding, with all safety arrangements removed.
7. *Locking arrangements.*—The gear which locks the grenade till it is actually to be thrown should be such as can be removed and replaced by a gloved hand. The replacements should be possible without any necessity for re-adjusting components.
8. *Fragmentation.*—They should fragment into small pieces with a violent effect, but the pieces should not have killing power outside a radius of 20 yards from the burst. As great a proportion as possible of the grenade should be fragmentable and the minimum mechanism.

9. *Standard type.*—They must be capable of being thrown by hand or fired from a rifle without any alteration or addition for either purpose.
10. *Manufacture.*—They must be capable of being manufactured in large quantities, by indifferent workmen, in small, cheaply and badly-equipped works. No part should be difficult to make, and the whole should be easy to fit together accurately.
11. *Inspection.*—They must be easy to inspect.
12. *Transport.*—They must be boxed in suitable man loads and stand any form of transport, even when each grenade contains its detonator. They should be capable of being dropped from aeroplanes, boxed, but with the detonators in a separate case within the box, as a supply of ammunition to advanced troops, without damage to the mechanism.

### B.—*Gas Grenades* and C.—*Smoke Grenades.*

The same points apply, and in addition :—

13. *Structure.*—They must not leak and must give a good cloud of gas or smoke, but gas grenades should not " arm " so quickly as other natures. If a taped mechanism is adopted for all types, it is only a case of lengthening the tape.

### *Rifle Grenades.*

All the points for hand grenades apply and, in addition to 7, the locking arrangement must be such that it can be removed and replaced while the grenade is in the discharger cup.

Also

14. *Range.*—They must have a range of 30 to 300 yards. By this is meant that the range must be controllable between those limits. If more range can be given so much the better, but other conditions should not be sacrificed in order to obtain great range. At over 300 yards it is most unlikely that sufficient accuracy could be obtained. The idea in giving a maximum short range is that a man supplied with grenades and a discharger cup can cover all ground, from the next shell hole to him up to the limit of range of a rifle grenade. He can throw up to 30 yards and fire over the rest.
15. *Propellent.*—They should be propelled by blank cartridge. It would be an advantage if each grenade could have its own blank cartridge so attached to it that insertion of the grenade in the discharger cup is impossible till the cartridge is removed ; this, however, is not essential, but *every* grenade issued should have its own cartridge.

It will be seen by a careful study of the Service requirements that many limitations are imposed upon the designer. The approved grenade will be a compromise which will conform in general with the requirements of the General Staff as regards use, training and transport, and at the same time with those of the technical staff as regards production, inspection, etc. The No. 54 2-inch grenade, of which details are given on pp. 133, 134, represents the solution to the problem which has been adopted in our own Service.

## CHAPTER IV—SECTION 3

### FUZES, IGNITERS, AND BURSTERS

The simplest ignition device for a grenade is the plain safety fuze lighted from some external source. This simple fuze was used in the early grenades, and also in some more recent ones, when gunpowder was the filling. With the addition of the detonator necessary for use with high explosives, it figured in many of the improvised grenades of the Russo-Japanese and the European wars. The next step was to improve the means of lighting and do away with the slow-match or other device of that kind. This was achieved by providing the end of the fuze with a bead of match-head composition, which was ignited immediately before throwing by rubbing against matchbox composition on a brassard worn by the thrower. A further improvement was to use a tape coated with matchbox composition and permanently attached to the grenade or fuze; the fuze was ignited by pulling the free end of the tape, which caused the matchbox composition to be drawn across the bead of match-head composition.

The next form of lighter was the friction tube (Fig. 8); this consisted generally of a tube filled with friction composition, in which was embedded a roughened rod or two strands of wire twisted together. The rod or wire projected from one end of the tube and was formed with a loop; into the other end was fixed the fuze. On withdrawing the rod or wire sharply from the tube by means of the loop, the friction composition ignited and lighted the fuze. This type of igniter was used in the British grenades Nos. 6 and 7, and was employed by the Germans in nearly all their hand grenades.

The "Igniter, Safety Fuze, Percussion" and Nobel's Igniter have also been used as lighters in some improvised grenades. The former device incorporates a striker and spring, while the fuze end carries a percussion cap; ignition is caused by releasing the striker. A natural development of this is the striker mechanism of the Mills' and some other grenades, but these have the added advantage that the striker is not released until the grenade actually leaves the hand.

The Adams' striker mechanism used on the British grenades Nos. 27, 28, 29, 34 (Fig. 7) and 37 was intermediate between the Nobel's igniter and the spring-operated striker mechanisms. It had no spring, but the striker was driven on to the cap by a blow if the grenade was to be thrown, or set back automatically owing to its inertia when the grenade was fired from a rifle. Until required for use the striker was held by means of a safety-pin and shearing wire. On removing the safety-pin the shearing wire alone retained the striker; the force of the blow on the striker-head or the shock of discharge, as the case may be, was then sufficient to cut the wire and drive the striker on to the cap.

The mechanism employed to fire percussion grenades, although known as a fuze, actually contains no safety fuze; normally it consists of a striker, percussion cap, and detonator, and the flash from the cap passes direct to the detonator.

The simplest form of this fuze is that used in certain hand grenades (Fig. 5), and works on the direct-impact principle. The striker, held in place by a safety-pin and shear wire, projects from the nose of the grenade, and is generally fitted with a kind of mushroom head. The grenade, being provided with a tail or its equivalent, falls on its nose, and the striker is driven in on to the cap, cutting the shear wire and exploding the grenade.

Another type of percussion fuze is that used in a large number of rifle grenades (Fig. 3) for which the ordinary direct-impact type is unsuitable owing to the setback which would occur—or which would always be liable to occur—on firing. (The failure of the No. 22 grenade is an instance of this.)

The mechanism is therefore reversed, and the initial shock of discharge, if made use of at all, only serves to operate a safety release, of which the "inertia collar and retaining bolt" is an example.

The action of this type of mechanism is as follows —The safety pin having been removed and the grenade fired, the inertia of the releasing socket or collar causes it to set back, thus uncovering the retaining bolts. The latter, being a loose fit in their recesses, cannot then prevent the movement of the striker pellet, which, however, is still held away from the detonator by a light spring called a creep spring.

As the rifle rod performs the functions of a tail, the grenade strikes the ground nose first and is suddenly checked; the striker pellet, however, owing to its momentum, sets forward, ejects the retaining bolts and, overcoming the resistance of the creep spring, strikes the detonator cap. This type of mechanism is used in most grenades of the Hale pattern. A variation of this design employed a wind-vane in place of the inertia collar (Fig. 2). This wind-vane was mounted on the threaded neck of the base piece of the grenade, and when fully screwed up covered the heads of the retaining bolts. There was, of course, no set-back action on firing, but the relative current of air caused by the flight of the grenade caused the wind vane to screw back and uncover the retaining bolts. On impact the action of the mechanism was the same as in the previous example.

Another substitute for the inertia collar was used in the German 1913 pattern rifle grenade (Fig. 9). In this case the place of the retaining bolts was taken by a pellet of compressed powder and for the striker to be freed it was necessary for this pellet to be ignited and burned away. For this purpose a second, small, inertia pellet, cap, and fixed striker mechanism was fitted into the hollow base piece of the grenade. On firing, the small inertia pellet carrying the cap set back on to the fixed striker; the flash from this ignited the powder pellet which burned away during the flight of the grenade.

On impact the mechanism functioned as in previous examples.

On the German 1914 pattern rifle grenade (Fig. 10) a somewhat complicated mechanism combining a direct impact striker and inertia collar release was used. In this fuze the striker point was hinged to the striker stem, and previous to firing lay upon its side so that it could not possibly come in contact with the cap; this permitted the striker stem with its mushroom head to be fully "home" in the grenade. In this position it was held against the action of a spring by a locking ball, which was in turn kept in place by an inertia collar. On firing the grenade the inertia collar set back and uncovered the locking ball; the spring then asserted itself, ejected the locking ball and forced the whole striker outwards a short distance. A second spring contained in the striker stem at the same time caused the hinged striker point to assume an erect attitude and thus "arm" the grenade. On impact the striker was driven back into the grenade, but the point being now erect, entered the cap and exploded the charge.

In all the percussion fuze mechanisms hitherto described it was necessary for the grenades in which they are fitted to strike the ground nose first. This necessitated some form of tail, which increased the weight and made the grenade cumbersome and awkward. To get over these difficulties fuzes were designed which acted in whatever attitude the grenade fell, whether nose first, tail first, sideways, or in any intermediate position. Fuzes of this type are generally known as "allways" fuzes (Fig. 11).

They were used by the Germans in some of their stick hand grenades and in some of their light minenwerfer shells. This fuze is the percussion fuze for the 3-inch mortar bomb, and in the near future will probably be more widely used as a fuze for hand and rifle grenades. The "allways" type of fuze generally takes the form of two telescoping members kept apart by a light spring and carried in a chamber with concave ends. The action of the spring forcing the two telescoping members apart against the ends of the chamber causes them to occupy the centre or deepest portion.

One of the telescoping members is formed as a striker, the other carries a percussion cap, and between them is the spring, known as the creep spring. If, now, either of the two telescoping members is forced against the other the striker will enter the cap and fire it. This action is brought about as follows, according to the attitude in which the grenade strikes the ground.

Should the grenade strike nose first or tail first, the momentum of whichever is the uppermost of the two telescoping members will cause it to be driven down on to the undermost, compressing the creep spring and firing the cap.

Should the grenade fall on its side the momentum of the two telescoping members will force them to leave the centres of the concavities in which they rest and move towards the edges; this causes the members to telescope upon one another and fire the cap as before. In any intermediate attitude a combination of these two actions will take place. It is obvious that unless special steps were taken, the action of throwing or firing the grenades would cause sufficient shock to operate the mechanism and cause a premature. One method of preventing this is to employ a safety bolt or pin; this pin passes through holes in the telescoping members and locks them rigidly to one another. To one end of the pin is attached about a foot of tape and a small weight; this tape is normally wrapped round the fuze and holds the pin in place and is itself kept in position by a transit cover. On removing the transit cover and throwing or firing the grenade, the tape unwinds and in a few yards falls off, taking the pin with it. The grenade is then armed and explodes on impact.

In grenades such as the French V.B. and the German 1917, which are purely rifle grenades, and are fired by means of a ball cartridge, the bullet itself is made use of to effect the ignition of the time fuze. This it does in its passage through the bullet way of the grenade by impinging against a projecting striker, which is thus driven into the cap and ignites the time fuze, or against the cap itself.

PART I, CHAP. IV. FIG. 8.

IGNITERS.

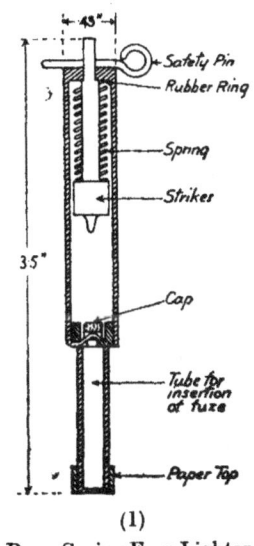

(1)
Brass Spring Fuze Lighter.

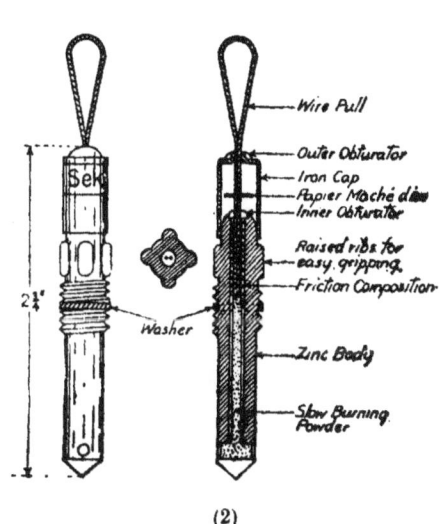

(2)
Friction Lighter.

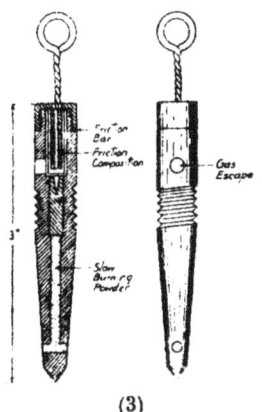

(3)
Friction Lighter.

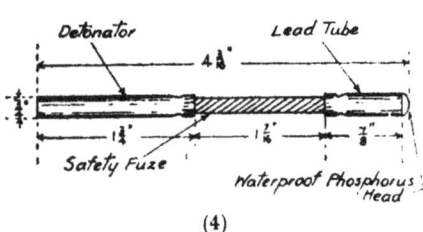

(4)
Match Head Fuze Lighter.

Part I, Chap. IV, Fig. 9.

GERMAN RIFLE BOMB, 1913.

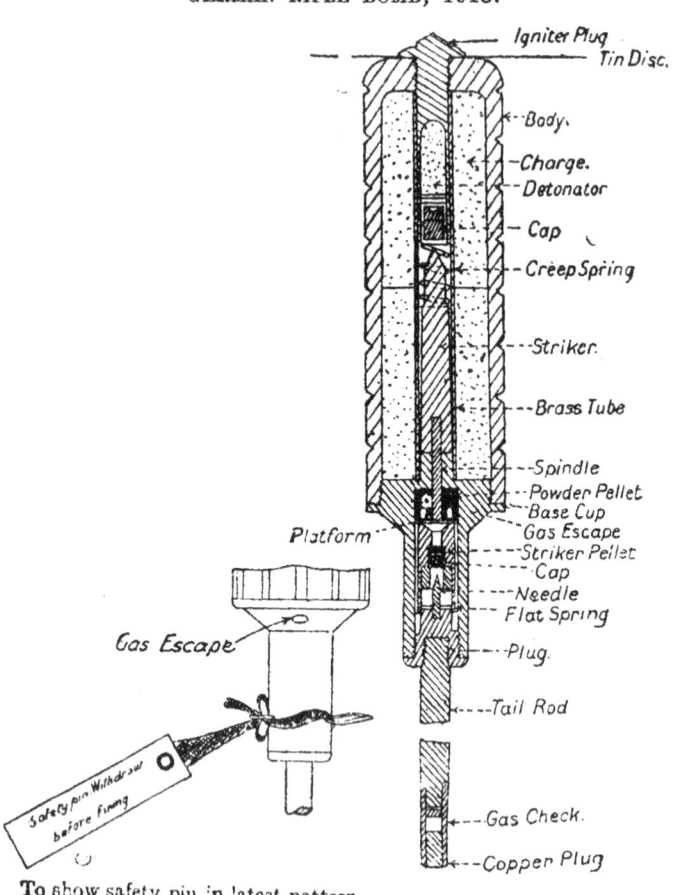

123

PART I, CHAP. IV, FIG. 10.

GERMAN RIFLE BOMB, 1914.

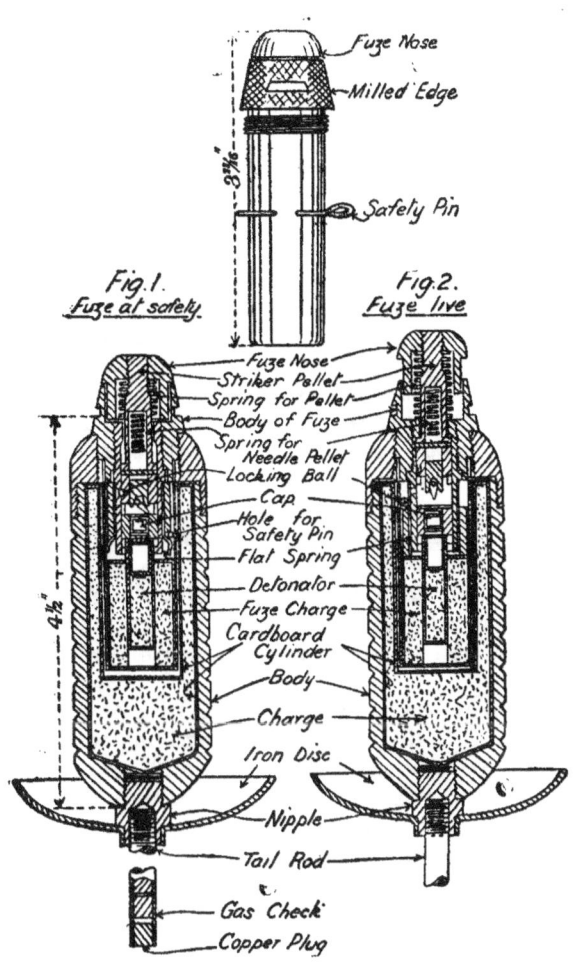

PART I, CHAP. IV, FIG. 11.

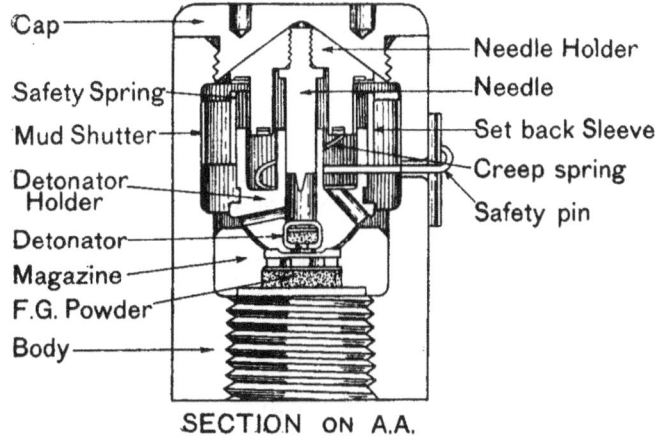

SECTION ON A.A.

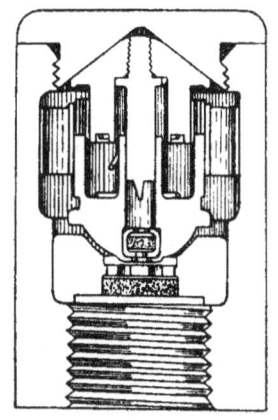

SECTION ON B.B.

## CHAPTER IV—SECTION 4

### Means of Projection

Although grenades were originally designed to be thrown by hand, it was not long before means were sought to obtain a greater range than was possible in a purely hand weapon. Small mortars were used at an early date for this purpose. The next step was to combine the mortar with the normal arm of the infantryman, i.e., the musket, in the form of what is known as the discharger. It is not known definitely when this device first made its appearance, but it was certainly not later than the latter half of the 17th century. Specimens dated 1743 exist, and are almost identical in form with those of the present day. Another pattern, dating from the time of James II, was combined with the butt of the musket (Fig. 12 (a) and (b)) instead of being a detachable fitting for the muzzle. When the musket was used as such the mouth of the discharger was closed by a hinged lid which served as the shoulder piece of the butt. When used for firing grenades the lid was opened and the musket reversed; in this position a single leg, hinged in front of the trigger guard, was swung out from a recess in the fore-end and served as a support. No further consideration seems to have been given to the problem until the Russo-Japanese war brought the grenade once more to the fore.

Hale produced a grenade designed to be fired from a rifle by means of a steel rod which was screwed into the base of the grenade and was a sliding fit in the bore of the rifle; the grenade was fired from the rifle by means of a blank cartridge. This method of projection was widely copied. In some cases, as in the German 1913 and 1914 grenades, a gas check was used to prevent the escape of the gases past the rod (see Fig. 10).

Various lengths of rods were used; those in the British Service varied from 6 inches to $17\frac{1}{4}$ inches for use with S.M.L.E. rifle. The general tendency was to obtain the maximum range by using the longest rod that the rifle would stand, and to obtain shorter ranges by one of three means—(a) By varying the angle of the rifle, (b) by attaching a disc to the grenade in order to increase the air resistance, as in the German 1914 grenade (see Fig. 10). or (c) by not inserting the whole of the rod into the barrel; a rubber ring fitting tightly on the rod served as an adjustable stop for this purpose. Generally speaking, rodded grenades were found to have certain disadvantages. They were not easy to carry; they were not as accurate as could be desired, and they were liable to cause ringed or bulged barrels. At the beginning of 1917 the Germans discontinued the manufacture of rodded grenades, as their inaccuracy was said to make them of doubtful value. A discharger and grenade similar to those already in use by the French was adopted at about the same time. In the British Service a discharger also came into use (Figs. 13 and 13A). With the French and German dischargers variation of range was obtained by altering the angle of projection; the British discharger was used as a constant angle of 45°, but had an adjustable gas port which permitted a greater or less portion of the gases to escape. The gas port type of discharge is possibly less accurate than the plain type, as portions of the charge are sometimes blown out through the port unburned and pressures may, therefore, be somewhat irregular. On the other hand, it does away with the clinometer, which is really necessary with the plain discharger if full value is to be obtained from its better grouping qualities. The British discharger is used with a special blank cartridge, whereas the French and German grenades are designed for use with an ordinary bulleted round; this latter feature simplifies the question of ammunition supply to a certain extent, but, on the other hand, involves the placing of the detonator to one side of the grenade, which may cause irregular fragmentation; it may also cause difficulties in training, as, owing to the fact that the bullet, after passing through the grenade, continues to practically its maximum range, firing of grenades must take place on ground giving the full danger area necessary for the bullet.

The firing of grenades from a rifle is, at best, something of a makeshift, and it has been suggested that the real solution to the problem should be sought in the direction of a small and very light mortar specially designed for the purpose. Accuracy and range would certainly be increased. The German Granatenwerfer (Fig. 14) is an example of this tendency, although it is much too heavy, and is otherwise little suited for use in open warfare. Rifles used for the firing of grenades frequently suffer to a greater or less extent, whether a discharger or a rod is used. Splitting of the woodwork owing to the recoil is the commonest failure.

In addition to the methods of projection already mentioned, various forms of catapult and spring gun have been used from time to time when better means were wanting.

### Discharger, Grenade 2-inch (Fig. 15)

Generally speaking, this discharger is a modification of the older 2½-inch pattern. Originally of the portless type intended for use with a clinometer, it was redesigned with a port and shutter, as it was found extremely difficult to obtain short ranges without some means of controlling the initial velocity.

The maximum range with the No. 54 grenade fired by means of a 30-grain ballistite blank cartridge is rather more than 300 yards. Two range scales are engraved on the edges of the gasport by which the shutter is set. One of these scales is for the H.E. grenade, the other for the smoke grenade. This is rendered necessary owing to the different weights and air resistances of the two types.

The method of attachment to the rifle is by means of a bridgepiece, which is semi-permanently attached to the nosecap of the rifle (in the case of the S.M.L.E., Mark III) by means of claw bolts, which engage in the cut-away portion near the muzzle. These bolts are clamped by means of feathered nuts which fit the slot in the handle of the bayonet; the latter, therefore, serves as a spanner, and no special tool is necessary. The bridgepiece is tapped for the attachment of the discharger barrel, and is cut away to afford clearance for the bayonet ring. The neck of the barrel is threaded to fit the bridgepiece, and is provided with an oil-hardened steel bush coned at the mouth. These bushes are a press fit into the bore of the neck, and are replaceable when worn.

In use, the discharger is screwed into the bridgepiece hand tight, so that the bush in the neck takes a seating on the muzzle of the rifle barrel. The hammering action of the recoil and the blast of the propellent gases are thus taken up by a replaceable component.

The coning of the mouth of the bush ensures that the gas blast from the rifle muzzle does not impinge upon a sharp corner. The hardening of the bush is to prevent loss of shape under the hammering action of the recoil.

The discharger may be attached and used without unfixing the bayonet.

PART I, CHAP. IV, FIG. 13.

CUP ATTACHMENT FOR FIRING MILLS GRENADE FROM A RIFLE.

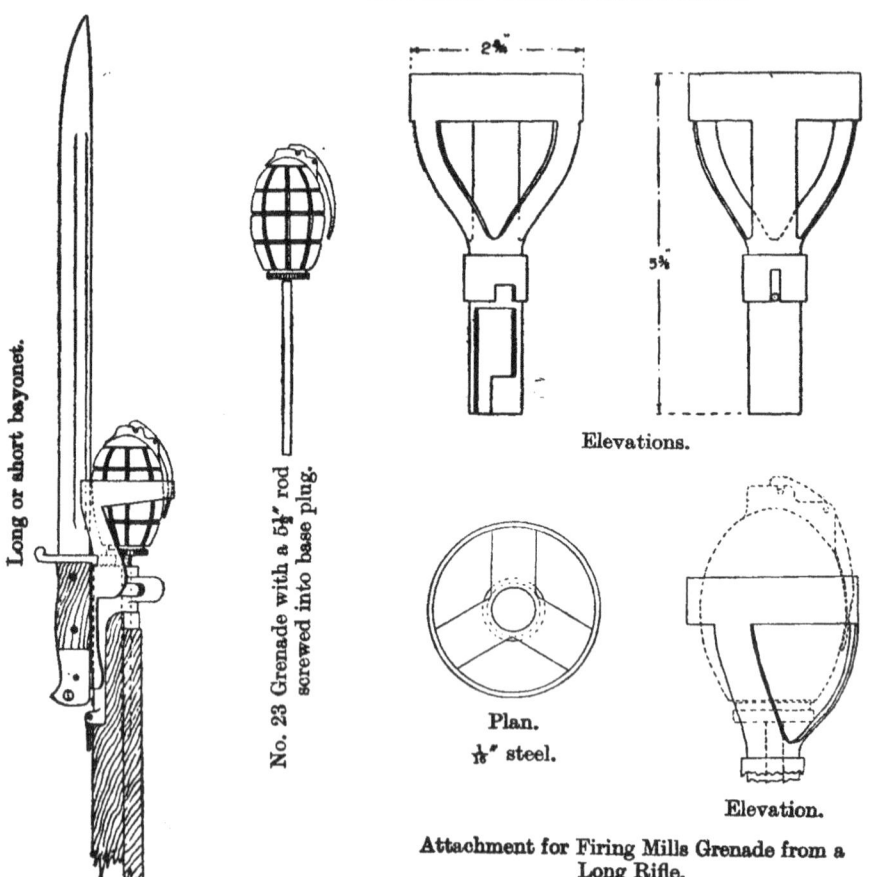

Elevations.

Plan.
$\tfrac{1}{16}''$ steel.

Elevation.

Attachment for Firing Mills Grenade from a Long Rifle.

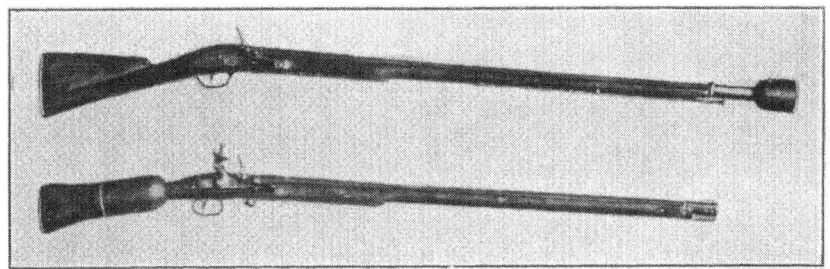

Part I, Chap. IV, Fig. 12 (a).

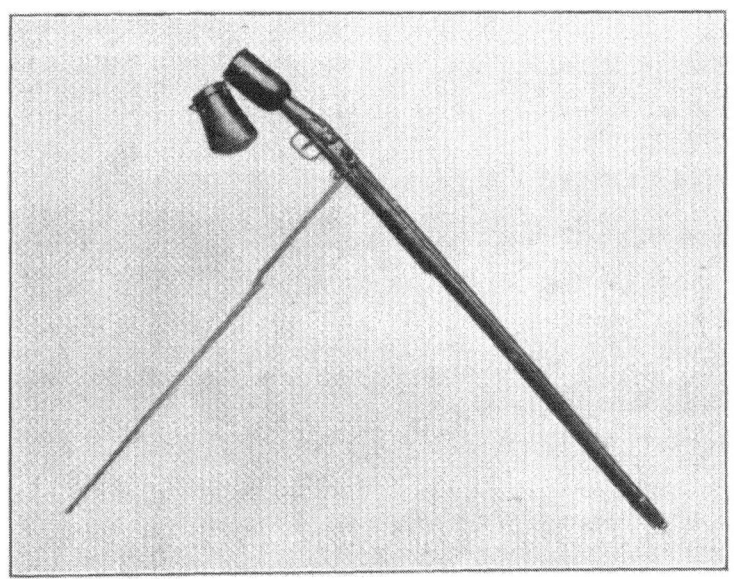

Part I, Chap. IV, Fig. 12 (b).

PART I, CHAP. IV, FIG. 13 (a).

DISCHARGER, BOMB, RIFLE NO. 1, MARK I. FOR ATTACHMENT TO RIFLES, SHORT, M.L.E.

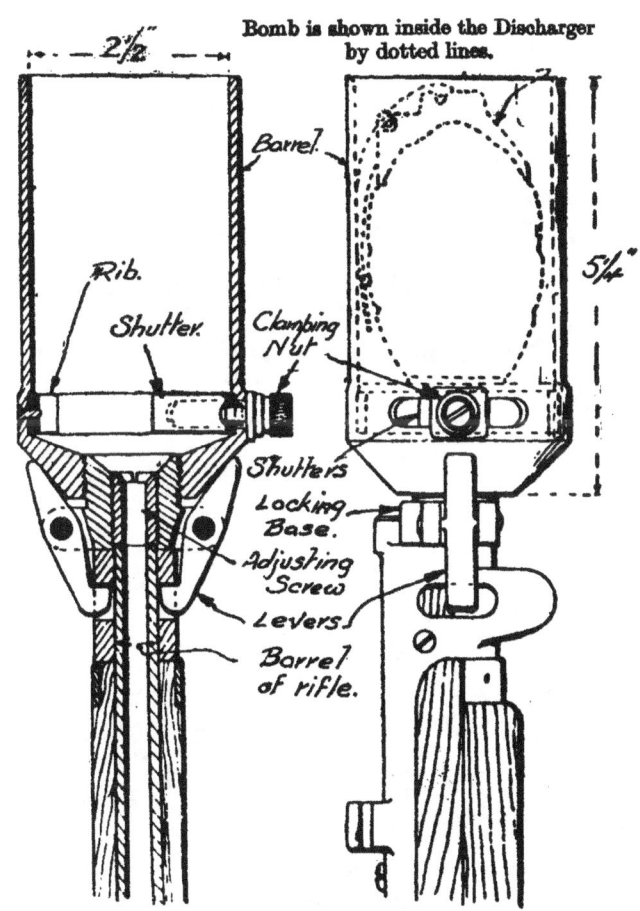

PART I, CHAP. IV, FIG. 14.

GRANATENWERFER.

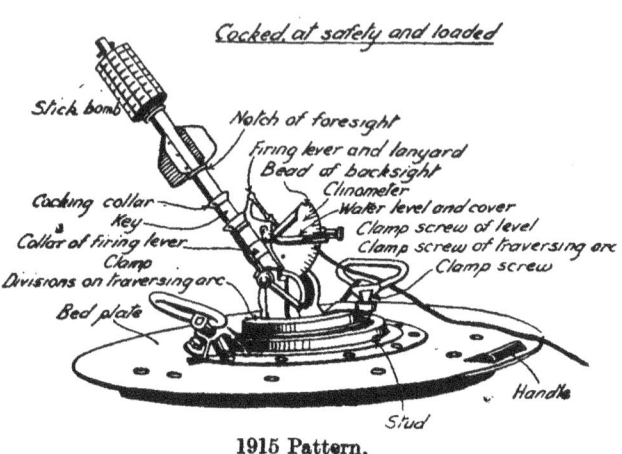

1915 Pattern.

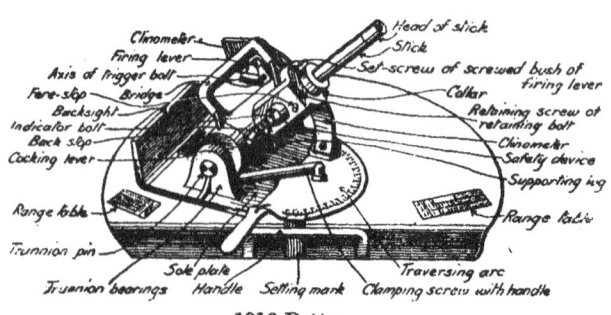

1916 Pattern.
Right-hand view.

PART I, CHAP. IV, FIG. 14 (a).

SECTION THROUGH "STICK" OF BOMB-THROWER SHOWING BOMB IN POSITION.

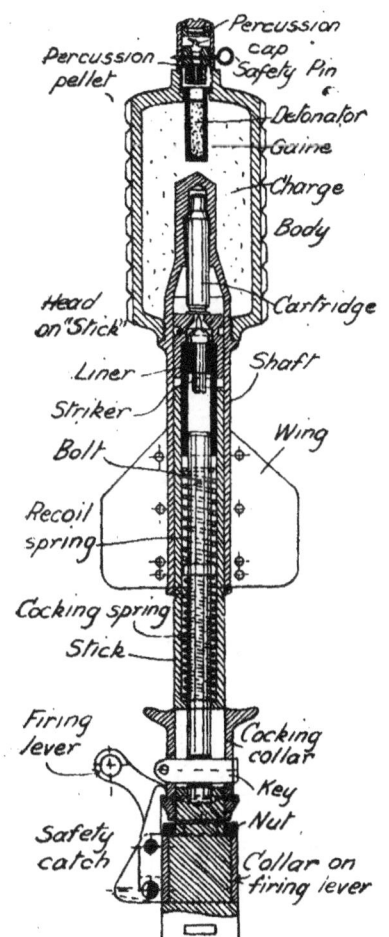

Part I, Chap. IV, Fig. 15.

Discharger, Grenade, 2-inch.

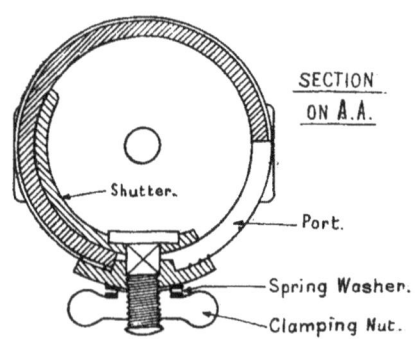

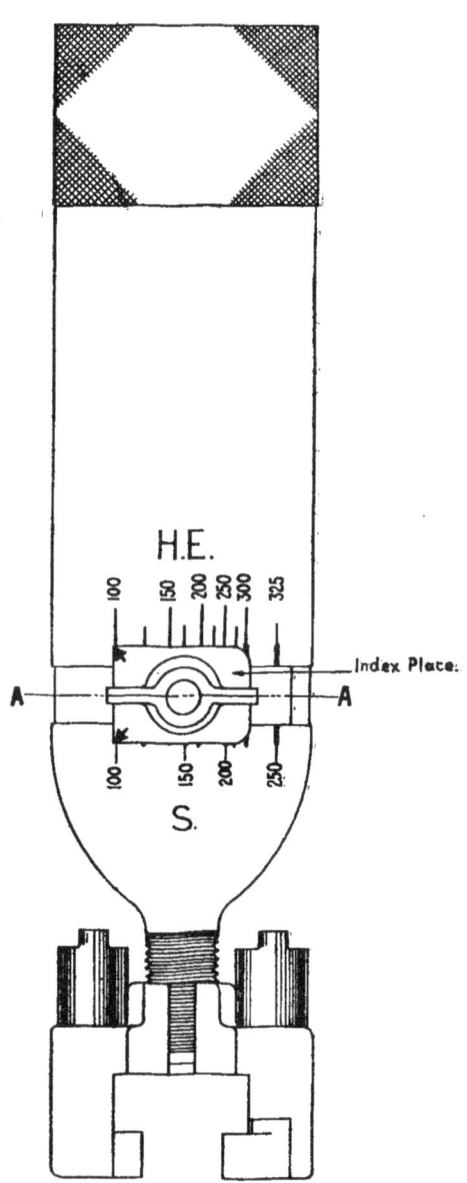

## CHAPTER IV—SECTION 5

## BRITISH AND FOREIGN GRENADES

### BRITISH

#### The No. 36 *Hand or Rifle Grenade* (Fig. 16)

This grenade is a direct descendant of the Nos. 5 and 23, or Mills grenade, and has undergone only slight modification in order to fit it for use with the discharger. The *body* is of cast iron with external segmenting grooves and is lacquered inside and out. It contains a *centrepiece* of aluminium alloy formed with two chambers to receive the striker and detonator respectively. The joints between centrepiece and body are sealed with a waterproofing composition. The *fillings* used are Ammonal, Alumatol, Amatol and Cilferite. These are put in through the *filling hole* which is closed by a brass screwed plug also sealed with waterproofing composition. The charge is detonated by means of an *igniter set* consisting of a No. 6 commercial detonator fitted with a 7-seconds length of Nobel's No. 16 safety fuze. The fuze is ignited by a ·22 rimfire cartridge cap which in turn is fitted into a type metal cap holder. The base of the cap is perforated to act as a gas escape. The cap is fired by means of a *striker mechanism* consisting of striker, spring, striker lever and safety pin.

The striker is formed with a head and neck. The head carries two striker points or nipples and is slotted to act as a gas escape. The striker spring is made of steel wire and is tinned as a protection against rust. When compressed in the cocked position it exerts a pressure of about 30 lbs. The striker lever is a steel pressing and is formed with two trunnions near one end. The safety pin consists of an ordinary split pin to which is fitted a split ring.

The mechanism is assembled in the cocked position and the grenade sealed with wax where the striker projects from the body. The base is closed by the steel base plug which in turn carries the steel gas check disc. The complete grenade weighs approximately $1\frac{1}{2}$ lbs.

*Action of the Mechanism.*—On removal of the safety pin the striker lever is kept in place either by the fingers gripping the grenade when throwing or by the sides of the discharger when firing. On the grenade being released by the hand or ejected from the discharger the lever is freed; the striker spring forces the striker downwards and fires the cap of the igniter set.

#### The No. 37 *W.P. Grenade* (Fig. 17)

The body of this grenade is of tinned sheet steel with soldered joints. A steel gas check disc is soldered to the base and a copper tube closed at one end is let into the top of the body to take the detonator and fuze. The filling of white phosphorus is inserted through a hole at the top of the body; this hole is afterwards sealed with a tinned disc soldered on.

White phosphorus is not, itself, explosive but on exposure to the air oxidizes with great rapidity producing quantities of thick white smoke. It is therefore intended primarily for the formation of smoke screens. Incidentally, it is strongly incendiary and can be used to cause injury by burning.

The grenade is burst open by means of a burster set consisting of a No. 6 commercial detonator, a 7-seconds length of safety fuze and a ·22 rimfire cartridge cap fixed into a cap holder similar to that used for the No. 36. To support the detonator when in the grenade a felt wad is fitted at the bottom of the detonator chamber. The cap is fired by means of an Adams striker mechanism of solid construction.

*Action of the Mechanism.*—On the safety pin being removed the striker pellet is kept in place by the shearing wire only. On firing the grenade the inertia of the striker pellet causes it to set back, cutting the shearing wire, and striking the cap. The explosion of the detonator in turn bursts the grenade open and scatters the contents.

If the grenade is to be thrown by hand instead of fired from the cup discharger the projecting end of the striker pellet is given a sharp blow. This cuts the shearing wire and drives the striker on to the cap.

### *The No. 28 Gas Grenade* (Fig. 18)

The body of this grenade was of cast iron, spherical in shape with a screwed opening for filling purposes and also to take the detonator sleeve. This detonator sleeve was also of cast iron and was screwed into the body. A screwed boss was formed at the mouth of the sleeve to take an Adams striker mechanism similar to that used with the No. 37, W.P. grenade. The burster set consisted of a cap, a 5-seconds length safety fuze, and a No. 8 commercial detonator.

*Action of the Grenade.*—This grenade was thrown by hand and the mechanism operated in the same way as that of the No. 37 W.P. grenade.

### *The No. 42 Signal Grenade* (Fig. 19)

#### *(Day Signal)*

This grenade is one of a series which differ from one another only in the nature of the signals contained within them. They are fired from the discharger. The body consists of a rolled paper tube closed at one end (the base) by a wooden plug. This plug is covered by a tinned sheet metal cap and is further strengthened at its centre by a steel disc let into a recess between the cap and the wood. Two holes are drilled through the cap and plug and carry 3 seconds' length of safety fuze. The outer ends of these fuzes are covered by waterproof paper discs. Above the wooden plug is placed a thick felt wad having a coating of mealed gun-powder gummed to its undersurface. This coating of gunpowder is in contact with the inner ends of the fuses and forms the opening charge of the grenade.

Five strands of quick-match are passed through a hole in the centre of the felt wad and are splayed out on both sides. Above the wad, and in contact with it, is packed the signal candle (yellow, purple, red or blue). In the case of the No. 42 grenade this consists of a rolled-paper case containing smoke composition, closed at each end by wooden plugs and having a central hole through which quick-match is passed up. (In the case of the blue candle the quick-match is passed up the side, not through the centre). One end of the candle is primed and covered with paper and the two wood plugs are connected by a binding wire.

One end of a 3-foot length of asbestos cord is attached to the middle of the candle so that when suspended it will hang horizontally. The remainder of the cord is coiled up on the top of the candle and the loose end attached to a parachute made of Japanese paper. A cardboard disc placed immediately above the candle protects the parachute from flames or sparks. The top of the grenade is closed by a tinned plate lid sealed with adhesive tape. A tinned plate ring is fixed to the body of the grenade near the top to serve as a stop for the grenade when placed in the discharger.

*Action of the Grenade.*—On firing the grenade the flash from the propellent ignites the fuses in the base and at the same time the grenade is driven from the discharger. After a delay of 3 seconds (by which time the grenade has reached its culminating point) the flame reaches the coating of mealed gunpowder on the felt wad and fires it. This blows off the lid from the grenade and ejects the contents. At the same time the strands of quick-match convey the flame from the gunpowder and light the smoke candle. After falling a short distance the parachute opens and supports the burning smoke candle at the end of the asbestos cord. The trail of smoke from the candle constitutes the signal.

### *The No. 55 Smoke Grenade (2-inch)* (Fig. 20)

With the earlier patterns of white phosphorus smoke grenades corrosion of the thin sheet metal bodies and leakage at the joints with consequent exposure of the phosphorus to the air were the cause of numerous fires. To overcome these defects as far as possible, more suitable material was used in the design now being described, and all joints were lapped over as well as being soldered.

In addition, the detonator sleeve passed through the grenade from end to end. The interior of this sleeve could thus be inspected for signs of corrosion or leakage.

The diameter of the body is 2 inches, so that the grenade can be fired from the same discharger as the No. 54 H.E. grenade.

The fuze, fuze cover and base plug are all similar to those used with the No. 54 and the action is the same.

The detonator constitutes the sole bursting charge of the grenade and, as in the case of the No. 54 grenade, is rimmed to prevent incorrect insertion in the grenade. In this case, however, the portion containing the fulminate is carried on a stem for the two-fold purpose of bringing the fulminate in closer proximity to the cap of the fuze and of placing it in the most effective position for opening the body of the grenade.

*Action of the Grenade.*—The grenade is manipulated in the same way as the No. 54 and the action of the fuze is the same. The firing of the detonator splits open the body of the grenade, scatters the filling of white phosphorus to a certain extent and accelerates its combustion. A dense cloud of white smoke is produced.

### *Signal Grenade (2-inch)* (Fig. 21)

This grenade is an example of a series differing one from the other only in the type and nature of the signal produced. The design generally follows that of the No. 42 series of 2½-inch grenades, except that a metal body is employed instead of the cardboard used in the No. 42 series. Chiefly on this account these grenades have a greatly improved resistance to damp.

### No. 54 H.E. Grenade (2-inch (Fig. 22)

#### *Weight 17 oz. (approx.)*

The design of this grenade was based on the experience of the Great War with a multitude of types, both of our own and foreign manufacture, and was produced to meet the requirements laid down by the general staff. The body of the grenade is of cast iron formed with two belts which are a close sliding fit in the discharger and serve as a seal for the propellant gases as well as a guide for the grenade. The diameter across the belts is 2 inches. The surfaces of the belts are tinned as a protection against rust. The wall thickness is so proportioned to the charge that small fragments only are produced and a thrower in the open is unlikely to be injured by his own grenade, provided he is more than 20 yards from the point of burst.

The base of the body is closed by a small brass plug, the top aperture—which is nearly the full diameter of the body—is internally threaded to receive the fuze mechanism.

Between the guide belts is the filling hole which is closed by a lead alloy screw-plug.

The fuze is of " allways " percussion type with tape-controlled safety pin. The fuze " gallery " is of lead alloy and has a screwed spigot by which it is attached to the body. The top of the gallery is closed by a cap. Passing through the bottom of the gallery and soldered thereto is a thin copper tube which, in the assembled grenade, forms the detonator sleeve. The gallery is screwed into the body over a broad rubber washer, which serves to seal the top opening; the detonator sleeve projects slightly through the bottom aperture, is turned over and soldered to the body.

The mechanism of the fuze consists of pellet, striker, creep spring and ball. The pellet is of brass recessed to receive the creep spring and striker. Into the bottom end of the pellet is pressed a perforated percussion cap. The striker has a double-pointed shank and conically cupped head; the shank is enlarged below the head and has a transverse hole through which the safety pin passes.

In the assembled grenade the safety pin passes through a hole in the gallery wall, across the top face of the pellet and through the transverse hole in the striker. So long as this pin is in place all relative movement between striker and pellet is prevented. To the outer end of the safety pin is affixed a length of tape and to the other end of this tape is attached a small flat lead weight. The tape being wrapped round the fuze gallery prevents the safety pin coming out and is itself retained in position (up to the moment of use) by a sheet steel cover. Indentations on this cover engage with a lip on the fuze gallery while its edges press firmly on the projecting rim of the rubber washer which serves as a seal between gallery and body. With the cover in place all entry of water into the fuze is prevented. As the base plug is also sealed with a rubber washer and the filling screw cemented in place the grenade is completely watertight and may even be immersed in water for days without ill effect.

The detonator of this grenade differs from the normal in that it is made with a rimmed base similar to that of a cartridge. The purpose of this rim is to prevent the detonator being inserted in the grenade upside down. The charge is 16 grains of fulminate of mercury.

The evolution of this rimmed detonator is of interest as the pattern originally used was of the usual form.

In the course of a trial by troops of an experimental pattern of smoke grenade an unheard of number of prematures and blinds were reported. On investigation it was found that in a large number of cases the detonators had been inserted into the grenades wrong way up, i.e., with the closed end opposite the cap of the "allways" fuze and the open end pointing downwards.

On firing the grenades from the discharger either the fulminate set back out of the detonator causing a premature at the mouth of the discharger or a blind resulted.

*Action of the Mechanism.*—The grenade is primed by removing the base plug (for loosening and tightening the bayonet may be used as a key by engaging the slot in the handle with a rib or feather on the base plug), inserting the detonator and replacing the base plug. The first action in throwing or firing the grenade is to remove the fuze cover by giving it a quarter turn in an anti-clockwise direction and lifting off: the wrapped tape is then exposed. In throwing, the grip of the throwers hand—without any particular thought on his part—keeps the tape in place; in firing, the sides of the discharger serve the same purpose. On leaving the hand or discharger the tape is free to unwind and the distance in which it does so depends upon its length.

With the normal foot-long tape unwinding is complete in about 10 ft. when thrown or in about 20 yards when fired; the tape then falls free, carrying with it the safety pin and arming the grenade. The striker and cap are then held apart by the creep spring only. On impact the fuze acts as described under "allways" fuzes.

## German

### *Stick Hand Grenade (Time): Hand Lighter* (Fig. 23A)

This is the best known of the German grenades used in the Great War. The body is of sheet metal $\frac{1}{16}$-inch thick, and has a hook fixed to the side for the purpose of hanging the grenade on the belt.

The charge of explosive is made up in a paper container or cartridge, and has a paper tube let into one end to serve as a detonator chamber.

The handle of the grenade is of wood and screws into a socket on one end of the body. Removal of the handle exposes the detonator chamber in the body and the detonator holder in the handle. The handle is bored out and carries an igniter consisting of a friction tube and a length of safety fuze. To the end of the friction wire is attached a string provided with a porcelain button. This string is coiled up in a recess in the extremity of the handle and is covered by a screwed cap. The complete grenade weighs about 1 lb. 13 ozs.

*Action of the Grenade.*—The grenade is prepared for use by inserting a detonator into the holder and screwing on the handle. Preparatory to throwing, the screwed cap is removed and the string grasped in one hand. On pulling the string sharply the friction tube is caused to act and lights the fuze. The grenade is then thrown and, after an interval depending upon the length of the fuze, bursts. Owing to the thinness of the metal body the effect of the burst is very local ; it is therefore possible to use the grenade in the open, under certain circumstances, without undue risk to the thrower.

*Stick Hand Grenade (Time) : Automatic Lighting* (Fig. 23B)

This grenade is similar to the ordinary stick grenade in general appearance and in the construction of the body. The handle, however, is of aluminium tube and contains an automatic fuze-lighting device consisting of a cylindrical weight which slides in the tube and is attached by means of a string to the friction wire of the lighter. The weight is normally prevented from coming out of the tube by an aluminium screw cap, which constitutes the safety device.

*Action of the Grenade.*—Before throwing, the screw cap is removed so that the weight is free to slide out of the tubular handle. On throwing, the weight leaves the tube with sufficient force to operate the friction lighter. The subsequent action of the grenade is the same as that of the ordinary grenade.

*Stick Hand Grenade : Percussion* (Fig. 23c)

Externally, this grenade closely resembles the pattern with hand lighter. The handle is of wood except for the upper portion, which is of metal, and contains a species of " allways " percussion fuze. The metal portion of the handle forms the fuze chamber, and is provided with conical end pieces.

The mobile portion of the fuze consists of the usual telescopic cap holder and striker pellets. The pellets are kept apart, and thereby held between the conical ends of the fuze chamber, by two locking pieces carried on the ends of a hair-pin-shaped steel spring. The two arms of this spring tend to spring apart, but are prevented from so doing by a tube or sleeve which slides over them. So long as this sleeve is in place the mechanism is locked. The sleeve also slides within the bore of the wooden handle, where it is retained by a screw cap similar to that used with the other stick grenades. This end of the sleeve is weighted.

*Action of the Grenade.*—As in the case of the automatic lighter, the screw cap is removed before throwing so that the weighted sleeve is free to slide out of the handle.

The grenade is thrown with an end-over-end spin. This causes the weighted sleeve to be flung out of the handle and free the hair-pin spring. The arms of the latter spring apart and withdraw the locking pieces. On striking the ground the momentum of the cap holder and striker pellets causes them to telescope one within the other. The striker enters the cap, which fires and flashes direct into the detonator. This action takes place in whatever attitude the grenade strikes the ground. In the case of it falling sideways the momentum of the pellets causes them to move from the deep centre of the fuze chamber towards the shallower edge (the ends being conical), and thereby to be forced inwards upon one another.

*Egg Hand Grenade* (Fig. 24A)

In order to obtain a greater range than was possible with the stick grenades, the so-called "Egg" grenades, weighing 11 ounces, were produced. The body was of cast-iron and was filled with a special powder which did not require a detonator. The powder was put in through a screwed hole at one end; this hole also received the igniter set. The igniter consisted of a friction tube and a length of safety fuze. A loop was formed on the end of the friction wire.

*Action of the Mechanism.*—The igniter was operated by pulling the wire loop either by hand or by means of a hook carried on a wrist strap. This ignited the fuze and the grenade was then thrown. A range of 50 to 60 yards could be obtained.

*Smoke or Gas Grenade* (Fig. 24B)

These two grenades were similar in appearance and construction and differed only in the nature of the filling. The body was formed of two hemispheres of tinned plate 0·04-inch thick, lap jointed and soldered. One end was provided with a filling hole, which was closed by a screw plug and lead washer, the other carried a tube which entered deeply into the body. This tube was closed at the inner end, and held a bursting charge of black powder. The mouth of the tube was screwed to receive the friction lighter. The weight of the gas grenade was about 1¾ lbs., and that of the smoke grenade about 2 lbs., of which about 21 to 23 ounces was the filling of smoke composition. The gas grenade was filled with methyl-sulphuric-chloride and a few small shot.

*Action of the Grenade.*—The friction lighter was operated and the grenade thrown in the ordinary way. A delay of 5 or 6 seconds was given by the fuze. The flame then reached the bursting charge of black powder and exploded the grenade. These grenades could also be fired from a rifle or thrown by a catapult.

*Rifle Grenade*, 1917 (Fig. 25)

This grenade is designed for use with a cup discharger and is fired by means of an ordinary bulleted cartridge. The maximum range is about 150 yards. The weight of the grenade is 750 grammes. The body is of cast iron and is made in two parts. A central sleeve or tube cast in one piece with the lower part of the body passes through the grenade from end to end. It is through this tube that the bullet passes on firing. The upper and lower parts of the body are held together by means of a nut screwed on to the outside of the central tube.

The filling of 50 grammes of Perdit is carried in the lower part of the body and is covered by a thick cardboard wad. The detonator sleeve passes through this wad into the space occupied by the filling. In the upper part of the body and at right angles to it, the central tube, is drilled a hole or chamber for the reception of the igniter. This consists of a percussion cap and a 5 seconds fuze. The cap is so fixed that it projects into the bore of the central tube and partially obstructs it.

*Action of the Grenade*—On firing, the bullet passes through the central tube and in doing so impinges upon the percussion cap of the igniter. The cap fires and lights the fuze. At the same time the gases following the bullet propel the grenade from the discharger. After a delay of 5 seconds the flame from the fuze reaches the detonator and explodes the grenade.

## FRENCH

The French explosive grenades are of three kinds, and differ entirely from one another in design. On the defensive, where cover for the thrower exists, a high explosive grenade, similar in form to the Mills, is employed. In offensive operations, where the thrower is likely to be in the open and therefore liable to be wounded by his own grenade, a special type giving a very small danger zone is used. The third type is purely a rifle grenade, and is fired from a discharger by means of a ball cartridge.

### *The Defensive Grenade* (Fig. 26)

The body of this grenade is of cast iron, serrated on the outside, and is generally similar in appearance to the Mills.

The safety lever is held by a split pin and ring of the usual type, and retains a plunger against the action of a coil spring.

The striker consists of a spring steel wire formed in the shape of a U, the outwardly projecting extremities of which are sharpened. This spring is normally compressed to such an extent that the ends cross over one another; in this position they are held by the lower end of the plunger passing between them. Opposite each of the pointed extremities of the spring is a percussion cap; from these caps a safety fuze leads to the detonator.

*Action of the Mechanism.*—Before throwing, the safety pin is removed, the lever being held close to the body in grasping the grenade. On the grenade leaving the hand the plunger spring is enabled to extend; the plunger is thus withdrawn from between the ends of the striker spring, which fly apart; the pointed extremities of this spring strike the caps and fire them, thus igniting the fuze, which in turn sets off the detonator and explodes the grenade.

### *The Offensive Grenade* (Fig. 27)

This grenade has a very local and chiefly blast effect; it is thus safe to the thrower when used in the open. It contains a smoke-producing explosive mixture. The body is of thin sheet metal and is oval in form. The open end is closed by a lead screw plug which forms a tubular fuze holder. The safety fuze is fixed into this holder and extends down into the centre of the explosive charge; to the end of the fuze is fixed a detonator. There is no detonator sleeve. The outer end of the fuze holder carries a four-point striker made of sheet metal and a brass collar; near the base of the brass collar holes are drilled to act as gas escapes. Sliding on the inside of the collar, but kept in place by indentations at the mouth, is a cap holder; the cap space of this holder is filled with friction composition. These portions constitute the striker mechanism. A transit cap covers the striker mechanism.

*Action of the Mechanism.*—The transit cap having been removed the mechanism is operated by giving a sharp blow to the head of the capholder; the resistance of the indentations is overcome and the holder is driven down into the collar. The fixed striker enters the friction composition, which fires; the flash passes downwards into the mouth of the fuze holder and ignites the fuze which, in turn, explodes the grenade.

### *The Rifle Grenade* (V.B.) (Fig. 28)

This grenade is fired from a discharger by means of a bulleted cartridge. The body is of cast iron, cylindrical in form, and having a tube passing through the centre from end to end to allow of the passage of the bullet on firing. A detonator and fuze sleeve is screwed into the top of one side of the grenade, between the central tube and the wall of the body. The external head of the fuze sleeve holds in place the lower limb of an "L"-shaped arm, which carries at the top of its upright limb a striker point. This striker point directly faces a percussion cap fixed in the head of the fuze sleeve. The upright limb is so placed that its upper end is slightly inclined over the mouth of the central tube of the grenade. On the opposite side of the central tube is a boss of the same size as the head of the fuze sleeve; these two together protect the striker arm from an accidental blow.

*Action of the Mechanism.*—On the grenade being fired the bullet from the propellent cartridge passes through the central tube and on emerging from the upper end impinges upon the striker arm, driving it into the percussion cap and igniting the safety fuze; the gases following the bullet drive the grenade from the discharger. The fuze sets off the detonator which, in turn, explodes the grenade.

PART I, CHAP. IV, FIG. 16.

RIFLE GRENADE, NO. 36, MARK I.

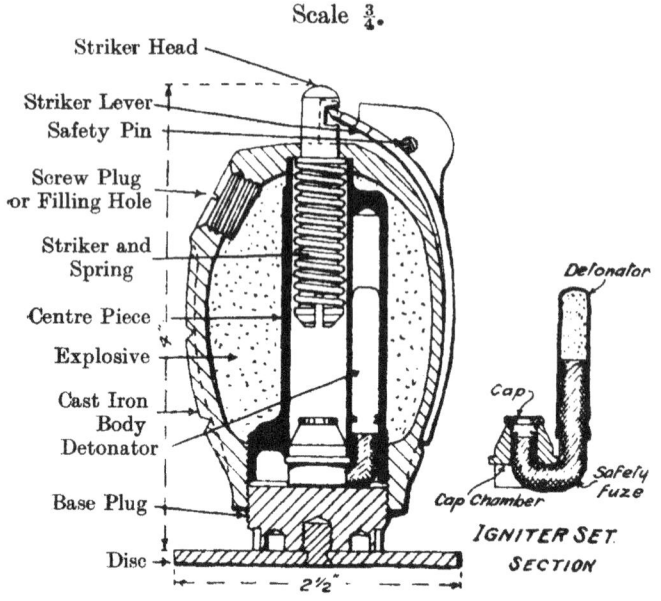

PART I, CHAP. IV, FIG. 17.

NO. 37, W.P. GRENADE.

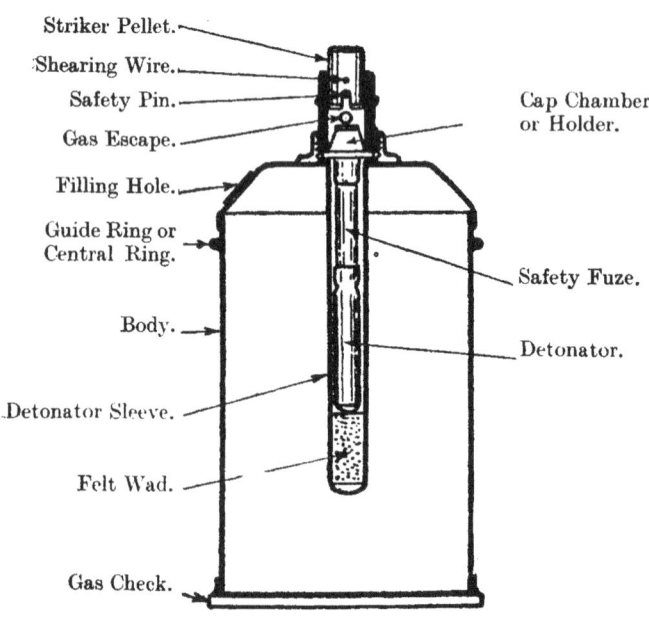

PART I, CHAP. IV, FIG. 18.

GAS GRENADE NO. 28.

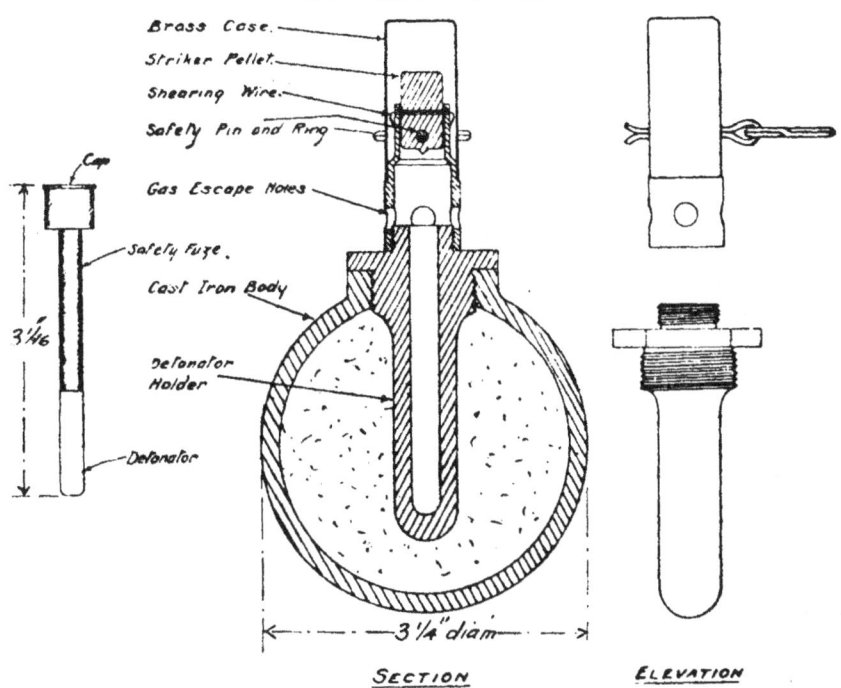

SECTION    ELEVATION

PART I, CHAP. IV, FIG. 19.

NO. 42 SIGNAL GRENADE.

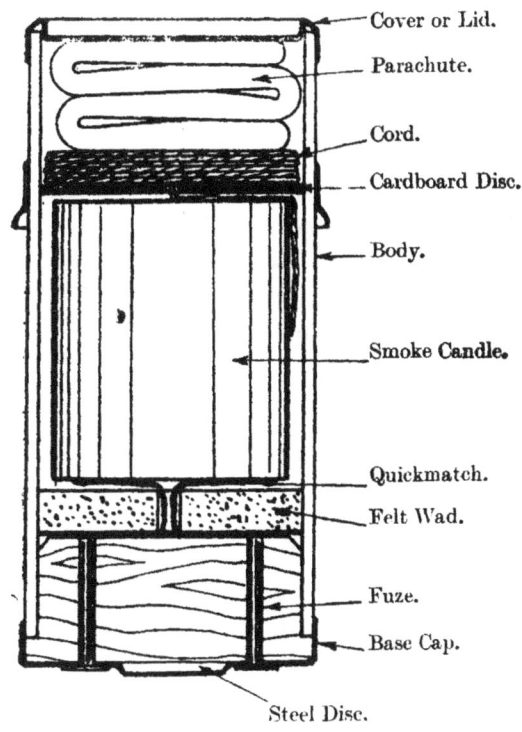

PART I, CHAP. IV, FIG. 20.

NO. 55 SMOKE GRENADE (2-INCH).

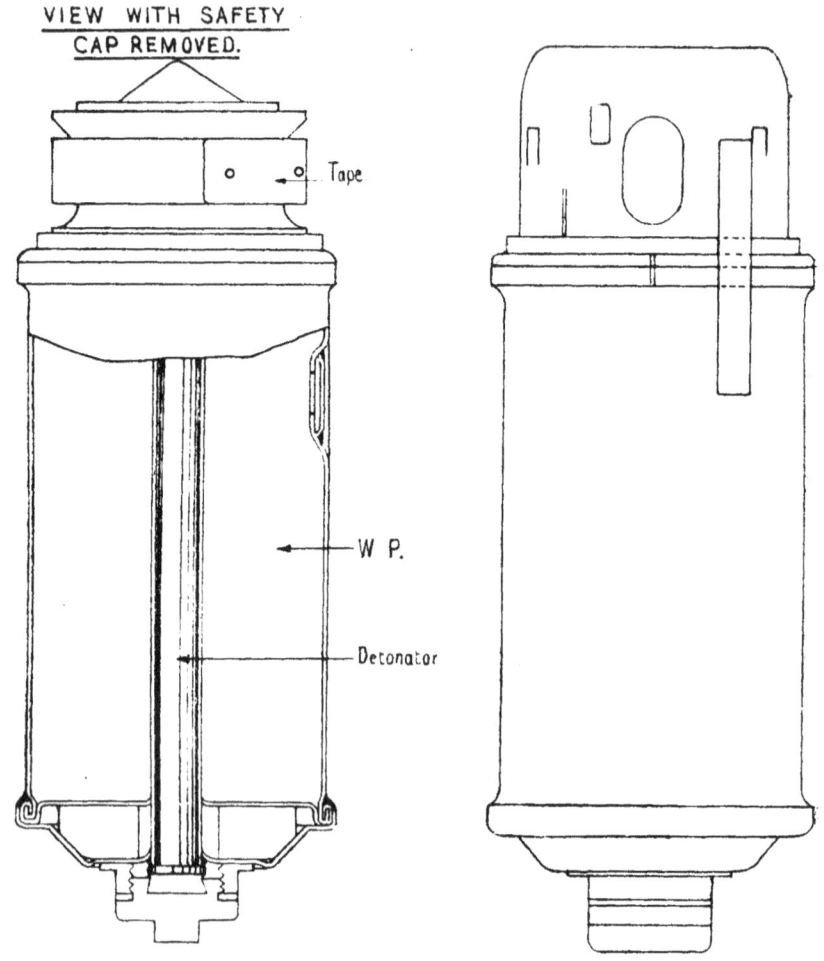

Part I, Chap. IV, Fig. 21.

Signal Grenade (2-inch).

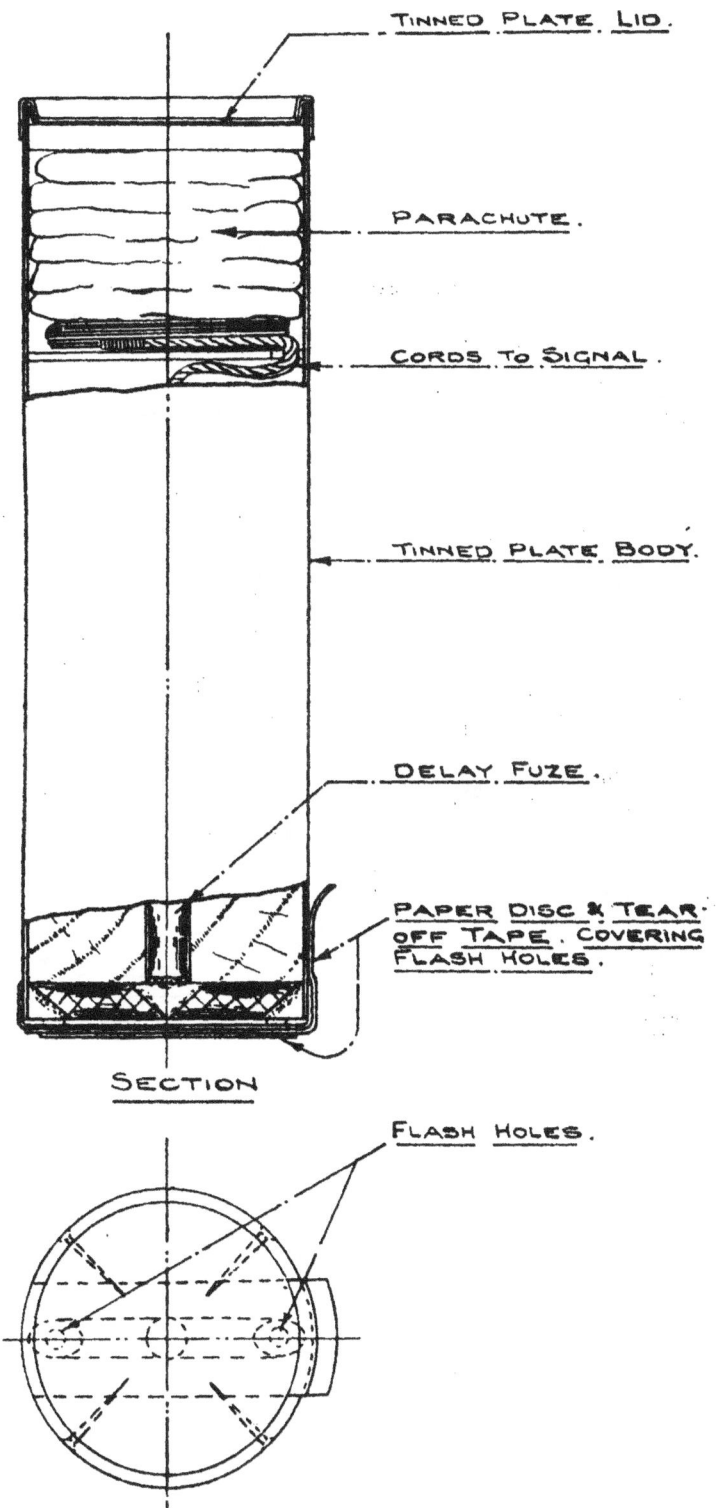

Part I, Chap. IV, Fig. 22.

No. 54 H.E. Grenade (2-inch).

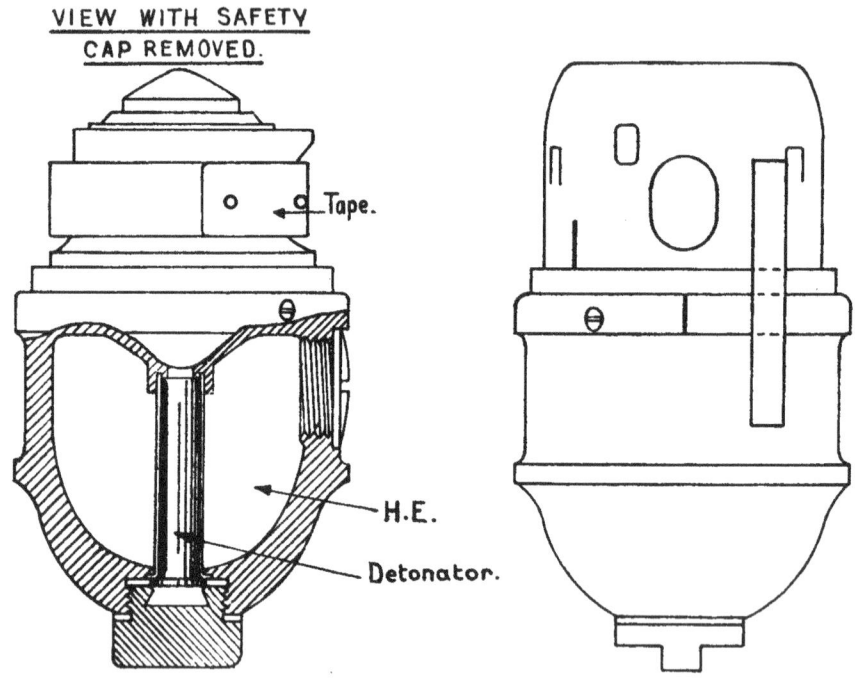

Part I, Chap. IV, Fig. 23 (a).

STICK HAND GRENADE (TIME).

(Cylindrical Hand Grenade with handle.)

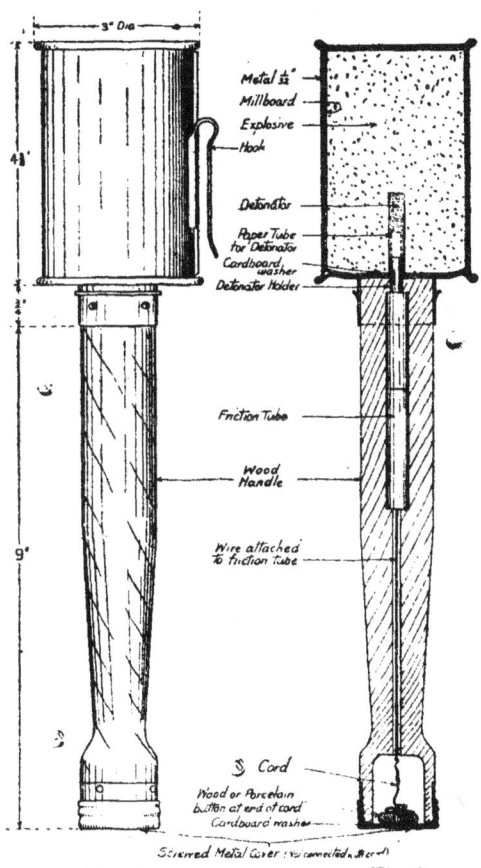

Fig. 1. Elevation.  Fig. 2. Section.

Part I, Chap. IV, Fig. 23 (b).

AUTOMATIC LIGHTING "STICK" HAND GRENADE.

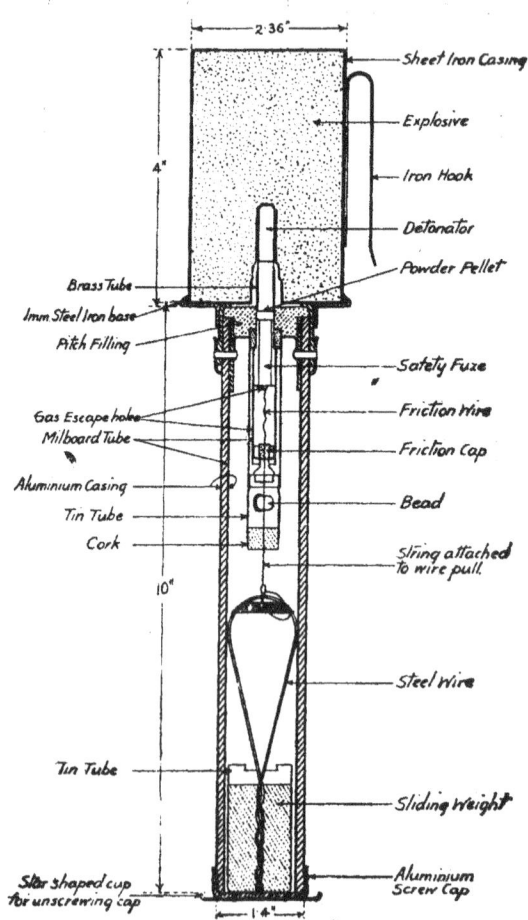

PART I, CHAP. IV, FIG. 23 (c).

STICK HAND GRENADE (PERCUSSION).

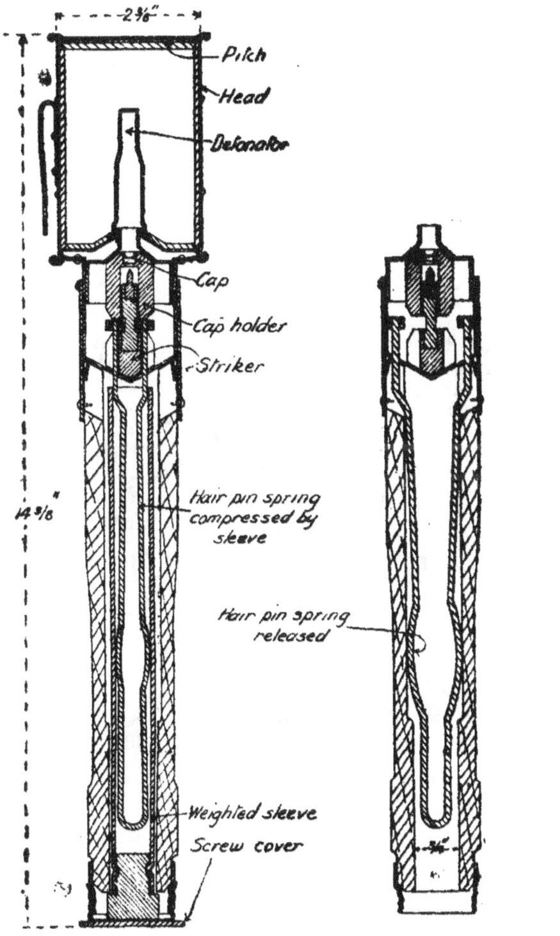

Section.

Section Showing Sleeve Withdrawn.

Part I, Chap. IV, Fig. 24 (a).

EGG HAND GRENADE (TIME).

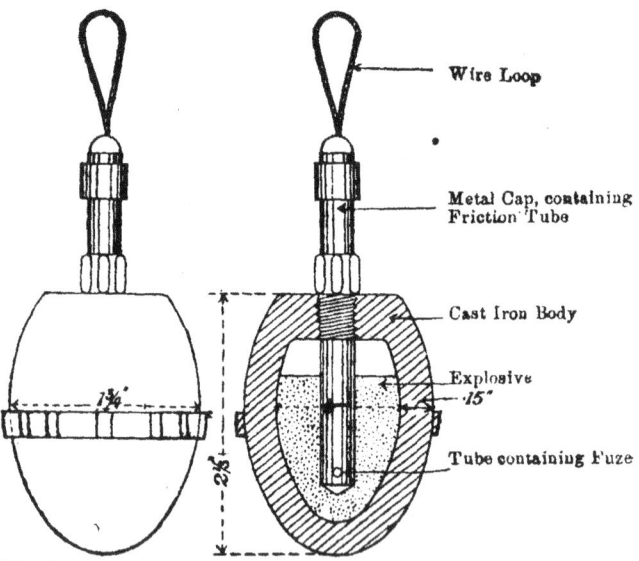

The form of the exterior differs slightly. Some Eggs have no segmented ring round their middle.

PART I, CHAP. IV, FIG. 24 (b).

SMOKE OR GAS HAND GRENADE (TIME).

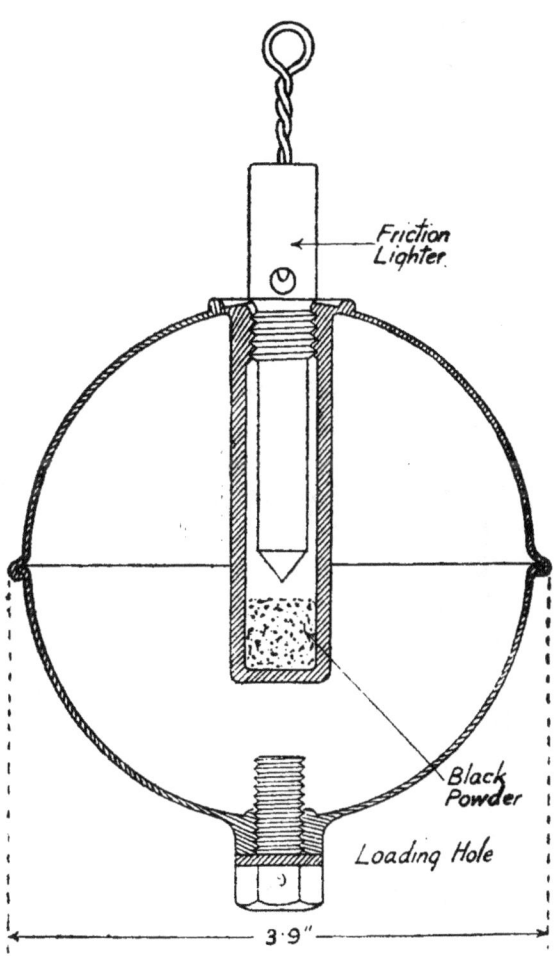

Part I, Chap. IV, Fig. 25.

RIFLE GRENADE 1917.

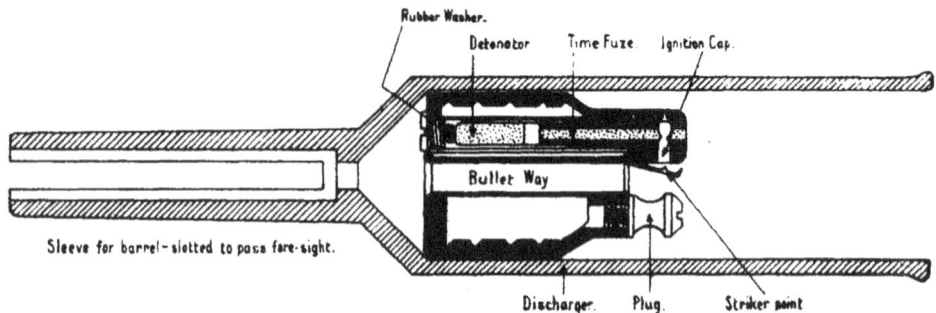

Part I, Chap. IV, Fig. 26.

DEFENSIVE GRENADE.

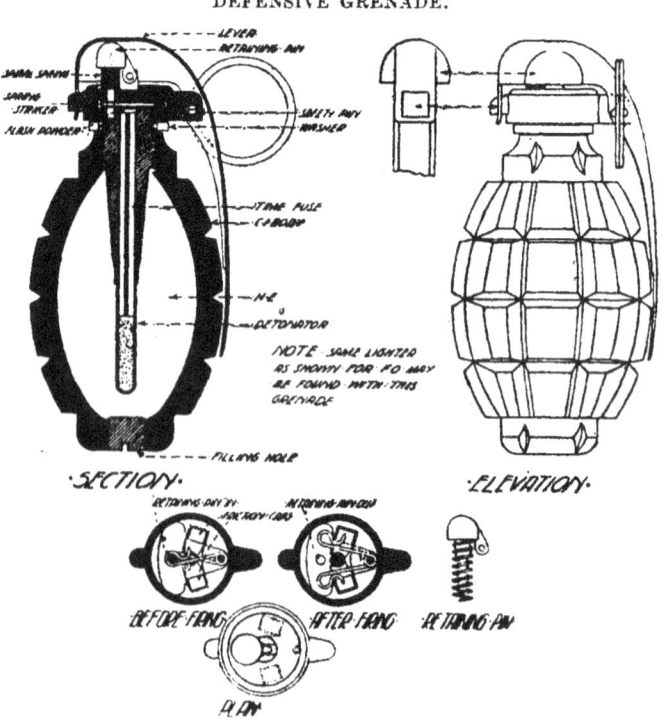

Part 1. Chap. IV, Fig. 27.

OFFENSIVE GRENADE.

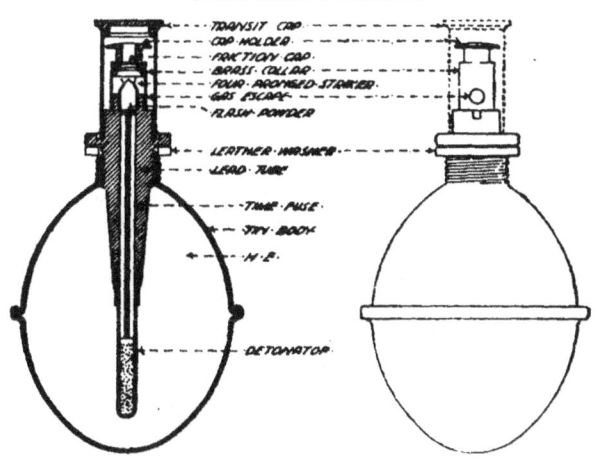

Part I, Chap. IV, Fig. 28.

RIFLE GRENADE V.B.

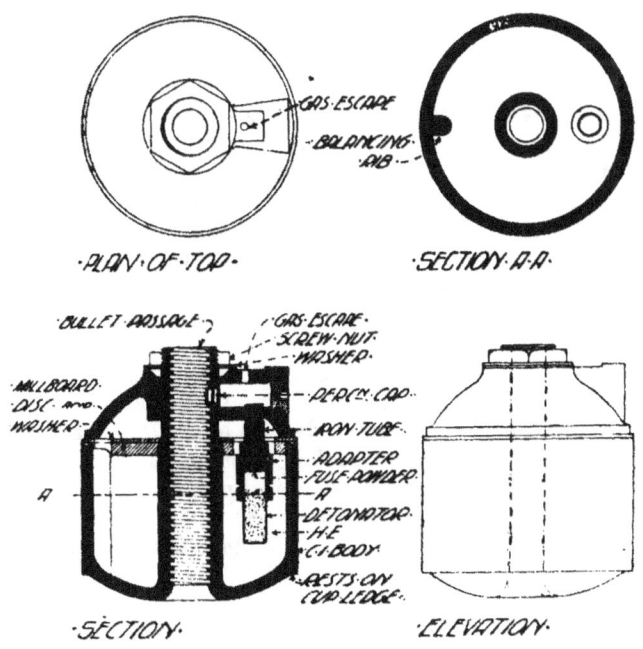

## CHAPTER V—SECTION I

## HISTORY OF THE MACHINE GUN

### INTRODUCTION

The necessity for automatic weapons has become such an accepted fact since the outbreak of the Great War that before considering details of the construction of some representative types, it will be as well to gain a more general idea of the experience which has born fruit in the development of the modern weapon. To this end a brief consideration of the evolution of design and some of the factors which influenced it will be of assistance.

The primary object of any automatic weapon is the economical development of fire-power. Tactics with a new weapon are necessarily in the main empirical, but subsequently tactics and design are interdependent, and therefore the design of a weapon must be influenced as much by the tactical needs which it is intended to satisfy as by considerations of purely mechanical efficiency. In war time the progress of design tends to be ruled by the needs of the moment, but during a prolonged period of peace the needs of the future are studied, and design, though at first responding to the lessons of war, eventually tends to reach a condition of comparative stability. This condition of stability may be upset either by a revolutionary departure in design or by some new method of destruction or defence due to science, compelling a reconsideration of tactics. But it remains true that in general the proper tactics for a new weapon are evolved by actual experience with that weapon.

In the following sketch of the evolution of automatic weapons no attempt has been made to trace their history in detail, nor has it been thought necessary to refer to or describe a greater number of the weapons themselves than is needful to illustrate the general trend of design.

### I.

The advent of fire-arms completed the ascendancy which infantry were already establishing by missile weapons over the old arm of decision, cavalry. Cavalry was still esteemed the most important instrument of battle till almost the end of the 15th century, when the fire-arm became really effective, and yet mobile enough for use in the field. To convey on one pair of wheels a number of fire-arms rather heavier than those carried by the soldier was a natural advance. Thus, at the battle of Ravenna in 1512, Pedro Navarro, who commanded the Spanish forces, drew up in front of his infantry 30 carts, on each of which several large arquebuses were mounted. The "orgue" or "organ gun" and the "ribaudequin" came into use on the Continent during the latter half of the 15th century. They consisted of a number of muskets mounted beside one another in a frame on a wheeled carriage, the latter having also a number of spears projecting in front of the barrels, and intended for the ultimate protection of the gunners from a frontal attack. The efficiency of fire-arms was long in reaching a stage at which they were capable of firing a continuous stream of shot, and it is interesting to note that an adaptation of the original idea of the organ gun was used in times as modern as the American Civil War, and even as lately as the Great War, when on certain occasions rifles were mounted in frames to sweep narrow approaches to the trenches, this being still a defensive use.

The introduction of fire-arms, like that of the crossbow, had evoked protests on the ground of inhumanity, and the same was the case with compound fire-arms, but this did not check their progress.

The earliest instance of the employment of a number of chambers revolving round a central axis, which marks the next stage in the development of mechanical fire-arms, occurs before the advent of the flintlock in the first half of the 16th century. A specimen of an arquebus of this period, with four chambers and a single barrel, each chamber having a separate flash-pan, may be seen in the Tower collection. This may perhaps be termed the progenitor of the modern light machine gun. In 1718 Puckle produced a gun fitted on a tripod mounting, which had a revolving block containing seven or nine chambers, either cylindrical, for ordinary warfare against Christians, or square, to shoot square bullets against Turks.

During the 18th century a compound arm was in use in France, consisting of a number of barrels fired by a single flintlock through a common touch-hole, and mounted in a single stock. Such arms were frequently constructed for the defence of fortifications, and had as many as six to ten barrels. Weapons of this type were made for the British Government as late as 1807 by the London gunsmith Nock, who was, incidentally, the originator of the so-called "Knox form," really "Nock's," which is still found in the barrel of the present Service rifle.

Up to the end of the 18th century designers could only hope to increase the fire power of any weapon by postponing the moment when it would be out of action during reloading—a defenceless period, the length of which depended on the number of barrels or chambers in the piece, each having to be separately charged from the muzzle and separately primed. Their efforts in this direction could not carry them further than an increase in the number of barrels employed and a reduction in the number of locks to a given number of barrels. Until breechloading was well advanced, the utility of the machine gun was limited almost entirely to defence or to preparation for offensive action.

## II.

The first effective breechloader used in large numbers by troops was the Dreyse needle gun, adopted by the Prussians in 1842. Twenty years after this came the first practical machine gun, invented by Dr. Gatling, of Chicago. In this gun a number of barrels were arranged round a central axis; behind the barrels was the reloading and ejecting mechanism, into which the cartridges were fed, falling by their own weight from a hopper or slide way above the gun. The whole of the reloading mechanism and the system of barrels were revolved, and the barrels loaded and fired successively, by turning a crank handle at the side of the gun.

The Gatling was used in the American Civil War, though it was never officially adopted by the Washington War Department, and it is interesting to note as indicative of the general attitude towards the new weapon that it was generally operated when brought into action by an employee of the Gatling Gun Company, who took the opportunity of demonstrating its powers.

For a long time the Gatling attracted little attention in Europe. The first machine gun to do so was the Montigny Mitrailleuse (Fr. mitraille = grape shot), which was introduced into the French Army on the eve of the Franco-Prussian War in 1870. This gun originated in Belgium about the year 1851, but was considerably improved before its adoption by the French Army. In an article contributed to the Royal United Service Institution in 1869, and entitled "Mitrailleuses and their Place in the Wars of the Future," Major Fosbery stated that in the preceding year he had been employed by the Government of India to report on the Montigny with a view to its adoption by the Indian Army, and gave a detailed description of its mechanism. The gun was a clumsy weapon and was mounted in much the same way as an ordinary field gun, which it somewhat resembled in form. It was of the multi-barreled type, and the cartridges were held in perforated iron plates. On loading the gun these plates were fitted into grooves on the breech block, which was then forced home and locked, each cartridge entering one of the chambers. The gun was fired by turning a handle which caused the barrels to be fired in succession. The secrecy observed by the French as to its design when it was adopted in 1870 on the eve of the war, and the journalistic accounts of this new and terrible weapon which was to ensure the success of their armies, hardly accord with the publicity given to it in the previous year.

The weapon was undoubtedly crude, but it was unfortunate for the progress of design that on this first occasion when the machine gun was officially on trial the tactics employed in its use were such that it was judged to be a failure. The French artillery had hitherto been organized in groups of three batteries, each of six guns. On mobilization one battery in each group was replaced by ten of the new mitrailleuses. Experience proved that the effect was equivalent to replacing six guns by a number of rifles, the extreme range of the mitrailleuse being the same as that of the Chassepôt rifle, rather more than a thousand yards. The new gun was as conspicuous as the light field gun with which it was linked, and yet was its inferior in range, power, and even mobility. Although this method of employment was general and it was regarded as an auxiliary to the field gun, there were occasions when it was used in co-operation with infantry. It was then that its true value was evident, and its use in this manner was attended by conspicuous success, but this fact does not seem to have been generally appreciated at the time.

Although in the war of 1870 the results of the use of the machine gun had been disappointing, its mechanical efficiency had become such that its possibilities could no longer be neglected. A few Gatlings were ordered for the British Army and Navy for trial in 1871 and a large number were sold to various other nations. Russia adopted this gun extensively, and it was used in the Russo-Turkish War which broke out in 1877, still, however, mainly as a defensive weapon. An instance of the use of auxiliary aiming marks for night firing occurs in this war at the Siege of Plevna, when the guns were used to cover bridges and other lines of approach at night time.

In common with other early machine guns, the Gatling suffered from the disadvantage that at the time of its introduction into Europe the solid drawn cartridge case had not been perfected, and consequently jams were frequent. The British Boxer cartridge, which was in use at that time was, in fact, very ill suited to machine guns.

The machine gun was first generally adopted by Britain for use in the Navy, as the result of experiments in connection with defence against the newly developed torpedo boat, which were carried out at Portsmouth in 1880. A variety of weapons were purchased in 1884 for this purpose, and included Gatlings, Gardners and Nordenfeldts. This latter gun showed a considerable advancement in design. It had four barrels arranged side by side and fed by hoppers; it was mounted on a solid conical mounting, and was operated and fired by moving a lever backwards and forwards. The mechanism was accessible, and if a stoppage occurred in connection with any one barrel, that barrel could be put out of action and fire maintained with the remainder. The barrels could be adjusted to fire simultaneously or in quick succession, and to diverge from or to converge towards the line of sight. The rate of fire was approximately the same as that of the Gatling (about 350 shots a minute) but a solid drawn brass case was used and jams were less frequent. The Gardner was operated by turning a drum-shaped mechanism mounted on a transverse axis in rear of the barrels. With all three of these machine guns accuracy of fire was greatly prejudiced by the torsion produced by the movement of the operating lever.

The first important land operations in which machine guns were used in our own service were those of the Zulu War of 1879, in which a few naval Gatlings were employed. Their fire power was invaluable in checking the momentum of a charge of savages, but their success was limited by imperfect mobility and a liability to jam at critical moments. The great value of a mobile and efficient automatic weapon, if such could be designed, in co-operation with infantry both in attack and defence was, however, foreseen by Lord Chelmsford, who was in command of the British forces at the time; but in spite of this and of its proved utility on many other occasions in savage warfare when naval weapons were available, the progress of the machine gun was for some time longer hampered by the original idea that it was a sort of field gun, and required a similar mounting.

It is not within the scope of this chapter to discuss the many varieties of mechanically operated weapons, and the various theories which were evolved for their employment, but it is worthy of note that the machine gun had now become a factor to be reckoned with and had commenced to make its influence felt upon tactics.

## III.

The year 1883 saw the commencement of a new stage in the evolution of the machine gun, since it was in this year that the first of the Maxim patents was taken out.

The Maxim gun, invented by the American engineer of that name, was a true invention and was all the more remarkable for its immediate success. The original design has been modified, improved, and adapted in later years, but the principle of all the various types at present in use remains the same. It was the first machine gun in which all the operations of extraction, ejection, feed and firing were performed automatically, and the soundness of its design was from the first unquestionable.

Maxim's system of operation, based on the utilization of recoil, was followed by another, that of utilizing a portion of the expanding gases evolved by the explosion of the charge to work the mechanism through the medium of some form of piston or lever. This system was adopted in the Hotchkiss and Colt guns, and has been successfully employed in the design of a great number of the lighter guns of more recent years. It has not, however, shown itself equally suitable for use in the heavier guns, in which accuracy and reliability are so essential for their employment in overhead fire. Yet, in spite of its increased efficiency, the machine gun was still for some time regarded with prejudice, owing partly to its poor record in 1870, but still more from the belief that it was now too complicated for use in the field, and that ammunition supply must be a very great difficulty. This was particularly the case in Germany; the first Maxims were not officially issued to the German Army until 1899.

The Maxim was first introduced into the British Army in 1891, and soon displaced the various weapons then in use. The tactics of its employment began slowly to evolve, but, although its value in savage warfare was proved and recognized against the Matabeles, in the Sudan and on the North-west Frontier, the enthusiasts who advocated its general adoption for civilized warfare met with considerable opposition.

In the South African War the machine gun was again on trial, but it was used in comparatively small numbers and with an imperfect realization of its tactical possibilities. It was not yet free from the tendency to jam at critical moments—a tendency, no doubt, greatly aggravated by the imperfect training of the gun teams. The old wheeled mounting was conspicuous and notably unsuitable for a country in which cover was generally scarce; in fact, the gun was not infrequently removed bodily from its carriage and brought into action some distance away on an improvised mounting, the wheeled carriage being left to draw the enemy fire. Although there were a few instances of its use for covering fire in attack, its success was more noticeable as a defensive weapon in block houses during the later stages of the war.

Besides the Maxim of rifle calibre referred to above, a heavier model, known as the "Pom-pom," was used by the Boers, and later also by the British. This was of 1·457-inch calibre and fired an explosive shell weighing 1 lb., the lowest weight allowed by the Convention of St. Petersburg. It was not capable of the sustained fire of the smaller gun, and required a carriage as conspicuous and almost as heavy as that of a field gun. It was therefore not a success at the time, but it may be noted that the design of machine guns of calibres larger than that of the rifle is receiving considerable attention at the present time as the result of the opening out of new fields for their employment.

The Russo-Japanese War, 1904–5, was the first war in which machine guns were used in large numbers by both sides. Either side grouped their machine guns in batteries of six or eight pieces, Russia using the Maxim and Japan the Hotchkiss. Russia started the war with her guns mounted on high-wheeled carriages, but these were later discarded in favour of a low tripod mounting, which proved more successful. The Japanese Hotchkiss appears to have been somewhat unreliable compared with the recoil-operated Maxim, and a Japanese report summarized in the Royal United Service Institution Journal (February, 1910) states that, on an average, at the battle of Hei-kou-tai cartridges jammed once in every 300 rounds. Yet on more than one occasion they annihilated battalions; their cross-fire was successfully used in defence, and they were of great value in attack when used for overhead fire. It is not surprising, therefore, that in Europe a keen and growing interest was manifest in the possibilities of the machine gun. In the German Army particularly, its future was foreseen and studied in the years which followed before the outbreak of the Great War.

With the increased understanding of its proper employment and capabilities, the machine gun began to be adapted to suit new tactical requirements. The need for greater mobility led to greater lightness of construction and the evolution of lighter and more efficient mountings was also influenced by the study of the use of ground and cover. The possibilities of indirect fire with machine guns were not unrealized, but the development of the necessary apparatus for their efficient employment in this connection did not progress very rapidly until it received the stimulus of necessity in the Great War.

## IV.

The conspicuous advantage in fire power obtained by the German Army in 1914 by the extensive use of the machine gun necessitated immediate steps being taken by the Allies to increase their own machine gun armament. In the British Army the number of guns of the Maxim type was increased as rapidly as possible, but their production could not keep pace with the expansion of the army. Consequently the Lewis gun was introduced, and the new units were armed with it in place of the Vickers, as through certain modifications and the lapse of the old patents the Maxim has now come to be called.

As the war progressed it was found that the mobility of the Vickers was insufficient to allow it to keep pace with infantry in the attack, and the need for a lighter form of automatic weapon which could be carried easily by a single man became evident. The purpose which such a gun was required to fulfil did not call for the capacity of the Vickers for sustained fire, nor ordinarily for use at long ranges, and the Lewis and Hotchkiss were adopted to fill the gap. The Lewis being the more accurate was adopted for infantry purposes, and the Hotchkiss, being more robust and better shaped for carrying in a bucket, for cavalry. The adoption of a lighter form of machine gun, in addition to the heavier type, was general throughout the armies engaged, but space does not permit of a consideration here of all the various weapons employed in the different armies. The more important mechanical features of a representative number of them are dealt with in succeeding sections.

At the end of the war the bulk of the fire power of the infantry unit was supplied by automatic weapons, and the principle of fire with movement had been developed to make the greatest possible use of them.

With the increase in the number of light machine guns to 2 to a platoon, the tendency is now for them to be pushed forward when the enemy has been located in order to pin him down while the rifle sections manœuvre. For this purpose the Lewis gun has two disadvantages, its weight is too great and it is too liable to stoppages. Consequently a new problem has arisen, that of the production of a weapon which shall be of lighter weight and increased reliability, and yet equal in accuracy and in volume of fire to the Lewis. In dealing with this problem the needs of all branches of the Service must be taken into account, since it is undesirable for many obvious reasons that a variety of weapons should be in use, each specially suitable for one particular branch.

## SECTION II.—THE CHARACTERISTICS OF AN EFFICIENT AUTOMATIC WEAPON

Before modern machine guns and light automatics can be considered in detail, a general understanding must be arrived at as to what characteristics are desirable in an efficient automatic weapon. The following tabulation of the conditions which the weapon should fulfil is arranged in order of desirability, but it must be understood that this order will be subject to considerable readjustment in the case of a weapon intended to satisfy particular tactical needs.

### CONDITIONS.

1. It should be capable of delivering an accurate fire and sustaining it over long periods.

2. It should be as light as possible in order to fulfil (a) of condition 3, taking into consideration the necessity of a certain weight to prevent unsteadiness when firing rapidly, even when the gun cannot be rigidly held.

3. It should be capable of :—

(a) going anywhere a man on foot can go.

(b) engaging any kind of target.

4. The mechanism should be so designed that :—

(a) the gun is always ready for " Rapid Firing."

(b) should occasions arise, the gun can fire single shots at will.

5. Its reliability under the worst service conditions should be beyond question. Its mechanism should be easy to teach, and its upkeep and preservation simple and economical.

6. Some cooling system should be provided to prevent undue heating of the barrel, as the higher the temperature of the barrel the less its resistance to wear.

7. Wear of barrel, due to the heat of combustion and the friction of the bullet, as well as the wear of the moving parts, should be negligible.

8. All parts should be accessible, particularly the barrel, to facilitate replacements.

9. It must be safe to handle, and the action should be positively locked to take the shock of discharge.

10. Its manufacture should be reasonably simple, and economical in cost and time.

11. Its rate of fire should be adjustable at the firer's will, between, say, 60 and 1200 rounds a minute.

12. It should fire the types of service S.A.A. in general use and be suitable for all branches of the service.

These conditions can now be examined more closely and their attendant advantages and disadvantages in practice compared and emphasized.

### *Condition 1.*

The greatest possible accuracy is required in order to fire over the heads of friendly troops to give them close support, and also for Barrage work; therefore, its accuracy should be as good as the ballistics of the ammunition will allow, and no appreciable " falling off," due to heat affecting the barrel (introducing " droop " or " bore enlargement "), should occur.

The ability to sustain accurate fire depends upon the rigidity of the gun and its mounting, its vibrations, the cooling system, the life of the barrel, the capacity of the feed arrangements, and, lastly, the ammunition.

### *Condition 2.*

The weight of an automatic weapon depends on the relative importance of the following three requirements :—

(a) Accuracy.

(b) Mobility.

(c) Reliability and durability.

The lighter a gun can be made, the better chance there is of condition 3 (a) being fulfilled, at the cost of accuracy and reliability.

Weight becomes necessary if the parts of the gun are to be strong enough to sustain fire for long periods.

The question of lightness is not so important for guns which are to be mounted in Tanks, and, in some cases, Aircraft, as for those used by Cavalry and Infantry.

### Condition 3.

(a) Weight, portability, balance and shape are factors affecting the fulfilment of this condition.

(b) The gun must be able to fire all types of ammunition (*e.g.*, armour-piercing, tracer, etc.) at all angles above and below the horizontal, when tilted to any degree of cant from the vertical, and from ground sloping to any degree in any direction ; and it must be reliable in operation even in the extreme case of being upside down if mounted in aircraft.

### Condition 4.

(a) Ability to open fire rapidly is a primary object of the gun, to which every other consideration must be subordinated. But it is not a good thing to keep a gun ready for immediate use over long periods, as certain springs are kept fully tensioned and are likely to deteriorate, and, furthermore, accidents may happen in spite of the careful application of safety catches.

(b) If single shots can be fired it is a distinct advantage, as they are useful for target indication by tracer ammunition, and for testing the energy of return springs and the setting of gas regulators.

The various methods employed to this end include slipping triggers which release the sear in time to prevent a second round being fired, and spacing the conveyor or otherwise arranging that the following round is not ready to be fed into the chamber as it would be in rapid fire.

### Condition 5.

There is no need to enlarge on this condition. The gun must be easy to teach, reliable, and such as to give the greatest confidence to the user. This confidence must not be shattered by frequent stoppages, due to :—

(a) Mechanical defects, overheating and faulty ammunition.

(b) External conditions, such as dust, sandstorms, gas, etc.

Few tests should be necessary in the field to ascertain the serviceability of the gun, and such tests should be conclusive without being elaborate.

The construction of the gun should enable it to withstand rough usage, and render it possible for a man with little training and little mechanical knowledge, to strip and clean all parts, with no fear of damage resulting or of the gun being wrongly assembled.

### Condition 6.

The ideal cooling system is that obtained with a machine gun mounted in aircraft, the barrel being exposed to meet the rush of the cold air encountered by the fast moving aeroplane. Rings fitted to the barrel to assist in radiation have been found useful in ground service guns (Madsen, Hotchkiss).

The system of air-cooling, however, is used in some guns which are fired in stationary positions, and is to a certain extent successful, but it is not ideal from the point of view of rapid, continuous fire, so that for such guns other means of keeping the barrel cool must be considered.

The water-jacket ranks next to air-cooling, especially if the water is mechanically circulated ; a pump worked by the action of the gun, has been tried and has given good results. Gravity tanks to ensure circulation, either by allowing the heated water to waste or return to the tank by convection can be used, but in all water-cooling systems, the water and necessary fittings materially add to the weight, and further, considerable trouble is occasioned when the temperature reaches freezing point. The need for replenishing the water, and the visibility of the steam generated, are also bad features in this system.

The induced-draught principle has been adapted to one gun, being brought about by the muzzle blast, which sucks in cool air through a casing which contains a radiator surrounding the barrel. This system is very effective, provided that the magazine capacity is small, so that frequent cessations of fire, necessitated by the changing of magazines, give opportunity for the barrel to cool.

## Condition 7.

The wear on a machine gun barrel in constant use is very great, due to the nature of its particular work, rapid firing. The friction of the bullet in passing along the bore and the very high temperature of the propellent gases, which reaches in many cases a point above that required to melt wrought iron, cause erosion of the metal. Cleaning also causes wear.

The temperature reached by the gases depends on the chemical composition of the propellent used, and its effect on the bore is accentuated by :—

(a) The high pressures developed in the chamber.
(b) The velocity with which the gases are projected through the bore.

The rush of gas at so high a temperature causes a slight volatilization and a "washing away" of the metal; to neutralize this, the barrel should be kept as cool as possible and where long bursts of continuous fire are essential, an efficient cooling system must be provided. It is well to remember that the hotter the barrel is the more readily it wears.

The wear due to the friction of the bullet could be considerably reduced by making the barrels of hardened steel, but manufacturing difficulties are great, the principal objections being the difficulty in obtaining suitable cutting and boring tools, the longer time necessary for machining operations, and the consequent increased cost. Constant lubrication of the bore reduces friction and various methods have been tried in this connection, including greasing the bullet and oiling the whole cartridge, but so far it has been found that the objections outweigh the gain. Grease and oil collect grit and are also liable to be rubbed off, while if the bore is to be lubricated at all it is essential that the lubrication should be consistent, otherwise accuracy suffers. A further disadvantage is met with if an oily lubricant is allowed to remain in the chamber, since undue stress may be thrown on the breech mechanism through the failure of the case to grip the walls of the chamber, while extraction may be rendered more difficult through the carbonization of the lubricant and the collection of grit.

The firing of "Tracer" and other special ammunition of similar design, is liable to accentuate wear, since these bullets contain chemical compositions the reaction of which, in some cases, commences before the bullet has left the barrel, and may have an injurious effect on the metal.

## Condition 8.

The wear and tear of service, apart from the great strain imposed on all parts of the gun during firing, causes damage, deterioration and wear, and consequently spare parts must be provided. To replace any breakage, the gun must be capable of being dismounted or stripped on the spot; it should, therefore, require but few tools of simple design to enable this to be done, and parts should be accessible so as to avoid the unnecessary removal of other parts. In water-jacketed guns, the design should render it unnecessary to remove the barrel—except for replacement—so as to obviate loss of water when in action.

The modern tendency is to arrange the components so that they interlock with each other; this has the advantage of doing away with a considerable number of nuts, pins and screws, which are apt to get loose. Any tendency towards inaccessibility must of course, be avoided, otherwise considerable delay may be introduced into the replacement of broken and damaged parts.

The greatest care must be taken that all components are made so as to ensure interchangeability, and, when the hot barrel is changed for a cool one, readjustment of the sights should not be necessary.

### Condition 9.

The design must be such that the gun cannot be made to fire unless the action is fully closed and the cartridge properly supported, *i.e.*, mechanically safe; even then it should only be able to fire through the operation of the trigger mechanism, the efficiency of which must be proof against heavy jarring.

Applied safety devices should be so designed that their application is by a single simple movement. The most efficient design, though not always practicable, is one which positively prevents any protrusion of the firing pin or striker, and the least efficient one which only prevents movement of the trigger and has no direct control over the firing pin or striker.

### Condition 10.

The problem of replacing the great wastage of machine guns and their components occurring in time of war, is one that calls for full consideration and elaborate provision. Large stocks of the parts liable to breakage have to be kept, and a considerable number of complete guns held in reserve. During hostilities it is obligatory upon manufacturers to supply the components at some definite rate and more often than not the call is made for an increased output. The predominant factor, therefore, is time, which renders it imperative that the design of all parts of a machine gun should be such that the number of operations, whether of machining or handling, is reduced to a minimum. The ultimate cost of a gun or component will largely depend upon the time taken to manufacture, the actual cost of material being but a small proportion of the total.

### Condition 11.

A very rapid rate of fire is necessary when engaging fleeting targets which may only be in view for a few seconds, such as are met with in a duel between aircraft. On the other hand, a slower rate of fire is normally required against a ground objective. When firing at a high rate of fire the gun overheats, the barrel wears quickly, and the cooling system becomes impaired. What is required, therefore, is a slow firing gun which can be rapidly and easily adjusted to a high rate of fire when circumstances demand.

The continuity of fire is always governed by the capacity of the magazine or conveyor, and the necessary time occupied in carrying out its exchange.

### Condition 12.

It is generally admitted that no special size of ammunition other than that in general use is permissible for machine guns, except for those intended for special purposes; and, further, in a machine gun barrel that is vented for gas operation, the vent should not prejudice the functioning of any special types of ammunition, nor should the entrance of the vent cause scrapings to come off the bullet or accumulate fouling.

It would also be an advantage if all magazines or conveyors used for small arms of the same calibre could be so designed as to be interchangeable, but this advantage is modified by the necessary smaller capacity of the conveyors of the lighter weapons.

It will be seen from the foregoing that it is not practicable to embody all the desired characteristics in one single gun, as some of them are contradictory. Further, the tactical requirements which a particular gun is intended to fulfil influence the relative importance of the above considerations in its design.

Having considered the characteristics of a supposed perfectly efficient weapon, all of which unfortunately cannot be incorporated in any one gun, it is necessary to investigate the various types of machine-guns that have been devised to fulfil some of these conditions, and the following is a summary of the general features found in them. No gun is likely to contain all these features and, naturally, details will vary in every case.

## SECTION III.—FEATURES OF DESIGN GENERALLY FOUND IN AUTOMATIC WEAPONS

(1) System of automatic operation.
(2) Feed arrangements.
(3) Method of supporting the base of the cartridge, which may be :—
   (a) By a definite locking of the action.
   (b) By the retention of the support by some form of "Inertia" until after maximum pressure has been reached.
(4) Safety arrangements, consisting of :—
   (a) Methods of ensuring mechanical safety when the gun is fired.
   (b) Means by which the gun can be rendered safe from accidental discharge.
(5) Cooling system.
(6) Means of storing energy.
(7) Trigger and firing arrangements.
(8) Extraction and ejection of the spent case.
(9) Sighting.
(10) Means of controlling the operation of the mechanism.
(11) Means of adjustment to regulate the rate of fire.

Before considering these features in detail, some indication must be given as to the various classes into which the different types of automatic weapons can be grouped. The design is as much dependent upon the tactical needs which the weapon is intended to serve as upon considerations of purely mechanical efficiency, and consequently questions of weight, calibre and tactical usage form a basis for the differentiation of the various types. The dividing lines cannot be considered inviolable, but generally the following four groups will suffice to include them all :—

1. Heavy machine guns—of a calibre above the average normal calibre for a rifle, say, 0·5 inch and upwards, and of necessity used with a stable mounting.
2. Machine guns proper—of normal rifle calibre and by reason of their tactical employment primarily designed for use with a stable mounting, such as the Mark IV tripod of the British Service.
3. Light machine guns—of normal rifle calibre, primarily designed for use with some form of light tripod or rest, and easily portable with the tripod or rest by a single man.
4. Automatic rifles—of normal rifle calibre or less, and primarily designed for use without any form of mounting or rest.

---

### Feature I.

#### Method of Operation

Since it is essential that the action of the mechanism should be automatic, some source of power from within the gun is necessary for its operation. The cycle of operations which must be faultlessly performed at a high speed after each round is fired, is as follows :—

1. Extraction and ejection of the empty case.
2. Positioning and feeding up of the succeeding round.
3. Re-setting the mechanism in position preliminary to firing.
4. Firing the next shot.

In old pattern machine guns, such as the Nordenfeldt, Gardner, and Gatling, the source of power was external to the gun. They were not, therefore, automatic in their action, and, consequently, have no place in the classification of automatic weapons.

In all modern designs, the necessary power is derived from the expansion of the propellent gases, and generally speaking it may be said that this power is utilized either directly by causing the expanding gases to come into contact with a reciprocating portion of the mechanism, or indirectly through the medium of recoil; in some cases the two methods are combined.

The various systems of operation may be grouped as follows :—

1. Direct operation by gas, *i.e.*, by means of diverted or trapped gas-pressure.
2. Operation by gas through the medium of loaded spring. In this system, diverted gas pressure acts on a short piston which operates the mechanism through the medium of a loaded spring.
3. Operation by recoil.
4. Operation by means of a combination of 1 and 3.

---

### 1. *Direct operation by gas*

The gas-piston system of operation is actuated by the use of a small portion of the propellent gases. This gas may be used in two ways :—

(a) By diverting, through a vent in the barrel, some of the resultant pressure.

(b) By trapping a portion of the gas on its escape at the muzzle.

The first method is the one usually adopted. The amount of gas energy required to be diverted is so small a quantity, that its diversion has no material effect on the bullet's velocity. The quantity necessary to overcome the resistance offered to the moving parts determines the position of the vent in the barrel. If the vent is placed near the chamber, it quickly becomes scored and enlarged, as, at this point, the pressure developed is extremely high. Should the vent be positioned near the muzzle, there is less time for the bullet to seal the bore after passing the vent, and consequently, the diverted pressure can only act on the mechanism for a shorter time. Again, if the vent is placed near the chamber there is more likelihood of the obturation being broken before the bullet leaves the bore than if it is placed nearer the muzzle, since the gas is diverted at an earlier stage. This must be taken into consideration in the design.

To prevent the vent interfering in any way with the bullet, it is essential that it should be placed wholly in a groove. Sharp angles must also be obviated, as they cause small portions of each bullet to be scraped off, which fouls the vent and gives rise to unsteadiness in the bullet's flight. For the same reason, the inner edges of the vent must be chamfered. This chamfering also tends to prevent gas erosion at this point.

(b) The second method of operation has not yet been incorporated in any machine-gun design, but it has been tried with success in some of the self-loading rifles. In these, action is brought about by providing a gas chamber on the outside of the muzzle. This chamber has a hole in prolongation of the bore, so that the closing of this hole during the passage of the bullet, momentarily traps the partially expanded gases. The resulting pressure either causes the chamber to move forward against a spring and so operate the mechanism through a connecting rod and links, or actuates a piston rearwards.

### 2. *Direct gas pressure acting on a piston which loads up operating springs*

In the Beardmore-Farquhar gun, which is operated on the above principle (a), the piston has only about half the travel necessary for the action to unlock, eject the spent case and insert a new round into the breech.

Between the piston and the action a spring is interposed which has no initial compression. When the piston moves back under the influence of the gas pressure the spring is compressed between the piston and the breech action. When the spring is loaded sufficiently, it starts the unlocking movement of the bolt and when this function is almost completed, the piston comes to the end of its stroke. To prevent the interposed spring from re-acting on the piston and so returning it to its original position, which would cause the spring to lengthen and so lose its energy, the piston is retained at the end of its backward stroke by a spring-operated catch. Thus the unexpended energy of the interposed spring is made to act in one direction *i.e.*, backwards, to give further movement and so to complete the full functioning of the action.

The catch for the piston is put into operation directly the interposing spring is compressed, and while the compression lasts the piston is still retained by it. The spring only remains in compression up to a point which is reached soon after the action is unlocked and the further backward movement has begun, when the spring extends itself to its fullest extent, and once more becomes floating.

During the piston's back stroke a long light spiral spring is compressed and acts as a retractor for the piston.

The chief points in this design are that the interposed spring ensures a delay in the unlocking of the bolt and that there is no direct connection between the piston and the action.

The principle objection is that it tends to be complicated owing to the number of parts employed.

*General remarks on gas-operated guns*

*Regulation.*—With all guns of this type, plenty of power is available and it is a very easy matter to arrange for the normal amount required, and for the adjustment of this amount, within limits, to ensure the smooth working of the gun. This adjustment is usually carried out by one of two methods :—

(a) By varying the size of the vent or its continuing channel.
(b) By altering the capacity of the chamber into which the gas expands.

With method (a) the actual diameter of the vent in the barrel is difficult to alter, but the control of the gas, after passing through the vent and before it meets the head of the piston, presents no difficulties. Should the area of the vent be smaller than that of the part of the channel which is adjustable, the degree of regulation will not be large. Adjustment, by having a number of vents in the barrel and so tapping off pressure at a position nearer or further away from the chamber, suggests itself, but would be impracticable, as it would entail closing all except the vent required. It may be noted that a gas port situated nearer to the breech is more prone to erosion, since gas temperature and pressure are higher, and it is subjected to the action of the gases for a longer period.

Method (b) is the better of the two as its range of regulation is large. The gas, after passing through the barrel vent, is allowed to expand in two directions—on to the head of the piston or into a chamber, the capacity of which is controllable by a regulating plunger. By adjusting the plunger the capacity of the chamber can be varied from nil to the full amount, and if placed so as to give the latter, only a small amount of pressure can act on the piston, as the gas enters the chamber and the pressure is reduced.

*Exhaust gases.*—No difficulties occur in getting rid of these. In types in which the piston is enclosed in a cylinder, the muzzle blast creates a suction in the bore which draws away the spent gases, and in those in which the piston or a pivoted lever is blown off a nozzle, the gases escape into the atmosphere as soon as the piston clears the nozzle.

*Fouling.*—The accumulation of gas deposit in and around the vents, gas chambers, cylinders and on the piston head is a constant source of trouble, and if steps are not taken to remove it as soon after firing as possible, it hardens and may have to be chipped off. Also the smooth working of the gun is affected and regulation has to be frequently resorted to.

*Cleaning and stripping.*—To facilitate the removal of fouling, frequent cleaning is necessary, so that ease in stripping is desirable.

*Balance and control.*—The accuracy of an automatic weapon depends, to a very great extent, on the steadiness with which it is supported or held during fire, and the greater the rate of fire the more difficult it becomes to maintain rigidity, since each succeeding round is fired before the gun has completely recovered from the shock imparted by the firing of the preceding round.

This effect may be called the disturbance of aim. The disturbance is noticeably more pronounced in some gas-operated guns than in others, and the reason for this cannot always be clearly defined since it is the result of the combined effects of a number of factors. These factors are :—

(1) The position of the centre of gravity and the method of supporting the weapon.
(2) The manner of holding, the firer's weight, etc.
(3) The total weight, and the weight of the moving parts.
(4) Movement due to pure recoil.
(5) The effect of the muzzle blast. This may tend to force the gun either forward or backward according to the design.
(6) Total resistance offered to the moving parts.
(7) Position of the gas vent in the barrel.
(8) Whether the piston works in a cylinder or is blown off a gas nozzle.
(9) Length of the piston's stroke.
(10) Type of recoil spring.
(11) The interior ballistics of the gun.

The ease of control of an individual gun depends on the balance of the above factors, and certain types are generally more difficult to hold than others. In practice wear has an effect, since such conditions as want of alignment in the working of the piston will have an unsteadying influence.

### 3. Recoil Systems

The mechanism in these systems is operated by one of two methods :—

(a) The backward thrust of the cartridge case.
(b) Recoil of the barrel.

(a) This method depends entirely on the momentum given to the action by the projected case, the movement of which must be delayed so that the bullet will have cleared the bore before an appreciable opening occurs between the breech and the barrel, otherwise the gas would escape, which would endanger the firer, lower the muzzle velocity and damage the mechanism.

When the cartridge case starts to move back, a portion of it is unsupported by the chamber, and, as the maximum pressure is reached when the cartridge case is in this position, it must have specially thick walls to prevent burst and separated cases, which are known as "separations." Also, to prevent the escape of gas round the outside of the moving case, its sides should be parallel, or nearly so. Parallel cases, however, tend to grip the walls of the chamber during the process of recoil upon which the operation of the gun depends. For ease of extraction, especially in non-automatic weapons, a cone-shaped case is essential, since with this form the friction between the case and the walls of the chamber is relieved as soon as any backward movement occurs.

As the gas pressure set up is equal in all directions, it follows that if the resistance offered by the breech action, due to its weight and other factors, is relatively high, compared with that offered by the weight and resistance during movement of the bullet, the respective velocity of the two will be in the inverse ratio of their total resistance.

The effect of retardation or opposition to the backward movement of the breech action can be brought about in several ways, such as by the use of heavy moving parts (an undesirable feature), by the introduction of friction, by utilizing the resistance of springs, or by otherwise arranging for some portion of the shock of discharge to be absorbed in a non-effective manner. A suitable combination of these methods is usually employed, and will be considered later under Feature 3. It must also be remembered in this connection that if the cartridge case on expansion is allowed to grip the walls of the chamber it will materially assist in the retardation of the backward movement, and this factor must be taken into consideration in the design of the gun. In the Schwarzlose machine gun, for example, the cartridge cases are oiled so as to prevent their gripping the walls of the chamber.

Owing to the small area of the base of the cartridge, and the maximum pressure being reached quickly, the action must be strong and consequently comparatively heavy, to withstand the shock. It is impossible with this system to arrange for a definite means of locking the breech, since the bolt or lock has to be free to move on the explosion of the charge.

In the case of water-cooled guns the prevention of water leakages is made easier by this system, since, the barrel being fixed, the necessary packing glands do not make contact with a sliding barrel. With a long barrel, high muzzle velocity and high gas pressure, however, a definitely locked action is more usually employed.

(b) Recoiling barrels can be sub-divided into two groups :—

    (1) Short barrel-recoil.
    (2) Long barrel-recoil.

N.B.—The terms " short " and " long " recoil are used in a relative sense.

(1) In guns of the short recoil type the backward movement of the barrel is much less than the length of the cartridge. This difference causes the extraction and ejection of the spent case and the placing of a new round in the chamber to present difficulties, and entails complication of design. The speed and distance of travel of that part of the mechanism— usually named the lock—which performs the above three operations, has to be accelerated, relatively to those of the barrel.

(2) With long recoil, the breech must move back sufficiently far to allow the feeding up of the new round. This is easily arranged by allowing the barrel to go forward while the lock is being retained until the round has entered, or partly entered, the chamber. Consequently the action is somewhat slower, affecting the rate of fire. In some cases, means are provided to introduce the cartridge before the barrel reaches the home position. The Madsen machine gun is an example embodying this idea.

It will be interesting here to discuss the means employed in machine guns, especially in those operated by recoil, to speed up the critical or normal rate of fire.

With recoil systems particularly, there is a limit to the number of rounds a minute that can be fired. This limit depends upon various factors, the chief one being the time taken to feed up the cartridge case and introduce it into the chamber. The action of this portion of the mechanism is always sluggish compared with the action of the other portions.

Should means be provided to facilitate the feeding up of the next cartridge, the rate of fire is thereby increased, and this speeding up can be done by incorporating a differential action into the mechanism. By this means the succeeding round is placed in the ready position before some of the other functions are completed, so that a cartridge is always waiting in position, ready to be received by the carrier in its action, thus ensuring a positive feed.

An example of this is found in the Vickers' machine gun ·303-inch, where the barrel recoils a short distance only, and in recoiling brings about preparation to feed. At the same time the recoil causes the lock to be accelerated, and before the lock's backward movement is completed, the barrel is made (by the aid of cams or eccentrics) to return to the firing position, reversing the feed action in doing so. This places the succeeding round in the ready position.

The question of unlocking the breech with recoiling barrels has to be carefully considered from the point of view of the bullet's exit. If the breech is opened too soon, obturation fails. Should it happen too late full functioning of the mechanism is prevented, because there is not sufficient energy and space left to give the required momentum to the moving parts, to carry out ejection and feed, etc.

Further, the whole force of recoil must be always assumed as acting with every round, therefore the fixing of the unlocking position is dependent upon the interior ballistics and design of the gun. Very little margin can be allowed for slight variations in recoil energy, which may arise from faulty charges or fluctuating resistance offered to the moving parts. The selection of suitable antagonistic springs needs careful consideration, since besides serving as a means of storing energy, they must be considered in relation to the other factors of resistance to recoil. In order to maintain the total resistance to recoil at a constant value in practice, it is also desirable that the springs should be adjustable.

---

## Feature 2.

### Feed Arrangements

The various types of conveyor used in automatic weapons may be grouped as follows :—

| Conveyor used | Guns used in |
|---|---|
| Belts, fabric or metallic | Maxim, Vickers, recoil-operated Browning, Colt, Schwarzlose, Parabellum. |
| Circular magazine | Lewis ·303-inch and ·5-inch, Gast, Beardmore-Farquhar, the early Maxim, Thompson. |
| Strips (stiff or folding) | Hotchkiss (except one of the new light models). |
| Box magazines (with platform and spring). | Madsen, Revelli, new pattern Hotchkiss, light and heavy Brownings. |
| Hoppers (circular or straight) | [Gardner, Gatling, Nordenfeldt.] |
| Combined belt and rotating wheel | Original Maxim. |

*Note.*—Both the Beardmore-Farquhar and the Thompson can be used with a small capacity box magazine as an alternative.

Belts may either be :—

(a) Disintegrating, or
(b) Continuous.

(a) Disintegrating belts are made of metal and are designed so that the cartridge itself forms the connecting swivel pin between adjacent links. When the cartridge is withdrawn from the link during the operation of feeding, the foremost link is freed and falls away. With these the initial loading is somewhat difficult, but the great advantage is that no portion of an expended belt can extend beyond the side of the gun, as would occur if a continuous belt were used, a condition that must not obtain in a gun mounted on aircraft; and, further, there is no limit, except capacity, to the length of the belt that can be made up. Care has, however, to be taken when the gun is mounted in aircraft that the expended links are deflected or caught in some manner, otherwise they might injure the firer or damage the more vulnerable portions of the machine.

(b) Continuous belts are made of fabric, and are of sufficient length to hold a suitable number of rounds, which number depends upon questions of weight, and on the number of rounds which may be required to be fired in a burst. Belts are usually made to carry 250 rounds, but the number varies with different guns. Their chief advantages are cheapness, freedom from distortion, lasting qualities from a Service point of view, and the small space into which they can be packed when empty. They are, however, affected by damp, which causes the pockets for the cartridges to shrink; also the pockets, owing to wear, eventually become enlarged and may allow the rounds to fall out.

Some types of these belts are fitted with metal strips which project beyond the edge of the belt as far as the nose of the bullet when the belt is filled. When the filled belts are packed in drums or boxes during transit, the fact that the ends of the strips are flush with the noses of the bullets prevents the rounds being jolted out of the belt against the side of the container, a tendency which is much increased when the case is cone-shaped.

When stripless belts are used with cone-shaped cases, light strips of wood are passed between the rounds at their bullet ends as they are packed in the container; this achieves the same object as the provision of projecting strips on the belt, in that the belt is prevented from slipping forward on the rounds during transit.

Continuous metallic belts which are used under certain conditions with the (British) Hotchkiss, have certain disadvantages. They are easily damaged, when the free insertion and withdrawal of the round are affected, and their flexibility suffers.

With all types of belts, in order to place the round in the chamber, some means must be provided to withdraw the cartridge from the belt and control it until so placed. The most successful method is that employed in the Maxim, where the round is pulled out of the belt, carried back, made to drop down in line with the chamber, and then carried forward, so entering it. This means cannot always be introduced, so that others, such as a claw which positions the round, or a rotating wheel or wheels, have been tried.

Circular magazines can be arranged to operate in both planes; vertically as in the Gast and Thompson; horizontally, as in the Lewis and Beardmore-Farquhar. This type of magazine is either rotated by a ratchet pawl, or is fitted with a clock spring to give self-rotation.

Strips have the advantage of being light and cheap to manufacture, but require working space on both sides of the gun for insertion and ejection; this space is governed by the number of rounds, and for the strip to contain a useful quantity will necessarily be large. Folding strips, however, enable this space to be reduced, a desirable feature in guns mounted in confined positions, as in tanks and cars.

Box magazines, though possessing certain advantages with regard to ease of feed, when arranged to work from the top of the gun have a tendency to interfere with the sighting, limit the field of view and render the gun conspicuous. If used in aircraft in this position they also suffer from the disadvantage of exposing a considerable surface to the resistance of the air.

The box magazine used in the Revelli machine gun (Italy) is of the multiple type, fitting underneath, and as each partition becomes empty, the whole is worked over more space by the action of the gun. This appears to be the best of this type, considering its large capacity.

The hopper type of feeding apparatus belongs to the early stages of the development of automatic weapons. The straight pattern consists briefly of a " guide-way " into which a number of rounds are fed by hand, and from which they drop into the feed-way of the gun in their turn as they reach the bottom. Another pattern consists of a circular container in which the rounds are moved round by rotating wheels, each round as it reaches the mouth of the hopper being allowed to drop by gravity into the feed way. The hopper is cumbersome, interferes with the sights, and during replenishment necessitates a degree of movement which seriously affects the ease of concealment of the gun.

The combined belt and rotating wheel type of conveyor was the first method adopted by Sir Hiram Maxim to bring about automatic feed, and can be best followed by reference to Patent Specification No. 3493 of the year 1883.

*General Remarks on Feed Arrangements*

1. The greatest care should be taken to see that belts, strips, magazines or other means of holding the cartridges in reserve, are correct in every way, as too much attention cannot be paid to them. More trouble resulting in stoppages, jams, lost opportunities, etc., is brought about by slight defects in these adjuncts passing unnoticed, than from any other source.

2. The efficiency of the gun is necessarily to a certain extent influenced by the quality

of the ammunition. Should this depart comparatively slightly from its correct dimensions, troubles in feed will result; these are most commonly due to long or short rounds. With rimmed cartridges, particularly in the case of certain guns, abnormally thick rims must be avoided.

3. A good way of testing the functioning of the extraction, ejection and feeding mechanism, is by loading and passing through the gun a few dummies, operating the gun by hand and observing the result.

4. It is necessary in some recoil operated guns to introduce a device which will prevent, or tend to prevent, stoppages due to the slow working of the feed arrangements, relative to the speed of the recoiling portions of the gun. Thus, in the Vickers, a gun of the Maxim type, a differential action is introduced, which ensures that a round is always ready for the loading action. This movement is brought about by cams on the crank forcing the barrel forward, and so positioning another round in the feed block while the lock is still moving backward, relative to the barrel, thereby giving a positive feed.

5. Whatever feed arrangements are conjoined to the method of loading, it is imperative for safety that the action of making the gun ready to fire does not place a round in the chamber, as a heated gun is liable to cause a cartridge to explode. This is not always possible, but under no circumstances, including stoppages, should it be possible for a round to remain in the chamber unsupported by the lock or bolt, except for a brief instant. It should also be possible, when the conveyor is removed, to see without difficulty whether a round has been left in the feedway, otherwise accidents may occur.

---

FEATURE 3.

The method of supporting the base of the cartridge in automatic weapons and of ensuring that the breech does not open before the gas pressure has dropped to a point consistent with safety and efficiency, is a very important feature and varies with different designs. These, however, can be divided into two distinct groups, as was indicated in the list of features of design on p. 159.

(a) Those in which the action is definitely locked and the cartridge rigidly supported until the bullet has left the barrel.

(b) Those in which the locking action cannot be called positive, but in which the support for the cartridge and the delay in the opening of the breech is effected by some other means, such as inertia assisted by springs or friction. This group can be described as working on the delayed action or hesitation system.

These two main groups will now be considered in detail.

*Group (a)*

The following methods of locking the breech are found in guns of this type:—

| Method of locking | Guns in which found |
|---|---|
| 1. Formation of a toggle joint | All Maxim types. |
| 2. Rotation of locking lugs | Lewis. |
| 3. Engagement and rotation of an interrupted screw | Hotchkiss (British). |
| 4. Link on bolt fitting into a locking recess in the body— | |
| (a) The link having a downward movement | Hotchkiss (French). |
| (b) The link having an upward movement | Browning (Light model) and Berthier. |

5. A stud on a pinion working in a groove or camway on the breech block, part of the groove being eccentric to, and part concentric with, the pinion. Locking of the action is effected by the passage of the stud from the eccentric to the concentric portion of the groove .. .. St. Etienne (French).
6. Breech block hinged at rear end, the front portion being raised for locking by the action of a stud working in a switch plate in the body.. .. Madsen.
7. Rising wedge or block locking the breech block to the barrel .. .. .. .. .. Browning (recoil operated).
8. Rear end of bolt dropping down in front of shoulders on the body .. .. .. .. Colt.
9. Bolt passing through slotted rings which, when rotated, form locking shoulders.. .. .. Gast.

The action of the foregoing methods of locking will now be described in greater detail.

### 1.—*The toggle joint*

Example : ·303 Vickers machine gun (Fig. 1 and Plate I)

In the firing position the centre of the crank pin (1) is below a straight line drawn through the centres of the side-lever axis (2) and the crankshaft (3), and the crank rests on the crank stops (4).

The shock of discharge being received by the lock is transmitted through the axes (2) and (1) to the crankshaft (3) which is rigidly supported relative to the barrel. The crank pin (1) being below the centre line, as described above, therefore tends to be forced downwards, but is prevented from moving by the crank bearing on the crank stops.

At the correct moment for unlocking, the crank is caused to rotate, the axis (1) lifted, and the toggle joint broken.

### 2. *Rotating locking lugs*

Example: Lewis Light Machine Gun (Fig. 2 and Plate II)

In the Lewis, which may be taken as typical of this design, the striker (1) is carried on a post (2) mounted on the rear end of the piston (3), and this post moves backwards and forwards in a slot (4) in the bolt (5). When the piston commences to travel forwards under the action of the return spring the striker post is lodged in the recess at the rear of the slot in the bolt, and its left side bears against the left side of the curved portion of the slot. The locking lugs (6) on the bolt, being engaged with the guide grooves of the body prevent it turning, and it is therefore carried forward until such time as the locking lugs are opposite the locking recesses. The striker post then travelling along the curved portion of the slot causes the bolt to turn, thus locking it to the body, and continuing along the straight portion of the slot carries the striker forward to fire the round.

During the return movement of the piston the reverse action takes place.

### 3. *Interrupted screw*

Example: Hotchkiss Light Machine Gun (British model) (Fig. 3 and Plate III)

With this method, which is also employed in the majority of B.L. Ordnance, two members of the mechanism, one of which is rigidly supported, carry male and female threads respectively. A portion of the thread is removed in each case, so that the one member can slide into the other, and then by a partial turn of one of them they can be locked together.

In Fig. 3, which shows a portion of the mechanism of the ·303 Hotchkiss (British), the breech block (1) enters the fermeture nut (2) by means of the gaps in their respective interrupted threads, a portion of which can be seen on (1). The fermeture nut is rigidly supported in the body and is caused to rotate through a small angle to engage the threads and so to support the breech block. This partial rotation is brought about by the action of the inclined groove (3) on the piston working on the boss (4).

### 4 (a). *Link and Locking Recess*

Example: Hotchkiss Machine Gun, Model 1914 (Fig. 4a and Plate IVA)

The underside of the hump (2) on the piston (3), during the completion of its forward movement, comes into contact with the hinged link (1) on the breech block (5), and forces the lower end of the link downwards into the recess (4), in which position it engages in front of resistance shoulders on the body. The breech block is then rigidly supported.

At the correct moment during the backward movement of the piston, the inclined surface in front of the recess (4) raises the link away from the resistance shoulders, thus leaving the breech block free to travel to the rear with the piston.

### 4 (b). *Link, with an upward movement to lock*

Example: Browning Light Machine Gun (Fig. 4b and Plate IVB)

In Fig. 4 (b) the mechanism of the ·303 Browning Light Machine Gun is shown in the firing or locked position.

The bolt-lock (1) and the bolt (2) are connected by the swivel pin (3). The bolt-lock is connected to the piston-slide (4) by means of the link (5) which swivels at (6) and (7). The shoulder (8) on the bolt-lock is, in the position shown, bearing against resistance shoulders in the body, and the bolt is thus rigidly supported.

When the piston commences its backward movement it carries with it the lower end of the link (5). The bolt is thus lowered from engagement with the resistance shoulders on the body and, as it is lowered, commences to move backward, slowly at first and then at the same speed as the piston when completely disengaged from the resistance shoulders.

During the forward movement of the piston the reverse action takes place.

### 5. *Stud on the pinion working in a camway on the breech block*

Example: St. Etienne Machine Gun (Fig. 5 and Plate V)

In Fig. 5 is shown the mechanism of the St. Etienne, in which the piston is blown forwards by the propellent gases, and actuates the mechanism by means of a rack and pinion.

In the figure the rack (1) on the piston engages with the teeth on the pinion (2). The stud (3) on the arm of the pinion works in the camway (4) on the breech block. The upper portion of this camway is eccentric to, and the lower portion concentric with, the path of the stud during the rotation of the pinion. In the firing position the piston is to the rear, and the stud (3) lies in the concentric portion of the cam-way. In this position the stud is below a line drawn through the centre of the pinion and parallel to the line of recoil, and as the pinion is rigidly supported in the body, pressure on the face of the breech block merely tends to force the stud further down in the camway. The action is thus locked to receive the shock of discharge.

During the forward movement of the piston the stud rises in the camway and, at the correct moment for unlocking the action, passes from the concentric to the eccentric portion of the camway. The continued rotation of the pinion then, through the medium of the stud, carries the breech block to the rear.

In Fig. 5 the stud is shown entering the concentric portion of the cam-way.

### 6. *Hinged Block*

#### Example: Madsen Light Machine Gun (Fig. 6 and Plate VI)

This is similar in principle to the Martini rifle action. The rear end of the breech block (1) is pivoted to the breech casing (2) by the axis pin (3). On the lower part of the breech block (1), at its front end, is a circular stud (4); this stud works in the guide grooves of the switch plate (5), the latter being fitted to the non-recoiling portion or receiver of the gun. The breech block, breech casing and barrel recoiling, it follows that the stud (4) and the guide grooves (5) control the movement, rising or falling, of the front part of the breech block. The movement of the stud is shown by the dotted line and arrows in the sketch.

### 7. *Rising Block or Wedge*

#### Example: 0·303 Browning (Recoil Operated). (Fig 7 and Plate VII)

In Fig. 7 is shown the locking mechanism of the ·303 Browning machine gun (water-cooled model). This gun is operated on the short recoil principle. The barrel recoils a short distance and in so doing unlocks the bolt and operates the accelerating mechanism which throws the bolt further to the rear against the action of the spiral " driving " spring. This spring reasserts itself to return the bolt and barrel and lock them together during the forward or loading action.

The breech block (1) moves in the guide grooves of the barrel extension (2). Fitting into this extension and capable of being moved vertically is the breech lock (3) which consists of a solid block fitted with a projecting guide pin (4).

In the firing position the breech lock is in its highest position and engages behind a shoulder on the bottom of the bolt, thus locking the bolt and barrel together.

At the correct moment during recoil the breech lock (3) is lowered from engagement with the bolt by its guide pin (4) striking the slanting surfaces of the right and left projections (8) of the lock frame (9).

During the forward movement of the recoiling portions of the mechanism, when the bolt has rejoined the barrel, the breech lock comes into contact with, and rides up, the breech lock cam (6) on the bottom plate (7), and thus engages behind the shoulder on the bottom of the bolt.

### 8. *Rear End of Bolt dropping down in Front of Shoulders formed in the Body*

#### Example: Colt Machine Gun (air-cooled). (Fig. 8 and Plate VIII)

Fig. 8 illustrates the locking action of the Colt air-cooled machine gun. This gun is operated by means of a pivoted radial lever being blown off a gas port underneath the barrel. Fig. 8 (*a*) shows the action locked in the firing position, and Fig. 8 (*b*) the bolt in its rearmost position. The radial gas lever, after having been blown off the gas port in the barrel, operates the gas lever extension (1), which in turn causes the slide (2) to travel in the direction of the arrow. The pin (3) which is attached to the slide (2), working in the slot (4), raises the breech bolt (5) clear of the resistance shoulders (6), and then withdraws the bolt (5) to the rear, as shown in Fig. 8 (*b*). The drawings give a far-side view of the slide (2) and resistance shoulder (6), which are duplicated; the flange (7) of the bolt (5) acts as a guide by working in between them.

### 9. *Rotating Rings*

#### Example: Gast Light Machine Gun (Fig. 9 and Plate IX)

In this gun two reciprocating barrels are used, the recoil of one supplying the motive power for the complete cycle of movement of the other, hence the description of unlocking and locking of one breech action will suffice.

Attached to the barrel extension (1) are rotatable discs (2) (only the near disc is shown in the figure) on the inner surface of which slots (3) are cut for the free passage of the flanges (4) of the breech bolt (5).

Backward movement of the breech bolt (5) on the explosion of the charge is prevented by the resistance shoulders (6) of the flanges (4) of the breech bolt (5) bearing against the periphery of the disc at the point shown by (6) as at this moment the slots (3) for the flanges (4) are not in alignment.

Rearward movement of the barrel extension caused by recoil brings the striking toe (7) of the disc (2) against the fixed abutment (8), thus causing the disc (2) to rotate through an angle of 45°, so aligning the slots with the resistance flanges (4), allowing free backward movement of the breech bolt (5).

During the forward movement, after the resistance flanges (4) are fully forward, the striking toes of the discs on the inner side (not shown in sketch), turn the discs through a similar angle in the opposite direction, thus locking the action.

*Group (b)*

### Delayed Action or Hesitation System

This system is employed in those guns which are operated by the projection of the spent case. (*See* Recoil Systems, p. 162.)

The various methods of ensuring this delay in the opening of the breech can be classified as follows :—

1. Those in which the principal delaying element is friction, introduced by some mechanical contrivance such as is found in the Revelli machine gun or the Thompson automatic rifle or sub-machine gun. In the former the opening of the breech is controlled by a rotating wedge which temporarily connects the breech block and barrel during the early stages of the recoil movement, and in the latter the shock of recoil is partially transmitted directly to the body through the medium of slipping inclined faces  The Villa Perosa also illustrates this feature, though it more correctly belongs to the inertia class, since the frictional element is less pronounced.

2. Those in which the principle delaying element is inertia assisted by the resistance to compression of springs, and other factors.

    In the Schwarzlose machine gun this delay is effected by means of a heavy breech action supported by an elbow joint and a strong spring.

    The Villa Perosa (mentioned in (1) above) together with the Thompson " sub-machine gun," fall into the automatic rifle class, though they are relatively short-barrelled weapons firing a pistol type cartridge.

1 (*a*).  The Revelli Machine Gun   (Fig. 10 and Plate X)

On firing, the pressure of the propellent gases, acting through the base of the cartridge, imparts a rearward movement to the breech block (1). At this stage the movement of the breech block (1) relative to the sleeve (2) is controlled by a wedge (3), which is capable of rotation about a fixed axis (4) at right angles to the axis of the bore. The rearward movement of the breech block (1) causes the wedge to rotate to the rear. In so doing the wedge (3), which passes through a slot in the sleeve (2), bearing against the latter at the shoulder (5), forces the sleeve (2), together with the barrel, rearwards. After a movement of about 4 mm. on the part of the moving portions, the wedge is entirely disengaged from the breech block (1), which is now free to continue its backward action under its acquired momentum. The wedge is maintained at its lowest position by its nose riding on the under surface of the breech block (1). Hence the sleeve (2) cannot move forward until the recess in the breech block (1) returns to within about 4 mm. of its initial position.

The axis (4) of the wedge is eccentric, and can be adjusted in one of three fixed positions in its bearings. This enables the amount of the jamming effect of the wedge on the moving parts to be increased or decreased to ensure the smooth working of the gun.

The return of the moving parts to the firing position is caused by a strong spring operating through a connecting rod (6), one end of which is hooked to a claw (7) on the bottom of the wedge, the other end being connected to an adjustable spring which, in its turn, is attached to the frame of the gun.

The claw (8) does not assist in the locking of the action, but is the medium by which automatic firing is controlled.

1 (b) Example : Thompson Automatic Rifle or " Sub-Machine Gun."

(Fig. 11 and Plate XI.)

On firing, the projection of the spent case drives back the bolt (1), which takes with it the H-piece (2). Resistance to the bolt's backward movement is at once encountered as the trunnions (3) of the H-piece (2) are engaged in the short 45° inclined slot (4) of the body (5), and so must lift the H-piece.

Resistance to the lifting of the H-piece (2) also occurs, as its front face in rising meets the rear face of an inclined slot of 70° (6) in the bolt, and further retardation results as the bridge (7) of the H-piece (2) in rising enters the jaws (8) of the actuator (9), which are set at 10° to the vertical.

If the above action is analysed, it will be seen that the backward movement of the bolt must necessarily be retarded by the fact that in the early stages the force of recoil is split up into several components acting in different directions, as follows :—

The inclined slot (6) in the bolt tends to force the H-piece downwards and backwards.

The rear face of the inclined groove (4) in the body reacts upwards and forwards (towards the barrel).

The inclined surface on the rear jaw of the actuator reacts downwards and forwards.

The direction of the resultant of all these components is upwards and backwards (*i.e.*, the direction of the resulting movement of the H-piece), but its value is small compared with that of the force of recoil in the original horizontal direction, and, further, it has to overcome the friction between the inclined surfaces engaged.

It is claimed by the inventor that this gun is so designed that owing to the rapidity with which the pressure in the bore rises to a maximum on firing, the bolt is supported by the adhesion of the inclined surfaces, until the pressure has again fallen to a much lower point, reached when the bullet leaves the barrel, and that not until then does the breech commence to open.

The determining factor is the state of lubrication of the surfaces, and it is essential for the successful application of this principle that this factor should remain constant. In this connection the method of lubrication of the cartridges by means of oil pads in the body (*see* Fig. 11 (10) ) should be noted.

2 Example : The Schwarzlose Machine Gun. (Fig. 12 and Plate XII.)

In this design resistance to the backward movement of the breech block is provided by the inertia of heavy moving parts, and by the resistance to compression of a strong return spring. This resistance is further assisted by attaching to the breech block the end of one arm of an elbow joint, the end of the other arm being pivoted to a fixed axis in the body.

The operation of the elbow joint is as follows (*see* Fig. 12). The backward movement of the breech block (2) causes the vertex (c) of the elbow joint to move through the arc C.C.i, and owing to the small angle between the crank link (3) and the crank (1) when the breech is closed, a large proportion of the force of recoil is transmitted to the fixed axis (4) of the crank and crank handle in the body. As the vertex (c) rises, however, and the angle between (3) and (1) increases, the resistance encountered by the breech block decreases, and although the main force of recoil has been absorbed as already described, the backward movement of the breech block is completed against the resistance of the return spring by the acquired momentum of the heavy moving parts of the mechanism. Additional resistance is met with in the backward movement in the cocking of the striker which is effected by a cammed toe on the crank link acting on a corresponding toe on the striker.

2 (b) The Villa Perosa is of little interest, except for the fact that it has two barrels which can be fired either independently or at the same time, and is capable of a very high rate of fire. The cartridge is supported by a strong spring and the inertia of the bolt, as in certain types of self-loading pistols. It was primarily designed for use in aircraft, and is not a "shoulder" weapon and, therefore, its classification as an automatic rifle is perhaps somewhat misleading.

---

## Feature 4.

### Safety arrangements

**A.—** *Methods of ensuring mechanical safety.*

In order that a gun may be mechanically safe to fire the mechanism must be so designed that :—

(1) the cap cannot be struck by the firing pin or striker before the breech is closed and the gun ready to fire ; and

(2) sufficient support, covering the limits allowable for slight variations in pressure from round to round, is given to the base of the cartridge case to ensure that the breech will not open before the bullet has left the barrel. Various methods of providing this necessary support of the base of the case are met with in different designs, ranging from those guns operated purely on the " inertia " system to those in which the action is positively locked at the moment of discharge. The object of introducing some mechanical means of giving additional support to the base of the cartridge, or of definitely preventing premature opening of the breech, is to increase the margin of safety obtainable.

The following examples illustrate some of the different methods employed to ensure safety in these respects :—

(a) *Lewis, Hotchkiss, Browning, Light Machine Guns.*

(1) In these guns the sequence of the action of the mechanism is so designed that it is mechanically impossible for the firing pin or striker to protrude through the face of the bolt before the breech is closed and locked.

(2) Gas pressure does not act on the piston until the bullet has passed the gas port and is near the muzzle, and, further, the unlocking of the action is not commenced until the piston has travelled a certain definite distance to the rear.

(b) *Colt Machine Gun.*

(1) The firing pin is driven forward to strike the cap by a hammer which is contained when cocked in the pistol grip unit.

The release of the hammer is controlled by two sears, the one being operated by the trigger and the other being disengaged automatically by a trip lever, operated by a groove in the slide, as the rear end of the bolt sinks in front of the recoil shoulders at the completion of the forward movement. Moreover, since the rear end of the bolt rises and falls in the rearward and forward movements respectively, the hammer, firing pin and cap are not in alignment except when the breech is closed and locked.

(2) The necessary delay in the opening of the breech is introduced in a similar manner to that described in the case of the other gas operated guns referred to in (*a*).

(c) *All Maxim types.*

(1) The cap is masked until the last moment before firing by a rising portion of the mechanism in which is the firing pin hole. In addition to this, the firing pin is retained by a sear until this is disengaged by the further travel of the side lever head after the toggle joint has been formed and the action locked. The firing pin is then either propelled forward by the lock spring to strike the cap or else held back by the engagement of the nose of the trigger in the bent of the tumbler until the firing lever is pressed.

(2) For a short distance the barrel and lock recoil together, and the toggle joint remains locked until it is broken by the tail of the crank striking the roller.

(d) *Browning Machine Gun (recoil-operated model).*

(1) In this design the trigger does not engage with the sear until the bolt is nearing the end of its forward movement and the breech is closed and locked. In addition to this, the action of the cocking lever which compresses the firing pin spring during the backward movement of the bolt is so arranged that the firing pin could not protrude through the face of the bolt until the forward movement was completed, even if it were not retained by the sear.

(2) During the initial stages of recoil the barrel and breech block remain locked together by the breech lock, which is caused to drop, and unlock the action, when its guide pin strikes the slanting surfaces of the projections on the lock frame.

(e) *Schwarzlose Machine Gun.*

(1) The breech block is propelled forwards by the return spring through the medium of the firing pin, of which a hinged extension engages in a bent on the breech block. The firing pin cannot reach its forward position in the breech block, and protrude through the face, until this extension is disengaged from the bent. When the firing lever is pressed the trigger bar is drawn to the rear, and in continuous fire the firing pin extension is disengaged from the bent on the breech block by its nose riding up the ramp on the trigger bar, when the forward movement is almost completed. When the firing lever is released the ramp on the trigger bar is outside the range of movement of the nose of the firing pin extension.

In addition to the above a stud on the breech lock link prevents the firing pin extension from rising from engagement with the bent until the breech block is almost home.

(2) The action of this gun is not locked. The necessary delay in the opening of the breech is dependent upon the inertia of the moving parts, the resistance of a strong return spring and the absorption of part of the force of recoil during its early stages through the medium of an elbow joint. The use of a relatively short barrel also assists in reducing the degree of delay which must necessarily be arranged for.

(f) *Revelli Machine Gun.*

(1) A stop is interposed between the firing pin and the face of the bolt which cannot be cleared until it comes opposite a recess provided in the body to receive it when both the bolt and barrel are fully home. The stop is forced aside into its recess by the firing pin itself in going forward to strike the cap. In addition to this a secondary sear is employed which holds up the firing pin until, when the breech block is almost home, the tail of the rotating wedge rises sufficiently to release it.

(2) The opening of the breech is definitely controlled by the action of a wedge rotating about a fixed axis, and connecting barrel-sleeve and breech block during the initial movement of recoil.

(g) *Madsen Light Machine Gun.*

(1) The breech block and breech block casing are driven forward by the return spring through the medium of a radial recoil arm pivoted in the body. When the forward movement is almost completed a cam on the recoil arm depresses a pawl, which in turn releases a sear from a radial hammer at the side of and on the same axis as the recoil arm. This hammer being actuated by its spring then flies forward and strikes an anvil in the end of the breech block casing, which transmits the blow to the firing pin. In addition to the above it is impossible for the cap to be struck before the breech is closed, since the cartridge is not fed into the breech by the breech block, and until the front of the breech block rises behind the cartridge to close the breech at a point $\frac{1}{2}$ inch before the forward movement is completed, the hammer, firing pin and cap are not in alignment.

(2) The barrel is joined at the breech end to the breech block casing which contains the hinged breech block, and these three members recoil together. The breech remains closed and locked for the first $\frac{1}{2}$ inch of the recoil movement, the front end of the breech block being then caused to rise, and to unmask the breech, by the action of its stud working in the groove in the switch plate.

B.—*Means by which the gun can be rendered safe from accidental discharge.*

This type of safety device is actuated by the firer, and when applied is intended to ensure that the gun cannot be fired accidentally, either during loading and unloading, remedying stoppages, etc., or at any other time when it may be required to render the gun temporarily inoperative. The following are some typical examples of the methods employed in different designs.

(a) Maxim type of action.—A catch which prevents movement of the firing lever, and which is applied automatically when the firing lever is released but has to be disengaged by the firer preliminary to firing.

(b) Lewis.—A notched "safety plate" which, when applied, retains the piston by fixing the position of the cocking handle.

(c) Hotchkiss (British model).—Control disc on the cocking handle which, when set in a certain position, prevents any movement of the trigger.

(d) Hotchkiss (Light model).—A lever embodied in the pistol grip which has to be pressed in order to render the trigger operative.

(e) Parabellum. Gast.—A supplementary safety trigger which, unless pressed, locks the main trigger.

## Feature 5.

### Cooling Systems

As already explained in the consideration of No. 7 of the conditions which an efficient automatic weapon should fulfil, the life of the barrel is to a large extent dependent on its temperature during firing. The maximum temperature which the barrel can reach is necessarily determined by the relative values of the following conditions :—

(a) The rate at which heat is absorbed.
(b) The heat capacity of the barrel.
(c) The rate at which the heat is lost by radiation and conduction.

The value of each of these conditions is governed by certain factors, which are inherent in the design of the gun and its ammunition, and these factors must receive careful consideration with due regard to the tactical needs which the gun is intended to satisfy.

These controlling factors are :—

Affecting (a).

(1) The propellent used.
(2) Lubrication or otherwise of the bore.
(3) The internal ballistics.
(4) The rate of fire.
(5) The number of rounds to be fired in any one burst.
(6) The intervals between bursts.
(7) The capacity of the magazine or conveyor.

Affecting (b).

The thickness of the walls of the barrel, it being assumed that all barrels are made of steel and that the specific heat is the same in all cases.

Affecting (c).

(1) The nature and area of the external radiating surface.
(2) The temperature and conductivity of the surrounding medium.

Having considered the factors which influence the temperature of the barrel during firing, automatic weapons can now be grouped under the systems employed in their design to control these factors.

Under this heading they fall into three main groups :—

(1) Those in which the barrel is simply exposed to the air.
(2) Those in which the barrel is cooled by an induced draught of air.
(3) Those in which the barrel is surrounded by a water jacket.

Arranged in these three main groups the different guns and their cooling systems are as follows :—

*Group*(1).

(a) Guns intended for use in aircraft. A thin barrel exposed to the rush of cold air made available by the speed of the machine.
(b) Recent light models of the Browning, Lewis, Hotchkiss, etc.—A thin barrel with no radiating rings.
(c) Colt, Madsen.*—A barrel of varying thickness but with numerous radiating rings.
(d) Hotchkiss (British* and French models).—A thick barrel with few, but large, radiating rings.

---

\* A second barrel always accompanies these guns in action, and their design enables a hot barrel to be quickly exchanged for a cool one.

*Group* 2.

*Lewis.*

In this cooling system a flanged radiator makes contact with the barrel throughout its entire length, the whole being enclosed in a cylindrical casing. The expansion of the propellent gases as they leave the muzzle is utilized to cause a draught of air to be drawn from the rear through the radiator casing between the flanges of the radiator. Cooling is also materially assisted by the large area of the fins and the high conductivity of the material (aluminium).

*Group* 3.

(*See previous remarks on* " *Condition* 6," *page* 156.)

Guns of the Maxim type, Schwarzlose, Browning, Revelli, Machine Guns.

*N.B.*—The advantages of this system of cooling are only applicable in the case of guns intended for sustained fire under ground service conditions.

The barrel for nearly the whole of its length is surrounded by water contained in a casing. To avoid the barrel becoming uncovered when the water level becomes low or when firing at extreme angles of elevation or depression, it is fitted as low as possible in the casing. To prevent leakages of water where the barrel passes through the ends of the casing, glands are necessary, and these are usually packed with asbestos fibre. A small air space is left in the top of the casing to give room for the expansion of the water if it should freeze, and also for the collection of steam on boiling.

Should extreme cold be anticipated, the freezing-point of the water can be lowered by the addition of some anti-freezing agent. Glycerine is generally used, but alcohol, sodium lactate, calcium chloride, or common salt solution will also serve. The use of any of these agents, however, is not without its disadvantages, the factors which have to be reckoned with being :—

(1) Possible corrosion of the metal either due to the agent itself or impurities.
(2) The effect on the mixture or solution of alternately heating and cooling it.
(3) The evolution of fumes when the temperature rises during firing, which may be dangerous if the gun is used in a confined space.

Continuous fire quickly causes the water to boil, about 600 rounds usually being sufficient, though this number naturally depends on the rate of fire and the original temperature of the water. Some way of escape for the steam generated is therefore necessary, and the usual method employed is to fit a pipe, with front and rear ports for the entry of the steam, along the top of the inside of the casing, one end of the pipe leading to a free exit through the casing wall.

To ensure that only the port above the water level is uncovered, and that steam, but not water, can escape at all angles of elevation or depression, a sleeve valve is fitted to the pipe which slides down by gravity to cover the lower port as the gun is elevated or depressed from the horizontal. This sleeve must be of a length not greater than the distance between the centres of the ports in the pipe, and not less than that required to avoid delay between the opening of one port and the closing of the other.

To assist concealment of the gun in action and to avoid as far as possible wastage of water, the escaping steam is led away by a flexible tube to some form of condenser.

Owing to the limited capacity of the casing, and the rate at which water is lost during firing in the form of steam, the water level sinks comparatively quickly, and if this is not attended to there is a danger of a portion of the barrel becoming uncovered and heating up excessively, particularly when firing at extreme angles of elevation or depression.

With most cooling systems of this type replenishment of the water becomes essential after about 2,000 rounds continuous fire.

## Feature 6,

### *Means of storing energy*

Whatever the method of operation, whether by recoil or gas, some of the energy derived from the expansion of the propellent gases must be conserved or stored in order that sufficient may be available to return the moving portions of the mechanism to their respective positions on firing, to feed up the succeeding round, and to close the breech. The amount so stored must be sufficient to ensure that these operations should be efficiently performed, irrespective of the influence of gravity, at all angles of elevation and depression. With a single exception the only method of storing this energy so far adopted is one which involves the use of a spring or springs, either by their compression, extension, or winding, during the backward movement of the mechanism, according to the type of spring used. The energy so stored is drawn upon by allowing the spring to reassert itself after the backward movement has been completed.

A hydraulic or pneumatic system or recuperator for doing this work could no doubt be devised, but the additional weight involved and other considerations would seem to make such a system impracticable.

The single exception referred to above is found in the Gast gun. In this design springs are dispensed with and the gun is operated by the alternate recoil of two reciprocating barrels, the recoil of the one being employed to perform those operations, in the case of the other, for which the energy is usually supplied by reacting springs. The employment of this unusual system of operation is, however, of doubtful advantage.

Where springs are used the choice of the type of spring, and the qualities which it must possess, depend on the mechanical features of the design of the gun. It is evident that resistance to shock and ability to maintain efficiency under conditions which involve considerable variations in temperature are important factors; in addition, the question of adjustability must receive careful consideration. This is necessary for two reasons. Firstly, the total resistance met with in the operation of the mechanism includes friction, the value of which for many reasons is liable to vary. Secondly, the available energy may also vary owing to such causes as the fouling of gas ports, variations in temperature, etc. Some balance between the amount of energy available to operate the mechanism, and the work which it has to do must be maintained. This can either be done by adjusting the springs or by regulating the actual amount of energy so available, for which in some designs provision is made. If from any cause the whole of the energy be not completely absorbed in overcoming inertia, friction, and the resistance of springs, the remainder will be received by the framework or body of the gun in the form of a shock, which will affect its stability. If the gun is under the physical control of the firer his aim will be affected and in any case accuracy will suffer. It is therefore evident that any adjustment for which provision is made in the design must be capable of being quickly and simply effected.

To avoid frequent adjustment of springs and regulators, some guns are fitted with shock-absorbers, of which a few examples are given below.

### *Pom-Pom (1 pdr. Q.F.), ·5 Browning*

In these guns the shock effect is minimized by incorporating in the mechanism an oil buffer, the piston of which is operated by the backward movement of the mechanism. In the Pom-Pom, an early heavy type of Maxim gun, this buffer also acts in the reverse direction and prevents a too violent forward movement, which would otherwise affect the stability of the gun.

### *Browning, Light and ·3-inch Models*

In the ·3-inch model the shock is received on a buffer plate. Behind this plate is a split brass cup into which fits a steel cone, and behind these again are a number of fibre discs, the whole being contained in a cylinder. When these component parts are forced together friction is introduced between the cup and cone, between the fibre discs and the walls of the cylinder, and between the exterior surface of the cup and the walls of the cylinder, thus producing a cushioning effect.

The same principle is adopted in the light model, with the modification that four pairs of truncated hollow metal cones, fitting into one another, are employed, the outer cone of each pair being split, and the place of the fibre disc is taken by a strong coil spring.

## FEATURE 7.

### Trigger and Firing Mechanism

The designs in this feature differ to a considerable degree in every type of machine gun, and their peculiarities are so numerous as to preclude their discussion except in principle.

In order to fire the first shot of a series, some part of the mechanism must be set in motion, and this motion must be checked in order to stop firing. This is usually effected by pressing and releasing a trigger or firing lever.

Continuity of fire is the primary object of an automatic gun, and this is controlled by means of the trigger and firing mechanism.

### Firing Mechanism

After the first shot of a series has been fired the trigger or firing lever must be kept pressed if the gun is to continue firing. For subsequent shots the mechanism of the gun operates independently and the forward projection of the striker or firing pin must, therefore, be timed automatically. This is usually effected in one of three ways:

1. Lewis light machine gun.—By attaching the striker to some member of the moving parts, the completion of whose forward movement is suitably timed.

2. Browning (light machine gun).—By employing a floating firing pin which is struck by a suitable portion of the mechanism on the completion of the forward movement.

3. Guns of the Maxim type.—By placing the firing pin in a moving frame called the lock, and providing a sear on which the firing pin is cocked, against the action of a spring, during the backward movement of the mechanism, and from which it is automatically released at the correct moment on the completion of the forward movement.

### Trigger Mechanism

1. Lewis, Browning light machine guns.—In these guns the trigger is connected with a pivoted sear, the nose of which holds back the piston until it is lowered by the partial rotation of the sear when the trigger is pressed. In the case of the Lewis the striker is mounted on the piston and in the Light Browning the forward movement of the piston causes a floating firing pin to be struck.

2. Madsen light machine gun.—This action is rather complicated, as it contains a double release, one by which the barrel is allowed to go forward after recoiling, the other by which the gun is fired. It is so arranged that the second release may be prevented after the firing of each shot, in order that single shots may be fired if required. For this purpose a thumb lever is provided which brings into operation an intermediate sear.

3. Beardmore-Farquhar light machine gun.—In the earlier design of this gun upwards of ten separate pieces of mechanism were employed in this connection. This large number was due partly to the round about way in which the release of the bolt had to be obtained, owing to the relative position of the trigger, and partly to the single shot action.

Simplicity of design and a small number of parts are very desirable in this feature, and from this point of view, the Lewis and Light Browning designs would seem to have an advantage over the others mentioned above.

## FEATURE 8.

### Extraction and Ejection of the Spent Case

The high chamber pressure developed on explosion causes the cartridge case to expand and grip the walls of the chamber. To render extraction easier, cartridges and chambers are coned, tapering towards the front, so that the slightest movement of the case to the rear will free it and make easy its complete withdrawal prior to ejection.

In all hand operated weapons, the first release of the case is effected by employing powerful leverage, this action being known as primary extraction. In automatic guns this first loosening of the case is not necessary, as extraction is effected by the gun itself, and plenty of power is available for it to be done in one movement. This, however, necessitates comparatively strong extractor jaws, since the case is virtually wrenched out of the chamber.

The usual methods of gripping the case are :—

1. By means of a single extractor or a pair, which are kept up to work by springs or by being made of spring steel.
2. By means of grooved slides which fit the base of the cartridge case.

Should the extractor or extractors rotate with the bolt or bolt head, one of the following two features will be found, to ensure that the round is retained under the control of the extractor's jaws :—

(a) A circular recess in the barrel face.
(b) A lip or projection extending outwards from the chamber.

If the extractor does not rotate, an inclined way is provided in the face of the breech, to enable it to ride away from the cartridge rim or groove. This prevents the extractor, under the influence of its spring, giving a side thrust to the cartridge as it lies in the chamber, any side thrust on the cartridge being detrimental to obturation. A single extractor is bad from this point of view.

The grooved slide or rising and falling extractor, as found in the guns of the Maxim types, is perhaps the best form. This type moves in a rectangular path, its action being as follows. The groove or extractor draws a round from the feed-block and holds it tightly while it is being carried back, lowered and carried forward into the chamber, where it remains ready for firing, while the extractor rises again to grip a new round in the feed-block. After firing, the spent case, still gripped by the grooves, is carried back, and when clear of the chamber, either falls off owing to its own weight, or is removed by more positive means.

Ejection should take place at such a point in the movement of the mechanism as to allow a bulleted round as well as a spent case to be cleared. The following methods have been adopted in the designs stated :—

(a) Guns of the Maxim type.—The case is either allowed to fall off the extractor by gravity, or is forced clear of the gun when the extractor rises on the completion of the forward movement of the block.

(b) Lewis, Parabellum.—The case is pushed off the face of the bolt by an ejector which is mechanically operated during the backward action.

(c) Hotchkiss, Revelli.—The case during the backward action comes into contact with a projection, which trips it out of the grip of the extractor.

(d) Madsen.—In this design the operations of extraction and ejection are both performed by one component, the action of which is as follows (see Fig. 6) :—

In the firing position the jaws (6) of the extractor (7) are out of engagement with the rim of the cartridge (8). On firing, the barrel (9), recoiling, takes with it the extractor and the pawl ejector (10). The lower end of the extractor coming into contact with the fixed ramp (11), causes the whole of it to rise, grip the rim of the case and to rotate, slightly withdrawing the case. This rotation is allowed for as the hole (12) for the axis pin (13) is enlarged as well as elongated.

To carry out ejection, the extractor must rotate further, and this is only possible by clearing the pawl ejector (10), which is effected by the stud riding up the ramp (15). When the pawl ejector is clear of the extractor, the lower part of the latter strikes the projection (16) the case then being ejected and deflected downward.

## Feature 9.

### Sighting

In order that fire may be brought to bear on any target the gun must be adjusted as follows :—

(a) For direction in the horizontal plane (*i.e.* for " line ").
(b) For angular elevation in the vertical plane, the angle of elevation depending on ballistic considerations.

The means provided for making these adjustments are described below.

(a) Adjustments for line are usually made with the aid of a foresight and backsight, which are attached to the gun with a certain definite distance between them, known as the sight radius.

The foresight is either of the blade or barleycorn pattern, so fitted to the muzzle end of the gun as to be adjustable laterally, for the correction of alignment during tests. For some guns, these sights are supplied in various heights, enabling the sighting to be adjusted for zero elevation.

The backsight is fitted with either a U or a V notch for use in conjunction with the blade and barleycorn foresight respectively, or with a circular aperture.

Should the distance between the backsight and the eye of the layer fall within a certain distance from the eye (about $10\frac{1}{2}$ inches with normal vision) a notch would appear blurred and would be difficult to use. The use of an aperture instead of a notch enables this difficulty to be overcome, and thus, in addition to certain other advantages, in certain cases enables a longer sighting radius to be employed. On service guns of the U.S.A., apertures of different sizes are made available by means of an adjustable disc.

(b) The angle required for a given range is obtained by raising the notch or aperture of the backsight through the necessary angular distance, the correct setting being obtained by moving an indicator (usually called the slide) on the stem of the sight, which is suitably graduated in units of measure. The sight itself may be designed on the tangent, radial, or arc principle. After setting the sight the gun is laid with the aid of the foresight.

The longer the sighting radius of these sights, the less the chance of small personal errors of the layer materially affecting the position of the mean point of impact.

Guns mounted in aircraft are not fitted with sights that can be set for varying elevations. A fixed setting is all that is needed, as their objectives are engaged at short range. In some 'planes a small or unit power telescope is fixed close to the seat of the pilot, who by directing his line of flight so that his target appears in the optical centre of the telescope thereby places his objective in the cone of fire.

Any lateral correction which may be necessary to counteract the effect of " drift " is generally ignored in the design of sights for automatic guns, but a correction is so made automatically, in the case of certain service guns of the U.S.A., in the setting of the backsight. In these guns the slide of the backsight moves up and down in an inclined slot, the angle which this slot makes with the vertical providing for an average drift correction.

The above forms of sight are in universal use, but, in addition, certain special sights, aids to adjustment, and accessories are met with.

1. Telescopic sights.—Optical sights of unit power or having a low magnification, of which the object is to increase the degree of accuracy of which the layer is capable.

2. Battlesights.—Fixed backsights giving an elevation suitable for use up to a certain limited range, and intended to ensure that the gun shall always be able to engage a target within that range without time being lost in the adjustment of the ordinary backsight.

3. Night sights.—Auxiliary sights which can be fitted to the existing fore and backsights in order to enable the gun to be laid at night. They are also used in conjunction with 4 (below).

4. Bar foresights.—These fittings are used for laying on aiming marks, etc., and also facilitate the use of switch angles when a change from one target to another is required. They consist of a graduated bar, fitted to the front end of the gun, on which a movable auxiliary foresight can be set to give definite angles of lateral deflection.

5. Direction dials.—These are fitted to the mounting and are graduated in degrees of angle. By the aid of a pointer, the gun may be swung through any desired angle from a given reference point.

6. Clinometers.—Adjustable spirit levels for giving the quadrant angle of elevation to the gun in indirect fire.

7. Clicking indicators.—Clicking devices incorporated in the elevation and traversing gears which give a lift or traverse of a small angle, ensuring regularity in searching and sweeping.

8. Windgauges.—A means of making some lateral allowance in degrees of angle for the effect of side wind on the bullet during flight. Wind gauges are found in the Schwarzlose, 0·30-inch (recoil-operated) Browning, and Colt machine guns. The first of these can be set for any allowance up to 3° right or left, while in the other two provision is made for any setting up to 20° right or left; in the latter the windgauge also fulfils the same purpose as the bar foresight described at (4) above.

9. Dial Sights.—(*See* Text Book of Gun carriages and Gun mountings, 1924, p. 224.)

These consist of a self-contained sight which, when set to the angle between the aiming point and the target, enables the correct line of fire to be obtained by laying on the aiming point. In more recent designs provision is made for giving both elevation and deflection to the gun. These sights are fitted to all mobile ordnance and are primarily intended for use in connection with indirect fire. When used on machine guns they are fitted to the gun itself or to some part of the mounting which moves with the gun. They were fitted to some machine guns used by the Germans in the late war, but their liability to damage through rough usage and their heavy cost are factors unfavourable to their general adoption.

10. Aiming Posts, Aiming Marks, Zero Posts.—These aids to laying are also made use of in indirect fire. They enable direction and elevation to be maintained when other methods are too slow or otherwise rendered impossible, *e.g.*, through smoke, dust, etc. Special lighting arrangements can be contained in or attached to the posts for night firing.

11. Anti-Aircraft sights.—These are sights specially designed for use on the ground against aircraft. They enable allowance to be made for the speed of travel of the target relative to the line of fire, and also enable the correct elevation for the range and altitude of the target to be given to the gun.

12. Aircraft Machine Gun sights.—These sights are designed for use in aircraft and allow for :—

(a) The speed of travel of both gun and target.
(b) The direction of the target's flight relative to the direction of fire.
(c) The direction of fire relative to the line of flight of the firer.

## Feature 10.

### Means of Controlling the Operation of the Mechanism

The mechanism should be so designed that by the setting of a simple contrivance, it it possible to arrange that, when the trigger or firing lever is pressed, the gun

(a) will continue to fire until the magazine or conveyor is emptied or the trigger is released.

(b) will only fire single shots, so that the trigger has to be released and pressed again for each subsequent shot.

(c) will not fire.

The mechanical means by which (a), (b) and (c) above are effected are respectively as follows:—

(a) By arranging that the pressure on the trigger or firing lever keeps the sear and bent (or their equivalents) out of engagement so long as it is maintained.

(b) By the use of tripped triggers or sears which must be allowed to re-engage before another shot can be fired.

(c) By locking the trigger or firing mechanism.

If such a device is not incorporated it may be possible to arrange for (b) by loading in a special manner or by spacing the rounds in the conveyor.

---

## Feature 11.

### Regulation of the Rate of Fire

In No. 11 of the conditions which an efficient automatic gun should fulfil, it was stated that some means of regulating the rate of fire was desirable. The design should allow of adjustment within limits being made by the firer, in order that the rate of fire may be varied to suit special conditions, as, otherwise, if only the maximum rate is available at all times, ammunition may be wasted and barrel wear will be greater. It is essential that this varying of the rate of fire should not affect the accuracy of the gun to an appreciable extent, though some alteration in the actual position of the mean point of impact cannot be avoided, since no gun or mounting can be perfectly rigid, and consequently the value of " jump " will vary.

It is possible to arrange for regulation by providing a means of controlling the speed at which the reciprocating portions of the mechanism can move, and this is the method employed in the following earlier devices :—

1. Maxim's patent of 1885, No. 1307.—In this the breech closing is retarded by a forward moving piston in a closed cylinder, liquid having to be displaced and forced through an adjustable opening in the piston.

2. Bloxham's patent of 1906, No. 27422, which was an integral part of the Maxim trigger mechanism. Its action depended on the displacement of liquid through an adjustable port in a piston, worked by the forward movement of the recoiling parts. This gear could be thrown in or out of action by a rotating member fitted to the firing lever.

3. Patent office specification 1907/13194.—In this case a fly-wheel was incorporated in the mechanism and was operated by the recoil. Regulation was effected by altering the weight of the flywheel by the addition or removal of weights.

None of the above has proved of definite practical value, but more recently this feature has been successfully incorporated in the following designs :—

1. The St. Etienne machine gun.
2. The ·5-inch Browning.
3. The Hotchkiss (recent light model).
4. The ·5-inch Vickers (Ground service model).

1. In this gun the backward movement of the mechanism causes a plunger to be driven into a cylinder containing liquid, which is thereby displaced. As the plunger is driven in a catch engages with the lock and delays its forward movement while the plunger is withdrawn and the liquid returns by way of a small channel to its original position. The size of this channel, and therefore the rate of fire, can be regulated by means of a valve which is under the control of the firer.

2. The principle of this device is similar to that mentioned in 1, and it also acts as a shock absorber. Regulation is effected by setting a pointer which indicates on a disc the approximate rate of fire obtained with a particular setting.

In 1 and 2 (above) the rate of fire is controlled by the firer, but in 3 and 4 (below), the regulating device merely serves to check the action of the mechanism and to lower the maximum rate of fire of which the gun would be capable were no such device fitted.

3. In this case the piston during its backward movement operates the lever of an escapement which times the duration of the engagement of a co-related sear with the piston at the end of the movement.

4. The primary object of the arrangement which reduces the rate of fire in this gun is to ensure positive feed. A pawl is fitted to the top cover of the breech casing, which, when the recoiling portions are fully back, retain the crank, and consequently the lock, until such time as the barrel has moved forward and the feed-block action has been thereby completed. In the Air Service model this device is not incorporated, as rapidity of fire is essential.

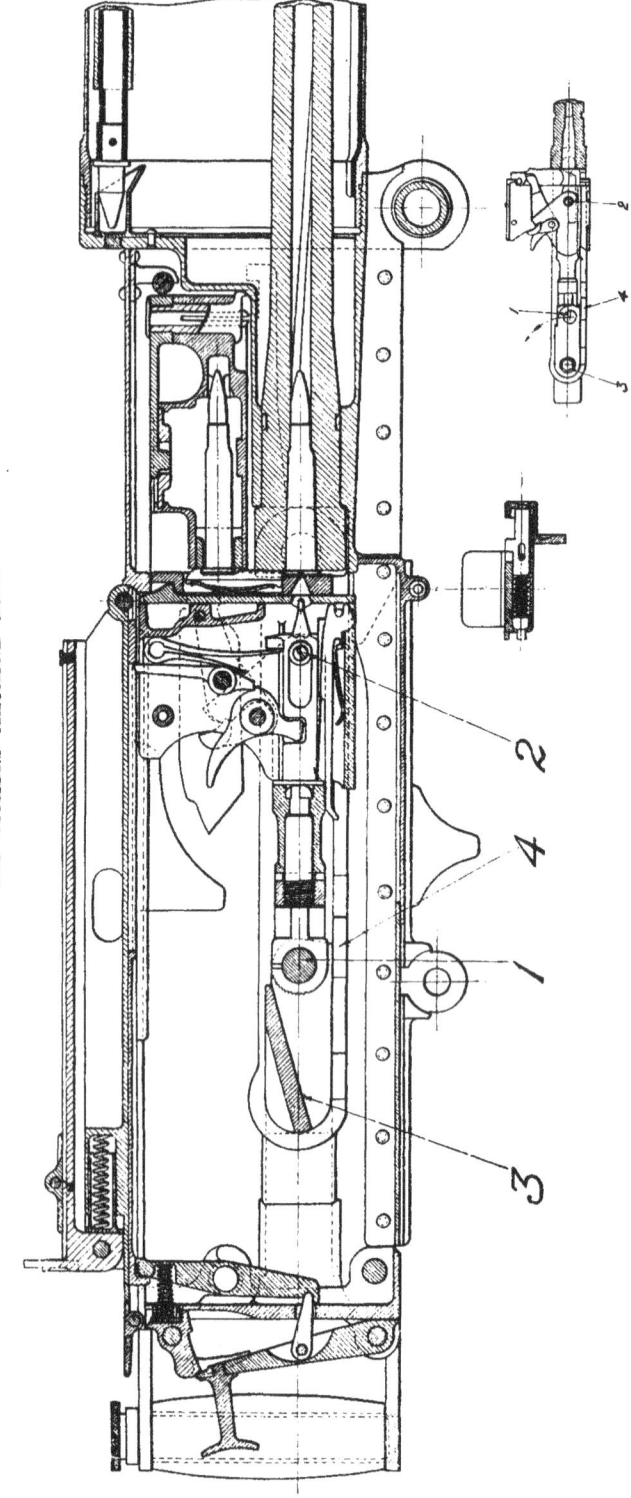

Part I, Chap. V, Fig. 1.
·303 Vickers Machine Gun.

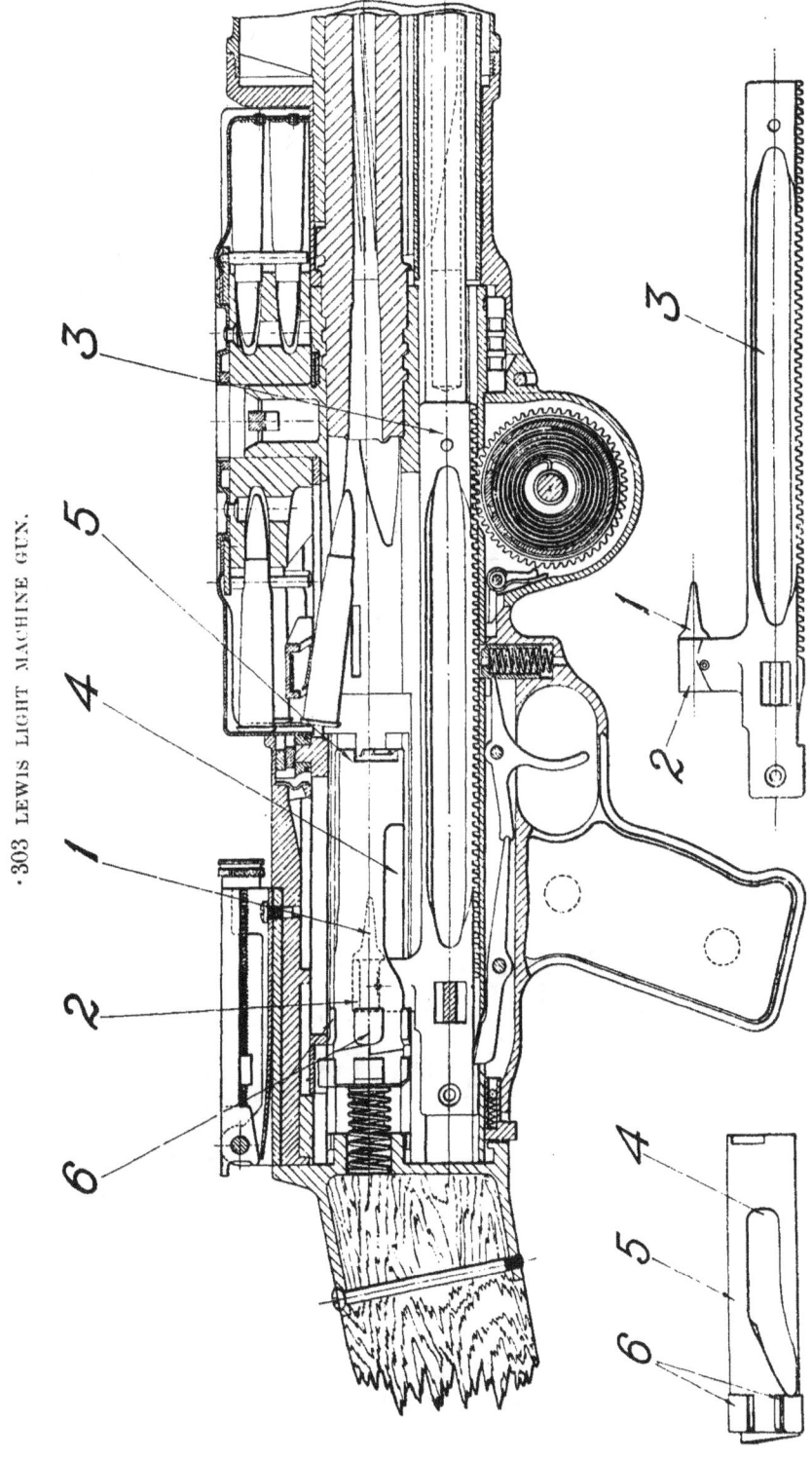

Part I, Chap. V, Fig. 2.

.303 Lewis light machine gun.

Part I, Chap. V, Fig. 3.

·303 HOTCHKISS LIGHT MACHINE GUN.

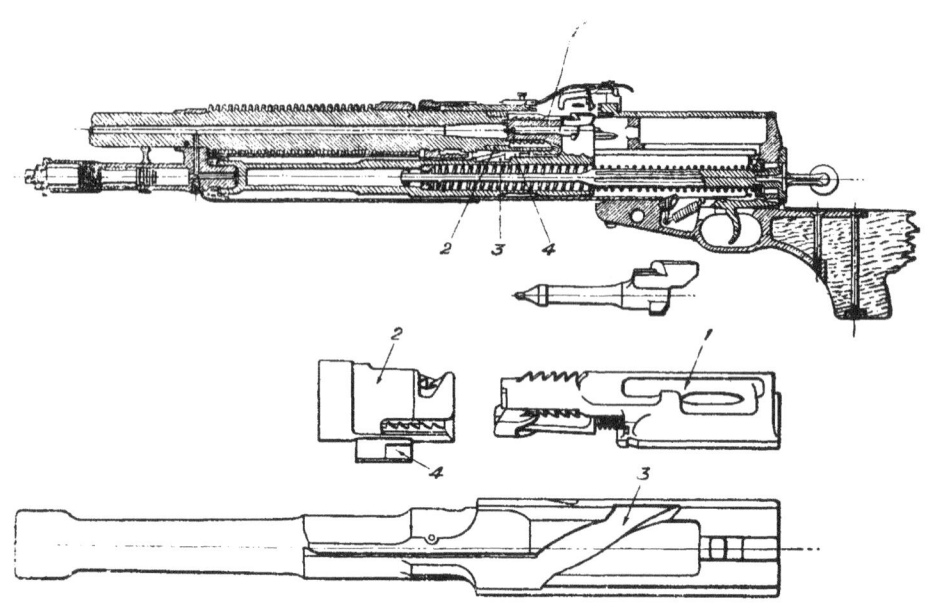

Part I, Chap. V, Fig. 4(a).

HOTCHKISS MACHINE GUN, MODEL 1914.

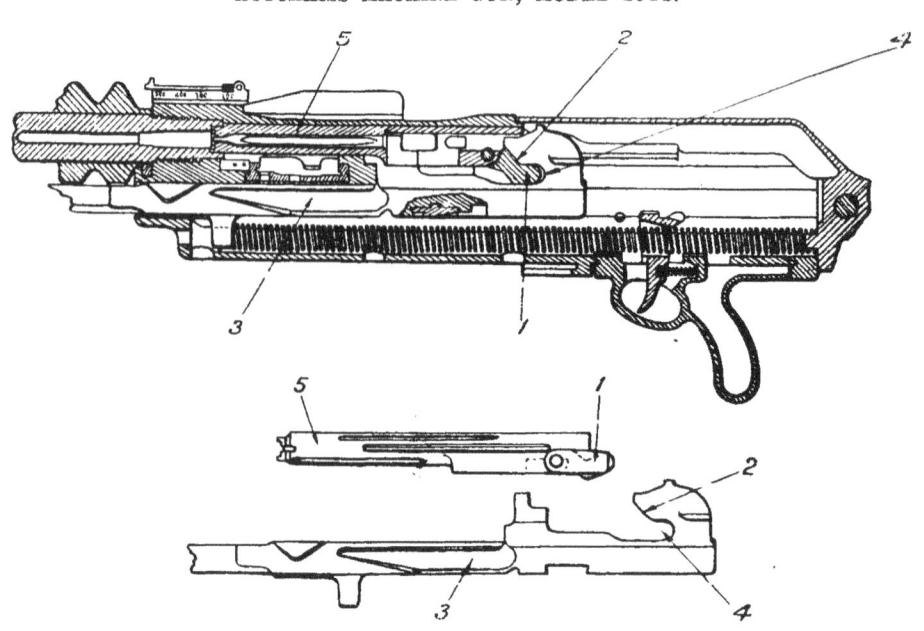

Part I, Chap. V, Fig. 4 (b).

BROWNING LIGHT MACHINE GUN.

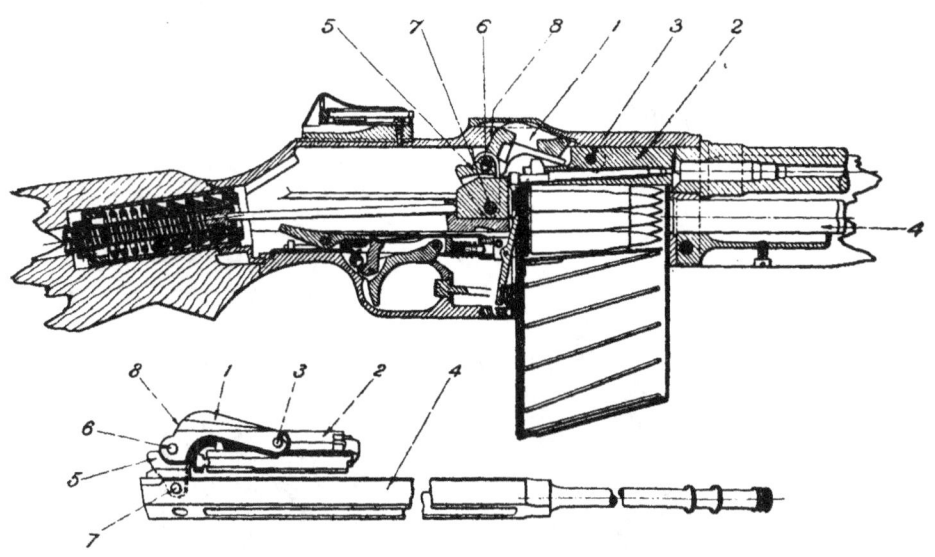

Part I, Chap. V, Fig. 5.

ST. ETIENNE MACHINE GUN.

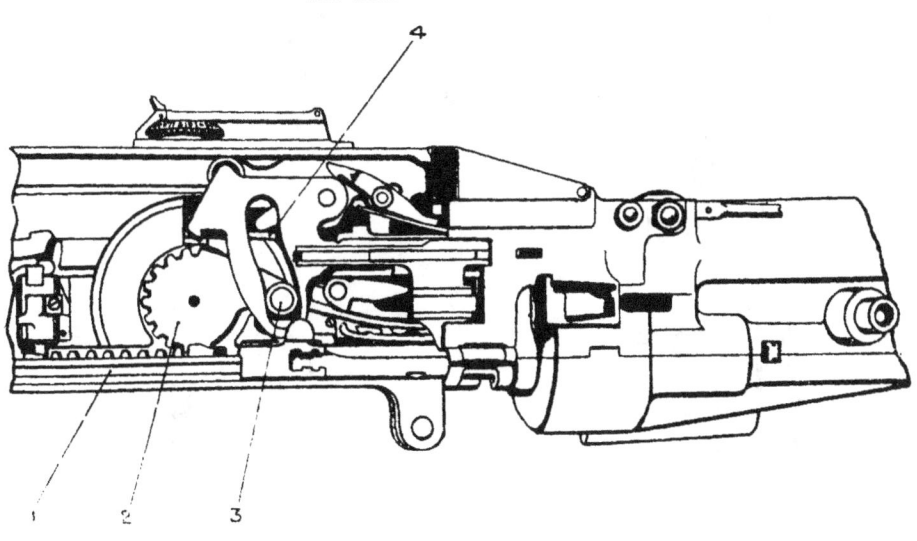

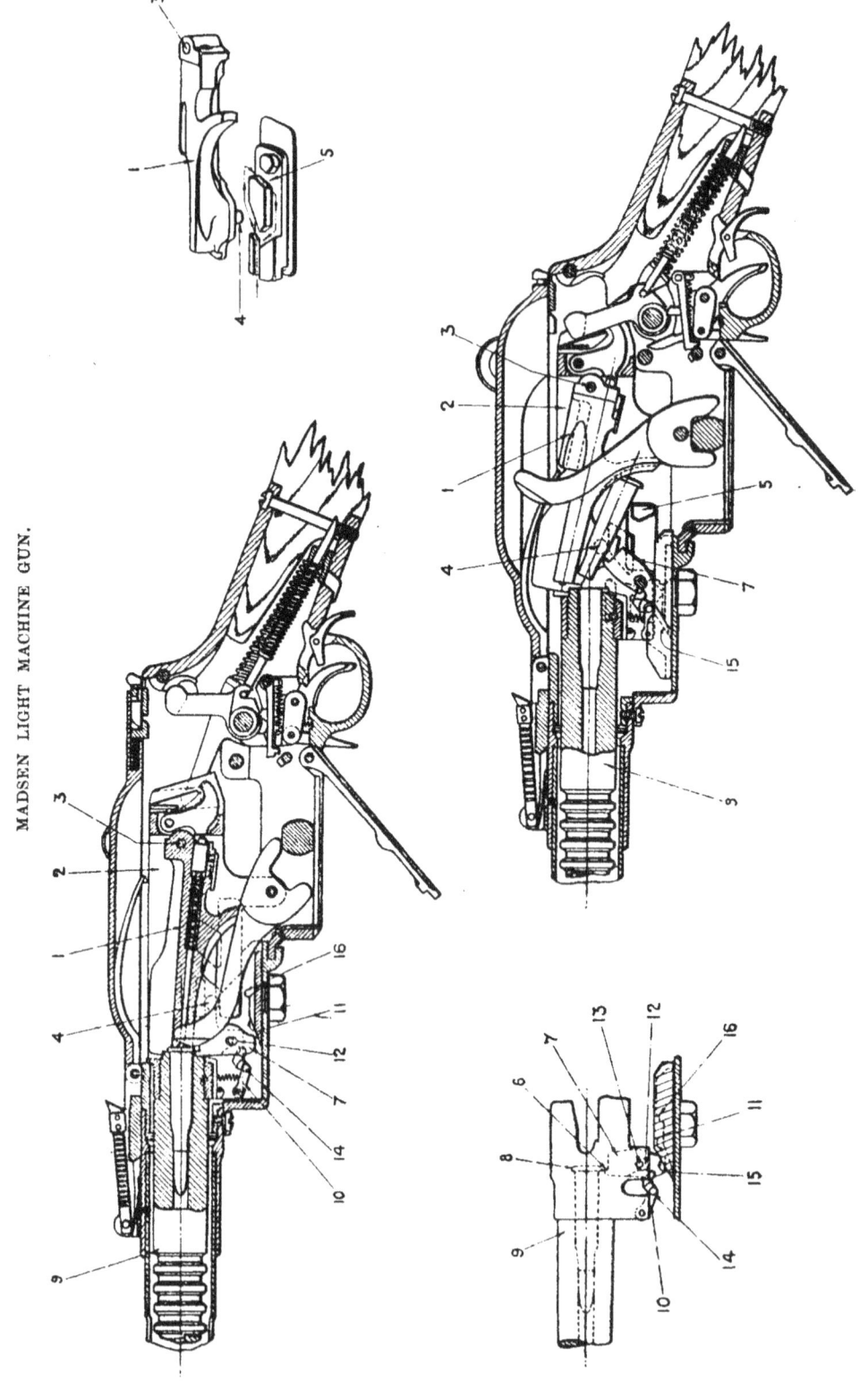

Part I, Chap. V, Fig. 6.
Madsen Light Machine Gun.

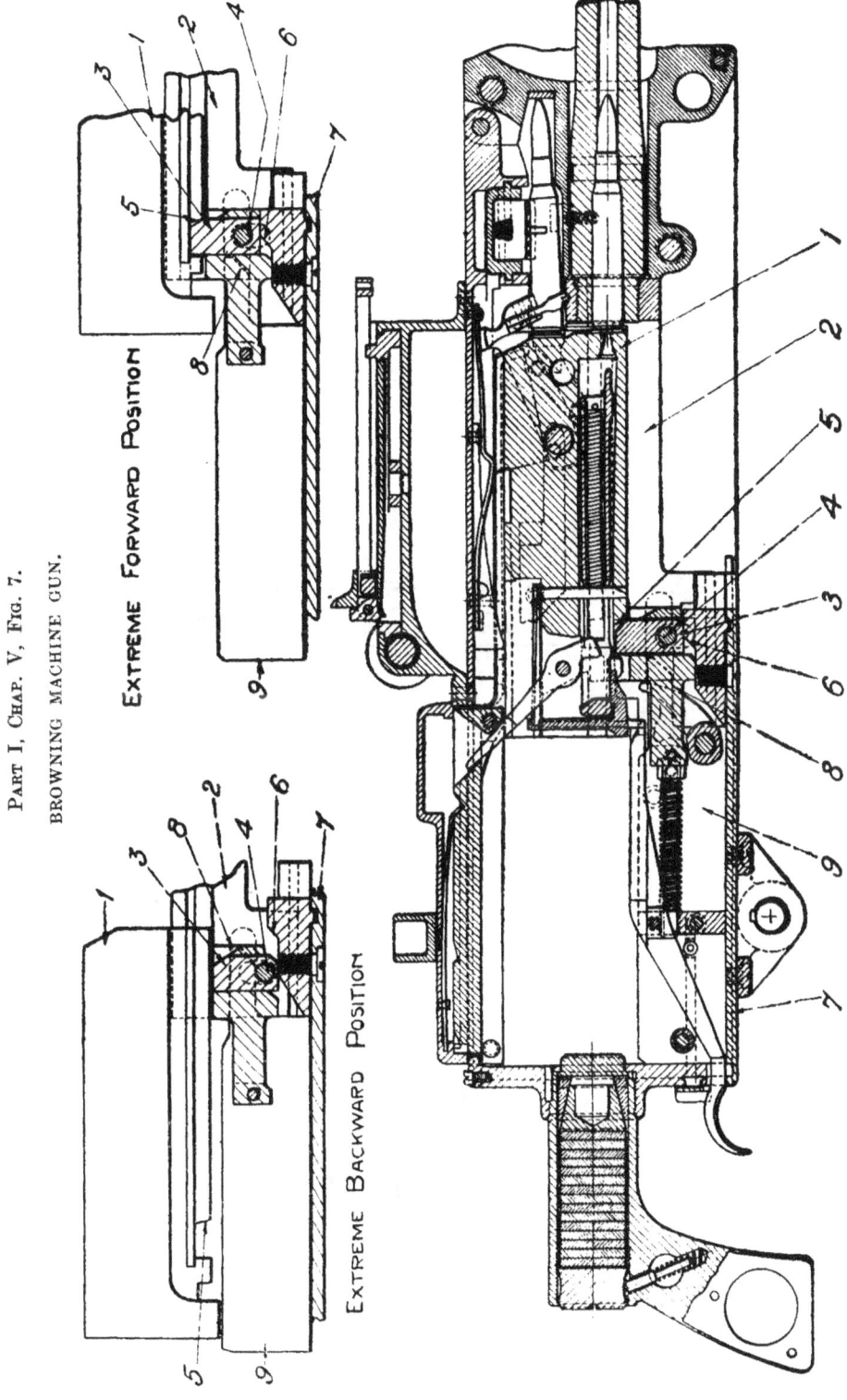

Part I, Chap. V, Fig. 7.
BROWNING MACHINE GUN.

PART I, CHAP. V, FIG 8.

COLT MACHINE GUN.

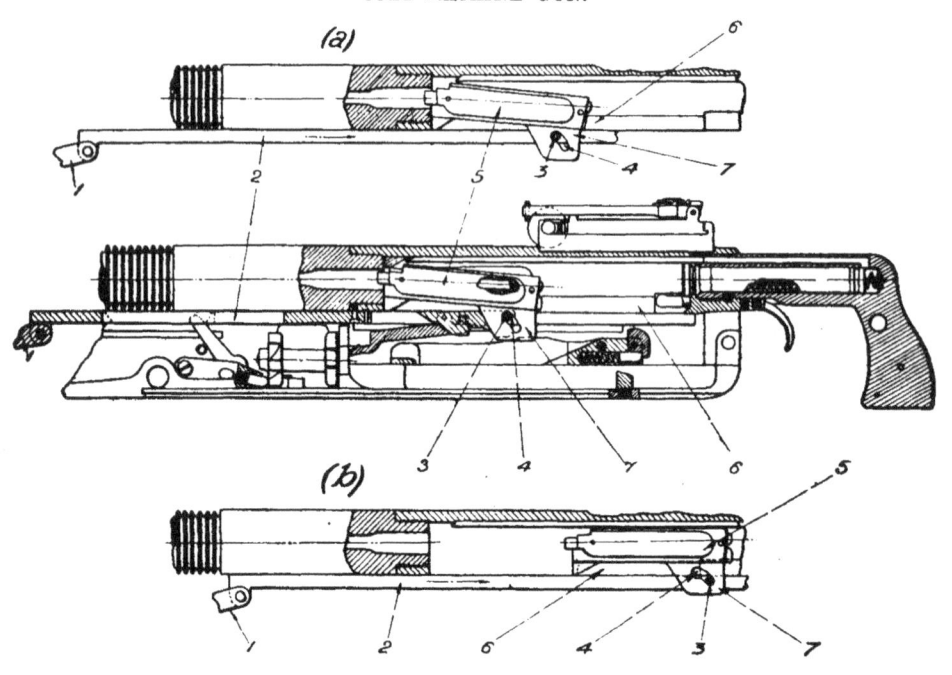

PART I, CHAP. V, FIG. 9.

GAS AND LIGHT MACHINE GUN.

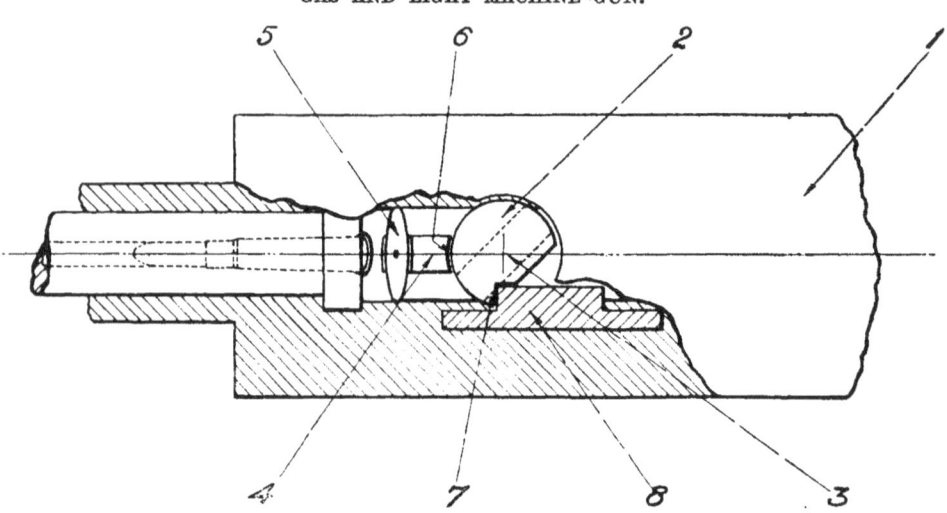

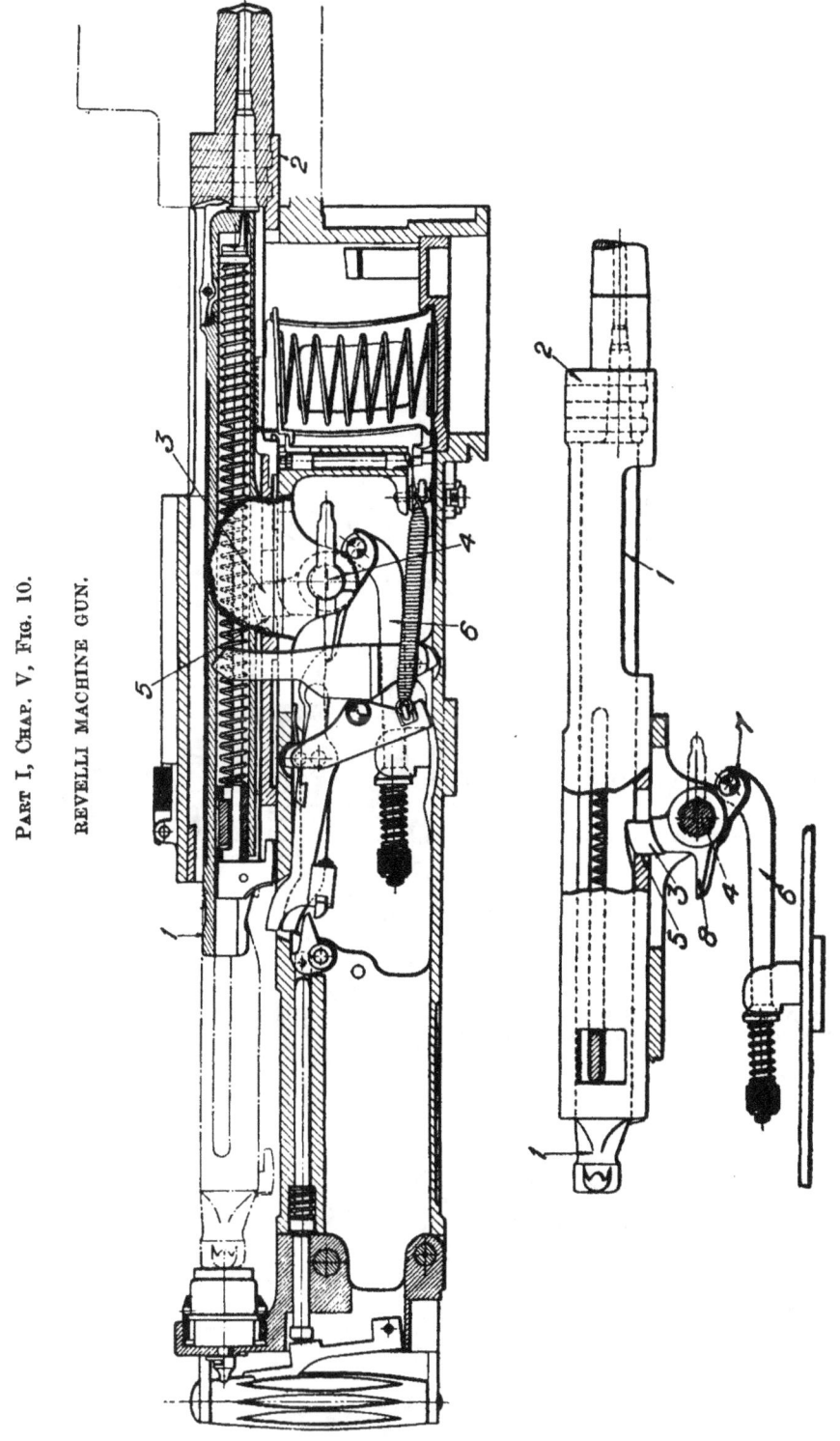

PART I, CHAP. V, FIG. 10.
REVELLI MACHINE GUN.

PART I, CHAP. V, FIG. 11.
THOMPSON SUB-MACHINE GUN.

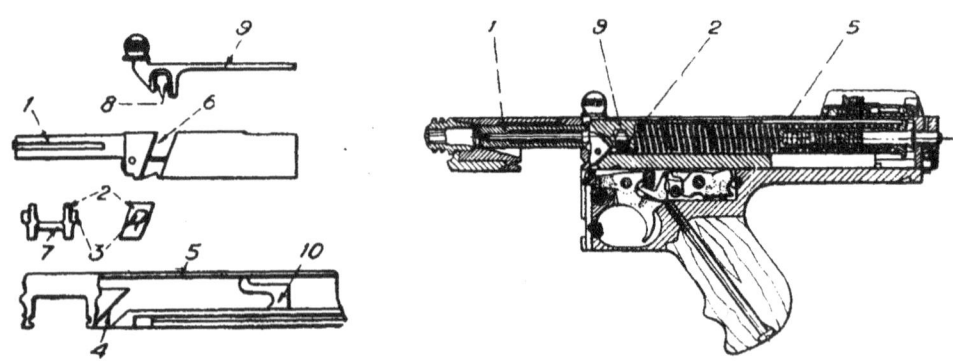

PART I, CHAP. V, FIG. 12.
SCHWARZLOSE MACHINE GUN.

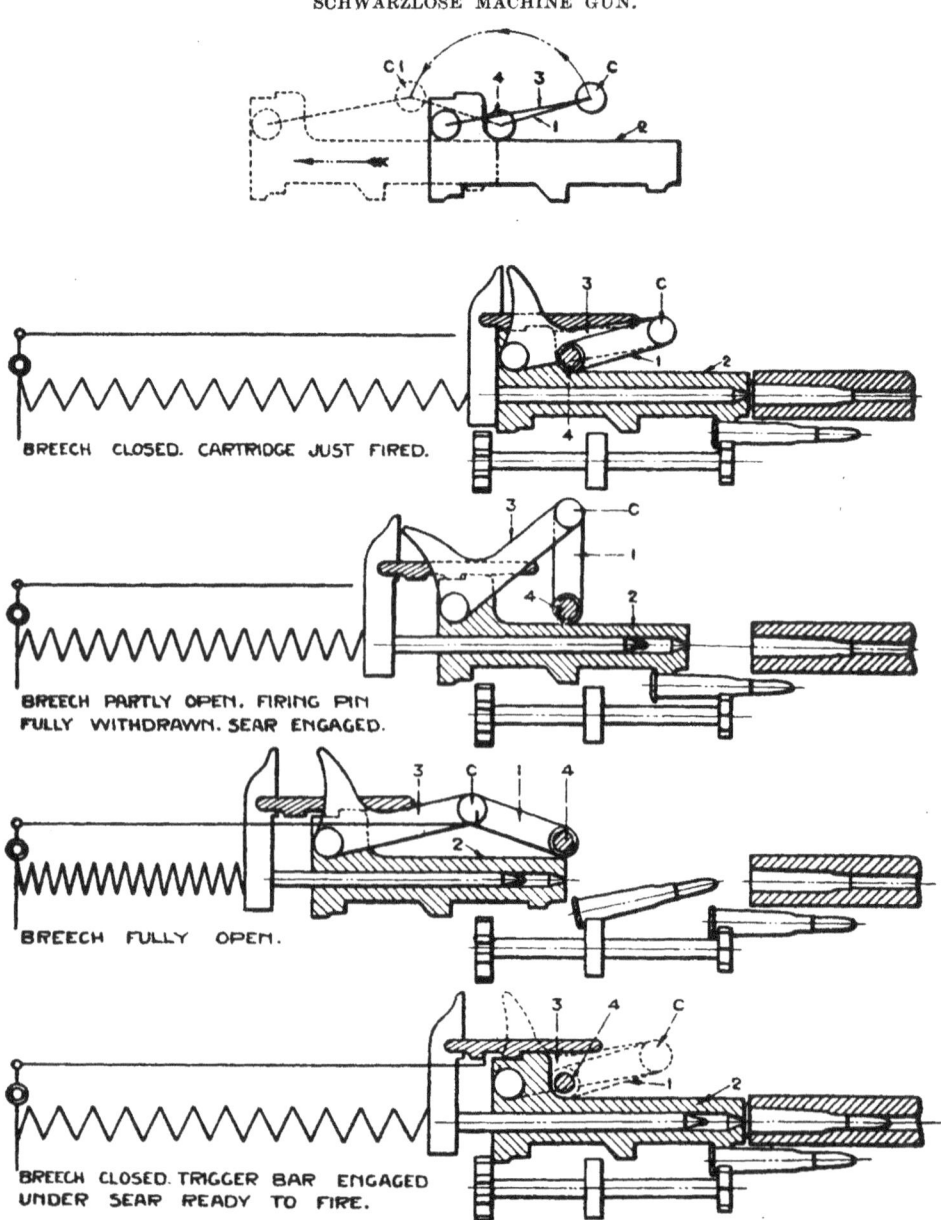

BREECH CLOSED. CARTRIDGE JUST FIRED.

BREECH PARTLY OPEN. FIRING PIN FULLY WITHDRAWN. SEAR ENGAGED.

BREECH FULLY OPEN.

BREECH CLOSED. TRIGGER BAR ENGAGED UNDER SEAR READY TO FIRE.

[*To face page* 192.

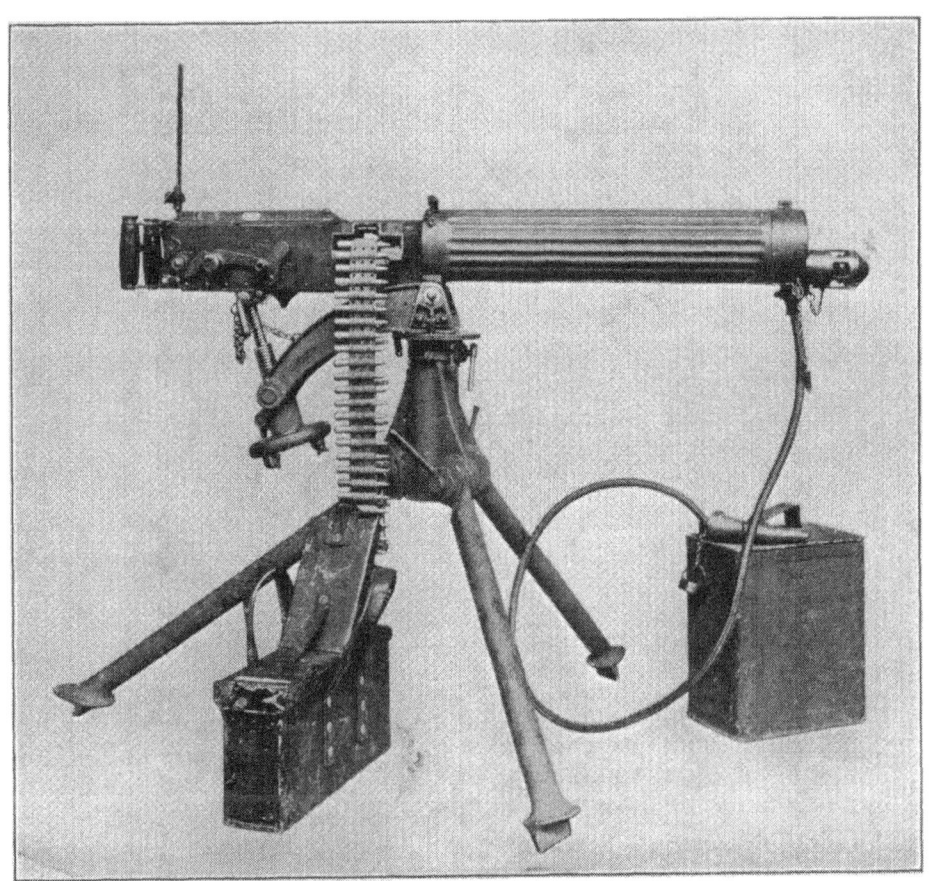

Part I, Chap. V, Plate I.
·303 Vickers Machine Gun (Great Britain).

Part I, Chap. V, Plate II.
Lewis Light Machine Gun (Great Britain).

PART I, CHAP. V, PLATE III.
HOTCHKISS LIGHT MACHINE GUN (GREAT BRITAIN).

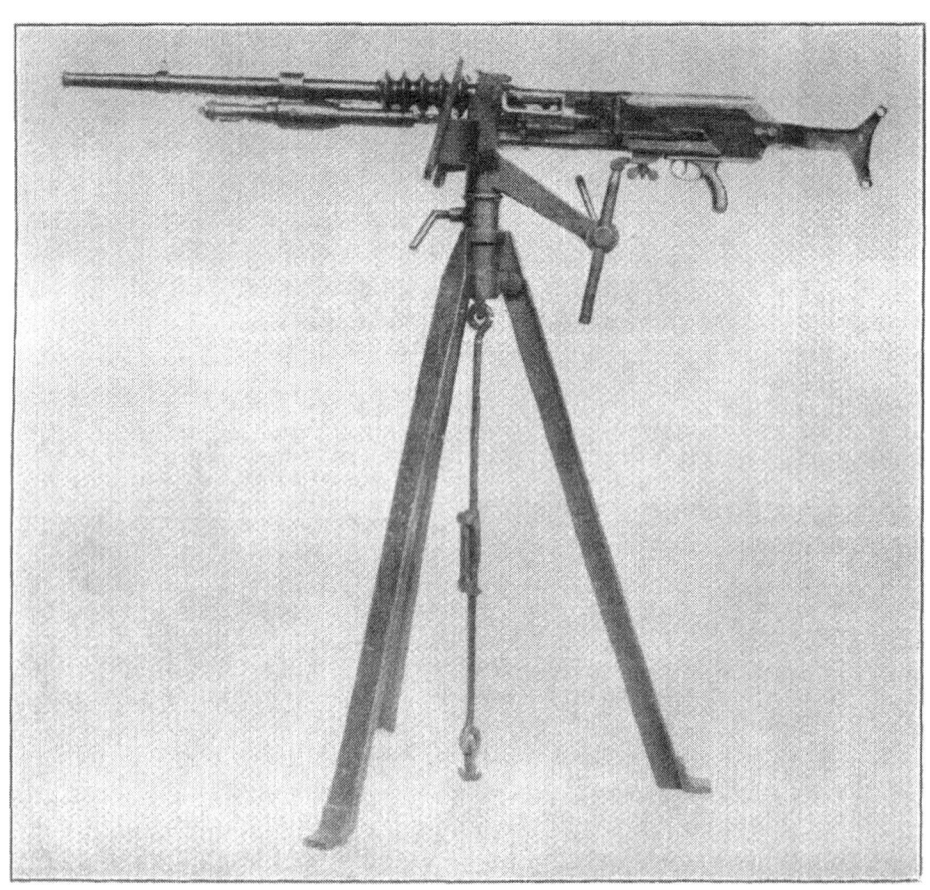

PART I, CHAP. V, PLATE IV (a).
HOTCHKISS MACHINE GUN, MODEL 1914 (FRANCE). (EXTEMPORISED MOUNTING FOR TEST PURPOSES.)

Part I, Chap. V, Plate IV (b).
Browning Light Machine Gun (U.S.A.).

Part I, Chap. V, Plate V.
St. Etienne Machine Gun (France).

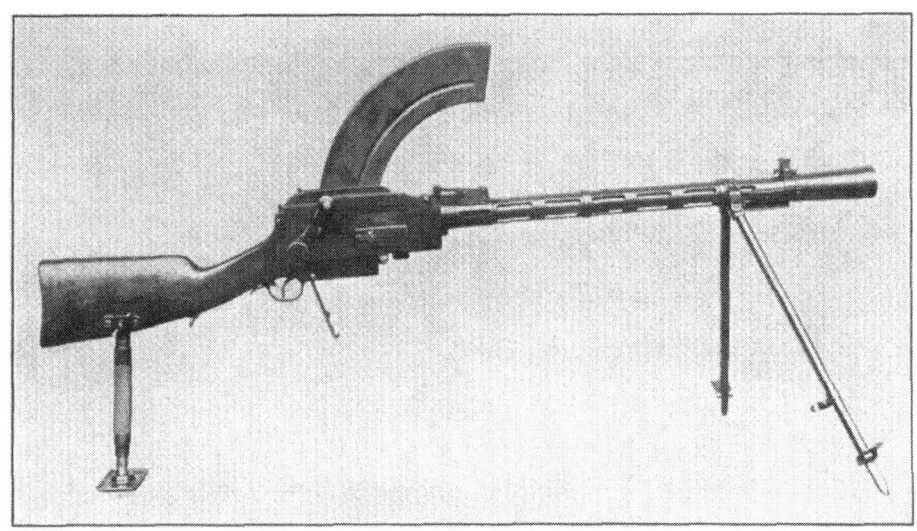

Part I, Chap. V, Plate VI.
Madsen Light Machine Gun (Denmark).

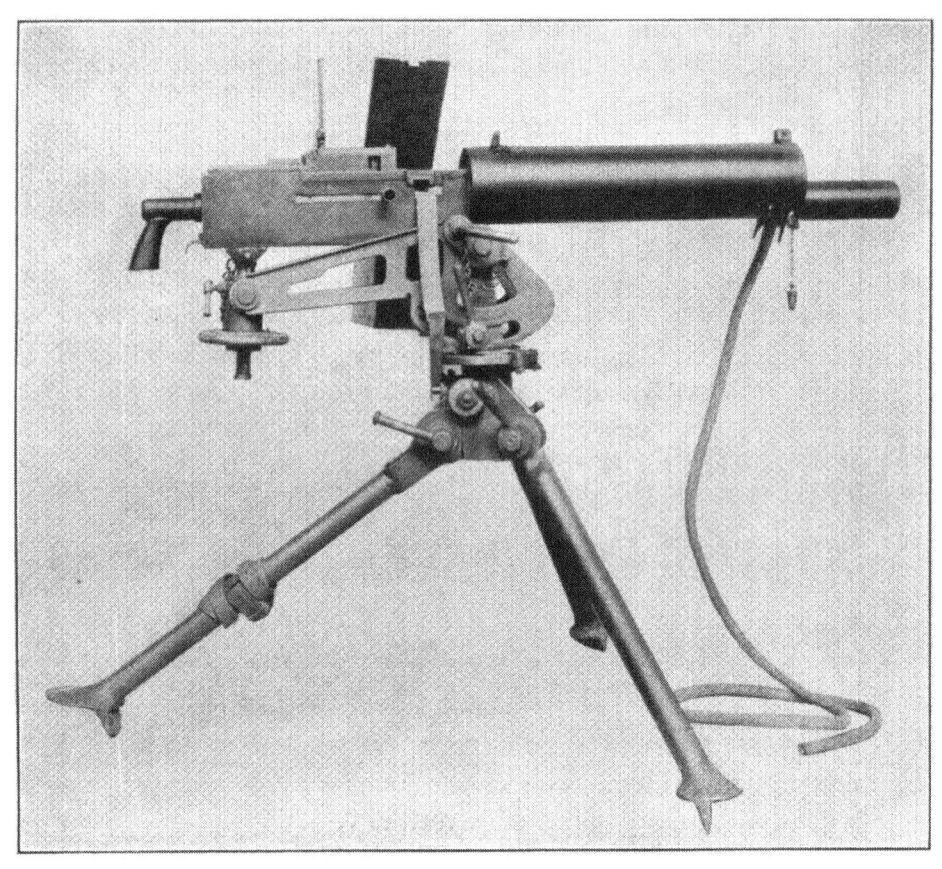

Part I, Chap. V, Plate VII.
Browning Machine Gun (U.S.A.).

Part I, Chap. V, Plate VIII.
Colt Machine Gun.

Part I, Chap. V, Plate IX.
Gast Light Machine Gun (Germany).

Part I, Chap. V, Plate X.
Revelli Machine Gun (Italy).

[*To face page* 193.

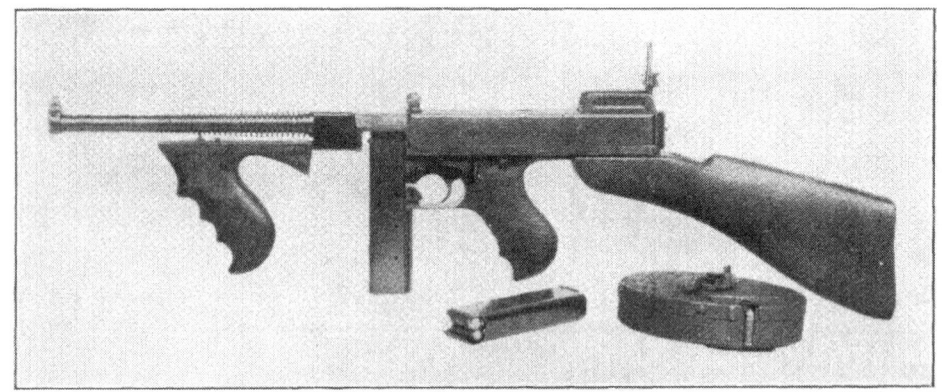

Part I, Chap. V, Plate XI.
Thompson Sub-Machine Gun.

Part I, Chap. V, Plate XII.
Schwarzlose Machine Gun (Austria).

## SECTION IV

## TESTS AND ADJUSTMENTS

If the efficiency of a gun is to be maintained under service conditions, certain parts must be made adjustable and replaceable in order that wear and damage may be made good. Certain prescribed tests must be applied to the gun periodically in order that wear and damage may be detected, and also, after the gun has been repaired or adjusted, to determine whether the repair or adjustment has been effective.

The tests usually carried out in service enable the state of the gun to be determined as regards the following details of its mechanism and performance :—

1. The protrusion and condition of the striker or firing pin.
2. The strength of springs.
3. The correct action of various operations performed by the mechanism, particularly as regards safety arrangements, locking or closing of the breech action, and the action of the firing mechanism.
4. The amount of cartridge head space.
5. The accuracy and sighting.
6. The state of the barrel.
7. The operations of extraction and ejection.
8. The efficiency of the feed arrangements.
9. The amount of friction between sliding surfaces met with in the operation of the mechanism.

Other features peculiar to the particular type of gun, and not covered by the above, necessitate special tests.

These tests will now be considered in detail, and the methods of adjustment, where this is possible, described with the test concerned.

### Test No. 1.—*For Protrusion, etc., of the Striker*

The distance the striker or firing pin point protrudes through the face of the bolt, or its equivalent, when striking the cap of the cartridge, together with the shape of the point, must be carefully inspected. Should the protrusion be insufficient or the point blunt, the cap will not be exploded. If on the other hand the protrusion is excessive or the point too sharp, the cap is likely to be pierced, and the consequent escape of gas at high pressure and temperature would have a detrimental effect on any metal with which it came into contact.

An effect similar to that produced by insufficient protrusion may be obtained, although the protrusion be correct, if the head space for the cartridge is excessive or the cartridge caps are deep set. Excessive protrusion may be caused by the effect of the continued hammering, during firing, on those parts which limit the forward movement of the striker or firing pin.

Method of Testing.—The normal method is by the use of high and low gauges whose limits differ slightly for each type of gun. This test can also be carried out by filling up the cap cavity of a gauge similar to a dummy cartridge with wax or some similar composition, inserting it in the chamber, and causing it to be struck in the manner usual on firing. The shape, depth, and position of the impression made by the striker is then examined.

Adjustments.—Firing pins in the case of automatic weapons are generally interchangeable and adjustments in the British Service are very rarely resorted to. In cases of faulty protrusion the substitution of another, serviceable, firing pin is usually sufficient, while if this fails the cause will usually be found to be in the wear or distortion of some other component effecting protrusion, and this component in its turn can be replaced by one which is serviceable.

*Test No. 2 for Strength of Springs*

The strength of a spring is always tested in service when it is in a state of readiness to commence work, *e.g.*, lock springs are tested when fully under compression and with the striker or firing pin in the cocked position.

The working life of a spring is considerably shortened should it be left in a state of tension or compression for any length of time. Too much stress cannot be laid upon the importance of releasing springs as far as possible when not in use.

When testing springs in the Service, a spring balance is used to register the amount of force exerted at the moment of testing. No particular test or the method of carrying it out can be mentioned here, as all springs vary in shape and act in different ways. Care must be exercised that in all cases the particular instructions given for the test are adhered to.

In some cases no method of testing springs in service is laid down, but before they are accepted for issue the manufacturing specifications demand that all steel for springs should be subjected to test by analysis. Also, after manufacture, all springs are subjected to a severe mechanical test, after which they must recover and return, within small limits, to their original shape.

*Test No. 3 for " Safety and Control "*

(1) The safety devices that are tested here are :—

(a) Those that act automatically.
(b) Those which have to be applied.

As regards (a) there is little likelihood of failure, and should this occur other faults would also be apparent. As regards (b) careful inspection is necessary to ensure that their application renders the gun safe under all conditions, even if subjected to a severe jar or fall when loaded.

(2) Testing for control implies ascertaining whether the action of the mechanism is correctly timed so that locking or breech closure is effected before a round can be fired, and that the firing mechanism, etc., operates at the correct moment.

Method of testing.

(1) The gun is loaded with dummies and operated by hand, the efficiency of (a) and (b) being checked by trial and inspection.
(2) Special tests are laid down for each type of gun, which include means of determining whether all the stops which limit the movement of rotating or straight components, give even contact and receive and distribute evenly any force which may be transmitted to them.

Adjustments.

Adjustments where permissible under this heading should only be carried out by a skilled armourer, and care must be taken that the adjustment of other portions of the mechanism is not thereby affected. The usual remedy is the substitution of serviceable for unserviceable parts.

In the case of weapons of which the components are interchangeable care must be taken that any adjustments made do not affect interchangeability.

*Test No. 4 for Cartridge Head Space*

1. With Rimmed Cases.—This type of case is positioned in the chamber by its rim butting on to the face of the breech, and in order to ensure easy closing of the breech the space between the face of the bolt (or its equivalent), when fully forward, and the face of the breech must not be less than the greatest thickness of rim likely to be encountered in any cartridge.

This space, known as cartridge head space, must, however, be limited, since if it is greater than the thickness of the rim, the case will be forced back on to the bolt face on firing, and that portion of it immediately in front of the rim will be thereby removed from the support of the walls of the chamber. If the extent of this unsupported portion is excessive the case may burst.

With the British Service cartridge, as issued, no case with a thickness of rim more than $\cdot 064$ inch, or less than $\cdot 058$ inch, is likely to be found, these being the high and low limits of the manufacturing tolerances. Gauges with rims $\cdot 064$ inch in thickness form the low limit when testing cartridge head space. Actions which close freely over this gauge will take all service cartridges. The amount of tolerance beyond this depends on several factors, and experiments have proved that the set back or case movement should not exceed $\cdot 02$ inch, otherwise separations may occur.

The maximum amount of cartridge head space allowable has been determined as $\cdot 074$ inch, and gauges of this dimension are used by armourers in carrying out this test. Even a cartridge with a rim thickness of $\cdot 058$ inch is then unlikely to receive insufficient support, as the set back is only $\cdot 074$ inch $- \cdot 058$ inch $= \cdot 016$ inch.

2. With Rimless Cases.—To ascertain the correct testing tolerance in cartridge head space in cartridges of the rimless type is more difficult than in the case of rimmed cartridges.

In this case the cartridge is positioned by its forward shoulder butting on the shoulder in the chamber, and consequently gauging must be from the shoulder in the chamber to the face of the breech action, assuming that the actual chamber length is correct. The magnitude of the variations in the distance between the forward shoulder of the cartridge and its base will naturally tend to be greater than the magnitude of the variations in the thickness of rim of rimmed cartridges. This factor and the amount of "crush up" which can occur with this type of cartridge, affect the degree of tolerance which must be allowed in the cartridge head space.

Method of Testing.—The points which require particular attention in carrying out this test, with the aid of the gauges described above, are briefly as follows :—

    i. The action and chamber should be thoroughly clean, and the latter free from distortion.

    ii. Where possible, the extractors should be removed.

    iii. The action should be operated by hand (when the gauge is in place), firmly but without undue force, and on no account should the mechanism be "fired" on to the gauge.

    iv. Where the body and barrel are not rigidly connected, (*e.g.*, Lewis or Hotchkiss), no longitudinal play should exist between them.

Sliding extractors of the Maxim type will not accommodate the high gauge, so that other means of determining the cartridge head space are employed. *See* "Handbook for the $\cdot 303$ in. Vickers Machine Gun, 1923" (Section 88).

Adjustments.

The method of adjusting cartridge head space varies with different types of guns as follows :—

    1. Maxim type.—Adjustment is effected by varying the effective length of the connecting rod, between the lock and crank, by means of the addition or removal of washers of varying thicknesses.

    2. Lewis, Madsen, Hotchkiss (French and Light models), Colt and similar types.— As a rule no adjustment is provided and the only remedy is the exchange of unserviceable for serviceable components.

    3. Browning (water cooled) machine gun.—The barrel can be positioned further from, or nearer to, the lock face by unscrewing it from, or screwing it further into the body. When the cartridge head space has been corrected the barrel is fixed by means of a spring catch which can fit into any one of several recesses on the barrel. This method allows for very fine adjustment, since there are

16 threads to the inch on the barrel, and 16 recesses for the spring catch. Thus the cartridge head space can be adjusted to $\frac{1}{16} \times \frac{1}{16} = \frac{1}{256}$ inch (·004 inch approximately).

4. Hotchkiss (British model).—The cartridge head space is affected by the screwing up of the locking nut which positions and locks the barrel, and by the condition, as regards wear, of the fermeture nut and its threads.

In some guns the substitution of a new barrel will affect the cartridge head space, and this is particularly so in the case of the Lewis, in which the condition of the projection on the rear face of the barrel is a determining factor. It should also be noted that it will usually be necessary to test for cartridge head space after the correction of such faults as the following :—

(a) Loose locking nut (e.g. Hotchkiss, British model).
(b) Badly fitting or worn fermeture nut (Hotchkiss, British model). (This necessitates exchange at the factory).
(c) Worn threads joining the barrel to the body.

*Tests Nos. 5 and 6.*

For sighting and the state of the barrel.—These two tests can be conveniently considered together, since the questions of accuracy and sighting are in many ways interdependent.

Method of Testing.—The methods employed in determining the state of the gun under these headings fall into two classes :—

1. Those which can be carried out in barracks or in a workshop.
2. Those which involve obtaining results from shooting on the range.

1. Test by examination.—The following points must be checked :—

(a) The condition of the sights as regards rigidity of fitting, damage and wear.
(b) Efficiency of the means of adjustment.
(c) The state of the barrel. This subject is dealt with in Chapter I, Section 5 ("Considerations affecting the accuracy of the Rifle") the same faults being common to all barrels.
(d) The positioning and fit of the barrel. This point is particularly important in guns such as the Hotchkiss (British model) where the barrel is frequently changed, and in recoil operated water-cooled guns, where the positioning of the barrel may be affected by packing glands.
(e) The fitting, rigidity and means of adjustment of the mounting.

2. Test by trial.—In this connection the gun is fired on the range by a skilled firer, in a still atmosphere and with reliable ammunition. The conditions of the test, as regards range, standards to be attained, and the number of rounds to be fired, vary with different guns, but are based on the capabilities of the particular type of gun when in good condition.

Before the actual test is commenced a few rounds must be fired to warm up the gun, and, when it is mounted, to allow the mounting to settle down. Aim is then taken in the usual manner at a large paper covered target ruled in squares, and a series of groups obtained. Each group must be obtained by continuous fire, and if stoppages occur, another series of rounds should be fired. From these groups the average mean point of impact, and the average figure of merit for the barrel are obtained, the method in each case being similar to that described under "Accuracy proof of Small Arm Ammunition," in Part II, Chapter V.

The position of the M.P.I. relatively to the point aimed at, provides a means of determining whether the gun is correctly sighted and the magnitude of any correction which may be necessary, while the figure of merit refers to the serviceability of the barrel.

It should, however, be noted in this connection that if it is suspected that the figure of merit is influenced by any unsteadiness in the mounting, the test should be repeated with a mounting which is known to be in good condition as to fit and rigidity, for purposes of comparison.

Adjustments.—Adjustments to the sighting are usually effected by fitting a higher or lower foresight, and by positioning it to the right or left as required. If this method is not possible allowances must be made in setting the backsight or in the selection of the point of aim, but it must be remembered, as regards elevation, that it does not follow that the allowance (in units of measure) which has to be made in setting the backsight, when firing at a certain range, will be constant at all other ranges.

### *Tests Nos. 7 and 8.—For the operations of extraction, ejection and feed*

The factors which may influence the efficient performance of these operations are as follows:—

(1) The strength of springs, and the general condition as regards wear and damage of the various components concerned.
(2) The timing of the various operations performed by the mechanism.
(3) The total resistance to movement met with in the operation of the gun, and the amount of friction introduced in the operations of extraction and feed.
(4) The amount of energy available to operate the gun.
(5) Abnormalities in the ammunition, as regards its dimensions and the quality of the metal of which the case is made.

Method of testing.—Certain faults can be detected by examination, by using good dummies and operating the gun by hand, but the only satisfactory method of testing the efficiency of the gun with regard to these operations is by observing the manner in which they are performed when the gun is actually firing. In this connection it is important to bear in mind that if, as in some guns, the cartridge case is lubricated, extraction may be rendered difficult by the accumulation of grit, etc., in the chamber, and also that excessive friction may be rendered evident by the irregular working of the gun or by a reduction in the rate of fire.

Adjustments.—The method of adjustment, apart from the replacement of worn or damaged parts and the elimination as far as possible of friction, consists of balancing the amount of energy available to operate the gun with the work which it has to do, and the manner in which this may be effected is discussed under Feature 6, page 177.

### *Test No. 9.—For friction*

It was stated under Feature 6 (page 177) that the friction which has to be overcome in the operation of the mechanism is a variable quantity, and that this necessitates some means of adjusting the balance between the resistance and the available energy. The smooth working of the gun may also be affected by the effect of gravity at extreme angles of elevation and depression, and it is evident that to keep this balance within the limits of control the value of friction must be kept down to a minimum.

The degree of resistance varies with the type of gun, but tends to be greater in the case of recoil-operated guns than in those which are gas-operated. In guns which are operated by the projection of the barrel, and which are water-cooled, additional friction is introduced by the packing glands which are necessary to prevent leakage of water; while in gas-operated guns, fouling is liable to accumulate on those parts which come into contact with the gases, and must be periodically removed.

Careful and regular cleaning will assist in reducing friction, and the question of lubrication must be carefully attended to. Over-attention in this respect is a good fault, but too much oil must not be applied, in order that the accumulation of grit may be avoided. The lubrication of parts such as a piston, which comes into contact with gases at a very high temperature, defeats its object, since the oil is quickly burnt and fouling increased. In some designs lubrication of the cartridge case is provided for, but in others oil finding its way into the chamber by being splashed on to the case will adversely affect the working of the gun.

A further consideration is the expansion of the various portions of the mechanism when the gun gets hot, and this must be allowed for when the manufacturing limits and tolerances are laid down, otherwise friction will tend to increase during firing, and frequent adjustments will be necessary.

Method of Testing.—It is only in the case of recoil operated, water-cooled guns working on the projected barrel principle that a test for friction must necessarily be laid down. The test described in the "Handbook for the ·303-inch Vickers Machine Gun (1923)," Section 87, is applicable in principle to all other guns of similar type, and consists of removing all springs affecting resistance and then determining with the aid of a spring balance the force which has to be exerted to bring about the movement of the recoiling parts.

Adjustments.—Friction may be rendered excessive by certain causes which are peculiar to certain types of gun, and, as in each case the cause must be discovered by investigation, the obvious remedy is its removal. This can usually be effected by lubrication, cleaning, or the replacement of worn or damaged parts.

The causes which should be particularly looked for in certain types of gun are as follows :—

(A) In recoil-operated guns of the projected-barrel type.
1. Tightness of the bearings through which the barrel slides.
2. A slight bulge in the barrel at or near one of the bearings.
3. Lack of parallelism between recoiling portions and the surfaces in contact with them.

(B) In guns operated by gas through the medium of a piston.
Gas fouling.

(C) In guns in which the barrel and body form two distinct groups (*i.e.*, nearly all those included in (B) above).
Want of alignment between the two groups, causing friction either during the stroke of the piston or (if it is fitted with a cup-shaped head to fit on to a gas nozzle) when it returns to its initial position.

The following causes are common to all guns and should always be looked for :—
(*a*) Dirt.
(*b*) Burred and damaged parts of the mechanism.
(*c*) Strained framework.
(*d*) Unequal wear of the bearing surfaces of bolt lugs, locking recesses and resistance shoulders, etc., which introduces side thrust, and so throws the mechanism out of line.

## SECTION V

### MOUNTINGS

A machine gun mounting may be defined as a convenient and suitable form of rest into or on to which the gun can be placed, and fired, and from which a result can be obtained which will satisfy the particular requirement for which the mounting is intended.

The large variety of form and type of mountings renders it desirable to classify them in the following groupings :—

1. Portable—for ground use, of light weight for convenient attachment and carriage on lighter types of gun on one man's person in action. These include :—
   (*a*) Monopods.
   (*b*) Bipods.
   (*c*) Light tripods.

2. Movable—for ground use, of heavier weight, capable of being carried as a separate unit on one man's person or readily transported over the ground with or without gun mounted thereon. These are employed for the heavier types of gun, and include :—

   (d) Heavy tripods.
   (e) Heavy tetrapods.
   (f) Sleighs.
   (g) Light-wheeled carriages.

3. Fixed—for ground use, usually of heavy weight, and employed in fixed positions in forts, earthworks, etc. These include :—

   (h) Cone—also for ships.
   (i) Overbank.
   (j) Parapet carriages.
   (k) Trench.

4. Aircraft—

   (l) For fixed guns.
   (m) For observer's guns.

5. Anti-aircraft—

   (n) Fixed posts—for ground use.
   (o) Tripods—for ground use.
   (p) Twin or multiple—for use on ships, etc.

6. Tank and armoured car—

   (q) Ball.
   (r) Gimbal.

The principal characteristics of each of the types of mountings mentioned in the foregoing groups are briefly as follows :—

*Group* 1—Portable.

(a) *Monopod*—single-legged rest for the fore part of the gun. In some forms it is provided with a grip which can be held by the left hand for steadying the gun during firing. Monopods are not generally favoured in view of their instability.

(b) *Bipods*—two-legged rests attached to the fore part of the gun. The legs are hinged, and splay outward to a stop when in use, and close up to the gun when not in use. Adjustment for height is occasionally arranged for also. In some cases provision is made for lateral traverse of the gun, whilst in others movement is obtainable in the action of the legs for this purpose.

Shoes either fixed or hinged are fitted at or near the bottom of the legs to prevent the latter from sinking into soft ground.

*Example.*—Mark III Field mount for the Lewis ·303-inch light machine gun.

(c) *Light Tripods*—are usually attached to the gun at a point not far in advance of the centre of gravity. Provision is made for lateral traverse of the gun, and also for the automatic opening of the legs from the closed-up position.

These mountings are more suitable than bipods for some types of gun, but bipods are in more general use.

*Example.*—Tripod for Hotchkiss ·303-inch machine gun.

*Group* 2—Movable.

(d) *Heavy Tripods* are well known, a good example being the service Mark IV for Maxim and Vickers ·303-inch guns. The legs are adjustable for height and for levelling the gun on uneven ground, and are of stout tubular construction with fixed ground shoes, the rear leg being longer than the two front legs in order to ensure stability against the kick of the gun. Saddles are sometimes fitted on the rear leg for the gunner, and to aid in stability.

The jointed end of the legs is connected to a body-piece having a socket or stem to take a crosshead or cradle for the gun. Provision is made on the body or crosshead, according to the design, for elevating and traversing purposes; graduated dials and a clinometer may also be embodied.

Mountings of this type must ensure a steady support for the gun, more especially when the latter is employed for indirect or overhead fire.

(e) *Tetrapods* are rarely used. A model has recently been produced in the United States of America, which tends to overcome the objections made to this type of mounting consequent upon the difficulty of obtaining a level base on uneven ground. In this model the two rear legs form one connected unit, the apex being horizontally jointed to the body of the mounting and capable of swing in a vertical plane at right angles to the gun, so enabling the legs to accommodate irregularities in the ground.

A compensating eccentric mechanism is introduced into the body of the mounting to enable the gun cradle to be adjusted to counteract tilt.

(f) *Sleighs*, constructed of thin pressed metal, and having an adjustable cradle with gears to admit of the necessary movements for elevating and traversing the gun. The rear end of the sleigh may be formed with arms which, together with two hinged front arms used as adjustable leg supports during firing, enable the mounting with gun to be carried in the form of a "stretcher" by two men.

Provision may be made on the sleigh for the carriage of spare barrels, belt boxes and tools.

*Example.*—German, as used in the Great War.

(g) *Light Wheeled Carriages.*—A mounting on two wheels with a shaft to enable the mounting with gun to be drawn by one or two men.

The mounting details follow normal design.

*Example.*—Russian, as used in the Great War.

*Group* 3—Fixed.

(h) *Cone.*—Is in the form of its designation with a divided flanged base arranged to fit in a holding down or deck ring around which it can be traversed and clamped in any desired position.

It is employed for guns of the Maxim and Vickers types and is provided with gun joints and elevating gear. It forms a very firm mounting, and for this reason is employed for the accuracy test of all machine guns of new manufacture and after factory repair.

(i) *Overbank.*—Is constructed on the parallel motion principle, and is adapted to fit into one of the other regular types of mounting such as the Mark IV service tripod or trench mountings. It could be mounted on a base of its own, if necessary. As its designation implies, it is for use in more or less fixed positions, and where it is desired to raise the gun temporarily over an earthwork or bank when required for action, and to swing it down under cover when not so required. A number were supplied for use in the British Service in the Great War, but they were not used to any appreciable extent.

(j) *Parapet Carriages* are wheeled carriages, having a long tubular shaft with horn-shaped handles, which can be rested upon a parapet and used for transport purposes. On the shaft is a rack along which the mounting portion can be raised or lowered as desired. In the earlier patterns one of the wheels is used as a capstan for raising or lowering the mounting portion. In the later model the wheels are much larger and are for transport purposes only.

(k) *Trench.*—At one time known as "Longfield" after the inventor, and as the designation implies were introduced specially for trench warfare during the Great War. They are simply constructed, mainly of angle and strip iron. The base portion is in the form of a sector on three short legs, the arc being fitted with a graduated plate. The upper portion which forms the gun mounting is pivoted to the apex of the sector and is capable of movement around the arc, to which it can be clamped in position as desired or traversed at will.

Additional and independent clamps are provided to form stops, which can be fixed in pre-determined positions on the arc to enable fire to be concentrated on selected targets. A cross-bar is fixed to the apex end of the sector for sand-bagging purposes.

The clamps can be fixed in daylight to form stops, so enabling fire to be concentrated on a selected location in the dark.

From the description it will be seen that the mounting enables the gun to pivot about its muzzle, and fire through a small loop hole.

*Group* 4—Aircraft.

(l) *For fixed guns.*—These are in the nature of brackets attached to the fuselage of an aeroplane, the brackets being adjustable for elevation and traverse for the initial " setting " of the gun or pair of guns mounted.

Provision for mounting a telescopic sight is usually made for guns mounted in pairs.

(m) *For observers' guns.*—The British Royal Air Force example is known as "The Scarfe Ring Mounting." The underlying principle of this type of mounting is freedom of movement and control of the gun by the gunner-observer in all accessible positions in azimuth. The base ring is fixed around the cockpit, and a second ring capable of an all-round traverse is fitted to it. To this second ring is pivoted a bow or hoop to which the gun mount is attached ; opposite the bow joints a quadrant is fitted to the same ring for the purpose of clamping the gun mount and bow in any desired position within its compass. The observer takes his position within the ring.

*Group* 5.—Anti-Aircraft.

(n) *Fixed posts.*—These take the form of an improvised post fitted with a suitable projecting joint to take the gun, and so enable the latter to be traversed and elevated, or they may be specially constructed on similar lines for use in fixed locations, such as camps or dumps.

(o) *Tripods.*—Of light construction for easy transport either by hand or on pack.

The British service pattern has hinged legs, with straps, which can be closed in to a central tubular post for compactness in transport. This post slides in the two brackets which form the joints for the legs and stays, and holes are drilled through it at suitable elevation intervals for a linch pin which rests upon the upper bracket, so transferring the weight of the gun, which is mounted in a holder fitting into the upper end of the post, to the legs.

The gun holders provided for the mounting described are for the Lewis and Hotchkiss ·303-inch guns respectively. They comprise the upper portions of the service bipod and tripod referred to as examples of (b) and (c) of Group 1.

(p) *Twin or multiple gun mountings.*—These may be employed where the concentrated fire of two or more guns is desired. It is obvious that under such conditions a substantially heavy mounting is required. For two guns the base may take any suitable form such as a metal cone, which can be secured to the deck of a ship.

The actual gun holder or cradle in which both guns are mounted must be capable of an all-round travel and elevation in azimuth.

Provision of a support for the gunner is necessary ; this may take the form of a strap in which he can hang back, the ends of the strap being attached to the gun cradle, or it may be a more rigid form of support.

Means are provided for firing both guns simultaneously or separately as required.

Sights may be conveniently mounted on the gun cradle between the two guns.

A multiple gun mounting of special design, known as the "Earpan," was tested by the army during the Great War. In this mounting provision was made for three Lewis guns, which were mounted in such a manner as to enable the guns either to converge or diverge from fixed centres, so as to decrease or increase the fire zone. The mounting was somewhat complicated, and the mechanism for simultaneous firing action rather difficult of adjustment.

Anti-aircraft gunnery being still in its infancy, further developments, both as regards guns, sights and mountings are to be anticipated.

*Group* 6.—Two types of mountings are used in British tanks and armoured cars :—

(*q*) Ball;

(*r*) Gimbal.

Both types are to be found in service armoured fighting vehicles, but the latter, which has been recently adopted, is considered more suitable for A.F.V. requirements.

They are mounted in the wall of the vehicle, in a sponson, or in a turret, according to the type of vehicle.

Finality has by no means been reached, either in design of gun or mounting, for either tanks or cars. In view of possible developments, it would be of little value to enter into further detailed descriptions here.

---

SECTION VI.

BLANK FIRING ATTACHMENTS

Blank firing attachments are employed for the purpose of simulating automatic ball firing on manœuvres, and also for the general training of machine gunners.

The form which the attachments take will vary according to the type of gun and blank cartridges employed.

For convenience of supply, and for economy, it is usual to employ, wherever possible, blank cartridges of simple type in common use for the service rifle, but difficulties in feed, etc., in some types of guns may render the use of a special cartridge desirable, *e.g.*, one having a hollow wood bullet which will break up into small fragments on exit from the barrel.

In the Vickers ·303-inch recoil-operated gun a muzzle attachment, similar in principle to that for ball ammunition, is used in conjunction with a worn-out service barrel, which is specially bushed at the forward end of the chamber in order to restrict the opening of the mouth of the service rifle blank cartridge, and ensure more regular combustion and pressure for the automatic operation of the gun. The service ammunition belt is employed.

The back pressure available from the muzzle attachment during firing can be regulated within limits by an adjusting screw situated in the front of the fixed part of the attachment; the head of this screw enters a muzzle cup screwed to the recoiling barrel, and its depth can be varied.

For gas-operated guns in which the gas port is situated somewhere in the bore between the breech and the muzzle, the attachment takes the form of a cylinder which is screwed, or otherwise affixed, to the muzzle of the barrel. This cylinder has a bore smaller than that of the barrel, the size varying according to the power required to operate the gun. A bushed chamber (as in the Vickers) is also used to build up the pressure and ensure complete combustion.

In this system the choke of the cylinder checks the free exit of the gases and so causes pressure to be temporarily retained within the bore for action through the gas port.

When wood-bulleted blank cartridges are employed it is usual to serrate the choke in the muzzle cylinder in order to ensure the complete break-up of the bullet.

Magazines, or other forms of ammunition feed containers used for ball ammunition can usually be employed for wood-bulleted blank cartridges without alteration, but with unbulleted blank some modification may be necessary.

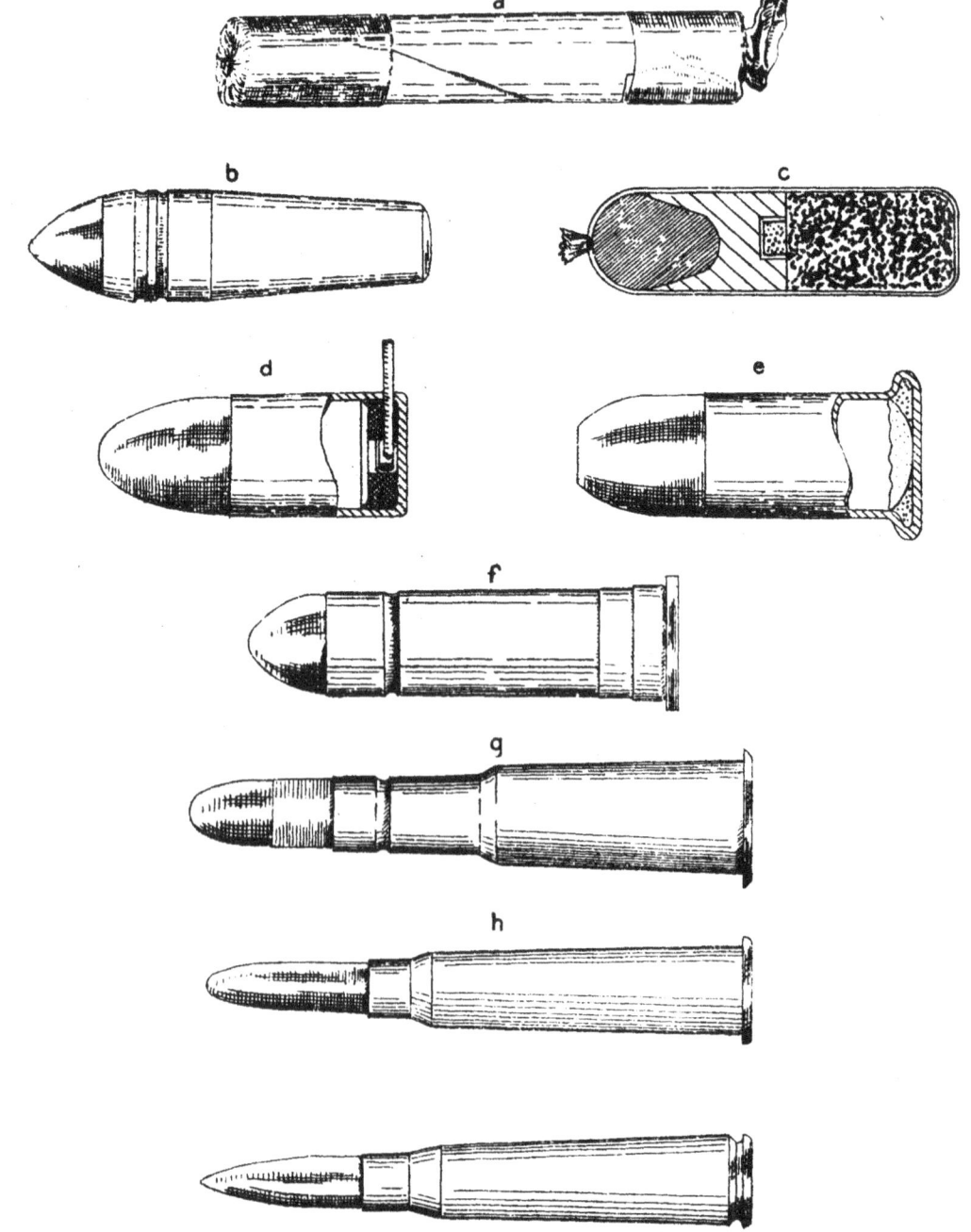

Fig. 1.

# TEXT-BOOK OF SMALL ARMS

## PART II

### AMMUNITION

#### CHAPTER I

#### HISTORY OF THE DEVELOPMENT OF THE SMALL ARMS CARTRIDGE

In the early day of firearms the components of the round were carried separately, with the powder in horns or flasks and sometimes in small tubes hung from cross-belts and known as bandoliers. Early in the seventeenth century Gustavus Adolphus, King of Sweden, ordered the powder and ball to be carried made up together in the form of a cartridge. This method remained in use in much the same general form and was applied to a variety of military weapons for many years, the cartridge usually consisting of a cylinder of paper, one end of which was torn open for the purpose of loading (Fig. Ia.)

The means of ignition was extraneous to the cartridge, and was first the flint and later the percussion system. The credit for the introduction of the percussion system in our Service goes to an Aberdeenshire divine, The Rev. Alex. Forsyth, who, at the request of the authorities, temporarily laid aside his parochial duties in order to conduct the investigation that culminated in the incorporation of his system in the Brunswick rifle in 1838. This system took the form of a tube filled with a composition containing mercury fulminate, which passed through the wall of the barrel at the breech and fired when struck by the falling hammer or cock. English and continental experimentalists produced such systems as the Westley-Richards detonating pellet and the Manton detonating tube, and in France a species of amorce or paper cap was employed with a certain degree of success.

The history of the evolution of rifle ammunition is so intimately associated with that of the arm that a detailed description of the development of the former would involve a good deal of unnecessary repetition of what has already been dealt with in the history of the latter. Prior to the introduction of the breech-loading principle, efforts were principally directed to the solution of the problems of difficult loading in a foul rifle and the elimination of loss due to windage, that is to say, the failure of a projectile that was an easy fit during loading to expand on firing and seal the bore. The real solution of the problem was found by Metford, who employed a bullet which " set up " or expanded under the influence of the gas pressure on discharge without the employment of a base cup or plug, which was " patched " or wrapped in paper over the cylindrical portion, and which was used in conjunction with a broad and shallow grooved rifling. This, however, was not until the advocates of the breech-loading principle were gaining ground and its advantages had begun to approach realization.

Pauly, as early as 1805, had produced a breech-loader and cartridges containing their own means of ignition, and the advent of an arm of this type in a continental army in 1842 marks one of the most important epochs in firearm history. The Dreyse needle gun introduced into the Prussian army in that year fired a cartridge having the powder and projectile separated by a wad, in the rear of which the cap was situated. Firing was effected by a long, sharp needle or firing pin which penetrated the cartridge envelope and passed through the powder charge (Fig. Ic.) Other inventors with varying fortune brought forward other systems, and of those intended for military purposes Sharp in America, and Westley-Richards in this country, in 1848 and 1861 respectively, produced weapons which were extensively used, although both adhered to the old system of the nipple and loose percussion cap (Fig. I*b*). In 1854, Morse, an American engineer, put forward a rifle and ammunition which were far ahead of their time, the weapon being a magazine arm firing cartridges with metallic cases containing their own means of ignition; but his efforts are understood to have been firmly suppressed by the authorities of that day as being extravagant and dangerous. About 1866 the French, following the Prussian lead, decided to adopt the *chassepôt*, firing a cartridge somewhat similar to that of the Dreyse except that the cap was in a wad toward the rear instead of in the front of the charge, was surrounded by powder and had the detonating composition facing the striker. About this date Great Britain, which had so far held herself aloof from the breech loader, decided to convert the existing ·577-inch Enfield into a breech-loading arm on the Snider principle.

The early continental breech-loading arms were but a qualified success, for although much was gained in rapidity in loading during a brief period, the extensive gas escape from the breech, and the fouling produced, usually resulted in failure to operate. This trouble was spared to the British who had bided their time until a system had been produced that relieved the actual breech mechanism of the duties of obturation, as the sealing of the gas is termed, and threw this responsibility upon the cartridge, which, of course, was closely supported.

The body of the final Snider cartridge was made of thin, coiled, brass with an iron disc forming the base and flange. It was developed by Colonel Boxer, of the Royal Laboratory, Woolwich, and was constructed on the central fire principle. Unlike the early types discussed, in which the cartridge was of paper and any fragments remaining were ejected by the bullet of the next round, this cartridge was self-sealing and the empty case was removed after firing. It had additional advantages, such as greater capacity to withstand rough treatment and exposure in inclement weather, though in these respects its capacity is not to be compared to that of its successors (Fig. I*f*.)

In referring to the introduction of the central fire cartridge usually attributed to Pottet and Schneider, it is only fair to draw attention to the efforts of the English gunmakers, Daw and Lancaster, although in company with Needham their attention was chiefly directed to the development of this class of cartridge for use in shot guns. Needham's cartridge was not, however, in the modern sense a central-fire cartridge. Daw patented his cartridge in 1861. Most of these designs were weak in the region of the base even for use in shot guns, where the chamber pressure seldom exceeds 3 tons to the square inch; but in justice it must be said that the support provided by the breech closure of the "drop down" arms of that day often became imperfect after much use.

The Snider was followed in 1871 by the Martini-Henry, in which a calibre of ·450-inch took the place of ·577-inch. The cartridge was bottle-necked, and the bullet not entirely dependent upon expansion for sealing the bore. The length of the bullet was 2·8 calibres as against 1·8 in the case of the Snider. The ammunition, so far as the general construction of the case was concerned, followed on Boxer's lines until about 1882, when the advantages of the solid-drawn case over the coiled brass or "built-up" type became manifest, particularly for use in machine guns where more violent conditions of loading and extraction usually exist (Fig. I*g*). It is of interest to note that the first manufacture in bulk of this new type of cartridge case on behalf of the Government is associated with the well-known name of Kynoch.

In 1886 the British authorities had produced and were considering the adoption of another design of Martini, of reduced calibre ·402-inch, but the Enfield-Martini, as it was called, was never issued owing to the need which arose for keeping pace with the development of magazine arms in foreign countries. In America the Spencer repeater of 1860 and the Winchester of 1866 (a development of the Henry) were working propositions, both using cartridges of the rim-fire type with calibres of ·56-inch and ·44-inch respectively. Switzerland in 1867 adopted the Vetterli, using a similar cartridge of ·41-inch calibre. Germany in 1884–85 converted her single-shot Mauser into a magazine rifle, using a central-fire cartridge, and in about the year 1886 France replaced the single-shot Gras by the magazine Lebel, also a central-fire rifle. All these early rifles employed magazines of tubular form either in the butt or under the barrel; hence the blunt form of bullet usually employed as a precaution against accidental explosions in the magazine.

The Remington rifle, calibre 11 mm. (about ·433 inch), was still popular, but the general inclination was in the direction of smaller bores. France and Portugal chose 8 mm. (·315-inch). Germany in 1888 adopted a Mannlicher with a calibre of 7·9 mm. (·311-inch), the calibre she still retains in her Mauser. This diminution in calibre, which became general, can be attributed in great measure to the foresight and enterprise of Major Rubin, a Swiss officer, who is entitled to the credit of developing a rifle of 7·5 mm. (·295-inch) calibre, using cartridges which, in many important respects, laid the foundation of modern practice. He employed a compound bullet and explored the possibilities of the rimless case, having first, in order to obtain ample chamber capacity combined with limited length, employed a split collar in order to bridge over the differences in diameter of bullet and case. He finally adopted a necked cartridge of solid drawn brass with a charge of compressed powder, an example followed by Great Britain on the introduction of the ·303-inch Lee-Metford in December, 1887. The military advantages of the small calibre prompted certain powers to adopt 7 mm. (·275-inch) and the 6·5 mm. (·256-inch), and a cartridge of this latter calibre adopted by Roumania about 1893 gained great popularity with sportsmen. A still smaller calibre was reached by the Lee straight-pull rifle adopted by the United States Navy about 1898, which possessed a calibre of 6 mm. (·236-inch), but in spite of much care and attention this cartridge's history is a record of failure.

Space does not permit of our following the ramifications of design since the adoption of small calibres. It is sufficient to say that, although low trajectories and light ammunition are as eagerly sought to-day as ever, experience indicates most clearly that so far as appears at the present time the existing limit in calibres is likely to obtain for many years to come. To-day we have calibres ranging from ·315-inch down to ·256-inch, bullet weights averaging less than 200 and nearer 160 to 170 grains, and muzzle velocities mostly in the region of 2,500 to 2,700 f.s. The claims of the central fire cartridge are firmly established, and, with the exception of the French " Balle D," compound bullets (called in America " metal-patched " bullets) are in universal use. Steel is tending to replace copper and cupro-nickel as the envelope material.

One of the most important of modern advances is in the direction of stream-lined or " boat-tailed " bullets, as they are popularly styled in America. The French " Balle D " is in contour a somewhat complicated specimen of the class, and while America has explored the possibilities of stream-lining with considerable success, Switzerland has in actual use in her service what may be regarded as the most up-to-date and successful example of the type. An examination of certain of Whitworth's projectiles is evidence of an early recognition of the ballistic advantages derivable from the employment of the rear taper. Match rifle shots who limit themselves to service standards as regards calibre and bullet weight, and use weapons, sights and methods of considerable refinement, interest themselves in exploring the possibilities offered by suggested improvements of the cartridge and barrel. The Bisley Meeting of 1922 saw them using a stream-line bullet of 174 grains weight emerging from the rifle at a muzzle velocity of 2,850 f.s. This ammunition was produced by Nobel's specially for the occasion.

While this chapter has dealt almost exclusively with service arms, interest and profit may be derived from glancing briefly at the effect of military small-bore evolution on sporting weapons. It was not long before the merits of the ·303-inch and ·256-inch cartridges became known to sportsmen, and all but the more conservative quickly became conscious of the advantages conferred by absence of smoke, light weight, and error-absorbing qualities of low trajectory at ranges outside the scope of the most efficient black-powder expresses. These advantages were no less appreciated by the more progressive rifle makers, who, conscious also of limitations as regards stopping power against dangerous game, sought to make good this palpable deficiency by means other than bullet design. In place of the old and well-tried 8, 10 and 12-bore rifles, ·577-inch, ·500-inch,

·450-inch, ·400-inch and ·360-inch expresses, now stands a range of weapons using cartridges which combine the advantages of absence of smoke, sustainable recoil, and lowness of trajectory over sporting ranges, with adequate bullet weight and enormous smashing power.

Cartridges suitable for big game of all varieties, ranging in calibre from the ·600 cordite with its 900 grain metal-patched bullet and 7,600 ft.-lbs. muzzle energy, down to such as the ·240-inch Holland "Apex" with its 100-grain projectile and muzzle energy of 2,000 ft.-lbs., are in use; and well-balanced and handy magazine rifles can be obtained for powerful cartridges such as Gibbs's ·505-inch, with its 6,430 ft.-lbs. muzzle energy. Most varieties of such cartridges are efficient, many extremely so, and the only objection that can be raised is perhaps that they are too numerous.

A cartridge of American origin that deserves mention is the ·22-inch Savage "Hi-Power," which fires a "metal-patched" bullet of 70 grains weight with a muzzle velocity of 2,800 f.s., and is the smallest representative of the high-power class. In spite of the recorded performances of cartridges of this type, it is considered that their use, except in an extremely circumscribed rôle and by very experienced hands, is likely to lead to disappointment. Military cartridges are designed to serve a definite purpose, whereas the uses of a sporting cartridge are manifold. It is an unfortunate fact that a great deal of misconception exists regarding the correct interpretation of ballistic data, with the result that disappointment is often due simply and solely to very gross over-taxation of the capabilities of even the most efficient small-calibre projectile.

Before these general remarks give place to the more detailed aspect of modern cartridge design, a few brief notes on cartridges containing their own means of ignition, other than central fire, may be of interest. The rim-fire system (Fig. I$e$) has been referred to in connection with the ammunition for the Spencer and Winchester carbines. Its use is now almost exclusively confined to the popular ·22-inch cartridges, and being ill-designed to withstand high pressures it is never likely to be the subject of wider application. Another fault lies in the large quantity of priming composition that has to be employed, which often produces serious corrosion in the absence of opportunities for the exercise of extreme care and attention in cleaning the barrel after use.

The remaining type is the now almost extinct pin-fire (Fig. I$d$), in which a cap situated inside the base is fired by the action of a pin or wire rod which passes radially from the wall near the base and is driven in by a falling hammer. This system was introduced by Lefaucheux, who, about 1836, produced such cartridges for use in a double-barrel "drop-down" action gun of his own design. Pin fire guns and revolvers are still met with, but they are fast falling into disuse. Many such revolvers are of the commonest quality, with calibres ranging from 5 to 15 mm. Cartridges of this type are somewhat dangerous to use and to carry on account of their projecting pins, and as they require to be loaded in a certain radial position they are difficult to manipulate, and are inapplicable to magazine weapons.

## CHAPTER II

### NOTES ON THE DESIGN OF THE MODERN MILITARY CARTRIDGE

In turning to the subject of the characteristics and design of the modern military cartridge, it is necessary to emphasize the fact that no matter how much skill and care is bestowed upon the design and manufacture of weapons, this must in the main be wasted unless the ammunition receives correspondingly careful treatment. The designer of military cartridges is confronted with problems and surrounded with restrictions which can be safely disregarded in many cases by his sporting rifle contemporary. He is required o keep down weight and bulk and yet to provide a cartridge that will not only produce allistics of a high order with regularity, but also withstand ill-treatment and exposure and still function correctly in magazine arms likewise subject to the severe conditions of service use.

The requirements of the cartridge may therefore be detailed as follows :—

(i) Moderate bulk and weight.
(ii) Ability to withstand rough usage and exposure.
(iii) Safety in store, transport and use.
(iv) A high velocity well sustained at long and distant ranges, obtained otherwise than at the expense of handiness or sustainable recoil.
(v) Ability to function correctly in rapid and sustained fire from magazine and automatic arms.
(vi) Capability of being rapidly and cheaply produced in times of emergency on the class of machinery, from the materials and by the type of labour then available.
(vii) Moderate chamber pressure well sustained.

The elements of the cartridge are :—

(a) The bullet.
(b) The cartridge case or "shell," as it is called in the United States.
(c) The propellent charge.
(d) The cap or primer.
(e) The wad, lubrication, etc.

(a) *The Bullet.*

With the exception of the French "Balle D," which is a solid bullet of copper-zinc alloy, the bullets of modern cartridges are "compound" (Fig. 1A), consisting of a core of

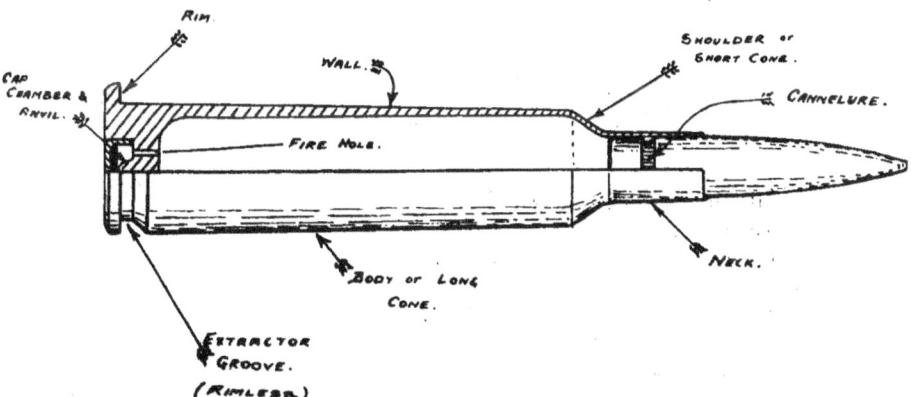

Fig. 1.—Diagram giving named of various portions of a rifle cartridge.

lead covered with a jacket or envelope of harder material, usually cupro-nickel, or soft steel coated thinly with some such anti-friction metal or alloy. Lead bullets of the old type are useless to-day as, apart from considerations of penetration, high velocity would

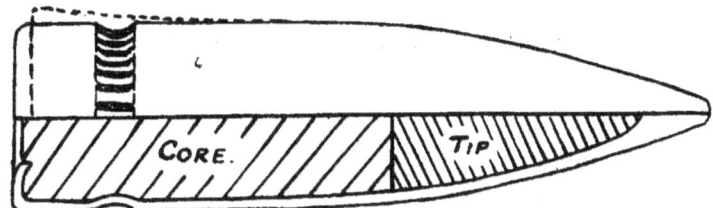

Fig. 1A.—Typical compound bullet—cannelured. (The dotted line is intended to illustrate "set up," but the condition is intentionally exaggerated.)

cause the soft metal to strip instead of following the spiral of the rifling. At the same time a certain degree of plasticity is requisite in order that " windage " may be avoided. Thus a lead core with an exposed base, combined with an envelope of harder material thinning off to the rear, provides the desired compromise.

The present British Service Mk. VII bullet is an example of the manner in which the compound projectile enables weight to be reduced without a corresponding reduction in length. In this bullet the lead-alloy core ends short of the point, the remaining space being filled with a tip of aluminium, fibre or paper. The weight is 174 grains as against about 196 grains for a bullet of equal bulk but with a solid lead-alloy core.

One of the principal troubles encountered by the user of small bore arms is metallic fouling or the deposit of portions of the envelope material on the surface of the bore. This is probably due to a condition akin to " seizing " in a bearing, and when once commenced is cumulative ; it sooner or later impairs and may even destroy the accuracy of an otherwise perfect barrel. Traces of cupro-nickel on the surface of the bore in the form of a thin plating or film do not cause trouble, but it is the formation of a lumpy deposit that is fatal to accuracy. This trouble, like the " leading " of former days, may be caused by roughness or a toolmark in the bore, or by a tight band or other influence restraining the free expansion of a hot barrel, while a barrel that has once been badly nickelled is usually more susceptible to a recurrence of the malady. Of the various means of removing such fouling the chemical are usually preferable to the mechanical. The first usually consisting in the dissolving away of the deposit with solutions containing ammonium per-

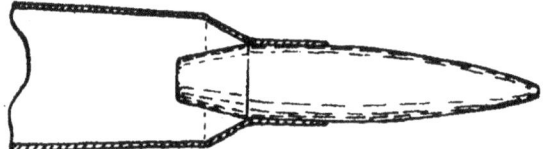

Fig. 2.—Typical stream-lined bullet in neck of case.

sulphate, etc., while the second is the use of an abrasive, which, if capable of abrading steel, necessarily removes part of the substance of the barrel.

Tin-plating has been tried as a preventive, but suffers from the serious objection that the bullet and case become almost inseparable. Generally speaking, the use of envelopes of soft steel thinly plated with cupro-nickel may be regarded as the best means of avoiding serious trouble from metallic fouling.

In the concluding portion of the history of the cartridge reference was made to the steady growth of opinion in favour of stream-lined projectiles, and while the purely ballistic aspect of the matter is outside the scope of this chapter, there are points connected with the design of this class of projectile that deserve special notice. It has been stated that the sealing of the bore is affected in some measure by the " set up " or increase in diameter of the rear part of the flat based bullet caused by the gas pressure as it is driven through the " leed " or conical entrance into the rifled portion of the bore, to which end a certain plasticity is purposely retained. Such conditions cannot exist in the taper-tailed bullet, firstly, on account of the fact that the pressure no longer acts on the reduced area of the base alone, and, secondly, because "set up" would produce deformation and vitiate the ballistic advantages sought.

There are two further points worthy of mention, one being that the existence of the taper necessitates the bullet being loaded deeper into the case (Fig. 2), thus usurping valuable charge space and rendering the use of a wad almost impossible, the other being the extreme importance and increased difficulty in ensuring exact concentricity of bullet and case where the length of bullet parallel is much restricted.

This latter point is of no small moment, since a bullet of this type, if untruly centred at the start, is almost certain to be more or less unstable in flight. Thus, in order to obtain the advantages possessed by stream-lining, the correct positioning of the loaded bullet in relation to the position and form of the " leed " requires the utmost care.

Misapprehension has been created by the considerable looseness with which the adjective " explosive " has been applied to rifle bullets. It is true that projectiles containing an explosive substance acting on impact have had restricted use in the larger black powder sporting rifles. The wounding power of projectiles is dealt with in Part III, Chapter X, but it may be stated here that when additional stopping power is obtained by exposing the lead core, by hollowing the nose of a compound bullet, or by weakening the restraining envelope in some other manner, the bullet is properly described as an expanding bullet (Fig. 3). Its destructive effect on bone and tissue is greater than

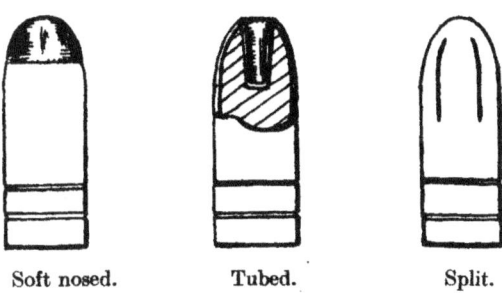

Soft nosed.   Tubed.   Split.
FIG. 3.—TYPICAL EXPANDING BULLETS.

that caused by a solid bullet, and quite different from that of a bullet opened by the explosion of an internal charge.

### (b) The Cartridge Case

Brass, usually an alloy of 70 per cent. copper to 30 per cent. zinc, is in universal favour as the material for cases, being strong, sufficiently ductile, non-rusting, well suited to drawing operations in manufacture, and of reasonable weight. Steel has been tried for the purpose, and possesses the advantages of cheapness of the raw material, strength, and lightness ; against which must be placed such considerations as greater wear of tools, increased liability to corrosion due to damp, and resistance to extraction even when plated with a thin coating of copper or similar non-ferrous metal.

Since it is obviously necessary to combine a maximum ballistic value with minimum weight, it is desirable to waste as little weight as possible in the case, but as this component is required to resist considerable stresses during loading, firing and extraction, it is clear that adequate strength is vital to efficiency. There must be sufficient metal in the head to sustain the severe back-thrust that occurs and is particularly marked when the chamber is not entirely free from lubricant. If the distribution of metal in the walls be too sparing, the extension due to longitudinal stresses set up on firing and in extraction may result in fracture or a " separation," as it is termed.

The bodies of cartridge cases are slightly coned in order to reduce to a minimum the distance through which it is requisite to withdraw the fired cartridge in order to loosen it in the chamber. This loosening is termed " primary extraction," and is performed, as a rule, by means of a powerful cam action brought into play by the initial movement of opening the breech.

Cases are bottle-necked in varying degrees in order to avoid the drawbacks due to too long a cartridge, and at the same time to provide ample capacity for the charge. The extent to which cartridges are necked down depends upon a variety of circumstances, but, in general, it may be said that a very " fat " cartridge, such as the French, is not well suited to box magazines. The precise angle of the shoulder is not a matter of great importance unless the cartridge is dependent on this point for correct positioning in the rifle chamber.

The design of the head of the case is an important feature, and the principal types are as follows :—

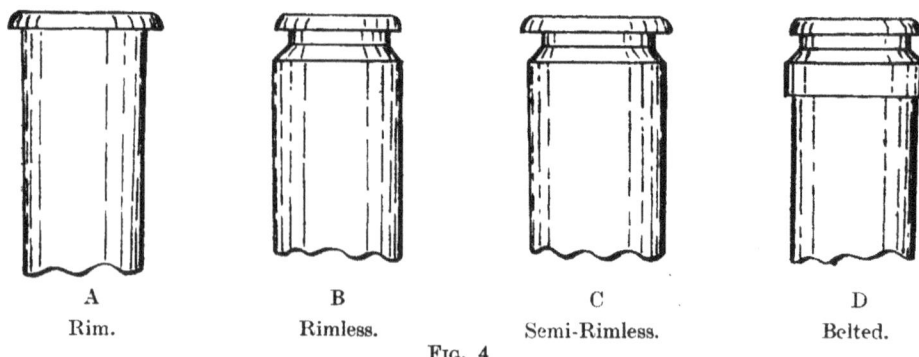

A
Rim.

B
Rimless.

C
Semi-Rimless.

D
Belted.

FIG. 4

*(a) Rim* (Fig. 4A)

As used by Great Britain, France, Roumania, Russia, Holland, etc., the cartridge is positioned in the chamber by the front of the rim or flange bedding on the face of the barrel, thus avoiding failures due to wear of chamber or to short-length cartridge cases. The projecting rim is apt to catch in the rims of the other cartridges or in other projections during " feed " from the magazine into the chamber.

*(b) Rimless* (Fig. 4B)

As used by United States, Germany, Switzerland, etc., the cartridge is positioned in the rifle chamber by its shoulder butting against the chamber shoulder, which acts as a stop. It is slightly more expensive to make than the rim type, but packs, " feeds," and extracts very satisfactorily. The extractor claw engages in the groove, but the latter must not be too deep or the head will be unduly weak.

*(c) Semi-Rimless* (Fig. 4c)

As used by Japan. This form is intended to provide the certainty of positioning of the rimmed case without the disadvantages of a prominent rim. The success of this attempted compromise is questionable.

*(d) Belted or Accles Type* (Fig. 4D)

This is another attempt at compromise which is used by Messrs. Holland for express rifle cartridges, but is not regarded as particularly suited to service requirements.

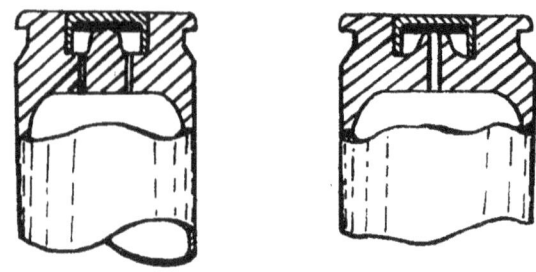

FIG. 5.—Fire holes.

The anvil or nipple against which the cap crown is driven by the firing pin usually takes the form of a projection from the bottom of the recess in the centre of the head termed the cap chamber, channels or fire holes being provided to conduct the cap flash to the

PT. II . CHAP. II.  To face p. 213.

## PROPELLENTS

- A . CORD, *also partially consumed.*
- B . TUBE, *also partially consumed.*
- C . STRIP. (AXITE).
- D . CUBE.
- E   FLAKE.
- F   CYLINDER.
- G . PERFORATED CYLINDER, *also partially consumed.*
- H . DISC, *also partially consumed.*
- I . PERFORATED DISC, *also partially consumed.*
- J . MULTI- PERFORATED DISC, *also partially consumed & broken up into* "SLIVERS".

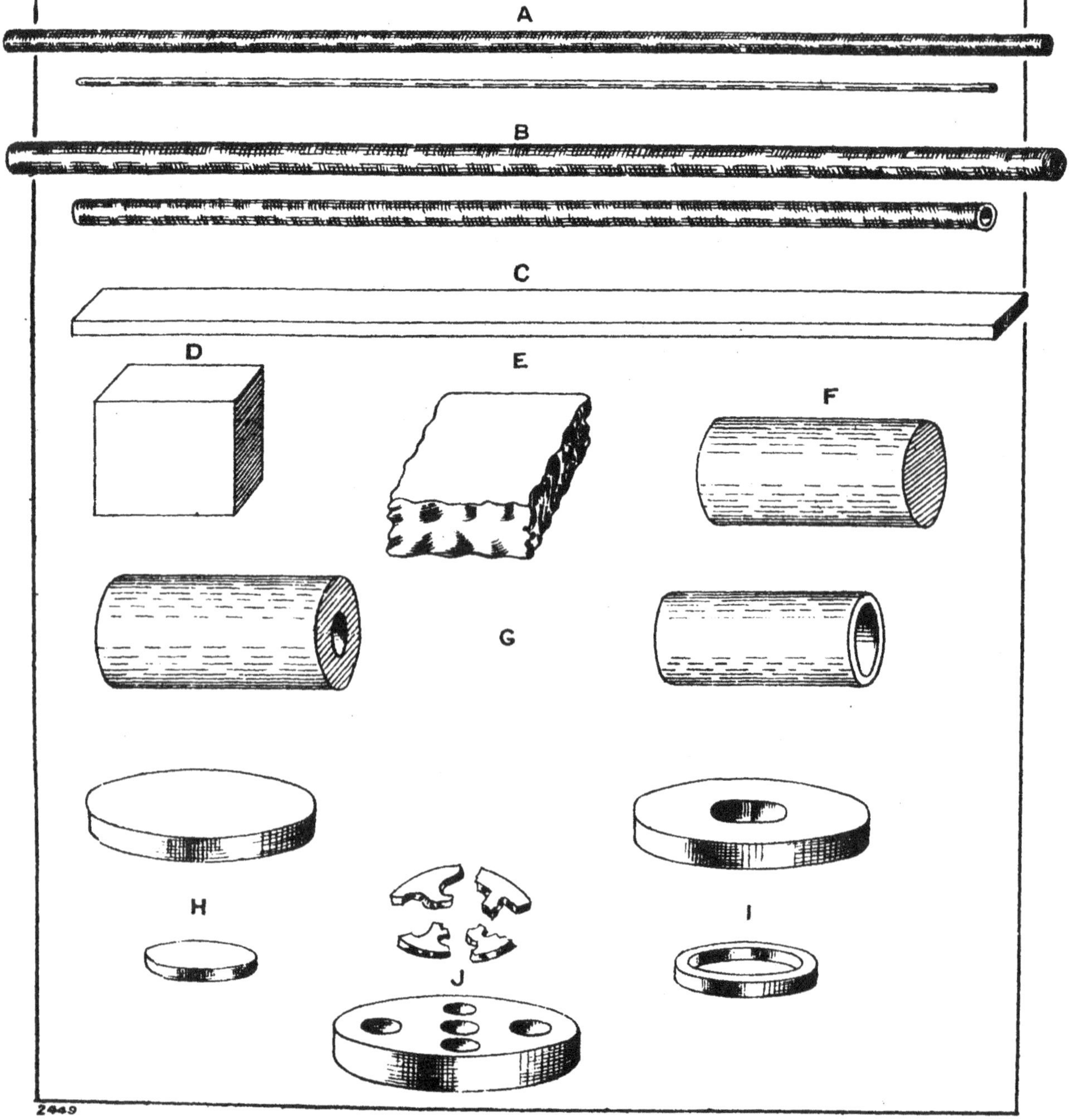

**FIG. 6.**

Malby & Sons. Lith.

propellent charge. Anvils are sometimes separate components, but more usually they are formed integral with the case. The contour of the anvil is important, and there is a growing tendency to use a flatter or more blunted form in order to obtain greater certainty of action in the event of an eccentric blow from the striker.

Fire holes should be of adequate size, and should be located well within the cap walls and not beneath them. In some cartridges a single fire hole passes axially through the anvil, but experience indicates that two holes at the root of the anvil are by far the most satisfactory design.

The diameter of the cap chamber is subject to certain limitations, particularly in the rimless design of case, for if of large diameter and combined with a deep extractor groove, the annulus of metal separating the two may be insufficient to withstand the back-thrust without undesirable deformation.

(c) *The Propellent Charge*

The subject of small arm propellents is dealt with in detail in Chapter III, and the present chapter will only contain a few general remarks.

The essential characteristics of a service propellent are :—

(a) It must be capable of being produced rapidly and economically from ingredients that are reasonable in cost and readily obtainable in time of war.

(b) It must, if correctly loaded, give regular and even pressures and velocities under varying conditions of storage, and must be immune from decomposition or spontaneous ignition, *i.e.*, be "stable."

(c) It must be easy and safe to load, non-hygroscopic, and free from products of combustion that are difficult to remove and deleterious to the barrel.

Modern propellents can be divided into two classes :—

(a) Nitroglycerine and (b) nitrocellulose. In the British Service the representative of (a) is cordite and of (b) Du Pont No. 16, cartridges loaded with (b) being distinguished by the addition of the suffix "Z" to the mark.

Propellents of both classes are usually manufactured in geometrical form as cords, strips, tubes, cylinders, cubes, flakes, or discs ; solid, perforated, or in the larger sizes, multi-perforated. (*See* Fig. 6). The advantage of geometrical form is the regular surface exposed to the action of the igniting flame and the extent to which the subsequent rate of combustion can be controlled.

The methods of loading the charge into the cases are dealt with in Chapter IV. Cordite charges require to be inserted into the cartridge case prior to the formation of the neck, whereas what may be termed granular propellents can be inserted subsequently to this operation, and simple bulk loading by means of charge plates can be employed.

The importance of stability cannot be overrated, particularly as British Service ammunition is stored and used in a greater variety of climates than that of almost any other Power. Further, present-day conditions of use often require that cartridges shall remain for some time in the chambers of intensely hot weapons. Earlier varieties of smokeless propellent were very apt to resent such treatment, and rifles burst from this cause were far from being unknown. Examination of these weapons often provided evidence of a condition of detonation or at least of a rise in pressure of abnormal rapidity.

It may not be out of place to give something in the nature of a homely illustration of the normal combustion of a propellent in a rifle compared with detonation. An ordinary door stands open, and by placing one finger against it and applying gentle pressure the door begins to close, and after it is once "on the move," its inertia being overcome, little effort is required to close it completely. If, instead, we take a running kick at one of the thin panels, it is very probable that the panel will fly to matchwood and equally possible that the door will fail to close. If a propellent is ignited at its surface it will liberate gas at a certain rate, and this gas acting upon the projectile will first overcome its inertia and then force it up the bore with increasing velocity ; the door is easily closed. If, on the other hand, our explosive detonates, the maximum force is almost instantaneously exerted, and we have the shattered panel. In the first case we have a gradual increase in gas volume and pressure, while in the second the structure of the explosive is broken down and the constituent gases seek suddenly to resume their normal bulk, the result being a shattering blow instead of the steady " push " which, though acting in all directions, finds an outlet by the path of least resistance.

(d) *The Cap (see Fig. 7)*

It is impossible to lay too great stress on the importance of a good, regular and powerful cap. Unlike the old black powders, the horny surface of the modern propellent requires a flame of some considerable duration and heat to provide that certainty of ignition and regularity of inflammation which is so essential to accuracy.

A composition much in favour consists of a mixture of mercury fulminate, potassium chlorate, antimony sulphide, sulphur and mealed powder, but some makers prefer to omit the last two ingredients, and occasionally mercury fulminate is also absent. The proportions of the ingredients vary considerably with different firms, each of which claims the advantage of certain qualities, such as increased regularity, stability and absence of any products of combustion injurious to the bore. The charge of composition, usually about $0 \cdot 6$ of a grain, is consolidated in the small metal cup or shell of the cap by direct pressure. The cap shell is formed of copper or brass, and although it must be sufficiently ductile to be indented by a moderate blow from the striker and expand to form a gas-tight joint with the case, it must also possess ample resistance to perforation.

The efficiency of a cap cannot altogether be judged by the size or brightness of the flame produced as shown by photography. It lies rather in the heat generated, nor should it be forgotten that since many photographic plates are but slightly sensitive to the influence of the heat rays, considerable caution needs to be observed when igniting values are being assessed from photographs.

(e) *Wads and Lubrication.*

The purpose of the wad is to provide some form of gas-tight seal as the bullet passes from the neck of the cartridge into the bore. In former times wads were required to clean and in some instances to lubricate the bore, but since very little residue results from the combustion of most modern smokeless powders, the rôle of the wad has changed with the propellent. The Martini-Henry cartridge had a cupped wax wad situated between two card wads in rear of the bullet. Cordite loaded cartridges contain glaze-board wads which are inserted before the case is necked, but in the case of those loaded with nitro-cellulose, a wad of similar material is inserted by a special method subsequent to necking. With stream-lined bullets the use of a wad is increasingly difficult owing to the increased distance the bullet is inserted into the case. This, coupled with the fact that the stream-lined projectile is non-deformable (*i.e.*, does not set up at the base when fired), constitutes a somewhat important consideration that must not be overlooked.

Although lubricated wads have fallen into disuse, the bullets of most modern small-bore cartridges are lubricated either—

(a) Internally, that is by the insertion of a lubricant such as beeswax in a " cannelure " or groove which encircles the bullet at a short distance from the base, and is covered by the neck of the cartridge case, or

(b) Externally, that is, by the application of the lubricant to the exposed surface of the bullet.

The purposes of lubrication are :—

(i) Reduction of friction.

(ii) Exclusion of moisture from the propellent charge.

(iii) Prevention of oxidation in the case of unplated steel or iron envelopes.

PT II. CHAP. II.   To face p.214.

1. Detonating Tube.
2. Amorce or Paper Cap.
3. Musket Cap.
4. Gun Cap.
5. Typical Cap for Rifle Cartridge.
6. Gun Cap on Nipple of Muzzle-loader.
7. Effect of Firing-pin on Cap.

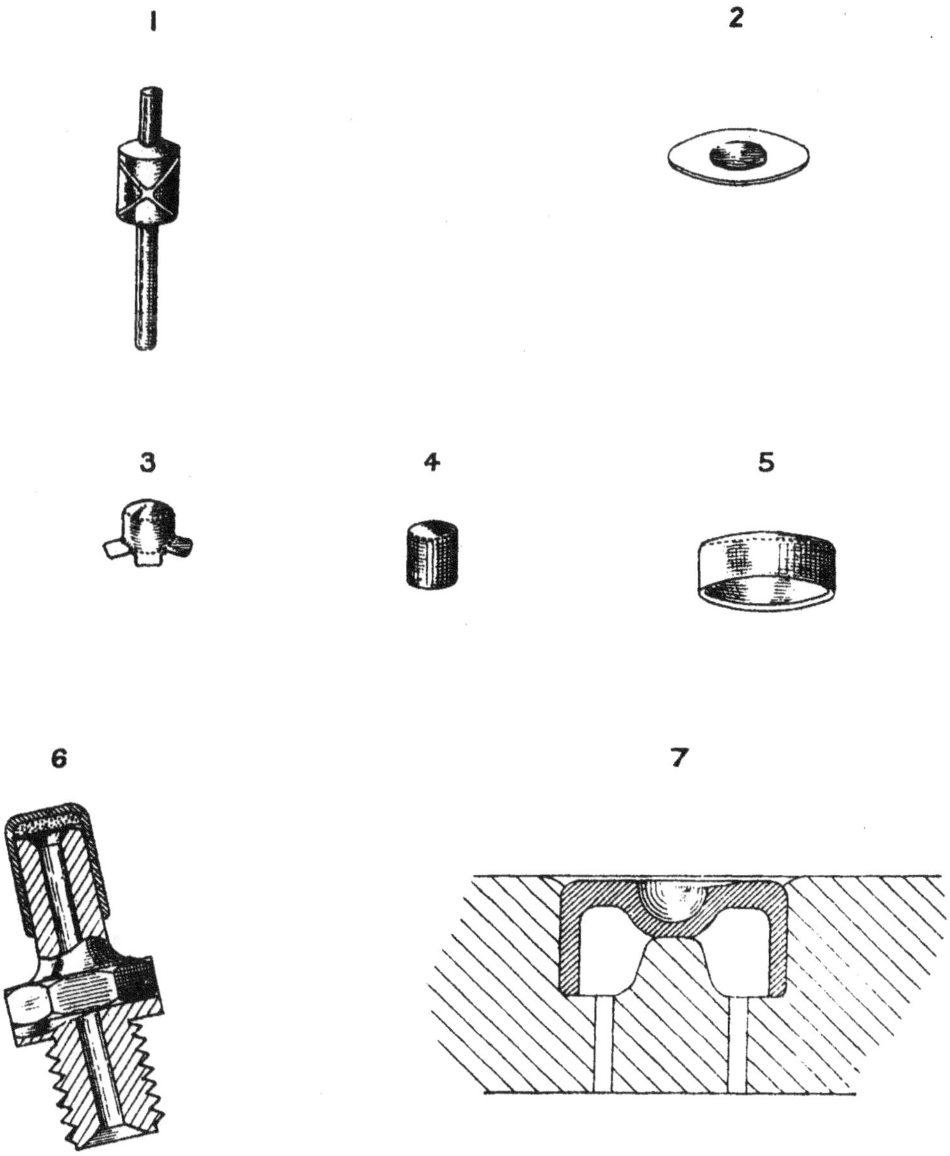

FIG. 7.

British service cartridges are treated by method (a), although until recently those loaded with nitro-cellulose were not lubricated, but were waterproofed with varnish at the junction of case neck and bullet. Stream-lined bullets cannot well be cannelured, and in consequence lubrication must be effected by method (b) or dispensed with entirely. Externally lubricated bullets, however, are objectionable for the reasons that they are apt to develop verdigris, are dirty, particularly in hot climates, and collect dust and sand.

*(f) Chargers and Clips.*

In order to accelerate the loading of magazine arms the cartridges are usually issued in chargers or clips holding from three to six, but usually five, rounds.

The design of these components does not come strictly within the province of ammunition design, since in great measure they are more influenced by the mechanism of the arm than by the details of the cartridge. The distinction drawn between chargers and clips is that in the case of the former the cartridges are forced from the charger into the weapon in one movement and the charger falls away, while in the case of the latter both clip and cartridges enter the magazine and the clip is retained in it until the last round is expended.

The British charger holds 5 rounds, the most generally popular number, the Swedish charger holds 6, the Italian clip 6, and the clip for the French carbine, 3 rounds.

Since, as already mentioned, rimmed cartridges do not lie well together on account of their projecting flanges, these flanges need to be overlapped or staggered, and the rounds kept in alignment by relatively deep walls to charger and clip. The rimless form of case on the whole is favourable to the provision of a neat and light design, but opinion is still somewhat divided on its general superiority.

---

# CHAPTER III
## EXPLOSIVES

### 1.—General Remarks

Explosives are substances which when suitably activated, as, for example, by the action of heat, exert a sudden intense pressure on their surroundings. This pressure is developed by the extremely rapid conversion of the explosive into gaseous products with liberation of heat.

The majority of explosives contain combustible portions consisting of carbon and hydrogen, like petroleum, but instead of having to draw the oxygen required for their combustion from the air, they have it already available in the explosive in a condensed form. For example, ordinary gunpowder is a mixture of the combustible substances, sulphur and charcoal, with a third ingredient, saltpetre, which contains the supply of oxygen necessary for their combustion. In other explosives, such as nitroglycerine or lyddite, the association of the oxygen with the combustible portions is much closer than in gunpowder, each being contained within the same ultimate particles or molecules.

Military explosives may be divided into four classes according to the purpose for which they are used :—

(1) Initiating agents.
(2) Propellents.
(3) High explosives.
(4) Gunpowder, pyrotechnics, and other compositions which cannot usually be detonated.

The explosives in class (1) are used in percussion caps or detonators for initiating explosion in the explosives of the other two groups. They are very sensitive to shock and friction, so that they can be readily brought to ignition by such means.

Class (2) comprises explosives such as cordite and nitrocellulose powder, which are used for the projection of shell from guns. Their function is to provide a relatively well-sustained, but not excessive pressure on the base of the projectile. They are always initiated by a hot flash, so that the relatively slow process of combustion from layer to layer, but not detonation, ensues.

The function of the explosives of class (3) is to provide a sudden very intense pressure so that shattering and disruptive effects are produced. Their initiation is generally effected by a detonator so as to bring the explosive to detonation, a wave-like mode of decomposition proceeding through the mass at an extremely high velocity and giving rise to intense local pressures. The explosives of this class are used in the Service as bursting charges for shells, torpedoes, etc., and for blasting, and include guncotton, dynamite, amatol and lyddite.

Class (4) includes a large number of compositions which, like gunpowder, are intended to burn without violent explosive effect. They are used in time fuzes, signals, illuminating and smoke-producing ammunition, etc.

## 2.—HISTORICAL OUTLINE

The early history of explosives is very indefinite and shrouded in fiction. Roger Bacon knew of a gunpowder mixture as early as the year 1242, but the real development of this explosive occurred only after the invention of firearms in the 14th century by Berthold Schwartz. Gunpowder appears to have been first produced in England for the purposes of war in the reign of Edward III, guns having been used against the French in the battle of Crecy, 1346 ; but it was not until Elizabeth's reign that its manufacture was properly established in this country. About 1561 George Evelyn, who appears to have learnt the art of manufacture in Flanders, set up mills at Long Ditton and Godstone, while powder works which later became Government factories were established about the same time at Faversham and Waltham Abbey.

At first gunpowder was used in a finely-divided state, but its behaviour was very erratic, and it was liable to foul the gun. These defects were subsequently overcome by forming it into grains of various sizes, but other disadvantages, such as the excessive amount of smoke produced on firing, were inherent in its composition. For want of a better explosive, gunpowder remained the principal military propellent and disruptive explosive until the latter part of the 19th century, when smokeless propellents and high explosives such as guncotton and lyddite were gradually introduced. Advances were, however, made during the early part of that century by the introduction of percussion caps, which became possible through the discovery of chlorate explosives in 1786 by Berthollet, and of mercury fulminate in 1799 by Howard.

In 1845 Schönbein discovered guncotton (nitrated cellulose), the use of which as a smokeless propellent for small arms he successfully demonstrated at Woolwich during the following year. Manufacture was started at Faversham, but before a year had elapsed a disastrous explosion wrecked the works. Similar explosions occurred in France, and as a result manufacture ceased in this country for sixteen years. Experiments which met with partial success were continued by General von Lenk in Austria, but it remained for Frederick Abel, who in 1863 took up the study of guncotton at Waltham Abbey, to devise the first really satisfactory method of manufacture and purification. The product was relatively free from the danger of spontaneous ignition, but was used only for torpedoes and mines.

Many attempts were made to use guncotton as a substitute for gunpowder for military propulsive purposes. To control the rate of combustion von Lenk prepared charges in the form of bobbins of tightly-wound guncotton yarn, while Abel compressed finely-pulped guncotton into blocks. Both products were much too porous to be reliable. Successful shot-gun powders made from guncotton or allied substances were brought out by Schultze in 1867, and by the E.C. Powder Company in 1882, but these were too quick for use in rifled arms.

A method of treating nitrocellulose was eventually devised which rendered it suitable for application as a military propellant. This consisted of an incorporation with suitable solvents which destroyed the fibrous character of the nitrocellulose, and on evaporation of the solvent left a compact material devoid of pores, and capable, therefore, of burning only from the exterior, layer by layer, to the centre. The gelatinized material, before the evaporation of the solvent, could, while still plastic, be rolled or pressed into sheets, cords or other desired forms, which retained their shape when dried. The first smokeless military rifle powder was that invented in 1884 by the French chemist Vieille for the Lebel rifle. Since that date a large number of similar nitrocellulose propellants have been introduced.

Nitroglycerine was discovered in 1846 by Sobrero at Turin. For many years its only application was in medicine for the treatment of angina pectoris. In 1859 Alfred Nobel, a Swedish engineer, discovered means for detonating it, and three years later commenced its manufacture in Sweden for blasting purposes. The explosive was at first used in the liquid condition, but so many accidents occurred that it became necessary to devise means for rendering it safe during transport and use. In 1867 Nobel produced "dynamite," in which the nitroglycerine is absorbed within the pores of Kieselguhr, an infusorial earth. Later he discovered that by dissolving 8 per cent. of nitrocellulose in nitroglycerine it is converted to the powerful gelatinous explosive, "blasting gelatine." In 1886 he made the further discovery that if the proportion of nitrocellulose was increased until it about equalled that of the nitroglycerine, and the materials were incorporated by malaxation or rolling between hot rollers, a horn-like product capable of being cut up or granulated into any desired form or size, and suitable for use as a propellant, resulted. This explosive was patented in 1888 under the name of "Ballistite," and is still employed in some countries. Camphor was originally included in the composition to facilitate the blending of the nitroglycerine and nitrocellulose, and to reduce the rapidity of explosion of the product, but the use of this ingredient was subsequently abandoned owing to its tendency to evaporate during storage and thus produce an alteration in composition and ballistics of the propellant.

To obtain a smokeless propellant which would remain constant in its ballistics, the Explosives Committee, of which Sir Frederick Abel was President, carried out a long series of experiments. The final result of their labours was the introduction of cordite into the British Services in 1888. This propellant consists essentially of nitroglycerine and guncotton gelatinized and blended together by the aid of a solvent (acetone). A small proportion of mineral jelly is incorporated into the mixture; originally introduced to prevent metallic fouling of the rifle, experience has shown mineral jelly to operate beneficially in cooling the propellant and in helping to maintain its uniform stability under varied climatic conditions. The name "Cordite" was chosen for this propellant owing to the cord-like form in which it is manufactured.

This explosive has proved, after years of use, to be satisfactory in its stability both chemically and ballistically. Its main defect is the erosion it produces, particularly in heavy ordnance, owing chiefly to the great heat it develops. To overcome this defect, a cooler propellant was worked out. This propellant, which is known as Cordite M.D. (modified cordite) contains the same three ingredients, but in different proportions. Experience with Cordite M.D. has shown it to give rise to more erosion than certain other propellants, but it remains still one of the best rifle powders.

During the late war certain other propellants were introduced into the British Service. Acetone, the solvent which is used in the manufacture of cordite, could not be obtained in sufficient quantity, and a modified composition known as cordite R.D.B. was introduced. This could be made with a mixture of ether and alcohol, solvents which were readily available, but the propellant was never used for small arms. The British factories were also incapable of turning out the enormous quantities of propellent required during the war, and it became necessary to purchase supplies from America. As the American factories were not equipped for the manufacture of cordite, but for nitrocellulose powders, and as these were found to shoot satisfactorily in British guns, enormous quantities of nitrocellulose powders were introduced into the British Service. The American rifle propellent, originally known as Dupont No. 16 powder, is now termed N.C.Z. in the British Service.

### 3.—General Characteristics of Small Arm Propellents

*Composition.*—Practically all modern smokeless propellents contain nitrocellulose in some form or other but it is customary to classify them into the two groups : nitroglycerine powders and nitrocellulose powders, according to whether they do or do not contain a proportion of nitroglycerine. Other ingredients are added for the sake of rendering the propellent chemically stable, reducing the amount of heat evolved on explosion, moderating the rate of burning, waterproofing, and for preventing electrification and sticking together of the grains. These ingredients will be referred to in the following sections. The composition of the propellents used in the British Service, together with certain explosion constants, are given in the following table.

*Composition and Explosion Constants of Different Propellents**

| Ingredient. | Cordite Mark I. | Cordite M.D. | Cordite R.D.B. | N.C.T. (American Smokeless Powder for Ordnance). | N.C.Z. (American I.M.R. or Dupont No. 16). |
|---|---|---|---|---|---|
| Nitroglycerine | 58 | 30 | 42 | — | — |
| Nitrocellulose | 37 | 65 | 52 | 99·4 | 92·4 |
| Mineral Jelly | 5 | 5 | 6 | — | — |
| Dinitrotoluene | — | — | — | — | 6·5 |
| Diphenylamine | — | — | — | 0·6 | 0·6 |
| Graphite | — | — | — | — | 0·5 |
| Nitrogen in Nitrocellulose (average) | 13·0 | 13·0 | 12·2 | 12·6 | 13 { 35 per cent. of 12·6, 65 per cent of 13·2 |
| †Heat of explosion (gram calories per gram ; water gaseous) | 1,114 | 939 | 904 | 777 | 815 |
| †Volume of gases produced on explosion c.c. per gram ; water gaseous) | 886 | 933 | 959 | 961 | 930 |

\* The figures given under composition are average throughout.

† Values obtained when propellent is fired at density of loading of 0·1.

*Physical properties.*—Most smokeless small arms propellents are manufactured in the form either of cords, strips or tubes cut approximately to the length of the cartridge case, or as small grains in the shape of square or round flakes, cubes or short cylinders or tubes. Shot-gun powders are for the most part in the form of irregular or rounded grains, but are sometimes made in very thin flakes. The texture of rifle powders is non-porous, so that burning takes place only from the surface; on the other hand sporting powders, which are required to burn much faster, are frequently made to be slightly porous so that the burning takes place almost simultaneously throughout the bulk.

Pure nitrocellulose powders have a horn-like consistency, and are greyish or yellow in colour; sometimes they are coated with graphite, which gives them a dark silvery-grey appearance. Nitroglycerine powders are softer than small arm nitrocellulose powders, and may vary in colour from light yellow to deep brown.

Nitroglycerine powders such as cordite or ballistite have only a very small tendency to take up moisture from the air. Nitrocellulose powders, however, are hygroscopic, and for this reason may alter in ballistics if left exposed to the air; special precautions are necessary, when loading them, to keep a correct degree of humidity in the filling rooms. Numerous efforts have been made to overcome this defect of nitrocellulose powders, but, so far, only a limited success has been achieved. For this purpose, the practice of treating the powder grains with substances such as dinitrotoluene or derivatives of urea, which are, absorbed into the surface layers, has in recent years been introduced. The effect of this is, however, only to reduce the rate at which moisture is taken up, but not the total amount absorbed. Coating the grains with graphite also assists slightly in protecting the surface from moisture, but the principal objects of this treatment are to prevent electrification of the grains, which occurs in its absence, and to prevent the grains sticking together during storage, and more especially during the rise of pressure in the cartridge when it is fired.

As far as experiments have hitherto shown, most rifle powders are fairly insensitive to shock, and are not exploded by the impact or passage through them of rifle bullets. The ignition or firing of one round inside a box of rifle ammunition has been found in repeated trials not to cause the explosion of the remaining rounds.

*Stability.*—Stability, both chemical and ballistic, under all conditions of climate, storage and use, is undoubtedly one of the most important properties of any explosive. With black powders, so long as they are kept dry, there was no question of want of stability. A quantity of gunpowder which was discovered walled up in Durham Castle, and must have lain there for centuries, was found to be in perfect condition. The nitrocellulose or nitroglycerine used in modern smokeless propellents, however carefully manufactured, decomposes very slowly at the ordinary temperature, and the products formed unfortunately promote the further decomposition of the explosive. The small size of rifle propellents, which permits of the escape of these products into the air, is undoubtedly an important factor which increases the chemical stability, but to obtain propellents satisfactory in this respect it is necessary to include in their composition small quantities of stabilizers, substances such as mineral jelly, chalk or diphenylamine, which convert the harmful products into innocuous substances. Generally speaking, the British rifle propellents are highly satisfactory in regard to chemical stability.

*Products of Combustion.*—Most smokeless propellents are composed of substances which when burnt are completely converted into gas. The products of their combustion do not vary very much, and are composed principally of a mixture of carbon monoxide, carbon dioxide, hydrogen, nitrogen and water vapour. When fired in a rifle, there are generally small residues of unburnt carbonaceous matter which, together with a certain amount of condensed water vapour, may produce a small quantity of smoke. The carbon monoxide and hydrogen, both of which are inflammable gases, generally take fire on issuing from the muzzle and give rise to the muzzle-flash. Under certain conditions of firing, however, these gases do not ignite and the carbon monoxide, which is very poisonous, may have injurious effects on the gunners should they be firing from a badly ventilated enclosure. Failure of the gases to ignite is more apt to occur with cool nitrocellulose powders than with nitroglycerine propellents.

*Ballistics.*—The introduction of magazine rifles and quickfiring guns necessitated, not only the employment of a powder producing little, if any, smoke, but also, to enable the weapons to develop their full effect, one giving much higher velocities than those obtainable with the old black powder, without exceeding the permissible limits of pressure in the bore. These improved ballistics became possible with the new powders, owing, to a great extent, to their colloidal and non-porous structure. Ignition having taken place on the surface of the grain or cord, combustion can only proceed by successive layers, with the result that, although a much larger total volume of gas, and therefore greater velocity of the projectile, is now developed than formerly, this development of gas may be made to take place gradually during the whole time of the passage of the projectile down the bore, with a correspondingly more uniform distribution of pressure. The total propelling force is naturally greater than formerly, but, as it is more sustained, the maximum pressure is not correspondingly increased. It follows, therefore, that for equal velocities much smaller charges are required than was the case when gunpowder was used, and the chamber pressures are lower, also for the same or lower chamber pressures higher velocities are obtained.

By increasing the thickness of the flakes or tubes or the diameters of the cords or cylinders, the surface of ignition for a given weight is decreased with a corresponding decrease in the initial development of gas and consequently of initial pressure, whereas the total time of combustion is increased. The thicker the flake tube or cord, the slower burning the powder and the larger the gun in which it can be advantageously used.

The rate of combustion of modern propellents is also regulated by the addition of the so-called "moderants," which both cool the powder and reduce its rate of burning. In the case of cordite the moderant, mineral jelly, is uniformly distributed throughout the explosive, but in the case of the nitrocellulose propellents, N.C.Z. and German rifle propellent, the moderants, dinitrotoluene and urea derivatives respectively, which, as mentioned before, are also used for waterproofing, are contained in the surface layers of the grains. The result is that the surface layers of these nitrocellulose propellents are less explosive and burn more slowly than the internal portions, and that the pressure developed in the chamber is correspondingly reduced; but, on the other hand, through the faster burning of the interior, the pressure is more maintained down the bore than with cordite. Propellents of this type are termed "Progressive."

From a ballistic point of view, the form of the propellent is important. A grained propellent can be more thoroughly blended than one in the form of cords, and for this reason should give more uniform ballistics. This advantage is, unfortunately, counterbalanced by a less uniform ignition of all the grains by the flash from the percussion cap. Chopped tube has an advantage in this respect over flake in that it allows the igniting flash to penetrate more easily through the bulk. Flake has the further disadvantage of tending at times to bunch together in the choke of the rifle, and thus to give excessive chamber pressures, while, on the other hand, several grains may be pressed together during firing so that they burn as a single thick grain.

*Effects on the Rifle.*—The principal effect produced by the propellent on the rifle is the gradual wearing away of the steel at the forward end of the chamber and in the adjacent bore. This is termed erosion. It is not a very serious trouble in ordinary rifles, which are unlikely to be used for a great number of rounds, but in machine guns it is one of the principal factors determining the life of the barrel.

The different propellents vary very considerably in the amount of erosion produced; in the case of cordite Mark I, the effect was so great that it was necessary to find another propellent. The amount of erosion is found to be closely connected with the quantity of heat evolved on explosion, figures for which are given for several of the most important propellents in the table on page 218. Nitrocellulose powders are the best in this respect, while cordite M.D. and R.D.B. occupy intermediate positions between these and cordite Mark I.

A further effect, which is traceable to the explosives used, is the tendency of the barrels of small arms to become rusty after firing. This is not accentuated by the gases produced by the propellent, but is due to the products of combustion of the cap compositions. These compositions almost always contain chlorate of potash, which on firing is converted into potassium chloride. Particles of this salt settle in the bore, where they absorb moisture and cause rusting of the adjacent steel. Since potassium chloride is readily soluble in water, the simple expedient of rinsing out and drying the barrel after firing is sufficient to obviate its effects.

### 4.—Manufacture of Small Arm Propellents

The manufacture of practically all natures of smokeless small arm propellents is generally conducted on the same broad lines, and the principal operations may be summarized as follows :—

1. Manufacture of a stable nitrocellulose.
2. Incorporation of the nitrocellulose with other ingredients together with a solvent to form a plastic colloid.
3. Formation of the plastic colloid into shapes of definite dimensions.
4. Elimination of the solvent.

*Manufacture of Nitrocellulose.*—Cellulose derived from a great many types of plant and tree has been used for the manufacture of nitrocellulose, cotton being that most frequently used. The material is, if necessary, extracted with solvents to free it from oil, boiled with alkali and lightly bleached. It is subsequently picked over by hand to separate pieces of wood, wire, string, etc., and passed through various machines which open out the fibres, further purify and ultimately dry it. These precautions are necessary since certain foreign matter has a bad effect on the stability of the resulting propellent.

The dry cellulose is next steeped in a mixture of nitric and sulphuric acids, mechanical stirring being used in some factories where short-fibred material is employed. The strength of the acids, time and temperature of nitration is varied according to the degree of nitration required in the product. At the end of the nitration the spent acids are removed from the nitrocellulose either by centrifugal wringers followed by drowning in a large volume of water, or by an ingenious displacement process invented by the Messrs. Thomson. In the latter process water is gradually run on top of the acids and nitrocellulose, while the acid is drawn off at an equal rate from the bottom of the nitrating vessel, comparatively slight mixing of acid and water taking place.

The nitrocellulose, which contains a considerable quantity of acid, is then boiled with water in large wooden vats, which are sometimes lined with lead, for about 40 hours. During the boiling the water is changed several times to get rid of acid and impurities, but it is important that the first boiling should be done with water which is actually acid since this destroys certain unstable impurities in the nitrocellulose more rapidly than if neutral or alkaline water were used. If the water is hard, a certain amount of chalk is deposited on the nitrocellulose during the later boilings; it may, however, be necessary to make a special addition of chalk or other alkali.

The boiled nitrocellulose is then finely pulped in machines similar to those used in the paper industry. During this process a certain amount of acid which was trapped inside the fibres escapes. The pulp is washed with several changes of water, and in some factories subjected to a further boiling. It is finally passed through traps and electromagnets to remove foreign matter and either centrifuged or pressed to remove excess of water.

*Manufacture of N.C.Z.*—This propellent is made from two nitrocelluloses of different nitrogen contents, one of which alone is gelatinized in the final powder. In the first instance, the nitrocelluloses are carefully blended in a pulped condition by stirring under water. The mixture is wrung so as to leave about 30 per cent. of water in it and then dehydrated by placing it in a cylinder in which it is first squeezed and then treated with alcohol, which is run into the top of the cylinder, gradually displacing the water and finally completely dehydrating the nitrocellulose. The block of nitrocellulose containing a definite proportion of alcohol is then broken up in a rotating drum and placed in an incorporating machine. This consists of a covered metal box, in which a double worm revolves, giving a kneading motion to the material in the mixer. At this stage a proportion of ether, together with a small quantity of graphite and diphenylamine is added. After mixing for one hour, the nitrocellulose mixture, which has been partly gelatinized by the solvents, is transferred to a press in which it is subjected to a pressure of 1,800 lb. per square inch.

The block so formed is then extruded as cords through a die containing many holes, the motion helping to effect further gelatinization. The cords are reformed into a block by pressure and finally extruded through a multi holed die which forms this into a number of tubular cords. Each cord is collected separately and is subsequently run through a cutting machine, by which it is cut into short lengths. The chopped grain is partly dried in a slow current of warm air from which the solvent is recovered, then immersed in water which displaces most of the residual alcohol. The grains are then placed in a tumbling machine under warm water, together with the requisite quantity of dinitrotoluene, and worked until the whole of this substance has been absorbed into the surface layers. The propellent is finally dried in air, until it contains the correct amount of moisture, and blended in large batches.

It is noteworthy that the gelatinizing solvent used for nitrocellulose powders is ether-alcohol and not acetone, the product from the latter solvent being much too brittle.

*Cordite Mk. I and Cordite M.D.*—These propellents contain nitroglycerine which is manufactured by nitrating glycerine, in approximately 1500 lb. batches, in a special nitrator which contains no pipes with cocks, and which is provided with cooling coils (brine) and air under pressure for agitation. The mixture of concentrated sulphuric acid and nitric acid is run into the nitrator, and the glycerine charge introduced beneath the surface of the acid. The temperature of nitration is kept at 10° C. by adjusting the speed of addition of the glycerine and the brine flow through the coils. The time of nitration is about 1 hour.

The nitroglycerine which separates on the surface of the waste acid is displaced into a preliminary washing tank and then subjected to a thorough washing treatment and finally dried by a filtration method.

The nitrocellulose used in cordite is known as guncotton, and contains 13 per cent. of nitrogen, almost the maximum quantity which can be introduced from the point of view of chemical stability. Although it is blended intimately with the nitroglycerine in the finished explosive, guncotton will not dissolve in nitroglycerine under ordinary conditions.

In the manufacture of cordite, the guncotton is first dried in a current of warm air in stoves, this operation having to be done with extreme care on account of the tendency of dry guncotton to fly about as dust and of its highly sensitive character. The correct quantities of dry guncotton and nitroglycerine are first blended to some extent together by hand, and are then introduced into an incorporating machine (described above), together with the requisite quantity of acetone. For the manufacture of Mark I about 23 per cent., and for M.D. about 40 per cent., of solvent is required. After working for about three and a half hours, when most of the guncotton has been gelatinized and blended with the nitroglycerine, the mineral jelly is added, and the working continued for an equal time. The resulting mass, which has a dough-like consistency, is pressed through a die and formed into cords. These are dried in stoves at 110° F., until practically the whole of the solvent has been evaporated.

*Sporting Ballistite.*—This powder consists of approximately three parts of nitrocellulose blended with two parts of nitroglycerine. Unlike cordite, the nitrocellulose used contains only about 12·6 per cent of nitrogen, and is completely soluble in nitroglycerine. Instead of drying the nitrocellulose, it is mixed under water with the nitroglycerine by air-blowing. During this process the nitroglycerine is rapidly absorbed by the nitrocellulose, causing partial gelatinization. Excess of water is squeezed out and the resulting mass passed between rollers heated to from 60° to 80° C. The heat causes partial gelatinization and forms the explosive into a sheet which is folded up and passed through the rolls a sufficient number of times to complete the gelatinization. The resultant colloid is then rolled to a definite thickness, cut into square flakes, and finally coated with graphite. Various substance have been added from time to time to improve the stability. Occasionally a small percentage of acetone, subsequently evaporated off, is added to facilitate gelatinization.

# CHAPTER IV

## SMALL ARM CARTRIDGE MANUFACTURE

### 1.—General Considerations.

The manufacture of small arm ammunition is a very highly specialized industry. The processes employed are numerous, and their full understanding involves a scientific knowledge of physics, chemistry and metallurgy, requiring a lifelong study. Even where no attempt is made to cover the whole field of the various types of sporting and military rifle and shot gun ammunition, and consideration is confined to one single design of cartridge, its satisfactory manufacture requires special technical knowledge which is not available in text books, and which can only be acquired by actual experience over a long period in an ammunition factory.

Apart from the necessity for scientific and technical knowledge, high quality ammunition requires extreme regularity of manufacture, and as cost is a matter of great importance, this regularity of manufacture must be obtained by mass production methods. As the science of cartridge manufacture has developed, appropriate machinery for each operation has been designed, and from time to time improved, until now cheap mass production of high class ammunition necessitates an up-to-date factory suitably laid out, and equipped with special modern machinery, each item of which is peculiar to the operation it has to perform, while the necessity for strict interchangeability of components is ensured by a careful organization controlling the sequence of operations, by the maintenance of all tools and machinery in a high state of efficiency, and by the ruthless elimination of faulty material during manufacture.

These general considerations governing cartridge manufacture have been put forward in some detail because they have an important bearing on the capacity of any nation to manufacture small arm ammunition in sufficient quantities in time of war. The rate at which ammunition was required during the Great War was so much in excess of peace time requirements, that it is evident that it is quite impracticable during peace time to maintain sufficient factories for a large war. On the other hand, it is clear that the manufacture of small arm ammunition is a serious enterprise, and that any attempt to increase capacity very rapidly must be attended by considerable risk. Not only, therefore, is it essential that there should be maintained in peace time sufficient manufacturing capacity to carry on during the early stages of a war, before new factories can be equipped, but one of the clearest lessons of the Great War is that it is vital that the difficulties of manufacture should be appreciated, and that as large a nucleus as possible of staff possessing practical knowledge of manufacture should be retained in the peace time factories available for purposes of rapid expansion.

The various types of military small arm cartridges differ considerably in design, and a description of the methods of manufacture of every one of them would occupy many volumes. Even in the case of one single design of cartridge, different factories use slightly different methods, depending on their own experience and local facilities, and although the various methods may not differ very largely, it is not possible to give a full description covering all variations in details of manufacture. So far as the aspect of the question mentioned above is concerned, this country is most interested in the manufacture of the ·303-inch Mark VII British rifle cartridge, and it is proposed, therefore, to limit this chapter to a consideration of this cartridge, dealing rather with the scientific problems involved, and confining the description of the processes employed to one typical method of manufacture.

The Mark VII cartridge has a pointed bullet weighing 174 grains with a lead core enclosed in a drawn cupro-nickel envelope. The point of the bullet is struck to a radius of about eight calibres, and in the pointed end of the envelope in front of the lead is an aluminium tip. This is required partly to bring the bullet to the correct weight for the required ballistics, and partly to balance the bullet suitably for accurate shooting. Near the rear end of the bullet there is a cannelure, which is filled with beeswax, provided partly for lubrication and partly to waterproof the joint between the bullet and case. The bullet is secured into the case by three indents made in the case, the metal of the case being pressed into the cannelure.

The case is of solid drawn brass, and has a rim or head at the base end by which the cartridge is positioned in the gun and extracted. The cap chamber is recessed into the back of the case, and has a fixed anvil, two fire-holes being provided.

The cap, which is made of copper, is pressed into the case and rivetted in place, and the joint between the cap and case is varnished.

The cap composition is pressed into the cap, and a disc of varnished tinfoil is placed over it, the whole being waterproofed with a coating of shellac varnish.

There are for this cartridge two alternative types of charge, cordite and nitrocellulose powder. The cordite used is in the form of tubular sticks of the type and size known as M.D.T. 5-2. The nitrocellulose powder is in the form of chopped tube, and is of the "progressive" type, that is to say, it is impregnated to a certain depth with dinitrotoluene which has the effect of slowing the rate of burning during the early stages of combustion.

In both cases a glazed board disc is provided as a wad and placed between the propellent and the bullet. The material of which the wad is made has to be carefully selected, and must be chemically neutral, since the presence of either acid or alkali is liable to affect the quality of the cordite in course of time.

## 2.—General Factory Organization

The manufacture of small arm ammunition divides naturally into four sets of operations which can be carried out in separate factories. The first of these is the foundry where the brass and cupro-nickel ingots are cast. Then there is the rolling mill where the ingots are rolled into strip. The third is the component factory where the cartridge cases and bullets and caps are made, and lastly there is the filling factory where the caps are primed, and the cartridges are loaded.

For purposes of ease of transport it is desirable that the foundry and rolling mill should be near to each other, and where the component factory is any considerable distance away, it is usually convenient to do the first component operation, which is known as "cupping," in or near the rolling mill, as it is easier to transport the cups from which the cartridge cases and bullet envelopes are made, than to take the more cumbersome strips of metal down to the component factory, and afterwards bring back the remains of the strip or "webbing," which is used again as scrap metal in the foundry.

The layout of the component factory should be such that the moving of components backwards and forwards is reduced to a minimum. Where the rate of manufacture is such that each operation requires several machines, these should naturally be alike for simplicity in tool making, and it is preferable to put all machines doing one operation together in one bay of the shop, which facilitates supervision, and to some extent prevents mixing of components in different stages of manufacture.

A considerable amount of annealing and pickling takes place in connection with component manufacture, and the production of fumes from the annealing ovens cannot be avoided, whilst the pickling processes involve the use of acids. It is advisable, therefore, to arrange the annealing, pickling, and washing processes in a chamber separate from the rest of the component factory, and with outside access for delivery of fuel and materials. As, however, a cartridge case has to be annealed several times during manufacture, there must be easy access to the annealing room from the component factory, and the machines for those operations which immediately precede or follow an annealing process should be grouped conveniently near to the annealing room.

The work in the filling factory is of course carried out under Danger Building conditions. According to the Danger Building regulations the employees have to pass into the building over a barrier, the factory side of which is known as the " clean " side. Inside the barrier suitable clothing and rubber shoes have to be worn, and the possession of matches or other means of obtaining a light, tobacco, knives or steel implements is forbidden. No naked light is allowed inside the factory, nor anything liable to cause a spark. A full description of Danger Building regulations cannot be given here, but further particulars can be obtained from " Magazine Regulations, 1922."

Some means must be provided for controlling the hygroscopic conditions of a loading field to suit the powder which is being loaded. A rifle range is also required within easy reach of the filling factory, equipped for taking velocities and pressures for the assessment of the charge when loading.

As is always the case in a repetition factory turning out very large quantities of similar articles, the manufacture of which is highly organized, the nucleus both of the component and filling factories may be said to be the tool room. In the case of small arm ammunition regularity of manufacture can only be obtained by tools very accurately made and strictly interchangeable, and the tools and gauges are themselves made in quantities in a repetition manner, strictly to drawing, and are submitted to an independent inspection.

Cartridge components are all carefully examined in an inspection room set apart for the purpose. The object of this examination is to eliminate components which are incorrectly made, either because a tool has broken or got out of adjustment, or because one of the operations has been omitted. Where a faulty tool is causing bad work, it is not practicable however to wait for it to be found out in the inspection room, as in the meantime it may have spoilt some thousands of components, and therefore to every batch of machines there is allotted an itinerant examiner whose function is to take components straight off the tools and examine them, and who stops the machine immediately, and has it adjusted, if it is found to be making faulty work.

Owing to the narrow limits to which the various components are made, and to the numbers involved, small arm ammunition gauges wear out very rapidly. For this reason a careful organization is necessary for the checking and standardization of gauges, and in the case of the gauges employed by the itinerant examiners, which are in constant use, it is customary to provide two sets, which are used on alternate days, the gauges being checked on the days that they are out of commission.

### 3.—Case Manufacture

The two most important qualifications in a cartridge case are that the brass should be of really good quality, and that after manufacture the various parts of the case should be of suitable hardness.

The cartridge case is made for various reasons slightly smaller than the rifle chamber into which it fits. When the charge is fired the internal pressure expands the case, and if the brass were dead soft it would not spring back after expansion sufficiently to prevent it clinging to the walls of the chamber. This would lead to difficulties in extraction, and the case would be liable to be torn asunder. If, however, the brass is harder it springs back to a greater extent after expansion, and the extraction difficulty is avoided, but hard brass is more likely to split or crack during expansion than soft, and extreme hardness is not permissible in a cartridge case.

The most suitable hardness is not the same at all parts of the case. Generally speaking, the case should be hardest at the base end, and get gradually softer towards the front. It is softness at the rear end which is chiefly responsible for extraction difficulties, while if the front end of the case is too hard, in which case there is greater likelihood of internal stress in the metal, the brass is liable to split spontaneously in course of time, especially in tropical climates. The hardness about the centre should be intermediate between that of the front and back, and an unsuitable distribution in the hardness may lead to the fracture of the case during extraction. A good deal of careful research into this question has been carried out, and the ideal hardness for different portions of a cartridge case has been generally determined.

The apparatus most commonly used for making tests on the hardness of cartridge cases or bullet envelopes is a special form of Brinell hardness testing machine. The process consists in measuring the diameter of the depression caused by a hardened steel ball of given size after the application of a known weight. The Brinell hardness number is the figure obtained by dividing the load in kilograms by the area of the depression in square millimetres. As the Brinell hardness number obtained with any given material depends on the size of the ball and the weight used, comparative results can only be obtained where one combination of ball and weight is used throughout. In the case of small-arm ammunition, where the materials to be tested are thin, it has been found convenient to use a 1-millimetre ball and a weight of 10 kilograms.

The ideal hardness of a cartridge case as indicated by this scale is as follows :—

| Distance from Base in inches. | Brinell Hardness Figure. |
|---|---|
| 0·15 | 135–145 |
| 0·2 | 135–145 |
| 0·3 | 135–145 |
| 0·7 | 135–145 |
| 1·1 | 125–135 |
| 1·5 | 110–120 |
| At neck | 90–95 |

The process of making a cartridge case consists in punching a disc out of a strip of brass, and pressing this into the form of a shallow cup. This cup is then pressed or "drawn" by a punch through a hollow die, so as to make the walls thinner and the cup deeper, and this process is repeated until the original disc becomes a long thin tube closed at one end. The end is then pressed, and the cap recess punched, and the whole case formed by other punches and dies, till it takes the shape of a cartridge case.

Now brass is a metal which when pressed or rolled or otherwise cold worked and distorted becomes harder and more brittle, until finally it becomes impossible to work it any more. It must then be softened, by the process of annealing. This consists in heating it at a suitable temperature and afterwards either quenching it in water, or allowing it to cool slowly.

The ultimate hardness of the various parts of a cartridge case is regulated by the arrangement of the operations, which are so schemed that the amount of distortion to which each part is subjected, subsequent to the final annealing, produces the amount of hardness required. In this connection the diagram given in Fig. 1 is interesting. This shows the ascertained hardness of some sample cartridge cases tested at various stages of manufacture. This diagram will not be thoroughly understood until the description of the operations which follows has been studied, but it is clear that previous to annealing the brass reaches a hardness figure somewhere about 160, and that after annealing this falls to about 65.

It is probable that the ideal hardness of the finished cartridge case at the various points cannot be obtained by this method with great exactitude, except at great expense, but with careful control of the operations and of the annealing the desired result will be attained within reasonable limits. It will, however, be clear that variations in the annealing have some effect on the quality of the finished cartridge case, and that careful annealing is therefore important.

The amount of softening that is obtained in a given time is greater if the annealing is carried out at a higher temperature, but a temperature above 850° C. is not permissible as the brass may become "burnt" and its quality destroyed. The following table

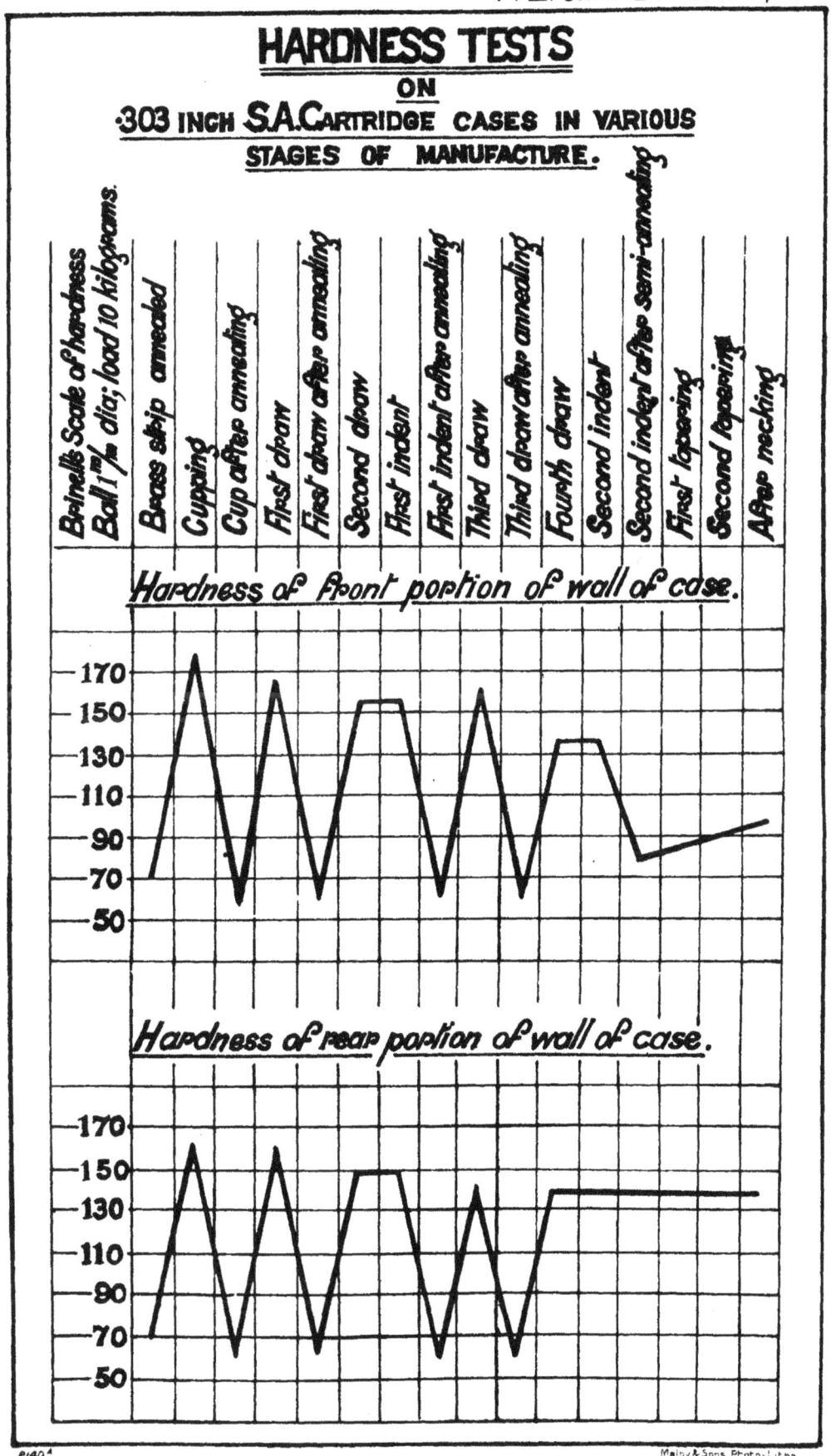

FIG. 1.

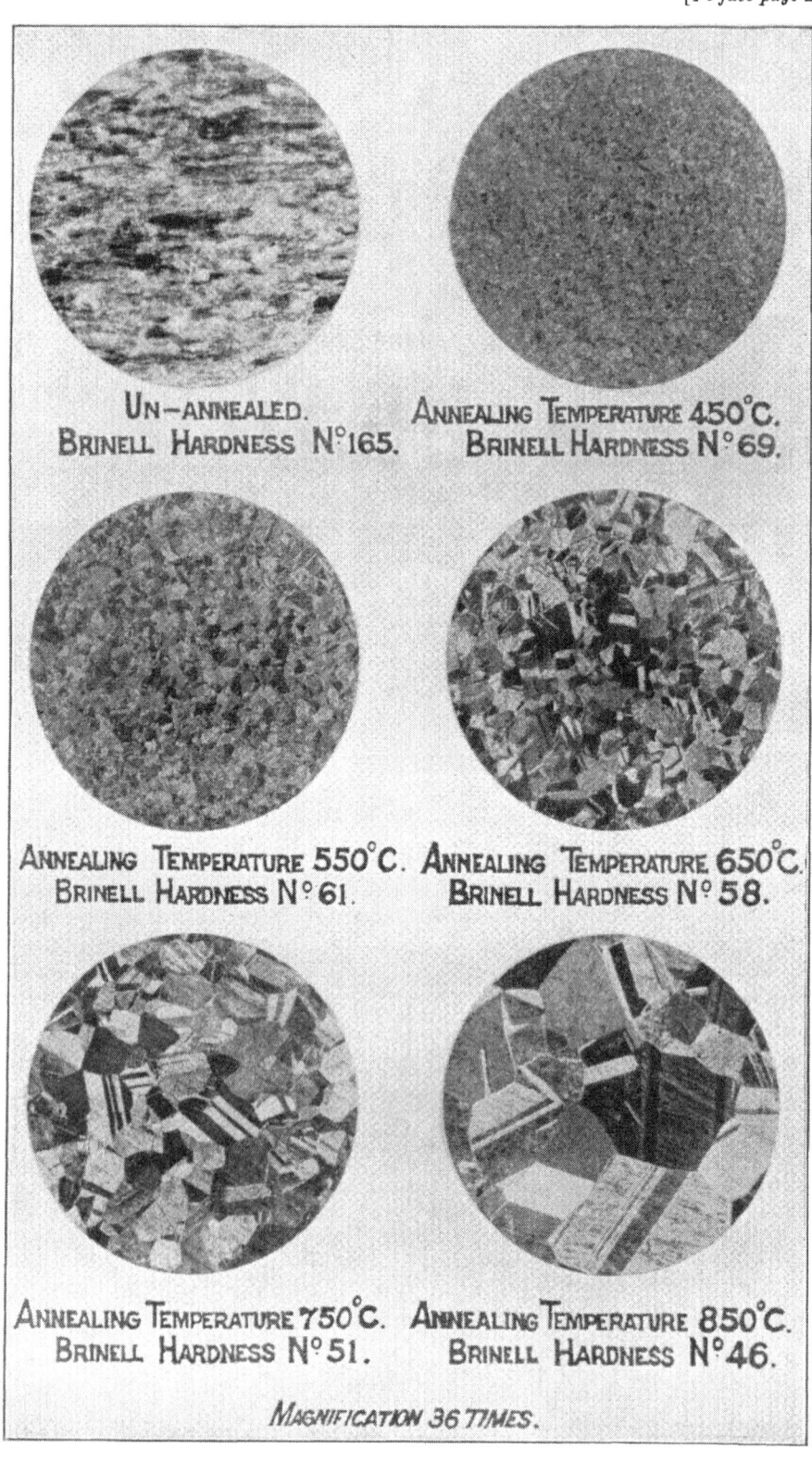

Part II, Chap. IV, Fig. 2.

gives the actual results of some tests which have been carried out on samples of hard rolled cartridge brass, showing the effect of annealing for a given time at different temperatures:—

| Annealing Temperature. Degrees Centigrade. | Yield Point. Tons per square inch. | Tensile Stress. Tons per square inch. | Elongation. Per cent. on 2 inches. | Contraction of Area. Per cent. | Brinell Hardness Number. 1 mm. ball, 10 kilo. load. |
|---|---|---|---|---|---|
| Unannealed | 36·80 | 39·10 | 15 | 67 | 165 |
| 450 | 10·77 | 23·65 | 54 | 60 | 69 |
| 550 | 8·62 | 21·88 | 63 | 61 | 61 |
| 650 | 8·02 | 20·95 | 66 | 64 | 58 |
| 750 | 6·70 | 19·94 | 74 | 63 | 51 |
| 850 | 6·05 | 19·25 | 74 | 57 | 46 |

*Note.*—The unannealed sample was taken from strip after rolling down to 48·5 per cent. of its original thickness.

The results of the tests on the brass annealed at 650° Centigrade may be considered satisfactory, and these are the conditions aimed at. The time required, however, varies to some extent with the weight of the brass being annealed, and small thin pieces, such as cartridge cases in the later stages of manufacture, are annealed at a lower temperature than thick heavy pieces such as brass strips.

Owing to the importance of the annealing operations, the temperatures of the annealing ovens are most carefully watched, and test pieces are taken from each batch annealed. These are sometimes submitted to a Brinell test, but generally the test piece is polished and etched, and its micro-structure examined. This method of examination affords a ready means of judging the true condition of the metal. In Fig. 2 are given photos of the microstructure of the brass samples mentioned in the tests above, both before and after annealing. It will be seen that there is a considerable difference in the crystalline structure of the brass when annealed in different ways, and that when the annealing temperature is high the size of the crystals is larger.

The brass which is used for making cartridge cases must be of suitable composition and good quality. However carefully the cases may be made and however correctly they may be annealed, unless they are made of good and suitable brass they may fail under the very severe strains to which they are subjected when the cartridge is fired.

The brass used for cartridge cases should contain from 68 to 74 per cent. of copper, and from 32 to 26 per cent. of zinc. It must not contain more than—

|  | Per cent. |  | Per cent. |
|---|---|---|---|
| Nickel | 0·2 | Cadmium | 0·05 |
| Iron | 0·15 | Bismuth | 0·008 |
| Lead | 0·1 | Antimony | Nil |
| Arsenic | 0·05 | Tin | Nil |

and not more than a trace of any other impurity.

A copper content not below 70 per cent. is generally preferred, and if scrap brass is included, only clean process scrap should be used to an extent not greater than 45 per cent. of the whole.

The metal is melted in a crucible usually made of plumbago. The scrap brass up to the percentage it has been decided to use is first put into the crucible and melted, a chemical analysis of the scrap being taken. Virgin copper and spelter are then weighed out in proportions calculated from the analysis of the scrap so as to produce the correct composition of the whole alloy. The copper is laid on the top of the crucible to get well heated, while the scrap brass is melting, and after the brass is melted the copper is added by degrees, and when this is all melted the spelter is put in. This melts quickly on insertion, and the whole is then thoroughly stirred with an iron bar, so as to make a homogeneous alloy. A little borax is added as a flux to facilitate the collection of impurities on the surface, and the molten metal is covered with a layer of charcoal and heated up to a temperature of about 1,180° Centigrade.

The crucible is then withdrawn from the furnace, the dross skimmed off, and the molten metal is poured into metal moulds. These moulds are in two pieces in the form of channels. The inner surfaces of these are oiled, and they are then put together and secured by links, and stood up on end. After the metal has cooled the links are knocked off the moulds, and the ingots taken out. The top or "pipe" end of each ingot is cut off, so that all spongy metal is discarded, and the ingots are scraped and filed, to remove any carbonised oil or other impurity, and visually examined for flaws or other defects.

The size of the ingot is determined by the required dimensions of the finished strip, which in turn are chosen so that the brass pieces from which the cartridge cases are formed may be punched out with the minimum of waste. A convenient section for the ingot is $3 \cdot 2$ inches wide by $1 \cdot 5$ inches thick. A chemical analysis is taken of each pouring, and the ingots are bonded until this is known to be satisfactory. The finished ingot should have a Brinell hardness of about 58.

The ingots are then taken into the rolling mill and are passed through the rolls, generally four times till they are reduced to a thickness of $\cdot 7$ inches. They are then annealed. The annealing oven in this instance consists of a large muffle in which the bars are placed until the requisite amount of annealing has taken place. The bars are then quenched in water, pickled in dilute sulphuric acid to remove scale, rinsed and washed in soapy water.

The bars are then again passed through the rolls, usually four times, until they are reduced to a thickness of $\cdot 33$ inch. As the bars are by this time becoming too long for convenient handling, they are cut in half. They are then again annealed, pickled in sulphuric acid, and washed as before.

The strips are then rolled about three times until they are reduced to a thickness of $\cdot 16$ inch, and they are again annealed as before, except that this time they are allowed to cool slowly and are not quenched in water, and after the pickling process the strips are visually examined all over and patches of dross or scaly metal removed with a scraper.

The strips are then finally rolled to the finished size, $\cdot 14$ inch thick, being passed through the rolls twice. The ends are then squared, and the strips again annealed, being allowed to cool slowly. On this occasion the strips are first pickled in the dilute sulphuric acid and rinsed, and they are then immersed in a mixture of nitric acid and sulphuric acid and rinsed and washed. The brass strips are then complete and ready for the cupping machines.

The cupping machines are powerful punches in which discs are cut out of the brass strip and pressed into the form of cups. Fig. 3 shows diagrammatically how this is done. A is the brass strip, B is a hollow cutting punch, and C the die. As the punch B descends into the die C it cuts out a disc D. In the centre of the hollow punch B is a separately actuated punch E, which then descends and presses the disc D downwards, and forces it through the die C, from which it emerges in the form of a cup. The first section shows the strip before the disc is cut out; the second section shows the punch E beginning to descend after the disc has been cut; and the third section shows the cup formed.

If the strip is at all thicker or is harder on one side than another, there is a tendency for the punch E to be bent sideways, and the cup is then made thicker on one side than on the other. This is very objectionable, as the fault is carried on through all the stages of manufacture and appears in the finished case.

In order to waste as little of the brass strip as possible, the positions of the discs are staggered. This necessitates the use of multiple punches, as the strip must be fed forward in a straight line. It is usual to use four sets of punches and make four cups at one stroke, and the discs are therefore not punched on the centre line of the strip. For these reasons it is essential that the strip should be of even thickness all over, and it is particularly

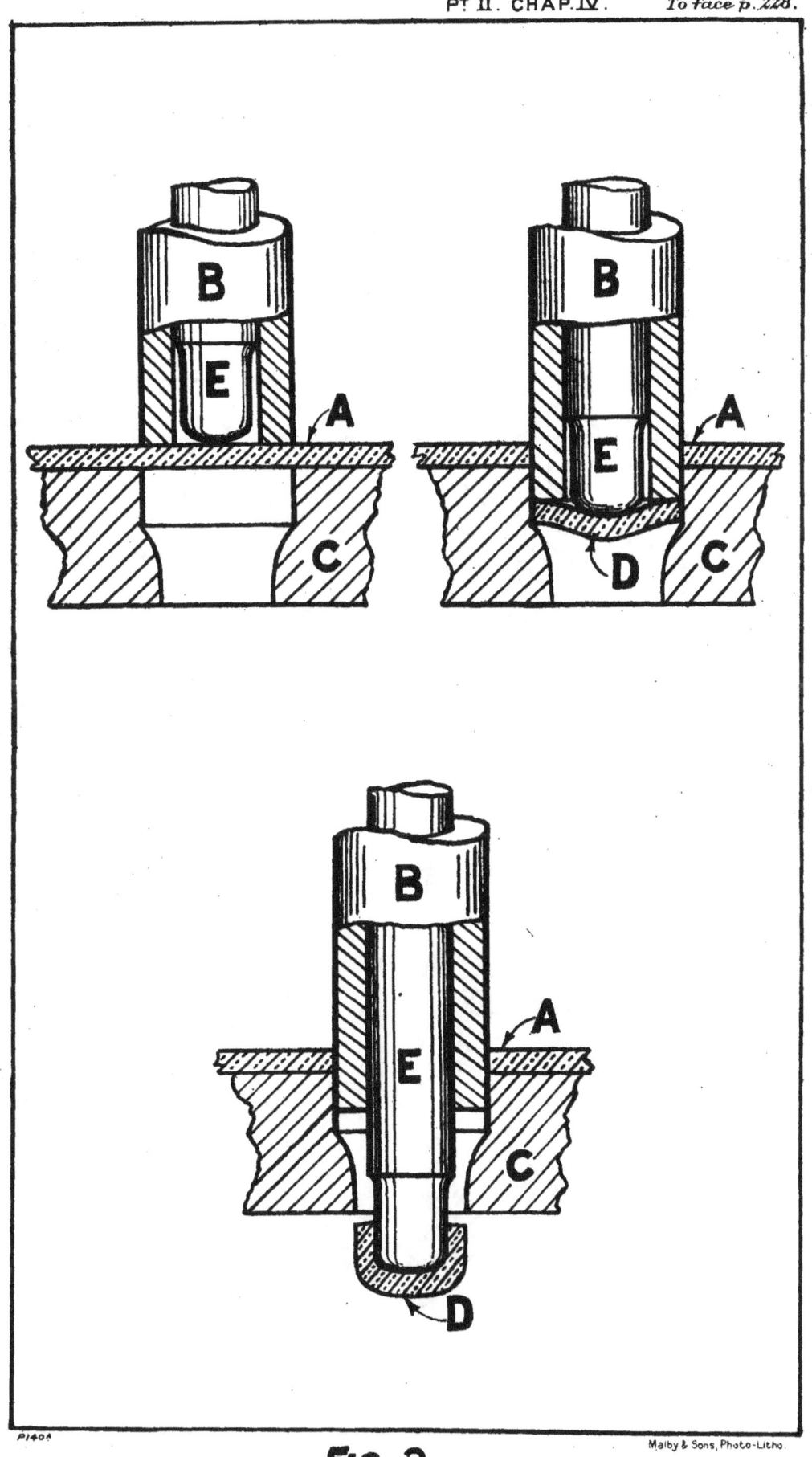

FIG. 3.

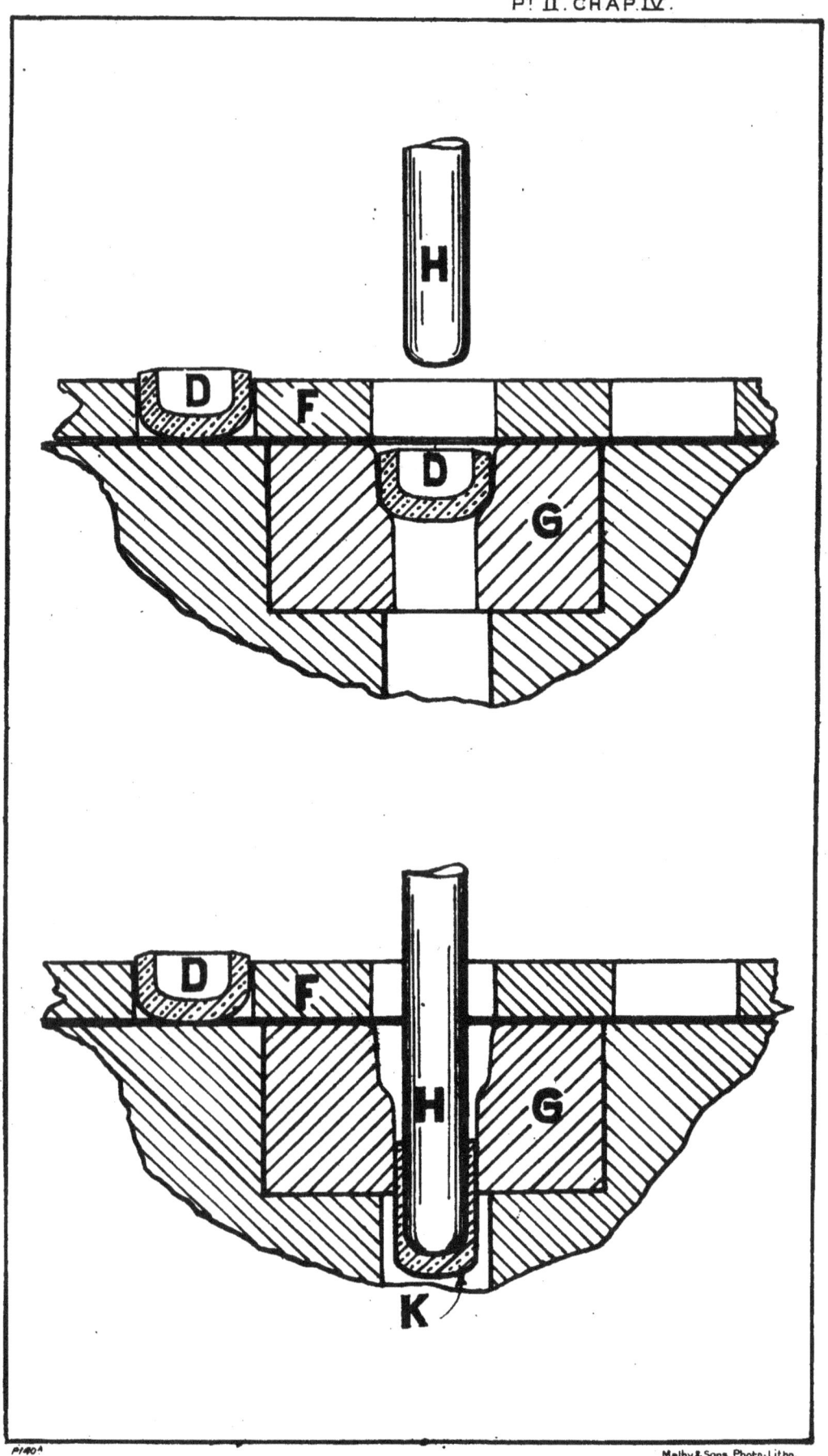

FIG. 4.

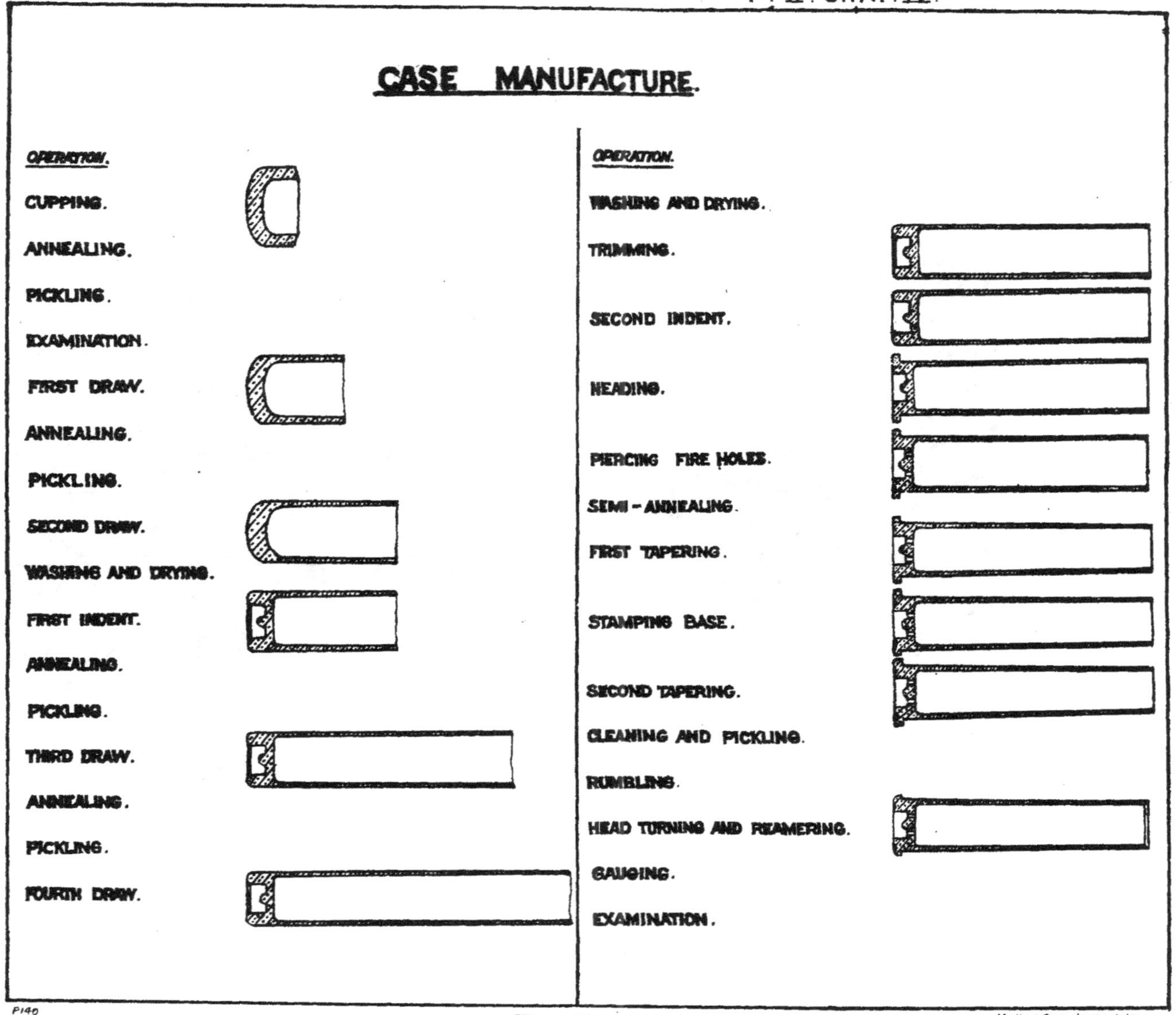
CASE MANUFACTURE.
OPERATION.
CUPPING.
ANNEALING.
PICKLING.
EXAMINATION.
FIRST DRAW.
ANNEALING.
PICKLING.
SECOND DRAW.
WASHING AND DRYING.
FIRST INDENT.
ANNEALING.
PICKLING.
THIRD DRAW.
ANNEALING.
PICKLING.
FOURTH DRAW.
OPERATION.
WASHING AND DRYING.
TRIMMING.
SECOND INDENT.
HEADING.
PIERCING FIRE HOLES.
SEMI-ANNEALING.
FIRST TAPERING.
STAMPING BASE.
SECOND TAPERING.
CLEANING AND PICKLING.
RUMBLING.
HEAD TURNING AND REAMERING.
GAUGING.
EXAMINATION.

P140

FIG. 5.

Malby & Sons, Photo-Litho

PT II. CHAP. IV. To face p. 229.

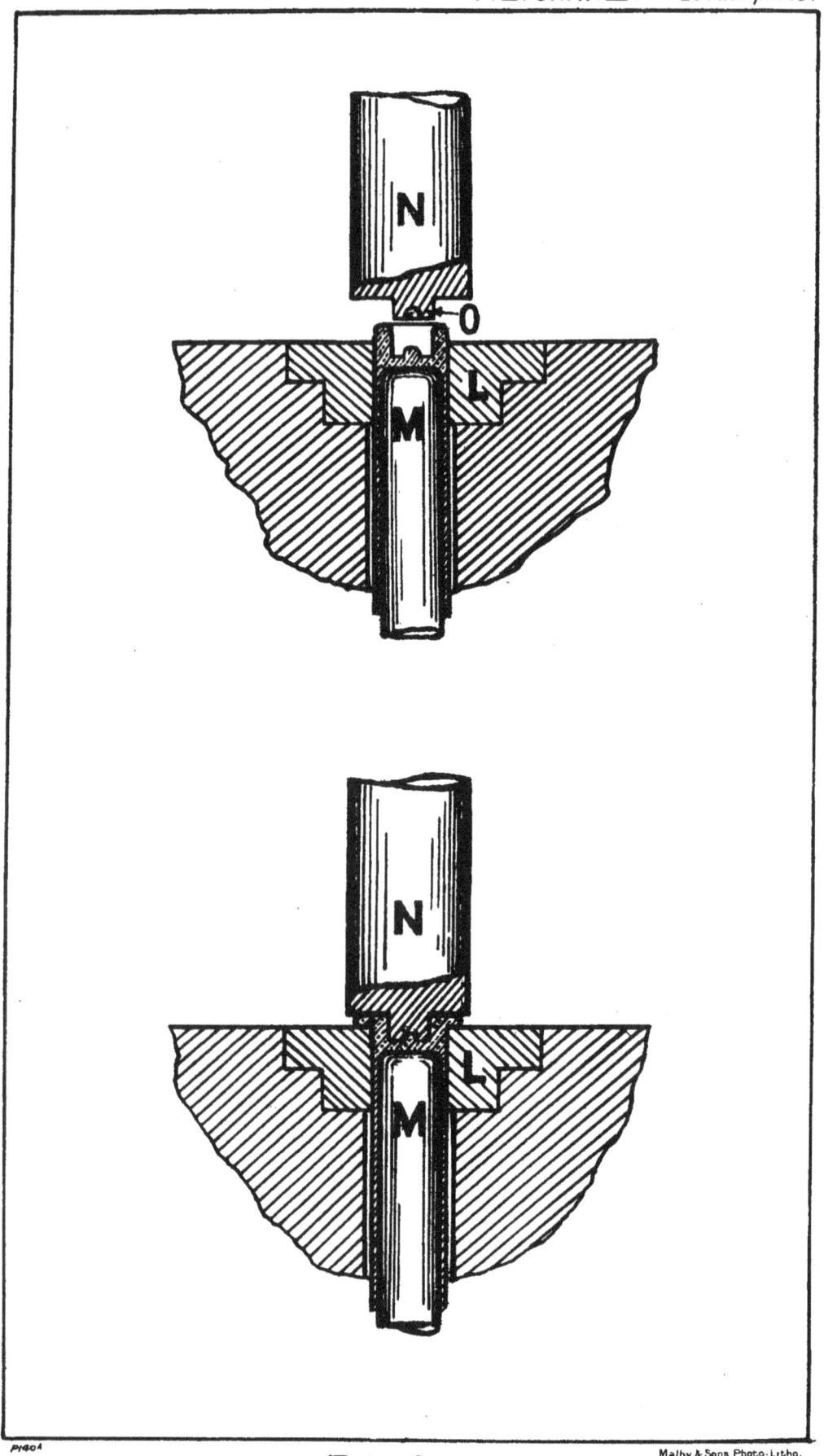

FIG. 6.

important that it should be the same thickness at the two sides as it is in the centre, and it must be equally annealed all over.

The cups have then to be annealed. Where, as in this and subsequent operations, the brass consists of a large number of small pieces, the annealing apparatus generally takes the form of a travelling tray, or rotating drum, which passes the pieces through the oven in a given time, the oven being maintained at a fixed temperature. After annealing, the cups are "pickled" or cleaned by immersion in dilute sulphuric acid, rinsed in water, washed in soapy water, and sent to the viewing room, where they are examined, and any which are eccentric or show cracks or other flaws are eliminated.

The cups have next to be drawn out into deeper and thinner tubes, until they take a shape approximating to that of a cartridge case. The process of "drawing" consists in pressing the cups through a hollow die by means of a punch. The punch is made smaller than the internal diameter of the cup, so that some of the metal of the crown of the cup is transferred to the walls. The die is also made smaller in diameter than the cup, and the difference in diameter between the punch and the die is such that the annular space between them is thinner than the walls of the cup. The effect of this is that the cup is reduced in diameter and the walls are made thinner, and the surplus metal goes to make the cup deeper. The process, which is shown diagrammatically in Fig. 4, is as follows:—The cups D D are fed into holes in a rotating disc F, which carries them in turn over the die G, into which they fall. The punch H then descends, and presses them through the die, from which they emerge in the form K. The diagram illustrates the first drawing operation, the first section showing a cup in position in the die before the punch has descended, while the second section shows the punch descending.

The amount of drawing which can be carried out at each operation is dependent on the extent to which the brass can be distorted without becoming too hard. It is usual to draw a cartridge case four times, and the approximate shape after each operation is shown in Fig. 5. After each draw except the last the cases are annealed and pickled in dilute sulphuric acid and washed in soapy water.

It will be realized that after drawing, although the walls of the case are fairly hard, practically no work has been done on the crown of the case, which is consequently comparatively soft. Advantage is taken of this fact to carry out the process known as first indenting before the case is annealed after the second draw. This process consists in pressing the crown of the case between two punches, the outer one of which is shaped so as to impress into the metal the recess for the cap chamber and anvil. It is very important that there should be no grease or liquid on the case or punch during this operation, since any liquid trapped in a closed recess such as the cap chamber would act hydrostatically, and would deform the chamber. For this reason after the second draw and before the first indenting the cases are washed in soda water and dried in hot air.

When cases are drawn the edges of the walls become uneven in thickness and ragged, and the fourth draw is arranged therefore to make the cases longer than required. After this operation the cases are washed and passed through boiling water, owing to the heat of which they dry automatically, and they are then trimmed to length, the excess metal being cut off in an automatic lathe.

Here again advantage is taken of the comparative softness of the head to carry out at this stage the second indenting, the heading, and the piercing of the fire holes. The second indenting is similar to the first, but is the finishing operation, and carries the cap chamber down to the correct depth, and forms it accurately to size.

The process of forming the head is shown diagrammatically in Fig. 6. The case is supported inside a die L by a central pillar M passing up the inside of the case. The head is formed by the punch N which presses the metal of the head on to the face of the die L. The travel of the punch is accurately regulated so that the head is left exactly the correct thickness, and the height of the pillar M is also carefully adjusted, as this controls the relative position of the inside of the case and the underside of the head. In order to prevent the cap chamber from becoming distorted during the heading process the punch N is provided with a register O, which enters the cap chamber and maintains it at the correct diameter, and at the same time ensures that the anvil is left at the proper depth below the base of the cartridge case. Two sections are given showing the case before and after the heading has taken place.

The piercing of the fire holes is carried out by a punch fitted with two needles, the case being supported inside.

After the fourth draw and subsequent indenting and heading operations the amount of work to be done on the base end of the cartridge case is very small, and if the cases were annealed at this stage they would ultimately be left too soft. There is, therefore, no annealing at the base end after that carried out subsequent to the third draw, and the amount of work which has been done at the end of the third draw has to be carefully adjusted, so that the further work to be done at the base end of the case just brings the brass up to the required hardness. On the other hand the metal requires to be softened at the front end of the case, because there is a good deal of forming still to be done, partly because the finished case is of tapered form, and the diameter at the front end has still to be reduced considerably, and partly because provision must be made for the necking and bulleting operations, which are carried out in the filling factory. The cases are accordingly now annealed at the front end only. This is done by passing them slowly in rotating carriers through a series of gas jets. The distance from the mouth to which the case is annealed is important, and is controlled by adjusting the height of the gas jets.

After the semi-annealing process the cases are tapered. This is done by hollow conical punches, which press the cases inwards and reduce their diameter. The tapering is carried out in two operations, the first being a roughing operation, while the second finishes the case to the exact size. The makers' initials, the year of manufacture, and the figure " VII " are also stamped with a punch on the base of the case, which is then cleaned in strong soda water and dipped in dilute sulphuric acid, rinsed in water and dried by rumbling in sawdust.

The heads of the cases are then turned down to the correct diameter in an automatic lathe and bevelled, and at the same time the mouth of the case is reamered.

The cases are then complete and are sent to the viewing room, where they are gauged and examined with a view to the elimination of those containing mechanical defects. The gauging is carried out in an automatic gauging machine which verifies the length, diameter of head, thickness of head, depth of anvil and diameter of cap chamber. The absence of fire holes is detected by traying up the cases and passing them in turn over a lamp. Other defects are found by careful visual examination.

A few samples are also taken from each batch and sent into the filling factory, and loaded up into complete cartridges and fired, to ascertain their general performance as cases. If the result of this firing proof is satisfactory that batch of cases is passed as complete and sent to the filling factory for loading.

#### 4.—Bullet Manufacture

The essence of a good bullet is symmetry. Of all manufacturing errors which affect the accuracy of shooting, want of symmetry in the bullet is by far the most serious. If the centre of mass of a bullet is not truly on the axis of the bullet it travels up the barrel of the rifle in a spiral, and the bullet departs on the line on which its centre of mass happens to be travelling at the moment the bullet leaves the barrel. If the centre of mass of the bullet is only one-thousandth of an inch away from the axis of the barrel, it will cause an error of between thirteen and fourteen inches at a range of 600 yards, and the size of the target obtained from a series of shots with bullets this amount out of truth at the muzzle will be increased because of this eccentricity by some twenty-seven inches. At the high velocity at which the bullet leaves the rifle the centrifugal force due to the centre of mass of the bullet being eccentric to the axis of the rifle is very large (about 2 lbs. a thousandth of an inch), and this causes the bullet to cut deeper into the rifling on one side and increases the eccentricity, so that a very small error in symmetry of the bullet may mean a much larger amount of eccentricity at the muzzle. The extent to which this happens depends on the hardness of the envelope and other factors, and cannot be calculated, but it can be seen that extreme concentricity of all parts of the bullet is very desirable. "Thick and thin" cups are, therefore, a much more serious fault in the bullet envelope than in the cartridge case, and great regularity in the thickness and hardness of the cupro-nickel strip is essential.

Good quality cupro-nickel is also necessary for the production of good bullets. If the cupro-nickel is too hard or brittle, the bullet envelopes will split and break up in flight, and if the quality of the metal is bad there will be trouble with metallic fouling.

Cupro-nickel is difficult to produce, one of the principal difficulties being the prevention of oxidation. If the ingots are not correctly made the metal is spoilt, and great care is taken in testing the ingots before any attempt is made to roll them into strips. When making the cartridge case an analysis of the brass is taken on casting, and thereafter the annealing is the important operation, and each batch of metal is identified by its annealing record, but in the case of cupro-nickel it is the practice to take a sample of the ingot and put this through the whole of the manufacturing processes, and make up bullets which are loaded into cases and actually fired at the range to see how the metal behaves as a bullet, and thereafter that batch of metal is identified by the records of its trials at the casting stage.

The annealing of cupro-nickel also requires great care. The temperature should be somewhere between 600° and 680° C., and on no account should be allowed to go above 700° C., as at this temperature the carbon in the alloy, which is always present with the nickel, changes into the graphitic state, and the alloy becomes intensely brittle and the metal is spoilt. The prevention of oxidation during annealing is also important.

Cupro-nickel, which after rolling reaches a Brinell hardness in the region of 180, should, after annealing, have a hardness of from 80–90. After each annealing samples are taken and tested mechanically and the microstructure examined.

Although cupro-nickel is difficult to make, and has to be annealed with great care, it is an alloy which, if of good quality and well annealed, will stand a great deal of manipulation, and after the final annealing of the strip, bullet envelopes are carried through all the processes of manufacture without further annealing.

When cupro-nickel is melted, a little scrap is first put into the bottom of the plumbago crucible, and then all the nickel and as much copper as the crucible will hold. These are then all melted down together, when the remainder of the scrap metal and the copper are added, and the whole is melted at a temperature of 1,300° C., the alloy being carefully stirred with a plumbago stirrer. An iron bar must not be used. A little charcoal is present throughout to reduce oxidation, and when all is molten a small quantity of manganese copper is put in and a little borax, and the whole surface of the liquid is covered with a layer of charcoal. The whole is then raised to a temperature of about 1,400° C. for 10 minutes. The metal is then poured into the moulds. Throughout the melting and during pouring every care is taken to prevent access of air to the metal.

As in the case of the brass, the section of the ingot is decided by the thickness and width of strip which is convenient for cupping, and in the case of the cupro-nickel ingot a convenient section is 4 inches wide by 1 inch thick. The ingots are taken out of the mould and the "pipe" end cut off and samples are taken for analysis and for testing as bullets, and the ingots are scraped and filed and visually examined for flaws, etc.

The ingots are first rolled, usually about six times, down to a thickness of ·5 inch, after which the strips are annealed. The first annealing is carried out in a muffle in the manner described in the case of the brass strip, and on the completion of the heating process the strips are plunged into water. The strips are then cleaned in dilute sulphuric acid and washed, and they are then rolled again, usually 15 times, down to the finished size of ·03 inches. When finished the strips should be of correct thickness all over within a thousandth of an inch above or below. After the ends are cut off square, the strips are annealed again. On this occasion they are rolled up in coils and put in metal pans, closed with fireclay to prevent access of air to the cupro-nickel, and after heating they are allowed to cool slowly. After cooling the strips are cleaned by dipping in a mixture of sulphuric and nitric acid, washed in hot water and dried in sawdust. They are then passed through a pair of roller cutters, which cut off ragged edges and finish the strips to the correct width.

The first operation in the manufacture of the bullet envelope from the cupro-nickel strip is "cupping," the process being exactly the same as that described in the case of the brass cups for the cartridge cases. In this case, however, the cups are not annealed, but are washed in soda water and rumbled, after which they are sent to the viewing room and examined. The cups are then drawn three times, the approximate shape after each operation being shown in Fig. 7.

After the third draw the ragged edge is trimmed, and the envelope cut to length. The point is then formed in two operations by punches, which press the end of the cup into a shaped die. After pointing, the envelope is cleaned in soda water, rumbled, and dried, and sent to the viewing room, where any faulty envelopes are removed.

The lead used for the cores of the bullets is specified to contain 98 per cent. of lead and 2 per cent. of antimony, the antimony being added to increase the hardness. The most convenient way to obtain the correct alloy is to melt up pure lead with lead containing, say, 12 to 15 per cent. of antimony, the proportions being chosen according to the results of analysis. The lead is used in the form of wire made in an extruding press. This consists of a cylindrical chamber into which fits a plunger, up the centre of which there is a hole. In this hole is fitted a die of the appropriate diameter for the wire which is being manufactured. The lead-alloy ingots are melted and the molten metal is poured into the chamber, where it is allowed to cool till it begins to solidify. The hydraulic pressure is then applied to the plunger, and as this descends reducing the volume of the chamber, the lead is squirted through the die in the form of wire, which is coiled up on a reel.

In the core-making machines this wire is fed into a hole in a bar, which then passes sideways, cutting off the requisite quantity of lead for a core. The bar then places the lead core opposite a shaped die into which it is pressed by a punch, and after the withdrawal of the punch the bar moves back to its original position and the shaped core is ejected from the die. In order that all the cores may be fully up to size and shape, the bar cuts off each time slightly more lead than is required, and a hole is provided in the side of the die, into which the excess is extruded under the pressure of the punch, this excess being cut off in turn when the core is ejected from the die. The finished cores are then sent to the viewing room and examined, damaged or faulty cores being discarded.

The aluminium tips are made generally in the same way as the lead cores. In this case, however, the aluminium is cast into billets, of the same shape as the chamber of the press, and the aluminium wire is extruded from these billets, which are heated and put into the press red hot.

The process of cutting off and forming the tips from the aluminium wire is exactly the same as that described in the case of the lead cores. After forming, the aluminium tips are sent to the viewing room.

After examination the components of the bullet are assembled. To do this the envelopes are arranged in trays, and first the aluminium tips and afterwards the lead cores are inserted by hand. The assembled bullets are then passed through a machine, in which they are fed in turn into a die, in which the lead is pressed into the envelope, and the edge of the envelope turned inwards by a punch, the bullet taking the form indicated in Fig. 7.

The next operation is the final pressing and marking. The marking consists in stamping the initials or trade mark of the manufacturer on the base of the bullet. This is useful, not only as an indication of the origin of the bullet, but also because it ensures that the lead is thoroughly pressed home, since, unless considerable pressure is applied, the lead fails to flow into the punch, and the marking is absent or indistinct, and the fact that there has been insufficient pressure is therefore evident on examination.

The bullets are next cannelured. This is done by rolling them between a rotating disc and a fixed guide. The disc is provided with a raised rim, which spins the cannelure into the bullet. This operation tends to distort the bullet somewhat and alter its diameter, and therefore after canneluring, the bullets are pressed through a ring die which rectifies them and sizes them to the correct diameter.

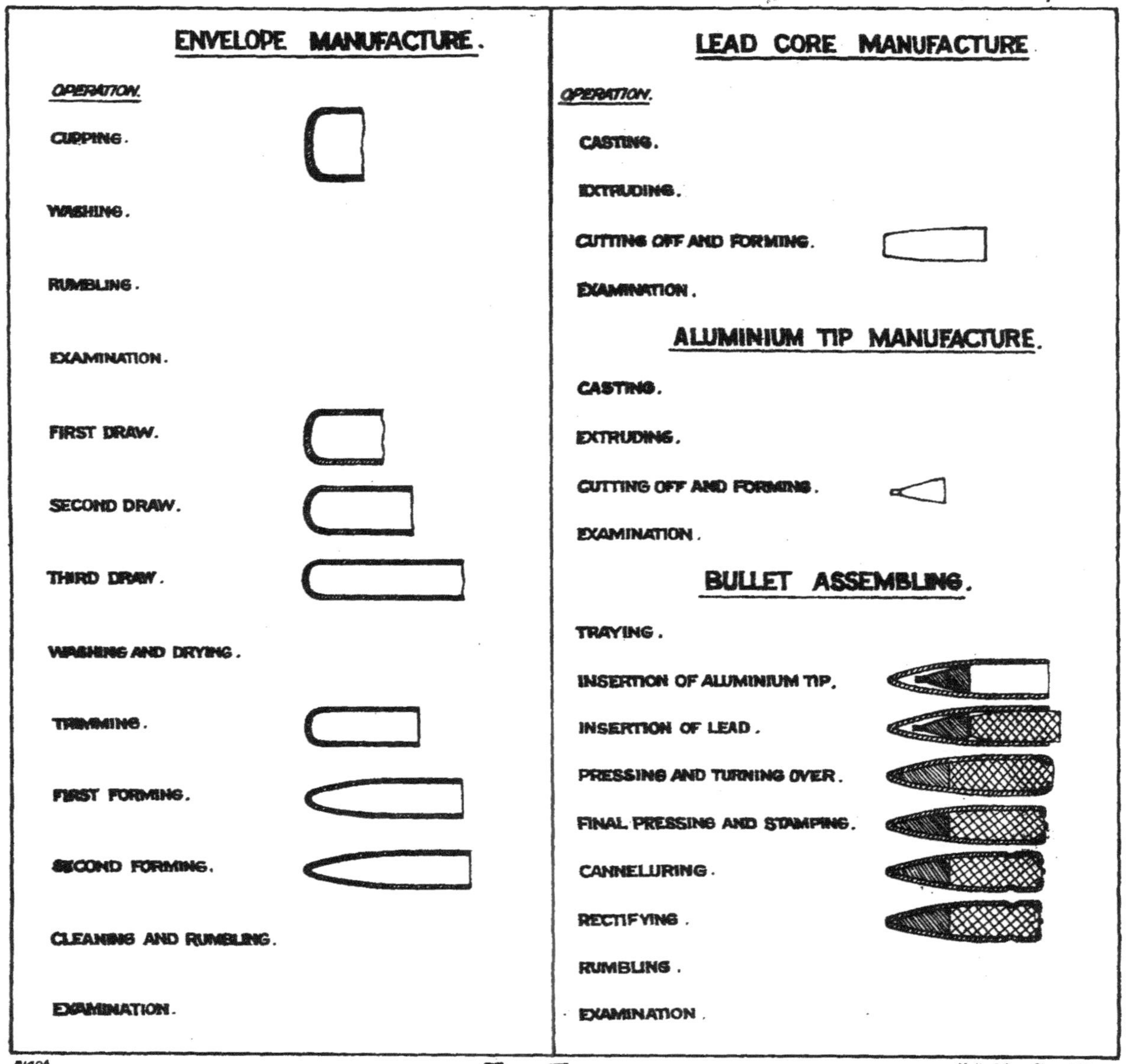

PT. II. CHAP. IV. To face p.232.

## ENVELOPE MANUFACTURE.

OPERATION.

CUPPING.

WASHING.

RUMBLING.

EXAMINATION.

FIRST DRAW.

SECOND DRAW.

THIRD DRAW.

WASHING AND DRYING.

TRIMMING.

FIRST FORMING.

SECOND FORMING.

CLEANING AND RUMBLING.

EXAMINATION.

## LEAD CORE MANUFACTURE.

OPERATION.

CASTING.

EXTRUDING.

CUTTING OFF AND FORMING.

EXAMINATION.

## ALUMINIUM TIP MANUFACTURE.

CASTING.

EXTRUDING.

CUTTING OFF AND FORMING.

EXAMINATION.

## BULLET ASSEMBLING.

TRAYING.

INSERTION OF ALUMINIUM TIP.

INSERTION OF LEAD.

PRESSING AND TURNING OVER.

FINAL PRESSING AND STAMPING.

CANNELURING.

RECTIFYING.

RUMBLING.

EXAMINATION.

FIG. 7.

The bullets are then cleaned by rumbling in dry cross-cut sawdust, and thoroughly inspected, all faulty bullets being discarded. A few samples from each batch of bullets are also taken and loaded up into cartridge cases and fired at the range to ascertain their general performance and accuracy. If they are satisfactory the batch of bullets is passed as complete and ready for the loading field.

## 5.—Cap Manufacture and Filling

Caps must be made accurate to diameter or they will allow an escape of gas between the cap and the cartridge case, known as a blowback, when the cartridge is fired. The material used must also be of good quality, or the caps will be liable to be fractured by the striker of the weapon in which they are fired. The actual process of making a cap is very simple, and consists in punching a cup out of a strip of copper exactly in the same way as the cups for the cartridge cases are made.

By far the most important part of the cap is the priming. This consists of six-tenths of a grain of the following composition :—

Eight parts by weight of fulminate of mercury.
Fourteen parts ,, of chlorate of potash.
Eighteen parts ,, of sulphide of antimony.
One part ,, of sulphur.
and One part ,, of mealed powder.

It is essential that the ingredients of this composition should be of good and regular quality, and that they should be properly mixed, and that the quantity inserted in each cap should not vary. Irregularity in the priming causes variations in the flash obtained, resulting in uneven ballistics and hangfires, while indifferent quality leads to instability and deterioration with resulting missfires. It is also of great importance that the composition should be well varnished to protect it from moisture, failing which deterioration will inevitably cause missfires in time.

Samples are taken of all the ingredients on delivery, and these are carefully analysed for purity. In the case of the chlorate of potash care is taken to see that the material contains not more than $0 \cdot 02$ per cent. of potassium bromate. The presence of bromate is believed to cause instability, and although this is uncertain, it is known that potassium chlorate containing potassium bromate in the presence of sulphur and moisture forms acids which do considerable damage to a rifle barrel, and bromates are therefore objectionable.

The sulphur used must be free from acid, as this causes instability of the composition.

Sulphide of antimony contains grit, which is the abrasive on which the sensitiveness of the composition depends. Antimony sulphide for use in caps should not contain more than $0 \cdot 25$ per cent. of its weight insoluble in acids, and of this residue not more than 80 per cent., or $0 \cdot 2$ per cent. of the whole, should be of a gritty nature.

The sensitiveness of the composition depends not only on the amount of grit, but also on its nature, and more particularly on the fineness to which it is ground. In practice a percentage of grit is adopted which is found to give the best results when ground in a certain manner, and the supplies of antimony sulphide are blended according to their analysis to give this percentage of grit, and the sulphide is always ground in the manner appropriate to that percentage.

The process of mixing the composition is carried out in the following manner. The fulminate of mercury is passed under water through a 30 I.M.M. sieve into a cambric bag, which is then hung up and the water squeezed out of it. It is then laid on a gunmetal table heated to 150° F. by hot water, and dried. When dry the requisite quantity is weighed out into a papier-maché pot.

The chlorate of potash is pulverized in a porcelain rumbler with gunmetal balls, 25 lbs. of chlorate being put in with 40 lbs. weight of gunmetal. When the chlorate has become finely divided it is passed through a 100 I.M.M. sieve, and that portion of it which will pass through this sieve, but which will not pass through a 170 I.M.M. sieve, is weighed out into a second papier-maché pot.

The sulphur and mealed powder are also passed through a 100 I.M.M. sieve and weighed out into a third pot.

The antimony sulphide is blended, crushed and ground and sieved. That portion of it which will pass through 100 I.M.M. sieve but which will not pass through a 170 I.M.M. sieve, is weighed out into a fourth pot.

All four pots are then taken into the mixing room.

The mixing is done in a conical shaped sateen bag suspended in a leather chamber so that there is a space all round the bag. The lower portion of the leather chamber is funnel shaped and is supplied with discharging openings, and underneath is placed a moveable tray, on which are placed the containers into which the mixed composition will be transferred. The whole apparatus, the general arrangement of which is shown in Fig. 8, is enclosed in a semi-circular armoured partition, so arranged that should an explosion occur the force of it would be directed towards the outside of the building. To the apex of the cone of the sateen bag is attached a cord, and the mixing is carried out by raising and lowering the cord, which turns the ingredients of the cap composition over and over inside the bag, the mixing being further assisted by strings of rubber discs and beads sewn inside the bag. The cord is actuated by a water motor placed behind the armoured screen, and arranged to work for three minutes at a time at the rate of 60 strokes a minute. All manipulation of the apparatus is also arranged to be carried out from behind the screen.

The procedure is as follows. The leather chamber and bag are first carefully cleaned, so that there is no risk of composition being left in the joints, and the apparatus is put in place, with a dust catcher on the movable tray underneath to catch any minute quantities of the ingredients which may be spilled. The operator then puts the antimony sulphide in the bag. He then puts in the chlorate of potash through a 70 I.M.M. mesh sieve, brushing it through if necessary with a soft brush. The sulphur and mealed powder are next inserted through another 70 I.M.M. mesh sieve, in the same manner, and finally the fulminate is gently poured in. When all the ingredients are in the bag, the operator retires from the room, and starts the water motor from outside the building. When the motor stops at the end of the three minutes, the operator again enters the building, and removes the dust-catcher from under the chamber by means of a handle at the back of the screen, and replaces it with the papier-maché containers, into which he intends to put the mixed composition, moving these into place also from the back of the screen. He then raises the lever attached to the cord, so that the sateen bag is completely inverted, and the composition falls into the leather chamber, which is shaken so that all the composition falls through into the containers below, which are then removed from under the leather chamber. All this manipulation is carried out from the back of the screen, and the operator then comes round to the front and takes the containers away and sends them to the cap filling room.

For purposes of filling the caps are trayed up in a frame consisting of a brass plate perforated with holes, arranged in rows, and each of the correct size to hold a cap, the brass-plate being backed by a plain steel plate. This frame is placed under a loading plate, which is of gunmetal and contains holes at corresponding centres, these holes being of such a diameter and length that each holds the exact charge of composition. Between the loading plate and the cap frame is a shutter plate also perforated with holes at corresponding centres. This shutter plate is arranged to be movable, and, while the loading plate is being charged, the shutter plate is adjusted so that the bottoms of the holes in the loading plate are closed. Across the top of the loading plate is arranged a trough which can be traversed over the plate by a screw worked from behind an armoured screen, and in this trough there are distributing vanes, which are moved backwards and forwards by a cam, as the trough travels over the plate. A general arrangement plan and diagrammatic sections of the apparatus are shown in Fig. 9.

The operator pours the required quantity of the composition into the trough and then

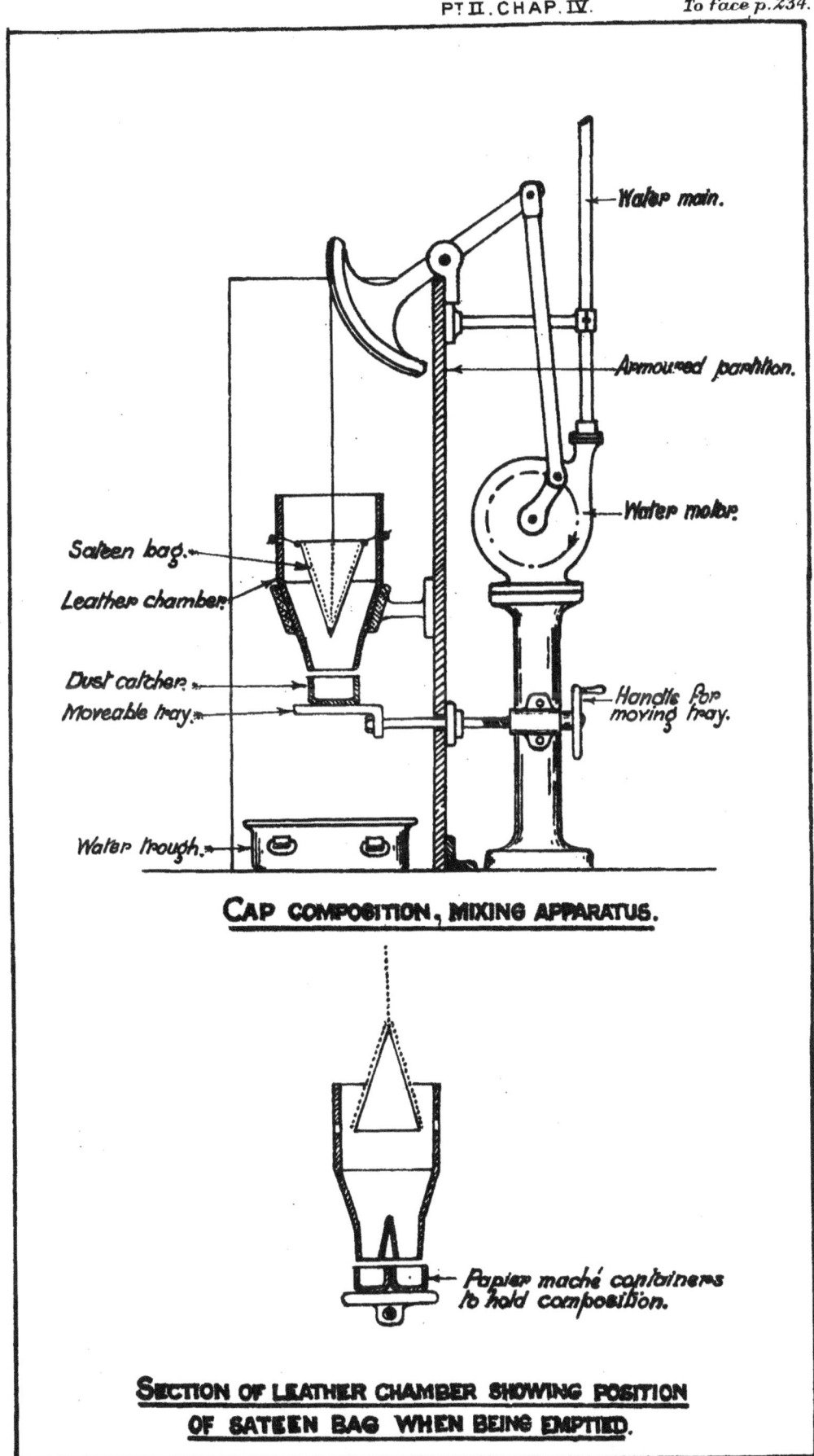

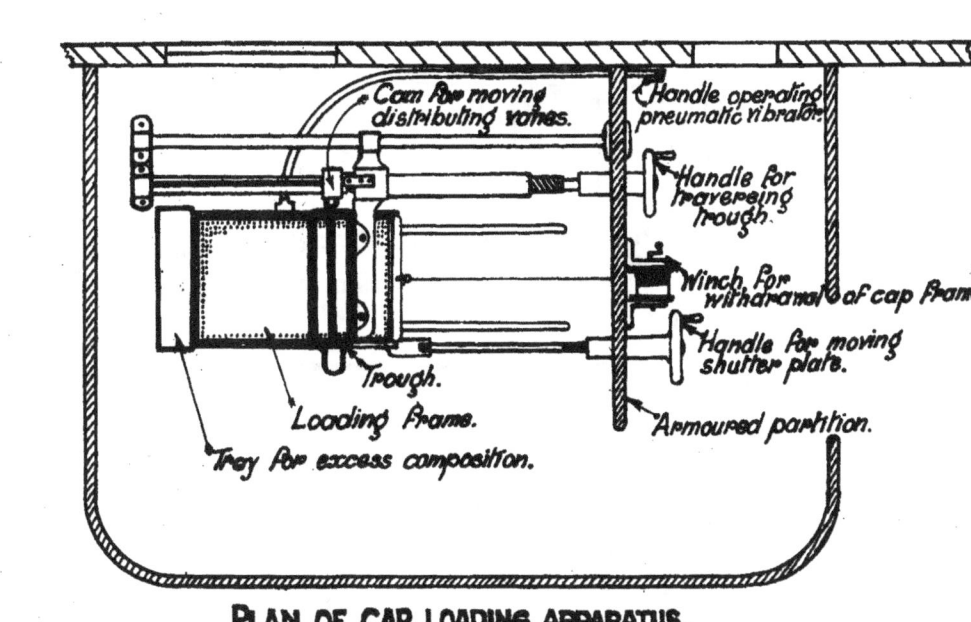

PLAN OF CAP LOADING APPARATUS.

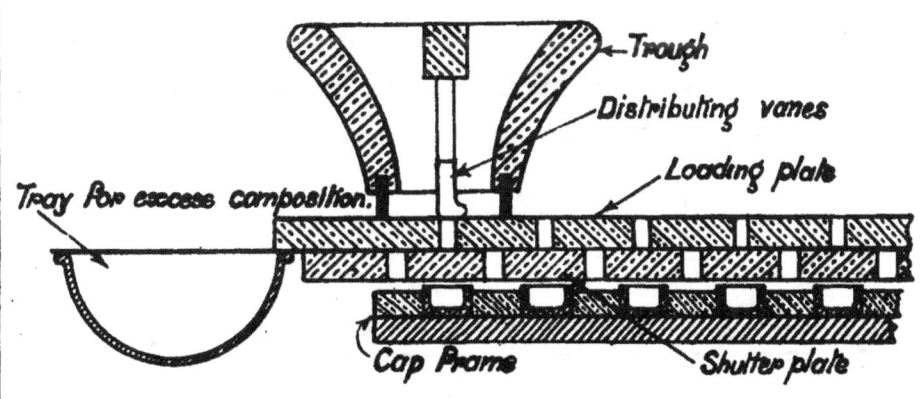

SECTION SHOWING LOADING FRAME PREVIOUS TO FILLING.

SECTION SHOWING LOADING FRAME AFTER MOVING SHUTTER PLATE.

FIG. 9.

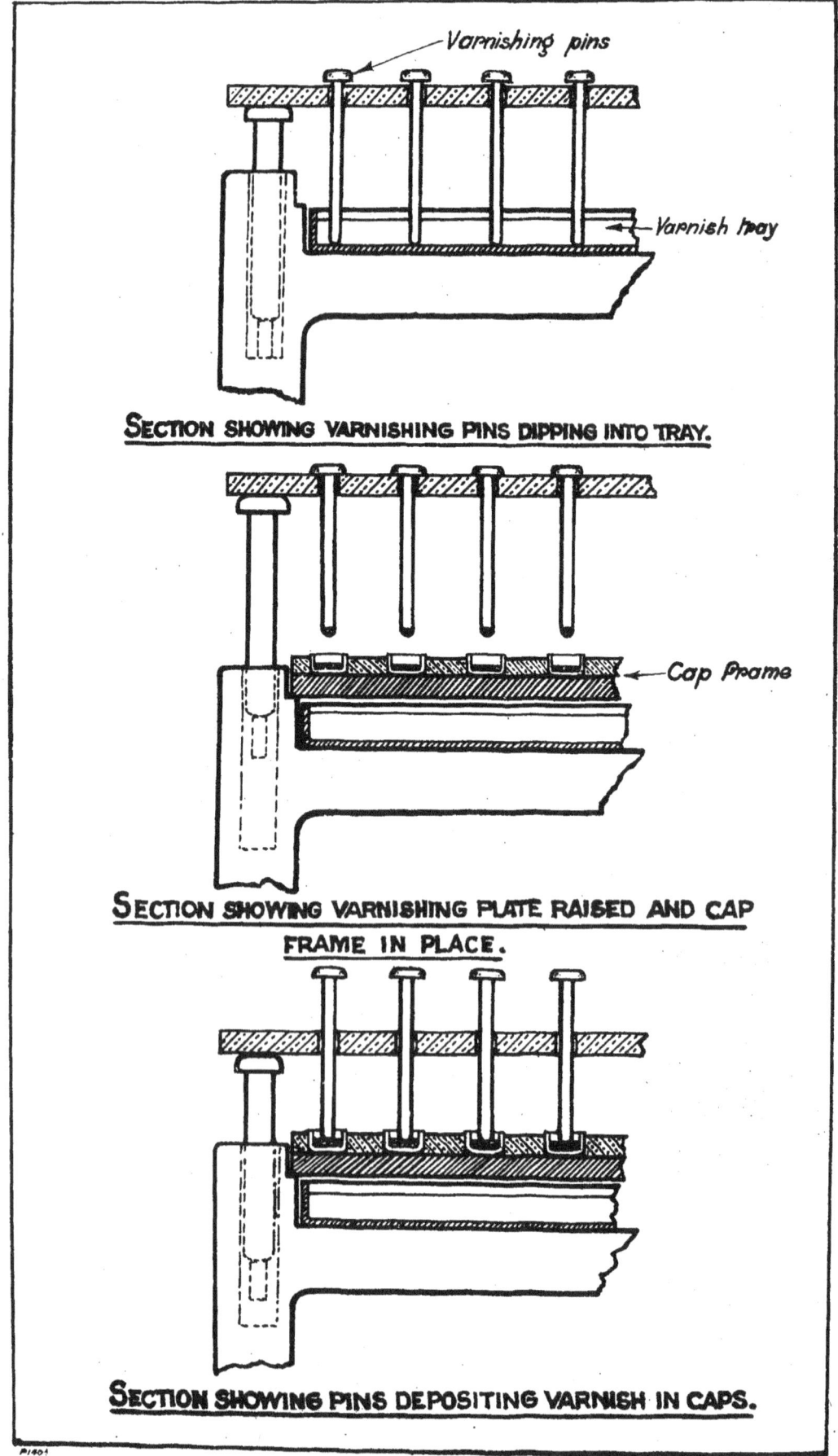

FIG 10

retires behind the screen, and traverses the trough over the loading plate twice. The distributing vanes sweep the composition over the holes in the loading plate so that these are filled and any excess composition is swept off and deposited in a tray placed at the end of the loading plate. The shutter plate is then moved also from behind the screen, so that the holes correspond with those in the loading frame, and the composition can then fall down into the caps. In order to ensure that all the composition finds its way into the caps, the apparatus is then vibrated by a pneumatic vibrator, and the cap frame is then withdrawn from under the loading frame.

The operator then comes from behind the screen and takes the cap frame containing the filled caps, and puts this in a press. This press contains a row of punches which press one row of caps at a time, the cap frame being fed forward after each stroke of the press. The pressure used is about 600 lbs. on each cap.

After pressing the cap frame is put in a chamber and any loose composition remaining in the caps is blown out by pneumatic jets.

Tinfoil discs are then inserted into the caps on the top of the pressed composition. There are several ways of doing this. One method is to cut the discs out of a sheet of foil by punches, placed at the same centres as the holes in the cap frames. The punches cut the discs on a perforated die plate, and then descend pressing the discs into the caps which are placed in the frame underneath. Where brass caps are used the tinfoil discs are sometimes cut on the edge of the cap itself.

After the insertion of the tinfoil discs the cap frames are again put into the presses, and the foil pressed on to the composition, the pressure used being about half that employed when pressing the composition.

The caps are then varnished. The varnish used is best orange shellac, 2 lbs. 2 ozs. of shellac being dissolved in one gallon of methylated spirits. The varnish is placed in a tray, over which is a plate perforated with holes at the same centres as the cap frame, and capable of being raised and lowered. In each of these holes a blunt ended brass pin rests (see Fig. 10). The plate is first lowered so that the pins dip in the varnish, and as the plate is raised each pin picks up a drop of varnish. The cap frame is then inserted under the varnishing plate, which is again lowered, so that each pin rests on the inside of a cap and deposits its drop of varnish.

The cap frames are then placed on a shelf till the varnish is dry, when they are put into a machine and turned over and shaken, until the caps all fall into a tray.

The finished caps are stored in batches of about 5,000 caps, each mixing of composition being treated as a batch. Samples are taken from each batch, and loaded into cartridge cases and fired. The flash from each cap should be capable of marking a card placed $8\frac{1}{2}$ inches away from the cap. If the firing trial on the samples is satisfactory, the batch of caps is sent to the loading field.

### 6.—Loading

The sequence of operations when loading cartridges varies to suit the different propellents used. For instance, where cordite is used in the form of sticks, the necking of the cartridge case is carried out in the filling factory after the insertion of the cordite and wad, whereas where cartridges are loaded with a granular propellent, the case is usually necked, or semi-necked, in the component factory. The cases can then be annealed after necking, and internal stresses in the necks, which are apt to cause splits, are largely avoided.

In either case the first loading operation consists in assembling the primed caps in the cases. Each cap is put into the cap chamber by hand and the cases are passed through the capping machines. In this machine the cases are carried cap upwards in a circular disc under two punches in turn. The first of these sets the cap in part of the way and straightens it, and the second presses it into the correct depth and rivets or "rings" it in. After examination the capped cases are trayed up into plates containing recessed holes, so arranged that the cartridge heads fit the holes and come flush with the face of the plate. The plates are then rubbed all over with purple varnish, which is then rubbed off again, leaving only the grooves round the caps full of varnish. The cases are then ready for loading.

Small arm cordite is usually loaded in the form of tubular sticks, and in this case consists of a number of tubular strands, usually about 44, grouped together into a rope, which is delivered on reels each holding from 50 to 60 lbs. of propellent. The number of strands chosen is such that the charge is formed by inserting the rope in the cartridge case, and cutting off the required quantity. The charge is controlled by regulating the length cut off for each cartridge, and the cordite should be so blended and reeled that some length within the allowable limits of 1·55–1·6 inches produces a charge which will give an average velocity between 2,340 and 2,420 feet a second, with an average pressure not exceeding 19·5 tons. (Further particulars in regard to velocities and pressures will be found in Chapter V.) Before loading, the exact charge for each reel is assessed by loading a few cartridges and measuring the velocity and pressure, and fixing the length of charge to give the required ballistics accordingly. Where there is reason to suspect that the ballistic qualities of the reel are irregular, an assessment may be made at more frequent intervals, and a portion of a reel only used before another assessment is made.

Sometimes the cordite is unreeled and laid out on a tray, about half the cordite on a reel being laid out at a time. The cordite is placed in a hut, in the wall of which is a trumpet shaped hole, through which the cordite rope passes into the loading room. The loading mechanism consists of a horizontally reciprocating head, containing three tools, across the face of which the cartridge cases, which are fed into a rotating disc carrier, are carried in turn. The first of these tools is a case punch, which slightly opens the mouth of the case. The second tool consists of an arrangement for gripping the cordite and carrying it forward to the extent of the travel of the reciprocating head, which is adjusted so as to feed the required length of cordite at each stroke. As each length of cordite is fed into the case a rotating knife cuts it off, and the case then passes to the third tool, which is a punch which presses the cut off strands into the case. The machine is arranged so that if there is a failure in the supply of cordite, and the punch can go right into the case, the machine is automatically stopped.

As a precautionary measure the cordite is arranged to pass through a ring, to which is attached by means of an inflammable cord a string supporting a guillotine across the face of the hole in the wall. Should by chance a fire start in the cordite, this burns the cord and releases the guillotine and thus isolates the bulk of the cordite in the hut before the fire can reach it.

As the cartridges come out of the loading machine they are placed in trays ready for wadding. The wads, which have previously been stamped out of sheets, are trayed up into a plate containing spaced recesses, each of which holds a single wad. This plate is then placed under a set of hollow punches connected to a vacuum pump, and on to the faces of which the wads are transferred by suction. The punches are then pressed down into the tray of loaded cases and the vacuum turned off, so that the wads are left in the cases.

The bullets are first lubricated before loading in a machine in which a disc runs in a bath of melted beeswax and deposits the wax in the bullet cannelure. The bullets are then trayed up and transferred into the cases, where they rest loosely on the top of the wads.

The bulleted cases are then fed into the necking machine, which necks the cases so that they are a close fit round the bullets and presses the bullets in the correct distance. The cartridges are next put through the indenting and coning machine, where first the indents which hold the bullet in are made in the case, and secondly the mouth of the case is coned inwards into the bullet.

The cartridges are then done up in bags and rumbled in turpentine to clean them, and they are then finished and ready for issue.

One of the methods of loading a granular propellent into the cases is as follows. The cases are trayed up and filled from a loading frame. The loading frame consists of two plates connected together by a number of tubes spaced to correspond with the holes in the tray holding the cartridge cases. The upper plate is in the form of a dish, and the tubes are of such a diameter and length that they will each just hold a charge. The volume of the charge is chosen according to the results of firing trials on a few sample rounds loaded up with the batch of powder in question, and the volume of the tubes in the loading plate is controlled accordingly by adjusting the distance between the two plates, by means of adjustment screws, placed at the four corners of the frame  The loading frame is placed on the top of a shutter plate containing holes spaced to the same centres as the tubes, and under which the trayed-up cartridges cases are placed, so that the mouth of a case comes opposite each hole. The loading frame is so placed that the tubes do not coincide with the holes in the shutter frame, and a weighed quantity of powder in excess of what is required, is poured into the dish, and distributed into the tubes by means of a soft brush until each tube is exactly filled. The surplus powder is then swept off the surface of the upper plate, and the loading frame is moved so that the tubes are opposite the holes in the shutter frame, and the powder then falls through into the cartridge cases.

The trayed-up cartridge cases are then placed under another perforated frame, in the holes in which are placed loose pins. This frame is lowered so that the pins can enter the cartridge cases, when it can immediately be seen if any case does not contain a charge, or contains only a portion of a charge. The loaded cases are then transferred in their tray under a series of punches which insert the wads in a cupped form into the neck of the case.

The bullets which are trayed-up, just as in the case of the ammunition loaded with stick cordite, are then inserted into the necks of the cases, and the necks are then finally closed in round them, coned and indented, after which the cartridges are rumbled in turpentine and issued.

---

## CHAPTER V

### PROOF OF SMALL ARM AMMUNITION, WITH SPECIAL REFERENCE TO ·303-inch MARK VII

#### 1.—SELECTION OF PROOF

The inspection of small arm ammunition after the completion of manufacture consists of two entirely distinct processes. The first of these is the assessment of its general shooting qualities, and involves the firing under various conditions of a small percentage of the rounds under review, carefully selected so as to form a representative sample of the whole. This process is generally referred to as "Proof."

The second process, which is dealt with in the next chapter, usually consists in the detailed inspection of each individual cartridge in order to remove isolated rounds which have been faultily manufactured.

The firing trials, or proof, of small arm ammunition vary slightly with the design of the ammunition, and the details to some extent depend upon the weapons in which it is intended that the ammunition shall be used. The following description, however, which refers to the proof of ·303-inch Mark VII ammunition, may be taken as typical.

The first point which it is essential to bear in mind is that at most only a small percentage of the total consignment under review can be fired at proof, and any serious failure at proof of that small percentage may involve the rejection of the whole consignment. It is therefore important that the rounds selected for proof should be really representative of the whole consignment, and they are in consequence selected by an expert "Prooftaker," a few here, and a couple there from as many boxes of the consignment as possible, and in order to prevent possible rejection of the whole batch owing to the faulty manufacture of an isolated round, the ammunition thus selected for proof is submitted to an individual inspection of each round, and any round found faulty is eliminated before the firing actually commences.

The normal and convenient size of a consignment is 200,000 rounds, and from such a consignment cartridges are selected for proof at the rate of three per thousand. If the batch contains more or less than this number, the quantity selected for proof varies accordingly, but however small the consignment, the number selected for proof is never less than 450 as this is the minimum quantity permissible if a comprehensive proof is to be carried out.

Speaking generally, the proof is designed to ensure that the ammunition is in all respects fit for use in all types of weapon in which it may be fired, and the firing trials are arranged to investigate the following features : Accuracy of shooting, velocity at the muzzle, pressure in the chamber, functioning in all types of service weapons, both new and worn, or out of adjustment within certain allowable limits, and freedom from defects in the cap, case or bullet.

## 2.—Accuracy Trials

By " accuracy " is meant the capacity of a series of shots, all fired in exactly the same direction, to group closely on a target.

Accuracy depends on the regularity of construction of the ammunition, and is affected by innumerable details in the design and manufacture. The chief causes of inaccuracy are variations in the loading of individual rounds and want of symmetry in the bullets.

Variations in loading either in the weight of the charge, or in the quality of the propellent, or in the loading of the cap, which in turn affects the rate of burning of the propellent, are liable to cause differences between round and round in the velocity of the bullet at the muzzle, and where variations in muzzle velocity occur the bullets have different trajectories, and their points of impact on the target vary accordingly.

Want of symmetry in the bullets, sometimes in shape, but more commonly in the thickness of the bullet envelope, causes the bullets to oscillate in flight, and is probably the principal cause of inaccuracy. Not only is a bullet which is oscillating during flight uncertain as to its point of impact on the target, but where the centre of gravity of the bullet is not truly central in the barrel, the bullet leaves the muzzle on a line of departure which is not strictly along the axis of the barrel.

Accuracy proof is carried out by firing series of rounds from rifles which are clamped to a mounting or rest so arranged that they always shoot in one direction. In the case of ·303-inch Mark VII ammunition, the shooting is carried out at 600 yards range from S.M.L.E. rifles mounted in rests which are usually described as " Fixed rests." This means that whilst the rifles are free to recoil straight to the rear, they are so housed that they cannot jump sideways or up and down, and after each shot the rifle can be readily restored to its exact original position.

There is a good deal more in the selection of a suitable rest than appears at first sight. A S.M.L.E. rifle barrel when being fired vibrates, so that the barrel is rising when the bullet leaves the muzzle. For this reason the angle of departure of a low velocity shot is greater than that of one with high velocity. The high velocity shot, however, has a flatter trajectory, and there is a range, which may be called the compensation range for that particular barrel, at which no matter what the muzzle velocity may be, within reasonable limits, all the shots hit the target theoretically at the same point. For a further study of this subject the theory of rifle compensation in Part I, Chapter I, Section 5, should be consulted, but for the present purpose it is sufficient to realize that if the rifle happens to be compensated for a range not far different from the range at which the accuracy trials are being carried out, the rifle will mask inaccuracy in the ammunition due to variations in velocity. For this reason it is essential for purposes of accuracy trials that the rest shall be of such a design that it holds the rifle barrel rigid, and prevents vibration during firing.

It is also desirable that the rest should clamp the barrel itself, and that the stock should be removed, and thus any warping of the stock which may occur as the barrel warms up is eliminated.

Barrels capable of accurate shooting are selected for these trials, and these are tested every morning by firing ten rounds from a batch of ammunition known to be accurate, and any barrel which fails to shoot this ammunition satisfactorily is discarded.

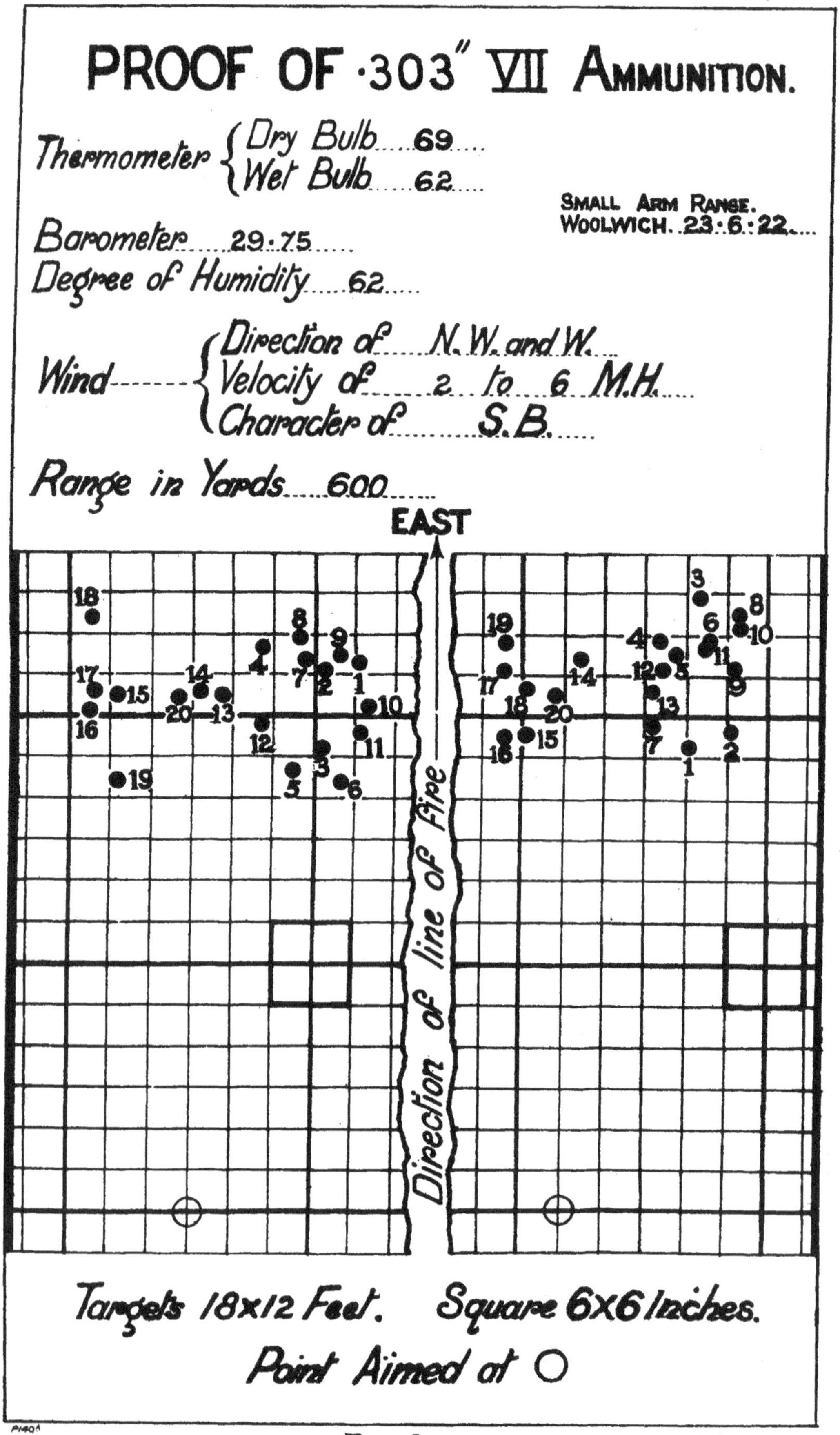

PT II. CHAP. V.   To face p. 239.

# PROOF OF ·303" VII AMMUNITION.

Thermometer { Dry Bulb ..... 69
             { Wet Bulb ..... 62

SMALL ARM RANGE.
WOOLWICH. 23:6:22.

Barometer ..... 29·75
Degree of Humidity ..... 62

Wind { Direction of ..... N.W. and W.
     { Velocity of ..... 2 to 6 M.H.
     { Character of ..... S.B.

Range in Yards ..... 600

EAST

Direction of line of Fire

Targets 18×12 Feet.   Square 6×6 Inches.

Point Aimed at ○

FIG. I.

The cartridges are fired in groups or "targets" of twenty rounds. The number of targets fired from each consignment varies, but is normally between eight and twelve. Two or more guns are used, and each pair of guns is fired simultaneously, so that both bullets are subject to the same external conditions, wind, etc. Further, whenever possible two batches of ammunition are tested concurrently, and in this case no two successive targets from one consignment are fired from the same rifle. This method of firing ensures that any barrel peculiarity which may develop during proof is more readily detected.

It will be realised that the exact portion of the target which is hit is not the main consideration, and that it is the grouping of the shots which is important, but if two barrels are laid alike and when firing the same ammunition simultaneously produce groups in corresponding portions of their respective targets, it is clear that any wide variation in the mean points of impact obtained when two batches of ammunition are being tested concurrently indicates a variation between the two batches, and thus any abnormality in general trajectory of any particular batch can be detected.

The target used is divided into 6-inch squares, and as each shot strikes the target its position is recorded on a sheet of paper, also ruled in squares, each shot being numbered in the record. The diagram Fig. 1 shows an example of such a record.

The mean point of impact is the mathematical centre of the various shots, and is obtained as follows. A vertical base line is drawn to one side or the other of the group of shots, and the distance of each shot is measured from this base line, and the average distance calculated. The mean point of impact will be somewhere on a vertical line drawn this average distance away from the base line. Similarly, a horizontal line can be drawn, which is at the average distance of all the shots from a horizontal base line drawn either above or below the group of shots. The mean point of impact is at the point where the vertical and horizontal lines thus found intersect. The accuracy of the shooting is assessed by calculating the average distance in inches of the twenty shots from the mean point of impact of the group, the figure thus obtained being termed the "Figure of merit." The figure of merit is, of course, increased as the shots are more scattered, and a high figure of merit means bad accuracy. The accuracy of a consignment of ammunition is the average figure of merit of all the targets fired.

Wind is the most important disturbing factor when testing for accuracy, and accuracy trials should not be carried out at all on an open range if more than a gentle breeze is blowing. The effect of wind is clearly shown in the accuracy targets given in Fig. 1, and the reason for numbering the shots becomes obvious, for it is clear that after shot 13 in both groups, the whole of the grouping went to the left, and the wind, which in direction was from the left, must have fallen.

A consignment of ammunition is considered to have passed the accuracy test if the average figure of merit of all the targets does not exceed 8 inches, but an allowance is made in the case of targets which have obviously been affected by wind.

If in any target a round is 3 feet or more away from the mean point of impact, it is considered a miss. If a miss is recorded during accuracy proof, a second accuracy proof is taken, and if another miss occurs at this second proof, the ammunition is considered to have failed at proof and is rejected, even though the actual figure of merit may not be greater than 8 inches. Should more than one miss occur at first proof, the ammunition is rejected without a second proof.

### 3.—Velocity and Pressure Trials and the Use of Standard Ammunition

The sighting of the various weapons is based on the performance of rounds arranged to shoot with a definite velocity at the muzzle, and it is essential, therefore, that the average velocity of the rounds forming each batch of ammunition should be within reasonable limits similar to that for which the weapons are sighted. The loss of accuracy occasioned by variations in the velocity between round and round has already been pointed out.

Actual velocity at the muzzle would be very difficult to measure, but it may be assumed that the average velocity over a short distance near the muzzle bears a sufficiently constant ratio to the actual muzzle velocity. In the case of the ·303-inch Mark VII cartridge, measurement of velocity is made by observing the time of flight from the muzzle to a distance of 180 feet. The average velocity over this distance is assumed to be the actual velocity at a point 90 feet from the muzzle, and the requirements are that at this point the mean velocity shall be between 2,340 feet and 2,420 feet per second, and that the average difference of the individual velocities of each of not less than 10 rounds from the mean velocity of all the rounds shall not exceed 30 feet per second.

Velocity measurements are always taken in a fairly new rifle barrel, housed in a fixed rest, and if necessary the cartridges are raised to a temperature of 60° F. The velocity is recorded on a Boulengé chronograph, the actual method of procedure being described in Part III, Chapter V.

The main object of measuring the pressure of the gases in the chamber is to ascertain that the ammunition is so loaded that it does not generate pressures dangerous to the weapons. Pressures are most commonly measured by recording the amount of compression obtained on a copper pad placed behind the cartridge in a rifle chamber, the cartridge case being oiled to reduce friction. The apparatus used is known as the "Ordnance Factory base pressure gauge," for further particulars of which *see* Part III, Chapter VI.

It is the characteristic of some propellent powders, particularly cordite, that the pressure generated by a given charge increases as the temperature rises, and it is important therefore, where ammunition may be used in the tropics, to test it for pressure at the highest temperature which it may attain in use, and in any case provision must be made for a cartridge which may be left for some time in a rifle chamber which is heated from having fired previous rounds. For this reason the ·303-inch Mark VII cartridge is tested for pressure at a temperature of 120° F., the cartridges being heated in an oven at this temperature for 2 hours. In order to pass inspection the mean recorded pressure of all rounds fired at this temperature must not exceed $19\frac{1}{2}$ tons to the square inch, whilst the pressure of a single round must not exceed $20\frac{1}{4}$ tons per square inch. The proof is usually taken on five rounds, but if the pressures are running very close to the high limit a greater number of rounds are fired.

Both the velocity and pressure obtained with any cartridge vary with the atmospheric conditions, and in order to obtain consistent results it is necessary to eliminate these variations. This is done by firing both for velocity and pressure in comparison with ammunition of known quality, which is kept for standardization purposes, and which is known as "Standard ammunition." The use of this standard ammunition also ensures that the whole of the apparatus used is in adjustment, and is giving dependable results.

Standard ammunition is very carefully made up with specially selected components. Care is taken that the bullets are of uniform weight and diameter, and that the cases are of uniform capacity. The charges are very carefully weighed, and are taken from a batch of propellent which is known to be regular throughout, and which has been stored until its ballistic qualities have settled down and become constant. The loading is carried out with the greatest care, and in fact everything is done to ensure that the cartridges composing any one batch of standard ammunition are as far as possible identical in all respects. The outstanding requirement is uniformity, and the absolute values of the velocity and pressure obtained are of comparatively minor importance, provided that the velocities and pressures throughout the batch are as nearly as possible alike, and that their values are accurately known.

The determination of the ballistics of ·303-inch Mark VII standard ammunition is a matter of great importance, as standard ammunition is used all over the world wherever ·303-inch Mark VII ammunition is manufactured, and it is the only means by which the ballistics of this ammunition are standardized, wherever it is made.

The actual assessment of the values of velocity and pressure of standard ammunition is carried out as follows :—Ten rounds are fired for velocity and ten for pressure, from each of four good barrels on each of two days. One day's firing is carried out in the morning and the other in the afternoon. All the cartridges are carefully heated to a temperature of 60° F. for the velocity trials, and to a temperature of 120° F. for the pressure trials. The firing is done with the greatest possible care, and precautions are taken to ensure that the conditions of firing throughout the whole series remain constant. If any single round from any of the series gives abnormal results, the whole series of ten rounds is discarded.

When the trials are complete, the results of each of the eight series of ten shots are averaged, and a mean result obtained for each series. These eight means are again averaged and one mean figure thus obtained for the whole of the rounds forming the eight series. This figure is taken as the assessed value either of the velocity or the pressure, as the case may be, of that particular batch of standard ammunition.

When carrying out pressure and velocity trials at routine proof, standard ammunition is used as follows :—At the beginning of each day's work five rounds of standard ammunition are fired from each barrel which is going to be used that day for velocity or pressure work. The difference between the results so obtained and the assessed value of the standard ammunition represents the variation from the normal due to the atmospheric conditions, and to the use of that particular barrel. Accordingly a correction of this amount is applied to all recorded velocities, or pressures, as the case may be, obtained in that particular barrel on that day.

By way of illustration the following is an "Abstract of results" of velocity and pressure of a certain consignment of ammunition, in comparison with "Standard."

| Nature of Cartridge. | Number of Barrel. | Rounds Fired. | Velocity. Feet per second. | | Uncorrected Pressure at 120° F. | | | |
|---|---|---|---|---|---|---|---|---|
| | | | Dry 60° Fahr. | Difference from Mean | Compression of Copper before Firing. | Compression of Copper after Firing. | Equivalent in tons per square inch. | Corrected Pressure. |
| Cordite Standard 23.11.21. Value at 60° dry, 2,370 feet-seconds. At 120° = 19·1 tons. | 888 | 98 | 2,363 | −12 | 0·114 | 0·140 | 19·5 | — |
| | — | — | 2,380 | + 5 | 0·114 | 0·130 | 18·7 | — |
| | — | — | 2,383 | + 8 | 0·114 | 0·136 | 19·2 | — |
| | — | — | 2,372 | − 3 | 0·114 | 0·133 | 18·9 | — |
| | — | — | 2,375 | 0 | 0·114 | 0·132 | 18·8 | — |
| | | Mean | 2,375 | M.D.6 | — | 0·134 | 19·0 | — |
| | | *Correction − 5* | | | *Correction + 0·1* | | | |
| ·303-inch Mark VII Cordite X. 1.2.22. | 888 | 103 | 2,348 | −16 | — | — | — | — |
| | — | — | 2,354 | −10 | — | — | — | — |
| | — | — | 2,384 | +20 | — | — | — | — |
| | — | — | 2,357 | − 7 | — | — | — | — |
| | — | — | 2,364 | 0 | — | — | — | — |
| | — | — | 2,376 | +12 | 0·114 | 0·136 | 19·2 | 19·3 |
| | — | — | 2,389 | +25 | 0·114 | 0·149 | 20·3 | 20·4* |
| | — | — | 2,362 | − 2 | 0·114 | 0·138 | 19·3 | 19·4 |
| | — | — | 2,327 | −37 | 0·114 | 0·133 | 18·9 | 19·0 |
| | — | — | 2,379 | +15 | 0·114 | 0·133 | 18·9 | 19·0 |
| | | Mean | 2,364 | M.D.14 | — | — | — | — |
| | | Correction − 5 | | | Mean corrected pressure = 19·4 tons | | | |
| | | Mean corrected velocity = 2,359 feet-seconds. | | | | | | |

From this abstract of results it will be seen that the assessed value of the standard in use is velocity 2,370 feet per second and pressure 19·1 tons, and that on that particular day on which the firing took place the results obtained from the standard rounds fired at the beginning of the trials were velocity 2,375 feet per second and pressure 19 tons. This means that from every velocity reading taken in that barrel during the day 5 feet per second must be deducted, and to every reading taken in the pressure barrel in which the above standard rounds were fired 0·1 ton must be added.

The table also records the results obtained from a consignment of ammunition fired on the same day, and shows that the mean velocity obtained was 2,364 feet per second, or after correction 2,359 feet per second. It is also clear from the table how the mean difference is obtained, which in this case comes out at 14 feet per second. These figures are well within the specification requirements.

In the case of the pressures, however, it will be seen that after correction the mean pressure comes out at 19·4 tons, which is just within the limits allowable, while the round marked * comes out at 20·4 tons, which is above the limit allowed for a single round.

#### 4.—Freedom from Defects and Functioning Trials

Any definite defect in the ammunition, such as a cap without any fulminate in it, or a cap chamber with no fireholes in it, would cause a failure under all circumstances, and provided the numbers fired were sufficient to bring it to light, such a defect would appear either in the velocity, pressure, or accuracy trials already described. The numbers of rounds fired in these trials, however, are comparatively small, and not sufficient to discover such defects with certainty, for although there might be no cartridges containing these defects among the small number of rounds fired, they might still occur with sufficient frequency seriously to impair the quality of the ammunition. It is only by firing a comparatively large number of rounds that it can be ensured that these defects, if they do exist, are so rare as to be unimportant.

Even if a greater number of rounds were fired, the ammunition, though it might still give satisfactory results in good barrels and well-adjusted rifles, such as are used for accuracy and velocity trials, might not behave equally well under the more strenuous conditions of machine-gun fire or in weapons which are worn out, because there are many cartridge defects which are latent only, and do not cause failures unless the cartridges are fired under unfavourable conditions. If these latent defects exist they must be discovered, and this can only be done by making the conditions at least as unfavourable as those which may be expected in the Service.

It is therefore necessary in the case of Mark VII ammunition that, beyond the rounds fired for velocity pressure and accuracy proof, any ammunition fired to detect defects in the cartridges themselves should not only be proved in rifles, but also in Vickers, Lewis and Hotchkiss machine guns, under such conditions of adjustment and wear as will develop any latent tendency to failure which may happen to exist. The tests for cartridge defects, therefore, are closely allied to those for functioning in the various weapons, and in practice are combined into one series of trials.

The more common forms of failure to which small arm cartridges are liable may be classified as defects in the cap, case and bullet respectively.

The principal cap defects are missfires, failure on the part of the cap to ignite the charge, hangfires, pierced caps, blowbacks, and caps out.

Missfires, that is to say, failure to fire on the part of the cap itself, may occur when there is insufficient composition in the cap, or when there is too much protective varnish over the composition. Missfires may also be due to deteriorated or broken cap composition.

Sometimes a missfire may be caused by the striker failing to drive the cap on to the anvil in the cap chamber, either because the anvil is sunk too far into the case, or because the striker has not sufficient protrusion, and in both cases the tendency to missfire is increased when the head space in the gun is large, allowing a clearance between the base of the cartridge and the breech face. As wear occurs the striker protrusion, both of the rifle and the Vickers gun, gets shorter, and in order therefore to show up any tendency to missfires which may exist due to low anvils, proof is carried out in both these weapons adjusted so that the striker protrusions are the minimum allowable.

Eccentricity in the cap chamber or firing pin also increases the tendency to missfires, particularly if combined with a low anvil, as owing to the domed shape of the anvil the striker does not impinge on the highest point of an eccentric anvil. Want of concentricity in the cap chamber is not, however, likely to cause a missfire unless it is clearly visible to the eye, and individual cartridges with eccentric cap chambers are eliminated during examination, and a special proof for eccentric caps is, therefore, not necessary.

Sometimes the cap, although firing itself, fails to ignite the charge. This may be due to weakness of the cap composition or faulty propellent, or may be caused by mechanical defects in the case, such as absence of fireholes, or by a wad covering the fireholes. Such a fault would cause a failure under all circumstances, and might occur in any proof, and no special proof is therefore required.

A hangfire is said to occur when an abnormal period of time elapses between the striking of the cap and the ignition of the charge. Hangfires may be, and with new ammunition usually are, caused by weak cap composition, which does not give a powerful flash in the propellent, and thus fails to ignite the charge thoroughly. Hangfires may also be due to deteriorated or damp propellent.

The presence of hangfires in a batch of small arm ammunition is a very serious fault. When shooting there is a natural tendency to cease to maintain the aim after the trigger is pressed, particularly in rapid fire, and if the bullet does not leave the barrel for an appreciable time afterwards, a miss is certain to be the result, and no accurate shooting is possible where hangfires are frequent. When ammunition is used in aircraft, where shots are fired between the blades of a revolving propeller, a hangfire is even more serious. In this case the gun is actuated by a timing gear which prevents it from firing except at periods when the propeller blades are not passing the muzzle. Should a hangfire occur the bullet is quite likely to hit the propeller blade and cause a serious accident.

If during any other proof a hangfire occurs which is sufficiently bad to be audible, it is noted, but the general tendency to hangfires is more carefully investigated in a special proof, which will be described later.

Pierced caps may be of two kinds. In one case the striker is driven right through the crown of the cap, and in the other that portion of the cap which has been struck by the striker is blown outwards. So far as the cartridge itself is concerned, the chief cause of pierced caps is the bad quality of the metal of which the cap is made, though the tendency to this trouble is increased where there is a large head space in the gun or high striker protrusion. Pierced caps are often due to a combination of circumstances, and sometimes their cause is somewhat obscure, but it is known that a heavy striker blow or a worn firing hole bush may give pierced caps, and even the rigidity of the gun mounting may have an effect.

For reasons connected with the timing of the various elements of the firing mechanism of the Vickers gun a greater protrusion of striker is adopted than in the case of either the rifle or Lewis gun. On the other hand, the Hotchkiss gun has a striker protrusion which, though not excessive when new, quickly increases with wear. Proof carried out, therefore, either in a new Vickers gun or in a worn Hotchkiss will show up any tendency to pierced caps which may exist.

A blowback is caused by the powder gas, which gets through between the cap and the walls of the cap chamber. It is due to the cap or cap chamber not being round, or to the cap being a bad fit in the cap chamber. Any tendency to blowbacks may be brought to light by an eccentric striker blow.

When a pierced cap or a blowback occurs in a rifle, the result is generally most unpleasant to the firer, and may be dangerous, owing to the fact that the hot gases may be blown into his face. Where the trouble is sufficiently serious and the quantity of powder gas is considerable, damage may be caused to the striker, and these defects are, therefore objectionable both in rifles and in machine guns.

Caps come out of the case when they are a loose fit in the cap chamber, but it has now been the practice for some time to rivet all caps in place, and where this is properly done caps practically never come out.

It is particularly undesirable that caps should come out in machine guns, especially in the Hotchkiss gun, as the loose caps are liable to get into and damage the mechanism. Unfortunately cartridges are more prone to develop this fault in the Hotchkiss gun than in any other weapon, owing to the fact that the after part of the cartridge case is not well supported, and if the case is at all soft it tends to expand under the pressure of firing and the cap is thus loosened.

Case defects may be classified under the following headings : bursts, separations, complete or partial, split necks, cases hard to extract.

A burst is probably the most serious fault to which a cartridge case can be subject. The case is usually said to have burst when it is fractured on firing somewhere between the base and the shoulder. Bursts are more likely to occur when the cartridge is fired in a worn gun with a large chamber, and the proof to detect the likelihood of bursts is fired under these conditions. A burst in or near the base is more likely to lead to an accident than one further forward, but whenever a burst occurs it is considered sufficiently serious to warrant rejection of all ammunition liable to this defect.

Bursts are nearly always due to the use of brass of unsuitable quality. If cases are drawn so that they are left too hard, or if they are too light and the walls too thin, the tendency to burst is increased, but bursts from these causes alone are rare. Cartridge cases may burst because they contain hair cracks or tool marks, but such defects can generally be seen and the cartridges removed during visual examination. Similarly individual dangerous cases made from cups punched from bad pieces of strip, where dross or scale has been rolled into the metal, usually show a flaky surface, and can be eliminated during visual examination, and in consequence such faulty cases should not be selected for proof at all. It is possible that on rare occasions a burst may occur on proof, due to an internal tool mark which is not visible on examination, but the great majority of bursts which occur on proof are due to the general bad quality of the metal of which the cases are made, and where a burst occurs due to an internal tool mark an examination of the fired case nearly always discloses the cause.

A separation consists of a circumferential fracture of the case. It differs from a burst in that the fracture is caused by the extraction of the case and not by the explosion. A separation is usually due to improper annealing, the middle of the case where the separation usually occurs being left too soft. When this happens the expansion to which the case is subject during explosion all occurs at the soft place, and leaves the metal at this point very thin and weak, and the case may then be fractured during extraction. Separations may be complete or partial. Where, however, the metal is stretched, but an actual fracture has not taken place, the defect is known as " Stretched metal."

Separations are more likely to occur in a weapon with a worn chamber and large head space, and other things being equal, ammunition is more likely to give separations in machine guns than in rifles, as owing to the rapidity of fire the stresses applied to the cartridge are more severe. The recoil mechanism of the Vickers gun appears to give a more violent wrench to the fired case than that of the Lewis gun, and the greatest tendency to separations occurs in a worn Vickers gun with a large head space.

If the separation is complete, the front part of the case is left in the chamber. For this reason the defect is particularly serious in a machine gun, because the next round is forced hard into the chamber but cannot get right home, and the machine gun is thus jammed, and there is usually a good deal of difficulty in clearing the gun. When the gun is in an inaccessible position the defect is even more serious, and hence separations are probably the gravest of all faults in ammunition used in machine guns in aircraft.

Any fracture of the case such as a burst or separation, indicating as it nearly always does that the rest of the ammunition is liable to the same defect, is so serious that it necessarily entails rejection. But this is not all. An epidemic of bursts follows a batch of metal rather than a consignment of ammunition, and similarly a tendency to separations coincides with a batch of cases as delivered from the component factory. Owing to the mixing of components which may take place in a filling factory, the cases made from any particular batch of brass will be found mixed with others in several consecutive consignments of ammunition. When therefore a burst or separation occurs during proof, special large proofs are taken from other batches of ammunition chronologically previous or subsequent to the consignment in question, even if these have already passed proof. These special proofs are fired under conditions likely to bring the fault to light until three clear consignments are found at each end of the series free from the defect. These batches are allowed to pass, but all others of the series are rejected, including any consignment which shows no fracture, but which is between others where fractures have occurred, as it is recognized that such a consignment almost certainly contains faulty cases liable to give trouble.

Split necks are usually caused by faulty treatment during manufacture which leaves the metal of the neck in a state of excessive internal stress. The fault becomes more pronounced as the ammunition gets older, especially if it is stored in the tropics. Split necks may show themselves before or after firing. Split necks occurring before firing are far the most important, but are practically unknown in new work. Split necks after firing are not so serious, but they are an indication that in course of time the cases are likely to split before firing, and split necks after firing occurring in the proof of newly made ammunition would entail rejection.

Cases which are hard to extract after firing are objectionable in a rifle, as they prevent rapid firing. The defect is usually due to the case being too soft in the rear half. Extraction difficulties are more likely to occur in a new barrel, and are increased when the chamber is hot. Tests for "Hard to extract," therefore, should involve fairly rapid fire.

The chief defects to which bullets are subject are: Tendency to break up on firing, failure to take the rifling, and metallic fouling.

When a bullet breaks up on firing, either the lead core is driven through the envelope, which is stripped completely off the core, or the envelope itself breaks up. In either case the defect is due either to faulty cupro-nickel, or to excessive hardness of the envelope. The fault is a rare one, but where it occurs it entails rejection of the consignment.

Failure to take the rifling is more common, and is generally due to the envelope being too hard, particularly at the rear end. When it occurs, the bullet quickly turns sideways on, and its flight becomes very erratic and its range is reduced. Bullets with a tendency to this defect will develop it much more readily in a worn barrel, particularly one with a worn leed, but there is no simple rule to decide how many rounds can be fired in a barrel before it occurs. To some extent it depends on the rate at which the rounds have been fired, and a given number of rounds fired "rapid" in a Vickers gun will wear the barrel nearer to the critical point than the same number of rounds fired in a rifle. On the other hand, a rifle barrel in peace time is worn out more by cleaning than by firing, and it is accordingly necessary to prove for this defect both in rifles and machine guns. It is difficult to say which type of machine gun is more prone to show up failure to take the rifling, but if the bullets tend to be hard it is probable that the Vickers gun will fire the greater number of rounds with safety.

When means are provided to prevent the powder gas getting round the bullet before it is set up, failure to take the rifling is largely eliminated, and for this reason all ammunition, both cordite and N.C. loaded, is now made with a wad. Failure to take the rifling is a serious defect both in rifles and machine guns, but particularly in the latter during barrage fire.

Metallic fouling is due to bad quality cupro-nickel, or is caused by insufficient hardness of the envelope. It is more prone to occur in a rough barrel than in a well-polished one, and any tendency to metallic fouling is increased in a hot barrel, and even by hot weather. If, however, the defect is sufficiently serious to be of importance, its presence will become evident during the accuracy trial, and no special proof for metallic fouling is therefore necessary.

In order to detect the possible existence of any of these defects, and to verify that the ammunition functions satisfactorily in the various weapons the following trials are fired: A hangfire proof, a proof to detect cartridges which are hard to extract, a proof into sawdust for recovery of bullets in order to ascertain whether they have taken the rifling properly, and a general proof in various weapons, known as "Casualty proof."

The number of rounds fired in each of these trials varies, and to some extent depends on the size of the consignment of ammunition under inspection. Some defects show up in more than one proof, and where this is so, the firing of large numbers of rounds at each proof is unnecessary. Throughout, the dominant consideration governing the quantities fired at each proof is the necessity of firing sufficient rounds to ensure with reasonable certainty that the number of cartridges in any consignment containing each or any particular defect forms such a small percentage of the whole consignment as to be unimportant.

*Hangfire Proof.*—This consists in firing a certain number of rounds in a new barrel, and measuring the period of time which elapses between the striking of the cap and the bullet leaving the muzzle. The time taken should not vary more than one two-hundredth of a second.

*Hard-to-extract Proof.*—In this trial cartridges are fired from a rifle with a new barrel as rapidly as possible in order to heat the chamber. If any round is found hard to extract by hand, the pressure required on the bolt handle to open the bolt is measured on a spring balance, and if it exceeds 25 lbs. the round is considered to be too hard to extract. The test is a very rough and ready one, but gives a fairly reliable indication of the general qualities of the consignment as regards this particular fault.

*Sawdust Proof.*—In this proof the rounds are fired from a rifle with an old barrel which has fired at least 5,000 rounds. The bullets are fired through screens into sawdust, and any bullets which break up, or tip, or fly sideways on, can be detected. The bullets are then recovered from the sawdust and the engraving of the rifling examined. The trial is particularly directed towards detecting bullets breaking up, or failing to take the rifling when fired in a rifle, but owing to the fact that a worn barrel is used, the proof may also bring to light missfires, blowbacks, bursts or split necks.

*Casualty Proof.*—The casualty proof is carried out in various weapons adjusted so as to bring to light the different faults to which the ammunition is most liable in the respective weapons. It consists of five separate firing trials, one each in the rifle, Lewis and Hotchkiss guns, and two in the Vickers gun.

In the rifle proof a weapon is employed which has fired at least 5,000 rounds, and has an old action with a large chamber and head space and a worn firing pin hole and eccentric striker. The striker is also set to give a small protrusion.

The Lewis gun used is fitted with a barrel which has fired at least 6,000 rounds, and a gun with a striker protrusion at the low limit is selected. Casualty proof with the Lewis gun is carried out through screens.

The gun used for the Hotchkiss casualty-proof is fitted with a new barrel and long striker protrusion.

In one of the Vickers gun proofs an old barrel that has fired at least 7,000 rounds is used, with an old action with large head space, and the firing takes place through screens. In the other Vickers gun proof the gun is adjusted to have a long striker protrusion and a small head space.

The casualty proofs in the rifle and Lewis gun are chiefly intended to detect missfires, blowbacks, bursts, and split necks; that in the Lewis gun being also arranged to discover bullets breaking up, or failing to take the rifling.

The proof in the Hotchkiss gun is directed entirely to discovering pierced caps, blowbacks, or caps out.

Of the two Vickers gun trials, one may bring to light missfires, blowbacks, bursts, separations, split necks or bullets breaking up, or bullets failing to take the rifling, while the other is intended solely for the purpose of disclosing a tendency to pierced caps.

## CHAPTER VI

## INSPECTION OF ·303-INCH MARK VII AMMUNITION

### 1.—NECESSITY FOR COMPLETE INSPECTION

Small arm cartridges after passing proof satisfactorily are submitted to a thorough inspection, the details of which depend to some extent on the design of the cartridge. As a rule every single cartridge is submitted to an individual inspection, but this procedure is not universal, and in some countries small arm ammunition is inspected on a percentage basis.

Wherever possible a complete inspection is very desirable. At first sight this may appear to be more costly, but it is not really so, and in any case it cannot be too strongly emphasized that the inspection of small arm ammunition must primarily be regulated by considerations of quality and not of cost.

Quite apart from the failure of ammunition on proof involving the wholesale rejection of complete batches, ammunition as turned out from the loading field may occasionally contain up to 5 per cent. of faulty cartridges. Such a percentage of faulty cartridges is quite inadmissible, and would lower the value of the ammunition for military purposes to an extent out of all proportion to the cost of a complete inspection, and might make it practically unusable.

Now it is obvious that should any batch of ammunition contain, say, 5 per cent of faulty cartridges, and the inspection be carried out on a 20 per cent basis then, even if all faulty cartridges were removed from the 20 per cent examined, the remaining 80 per cent unexamined would still contain a quantity of unserviceable rounds equivalent to 4 per cent of the whole. Inspection on a percentage basis, therefore, is only permissible on the understanding that if more than an insignificant quantity of bad cartridges are found the whole supply is rejected, and where this is the case the manufacturer is forced, in self-protection, to carry out a complete inspection himself, and must charge accordingly, so that the expenditure is thus incurred in any case.

It so happens that some of the most serious faults which might be dangerous or even fatal to the firer, or which would put the weapon out of action, are sporadic ones, which may not occur in many millions of rounds, and which then for a few rounds become epidemic ; and it is only a complete inspection which will find such faults.

Complete inspection, therefore, should be looked upon more in the light of a manufacturing process, essential for the production of high-quality ammunition, but one which for obvious reasons should preferably be carried out by the purchaser.

Even with a complete inspection the elimination of every faulty round from a batch of ammunition is a much more difficult matter than is generally supposed. It is notorious that, however careful proof-readers may be, and however many times they may scrutinize a book, it is not possible to guarantee that every misprint is eliminated. Bad cartridges must be looked upon in the same light as misprints, and it must be recognised that it is not possible to eliminate every one of them, but experience has shown that, provided the inspection is really carefully carried out, more than one examination of cartridges is not necessary, since after one examination the number of bad rounds left in the ammunition is insignificant.

For these and other reasons it has for many years been the practice in Great Britain to inspect small arm ammunition on a 100 per cent basis, and the inspection which is applied to the ·303-inch Mark VII cartridge, which involves weighing, gauging and visual examination, and which is described below, is carried out on these lines. It will be seen, however, that the whole organization is arranged so as to provide, where possible, a check examination on a small percentage of the rounds, not so much with a view to a second inspection, but rather with the object of keeping a close watch on the quality of the work done by the various operators.

The inspection of the cartridge components is carried out during manufacture in the factory, and has already been referred to in Chapter IV, and inspection after manufacture is generally confined to the complete cartridge. It is, however, usual to break down a few rounds from each batch in order to examine the quality of the internal manufacture, and from time to time a chemical analysis of the components is carried out.

After examination small arm cartridges for military purposes are usually packed ready for use, either clipped up in their chargers and in bandoliers, or arranged for use in machine guns, and, as they may be stored in damp climates, they are generally sealed up in tin-lined boxes. This should be done as soon as possible, before there is any chance of damp getting into the ammunition. It is convenient to pack and seal up the ammunition in the boxes in which it will ultimately be stored immediately after the examination is complete, in the same building in which it is inspected, as it is difficult to transport small arm ammunition in large quantities unless it has been packed.

Assessment of quantities is an essential part of inspection. Counting by individual cartridges would be tedious, and can be carried out in a much simpler manner by counting the ammunition after it has been packed in the boxes, each of which contains a definite number of rounds, and for this reason also it is convenient to deal with the packing operations at the same time that the inspection is carried out.

The inspection and packing of ·303-inch Mark VII ammunition thus consists of six processes :—

(i) Breaking down and chemical analysis.
(ii) Weighing.
(iii) Gauging.
(iv) Visual examination.
(v) Packing.
(vi) Assessment of quantities.

## 2.—Break Down and Chemical Analysis

A few rounds of each consignment, usually taken from those selected for proof, are broken down and the individual components weighed in order to ascertain whether the components are all within the limits allowed, and more particularly to see that the variations between round and round in the weight of any component are not excessive.

The allowable variations in weight in grains of the components of the ·303-inch Mark VII cartridge are as follows :—

|  | Normal weight. | Maximum weight. | Minimum weight. |
|---|---|---|---|
| Case | 170 | 175 | 165 |
| Cap | 5·5 | 5·5 | 5·5 |
| Bullet | 174 | 176 | 172 |
| Charge (Cordite) about | 37·5 | 37·5 | 37·5 |
| Total | 387 | 394 | 380 |

Note.—The normal weight of the nitrocellulose charge is about 41 grains.

Opportunity is taken when breaking down these rounds to ascertain that the bullets have been properly lubricated, that the varnishing of the cap has been adequately carried out, and that the correct propellent has been used. The pull required to extract the bullet is also measured. This should be not less than 60 lbs.

It is important that the propellent used should be chemically stable, and the manufacturer is required to declare from what batch of propellent any consignment of ammunition is made. Whenever manufacture from a new batch of propellent commences, a few cartridges are broken down and the propellent tested for stability, and occasionally the various components are chemically analysed.

The stability test for the propellent is known as the "Abel heat test," and is based on the fact that the rate at which the propellent will give off nitrogen peroxide when heated in air is an indication of its instability. The propellent is ground and sieved and placed in a test tube heated in a bath of water, the temperature maintained in the case of newly manufactured propellents being 180° F. In the test tube is suspended a paper impregnated with potassium iodide and starch, the upper half of which is moistened with a mixture of glycerine and water. The nitrogen peroxide causes a brown line to appear at the junction of the wet and dry portions of the paper, and the test is complete when this line attains a certain tint. If the stability of the propellent is satisfactory, the nitrogen peroxide should be evolved so slowly that the time taken for the development of the correct tint after the commencement of the heating is not less than 30 minutes. The test is a comparative one, and accurate results can only be ensured if standard conditions are maintained.

### 3.—Weighing

Some cartridges contain no charge, while others contain a second bullet in addition to the one fixed in the mouth of the case. This second bullet may be found in a case in which there is no charge or in one in which there is a considerable portion of a charge. "No charge" and "double-bulleted" cartridges only occur very rarely, but they may both be extremely dangerous.

If a cartridge containing no charge at all is fired in a rifle, the cap is not usually powerful enough to put the bullet into the bore, but only a very small portion of a charge need be present to drive the bullet well into the bore, and unless the amount of the charge is sufficient to eject it, it lodges as an obstruction. If this is not noticed, a very high pressure is generated when the next round is fired, and generally the gun is wrecked.

A double-bulleted cartridge containing no charge, or only a small portion of a charge, behaves in a very similar manner, but a double-bulleted cartridge containing a considerable charge, produces an abnormally heavy pressure, which may damage the weapon. There is also the further danger with such a double-bulleted cartridge that one or both of the bullets may be left in the barrel and cause disaster when the next round is fired.

Every round of each consignment is weighed in an automatic weighing machine, with the object of eliminating cartridges containing no charge, which are too light, or double-bulleted rounds, which are too heavy. The type of weighing machine which is used for the purpose is mechanically operated and forms part of a combined weighing and gauging machine. The cartridges are fed into it down a guide, and are weighed in turn, those that are too heavy or too light being diverted into a "reject bag," and those within the limits to which the machine is set being conducted down to the mechanically-operated gauging machine.

Sometimes a portion of a bullet or a bit of metal may find its way into the cartridge during manufacture, but unless this weighs about 80 to 100 grains, it is not likely to cause a dangerous obstruction. Double-bulleted cartridges which are dangerous are therefore considerably above the normal weight, and it is fairly easy to eliminate them. The weighing machines are in practice set to throw out any cartridge which is more than 70 grains heavier than the lightest cartridge allowed or, say, about 60 grains heavier than the normal.

Cartridges with no charge are more difficult to eliminate owing to the light weight of the charge itself, and to the variations which occur in the weight of the components.

From the figures given above, it is clear that no cartridge can be rejected for being too light whose weight is not below 380 grains, while it is possible for a cartridge of this weight to have a case and bullet up to or even above the limits of weight allowed and a charge of about 23 grains or less.

Actually some slight variation also occurs in the charge required to give correct ballistics, and in the weight of the cap, so that the variations in weight to be dealt with are somewhat greater than the above figures show.

Advantage, however, is taken of the fact that there is a tendency for components to run to one weight either above or below the normal, and it is possible therefore to weigh to closer limits by taking the average weight of the batch of cartridges and setting the weighing machines accordingly. On the other hand, some allowance has to be made for the uncertainty of the machine, which may throw out or retain incorrectly cartridges which are very near the limit. For this reason, and also in order to provide for cartridges which occasionally fall off the scale pan into the " reject " bag by vibration, all rejected cartridges are weighed a second time, the machine being set to reject at a weight 2 or 3 grains lighter than at the first weighing.

The actual procedure is as follows. One hundred cartridges are taken and weighed, and the average weight of the batch thus obtained. Assuming this average to come out at, say, 387 grains, all the machines would be set for the first weighing to throw out light cartridges at 382 grains, and heavy cartridges at 452 grains. Any rejected at this first weighing are again carefully weighed in a machine set to throw out light cartridges at 380 grains. Those rejected a second time are weighed on a chemical balance by hand, and the heavy ones sorted from the light, and the causes of rejection analysed, the cartridges being broken down and the components separately weighed if necessary.

If care is taken, rounds with low charges of less than 15 grains, or cartridges more than about 60 grains heavier than the normal, can be discarded almost with certainty. A charge of 15 grains is not dangerous to the weapon, as it is sufficiently powerful to ensure that the bullet is not left in the bore.

### 4.—Gauging

For purposes of manufacture both high and low limits are set for most of the principal dimensions of the cartridge, and it is important that the ammunition is made within these limits. In most cases, however, where cartridges are made smaller than the limits the result is a failure on proof. For instance, cartridges that are too small are liable to give bursts, particularly in a chamber which is worn up to the largest allowable limits. Again, bullets of small diameter may fail to take the rifling and the bullet therefore does not fly correctly, and turns sideways on. Where the low limit on any dimension of a cartridge is decided solely by some such consideration, and the performance of the ammunition on proof indicates that the particular fault is absent or so rare as to be insignificant, then it is not necessary to gauge each individual cartridge to ascertain that it is not too small at that particular point.

There are, however, some exceptions to this rule. For instance, cartridges must not only not be too long, because they would jam in the magazine of a rifle, but they must also not be too short, as in this case they would give trouble in the drum of a Lewis gun, and it is necessary, therefore, to gauge every cartridge to see that it is neither too long nor too short.

Cartridges with thin heads are also liable to give trouble. In a gun with a big head space the result may be a missfire, and such cartridges are liable to droop when held in the lock of the Vickers gun, and the bullet, instead of entering the chamber, hits the end of the barrel. In both these cases, however, unless the heads are so thin that the defect is clearly visible without measurement, failure is not likely to occur, and it is not necessary, therefore, to gauge for thin heads, this fault being eliminated on visual examination.

Accordingly, every cartridge is gauged to verify that it is not too large to enter the chamber of the gun, and that it is neither too long nor too short.

The high limits of size of the cartridge are slightly smaller than the size of the chamber. This is necessary to ensure that the cartridge will enter the chamber easily. At the mouth of the case the cartridge should be a close fit in the chamber, but it is not desirable to fit so tightly elsewhere. Provision must also be made for wear of gauges, and the smaller the size to which it is permissible to make the high limits of the cartridge,

Part II, Chap. VI, Fig. 1.

[To face page 251.

Part II, Chap. VI, Fig. 2.

Pt II. CHAP. VI. *To face p.251.*

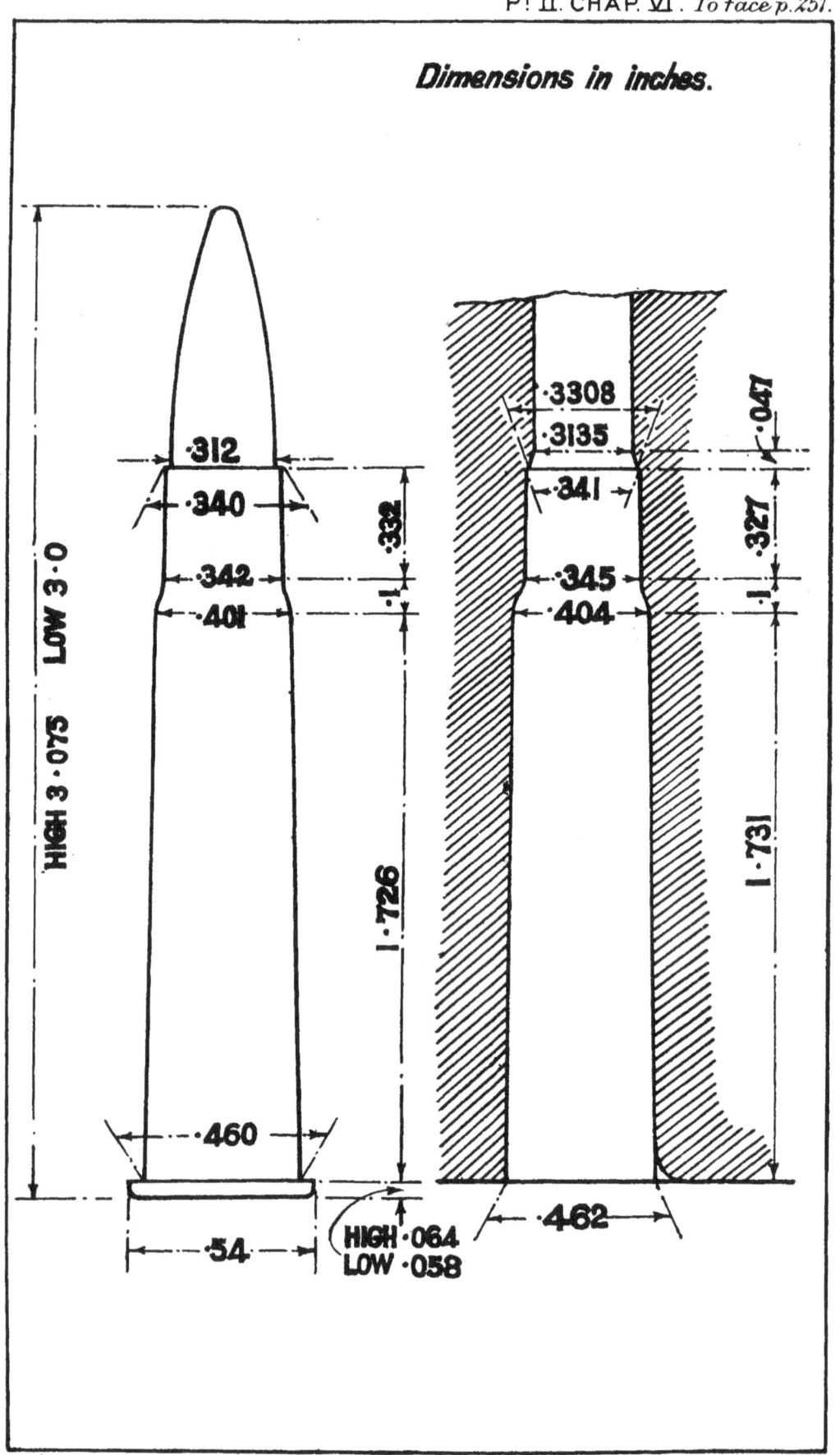

**FIG. 3.**

the greater is the amount of wear allowable on the gauge before it passes cartridges too large for the chamber. On the other hand, cartridges must not be made too small for the reasons already given, and the tolerances between the high limits of the cartridge and the chamber dimensions are a compromise, settled largely by experience, so as to allow the greatest possible amount of wear in the gauge without making the cartridge so small at any place that it is likely to fail on firing, and at the same time to leave provision for a reasonable variation in size between round and round in order to facilitate manufacture.

The cartridges are gauged in an automatic gauging machine, divided into two parts. One portion, which is known as the "Body" or "Socket" gauging machine, measures the complete cartridge all over, and verifies that it is not too large to enter the chamber of the gun. The other portion measures the overall length and eliminates cartridges which are either too long or too short.

The body gauging machine is arranged to receive the cartridges direct from the weighing machine, and feeds them in turn into the gauge. If they enter the gauge easily, a slider actuated by a spring then passes across the mouth of the gauge, and moves a directing shutter in a chamber into which the cartridge subsequently passes, and the cartridge is then conducted down into the length gauging machine. If, however, the cartridge is too large to enter the gauge, or enters it with such difficulty that the slider spring is not strong enough to force the slider across the gauge, then the slider fails to move the directing shutter and the cartridge is diverted into a "reject" box.

The length gauging machine consists of a rotating disc containing radial slots, into which the cartridges are fed from a hopper. The length of the slots is controlled by a central fixed gauge, so arranged that the slot lengthens as the disc rotates. Long cartridges are carried further round the disc before they fall through the slots than short ones, and thus short, correct and long cartridges are sorted into three separate boxes.

A picture of the combined weighing and gauging machine is shown in Fig. 1, while Fig. 2 is a view of a section of a small arm cartridge inspection shop, showing a group of these machines in action. Such a section, working day shift only, would handle about 3 million cartridges per week.

Cartridges which are correct to size may sometimes be unfairly rejected, because they have a piece of beeswax or grit adhering to them, which prevents them from entering the gauge. The gauges also become fouled with dirt and require cleaning out from time to time, and may therefore reject correct ammunition. Cartridges which fail to pass in the gauging machines, therefore, are gauged a second time. In order to give every cartridge an equal chance of passing, and to make quite certain that correct cartridges are not rejected, it is usual to carry out the second gauging in machines set apart for the purpose and fitted with gauges which are nearing the end of their useful life, and which will pass cartridges slightly oversize. Sometimes the second gauging is done by hand, but machine gauging is preferable, as it is more reliable. The hand socket gauges are ground accurately to length and arranged to slide on a surface table under a crossbar. If the cartridge is too big to go right home into the gauge, it catches against the crossbar and is then rejected.

Cartridges which fail to pass at the second gauging are carefully examined in investigation gauges, arranged to measure individual parts, in order to ascertain the reason for their failure. The complete gauging of all the individual parts of the cartridge requires quite a large number of gauges, which, however, cannot be described here in detail.

The maximum allowable limits of size of the ·303-inch cartridge and the dimensions of the standard ·303-inch rifle chamber are given in Fig. 3, and by comparing these the tolerances at the various points can be readily ascertained. It will be seen that they vary from ·003 inch downwards, and the allowable wear on the gauge must therefore be somewhat less than this.

Owing to the very small tolerances permissible, all small arm ammunition gauges have to be made with extreme precision. By far the most important of small arm ammunition gauges are the machine and hand socket gauges, and in these the accuracy aimed at is two ten-thousandths of an inch in diameters, and half a thousandth of an inch in lengths, and owing to their complicated shape and small internal diameters these full form or socket gauges are very difficult to make.

The accuracy of these gauges at all the essential points is checked by means of test plugs, very accurately ground to the correct diameter for the particular point to be measured, and so shaped that when the plug has entered the correct distance into the gauge a shoulder on the plug is flush with the face of the gauge. Owing to the very gradual taper of most parts of the gauge, a very small excess of diameter allows the plug to enter appreciably further into the gauge, and the shoulder falls below the face of the gauge, whereas if the gauge is small the test plug cannot go in far enough. Fig. 4 shows a set of plugs and a section of a gauge with one of the plugs in position. In addition to the test plugs for each of the important diameters of the gauge, a full form plug, marked FF on Fig. 4, is used. This is made to the shape of the complete cartridge, and ensures that all the various parts of the gauge are properly concentric round the same axis.

As is always the case where very large numbers are involved, the question of the standardization and the control of the wear of all gauges are matters of great importance.

Standardization is generally obtained by frequent reference of all check gauges used for measuring the working gauges to a "standard" or "reference" gauge, so that any alteration in the checks, due to wear or other causes, can be readily detected.

The checks for the full form cartridge gauges, if in constant use, wear very quickly, and it is necessary to check them frequently in the reference gauge. The checking of gauges is done by a highly skilled man, who, if he finds any check gauge has worn slightly and enters too far into the reference gauge, compensates for this by allowing the check to enter a similar amount into the gauge which he is checking, before declaring it to be too large. This use of worn checks, however, can only be allowed to a very small extent, and as soon as any serious wear occurs, the check must be discarded and replaced.

The reference gauge itself also wears in time, and unless this wear is watched, and the reference gauge replaced when necessary, loss of standardization results. To prevent this a set of very accurately made checks and reference gauges is kept. This set of check gauges is very seldom used and practically never wears out, as it is kept solely for the purpose of checking the reference gauges or any important set of check gauges, such as may be made for use as standards in other parts of the world. This set of standard check gauges therefore, not only prevents loss of standardization over a period of time, but also ensures that all similar small arm cartridge gauges throughout the world are made to one fundamental unit of length.

The use of the working gauges after excessive wear has taken place is only prevented by constant examination while in use, and each type of gauge requires different treatment, depending on its design and construction. The full form socket gauges in the automatic machines require very carefully watching in this respect. In these gauges the actual measurement of the cartridges takes place between the gauge and the lower surface of the slider, and the gauge is fitted into the machine so that its relative position in regard to the slider can be accurately adjusted. The adjustment is made so that a steel disc of the correct thickness fitted into the head recess in the gauge will allow the slider to pass over the gauge, while a second disc, slightly thicker, prevents it from passing. From time to time, as the gauge or the slider wears, the position of the gauge is adjusted so that the head space remains within the limits set by the discs.

Although the dimensions governing the allowable thickness of the cartridge head can thus be kept fairly constant throughout the life of the gauge, there are other parts of the gauge where wear occurs, and where adjustment is not possible. The gauges are allowed to wear at these points up to certain dimensions based on experience, and dependent on the largest size of cartridge which will function properly in a rifle chamber, and when any gauge wears up to these dimensions it is discarded. The progress of wear is measured by means of check plugs generally similar to those employed when a new gauge is being examined, but made to the dimensions to which the auges are allowed to wear. Socket

PT II. CHAP. VI.   To face p. 252.

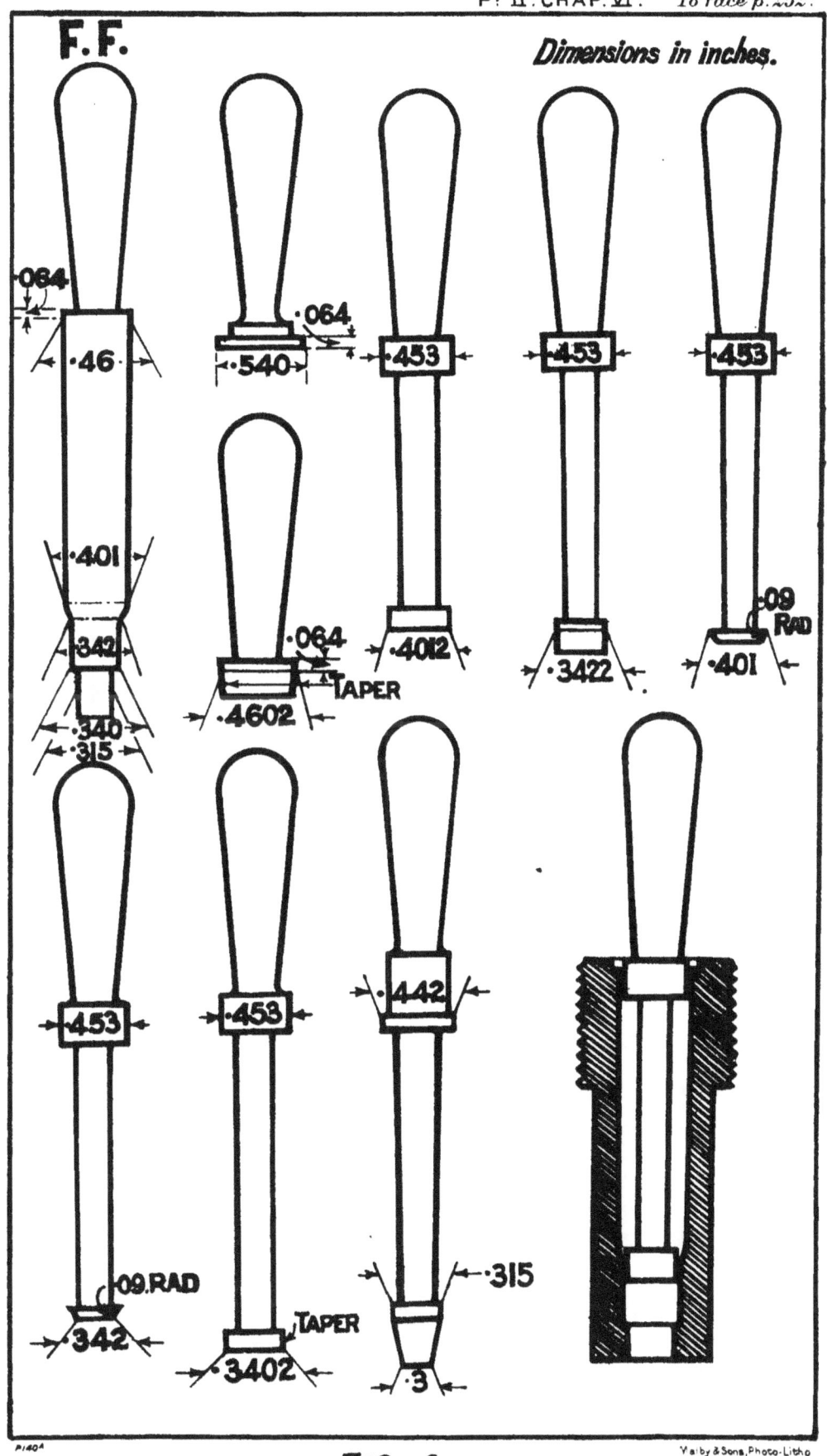

FIG. 4.

gauges which are in constant use day and night are checked both for adjustment and wear daily.

The life of a machine socket gauge varies considerably. Ammunition which is gritty or sticky with beeswax, necessitates frequent cleaning of the gauge, which wears the gauge out very quickly, and when the ammunition is dirty the gauge will not deal with nearly so many rounds as when the cartridges are clean. On the average the life of a socket gauge varies between half a million and a million and a half rounds.

### 5.—Visual Examination

All cartridges which successfully pass the weighing machines and gauges are submitted to a careful visual examination in a good light.

It is astonishing how many different defects may be picked up on visual examination. The examiner has to look at the bases of all cartridges to see that each one is properly fitted with a cap, and that the cap has not been inverted or otherwise faultily inserted. He must see that the cap is not scored or scaly, and that it is neither too high nor too low in the cap chamber, as a high cap may be due to a shallow cap chamber and high anvil, which will make the cartridge dangerous, while a low cap may denote either that the cap has been pressed in too far, or that the anvil is low, in which case the cap is liable to miss-fire. The examiner must also verify that the cap chamber is properly made and central, and that the ringing in of the cap is adequately done and concentric with the cap. He must also see that the cartridge is marked with the correct base mark, and that there is sufficient but not too much bevel, at the back of the head. The correctness of this bevel is important, as smooth feeding in the Vickers gun depends upon it. It is also necessary to see that the head is not thin, and that it is truly circular and concentric with the body of the case.

The examiner must also examine the body of the case for scores, hair cracks, etc., and for scaly metal. He may also find occasional rounds with split necks, or other splits and cracks, and there may be rounds with shoulders crushed up in the necking process, or otherwise distorted. He must also eliminate bullets with scored or scaly or split envelopes, and those which have blunt noses or have been in any way damaged.

All these defects, and others, the examiner must remember, and although he may look specially for scores and fine cracks and one or two similar important defects, it is essential, if he is to examine cartridges with reasonable rapidity, that he should know so thoroughly what a correct cartridge should look like, that any one in the least degree abnormal catches his eye at once. For this reason it will be recognised that effective cartridge examination requires a certain amount of experience and a good deal of practice.

The procedure adopted is that the ordinary examiner puts aside as doubtful every cartridge that appears to be in the least degree defective, and the cartridges thus put aside are re-examined more carefully by special or "final" examiners who have more experience.

In order to make certain that the examination is sufficiently thorough, the cartridges accepted by each ordinary examiner are re-examined from time to time at uncertain intervals by one of the final examiners.

### 6.—Packing

After examination the cartridges are packed into boxes in one of four different ways. For use in rifles they are packed either in chargers and bandoliers, or in chargers and charger cases. Ammunition for use in machine guns is packed in cartons, or in machine gun belts. Packing for machine guns in cartons has superseded the old method of packing in paper bundles, but ammunition packed in cartons is still described as "bundle packed ammunition."

In all cases small arm ammunition is packed in wooden boxes fitted with a tin lining. These boxes are usually strengthened with hardwood ends or battens, and are provided with a lid, secured either by a pin or catch, so that they can be opened quickly. They are carried by rope or web handles attached to the ends of the boxes, and the tin linings are arranged with an opening, closed by a soldered lid with a tear-off handle, the lid and the tin lining being painted black. Arrangements are made for sealing the boxes up after the ammunition has been packed into them. Boxes containing ammunition for use with rifles are stained dark brown, while machine gun ammunition is packed in boxes coloured green. The green boxes which contain carton packed ammunition are also provided with battens, one at each end, so that they can readily be distinguished by feel in the dark from boxes containing rifle ammunition.

When ammunition is packed up in bandoliers, it is first clipped up in chargers, five rounds in a charger. Then each of the five pockets of the bandolier is filled with two chargers put head and tail. The pockets are then closed up, and the bandolier which thus holds fifty rounds is folded up into a bundle. Twenty of these bundles, or 1,000 rounds, are packed into each tin-lined wooden box, the lining of which is then soldered down, and the soldered joint painted black. The box is then closed and sealed.

Packing for rifles in charger cases is now seldom used. The charger case consists of a cardboard box which contains 20 rounds clipped up in four chargers. This cardboard box is closed by a buckram band fixed with shellac. Fifty of these filled charger cases—or, again, 1,000 rounds—are packed in the same wooden box as is used for bandolier-packed ammunition.

Ammunition in cardboard boxes for machine guns is packed head and tail either 48 or 50 rounds in a box. The boxes are then closed by paper wrappers fastened with shellac. Twenty-six of these cardboard boxes are packed in each tin-lined wooden box, which thus holds either 1,248 or 1,300 rounds.

When ammunition is packed in machine-gun belts, each cartridge is put into the belt by hand, and the belt is then passed through a machine, which presses the cartridges firmly home in the pockets, and positions each cartridge correctly in the belt. Each belt is then examined to see that none of the pockets are split, and at the same time the cartridges are counted. Each belt should hold exactly 250 rounds. The belts are then packed in 500-round tin lined boxes, each box holding two belts in separate linings.

It is very important from a military point of view that it should be possible to tell at a glance what ammunition is in a box, without having to read an elaborate description on the label, and, as small arm ammunition is generally stored piled in huge stacks, the top of the box is seldom visible, and it is exceptional if more than one side can be seen. To meet these conditions, small arm ammunition for British service is labelled on all four sides with labels known as "distinguishing" labels. These labels are of simple but characteristic design and embody symbols denoting the various types of ammunition. The symbol for the Mark VII cartridge is a green grid. The labels are printed in distinctive colours, which indicate the group under which the particular ammunition falls in the Magazine Regulations. For instance, all Mark VII labels are printed green, indicating that the ammunition is in Group VI. The printing is also as large as space will allow, and in addition to the symbol gives information as to the type of ammunition contained in the box, the number of rounds, and the method of packing.

Attention is also given to the necessity of tracing the history of the ammunition during its life in the service. Every consignment of ammunition is identified by a batch number. This consists of a letter or letters denoting the manufacturer, followed by the date on which the ammunition was submitted for inspection. Thus, Government (or Royal Laboratory) manufacture submitted for inspection on Armistice Day would be called R.L. 11.11.18. The whole history of the ammunition, including its performance on proof, quality on examination, method of packing, quantity, etc., is recorded against that batch number or "make and date," and it is obviously of advantage that the ammunition while in the service should be capable of being identified with its make and date as long as possible. It is recognized that a tin lining may sometimes be taken out of its box and replaced in another box, and even bandoliers, cardboard boxes, etc., are taken out of their linings and made up into smaller packages. In order to provide for these contingencies, full particulars of the ammunition, including the make and date, are printed on a "descriptive" label. One of these is fixed on the top of the box in a recess, so that it will not rub off, and a second identical one is fixed on the closing lid of the tin lining. The same information is also printed on one pocket of every bandolier and on the paper wrapper closing each cardboard box, and the make and date is stencilled on each end of the wooden box.

In addition to the above labels, the Government group and division and explosive labels dealing with the regulations for storage and transport are fixed on one side of the box, and the box is sealed with a seal which indicates the station where the ammunition was packed.

It is customary to attach the labels with shellac varnish, as this is found to be more permanent than paste or other adhesive, and in tropical countries labels protected by shellac are not so readily attacked by white ants.

The boxes containing each batch of ammunition are numbered serially, and this serial number is stencilled on the top of the box, and the gross weight of each box is stencilled on one end.

### 7.—Assessment of Quantities

Although the counting of the cartridges after they are packed in boxes is simpler than counting each cartridge separately, it necessitates two precautions.

Firstly, it is essential that, as the ammunition is inspected in its uncounted state, cartridges from different batches should not be mixed with each other. This is provided for by the fact that the inspection shops are divided into compartments, and only one batch at a time is admitted to any compartment.

Secondly, it must be clear that each box holds its full complement of rounds. To make sure of this point, every bandolier or cardboard box after filling is weighed, and every ammunition box is weighed before and after packing, so that the net weight of the contents can be ascertained. Owing to the variations in the weights of individual rounds, the net weight of the contents of an ammunition box may vary more than the weight of a cartridge, and the weighing of the wooden boxes, therefore, only verifies that there are the full number of bandoliers or other packages in the box. The weighing of the bandoliers themselves, however, will disclose if there is a cartridge missing.

To prevent mistakes in the counting of the boxes, each box is stencilled with a serial number, but even so, it is quite possible to miss a box or give two or more boxes the same number. This serial number is, however, also used for recording the names of those who examined the ammunition, packed the box, etc. These particulars are chalked on each box, and every box has to be checked over and the particulars entered in a book, and any discrepancy in the numbering is therefore at once brought to light.

Cartridges left over and insufficient to make a complete box are counted individually. Rejects are also sorted and counted under their various classifications.

### 8.—Inspection of Packing Accessories

Inspection is not confined to the cartridges themselves. Faulty packing accessories may affect the serviceability of the ammunition, for instance badly made chargers or bandoliers might cause delay or inconvenience in loading, or leaky boxes might result in deterioration of the ammunition in storage. Steps are therefore taken to ascertain that all packing accessories are correct in every respect.

Thus chargers are gauged to see that they are of the correct dimensions both to hold the cartridges and to enter the gun properly, and are examined to see that they have been correctly made, that the springs have been hardened, and that the charger is suitably blackened.

Bandoliers are gauged, particularly as to the size of the pockets, and the stitching is carefully examined.

Belts are gauged to see that the pockets are of the correct size, square across the belt, and the proper distance apart. The pockets are also counted and the belt generally examined. It is usual to carry out a functioning trial on a small percentage of each consignment. This consists in filling the belts with ammunition and firing in a Vickers gun.

Boxes are gauged for size, and examined to see that they are correctly constructed, and have not been made of faulty timber, and are properly stained. The handles are also tested and the fastenings looked at. The tin linings are removed, gauged and examined, and tested for leakage, and painted with a black preserving compound.

Other packing accessories are dealt with according to requirements, but space will not permit of a detailed description.

## 9.—Storage

Ordinary small arm ball cartridges are classified as safety cartridges, and in the magazine regulations are placed in Group VI. This means that they must not be stored in magazines, but may be kept in almost any other kind of building, provided that it is not subject to undue fire risk, and is used exclusively for such storage. The building must also be capable of being locked up, and safe custody of the ammunition must be provided for. For further particulars of storage conditions see "Magazine Regulations, 1922."

When selecting a building for the storage of small arm ammunition it is most important that it should be perfectly dry, and that the floors should be strong. It is also preferable that the store should be so arranged that the ammunition is not exposed to the direct rays of the sun. In the climate of the British Isles artificial heating is not necessary, and buildings specially designed for the storage of small arm ammunition in home depots are not usually artificially heated. If, however, heating is provided, overhead pipes are preferable, and where radiators are installed care should be taken to see that the ammunition is not stacked too close to them.

A building 100 feet square and 14 feet high, which is a suitable height for the purpose, would, after making reasonable allowance for gangways, etc., accommodate one hundred and twenty million rounds of boxed Mark VII ammunition. Under these circumstances the weight on the floor would in places be as much as three-quarters of a ton per square foot.

The harmful effect of damp is, of course, greatly increased if the tin linings of the boxes through rust or from other causes cease to be airtight, and a percentage of the linings are examined annually in all stations both at home and abroad, in order that any tendency to rusting may at once be discovered. In home stations the life of small arm ammunition, provided it is kept dry, is generally found to be such that the annual turnover caused by training and practice requirements is sufficient to prevent accumulation of unserviceable stocks, but in the case of stations overseas, where atmospheric conditions and fluctuating temperatures tend to deteriorate the ammunition more rapidly, a periodic examination of the ammunition itself is necessary. It is customary therefore to begin to take firing proofs as soon as ammunition stored in the tropics becomes five years old, or seven years old in other stations abroad, and it is not usual to retain ammunition in tropical stations after it has been manufactured eight years, or ten years in other stations overseas. For more detailed particulars of the regulations governing the periodic examination of small arm ammunition in the service see Pamphlet VI, Regulations for Army Ordnance Services, Part II.

Pt. II. CHAP. VII.   *To face p. 256.*

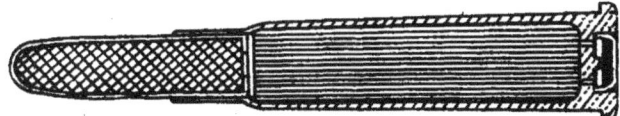
CARTRIDGE S.A. BALL
·303 INCH  MARK VI

CARTRIDGE S.A.
ARMOUR PIERCING
·303 INCH  W. MARK I

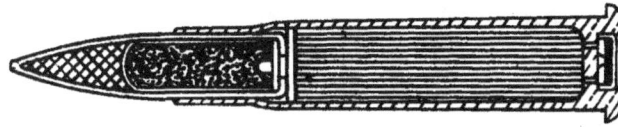
CARTRIDGE S.A. TRACER
·303 INCH  G. MARK I

CARTRIDGE S.A. RIFLE
GRENADE ·303 INCH
CORDITE  H. MARK II

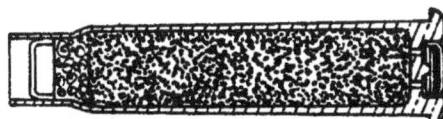
CARTRIDGE S.A. RIFLE
GRENADE ·303 INCH
BALLISTITE  H. MARK IZ

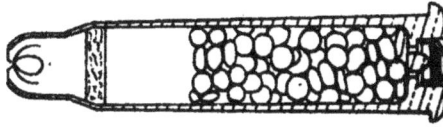
CARTRIDGE S.A. BLANK
·303 INCH  L. MARK V

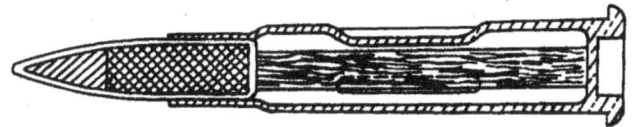
CARTRIDGE S.A. DRILL
·303 INCH  D. MARK VI

*Fig. 1.*

PT II. CHAP. VII.   *To face p. 257.*

 CARTRIDGE S.A. BALL REVOLVER
·455 INCH   MARK II

 CARTRIDGE S.A. BALL
PISTOL SELF LOADING
·455 INCH MARK I

 CARTRIDGE S.A. BLANK REVOLVER
·455 INCH L. MARK II T

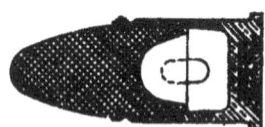 CARTRIDGE S.A. DRILL REVOLVER
·455 INCH  D. MARK I

 CARTRIDGE RIM FIRE
·22 INCH MARK I

FIG. 2.

# CHAPTER VII

## MISCELLANEOUS MILITARY AMMUNITION

In addition to the ·303-inch Mark VII rifle and machine-gun ammunition, the manufacture, proof and inspection of which have been described in the preceding three chapters, there are several other types of small arm cartridge in use in the British Government service, and it is proposed in this chapter to deal with the most important of these, calling attention, so far as space will permit, to any peculiarities they may possess, either in construction or performance.

These special cartridges may be divided into three main categories, those suitable for use in the various ·303-inch service weapons, pistol cartridges, and miniature ammunition for training purposes. Of the special cartridges suitable for use in the ·303-inch service weapons, the most important are the Mark VI cartridge, the armour-piercing, tracer, rifle grenade, blank and drill cartridges. Some of the types for use in ·303-inch weapons, notably the Mark VII, the armour-piercing, and the tracer cartridges, are also made of special quality for use by the Royal Air Force.

Fig. 1 shows sections of the various cartridges suitable for use in ·303-inch weapons, while in Fig. 2 are given sections of the various types of pistol and miniature ammunition.

### 1.—The ·303-inch Mark VI cartridge

This cartridge was the immediate predecessor of the present service cartridge, and is now virtually obsolete for army purposes, but it is still used by certain colonial troops and other forces in the Empire. It has a case identical with that of the Mark VII cartridge, but differs from the latest service model chiefly in that it has a round-nosed bullet weighing 215 grains. The propellent used is Mark I cordite, size $3\frac{3}{4}$. Nitrocellulose is not used in this cartridge. The normal charge is 30 grains, and the average velocity obtained in the long rifle is 1,970 feet per second measured over 180 feet, with a pressure of 17·5 tons.

This ammunition was designed for shooting in the long M.L.E. rifle, which is sighted suitably for the ballistics of this cartridge. It is not suited for shooting in the S.M.L.E. or short rifle, although the bore and the chamber size are the same, as the short rifle is sighted for the higher velocity Mark VII cartridge. Owing partly to the reduced efficiency of the blunt-nosed bullet, and partly to the lower initial velocity, the Mark VI cartridge has not so flat a trajectory as the Mark VII, and its danger zone is consequently smaller.

The methods of manufacture and inspection are generally the same as in the case of Mark VII. In some ways, however, the cartridge is easier to make than the Mark VII, because the lower pressure and velocity give less tendency to bursts, metallic fouling, etc.; in fact, the Mark VI cartridge is not designed quite so near to the limits of strength of the materials employed.

As it was designed at a time when the machine gun did not play so important a part in warfare as it does to-day, the cartridge has always been packed in chargers and charger cases. The characteristic symbol on the distinguishing label is the same as that of the Mark VII, except that the numeral VI is overprinted in black.

### 2.—The ·303-inch Armour-piercing Cartridge

Modern requirements include a cartridge having a bullet with greater penetrating power than the ordinary service bullet, as it is not possible with a calibre of ·303 inch to produce ordinary ball ammunition which will penetrate loophole plates or the armour carried by tanks, armoured cars, aircraft, etc.

It is found that the greatest penetration is obtained with a hardened steel projectile, but it is not feasible to make the bullet in this form, as it would not take the rifling, and the rifle barrel would be very quickly destroyed. The design therefore takes the form of a hardened steel core enclosed in a bullet envelope, and to enable the bullet to set up into the rifling, a lead sleeve is provided between the core and the envelope. Even so, the set-up obtained is very small, and, although the bullets are made larger in diameter than the ordinary service bullet, the accuracy is not good. Extreme accuracy, however, is not essential, as there is little advantage in the use of an armour-piercing bullet at greater ranges than about 400 yards.

The penetrating power of an armour-piercing bullet depends almost entirely on the striking energy in the core. Both the weight of the core and the velocity should therefore be as great as possible. The weight of the core can only be increased by making the core diameter as big as possible, and this means that the thickness of the lead sleeve has to be cut down to the lowest practicable limits. High velocity also involves high pressure, and an armour-piercing cartridge is therefore bound to put great strains on the rifle, and owing to the hardness and rigidity of the bullet, must cause excessive barrel wear. As, however, the total number of armour-piercing cartridges fired from any one rifle during its life is not likely to be great, the cumulative destructive effect is not likely to be serious.

The shape of the core to some extent affects its penetrating powers, a pointed core giving better results at normal impact, and a flat-nosed core at the more oblique angles. The shape of the core point is therefore chosen for the best all round performance. At the longer ranges the ballistic efficiency is of importance, and the external shape of the bullet is arranged so that the velocity is reduced as little as possible by the resistance of the air.

The service armour-piercing bullet has the same external shape as the Mark VII bullet. It consists of an outer envelope composed of steel coated with cupro-nickel, which contains a hardened steel core ·25 inch in diameter, with a thin lead sleeve between the core and the envelope. The diameter of the bullet varies from ·312 to ·314 inch. This bullet is used in conjunction with an ordinary Mark VII case loaded so as to give a nominal muzzle velocity of 2,500 feet per second. With the charge necessary to obtain this velocity the pressure is about 20·5 tons. This ammunition is not accurate, and a figure of merit of 10 inches when proved for accuracy at 600 yards is considered a good performance. With well made cores this bullet should penetrate a 10 millimetre armour plate of good quality at 100 yards practically every time, and the ammunition would be accepted on proof if 70 per cent. of the bullets penetrated under these conditions.

This cartridge is known as Cartridge, S.A. Armour Piercing, ·303 inch, W, Mark I, and for purposes of distinction the annulus of the cap is coated with green varnish. The base mark is W I, but early issues of this type of cartridge had base marks VII W, or VII W Z. There is also an earlier type of armour-piercing cartridge, which may be found in the service, which had a base mark VII P. This cartridge differed from the W, Mark I, in that it had a lighter core, and the velocity was not so high.

The armour-piercing cartridge is packed in cartons holding 48 or 50 rounds, 26 cartons being packed in each tin-lined wooden box, which thus holds 1,248 or 1,300 rounds, as the case may be. The characteristic symbol is a green disc on a white ground overprinted W I in black.

### 3.—Tracer Ammunition

"Tracer" cartridges are so called because they leave a visible wake or "trace" behind them so that the trajectory can be seen. They are used to ascertain visually where the bullets are going, with the object of correcting the aim, and it is therefore essential that tracer ammunition should have approximately the same trajectory as other service bullets. Unfortunately the trajectories of the ordinary service and armour-piercing bullets differ somewhat, owing to the greater velocity of the armour-piercing bullet, and it is not possible to imitate the trajectories of both exactly.

The trace is obtained by filling the bullet with a composition which burns during flight, and the result is that from the time the bullet leaves the muzzle until the composition is completely consumed the bullet is continually becoming lighter. The shape of its trajectory is therefore not the same as that of an ordinary bullet whose weight does not alter. Nevertheless, it is possible to make a tracer bullet which up to a range of 600 yards has a trajectory sufficiently close for practical purposes to those of both the ordinary service and armour-piercing bullets, the trajectories of which do not differ much up to this range. Beyond this range the path of the tracer bullet diverges considerably, but this is of no great consequence as the bullet does not trace much further than this.

The British Service tracer cartridge was originally known as the S.P.G., but its official name is Cartridge, S.A. Tracer, ·303-inch, G, Mark I. It has a bullet containing a cavity open at the base end, and fitted with a mixture of barium peroxide and magnesium powder. The bullet is loaded into a ·303-inch Mark VII case, the tracing composition being ignited by the flash of the propellent. Earlier issues of this ammunition were loaded with nitrocellulose powder, but a cordite charge is now more generally used.

For purposes of distinction the annulus of the cap is coated with bright red varnish, and the base mark on all cartridges issued subsequent to December, 1927, is G I. On issues previous to 1928 the base mark is VII G.

This ammunition may be used in all types of ·303-inch weapon, and functions equally well in all. So far it has not been found possible to manufacture it so that every bullet traces with certainty, but more than 70 per cent. should trace for at least 300 yards and over 50 per cent. should trace to about 600 or 800 yards. When fired for accuracy at 500 yards, from a fixed rest, 75 per cent. of the shots should be within a 5-ft. circle, a performance which does not compare favourably from the accuracy point of view with ordinary ball ammunition.

This tracer bullet is very destructive to the weapon in which it is fired. The barrel becomes rapidly eroded, and there is a tendency for the bore to become coated with a hard metallic deposit, which it is not easy to remove if it is left in for more than a short time. The best way of getting rid of this deposit is to fill the rifle barrel with a strong solution of ammonia and leave it to soak overnight. When firing S.P.G. cartridges it is desirable to fire ordinary or armour-piercing ammunition at the same time, using alternate rounds of the different types, as this has a decided effect in reducing the deposit. The use of S.P.G. interspersed with other ammunition in this way is usual in a machine gun.

·303-inch tracer cartridges are packed in 48 or 50-round cartons, 26 cartons being packed in each tin-lined wooden box. The characteristic symbol is a green triangle on a white ground overprinted G I in black.

These cartridges are classified for storage purposes as safety cartridges in group VI.

### 4.—SPECIAL AMMUNITION FOR THE ROYAL AIR FORCE

Although it is not correct to say that the ammunition referred to under this heading is specially designed, all types of service ammunition which are used by the Royal Air Force in controlled guns are made in two grades, and in the case of the Air Force or "Red label" ammunition, as this grade is called from its characteristic label, particular attention is paid to immunity from hangfires, missfires and functioning defects in machine guns.

The term "controlled gun" refers to a machine gun mounted in an aeroplane, and arranged to fire between the blades of the propeller along the axis of the aeroplane only. Should a bullet hit a propeller the result would be most serious, and the gun is controlled, therefore, so that it can only fire when the propeller blades are not passing the muzzle. The controlling mechanism is hydraulic, and actuates the trigger of the gun by means of impulses or energy waves, generated by a cam connected with the propeller shaft, and transmitted to the gun through a liquid contained in a tube.

After the impulse is imparted to the liquid, a certain time elapses before the bullet passes the propeller, and during this time the propeller blades rotate to an extent depending on the rate of revolution. At best the time lag cannot be reduced much below twelve-thousandths of a second, and a propeller revolving at 800 revolutions per minute would during this period rotate about $57\frac{1}{2}°$, whilst at 1,600 revolutions per minute the rotation would be about 115° and so on. Supposing it is desired to time a gun so that it can be fired at any engine speed between 800 and 2,500 revolutions per minute, and assuming that owing to the width of the propeller blade no shot must pass closer than 15° from the centre line of any blade, then the cam would be so placed that at 800 revolutions per minute the bullet would pass 15° behind the propeller blade, while at 2,500 revolutions it would pass—

$$15 + \frac{1700}{800} \times 57\tfrac{1}{2} = 137\tfrac{1}{2}°$$

behind the same blade. At this speed it would be within $27\frac{1}{2}°$ of the danger zone of the next blade of a two-bladed propeller, while it is obvious that with a four-bladed propeller this variation in speed is not permissible.

Suppose now that the time lag is greater, the amount that the propeller moves at any given engine speed is increased proportionately. Thus if the time lag is twenty-thousandths of a second, and the gun is set to fire 15° behind the propeller blade at 800 revolutions, it will be entering the danger zone of the next blade at slightly over 2,000 revolutions per minute. It will be seen, therefore, that any increase in the time lag necessitates a reduction in the variation of engine speed which can be allowed during firing, and it is of vital importance that the time lag should be as small as possible and absolutely regular.

The time lag is due, firstly to the time taken for the impulse to travel up the tube of the controlling mechanism, secondly, to the time required for the striker to hit the cap, thirdly to the time necessary for the cap to ignite the charge, and lastly, to the time taken by the bullet in travelling up the barrel. Provided the mechanism and gun are in good order, the first two causes for delay should be fairly constant. The third is the most likely cause of excess time lag, while the last may affect the time taken to a small extent if there is any considerable variation in the velocities of the bullets. It will be seen, therefore, that hangfires are the chief cause of abnormal time lag, and that immunity from hangfires is of paramount importance.

The necessity for immunity from functioning defects in a machine gun is fairly obvious, as it will be realised that stoppages during an aerial engagement would be very serious, and owing to the inaccessibility of a machine gun in an aeroplane whilst in flight, any failure such as a separation might put the gun out of action until a landing could be effected.

Red label ammunition is accordingly most carefully inspected, and any which is not of really first-class quality is ruthlessly rejected. During proof the ammunition is required to pass a special test to verify that it is immune from hangfires. This test approximates very closely to the actual conditions obtaining in an aeroplane. A Vickers gun is used, controlled by an actual timing gear, actuated from a shaft driven by an electric motor. On the shaft in place of a propeller is mounted a steel disc on which are arranged two cardboard targets diametrically opposite to each other. The disc is rotated at a constant speed of 1,200 revolutions per minute, and the position and grouping of the shots is recorded on the targets. The gun is mounted so that the bullets pass at a radial distance of 14 inches from the centre of the disc, and it is required that all shots but one should be within an arc of 4 inches in length, and all shots within an arc of 6 inches. A 6-inch arc at this speed is equivalent to a variation in time lag of about three and a half-thousandths of a second, but in practice all shots but one are usually much closer than this. Special provision is made for one shot to be slightly ater than the remainder, because the action of the timing gear is such that the first shot of any series is sometimes 2 or 3 inches behind the others, and this abnormality is not the fault of the ammunition.

Actually all ammunition of the types in question manufactured under peace time conditions is up to Red Label standard of quality, and there is no real difference in quality between the two grades when the ammunition is new. Owing to the fact, however, that immunity from hangfires is largely dependent on caps, and that any deterioration in the caps might make such ammunition unsuitable for use in controlled guns, a proof of each batch of Red label ammunition is taken annually and tested for hangfires, whereas this is not necessary in the case of ground service ammunition. Any batch of Red Label ammunition which fails at this annual hangfire test is relegated to ground service.

For purposes of identification the bases of cartridges made for the Royal Air Force for use in controlled guns are stamped with all four figures of the date thus, "1918," whereas ammunition for ground service is stamped with the last two figures only thus, "18." Ammunition containing all four figures of the date will, however, be found among ordinary stocks, as batches made for the Royal Air Force but rejected for that service are from time to time accepted for ordinary stocks, provided their quality is suitable.

Red Label ammunition is packed in 48-round cartons, which are enclosed in a tin-lined wooden box. Two types of box are in use, one containing 26 cartons and thus holding 1,248 rounds, and the other containing two tin linings, each holding 8 cartons, or a total of 768 rounds per box. Ammunition for use in controlled guns is labelled with a special red label, characteristic to ammunition up to Red Label standard, and all labels on the box and linings, and the wrappers closing each carton, contain the inscription "Special for R.A.F.," all labels except the distinguishing labels being printed in red.

### 5.—Rifle Grenade Cartridges

Rifle grenades in the service may be divided into two classes, those that are fitted with rods, and those without. Rodded rifle grenades are arranged for firing direct from the rifle, the grenade being held by the rod, which is inserted into the barrel. Grenades which do not use rods for firing are supplied with a circular disc which fits into a discharger cup attached to the muzzle of the rifle.

Owing to the fact that the rifle grenade rod fills up a portion of the bore of the rifle, the space available for the expansion of the gases when the charge is fired is considerably less than when the grenade is fired from a discharger cup, and for this reason the same loading would not be satisfactory with both types of grenade. There are thus in the service two types of rifle-grenade cartridge, suitable for firing rodded grenades and grenades from discharger cups respectively.

The service cartridge for firing rodded grenades has an ordinary ·303-inch cartridge case and cap, the propellent used being cordite. In the case of a rodded grenade the space in which the propellent gases expand, though less than when a discharger cup is used, is a great deal more than it is when a bullet is fired from an ordinary cartridge. This necessitates some precautions to ensure that complete ignition takes place, because cordite is a propellent the rate of burning of which depends upon the pressure at which it is burnt, and where the space is excessive there is some difficulty in getting the initial pressure to make the cordite burn sufficiently rapidly. Unless, therefore, a very rapid propellent is used, there is some danger that all the cordite will not be burnt, and for this reason a quick burning size of cordite must be used. The charge consists of 43 grains of cordite sticks, each strand being ·033 inch in diameter, and, to ensure complete ignition, gun-cotton tufts, each weighing about one grain, are placed both at the top and bottom of this charge. A cardboard cupped wad of special design is also used, placed in the mouth of the case. For purposes of identification the cases of these cartridges are blackened all over.

The cartridges are tested for range, and to pass proof must throw a service No. 23 rifle grenade, fitted with a 6-inch rod, 90 yards when fired in a service short rifle at an angle of elevation of 40 degrees, and under these circumstances the mean variation in range must not exceed 5 yards. The cartridges are also put through a " casualty " proof, and must be free from any liability to give missfires, hangfires, burst cases, etc. Each round of every consignment is also weighed, gauged, and examined.

This cartridge should not be used with a discharger cup, as the propellent is not sufficiently rapid to secure complete ignition.

The service grenade cartridge for use with a discharger cup is generally similar in construction, but differs in the form of charge used. This consists of 30 grains of sporting ballistite. No guncotton tufts are employed, the charge being held in place by a cupped wad. The cartridge can be distinguished from that used with a rodded grenade, because the front half of the case only is blackened.

Trials for range are carried out with a service rifle and No. 36 rifle grenade, and the cartridge must be capable of obtaining a range of at least 200 yards with this grenade fired from a discharger cup with the valve closed. For this trial the rifle is elevated to an angle of 45 degrees, and the rifle butt is rested on the ground. The mean variation from the average range must not exceed 10 yards.

This cartridge must not be used with rodded grenades, as the ignition of the ballistite charge is much too rapid and dangerous pressures would be produced.

Both types of cartridge are issued to the service packed in the same box as the grenades with which they are to be used, the requisite number of cartridges being contained in a small tin box enclosed in the main package.

### 6.—The ·303-inch Blank Cartridge

The chief essentials in a blank cartridge are that it must make as much noise as possible, and must be perfectly safe in use. The chief danger in the use of a blank cartridge is that when fired some portion may be ejected from the barrel in the form of a dangerous missile. There is some risk in using a mock bullet or wad, therefore, unless it is certain that it will be pulverized without fail by the firing of the charge. There must also be no risk of confusion between blank and ball cartridges capable of functioning in the same gun. It is also necessary that blank ammunition should function satisfactorily in all types of service weapons, and if any such weapon functions automatically with ball cartridges, it is desirable that blank ammunition should, if possible, also cause it to function automatically.

The blank cartridge at present in use in the Service is known as Cartridge, S.A. Blank, ·303-inch, L, Mark V. In this cartridge the ordinary ·303-inch Mark VII case is used, but the base mark on the cartridge cases varies, as cases are frequently used which were originally manufactured for ball or other types of ammunition. The charge, which is very quick in action, consists of 10 grains of sliced cordite, each slice being ·2 inch in diameter and about ·0055 inch thick. Over the cordite in the neck of the case is placed a strawboard wad, and the mouth of the case is then crimped in. There is no bullet. This cartridge is intended chiefly for use in rifles, but by means of a special barrel and attachment it can be made to function automatically in a Vickers gun.

The rounds fired at proof are submitted to the usual tests to discover liability to case defects, and special care is taken to watch for any tendency to fragmentation of the crimped neck. A certain number of rounds are also fired to ascertain that all wads pulverize properly. For this purpose firing takes place at a screen placed ten yards from the muzzle, and the ammunition is not considered satisfactory if any charge or wad marks the screen.

All inspection and packing is carried out in a building into which no ball ammunition of any kind is allowed to enter. Blank cartridges are usually packed in bundles of ten, which in turn are packed in barrels, of which there are two standard sizes, holding 3,200 and 1,900 rounds respectively. All labels are printed in red on a blue ground, the characteristic symbol being the ·303-inch Mark VII grid printed in red and overprinted L V in black.

## 7.—Drill Ammunition

The ·303-inch " drill " cartridge is a dummy cartridge used for the training of troops. It was originally known as the " dummy drill " cartridge, but the word " dummy " has been dropped as applying more appropriately to the sample cartridges used for demonstration and instructional purposes, which are a complete dummy replica of the live cartridge they are supposed to represent. This the " drill " cartridge is not, and, in fact, one of its most important characteristics is that its appearance should be as different as possible from that of a live round. As the cartridge is used for training troops in the use of the various small arm weapons, it must conform to the standard ·303-inch gauging limits, and it must be of sufficiently robust construction to function continuously in the rifle or machine gun when worked by hand.

The latest type of ·303-inch service drill cartridge is the D Mark VI. The case is made in white metal, and to make it more distinctive from that of the Mark VII ball cartridge, it is provided with three longitudinal grooves which are painted red. The bullet is an ordinary Mark VII bullet secured into the case in the usual manner, and to prevent the bullet being driven into the case by constant use, it is supported inside the case by a plug of wood. There is no cap, and the recess forming the cap chamber in the neck of the case is left empty.

Drill cartridges are packed in bundles of 10, 1,400 rounds in a tin-lined wooden box. The symbol on the distinguishing label is the grid characteristic of the ·303-inch Mark VII ball cartridge overprinted D VI. All labels are printed in black on a cerise ground.

## 8.—Pistol Ammunition

Pistol ammunition is intended exclusively for use at short ranges, and its most important characteristic is stopping power. The minimum striking energy required to stop a man with certainty is about 60 foot pounds, and this is not sufficient unless the whole of the energy is absorbed in the target (see Part III, Chapter X). A small bullet which obtains all its energy by virtue of high velocity will not stop a man, therefore, if it passes right through him without losing much energy in the process, unless it happens to hit a vital part of the body, and pistol ammunition in consequence tends to have a heavy bullet and large calibre rather than a high velocity.

There are two types of service pistol cartridge, suitable for use in revolvers and self-loading pistols respectively. The former of these, which is known as " Cartridge S.A. ball revolver ·455-inch, Mark II," is designed for firing in the service type of Webley revolver of ·455-inch calibre, and is also suitable for the Smith & Wesson and later patterns of Colt revolvers of the same calibre.

This cartridge has a solid-drawn brass case of cylindrical form, with a head or rim by which the cartridge is positioned and extracted. The cap is generally of the same form as that used in service rifle ammunition, but is smaller, and contains 0·4 grains of composition. The bullet is of lead without an envelope, and to obtain the necessary hardness the lead is either mixed with tin in the proportions of twelve to one, or is alloyed with 1 per cent of antimony. The bullet has a rounded nose and hollow base, and weighs about 265 grains. Three cannelures are provided, lubricated with beeswax. The fixing of the bullet into the case is carried out by canneluring the case into one of the cannelures of the bullet, the mouth of the case being also coned into the bullet. The charge consists of about $5\frac{1}{2}$ grains weight of Mark I cordite in chopped form, each grain of cordite being ·01 inch in diameter and ·05 inch long.

Only a low standard of accuracy is obtained, and a figure of merit of four inches at a range of fifty yards is considered satisfactory. The velocity obtained is about 580 feet per second, but a variation of 30 feet above or below this is permitted, and no limit is placed on the mean deviation from the average velocity.

The danger from overheating from rapid fire, which occurs in a rifle or machine gun chamber, practically does not exist in the case of a revolver cylinder, and the pressure proof is therefore carried out at 60° F. Copper crushers compressed to 4½ tons are used, and the mean pressure should not exceed 5½ tons, while no single round should give a pressure exceeding 6 tons.

In addition to the usual casualty trials, this ammunition is tested for penetration, and the bullets should be capable of passing completely through three one-inch deal boards placed one inch apart, at a range of ten yards.

It will be realized that when a revolver is fired the recoil is very sudden, and the cartridge cases in the several chambers are all drawn sharply backwards. If the bullets, which have considerable inertia, are not very securely fixed, they tend to remain behind, and "creep" out of their cases. When the revolver is being reloaded every one or two shots, it is possible for a cartridge to remain in a chamber during the firing of several rounds. If during this process the bullet creeps forward it may finally protrude from the front of the chamber and prevent the cylinder rotating. Revolver ammunition therefore undergoes a special test for security of bullets. The rounds so tested are each placed in a chamber of a revolver, and retained there while 25 rounds are fired in the other chambers, and at the end of the test the bullets must not have moved out of their cases more than one-sixteenth of an inch.

Although the standard of accuracy is low, and considerable variations are allowed in velocity and pressure, the revolver cartridge is one which to give satisfactory results requires in some respects considerable regularity of manufacture. Owing to the construction of the revolver the first part of the travel of the bullet is in the smooth bore of the cylinder chamber, and when the bullet enters the barrel there must of necessity be a space between the cylinder and the barrel through which the powder gases can escape. These conditions involve the use of a powder which burns very rapidly, and some difficulty is experienced in controlling the pressure, since there is a tendency for isolated rounds to give abnormal pressures if the loading is uneven. Another most important factor affecting the pressure is the extent to which the bullet is secured into the case, and the pull required to extract the bullet from the case should not only be sufficient to prevent the "creeping" referred to above, but should also be regular. It is the practice to break down during inspection no less than 100 rounds from each batch of revolver ammunition, in order to ascertain whether the ammunition has been carefully loaded, and whether the charges are all within reasonable limits of the same weight. At the same time the force required to draw the bullets is measured. This should be not less than 110 nor more than 140 lbs.

Revolver ammunition is weighed, gauged and examined during inspection, generally in the same manner as ·303-inch Mark VII ammunition, except that the gauging is done by hand. Although the weighing process eliminates some faulty rounds, it is not so definite a safeguard as in the case of rifle ammunition, because the allowable variation in the weights of the components is such that it completely masks any variations in weight of charge, and "no charge" rounds cannot be eliminated with certainty. There is in fact no known method of ascertaining definitely that revolver cartridges with no charge do not find their way into the service, though the large number of rounds broken down from each consignment is to some extent a safeguard, as they afford a fairly good indication whether the loading has been carried out with care.

This ammunition is packed in cartons, each holding 12 rounds, these cartons being in turn packed in a tin-lined wooden box. The standard package is the Box A.S.A. H.9, holding 240 rounds, but during the Great War large quantities of revolver ammunition were packed in the rifle ammunition box (Box A.S.A. H.1), which holds 2,160 rounds of revolver ammunition. The characteristic symbol is a representation of a revolver chamber in green on a white ground, overprinted II in black.

The service cartridge for use in self-loading pistols is designed for the Webley & Scott self-loading weapon of ·455 inch bore. It has a solid drawn brass case of the rimless type, that is to say, it has no rim or head, and the cartridge is positioned and extracted by the extractor which fits into a circumferential groove near the base of the cartridge. The cap is identical with that of the revolver cartridge. The bullet has a nickel-plated copper envelope and a lead core, and weighs 224 grains. The normal charge is 7 grains of pistol cordite. No wad is provided.

The requirements as regards pressure, accuracy and penetration are the same as those of the ·455-inch revolver cartridge, but the velocity is higher. This should be about 700 feet per second at a distance of 30 feet from the muzzle. The cartridge is submitted to the usual casualty proof, and must function satisfactorily in the Webley & Scott self-loading pistol. Each cartridge is weighed, hand gauged and examined, and the cartridges are packed in seven round cartons in tin-lined wooden boxes, the normal package being the A.S.A. H.9 box, which holds 252 rounds. The characteristic symbol is a green Plimsoll mark on a white ground, overprinted I in black.

There is also in the service a certain quantity of ·320-inch ammunition for use in the Webley & Scott self-loading pistol, but this is not recognised as a Government store.

The only pistol blank ammunition in the service is that suitable for use in the ·455-inch revolver. This cartridge has a brass case similar to that used for the revolver ball cartridge. The cap is, however, primed with a special composition consisting of six parts of fulminate of mercury, six parts of chlorate of potash, and four parts of sulphide of antimony, one-quarter of a grain of this composition being used in each cap. The charge consists of eight grains of black powder, and over the charge are placed two felt wads, the neck of the case being then crimped over.

When fired at proof this cartridge must not mark a screen placed at a distance of seven yards from the pistol mouth.

Pistol blank is packed in twelve round cartons in the A.S.A. H.9 tin-lined box, which holds 396 rounds. The characteristic symbol on the distinguishing label is the same as that of the revolver ball ammunition, but overprinted L II$^T$ in black, all labels being printed in red on a blue ground.

There is also a revolver drill cartridge. This has a white-metal case provided with three longitudinal grooves coloured red. The bullet is generally similar to that of the revolver ball cartridge, but the lead contains 5 per cent. of antimony, this alloy being harder than that used in revolver ammunition. The bullet is secured into the case by canneluring. There is no cap, and the cap recess in the back of the case is filled up with a fibre pad secured by three punch marks in the metal of the case.

This cartridge is packed in bundles of twelve, in the A.S.A. H.9 box, which holds 276. The characteristic symbol is a representation of a revolver chamber, overprinted D. I., all labels being printed in black on a cerise ground.

There are no blank or drill cartridges for use in the self-loading pistol.

### 9.—Miniature Rifle Ammunition

The service miniature rifle cartridge is the ·22-inch rim fire Mark I. There is however no exact design for this cartridge, and the specification governing its manufacture is worded so as to allow of small variations in construction, sufficient to admit the use of similar trade patterns. The chief requirement is accuracy, and it is essential that the cartridge should function satisfactorily in the ·22-inch service short rifle.

The term " rim fire " signifies a cartridge without a cap, the flange or rim of the case being hollow and filled with cap composition. In a rim fire rifle the striker is placed eccentric to the axis of the barrel, and when the rifle is fired the striker pinches the hollow rim and thus fires the charge.

The case of the Mark I cartridge is solid drawn, and is usually made of copper, though the use of brass or cupro-nickel is permitted. The rim is usually primed with about 0·4 of a grain of cap composition, the ingredients of which are identical with those used in the ·303-inch Mark VII cartridge.

The bullet is made of an alloy of lead, and weighs 40 grains. Three cannelures are provided, usually lubricated with beeswax. The bullet is secured into the case by coning, indenting or crimping.

The charge may consist either of cordite, or rim neonite, or other nitrocellulose powder, the most common form of charge being about 1·2 grains of rim neonite. No wad is provided.

The accuracy of the ammunition must be such that when fired from a service ·22-inch short rifle mounted in a fixed rest it must be capable of putting 95 per cent. of the bullets into or cutting a ¾-inch circle at 25 yards, the rifle being cleaned not oftener than once in 60 rounds. The cartridges must also be free from hangfires, missfires, split cases and blowbacks, and must load and unload freely in the rifle.

This is the only service small arm cartridge the inspection of which is carried out on a percentage basis. Although in the case of other types of small-arm ammunition, a complete inspection of every round is essential, there are reasons why in the case of the rimfire cartridge this is not so. It is practically impossible to manufacture a round which is dangerous either to the firer or to the weapon, and under these circumstances the presence of an occasional defective round is not so serious as it would be in the case of ordinary ball ammunition, since the rim fire cartridge is used exclusively for training purposes. The cost of a complete inspection would also be prohibitive.

For service purposes these cartridges are packed head and tail in cardboard boxes, each holding 100 rounds. Ten of these cardboard boxes are enclosed in a tin lining with a tear-off lid, and ten such tin linings are in turn packed in a wooden box, which thus holds 10,000 rounds. The characteristic symbol on the distinguishing label is a green target on a white ground, overprinted with the figure I in black.

# PART III

## BALLISTICS

### CHAPTER I

#### INTERIOR BALLISTICS (DESCRIPTIVE)

Interior ballistics treat of the events which happen in the crowded instant between the release of the striker from the control of the sear, and the sound of the fired shot reaching the ear of the firer.

The happenings, taken one by one, would be quite simple if it were not for their extreme rapidity. It is desirable to start by forming some mental image of this rapidity. The whole period in a modern military rifle is about 0·01 second or one-hundredth part of a second. This is already such a small amount that fractions of it are confusing.

The unit of time required to space out the events is a ten-thousandth part of a second, which is about the same fraction of a second as a second is of three hours. Fortunately, the aeroplane propeller through which it is often necessary to shoot a machine gun without hitting the blades gives an easy scale of time to read. A propeller about 5 feet in diameter, running at about 1,700 revolutions a minute, takes just one of these time-units to move its tip forward half an inch, or to advance one degree out of the 360 degrees in the circle. To save errors in printing and in thinking, this unit of one ten-thousandth of a second will be called "1 t.t.s." where t.t. means ten thousandths. The whole period of a hundredth of a second may now be called "100 t.t.s."

This period of about 100 t.t.s. is composed of four smaller periods, of which the first and by far the longest is about 58 t.t.s. during which the striker of the S.M.L.E. rifle is flying forward under the impulse of the main spring to hit the cap and cause its explosion.

The second period is only some 2 t.t.s. during which the cap explodes and then ignites the charge of powder, and that charge develops enough gas to start the bullet out of the neck of the case and cause it to begin its journey to the muzzle.

The third period is about 12 t.t.s., and is entirely occupied with the travel of the bullet along the two feet of the barrel.

The fourth and last period is about 22 t.t.s., which is the time the sound of the shot leaving the muzzle takes to get back to the ear of the firer, although the distance from the muzzle to the ear is only some thirty inches.

These four periods added together make up the total of 94 t.t.s. or about a hundredth of a second. Before any details of the nature of powder gas are entered into, it is proposed to describe the events in each period in considerable detail and to remark on the chief items.

*Periods 1 and 2.*—The striker has to be so made that the blow it gives to the cap is hard enough to crush in the metal of the cap against the lump of brass left standing up inside the cap chamber. It is easy to obtain a fired case and saw it in half lengthwise. It will then be seen that the cap as seen in section is a small U-shaped piece of thin metal sitting in a recess formed out of the solid metal of the base. The lump standing up in the cap chamber is called the anvil, and on each side of it there is a small hole called the fire-hole leading into the powder chamber or the cartridge case proper. The striker blow has to dent the cap sharply enough to crush the cap composition up against the anvil so quickly that the resulting heat and friction are enough to detonate the cap composition just as a toy pistol detonates the little paper caps sold to children. The blow must be hard enough to do this with complete certainty, and yet not so hard as to pierce the metal of the cap. Such a blow is equivalent to $\frac{1}{4}$-lb. falling 2 feet and delivering $\frac{1}{2}$-ft. lb. of energy. A blow of $\frac{1}{8}$-ft. lb. will rarely if ever fire the service cap. The service bolt gives a blow of about $\frac{3}{4}$-ft. lb., and its speed on striking is about 20 f/s. The length of travel of the striker is about ·8 inch under a main spring of $11\frac{1}{2}$ lbs. thrust, giving a time of travel forward of 58 t.t.s. This time can be very easily measured on the Boulengé chronograph as it represents nearly $\frac{1}{2}$-inch fall on the long rod, where 1/100 inch can be read with certainty. It is long enough for a man to move his rifle and spoil his aim, but its actual length has no effect on the explosion.

With too hard a blow the cap may pierce and drive the striker to full or half-cock, which is disconcerting for the next shot. But with too weak a blow the explosion may be seriously delayed and also reduced so that a loss of velocity of over 100 f/s, may result. With a really defective cap and striker a hang-fire of as long as 15 seconds has been twice observed at Woolwich in some 50 million rounds fired for proof. In rapid fire from the shoulder at the rate of 60 rounds in 60 seconds, the firer has been injured, but not severely, several times by a cartridge exploding after he had opened the bolt. As a rule, however, ignition has no effect on ballistics provided that ordinary care is taken. Within ordinary limits the number and size of the fire holes, the shape of the anvil, the strength of the blow, and the nature and amount of the cap composition are matters of purely technical interest. The manufacturer by long years of practice supplies an article so good that only experiments lasting over years can hope to improve on it.

*Period 2.*—When the cap composition is fairly struck it explodes by what is called a detonation. The whole substance of the composition turns into gas at a very high temperature practically at once, perhaps in a hundredth part of a t.t.s. Except for our familiarity with the phenomenon it is an amazing occurrence. What was half a grain of solid heavy stuff at the same temperature as the air becomes a white hot mass of gas as hot as the electric arc lamp, and of such cubic capacity as to be capable of raising the space available for its reception, (*i.e.*, the small air spaces between the powder grains) to a pressure of some 10,000 lbs. or, say, 5 tons to the square inch. Owing, however, to a well-known property of gas, that its volume and consequently the pressure it exerts depends upon its temperature, the actual pressure produced is not much more than 2,000 lbs. to the square inch, because the brass case and the powder are so quick to cool it. At any rate the pressure inside the cartridge rises practically at once from 30 inches of mercury to something like 200 times as much, and fills the powder chamber with an intensely hot gas carrying with it certain small but hot pieces of solid from the varnish and other inert parts of the cap composition. It is not known how much the temperature, how much the pressure, and how much the solid hot particles each contribute to the lighting of the surface of the powder grains, but any cartridge maker can guarantee to light any powder which he uses.

When once the surface or some part of the surface of the powder is well alight, the powder charge itself begins to make gas and tends to raise the pressure in the powder-chamber proper. This " young gas," as it may be called, has a double office to perform. It raises the pressure in its chamber, and it loses its own heat and pressure very quickly in warming up the extreme surface of the brass case and of the rest of the powder. It also leaks out of every tiny crack it can find anywhere, and by its pressure it produces an enlargement of every orifice. It is, however, subject to the laws of every moving body having weight; it cannot produce motion except at the expense of time. The pressure it exerts against the base of the bullet and against the walls of the case only produce their effect gradually. The general effect is that the powder charge burns from its extreme surface inwards more and more violently, the cartridge case swells outwards more and more, and at last the hold of the neck of the case on the bullet is released enough to allow the bullet to move forward. In a rifle this period can only be estimated because the size of the forces at work can only be guessed. But as soon as the force is great enough to start the base of the bullet forward this period is over. An estimate of the length has already been given as 2 t.t.s.

*Period* 3.—This period begins at the end of the second period, that is, as soon as the base of the bullet begins to move forward. At once a very interesting conflict ensues. The base of the bullet moves, but there is no particular reason why the point should move also. The sides of the bullet can expand and they do so until they make firm contact with the walls of the mouth of the case or of the barrel. This phenomenon is called the set-up or upset of the bullet, and it is a vital point in all ordinary bullets which have a flat base. All pre-war bullets were made with flat bases in England, whether they were of plain lead or had a jacket of cupro-nickel covering the whole front, but open at the back, or whether they were of the sporting pattern, often called Dum-Dum, with solid base and lead exposed at the nose. Every one of these bullets (as also the old Snider bullet with its clay plug in the base) depends for its accuracy and proper behaviour on its power of setting-up and sealing any gas-escape before the young gas can creep or slide to any extent over the bullet's parallel portion and get a gas-leak fairly established for high pressure gas to pass through. This is very easily proved by cutting down a barrel from 24 inches in length to 3 inches, so that the nose of the bullet sticks out of the barrel. If such a barrel is fired at a few yards' range with an ordinary cartridge and the bullet is caught in sawdust (free from resin) the bullet is seen to be expanded in front of its base end to some ·350-inch diameter instead of ·310 as made. The base of the bullet is mushroomed quite symmetrically like a pat of butter hit on the top and squashed up. The cause is obvious, namely, that the gas pressure is so great at the breech end of the gun that the bullet would lose all shape but for the walls of the barrel holding it together. If, however, the barrel is as long as 6 inches the bullet is recovered in its normal shape, showing that the pressure is then insufficient to add to its upset.

The conclusion is that the fate of the bullet is settled in its first inch of travel. It either upsets and barricades the young gas or else the young gas wins the race and gets over and round the bullet in sufficient quantities to afford a passage for the high-pressure gas to pass over the bullet and constrict its circumference so that it passes down the bore, floating on or swimming in a bath of gas, without touching the walls of the barrel. Such deformation and squeezing out lengthwise of the bullet is often called " wire-drawing."

Between a perfect upset and complete wire-drawing all gradations are possible, but as a rule the gradations run from perfection to absolute keyholing, or bullet striking side-on, very quickly indeed. A really good flat-based bullet should be so soft at the base that if made ·306 in diameter it can be fired properly from a barrel ·307 or more in diameter. The addition of a wad of soft felt or a glaze board disc enables a harder bullet to be fired from a bore of large diameter by acting as a gas check, but any ordinary bullet can be made soft enough to shoot as above shown without a wad.

With streamline or armour-piercing or tracing bullets quite another problem is introduced. These bullets are so hard that they do not upset at all. They should, therefore, be made of diameter almost large enough to bottom in the grooves. The young gas gets over them, but is not able to wire-draw them. In due course they advance far enough into the rifling to barricade the high-pressure gas by their mere size and strength.

To return now to the question of time. The base of the bullet begins by starting forward. At once the size of the powder chamber is increased, and a competition begins between the rate of evolution of new gas, as the powder burns, and the rate of increase of the space available behind the bullet for the gas, as the bullet advances under the pressure of the gas behind it. During the time in which the gas is forming faster than the space increases, the pressure behind the bullet increases. It begins by being just enough to move the bullet, say, 5 tons to the square inch. It rises until the two rates are equal, and this is the instant of maximum pressure, something under 20 tons. A balance or equilibrium exists then for an instant, and then the bullet finally wins the competition and moves faster than fresh gas can be made, thus causing the pressure to fall off continuously until the muzzle is reached. The gas is then released into the air and its pressure falls off almost instantly to atmospheric pressure, accompanied by the familiar report of the explosion. During the period of competition a time of about 2 t.t.s. elapses, and whilst the competition is at its greatest intensity the pressure is rising at the prodigious rate of at least 100,000 tons a second. In Chapter III diagrams are given showing graphically the rise and fall of pressure as the bullet travels along the barrel.

The effect of the pressure rising at this rate inside the gun is very similar to that of an intensely sharp and heavy blow. The whole gun shudders from the impact of the rapidly moving particles of gas. The cartridge case is violently expanded against the walls of the chamber, and is thrust backwards against the face of the bolt, which in turn is thrust against the resistance shoulders on the body. As a result of this the barrel in particular is thrown into severe vibration, affecting, among other things, the sighting (*vide* Part I, Chapter I, Section 5). The expansion of the case is so rapid and strong that in the absence of grease in the chamber the case and barrel and bolt behave as a solid. This can be shown experimentally by cutting off the screw thread from a barrel leaving the barrel a loose fit in the body. A gun so mutilated can be fired without anything unusual happening. The barrel remains in the body, although the slightest trace of grease in the chamber will cause the barrel to be thrown forward some feet, and the cartridge case to blow off its base dangerously.

As before mentioned, the base of the normal flat-base bullet starts before the tip of the bullet, the lead in its cupro-nickel envelope flowing outwards to fill any vacant space. It is now that the office of the "leed" or the cone joining the parallel of the bore to the chamber is apparent. In a new barrel of a S.M.L.E. rifle this slope is about 1 in 100, or 35 minutes of angle. It is situated about $\frac{1}{4}$ inch in front of the bullet as it lies in the chamber ready to fire, and this $\frac{1}{4}$ inch constitutes the free travel of the bullet before it meets the metal of the rifling to be engraved by the lands. The shorter this free travel and the steeper the leed, the greater is the resistance to forward motion and the quicker and the higher the gas pressure rises. In a barrel which has fired some 5,000 rounds of Mark VII cordite and has not been worn out by the use of wire-gauze, the free travel is over 1 inch, and after 10,000 rounds it is over 3 inches. The maximum pressure is then some 2 tons lower, and the velocity perhaps 200 f/s. lower, but the accuracy is hardly impaired, except for the highest class of bullseye competition. If, however, the bullet is wrongly made and is too hard, such worn barrels will fail to rotate their bullets and be useless and dangerous to friends in front. There is at the Small Arms Range at Woolwich a S.M.L.E. barrel which has fired 30,000 rounds of Dupont loaded Mark VII. It is now occasionally put into the accuracy house for exhibition purposes, and it rarely fails to put its 20-shot group at 600 yards into a circle 18 inches in diameter with pre-war K.N. ammunition, which has a very soft base bullet. This barrel is not mentioned as a curiosity, but because of the extreme importance of the principles which it illustrates.

The upset and engraving of bullets are intimately associated with the subject called by ballistic writers "forcement." Together with "friction" and "cooling" it comprises the main difficulty of the mathematical treatment of the motion of a shot in the bore, in which the science of interior ballistics consists. The allowance to be made for forcement in a rifle has not yet been subjected to any simple rule, and, in view of the plastic nature of the bullet and the intensity and rapidity of the forces at work, this is hardly surprising.

After the moment of greatest pressure is past, the bullet continues to gather speed up to the muzzle, as a glance at the diagrams in the succeeding chapters will show. The pressure of the gas behind it is the efficient cause of the acceleration, and this pressure is diluted by the two factors of cooling and friction. These factors make themselves very evident during the firing of the third belt of 250 rounds in a Maxim gun starting from cold. The water jacket boils during the third belt and clouds of steam escape. The energy required to boil the water is abstracted from the gas, and is roughly as big as the energy imparted to the bullet. Part of this loss is occasioned by the friction of the bullet in the barrel, but the larger part is due to the flow of heat through the solid metal of the barrel into the water jacket. Barrel steel is a very good conductor of heat, and behaves to heat as a sieve does to water. The energy of the powder gas passes through the barrel into the water at about the same pace as it passes into the bullet. Out of every 100 ft. lbs. of energy contained in the charge, some 30 ft. lbs. appear in the bullet as useful energy, about the same quantity gets into the water jacket and merely boils the water. The rest, or nearly 40 ft. lbs., is blown out of the muzzle, and causes nothing but noise, and recoil, or kick.

Most of the research work which has been done on powders and has been published has been for the benefit of heavy ordnance. With the exception of R. H. Housman, of Kynoch's, who died young in 1905, and of F. W. Jones, now of Nobel's, and still in the prime of life, there are no authors outside the Government factories of Europe and America who have published anything bearing on the practical numerical solution of the problems of interior ballistics presented by rifles as rifles.

Many excellent books and articles on rifles have been published, especially in America, and such names as Dr. Mann, Townsend-Whelen, Caswell, A. C. Gould, are as familiar and as esteemed as those of Dr. Kelly, Fremantle, Ommundsen, and Robinson. But of hard figures and formulæ for velocity and pressure they give no details.

The literature on interior ballistics is quite extensive, but it is chiefly written from the artillery point of view, and consequently the methods of attacking the numerical problems involved require a good deal of modification before they can be applied to rifles. In particular, the phenomenon of the water boiling in the Maxim jacket shows that the rate of loss of heat in a rifle is very different from that in a heavy gun. The assumption that the expansion work of the gas occurs without loss of heat or "adiabatically," as the term goes, has therefore to be used very sparingly when dealing with rifles. In general, the detailed methods of calculating pressures and velocities which are suitable for ordnance require rather extensive modification for rifles, although the great body of the theory is common ground, and the ballistics of both weapons are governed by the same physical laws.

The three outstanding difficulties of interior ballistics are thus

(1) forcement,
(2) cooling,
(3) friction,

and as no experimental data separating them or giving them exact definition are available for rifles, recourse has to be made to other methods of attack.

*Period 4.*—This period, strictly speaking, is the time taken by the sound of the released gas at the muzzle to travel to the ear at the rate of about 1,100 f/s, or, say, 22 t.t.s. It includes, however, the time taken by the sound to pass from the drum of the ear to the consciousness. It is well to mention such a seemingly small matter because it is quite certain that during the whole interval from the pull of the trigger to the exit of the bullet the human element (even in a fine target shot) may cause an error, more especially when the man is not in perfect health and condition. It is only by prolonged introspective analysis that such errors can be located, but their occurrence is agreed to by all first-class shots. The cartridge is often blamed by the ordinary man for an unaccountable shot, but the human element must never be lost sight of.

During this fourth period a ballistic event occurs which is apt to escape notice. It is connected with recoil, and composes quite a large proportion of it. As soon as the bullet leaves the muzzle the confined gas rushes violently out of the bore, and in doing so gives a noticeable backward thrust on the gun. This thrust is utilised in the Maxim gun to assist recoil when blank charges are used or a barrel is worn. The so-called "muzzle attachment" is made fast to the mounting and catches some of the blast, so dragging the mounting forward against the recoil of the barrel, thus increasing the relative motions of barrel and mounting, or adding to the recoil. Conversely, the blast can be used to reduce the recoil felt by the firer by attaching a suitable fitment to the barrel. The blast then impinges on the recoil reducer and drags the barrel forward, away from the shoulder, materially lessening the feeling of recoil. At the same time, a blast of air or even hot gas is liable to be thrown back into the firer's face unless the angles of the front plate are nicely calculated.

These remarks conclude the analysis, so far as space permits, of the events which occur in the four periods of time. They are by no means exhaustive, but they may serve to present some useful mental pictures on which the student can himself embroider the details.

Before a description is attempted of the methods of calculating the pressure and velocity of a given combination of rifle and cartridge, a short account will be given of the principal experimental methods which have been employed by investigators to gather the numerical facts. It will be arranged in chronological order, as it is highly probable that these investigators confronted the difficulties and invented the appliances required in the most logical and natural order.

Roger Bacon, usually called Friar Bacon, and also known as Doctor Mirabilis (1214-1292), was born in the year before Magna Carta was signed at Runnymede, and lectured at the newly created universities of Oxford and Paris, where the ancient Greek learning was first introduced to modern Europe. To Bacon the discovery of gunpowder is undoubtedly due, as is shown conclusively by Lieut.-Col. H. W. L. Hime in his book *Gunpowder and Ammunition*, published in 1904.

Berthold Schwartz, a German monk, is often said to have discovered it about the year 1320. Following the opinion of Benjamin Robins, it is more likely that Schwartz, having pounded some gunpowder in a mortar, afterwards covered it with a stone. A spark accidentally flew into the mortar, and the explosion blew the stone to a distance, from which accident he was taught the simplest way of applying the power of gunpowder in war. The shape and name of "mortar" given to early guns very much corroborate this conjecture.

Leonardo da Vinci (1452-1519), the universal genius of Florence, asked himself questions on quite modern problems:

"What shape of powder ignites the quicker? What difference does it make if ignition takes place at one end or the other or the middle of the powder charge, and what if with the same weight of powder the grains are long or short, round or cubic?" He also explained correctly the cause of the report.

To his first two questions no practical importance attaches itself for rifles, but his third question can be answered by the two diagrams published by his fellow countryman, Col. Bianchi at Turin in 1914. He considers several shapes, such as long tubes, long strips, flake tubes chopped to two different lengths, long cords and spheres. Fig. 1, which is copied from one of his diagrams, shows the relation between the weight burnt and the thickness burnt. It will be seen that long tubes which burn, of course, on the inside as well as the outside, give a simple proportion, viz., when half the thickness is burnt then half the weight is burnt, and so on. At the other extreme are spheres. Practically half the weight is burnt when the diameter or thickness of each spherical grain is only reduced by 20 per cent., so that they give off much gas at first and then fade away slowly.

In Fig. 2 Col. Bianchi shows the relation between the thickness burnt and the change of rate of emission of gas from its initial rate. Here it is seen that long tubes give a constant rate, and spheres give a rapidly diminishing rate. Modern theory is in favour of a long tube or what is the same, a wide flat sheet, which is equivalent to a long tube cut down the side and opened out. When for some manufacturing or other reason this shape is not possible, the next best shape is, taken generally, flake or cord. Strip, or axite as it is often called, is avoided because it is said to pack so close that the flame cannot pass properly to light it. In rifles there is supposed to be an advantage in long tubes, but in practice the advantage is not particularly noticeable.

Benjamin Robins (1707-1751) died with his pen in his hand at Fort St. David, as Engineer General to the Honorable the East India Company. "This excellent person was born at Bath" and was withal a "most sprightly and agreeable companion." His *New Principles of Gunnery*, first printed in 1742, shows a grasp of the subject so sure and so much in advance of his time that for a hundred years and more his conclusions constituted an unexhausted mine of information on the science of weapons of war.

Three of his problems were:—

(1) the determination of the force of gunpowder.
(2) the resisting power of the air.
(3) the nature and advantage of rifled barrels.

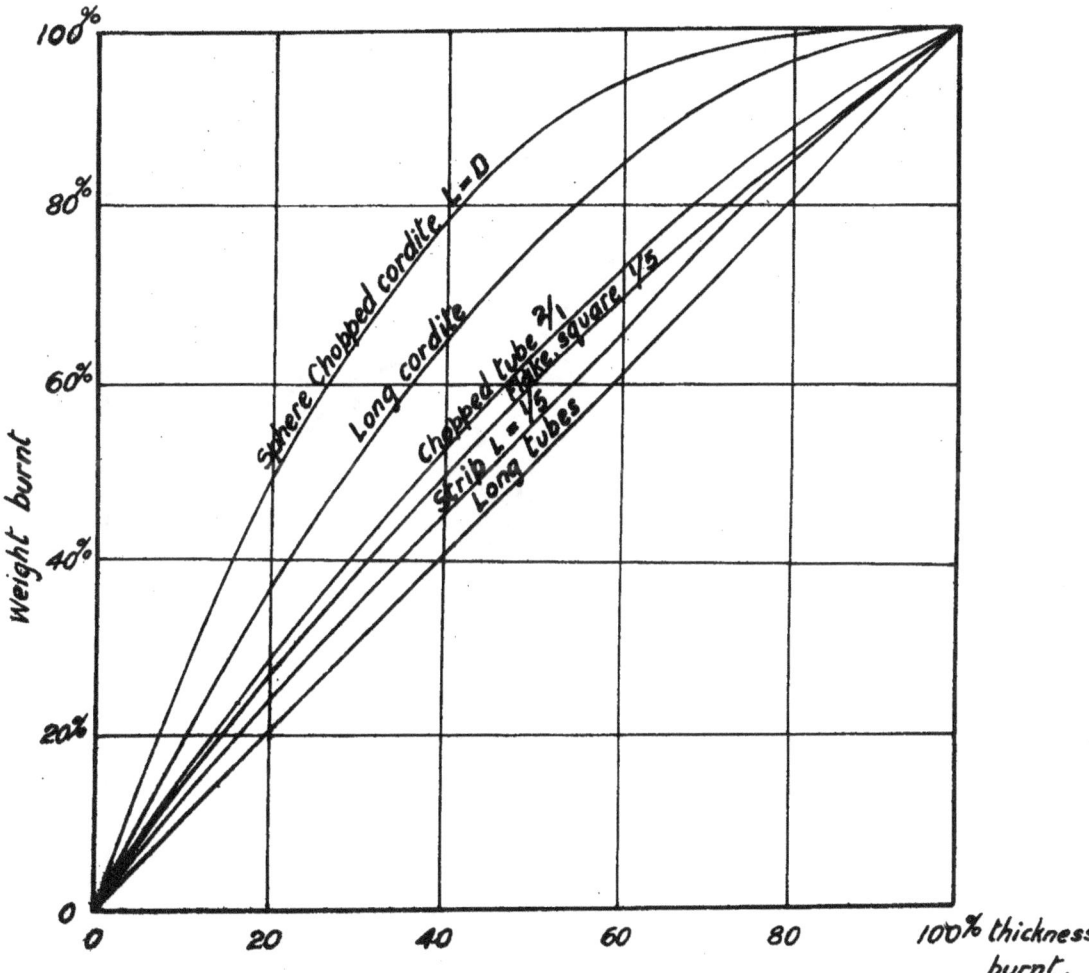

PART III. CHAP. I.

FIG. 1.

To face p. 272.

PART III. CHAP. I.

FIG. 2.

To face page 273.

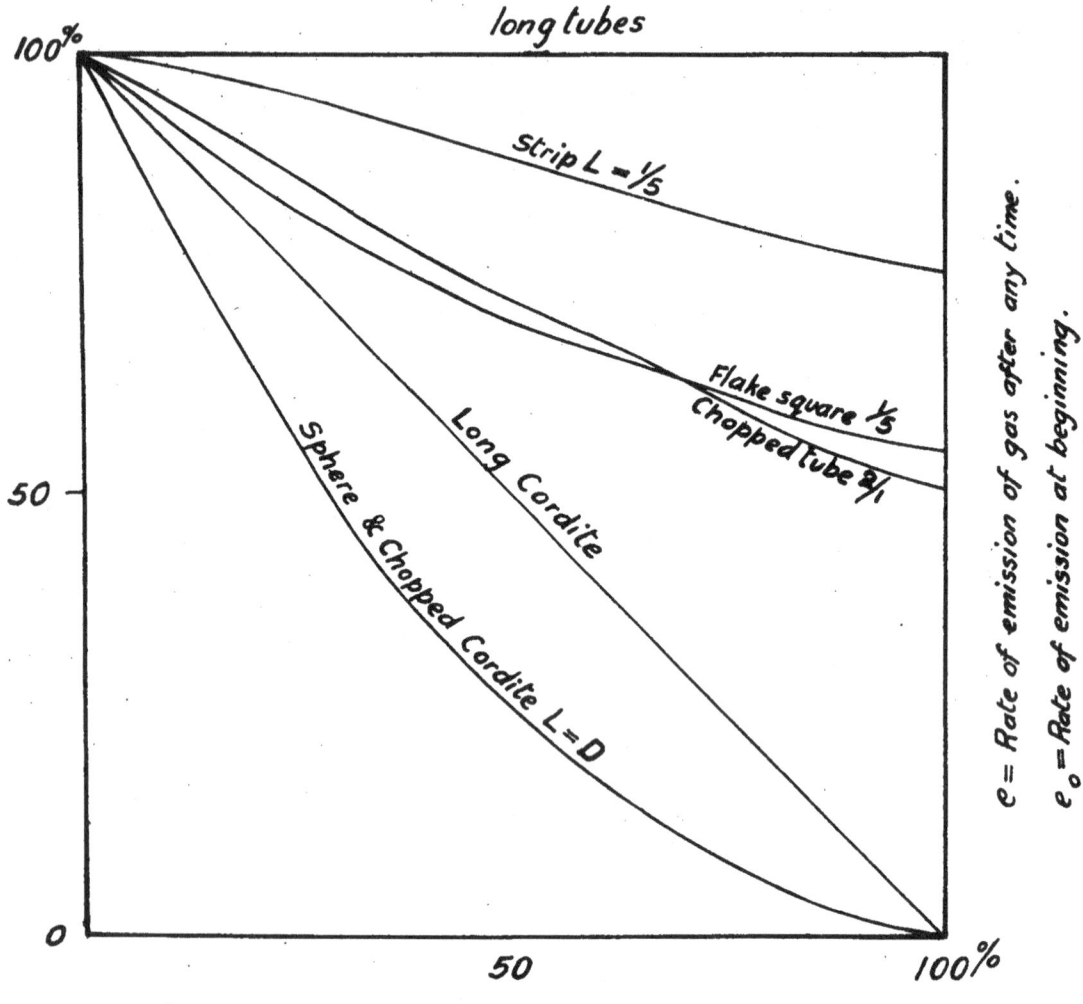

The first of these will be explained when the names of Noble and Abel are reached, and the third now requires no remark. The second, however, involved the invention of an instrument to determine precisely the velocity of a bullet at any distance from the muzzle.

Robins invented the ballistic pendulum for this purpose. It is described fully in Chapter VII. It consisted of a big block of wood hung from a large scaffolding. The bullet was fired into the wood and the recoil, in inches, of the wood was noted. Knowing the weight of the wood and of the bullet, and the length of chain suspending the wood as a pendulum, the velocity of the bullet on striking is easily computed with great precision.

Before this instrument was invented no one had any clear notion of the real muzzle velocity of any shot, and without proper information on this subject it is impossible to begin constructing a rational theory about the force of powder. The importance of this invention cannot be overrated, as it is the foundation of all our present knowledge. It is often said that Robins thought that the whole of the powder was burnt before the shot started to move. He never said so, but if he did he would not have been far wrong because the cannon powder of those days was generally "serpentine," which was as fine as sand and as soft as flour, as opposed to "musquet powder," which was corned or grained "to render it more convenient for the filling into small charges." Robins actually said that "this postulate, though not rigorously true, may yet be safely assumed in investigating the effects of powder."

In particular, Robins published the first indicator diagram, Fig. 3, of the pressure in a gun, where the curve K H N Q is a hyperbola with axes A.I. and A.B.

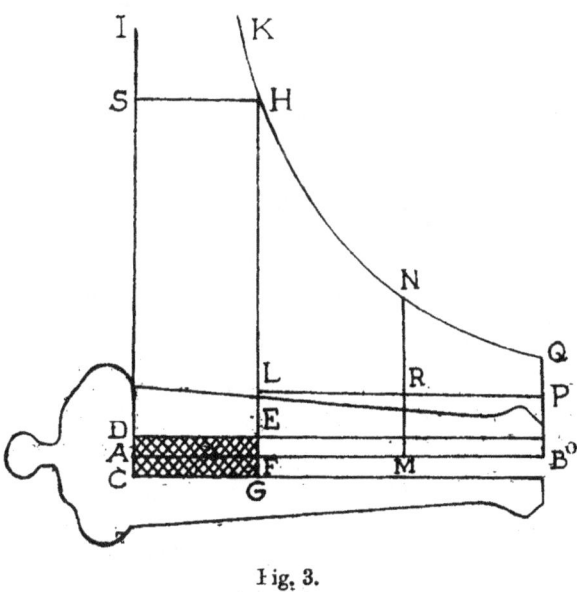

Fig. 3.

If, then, the powder charge D.G. is suddenly turned into gas of area S H F A and pressure L H, the advance of the shot towards the muzzle causes the point H to describe the curve H N Q as the line of gas level sinks from S H towards A B; keeping its area constant and its level parallel with S H or A B. We now know that things are not so simple, but till quite lately the idea was near enough to the facts to escape serious criticism. Even now it gives a good general picture where refinement is not aimed at.

The Chevalier d'Arcy, writing in 1760, described his method of arriving at the pressures inside the bore by noting the change of velocity as the barrel was cut down in length inch by inch. His actual writings have not been consulted, but his method is frequently used now with rifles, and is most useful. By its means we know that the rifle barrel has to be lengthened about 6 per cent. to increase the velocity by 1 per cent., and also that the gas pressure at the muzzle of the S.M.L.E. must be about 3 tons a square inch.

Charles Hutton (1727-1823) was a profesor at the Royal Military Academy at Woolwich. He experimented in "The Warren," in the Royal Arsenal, near the present main gate, now built over with offices. He shot down the Thames itself along Gallions Reach, when he wanted a long range, having observers on each bank. His "Tracts" in three volumes were printed in 1812, but his famous tracts XXXIV to XXXVII were printed by the Royal Society in 1784. In them he showed a systematic scheme for studying ballistics by firing for muzzle velocity into his large ballistic pendulum with different charges of powder and of shot, and also of observing the recoil of his gun by mounting it on another ballistic pendulum and noting the recoil distance of the gun. One fine experiment was with a 2-inch gun and a 1-lb. ball having the same sectional density of projectile (or $W/d^2$) as our Mark VII ·303-inch, viz., ·271. His gun was cut down by 4 inches at a time from 80-inch to 28-inch shot travel, and his charges were varied from 2 oz. to 16 oz., giving velocities from 774 f/s to 2,200 f/s. He found, until the powder charge was so large that powder was blown out unburnt, that increases of 5 per cent. in barrel length or 2 per cent. in powder charge gave 1 per cent. extra velocity. These are very near the figures now found with modern appliances for such powders as he used. His method is still a routine method for rifle work, and is likely to remain so.

In addition, he devised a method for proving powder. He had a weighed charge fired blank in a mortar suspended as a pendulum and noted the recoil. In doing this he did very much what modern practice is veering round to when the "vivacity" of a powder is required to be measured. It is judged by its shooting and not by its physical or chemical characteristics.

Benjamin Thomson, Count Rumford (1753-1814), an American citizen, who spent most of his life in Europe, was contemporary with Hutton, and communicated the result of his researches to the Royal Society in 1797. Besides working on the same lines as Hutton, he broke new ground by devising a way of determining the actual pressure of fired gunpowder. He made a small strong mortar of $\frac{1}{4}$-inch calibre capable of holding 28 grains of powder. On top of his powder he arranged a leather wad as a gas check, and on top of that he placed his weight. He fired the powder by the simple expedient of making the base of his mortar red hot, using ingenuity in so fashioning the shape of his base that the heat was conducted easily to the base of the powder-chamber. He varied the weight of his stopper till the explosion just lifted it off its seat. Knowing the area of his bore and his weight, he could obtain the actual pressure in pounds per square inch just as in the case of the safety valve of a steam boiler. His results enabled him to give a formula for the pressure of powder gas at any given density, and up to 5 tons per square inch, or a density of one third that of water, he was very correct. His method, with the addition of modern refinements, is still the standard method of measuring gas pressure.

General G. Piobert may be taken as the last of the ancients. In his *Traité d'Artillerie*, 1859, he mentions the chemist's work, and especially that of Gay-Lussac, and discussed the chemical nature of the gases formed by exploded gunpowder. His chief contribution lies in the suggestion, which is fairly true for modern smokeless powders, but which is certainly untrue for black gunpowder, that each grain composing the charge burns by parallel layers till the whole is consumed.

General T. J. Rodnam, of the Ordnance Department, United States Army, did a quantity of work on pressures between 1857 and 1861. His special claim to mention is that he first devised and used a crusher gauge to take pressures inside a gun, which he called an "indenting apparatus." In its modern form it is fully described in another chapter.

Captain Commandant P. Le Boulengé, of the Belgian Artillery, invented, about 1860, the electric chronograph which bears his name. It was in use in England in the Arsenal before 1870, and is still in daily use. It is fully described in Chapter V.

In 1857, to quote Sir Andrew Noble, Bunsen and Schischkoff published their very important researches on gunpowder. By the help of experiment they deduced from theoretical considerations the temperature of explosion, the maximum pressure in a closed vessel, and the total theoretical work which gunpowder is capable of performing on a projectile.

The last and the greatest of all these experimenters was Sir Andrew Noble, the head of the firm of Armstrong, who died in 1915 at an advanced age. He began experiments on a large and important scale before 1870, and in conjunction with Sir Frederic Abel (who died in 1902), he placed modern ballistics on a thoroughly sound experimental basis. The French school of thought includes the great names of Sarrau, Vieille and Charbonnier, and is justly famous for its severely logical and mathematical analysis of their own and the English experimenter's facts and figures.

A full description of Noble's work would require a book by itself. The briefest possible summary must suffice, as his own writings are available in all large libraries. His objects were :—

(1) To ascertain the products of combustion of propellents.
(2) To determine the law connecting the density of the gas and its pressure.
(3) To investigate the effect of changing the size and shape of the individual grains of the propellent.
(4) To determine the heat of explosion.
(5) To determine the work capable of being done by the explosion.

To effect these objects he used a strong bomb capable of holding several pounds of powder, which he exploded by electrical means. By immersing the bomb in a water bath he ascertained by the rise of temperature of the water the total heat given off in the explosion, which is a simple multiple of the energy in foot-pounds per grain locked up in the powder and capable of doing work when released by explosion. After the bomb had cooled down he drew off the cold gases and analysed their chemical nature and measured their volume. During the explosion he also measured accurately the rise of pressure and the rate at which it rose.

From these few details, obtained at the cost of enormous labour and expense, he calculated by rules of chemical arithmetic the actual temperature of the explosion, and found it in many instances as hot as the electric arc. From the temperature and the volume of the gases formed he also calculated the greatest pressure which the explosion should attain, and used this as a check on his mechanical pressure gauge. In addition, from his measured maximum pressure and the particulars of loading he calculated by another method the temperature of the explosion as another cross-check on his own results. In the end he calculated and defined the "force" of the powder as the pressure in tons per square inch obtained when modern smokeless powder is exploded in a vessel capable of holding 100 lbs. or other units of water, and actually containing 61 units by weight of powder.

With this brief outline the subject must be left. After Noble's work the reign of the chemist, physicist, and mathematician begins. For ordnance their help is indispensable, but for rifles their work is chiefly to give a scaffolding or skeleton of the idea on which rifle ballistics may some day be placed on as sound a footing as those of heavy guns.

## CHAPTER II

### EXTERIOR BALLISTICS (DESCRIPTIVE)

Exterior ballistics deal with the motion of the bullet from the instant it leaves the muzzle until it is at rest. This period may extend from the fraction of a second to perhaps a couple of minutes in the case of the most powerful small arms. For all ordinary purposes it rarely exceeds half a minute, which is about the limit of barrage fire at an elevation of 20 degrees.

The following brief and very general account of the subject is compiled from many sources, of which the principal are :—

(1) The lecture notes of Sir George Greenhill to the "advanced class" at Woolwich and his correspondence and conversation during the last twenty years.

(2) The original note-books of Bashforth and his printed works, supplemented by a few letters written to the compiler.

(3) *Balistique extérieur rationelle*, by P. Charbonnier, Encycl. Scien., Paris, 1907.

(4) *Handbook of Ballistics*. Vol. I, "Exterior Ballistics," second edition, by C. Cranz and K. Becker; Berlin, 1912, translated into English 1921. (H.M. Stationery Office. 30s. 479 pp.) One of the most complete textbooks on the subject which has ever been published in English.

The account has been arranged more or less in chronological order, and, except on a very few of the salient points, all details have been omitted. The chief headings are :—

(1) The division of the trajectory into its three most important sub-divisions, viz., beginning, middle and end, followed by remarks on the angle of elevation required to give the maximum range.

(2) The parabolic or unresisted trajectory and the recognition from the earliest times of the effect of a resisting medium.

(3) Early experiments on the resistance of the air, from Newton's time up to the introduction of rifled ordnance, with remarks on the change of resistance above the velocity of sound.

(4) **Bashforth's appointment** as professor at the Artillery College, Woolwich, in 1864, and his life work, including a description of his chronograph and methods.

(5) Ballistic tables compared to show the meaning of the coefficient of form and the ballistic coefficient or bullet value.

(6) The wind problem.

(7) The principle of the rigidity of the trajectory.

(8) The twist of rifling required, also drift, yaw, and the rotation of the earth.

1.—The Subdivision of the Trajectory into Three Parts

The whole period of flight may be regarded from three separate aspects, and in each aspect it may be divided, roughly, into three separate parts.

It may be regarded from the point of view (A) of its utility, i.e., the object for which it is employed by the firer, or it may be regarded (B) from the bullet's own point of view, i.e., as regards the chief disturbing influences during flight; and thirdly (C) from the point of view of "the ancients."

From the firer's point of view (A) the range is of the greatest consequence, and this may be either—

(A 1) Short range, allowing of direct fire by individuals without range-finders.

(A 2) Medium ranges, where errors of range-finding are appreciable even in collective fire.

(A 3) Long or indirect or curved fire, where accuracy of range-finding is of prime importance, and machine-gun fire supersedes rifle fire.

From the bullet's point of view (B) the three periods are—

(B 1) The first few score of yards where the bullet is recovering from the initial disturbances of the shock of discharge. During this period the bullet is liable to be a "tipper," and to make slightly oval holes in a card target.

(B 2) The normal period (after, perhaps, the first hundred yards of flight), during which the bullet travels very much in the way in which it is often assumed to be always travelling—that is, steadily, point first, with its spinning axis nearly coincident with the tangent to the trajectory. It is possible to calculate very closely the full particulars of this part of its flight by using the proper ballistic tables.

(B 3) The period in which "yaw" begins and develops, until at last the bullet flies almost broadside on. The earliest symptoms of "yaw" begin at medium ranges, but they assume no great importance during the first five seconds or so of flight. After about a mile or so the yaw is liable to become so pronounced that ordinary methods of calculating the trajectory fail entirely.

The third aspect in which it may be regarded (C) is that given by the standpoint of "the ancients," and perhaps this is the most illuminating of all. They had no precise knowledge of what we now call mathematics, but they had their own eyes and ears and logic to help them, and these adjuncts were as acute as our own.

They divided the whole trajectory into three parts as in the inset figure (a).

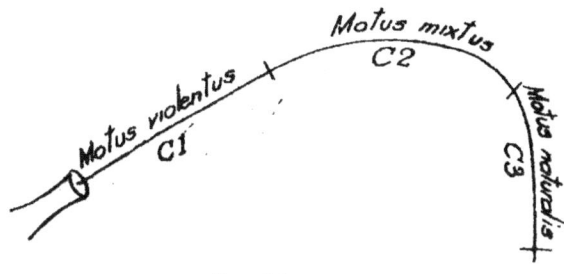

Fig. (a).

(C 1). *Motus violentus*, or the violent motion approximating to a straight line or to the vulgar idea of a so-called point blank range. This corresponds almost exactly to A 1. They never defined the actual distance in yards, but by implication they meant exactly what a modern gunmaker means when he sells a sporting rifle guaranteed to have a flat trajectory up to so many yards.

(C 2) *Motus mixtus*, which also by implication is the same as A 2 ; the ball is in a transitional stage as regards hitting the quarry as the diagrams show. In this stage errors of holding or of range judgment matter enormously.

(C 3). *Motus naturalis*, or the natural dead drop of any projectile at the end of its trajectory, ending in a vertical fall at a so-called "terminal velocity."

In the trajectory of any small arms weapon these three phases are easily distinguishable.

It is an observed fact that the greatest range of any hand gun occurs at an elevation of about 33 degrees, and not at the conventional 45 degrees. A revolver, a shot gun, a 0·22-inch rim fire and a service rifle all give their greatest range at about 33 degrees, and their shot all fall nearly vertically at any elevation which approaches or is higher than that angle. A rough but safe rule is that from 3 degrees elevation to that of the extreme range, the angle of descent is $2\frac{1}{2}$ times the angle of elevation.

It is now common knowledge that "Long Bertha" shelled Paris, using well over 50 degrees elevation because it was advantageous to get the shot out of the denser air near the ground as soon as possible. It was thus enabled to "make range" in the highest strata of the air, far above the top of Mount Everest, where the resistance is negligible because the barometer stands always at less than 2 or 3 inches of mercury. But the projectile of the hand gun has quite another problem to solve. It has to make range before its angle of descent becomes very steep, and in consequence its best angle is well below 45 degrees. The *motus naturalis* or terminal velocity of rifle bullets is very low, making the retardation due to air resistance very great in the early part of the flight if the velocity is high. The portion C 1 of the trajectory and the early part of C 2 need not therefore be given so much elevation because the C 3 part really makes very little range.

2.—The Parabolic or Unresisted Trajectory and the Recognition from the Earliest Times of the Effect of a Resisting Medium

The general shape of the trajectory is called parabolic because on a flat earth and in an unresisting medium the path of a projectile is actually a parabola, as was first proved by

Galileo in 1638. Very shortly afterwards he noticed that, owing to the curvature of the earth, the force of gravity did not act in parallel lines, but in lines converging to the centre of the earth. As a consequence, the usual path of a projectile fired on the earth in the absence of air would be a portion of an ellipse. If the shot were fired horizontally from an eminence at any ordinary velocity the elliptical path would intersect the earth's surface and the shot would remain on the earth. Should, however, the velocity be as high as 26,000 f/s, the shot would never return to earth, but would become a satellite with a circular orbit passing through its original firing point once in every seventeen revolutions. If the velocity was even greater than 26,000 f/s and less than 36,000 f/s, the orbit would be elliptical. Exactly at 36,000 f/s it would go off into space in a parabola, and at still higher velocity in a hyperbola. It would only describe a trajectory in a straight line when fired vertically upwards or downwards.

The inaccuracy of the guns of that early period and their low velocity masked the divergence of the real air trajectory from a parabola for fully a hundred years, although, of course, such men as Galileo and Newton were aware of the fore-shortening caused by the resistance of the air. This resistance is known almost instinctively by everyone, and also its practical effect. A child will crumple a piece of paper before using it as a missile, and the tighter the paper is crumpled the better it flies.

The most important property of a projectile is its terminal velocity, which was made very evident during air raids in the war. The falling bombs fell very hard and fast, yet the pieces of exploded shrapnel or a burning airship fell at very distinctly different speeds, and this because their sectional densities or terminal velocities were different. When the acceleration of gravity is equal to the retardation of the air the body arrives at the velocity of fall called its terminal velocity. The acceleration of gravity is always 32 f/s in each second, but the retardation of the air depends primarily on the weight of the body and its cross section. The heavier the body or the less its cross section the greater is its sectional density and power of getting through the air.*

The parabolic theory is, however, still useful as assigning limits for trajectories, especially with low velocities and heavy projectiles. For instance, the extreme range of a rifle grenade or trench howitzer bomb can be given by the following rule. The square of the muzzle velocity in f/s divided by 100 gives the extreme range in yards. Thus 100 f/s gives a maximum range of 100 yards and 1,000 f/s gives 10,000 yards.

Cranz, following Charbonnier (*Balistique Extérieur Rationelle*, Vol. I, Chap. 5), gives a list of eleven certainties in a trajectory arising from the parabolic theory. As a stand-by or check of rapid calculation they are worth recording :—

(1) The horizontal component of the velocity of a shot continually decreases.

(2) The angle of descent is greater than the angle of departure.

(3) The height of the vertex lies between the heights of the vertices of the two parabolas drawn (*a*) with muzzle velocity and angle of elevation, (*b*) with striking velocity and angle of descent.

---

* It is interesting to note that although Galileo dropped weights from the Leaning Tower of Pisa expressly to show that Aristotle was wrong in maintaining that a two pound weight fell twice as fast as a one pound weight, yet Aristotle never said anything of the sort. If Galileo had referred to Aristotle's "Physics" Book IV, Chapter viii, sect. 8 to 11, he would have found the Philosopher's brief remarks explaining as above that in a resisting medium the terminal velocity of a body depended upon its weight if its cross section was unaltered. These are so concisely worded that they are better consulted in the *Opera Omnia* of St. Thomas Aquinas, A.D. 1248 (Leonine edition, Tome ii, texts 71 and 74, pp. 183-187), where the Angelic Doctor devotes several pages of his commentary on Aristotle's Physics to a lucid exposition of the effect of weight, shape and cross section on terminal velocity, called then *motus naturalis*, or the constant velocity natural to a body falling freely though a resisting medium. Readers interested are referred to the columns of *The Tablet* headed "Literary Notes" for November, 23, 1912, and to the correspondence columns of *Nature* for January 22 and 29, 1914, for fuller details on Aristotle's observations and the mediæval commentary on them. They were overlaid and unknown till 1896, when they were incorporated into the philosophical lectures of Fr. David, O.F.M., at Louvain University.

(4) At a defined height on the ascending part the velocity is greater than at the same height on the descending part of the trajectory.

(5) The vertex is nearer to the point of strike than to the muzzle.

(6) If the descending branch of the trajectory is continued indefinitely below ground level, it becomes vertical after an infinite time and at an infinite depth below the level, yet the horizontal range to the tangent at infinity is itself finite. The velocity after this infinite time is exactly equal to the terminal velocity of the shot. The apparent paradox of finite velocity and horizontal range after infinite time is worth noticing.

(7) The point of minimum velocity of the shot is always in the descending branch of the trajectory.

(8) The point of greatest curvature in the trajectory is in the descending branch, and lies between the vertex and the point of minimum velocity.

(9) The vertical component of the velocity of a shot continually increases in the descending branch, and at any defined height is always greater in the ascending than in the descending branch.

(10) The time of flight in the descending is always greater than the time in the ascending branch of a horizontal trajectory.

(11) The distance measured along the trajectory from the muzzle to the vertex is always greater than the distance measured in the same way from the vertex to the point of strike on the horizontal plane.

3.—EARLY EXPERIMENTS FROM NEWTON'S TIME UP TO THE INTRODUCTION OF RIFLED ORDNANCE, WITH REMARKS ON THE VELOCITY OF SOUND IN AIR

Newton made the first quantitative experiments on air resistance by dropping balls, in 1710, from Wren's unfinished dome of St. Paul's Cathedral, and noting the time of falling of balls of different sectional densities. He concluded that the natural or square or quadratic law was true, so that 1 per cent. addition to the velocity gave 2 per cent. addition to the resistance. He argued that the faster the ball went, the more particles of air it met in a given time, accounting for 1 per cent. Further, that the faster the ball hit each air particle the more momentum was taken out of the ball, which gives the second per cent.

Hence the law that air resistance varies as the square of the velocity is called the natural or Newtonian law. Newton, however, is generally supposed to have failed to find a method for calculating trajectories in a medium resisting as the square of the velocity, and it was left to be produced as the result of a curious challenge narrated by Bashforth in his *Motion of Projectiles*. In 1718 Keill proposed the problem as a challenge to John Bernoulli, it being of course supposed, according to the fashion of those days, that the proposer knew how to solve his own problem. Bernoulli received the challenge early in February, 1718, and in a short time obtained a solution not only of the problem where the resistance varies as the square of the velocity (which was proposed), but also for a resistance varying according to any power of the velocity. Before publishing his solution, however, he deemed it just and proper to propose the same problem to Keill. This he did in May, annexing a time limit till September, with the addition that if at the expiration of the time no solution was received, silence would be taken as a tacit confession of incompetency. No solution was furnished by Keill, and Bernoulli argued that Keill in his difficulty would undoubtedly seek assistance from all the English mathematicians, including Newton. Bernoulli published his analysis in 1721, and the solution was rediscovered by Adams in 1866, when Bashforth was getting his first reliable measurements of the actual resistance of the air.

The first notable experiments at gun velocities were made by Benjamin Robins with his ballistic pendulum a few years after Bernoulli's publication. Robins found that Newton's law held true up to about 1,100 or 1,200 f/s but that at higher velocities the resistance seemed suddenly to increase about threefold. He rightly ascribed this sudden shift to the fact that the velocity of sound in air is about 1,100 f/s, so that at higher velocities the shot is passing through undisturbed air. Below 1,100 f/s the sound wave travels through the air faster than the shot, so that the air in front of the shot is not quiescent, but disturbed. This disturbance affects by some mechanical means the resisting powers of the air, reducing it very largely.

It is a well known fact that when the shot is moving faster than 1,100 f/s it carries a cracking noise along with it in its flight as though doing what probably it actually does, that is, breaking a way continually through the air. This is very noticeable when one is walking down a long rifle range (say a mile long) provided with an iron target. Near the firing point one bang is heard when the rifle is fired, followed after about 7 or 8 seconds by the sound of the bullet hitting the iron target. 300 yards from the rifle a sharp crack is heard as the bullet passes overhead, followed after nearly a second by the bang or thud of the gun, and then after 6 or 7 seconds by the sound on the target. At 600 yards the crack of the bullet is heard about a second after the flash or puff at the muzzle, then nearly two seconds after the flash the gun is heard, and after another five seconds the target is heard to be hit. But at 1,500 yards the bullet is not heard to pass, except perhaps as a hum, because its velocity is then less than 1,100 f/s. The flash is seen, and the gun's thud and the strike on the target are heard. This sudden change of effect at the velocity of sound is further exemplified by the failure of the "Maxim silencer" to produce its hoped-for result. This silencer is attached to the muzzle of the rifle, and is intended to act in the same way as the silencer of a motor car. It provides an expansion chamber for the muzzle blast, which is only released into the open air when its pressure is almost negligible and its power of making a noise has vanished. It is quite successful in silencing the gun entirely when the muzzle velocity is below 1,100 f/s, as is often the case with miniature rifles such as 0·220 rimfires, but with a military rifle it has but little effect. With any ordinary rifle it merely kills the sound of the muzzle blast and slightly reduces the recoil, but it has and can have no effect at all on the sound of the bullet hitting and splitting the air, which is the greatest source of noise with a modern rifle.

Firing at 1,000 yards at an earth stop butt, the sound of the bullet hitting a canvas target and going into the bank seems to be heard by the firer about 5 seconds after firing, but it appears that this sound is not really made by the bullet hitting the bank, but is the reflection of the bullet's crack from the bank. If a shot is purposely fired well over the bank, the same sound is heard and after the same interval. If a row of tree trunks or posts extends along the line of fire, the sound of the bullet's crack is heard reflected off each post like a boy rattling his stick along area railings. At a mile range a bullet is only heard to hit the ground with a tap.

It is often said that a bullet has two airs to penetrate, one air when above 1,100 f.s., and another, quite a different air, when below the velocity of sound. It would be more realistic to say there are three airs, one from 0 to 1,000 f.s., one from 1,400 f.s. and above, and a third joining the first and second extending from 1,000 f.s. to 1,400 f.s.

Ballistic tables have not been made on this supposition owing to certain analytical difficulties in calculation, but our ordinary English tables are really made for some seven or eight different natures of air joining into one another, but each resisting according to a different power of the velocity.

Robin's method of charting the resistance was to fire a series of groups of rounds with a definite weight of powder and shot at his ballistic pendulum, varying the range for each group, and noting the average of the striking velocity at each range. In this way he obtained a set of figures showing the loss of velocity along the range, whence he calculated from the known weight and diameter of his shot the retarding effect of the air. Hutton repeated and extended Robins's work and confirmed his results. Didion and Piobert in France, about 1840, and Count St. Robert in Italy continued the research, all of them using round shot.

#### 4.—Bashforth's Work with his Chronograph

To quote Bashforth, "On the institution of the Advanced Class for R.A. officers in 1864, there was no satisfactory work on ballistics, and no experiments made with elongated shot could be found which were consistent among themselves and therefore deserving of confidence." It accordingly became his duty as Professor of Applied Mathematics and Referee of the Ordnance Select Committee to recommend that research work should be instituted. This recommendation was approved by the Council of Military Education, and the Bashforth Chronograph had its first trial in November, 1865, with ten equidistant screens spaced 100 feet apart. This actual instrument is now in the South Kensington Science Museum, and was borrowed during the war for use at the Small Arms Range in Woolwich Arsenal within a few furlongs of where it was first used. It worked quite satisfactorily after its 50 years of storage as soon as it was cleaned and connected up. To a modern maker it looks a crude enough instrument, for which reason, perhaps, so eminent a ballistician as Dr. Cranz, of Berlin, expresses a doubt "whether the apparatus as constructed, can measure the time intervals with sufficient accuracy." A careful study of the original instrument and experience with its working showed that this suspicion, although reasonable, is not actually justified, as will be seen by the following brief description of it.

Fig. 1 shows the original picture of the machine, which is about five feet high. Fig. 2

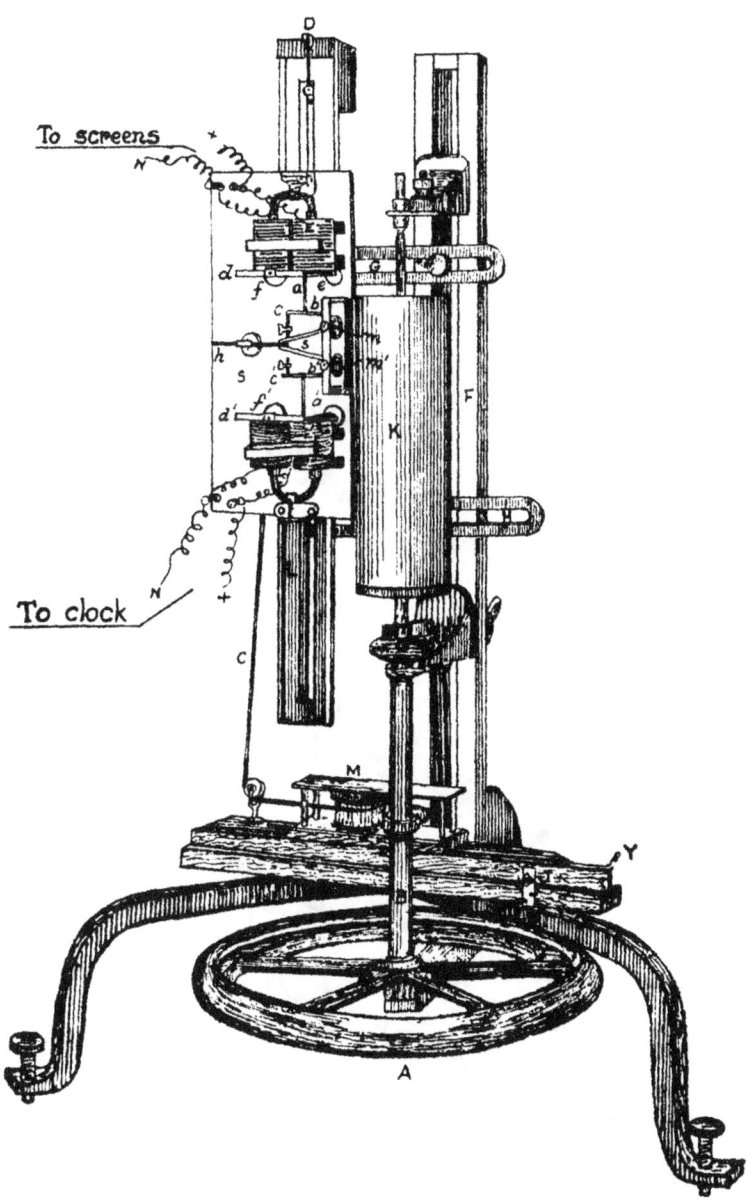

PART III, CHAP. II, FIG. 1.—THE BASHFORTH CHRONOGRAPH.

shows the arrangements of the screens. Bashforth's idea was to register the exact time to the nearest thousandth part of a second at which a shot passed through each of his equidistant screens. The screens were made of cotton threads about one inch apart, each

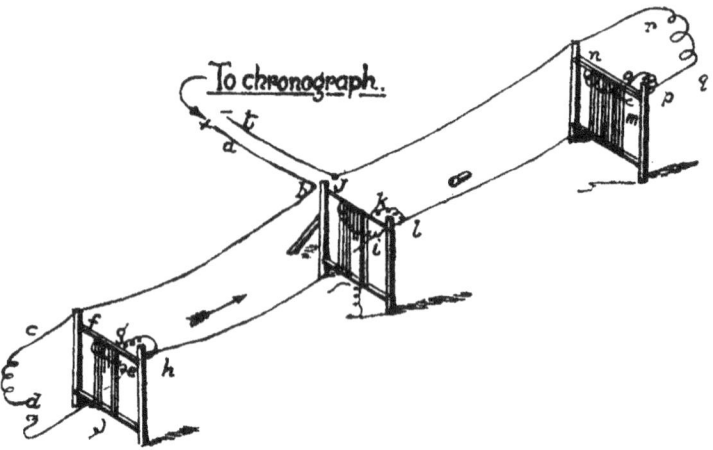

PART III, CHAP. II, FIG. 2.—SCREENS 150 FEET APART FOR BASHFORTH CHRONOGRAPH.

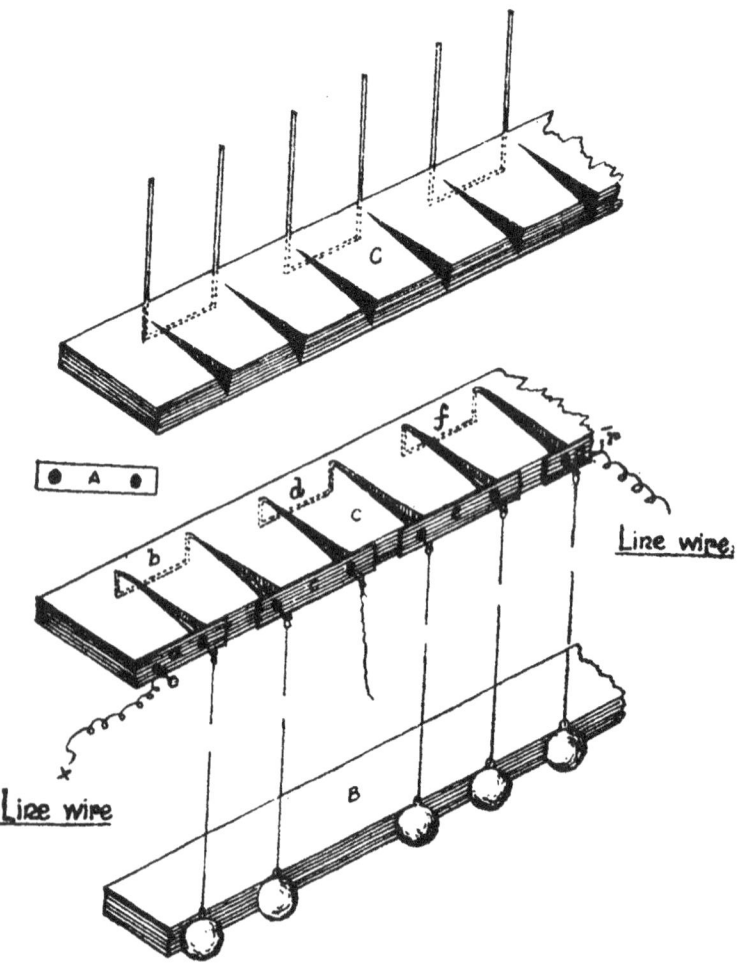

PART III, CHAP. II, FIG. 3.—DETAIL OF SCREENS SHOWING METHOD OF OBTAINING BREAK AND MAKE.

thread being kept taut by a small weight, so holding down a brass wire spring contact shown in Fig. 3. The passage of the shot cut a thread and released the contact. During the tiny fraction of a second which the contact took to fly across and remake contact the electric current was broken. This break affected the electro-magnet E (Fig. 1) actuating the recording pen m which marked on the revolving cylinder K, causing it to make a distinct "kick" mark at each make or break. He measured time by a very accurate clock actuating the magnet $E^1$ and the pen $m^1$ every half-second. The two pens, m and $m^1$, worked side by side on the cylinder K, which cylinder was kept in motion by a spin of the heavy flywheel A to which it was attached. At the same time the whole instrument board, S, was lowered by the unrolling of the string C passing over the pulley D. Thus after each shot there was a double spiral line traced on the cylinder, one line scaled in precise half-seconds and the other by screen breaks. The translation of this record into actual time intervals is a mere matter of common sense and of avoidance of or allowance for instrumental errors. When once this is properly effected, the history of the shot's flight and of its retardation is laid open to analysis.

The particular points which ought to be noticed in this apparently rough instrument are:—

(1) The recording drum is mounted on a *vertical* spindle, and is not driven by clockwork, but by the spin of its heavy flywheel. Consequently, it loses its spin very slowly and quite regularly by friction alone. The rate of loss of spin can be determined by inspecting the distance apart of the clock beats. The paper moved about 10 inches per second. The clock breaks could be read to ·001-inch or one ten-thousandth part of a second with normal diligence, as also the screen breaks. This regularity of the time record is impossible with any but a vertical spindle, dying down in speed by friction alone. Nearly every experimenter has overlooked this immense advantage of a vertical drum and flywheel.

(2) The make or break of the electro-magnet can be so arranged as to cause a blow on the rod carrying the pen, and this blow can be so arranged in speed and angle as to produce a right angle in the record line on the drum.

(3) The clock can be regulated to beat exact fractions of a second by taking the usual precautions of an astronomical clock, so that the true time scale is freed from certain of the known irregularities of a tuning fork.

(4) The spring of the brass wires in each screen can be relied upon to break contact regularly as soon as the cotton is broken, and the cotton can be made to break quite regularly. The modern screen is composed of metallic wires actually carrying current, and the irregularity of their electrical break is therefore obviated.

(5) The snapping over of the brass spring can be made to be very regular so that the current is restored regularly, and not by a succession of electric surges, as may happen when electrical relays are used to restore the current.

(6) The time of passing each screen is recorded by the interruption of the same galvanic current under precisely the same conditions.

Bashforth's method may be illustrated by part of the original manuscript notes made by him for round 479.

| Screen | Recorded Time | Correction | Corrected Time | First Difference | Second Difference |
|---|---|---|---|---|---|
| 1 | 73·856 | + 0·004 | 73·860 | | |
| | | | | 2·136 | |
| 2 | 76·000 | − 0·004 | 75·996 | | |
| | | | | 2·172 | |
| 3 | 78·180 | − 0·012 | 78·168 | | |
| | | | | 2·218 | 0·036 |
| 4 | 80·390 | − 0·004 | 80·386 | | |
| | | | | 2·254 | |
| 5 | 82·640 | 0 | 82·640 | | |
| | | | | 2·290 | |
| 6 | 84·910 | + 0·020 | 84·930 | | |

The muzzle velocity is about 2,300 f/s and $W/d^2 = 50/36$.

The distance apart of the screens was 150 feet, so that time is recorded over a series of equal distances. The recorded time is measured on a scale of about 100 units to 3 seconds, as the revolving cylinder is about six inches in diameter and turned about twenty times in a minute. His measuring micrometer was divided into 100 parts for each circumference, so that the whole six screens were included on about 11 divisions, or about one-ninth of a turn. After examining the recorded times he applied small corrections to them to make them run smoothly. The result is shown in the column headed "corrected time." The next column shows the difference of each record from the succeeding record, and these differences differ again, but, owing to the smoothing, by a constant amount, viz., $0 \cdot 036$. It is this last figure which is required, because it is itself a direct measure of the resistance of the air to the shot as the shot is passing between the third and the fourth screen, at the exact velocity it has at the half-way distance.

It is often said, even by responsible mathematicians, that Bashforth assumed that the resistance of the air varied as the cube or third power of the velocity. He actually made no such assumption, and his whole life work was based on the assumption that there was no simple law, and that his business was to tabulate the actual resistance in the most convenient way, and then to produce a set of tables in the form of a "ready reckoner" by which the trajectory of the shot could be figured out piecemeal. He noticed that if the actual resistance in lbs. of the air on the head of a shot was divided by the cube of the velocity of the shot at that particular instant, the resistance could be called $K V^3$, so that if he knew the value of K he knew the resistance. He also noticed that if a round was fired through a series of equidistant screens, and the resulting times of passage through each screen were arranged and differenced as above, the second difference was a simple factor of the K as above.

By firing a series of rounds, then, with one particular sort of shot, and varying the muzzle velocity, he obtained a set of values of K for each half-way velocity. By examining and cross-examining the records of each round with the assistance of the round of next higher and lower velocity he was enabled to criticize the time record of each screen and adjust his small corrections so that his second differences harmonized, the one with the other next to it. This second difference of $0 \cdot 036$ was his first attempt. On revision it was adjusted to some such figure as $0 \cdot 034$ or 6 per cent. less. In the end, when he had completed his firings from 2,600 f/s to 800 f/s muzzle velocity, it can be confidently stated that his final table of resistance was within one per cent. of the true answer for the actual shot and guns he used, under the conditions in which he fired them. An experimental determination of so complicated a thing as air resistance is, when carried out to the accuracy of one part in a hundred, an achievement to be justly proud of. It is equivalent to weighing a $\cdot 303$ bullet to within two grains, not by using a chemist's balance, but by actually firing it and timing its journey.

It is by no means to be supposed that Bashforth never made a mistake. He was clearly unaware that an electro-magnet takes so long as it does to become of full strength, and also to lose its pull again when the current is broken, although he knew it took some time. It is almost certain also that he supposed that the resistance of a pointed shot of one particular curvature could be determined from that of a standard shape by a simple multiplier, although he was fully aware that there was no simple relation between the resistance of a spherical and that of an elongated shot. His statement also in *The Motion of Projectiles*, 1872, p. 30, that air resistance is little affected by the more or less pointed apex of the shot, and depends chiefly on the curvature of the shoulder, was unfortunate, as it seriously affected English research for over 30 years.

The description of Bashforth's method is enough to show the general scheme of all methods for determining the resistance of the air, however widely different the instrument may be. From his general results it may be concluded that $p$, the resistance in pounds deadweight, varies :—

(1) as the square of the diameter of the shot,
(2) as the density of the air or weight per cubic foot,
(3) as some function of the shape and so-called steadiness of the shot, and
(4) as some function of the velocity of the shot.

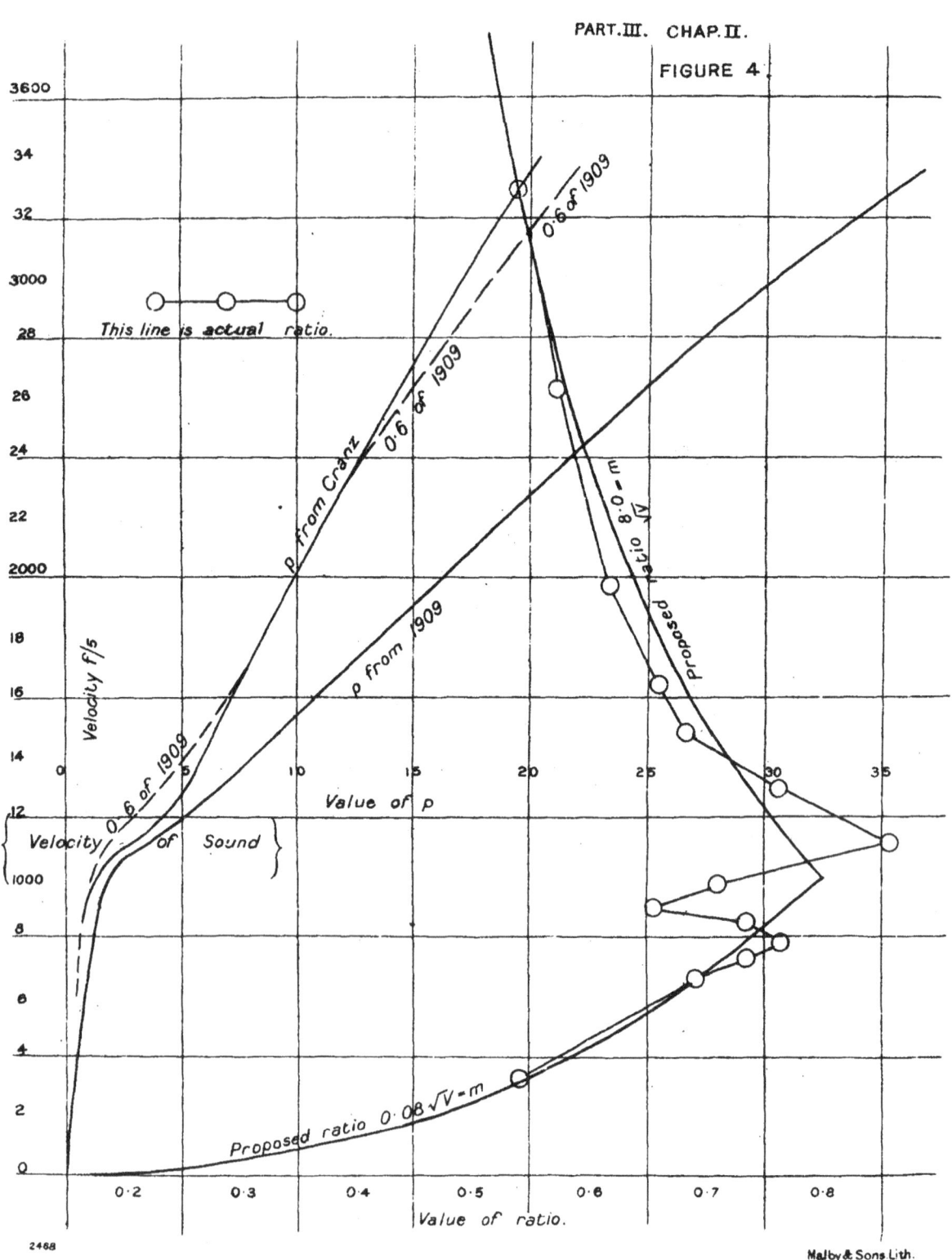

Cranz says that these deductions are conventional and seem reasonable, but concludes with the well-warranted remark that they are only adopted because there is nothing better to replace them. Here, of course, he refers to strict scientific principles, and in this conclusion all practical men must agree with him.

### 5.—Ballistic Tables and the Value of C

From the above remarks it will be seen that there is no necessary connection between a set of ballistic tables and the performance of an actual bullet. The bullet behaves according to its nature, while the tables are calculated for a bullet behaving in some definite and particular way. Between the tables and the actual performance of the bullet there is therefore no natural and necessary affinity.

In Fig. 4 is shown the actual resistance curves of the shot used for the last edition of this Text Book, and also the curve used for this edition. The old curve published in 1909 is founded on experiments carried out at Shoeburyness during the period 1904 to 1906 with what were then called modern artillery projectiles. The new curve is based partly on Cranz's work with the German service bullet just before the war and partly on experimental results obtained at the Woolwich Small Arms Range during the war.

There is a very marked difference between the two, and it can hardly escape notice that the multiplier to connect the two curves changes continuously with the velocity. To make it quite clear an extra curve has been added to give the relation between the two at each velocity. It is at once seen that if a modern bullet is fired at a velocity of 2,600 f/s, and its trajectory is studied by the aid of the old tables down to 2,000 f/s, a multiplier or "coefficient of form" (generally denoted by " n ") of about $0 \cdot 6$ on the average is required to connect the two. If, however, the trajectory is studied down to 400 f/s it is obvious that $0 \cdot 6$ is the wrong coefficient; but the exact value of the true coefficient or multiplier is not easily ascertained.

According to the theory of the tables it should suffice for entry to them to work out the sectional density of the bullet by dividing its weight in lbs. by the square of its diameter in inches, and then to divide this result by a factor of shape called above the coefficient of form. The result should be the "bullet value" or the ballistic coefficient, commonly denoted by the letter C. In any well-constructed tables the value of C should be the same for the same bullet always. To enable this to be the case the new tables have been calculated and published. It is hoped that they will serve all practical purposes up to about 1 or $1\frac{1}{2}$ miles without serious error. Beyond this range the effect of stream-lining and of yaw have not been studied long enough to enable a definite opinion to be formed. It should be observed that the accuracy of the value given to " C " is dependent upon the bullet moving steadily point foremost.

### 6.—The Wind Problem

There can be no doubt in the mind of any practical rifleman that the wind has a very serious disturbing influence on the direction of a bullet, and has some much smaller effect, but still an appreciable effect, on the range. The formulæ giving these effects are fairly simple, as also are the explanations of the reasons for them.

The formula for lateral effect is always near enough to the truth to be expressed by the following statement if the result is wanted in minutes of lateral deflection at ranges up to $\frac{3}{4}$-mile. The deflection is then proportional

(1) to the speed of cross wind,
(2) to the range,
(3) to the *delay* of the bullet over the range.

As regards (1), it is generally known that the cross speed of a wind is proportional to the trigonometric cosine of the angle of the wind, so that if a III o'clock wind is worth 8, a II o'clock is worth 7, and a I o'clock is worth 4, since the cosine of $90°$ is $1 \cdot 00$, of $60°$ is $\frac{7}{8}$ and of $30°$ is $\frac{1}{2}$.

As regards (2), this is an approximation, but it is so very near the exact truth that the error can never be demonstrated on the actual firing range.

As regards (3) this is a real fact, and it was demonstrated at Woolwich by a special set of experiments to convince a sceptical range officer. The delay is the difference in time of flight over a given range between the actual observed time and the time calculated from the muzzle velocity if the bullet had been fired in a vacuum.

Suppose that the time of flight of the service bullet with M.V. 2,400 is 1 second for 1,500 feet. Its time in a vacuum would then be 1,500 divided by 2,400 or 0·625 seconds. Its delay would be 1.00 — 0·625 or 0·375 seconds, and the wind deflection of other bullets for 500 yards would vary as 0·375 varies, but not as 1·00 varies or as 0·625 varies.

The special experiment framed to prove this point was selected by the sceptical officer himself. After making the calculations he pointed out that if this "delay" condition was true, the paradox would occur that the ·303 inch bullet would want the same wind allowance with the full charge of 40 grains and with the half charge of 20 grains, and would want more allowance with the three-quarter charge of 30 grains. Having framed his own conditions, he obtained permission to shoot out the experiment, and selected a day with a strong right wind worth about 4 feet at 500 yards. He fetched the whole range staff out to see the result, and proved himself to be in error. The full and the half charge struck 4 feet to the left and the three-quarter charge struck a good half yard further to the left, confirming his calculations, but contradicting his preconceived notions. The experiment was subsequently repeated on a day with a strong left wind, and calculation was again vindicated.

If the problem is considered in its most elementary form it can be seen that there are three separate steps to be made. Consider with a constant range :

(a) The firer in a train shooting at a stationary target.
(b) The firer stationary and shooting at a mark on a train.
(c) The firer in a train shooting at a mark on another train travelling parallel with it, in the same direction and at the same speed.

In case (a) the firer has to allow merely for the speed of his own train ; and reducing it to exact figures, suppose the range is 500 yards and M.V. 2,400, giving a vacuum time as before of 0·625 seconds and a real time of 1·00 seconds. Suppose the train to be moving at 40 f/s or at $\frac{1}{60}$ the initial speed of the bullet. The bullet will have a forward velocity of 2,400 f/s and a cross velocity of 40 f/s at the instant it leaves the gun. Its angle of projection will therefore be at a slope $\frac{1}{60}$ forward, and it will strike $\frac{1}{60}$ of the range forward or 25 feet.

In case (b) the train will have moved 40 feet in the 1·00 seconds, and the bullet will strike 40 feet behind.

In case (c) the bullet will strike 40 less 25, or 15 feet behind the mark.

Very clearly, then, the 40 feet is the wind speed multiplied by the actual time of flight, and the 25 feet is the wind speed multiplied by 0·625, so that the difference, that is to say, 15 feet, is the wind speed multiplied by the delay which is 0·375. This establishes conditions No. 1 and No. 3.

Condition No. 2, that the deflection is proportional to the range, is most easily established by calculating the deflection in feet at 200, 500, 1,000 and 1,500 yards, and reducing the results to minutes. The answers will be nearly exactly proportional to the range, and this is confirmed daily at target practice. The detailed proof of the proposition can be found written out from the technical point of view in an article by Greenhill and Hardcastle published in the *Journal of the United States Artillery*, May, 1903. It includes the effect of wind on range. From the arguments there used it can be seen that no ordinary wind can affect the range by an appreciable amount up to about half a mile. Half a minute of elevation is the utmost that can be expected. Beyond one mile the effect of wind on either deflection or on elevation may be safely taken as a trifle less than the actual movement of air during the actual time of flight. This is because the air delay in time is so very great compared to the vacuum time. It is as well to point out also that a wind which does not blow parallel to the horizontal plane lifts or lowers the shot on a vertical target by the same amount in feet (or minutes) as a lateral wind of the same strength blowing at the same angle deflects the bullet.

The problem of the effect of wind was first solved by Didion about 80 years ago for round shot, which always present the same cross section to air resistance. In his *Cours Élémentaire de Balistique*, Paris, 1859, p. 27, he defined the delay as above, and stated that the deflection in feet was equal to the space passed over by the current of air in the time represented by the delay. About the year 1890 Sir G. Greenhill, in discussing the graduation of naval sights with the late Captain F. Younghusband in Woolwich Arsenal, brought up the subject of wind deflection. Younghusband thereupon suggested the applicability of Didion's method to the solution of the problem in the case of elongated rifled projectiles. He pointed out that owing to the gyrostatic domination of the spin a properly rotated shot turns its spinning axis till it faces the resistance, so presenting like the round shot the same cross section always to the air resistance.

The bullet which has already been given as an instance has, when fired from the train, two resistances to encounter. There is the enormous hurricane of 1,630 miles per hour, or 2,400 f/s, caused by its own forward motion acting straight backwards and a side wind of 40 f/s (27 miles per hour), or $\frac{1}{60}$ of the hurricane speed, acting broadside. The two together act like one resistance at an angle of 1 in 60 backwards, and having a speed of 2,403 f/s, and to this resistance the spinning shot turns its nose and keeps it so turned, thus presenting the same cross section to the air resistance. This turning of the axis is accompanied by a certain amount of gyration, which may be even audible in a strong cross wind, but these gyrations are soon damped down and the shot steadies itself and goes to sleep like a top, with its nose pointing slightly up wind, just as a pilot in a strong tideway making a straight passage over the sea bottom keeps his keel inclined up stream.

### 7.—The Principle of the Rigidity of the Trajectory

If a fire hose is directed at a small angle, say two degrees, with the horizontal and kept still, the column of water shows clearly the curve of the trajectory of each drop, and this curve can be photographed. If now the hose is elevated to five degrees and photographed again to the same scale and compared with the first, it will then be seen that the curvature of the two jets is the same, and although their ranges differ, the two-degree jet will be nearly exactly over the five-degree jet if the photographs are superimposed and one of them tilted three degrees, with the nozzle as centre. This illustrates the exact meaning of the rigidity of the trajectory.

Trajectories of small angles of elevation of the same rifle and cartridge up to about ten degrees can be treated as if made of stiff wire, which, when tilted round the muzzle as centre, will lie one on the other up to the limit of their range. Thus if a ten-degree trajectory is drawn to scale and cut out of cardboard of the shape given by G C A, it can be pivoted at G, the muzzle end, and tilted to the position G D B, through the angle A G B. (*See* inset, Fig. b.) Then G E will be the range of the trajectory whose elevation is smaller than ten degrees by the amount of the angle A G B.

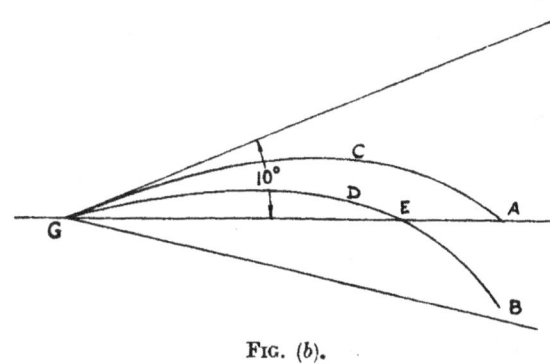

Fig. (b).

The original trajectory can be marked off to show the position of the bullet after each second of flight, and also its velocity at each second. These scales will be equally applicable to the tilted trajectories. The limit of 10° from the horizontal will be found small enough to exclude any serious errors with rifles and yet to include almost every practical firing problem before " yaw " becomes very serious.

In the numerical examples in Chapter IV this principle of rigidity is taken for granted, and except for short range aeroplane work all the examples are worked at angles of less than 10°.

### 8.—The Twist of Rifling, Drift, Yaw and the Rotation of the Earth

In 1879 Bashforth's successor at Woolwich, Professor Greenhill, found that the sharp spin given to howitzers set up such heavy torsional strains between the liner of the gun and the jacket that the gun sometimes wrung its own neck, and by shearing the copper vent automatically spiked itself. The spin was usually cut at one turn in 12 calibres, so that a 6-inch howitzer shell turned round once in each six feet advance (or one in 12 calibres), equivalent to 1 turn in $3\frac{1}{2}$ inches for the ·303 rifle. Thereupon he set himself to investigate the whole question of the spin of rifling required to produce steady flight against the resistance of the air, a subject which was then in a chaotic state. The mathematical argument was exceedingly complicated, but it resulted in a simple table of the minimum spin that is required to overcome instability.

As pointed out at the time by Captain J. P. Cundill, the hypotheses underlying the table are as follows :—

- (a) The projectile is of the form of a prolate spheroid, or egg-shaped, the only shape for which the " stream-lines " are known.
- (b) The medium through which it moves is frictionless and homogeneous (i.e., incompressible).
- (c) Gravity is neglected, and hence the curvature of the trajectory does not enter into the problem.

The validity of these assumptions has been a matter of considerable controversy, as they are obviously unsatisfied by any bullet flying in air ; but it is found that in the case of rifle bullets a reduction of spin in the proportion of 4 to 3 below the tabular value is a sufficient and reliable discount for the hypotheses, and gives enough and not too much spin for stability. The table can be obtained for all practical purposes from the following rule for solid lead bullets of specific gravity 10·9. "The length of the bullet in calibres multiplied by the spin required in calibres is 150."

Thus for a bullet 5 calibres long (or $1\frac{1}{2}$ inches for ·303) the greatest twist required is one turn in 30 calibres or 1 turn in 9 inches for ·303. When the density of the bullet is less than that of lead or the density of the resisting medium is greater than that of air, the spin should be increased as the square root of the ratio of the densities. Thus aluminium is four times as light as lead and requires double the spin, whereas water is 900 times as heavy as air so that the spin has to be increased 30-fold if the bullet is to be stable in water. Hollow shell such as the cupro-nickel envelope of a Mark VII bullet from which the lead has been melted require rather more spin than solid bullets, but so little more that the ·303 rifle will fire an empty envelope correctly. The actual calibre of the bullet is of no consequence, as the rule is expressed in calibres ; nor is the actual muzzle velocity, because the number of turns per second varies as the muzzle velocity.

In the steady or stable motion of a bullet the centre of gravity describes a very long helix of very small diameter ; in fact, the bullet must always be a slight " tipper " ; but the tipping is very small indeed, as the length of the helix of the ·303 is about 5 yards and its diameter less than one-hundredth of an inch. In actual practice Greenhill's figure 150 can be increased safely to 200 and still control the bullet. This reduction of spin required in practice is probably due to air being compressible, viscous, and possessed of friction.

In a later unpublished paper written in 1912 Sir George Greenhill investigated the stability of the shot whilst it was passing up the barrel. He pointed out that centrifugal wringing machines used in steam laundries or milk separators were driven either from above or below according to the weight and speed employed. The central spinning shaft for a slow moving centrifugal can well be taken from the roof with the centrifugal hanging from it, but if any fast spinning is required a roof suspension will smash itself to pieces, necessitating the shaft coming up through the floor for high speeds. The centrifugal pan then rides and spins quietly like a top asleep, which it really is. So, too, a shot as it starts slowly from the breech should be rotated by a driving band slightly in front of the centre of gravity, but as speed is picked up the rotating band should be moved (if such a thing were possible) to some position one or two calibres behind the centre of gravity if the shot is not to " fight " on its way to the muzzle and damage itself and the gun. With rifle bullets the rotating ring is really the whole of the parallel of the bullet and the fighting power of the bullet is so controlled by the enormous surface of contact that it has little chance to display itself. From the most recent experiments with stream-line bullets, which have a much shorter bearing surface, it seems probable that more attention will have to be given to this subject than in the past.

As a direct result of the spinning of the bullet in the air the phenomenon of " drift " is observed.

It was in 1746 that Benjamin Robins showed that the sole virtue of rifling lay in giving spin to the ball on a predetermined axis. No ball could be fired from a smooth bore without rubbing against the sides of the bore and acquiring a spin, the axis of which depended on the chance position of the last rubbing contact. He relates that he purposely bent a musket barrel a few degrees to the left 3 or 4 inches from the muzzle, and predicted that the bullet would incurvate to the right, and " this upon trial did most remarkably happen."

The following explanation of the effect of rifling and the phenomenon of drift, together with the plates, is reproduced from the *Text-book of Gunnery*, 1887, by Major G. Mackinlay, pages 133 and 242 :—

" Let us consider why a rifled projectile, which rotates about its longer axis, can be made to travel through the air with its point always approximately first. Suppose that the projectile has reached some position, D (Fig. 5), in its flight, where gravity has caused the axis to make a certain angle with the trajectory ; we may suppose the resultant resistance of the air to act along SR. But the effect of the couple R.GM in the case of a spinning shot, is not to raise the point of the projectile and give rotation round the centre of gravity, on an axis at right angles to the plane of the trajectory ; but it causes a slow movement of the point laterally, with service projectiles having right-handed rotation, to the right of an observer stationed behind the gun and who looks down the range ; with left-handed rotation the movement would be to the left. This is shown in the plan, and the projectile sets laterally across the original direction. As seen in the plan, the couple R.GM, instead of turning the point still more to the right, gives it a slow downward movement. The point is thus approximately kept down to the trajectory. The resultant resistance, as seen in the plan, also exercises a retarding effect, and as the direction of the projectile is inclined to this resultant, a lateral force is exerted, which causes a tolerably uniform lateral acceleration, giving rise to drift to the right. The plan of the trajectory is thus seen to be a curved line and not a straight one, and the trajectory itself has a double curvature, and is not contained in one plane.

" The gyroscope illustrates the stability of the direction of the axis of a spinning body and the tendency of a rifled elongated projectile to travel nearly point first in flight, and it also gives a reason for the slight turning movement of the axis of the projectile caused by the resistance of the air.

" A gyroscope (Fig. 6) for such a purpose consists of a carefully centred heavy model of an elongated projectile, some $2\frac{1}{2}$ inches long, free to revolve on pivots inside a brass ring, the axis of revolution being a diameter of the ring, which has externally two arms in its

290

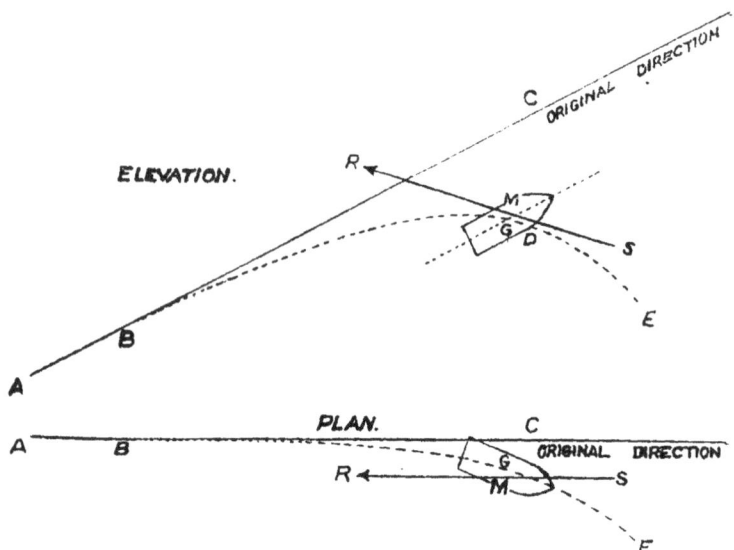

PART III. CHAP. II, FIG. 5.

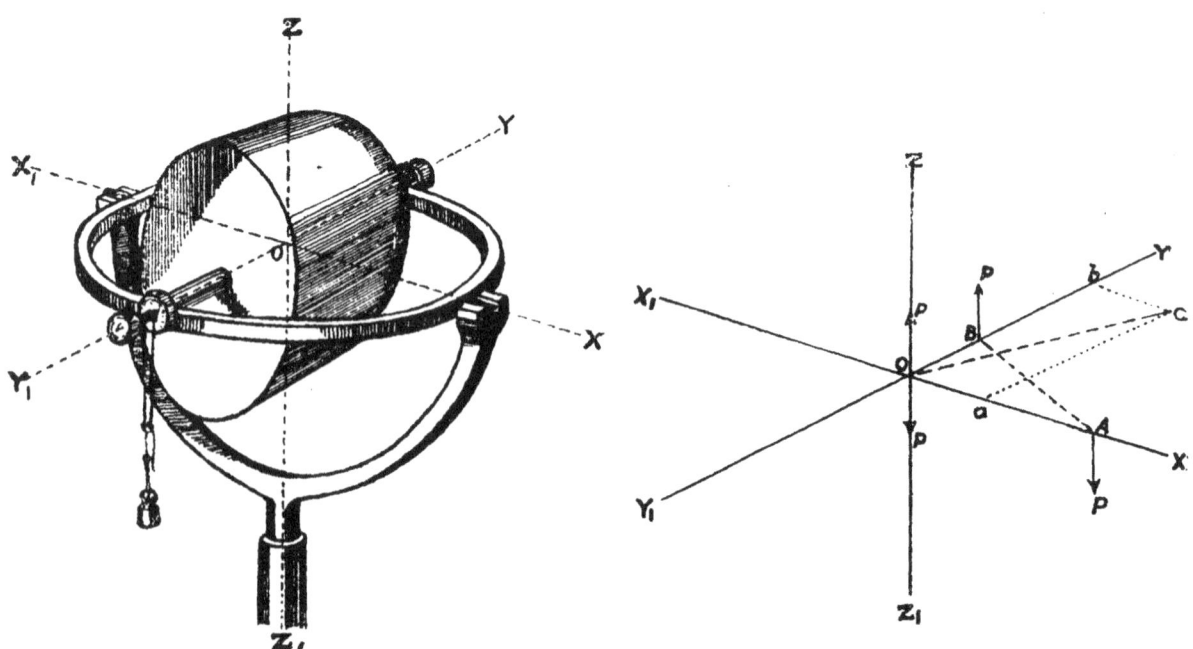

PART III. CHAP. II, FIG. 6.

PART III. CHAP. II, FIG. 7.

own plane, in prolongation of a diameter at right angles to the axis of the projectile. On these the ring can revolve in bearings in a vertical brass half ring, which has a stem under its middle fitting into a socket in a heavy stand. The half ring can revolve round its vertical axis.

"The centre of gravity of the model projectile is adjusted to be over the vertical stem by screwing the bearings of the pivots in or out together, and it will then remain in any position. All the three axes of rotation intersect in the centre of gravity O of the projectile, and movement of the axis in any direction is easy.

"If the gyroscope's projectile (not spun) is placed horizontally as in the figure, and a small weight is suspended from the brass ring behind it, the point of the projectile will immediately rise; and thus the effect of the couple caused by the resistance of the air in raising the point of a non-rotating elongated projectile in flight is imitated. But if the model projectile is caused to spin round its longer axis with a right-handed rotation, the addition of a small weight causes the axis to move its point round to the right (looking from behind it) with a slow movement in a horizontal plane, and thus the couple caused by the resistance of the air on a rotating projectile in flight is imitated. If one of the bearings on the brass ring is now pushed laterally in the direction of this motion, it does not increase the movement, but the axis turns in a plane at right angles and the point dips.

"In these last two cases a turning motion of the axis ensues in a plane at right angles to the couple, similar to the lateral movement of the axis of an elongated projectile in flight.

"This can be explained as follows:—

"Place the brass ring horizontal, and through O, the centre of gravity of the model projectile (Fig. 6), draw the three axes of revolution at right angles to each other, and call them OX, OY, OZ. Reproduce these lines in Fig. 7, which (for simplicity) is divested of the rings, projectile, etc., of Fig. 6. The original rotation round the axis $Y Y_1$ (Fig. 7) may be represented by a couple P.OA; with a certain length of arm OA: the couple caused by the weight, or by the resistance of the air in the case of a real projectile in flight, may be shown by P.BO, the force P being taken equal in both couples, but the arms different in length, since the couples are not equal to each other. Mark off on OY a distance $Ob = OA$, on OX make $Oa = OB$, and complete the parallelogram $ab$ with its diagonal Oc. The forces P and P at O neutralize each other as they are equal, and act along the same line in opposite directions, and there remain P at B, and P at A, constituting only one couple, causing rotation about a fresh axis Oc, which is the diagonal of the parallelogram $ab$, and at right angles to AB. This indicates that the axis of revolution slowly moves from the direction $Y_1 Y$ to Oc; and this turning movement will continue as long as both couples exist.

"The very slow continuous change of the inclination of the earth's axis, called the precession of the equinoxes, is a movement of a similar nature, which can also be illustrated by means of the gyroscope."

The actual amount of drift at ranges up to 1,100 yards with the ·303 Mark VII is known by actual trial on a large and careful scale to be less than one foot or one minute of angle. This is easily proved by setting the sights to the correct elevation and deflection to make central bull's-eyes at 1,000 yards in a flat calm. Aim is taken at a small bull on a large card at 25 yards. The shot holes are then found about 14 inches above the bull and less than ¼ inch to the right of the plumb-line through the bull.

At ranges up to about one mile, as pointed out by Metford, an inclination of the back sight to the right of about two degrees for a left-hand spiral corrects very well for the effect of drift. At ranges greater than about a mile the drift is more considerable, and may amount to as much as 100 yards at extreme ranges. A good working rule, then, is to multiply the tangent elevation in degrees by four and allow that number of minutes right deflection to correct for drift.

"Yaw" is a phenomenon observed more or less with all guns at long ranges. Firing over large stretches of sea-sand, the bullets make holes or marks in the sands which clearly indicate that they are not flying with their spinning axis parallel to the tangent of the trajectory. Divergences of as much as 45 degrees are by no means uncommon. The actual efficient cause of yaw is not completely known, and is not simple. It is certainly gyroscopic in nature, and it increases distinctly with the rate of bending of the trajectory measured in degrees turn per second. In the first few hundred yards this rate is about one degree per second, but a little after the vertex of the 10-degree trajectory the rate has increased to seven degrees per second. At the same time, the forward velocity of the shot has fallen to about a quarter of its original value, and the rate of spin of the bullet has fallen very much less. This results in the bullet having a very great amount of over-spin just at the time when the bullet requires to turn its nose down quickly so as to keep parallel with the trajectory.

The rate of diminution of spin is not known. The Mark VII starts with the prodigious spin of one-sixth of a million revolutions per minute, or 2,930 revolutions per second, and perhaps loses its spin at the rate of 5 or 10 per cent. per second. If it lost its spin at the rate of 10 per cent. per second, it would still be spinning at 1,380 r.p.s. after seven seconds, and at 2,000 r.p.s. if the loss was only 8 per cent. per second. If the loss was as low as $3\frac{1}{2}$ per cent. per second it would be spinning at 10 r.p.s. after 1,000 seconds, or rather over a quarter of an hour. The effect of yaw is to increase the air resistance enormously and to shorten the range.

The rotation of the earth during the time of flight of a bullet has a real effect on both its range and its direction, although these are both extremely small. In very powerful naval guns at very long ranges a correction may properly be made for its effect. Metford pointed out that a bullet in its flight behaves like Foucault's pendulum, so that if at the instant of firing the plane of the trajectory passes through a particular star near the horizon, it continues to pass through that star throughout its flight. The star moves 360 degrees in 24 hours, or a quarter of a minute of angle in one second of time. In a vacuum the effect may be considerable, but in actual air it may be neglected in comparison with other causes of error. Cranz works out numerically an example of a 12-inch gun with velocity 2,500 f/s fired due north in latitude 54 N. and at an angle of elevation of 40 degrees, giving a range of 37,000 yards. The effect is to shorten the range by 10 yards with 160 yards right deviation.

---

## CHAPTER III

### INTERIOR BALLISTICS (NUMERICAL)

The problem of designing a military cartridge is very much simpler than the corresponding problem for ordnance. In the latter case a complete and accurate knowledge of the behaviour of all propellants under all conditions is required because the uses to which ordnance is put are so multifarious. In the former case the conditions are so strictly limited that there is but little scope for radical changes from existing patterns.

In a cartridge for military rifles as opposed to sporting rifles the limits of size and weight are all-important, so that the problem is confined to a design of maximum efficiency. This at once settles one point, that the cartridge case must be as full of powder as possible, otherwise weight and space are wasted in carrying a cartridge case larger and heavier than is absolutely necessary.

The other chief limits may be summarized as follows :—

A.—(1) The size of a man's hand and the length of its effective movement.
    (2) The weight a man can carry.
    (3) The recoil a man can stand comfortably.

B.—(1) The strength of steel.
    (2) The power of steel to resist erosion.

C.—The temperature of explosion of possible powders.

D.—Effect of—
    (1) Air resistance on trajectory.
    (2) Length of bullet on air resistance.
    (3) Specific gravity of bullet material on length of bullet.

E.—Rounds per minute required for machine guns.

These general limits give rise to technical limits as follows :—

(a) The overall length of the complete cartridge is limited by $A_1$ and also by E, because the longer the cartridge the greater is the stroke of the reciprocating part of a machine gun and the fewer the strokes or rounds per minute.

(b) The weight of the cartridge and the gun is governed by $A_2$ and partly by E.

(c) The recoil is limited by $A_2$ and $A_3$, and is proportional to the weight of bullet and muzzle velocity, though its effect is diluted by the weight of the gun.

(d) The calibre is governed by the weight of the bullet and the air resistance $D_1$.

(e) The maximum pressure is governed by $B_2$ and C. It is not really dependent on $B_1$, as in practice steel is amply strong.

(f) The density of loading or ratio between weight of charge and weight of water, when either fills the case, is $0 \cdot 85$. All powders weigh in the solid about 400 grains per cubic inch. Water weighs 253 grains, but owing to the granulation of the powder 215 grains of powder bulk as large as 253 grains of water, giving a density of $0 \cdot 85$, varying slightly according to the granulation. Thus all the space available in the case is filled.

(g) The barrel length is about 2 feet, from $A_2$.

(h) The bullet length must be 4 or 5 calibres, from $D_1$ and $D_2$.

    Air resistance rapidly reduces bullet energy, and is seriously increased by a short bullet. (a) tends to curtail its length.

(i) The bullet's weight is affected by (c), (d), and $D_1$. In practice $D_3$ is of little consequence. Lead is rather too heavy for $D_1$ since a long point is difficult to construct without making the bullet too heavy, and steel is rather light for (a) and $D_2$.

The numerical calculations which require notice may be divided under six headings :—

1.—The principle of mechanical similitude, which is the most valuable stand-by for rifles.

2.—The practical method used by Housman at Kynoch's following Robins and Hutton.

3.—The adaptation by Lieut.-Colonel R. K. Hezlet, C.B.E., D.S.O., to rifles, of an ordnance method of calculation, specially for this chapter.

4.—A consideration of the energy contained in powder and the amount transmitted to the bullet.

5.—The articles of Mr. F. W. Jones published in *Arms and Explosives* during the last twenty years.

6.—The method recommended for general purposes by Hezlet in the new volume of *Encyclopædia Britannica*, 1922, called the monomial method.

## Section 1

### The Principle of Mechanical Similitude

(a) The general principle.

(b) The special variation of it for rifles due to Housman.

(a) An exact model to a reduced scale of any machine is the simplest illustration of mechanical similitude—say, a scale of 1 inch to a foot. All lengths on the original are reduced to one-twelfth; all areas are reduced to 1/144th, or one-twelfth squared, and all volumes or weights to 1/1,728th, or one-twelfth cubed. A rifle considered as a machine has the following parts:—

| | | | |
|---|---|---|---|
| Weight of powder charge .. .. .. .. .. | a weight | $w$ | lb. |
| Weight of bullet .. .. .. .. .. .. | a weight | $W$ | lb. |
| Capacity of cartridge case behind the bullet .. .. | a volume | $c$ | cubic inches. |
| Thickness of powder granulation .. .. .. | a length | $L$ | inches. |
| Length of shot travel to muzzle .. .. .. .. | a length | $u$ | inches. |
| Nominal diameter of bore or calibre .. .. .. | a length | $d$ | inches. |
| Weight of complete rifle .. .. .. .. .. | a weight | $W_1$ | lb. |

If a loaded rifle were to be sawn lengthwise exactly down the middle and a photograph taken of the section with a foot-rule lying beside the section and included in the picture, it would be an easy task to ascertain the calibre and all the other details. But if the foot-rule was not there, it would still be possible to ascertain the details, but in terms of the calibre and not in actual figures. The calibre could be assumed to be 1 inch, and from that all capacities, areas and lengths could be worked out, and all weights also if the specific gravity of the materials were known—that is, the weight per cubic inch of wood, metal, etc., employed. As soon as the actual calibre in inches was given, these weights, capacities, lengths, etc., could be given in actual ounces, cubic inches, etc.

If the actual calibre is $d$ inches, the figure worked out for an assumed 1-inch gun would have to be multiplied by $d^3$ for weights and capacities, and by $d$ for lengths. The indicator diagram of pressures and the velocity curve from breech to muzzle drawn on the picture would be unchanged, but the time and energy scales would vary with the calibre as in Fig. 1, although their shapes would remain the same. The various parts when given for the unit rifle of 1-inch calibre would be called

$$\frac{w}{d^3}, \frac{W}{d^3}, \frac{c}{d^3}, \frac{L}{d}, \frac{u}{d}, \frac{d}{d}=1, \frac{W_1}{d^3},$$

the time scale would be in terms of $t/d$, the energy scale in terms of $d^3$, and the pressure P, velocity V, and loading density $\Delta$ would remain unchanged. Similar rifles would be those whose various parts reduced to unit gun dimensions were the same. Then, provided that the friction, heat loss and forcement varied in proportion to the calibre, these guns would all have the same ballistics. This method is more suitable to ordnance than to rifles, as will now be explained.

(b) Housman's special problem at Kynoch's was to supply cordite cartridges for the various sporting rifles designed by gun-makers to suit their customers' requirements. For various reasons he only had one sort of cordite available for them all, known in the trade as rifle cordite, and in the Service as Mark I cordite, size $3\frac{3}{4}$, as made for ·303-inch Mark VI cartridge. All the rifle barrels used were 30 inches long or thereabouts. Mark VI cartridge contained about 60 strands of cordite. By making certain assumptions as to friction he could take one strand as his unit and, imagining the bullet to be made up also of 60 strands of lead stuck together, and similarly the capacity of the cartridge case and

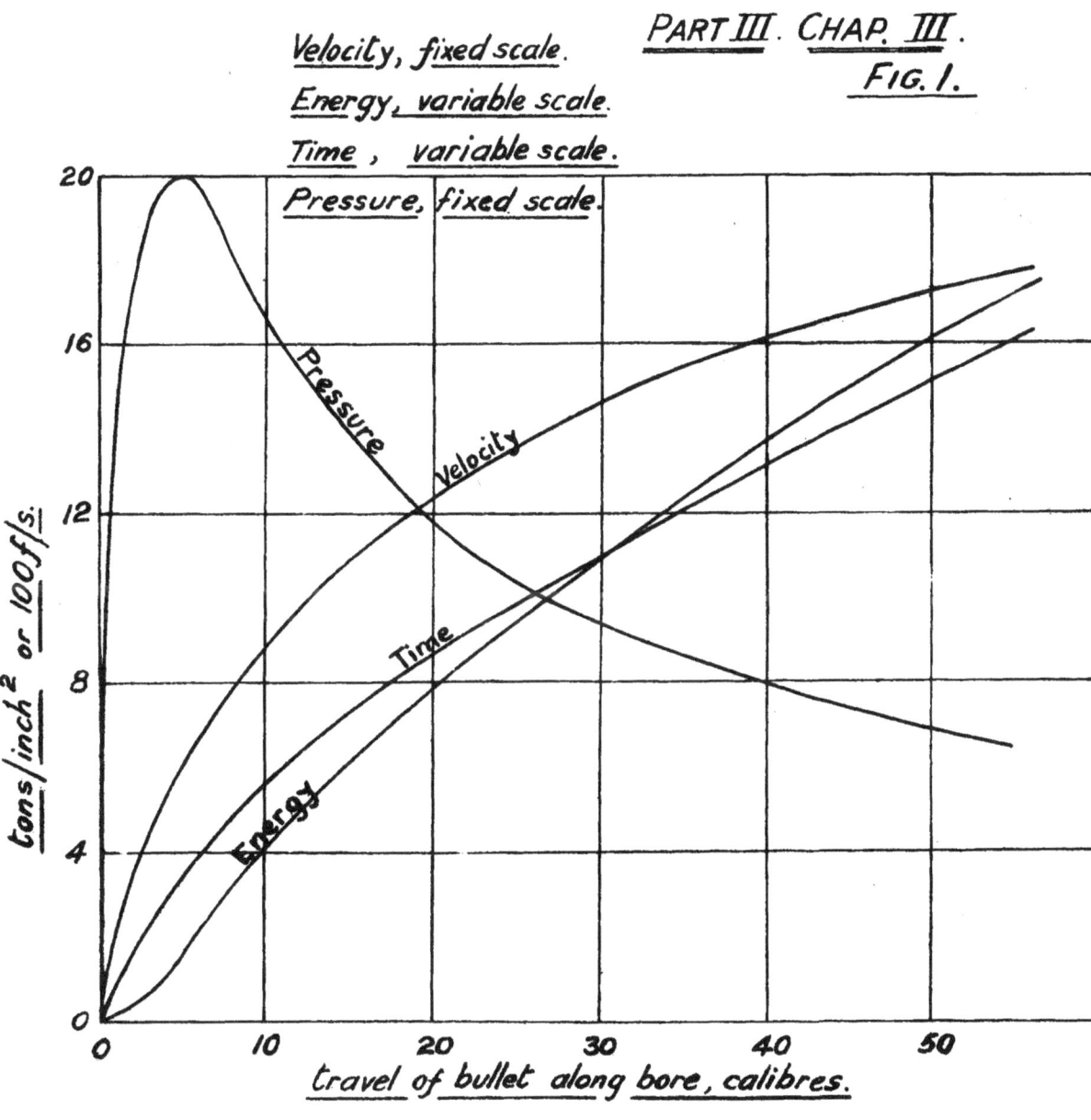

To face p. 294.

PART III. CHAP. III.
FIG. 1.

the section of the barrel, he could build up a cartridge for any calibre from ·256 to ·600 or higher to give exactly the same pressure and velocity as Mark VI by merely varying the number of strands of his elementary cartridge in the proportion of the area of his new bore to the area of the ·303. This area varies as $d^2$, so that if his new bore is ·606 he would require four times as many strands, etc. His barrel length remains the same and so does his loading density and his cordite thickness, together with his pressure, velocity, and time curves. The effect, then, of keeping to one powder and one approximate barrel length is to simplify calculation.

In addition, it so happens that sportsmen in general have decided that a bullet about $1\frac{1}{4}$ inches long is the best length whatever the calibre, making $W/d^2$ about ·3 if lead is used. The cause for this decision is not quite certain, but it is connected with the size of a man's hand, his ability to stand recoil, and also with air resistance.

The various parts of the cartridge and gun when expressed in terms of the unit rifle of 1-inch calibre are then

| $w/d^2$ | $W/d^2$ | $c/d^2$ | L | $u$ | $d$ | $W_1$ | P | V | $t$ |
|---------|---------|---------|---|-----|-----|-------|---|---|-----|
|         |         |         |   |     |     |       |   |   |     |

The units employed are :—

| | | |
|---|---|---|
| Weight | .. | pounds of 7,000 grains. |
| Capacity | .. | cubic inches. |
| Length | .. | inches. |
| Pressure | .. | tons per square inch. |
| Velocity | .. | feet per second. |
| Time | .. | seconds divided by 10,000 (t.ts). |
| Recoil (momentum) | .. | pounds weight multiplied by velocity, f/s, *i.e.*, foot-poundals. |
| Energy | .. | foot-pounds. |

The mechanical similitude used in this chapter is Housman's, and not the full similitude employed for ordnance or ordinary working models of machines.

In the sixth and final heading of this chapter it will be further simplified by using the constant density of 0·85, which defines the weight of powder charge as soon as the chamber is given. Also only one nature of powder at a time will be considered and one barrel length of 2 feet. This reduces the variables to three only, viz., $W/d^2$, $c/d^2$ and L, so that if the greatest pressure is also defined there are only two items to be evaluated, the muzzle velocity and the recoil. In this way the design of a military cartridge is reduced to the barest skeleton and vastly simplified.

## Section 2

### The Practical Method used by Housman

In 1903 Housman published in the *Kynoch Journal* the results of a very long series of experiments he had made using a 0·450 express rifle with 30-inch barrel having a chamber capacity of 0·415 c. in. or $c/d^2 = 2·05$. The powder was all from one batch of rifle cordite. The charges varied from 40 to 100 grains by 10 grains at a time, and each charge was shot with bullets of four different weights.

| Bullet weight (grs.) | 365 | 480 | 640 | 800 |
|---|---|---|---|---|
| $W/d^2$ | 0·258 | 0·339 | 0·452 | 0·565 |

Pressures and velocities were taken for all rounds, and the complete results were published in the form of curves on squared paper, one set of curves for pressure and one for velocity. The two sets have been assembled on to one sheet, and are given as Fig. 2. The usual load for this rifle is a 480-grain bullet and 70 grains of cordite, giving about 2,050 f/s and 15 tons pressure according to Fig. 2.

The following points should be noticed :—

(a) For each bullet the velocity varies almost exactly with the charge as the velocity curves are nearly straight lines.

(b) With very small charges the light bullets do not give as much velocity as the heavier ones.

(c) The curvature of the pressure lines is most pronounced, and is approximately given by the formula $\log P = $ a constant $+ 2 \cdot 5 \log w$ or $P = k\, w^{2 \cdot 5}$, indicating that 1 per cent. increase in the charge causes $2 \cdot 5$ per cent. increase in the pressure.

This very large experiment was analysed thoroughly during the war by the Inspection Department staff at Woolwich to ascertain the law connecting $P, V$, $w/d^2$ and $W/d^2$ with all else constant.

One result is given on Fig. 3, which was obtained by merely rearranging the curves of Fig. 2 and drawing the logical conclusions by adding the four curves on the network. This diagram purports to give the complete ballistics of the whole family of 30-inch rifles which have a value of $c/d^2 = 2 \cdot 05$ for their chamber capacity and use Mark I cordite, size $3\frac{3}{4}$. Provided that Housman's pressures are correct, and that change of calibre does not seriously affect the losses due to friction, etc., this diagram does what it purports to do. The horizontal scale is an even scale of $W/d^2$ from 0 to $0 \cdot 5$, the vertical scale is in logarithms, and shows pressure from 4 to 40 tons. It will be noticed that $W/d^2$ for all practical modern bullets lies between $0 \cdot 2$ and $0 \cdot 4$. The service $\cdot 303$ rifle also happens to have $c/d^2 = 2 \cdot 05$, so that this chart gives all possible information for it with this cordite. For instance, Mark VI has a velocity of 2,050 f/s. If the Mark VI line is followed up vertically till it comes a quarter way between 2,000 and 2,200, the pressure 14 tons is read off.

Curves of equal powder charge can be easily plotted from Fig. 1. The one drawn gives the apparent anomaly that a bullet of $W/d^2 = 0 \cdot 26$ gives a higher velocity than any other bullet heavier or lighter, and this is an experimental fact not peculiar to this chamber. The energy per grain of powder varies very rapidly with the pressure and bullet weight. For ordinary values of $W/d^2$ the total muzzle energy of a bullet depends almost entirely on the maximum chamber pressure. If the pressure is kept constant by varying the powder charge, the muzzle energy of the bullet used will be nearly constant whatever its weight.

The dotted curve of "all-burnt" (Fig. 3) refers to the completion of combustion of the powder. Any load that gives ballistics which are to the right and above this curve burns all the powder before the muzzle is reached. Loads which give results below and to the left of the curve blow out some of their powder unconsumed from the muzzle. The formula for "all-burnt" is always of the shape $V(W/d^2) =$ some number. The actual number is found by experiment, and varies for each nature of powder. For Mark VII $\cdot 303$ cordite, known as M.D.T. size 5-2, the figure is about 500, and for Mark VII $\cdot 303$ Z powder, Dupont No. 16, it is about 600. As $W/d^2$ for Mark VII is $\cdot 271$, the Z charge is all burnt when the velocity is about 2,200 f/s. and the cordite charge at 1,850 f/s., or less than half-way along the barrel.

Housman published his figures in 1903, and revised the pressures in the next issue owing to an error he had detected in his pressure gauge. He explained that all gauges were subject to two errors, one causing the gauge to over-record and the other to under-record. He promised to publish the reason and the remedy, but, although he lived for another two years, he failed to do so. The reason is now well known, but the remedy is not. In consequence, the shooting out of diagrams like Fig. 2, although easy and rapid, is not altogether a satisfactory method of solving the problem of rifle ballistics.

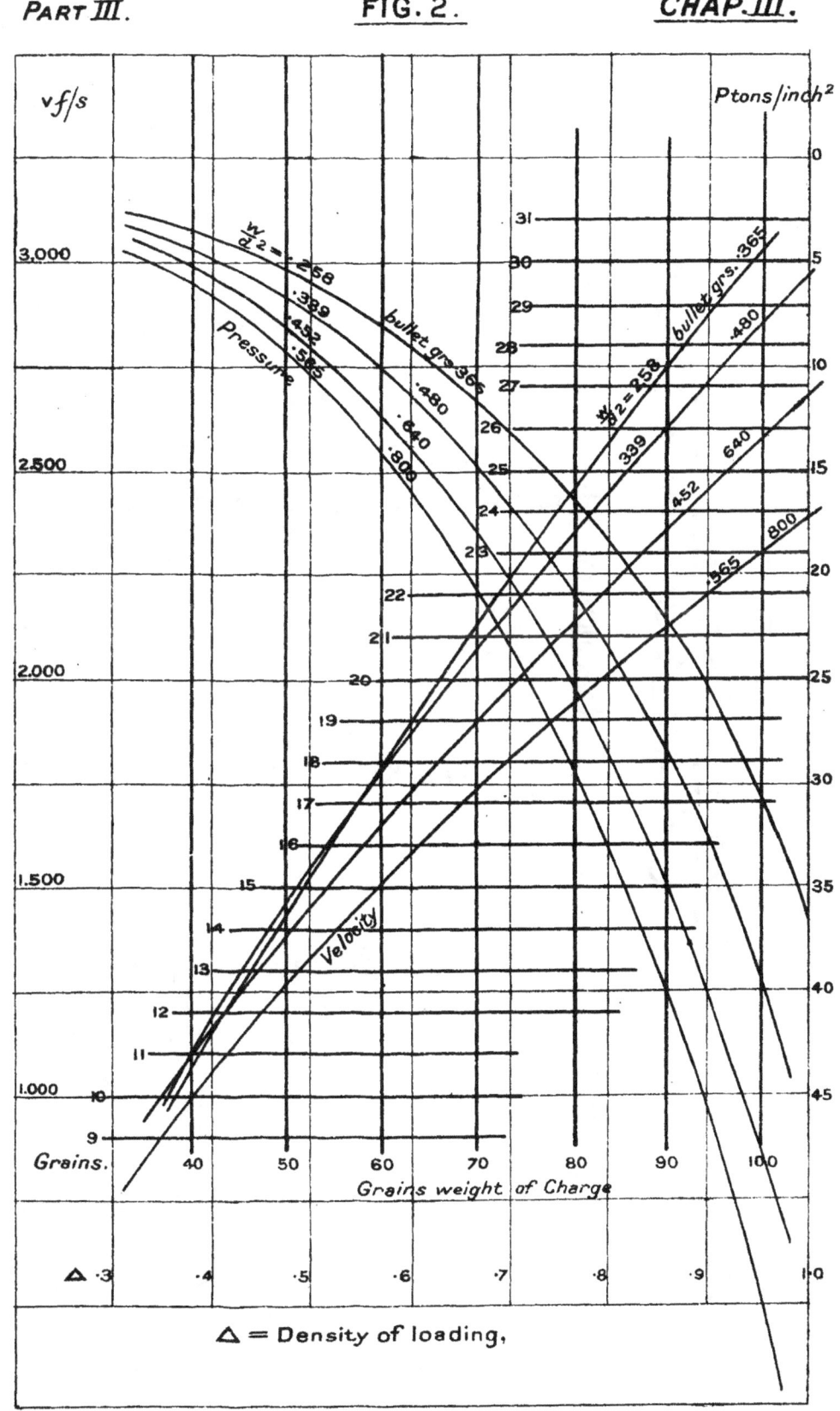

PART III.   CHAP. III. FIG. 3.

"HOUSMAN'S Experiment in 1903."
(published figures re-arranged)

Horizontal Scale is W/d² Sectional Density of bullet.
Slanting straight lines are Velocity.

To face p. 296.

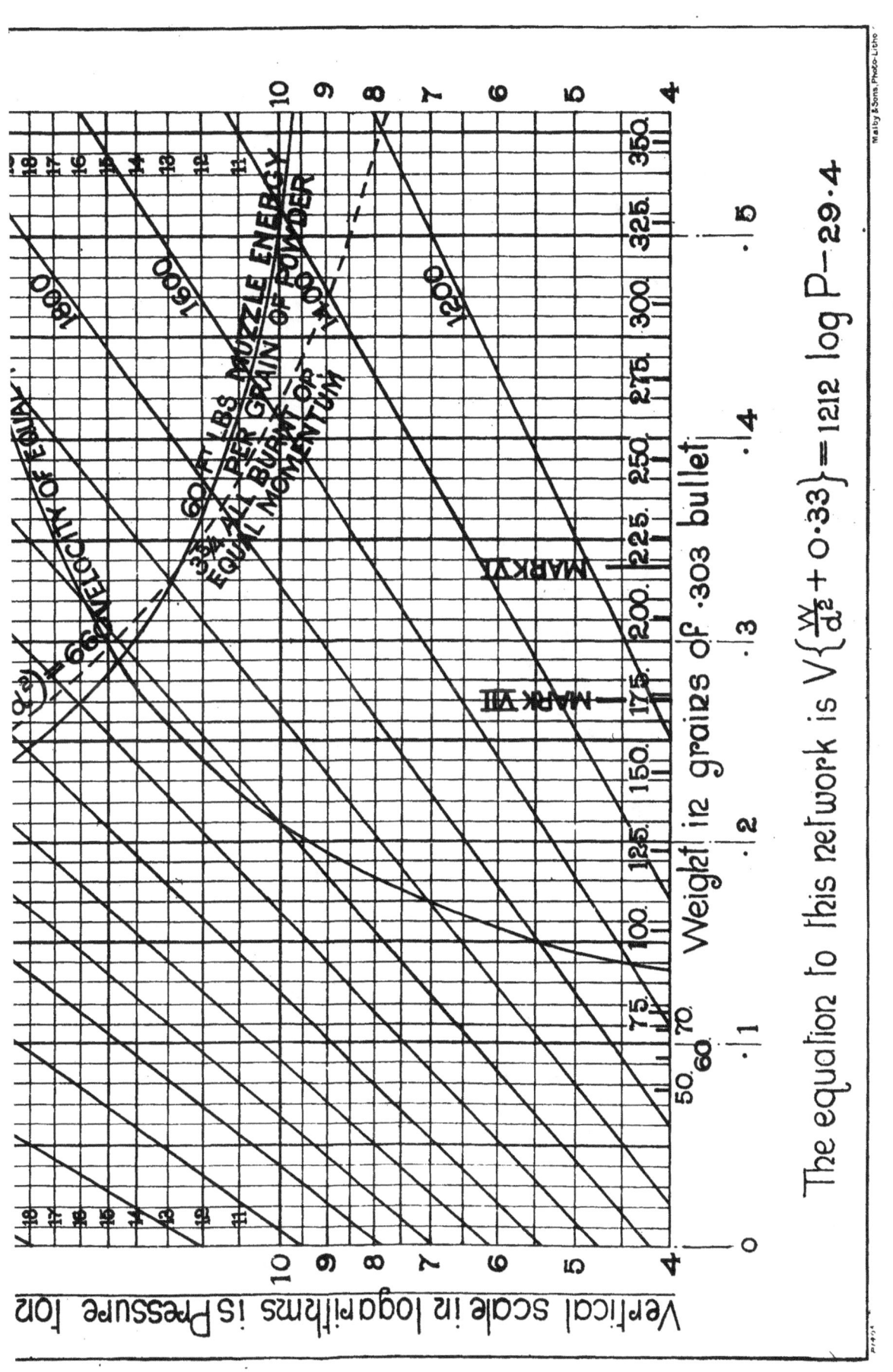

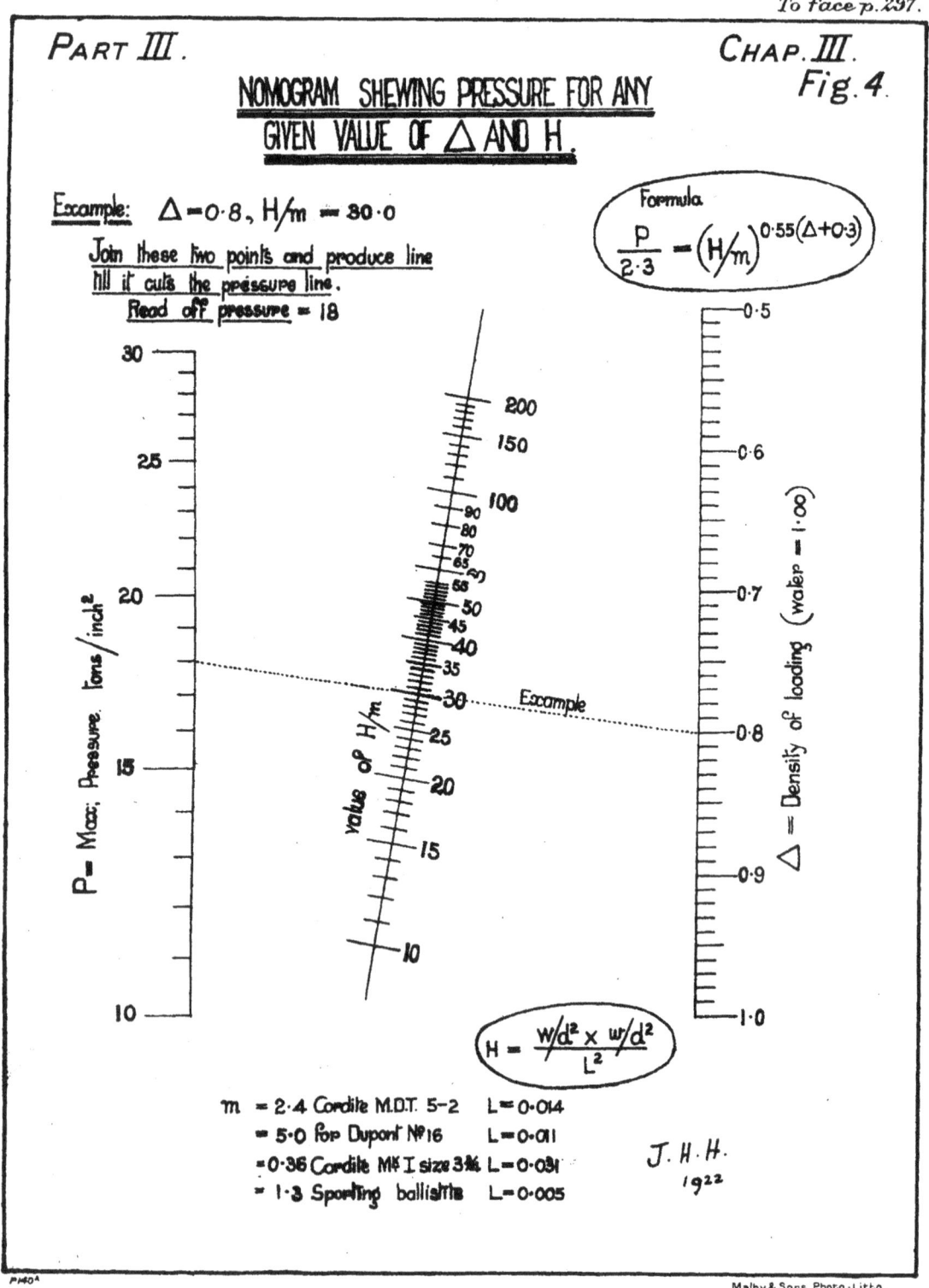

Fig. 4. Nomogram shewing pressure for any given value of Δ and H.

## Section 3

### Rational or Formal Methods of Calculation following Charbonnier and Hezlet's Methods

A complete scheme for calculating velocities and pressures in a gun is based on results obtained by experiment in closed vessels, chemical calculation, and mathematical analysis. For ordnance the present position of the problem is given in the *Revue d'Artillerie*, April and May, 1920, by Captain F. Desmazières. For rifles the position is still indeterminate, owing to the great difference in the proportion of available energy lost in heating, forcement and friction, which have not yet been closely analysed. The basis of the measurements is, however, the same, and it is due to Charbonnier. Taking the actual elements of loading, $w$, $W$, $c$, $L$ and $d$, he combines them into two expressions:—

(1) The density of loading or the ratio of the charge weight and chamber capacity, $w/c$, called hereafter $\Delta$.

(2) The parameter of loading or $\left(\dfrac{W/d^2}{L} \times \dfrac{w/d^2}{L}\right)$, which is called hereafter $H$.

Using these two expressions only, his work, so far as maximum pressure is concerned, has been reduced to a formula of the utmost simplicity:

$$\log P - 0\cdot 120 = 0\cdot 883 \{(\Delta + 0\cdot 3) \log H/k\},$$

where $k$ is an experimental factor to be found by one firing of the particular chemical powder under consideration.

This formula is not to be found in his writings, but it has been found by plotting his results, and it gives them with considerable precision. It is of a sufficiently flexible form to suit the requirements of a physical problem containing three experimental constants, as it is

(1) $$P/\alpha = \{H/m\}^{\beta(\Delta+\gamma)},$$

where $\alpha$, $\beta$, $\gamma$, and $m$ can be adjusted by trial and error. Tested for rifles, it gives results which do not fulfil the known results, but it is easily translated into the form

(2) $$P/2\cdot 3 = \{(H/m)\}^{0\cdot 55\,(\Delta+0\cdot 3)},$$

where $m = 2\cdot 4$ for M.D.T., size 5–2 $(L = 0\cdot 014)$
$\phantom{\text{where }m} = 5$ for Dupont Z $\phantom{\text{size 5-2 }}(L = 0\cdot 011)$
$\phantom{\text{where }m} = 0\cdot 36$ for cordite $3\frac{3}{4}$ $\phantom{\text{size}}(L = 0\cdot 031)$
$\phantom{\text{where }m} = 1\cdot 3$ for sporting ballistite $(L = 0\cdot 005)$.

It has been reduced to a nomogram (Fig. 4), and can be relied upon to give good average results and to avoid dangerous pressures.

Thus, for ·303 Mark VII cartridge with $37\frac{1}{2}$ grains M.D.T. and 174 grains bullet

$$W/d^2 = 0\cdot 271, \quad w/d^2 = 0\cdot 0585, \quad \text{and} \quad L = 0\cdot 014.$$
$$H = 81 \quad \text{and} \quad H/m = 33\cdot 5.$$

As the case holds 48 grains of water, $\Delta = 0\cdot 781$.
Whence $P = 18\cdot 5$.

For the velocity of rifles calculated from first principles it has been by no means easy to find a solution. Assistance was sought from Lieut.-Col. R. K. Hezlet, C.B.E., D.S.O.,

R.A., and on the basis of Sir J. Henderson's treatment of the interior ballistic problem (given in *Proc. R.S.*, series A, vol. 100, page 461) he gave a modified form of equation suitable for rifles by the introduction of an experimental constant.

His method determines the energy per grain to be found in the bullet, and may be summarised as follows:—

(1) From chemical considerations the indicated energy per grain can be written in terms of the temperature, as

$$\mathrm{I}\,.\,\mathrm{E/gr} = a + bt + ct^2.$$

(2) The pressure multiplied by the effective volume per lb. $= \mathrm{R}\,t$, where R depends upon the percentage volumes of the various sorts of gas resulting from the explosion.

(3) From the law of Hugoniot and Sebert (1882) that the rate of burning is proportional to the pressure, and, therefore, that the velocity of the shot before the completion of burning of the powder is proportional to the thickness of the layer of powder which has been burnt, it follows directly that at the moment of "all burnt" the indicated energy per grain varies directly as Charbonnier's H.

From these three fundamental theories Hezlet shows that the following equation is true for tubular cordite when the charge is completely consumed in the rifle,

$$\log\left(\frac{\text{air space to muzzle}}{\text{air space in chamber}}\right) = \left[\frac{b}{\mathrm{R}}\log t_1 + \frac{2\cdot 303\,\mathrm{R}}{2c}t_1\right] - \left[\frac{b}{\mathrm{R}}\log t_m + \frac{2c}{2\cdot 303\,\mathrm{R}}t_m\right],$$

where $t_1$ is the temperature at "all burnt," and $t_m$ is the temperature at the muzzle.

Now R, $b$, $c$ are constants; and $t_1$ is a function of the indicated energy per grain, and therefore, of his symbol $\mathrm{B}^1$.

So that the expression enclosed in the first square bracket is a function of $\mathrm{B}^1$, or $\phi\,(\mathrm{B})^1$, and the expression in the second square bracket is a function of the indicated energy per grain, or $\psi\,(h)$, and, therefore, of $\mathrm{B}^1$. He works out tables of $\phi\,(\mathrm{B})^1$ and $\psi\,(h)$, and over the region in which these tables are useful for rifles—

$$\phi\,(\mathrm{B})^1 = 7\cdot 222 + 0\cdot 0034\,\mathrm{B}^1$$

$$\psi\,(h) = 7\cdot 392 - 0\cdot 011\,\mathrm{I}\,.\,\mathrm{E/gr}.$$

Turning now to the purely numerical aspect of the subject, and calling air space to muzzle $v$ and air space in chamber $v_0$, we have

$$\log v/v_0 = \phi\,(\mathrm{B})^1 - \psi\,(h).$$

Here $v_0$ is the volume of the chamber less the volume of the solid explosive, or

$v_0 = c - w/398$, since there are 398 grains of cordite in 1 cubic inch of solid cordite.

$v = v_0 +$ volume of the rifled bore, including the grooves;

$= v_0 + 0\cdot 82\,d^2u$, where $0\cdot 82$ is used, instead of $0\cdot 785$ to allow for the grooves.

$\mathrm{B}^1 = \mathrm{B}/k$, where $k$ is an experimental factor;

$\mathrm{B} = 15700/\mathrm{H}$, where H is Charbonnier's parameter;

and $\psi\,(h)$ is already given as $7\cdot 392 - 0\cdot 011\,\mathrm{I.E/gr}$.

This I.E/gr. is much larger than the effective E/gr., as can be seen by looking at the indicator diagrams already given, and Hezlet calls it K times as great.

The formula then becomes

$$7\cdot 391 - 0\cdot 011 \text{ I.E/gr.} = 7\cdot 222 + 0\cdot 0034 \text{ B}/k - \log v/v_0;$$

or

$$\text{effective E/gr} = (\text{I.E/gr.}) \text{ K} = 15\cdot 46 - 0\cdot 309 \text{ B}/k + 91 \log v/v_0;$$

where K and $k$ are two experimental factors, to be found from the firing results for velocity of at least two different loads.

Hezlet proposed to find the value of $k$ from his pressure formula, leaving K to be found by firing for velocity, and this would be sound if we could rely on the pressures. But owing to the difficulty of obtaining rifle pressures with sufficient absolute accuracy he agrees to find K by a second firing for velocity with a different load.

For M.D.T. cordite as the result of many more than two firings

$$k = 1\cdot 30;$$
$$\text{K} = 1\cdot 39.$$

It has been noticed lately that better results are obtained by applying the K factor to the expression $\log v/v_0$. There is no justification for the change beyond the ascertained fact that it gives better results. The final formula which results, and which can be recommended for use, is

$$\text{Effective E/gr.} = 15\cdot 6 + 65 \log v/v_0 - 3110/\text{H};$$

and since

$$\text{E.E.gr.} = \frac{\text{WV}^2}{w,\ 2g,\ 2240},$$

$$\text{V} = 671 \sqrt{\frac{w}{\text{W}} \text{E.E/gr}},$$

which is the velocity formula for use in this system.

A numerical example for Mark VII $\cdot$303-inch will show the simplicity of the working. From the figures given for the pressure formula

$$\text{H} = 81\cdot 0.$$

The chamber is $0\cdot 190$ cubic inches.

$v = 0\cdot 190 - 37\cdot 5/398 = 0\cdot 096;$

$u = 23\cdot 5$ inches; $d = 0\cdot 303$ inch;

$0\cdot 82\ d^2\ u = 1\cdot 77$ cubic inch, $v = 1\cdot 86;$

$v/v_0 = 1\cdot 86/0\cdot 096 = 19\cdot 3$, so that the original air space undergoes $19\cdot 3$ expansions to the muzzle;

$\log v/v_0 = \log 19\cdot 3 = 1\cdot 286;$

$1\cdot 286 \times 65 = 83\cdot 6,\ 3110/\text{H} = 38\cdot 4;$

$\text{E.E/gr.} = 15\cdot 6 + 83\cdot 6 - 38\cdot 4 = 60\cdot 8;$

$$\text{V} = 671 \sqrt{\frac{37\cdot 5}{174} \times 60\cdot 8} = 671 \times 3\cdot 62 = 2427.$$

For Dupont Z powder the formula is

$$\text{E.E/gr.} = 15\cdot 6 + 65 \log v/v_0 - 7200/\text{H},$$

since the proportion of the gases produced is very similar. The difference in the figures 7,200 and 3,110 is due to the different rates of propagation of the flames through the powder.

## Section 4

### The Energy contained in the Powder and the Amount carried away by the Bullet

Black gunpowder was discovered by Roger Bacon. It was not invented. Modern smokeless powders are invented by the chemist to comply with certain conditions. The important numerical details which he has to consider are as below :—

(1) Total volume of gas measured at normal temperature and pressure (0° Cent. and 760 mm.) which is produced by a given weight of explosive. This is measured in cubic centimetres per gramme weight.

(2) The total units of heat (calories) produced by the combustion of a gramme of explosive. This amount happens to be almost precisely five times the foot-lbs. of energy per grain weight, and as the latter it will be given.

(3) The percentage volumes of the several sorts of gases produced.

(4) The temperature of combustion in degrees C.

(5) The "force" of the explosive as defined below.

The following table gives the figures for 1, 2, 4 and 5 for several powders, together with the percentage of nitroglycerine.

| Powder | 1 | 2 | 4 | 5 | N.G. |
|---|---|---|---|---|---|
| Gunpowder | 280 | 141 | 2,230 | 15 | 0 |
| Mk. I Cordite | 880 | 230 | 3,100 | 75 | 58 |
| M.D. Cordite | 930 | 195 | 2,700 | 70 | 30 |
| Dupont Z | 960 | 155 | 2,470 | 66 | 0 |
| Norwegian ballistite | 890 | 190 | 2,600 | 66 | 40 |
| Guncotton | 875 | 196 | 2,500 | 64 ? | 0 |
| M.D. Cordite without mineral jelly | 820 | 230 | 3,000 | 70 ? | 30 |

It will be noticed that the temperature of explosion does not depend upon the amount of nitroglycerine in the powder, but is associated with a high value of energy (column 2) and a low value of total gas (column 1). The effect of omitting the vaseline or mineral jelly from M.D. cordite shows the toning down power of suitable ingredients. A high temperature of combustion is one of the chief causes of erosion of the bore of rifles, and unfortunately it is always associated with the most desirable quality of a high energy per grain. Weight for weight M.D. cordite develops nearly 25 per cent. more energy than Dupont Z, and in a cartridge case of a limited size this is important.

The "force" of the powder is defined from the law of Noble and Abel that the maximum pressure in a closed vessel P tons/inch² is given by the formula :—

$$P = \frac{\text{force}}{\frac{1}{\Delta} - \frac{1}{1.58}} = \frac{f}{\frac{1}{\Delta} - 0.633},$$

where $\Delta$ is the density of loading and 1·58 is the specific gravity of modern powders. If $\Delta = 0.612$ or 612 grains of powder are put into a vessel capable of holding 1,000 grains of water, it is obvious that the denominator becomes 1·000 and that the "force" is the pressure of the powder when fired at a density of 0·612. From column (5) the pressure for any density can be evaluated.

The force is also defined chemically as VT/41,840, where V is the figure in column 1 and T is the absolute temperature or 273° greater than column (4).

By actually exploding powder at a known density of loading in a closed vessel and observing its maximum pressure, the force can be obtained by calculation from Noble and Abel's law. With a knowledge of the total volume of gas so produced, the temperature of explosion can be evaluated from VT/41,840.

T can also be evaluated from the percentage volumes of the various gases formed, as found by the chemist when examining the gases, in combination with the figures given in column (2).

The total energy per grain is obtained by immersing the closed vessel in a water bath before firing and leaving it there till cool. The rise of temperature of the water bath gives the heat evolved by firing.

By using all of these methods a cross-check is effected on the results of experiment and of calculation, which cross-check is of great value when dealing with such a complex.

In the end the chemist can produce a powder having a definite temperature of explosion, a definite energy content per grain, and a definite force and within limits can modify his powders to suit practical requirements.

As the meaning of these high temperatures is not very evident from figures alone, the following list of common temperatures is given:—

| | |
|---|---|
| Low red heat | 500°—600° C. |
| Copper melts at | 1083° C. |
| White heat | 1500° C. |
| Platinum melts at | 1755° C. |
| ,, boils at | 2400° C. |
| Electric arc | about 4000° C. |
| The temperature of the sun is estimated at | 5555° C. |

A detailed analysis of the distribution of the energy of the cordite charge in the ·303 Mark VII will now be given, condensed from an article by Major Hardcastle in the *R. A. Journal*, October, 1918.

The charge is $37\frac{1}{2}$ grains of M.D. cordite (tubular). Each grain contains 195 ft. lbs. of energy, so that the total energy in the charge is 7,320 ft. lbs. The bullet weighs 174 grains, and its M.V. is 2,440 f/s so that the bullet has a muzzle energy of 2,300 ft. lbs., leaving 5,020 ft. lbs. to be accounted for.

The following facts have been observed:—

(1) Allowing a small amount for loss of heat in radiation, it takes 600 rounds to bring the water in a Maxim jacket to boiling point. The capacity of the Maxim jacket is about 7 pints, or 8·8 lbs. of water. The weight of the brass and iron in the barrel and jacket is about 20 lbs., counting as 2 lbs. of water for an equal rise in temperature. So that about 11 lbs. of water have to be raised from 60° F. to 212° by 600 rounds. It requires 777 ft. lbs. of energy to raise 1 lb. of water 1° F., so that each round puts 2,140 ft. lbs. of energy into the water jacket.

(2) The momentum of recoil of a Mark VII cartridge is about 80 (lbs. × ft. secs). The Maxim barrel and recoiling parts weigh about 9 lbs., so that the velocity of recoil is about 9 f/s, giving an energy of recoil of about 11 ft. lbs.

(3) The energy of the powder gas as the bullet leaves the muzzle is about that due to half the charge and the whole M.V., say 254 ft. lbs.

(4) A Mark VII bullet with a small hole bored near the base in the envelope sometimes sprays out molten lead as it flies. From this fact it has been calculated that the heating of the bullet absorbs 260 ft. lbs., the 40 grains of the cupro-nickel envelope being heated to 550° F.

(5) As this 260 ft. lbs. is produced in the bullet by friction, it is only right to suppose that the barrel is heated by friction by a like quantity, viz., 260 ft. lbs.

(6) The energy of rotation of the bullet accounts for 35 ft. lbs.

(7) The heat of the ejected cartridge case accounts for 5 ft. lbs.

(8) This leaves 2,315 ft. lbs. to be accounted for in the energy of the muzzle blast.

Indicator diagrams have been drawn in Fig. 5 showing these tentative results.

(a) The diagram of effective pressure clearly contains only the bullet energy 2,300 ft. lbs.

(b) The indicator diagram contains all that can be called "work," viz. :—(a) + (2) + (3) + (4) + (5) + (6) or 3,120 ft. lbs.

(c) The diagram based on Noble and Abel's law with force = 70 and max. pressure = 19·0 tons, shows the greatest pressure that can possibly exist. Its area must therefore be greater than the sum of (a) added to (1) to (6) excluding (5), which is already accounted for in (1).

The following table results :—

|  | Total ft./lbs. | Per cent. | Ft./lbs. per grain. |
|---|---|---|---|
| Effective | 2,300 | 31·4 | 61·3 |
| Difference | 820 | 11·2 | 21·9 |
| Indicated | 3,120 | 42·6 | 83·2 |
| Difference | 1,880 | 25·7 | 50·2 |
| Noble and Abel | 5,000 | 68·3 | 133·4 |
| Difference | 2,320 | 31·7 | 61·8 |
| Total | 7,320 | 100·0 | 195·2 |

This table shows that roughly one-third of the total energy of the charge is employed usefully in the bullet. One-third is blown out of the muzzle, and one-third is used in heating the gun and water jacket, etc.

Besides the facts already enumerated in making the above estimate, it may be mentioned that Housman in 1903 made an estimate for the Mark VI cartridge, and in 1908 Cranz and Rothe in Germany made an estimate for the German rifle (see *Arms and Explosives*, September, 1917). Both these estimates are very much to the same effect.

As a further check the area of the indicator curve of Noble and Abel's law has been summed. It comes to 5,032 ft. lbs. The energy of the blast has also been calculated by the chemical equations, using an actual muzzle pressure of 2·8 tons inch. This gives 36·1 per cent. of energy blown out in the blast, as against 37·1 per cent. by the above table.

In the drawing of Fig. 5 advantage has been taken of the known fact that 10 per cent. shortening of the barrel reduces the muzzle velocity by 1·4 per cent. nearly.

Using the mathematical formula

$$\text{Effective muzzle pressure} = \frac{0 \cdot 14 \times \text{muzzle energy} \times 24}{\text{volume of rifled portion (cubic inches)}},$$

the effective pressure curve can be drawn from the muzzle backwards towards the breech as far as the 1·4 per cent. law holds.

The velocity for any barrel length is known by cutting down a barrel inch by inch and actually measuring the velocity with the shortened barrel. With a very short barrel some error is caused by the effect of the muzzle blast in adding to the velocity after the bullet has left the muzzle and this happens at about half velocity or a 2-inch shot travel.

Below half velocity (*see* Fig. 6) all the curves are merely estimated, and are therefore

Pᵀ III. PRESSURE–SPACE DIAGRAM OF ·303 VII M.D.T.        CHAP. III.
Fig. 5.

NOTE: a straight line V, is caused by a parabolic P curve.
  "    "    E,  "   "   "  constant  P   "

Difference
2320 ft lbs
blown out

J.H.H. 1922

Velocity V
Energy E
time
Noble and Abel
(uncooled closed vessel)
pressure 3120 ft lbs
indicated crusher gauge
effective pressure 2300 ft lbs
3900 + 1100 = 5000 ft lbs
1100 ft lbs
half velocity travel

Shot start — inch, Shot travel — Muzzle

pressure in tons/inch²
time in ft·s or 1/10,000 second
velocity in 100 f/s
energy in 100 ft lbs

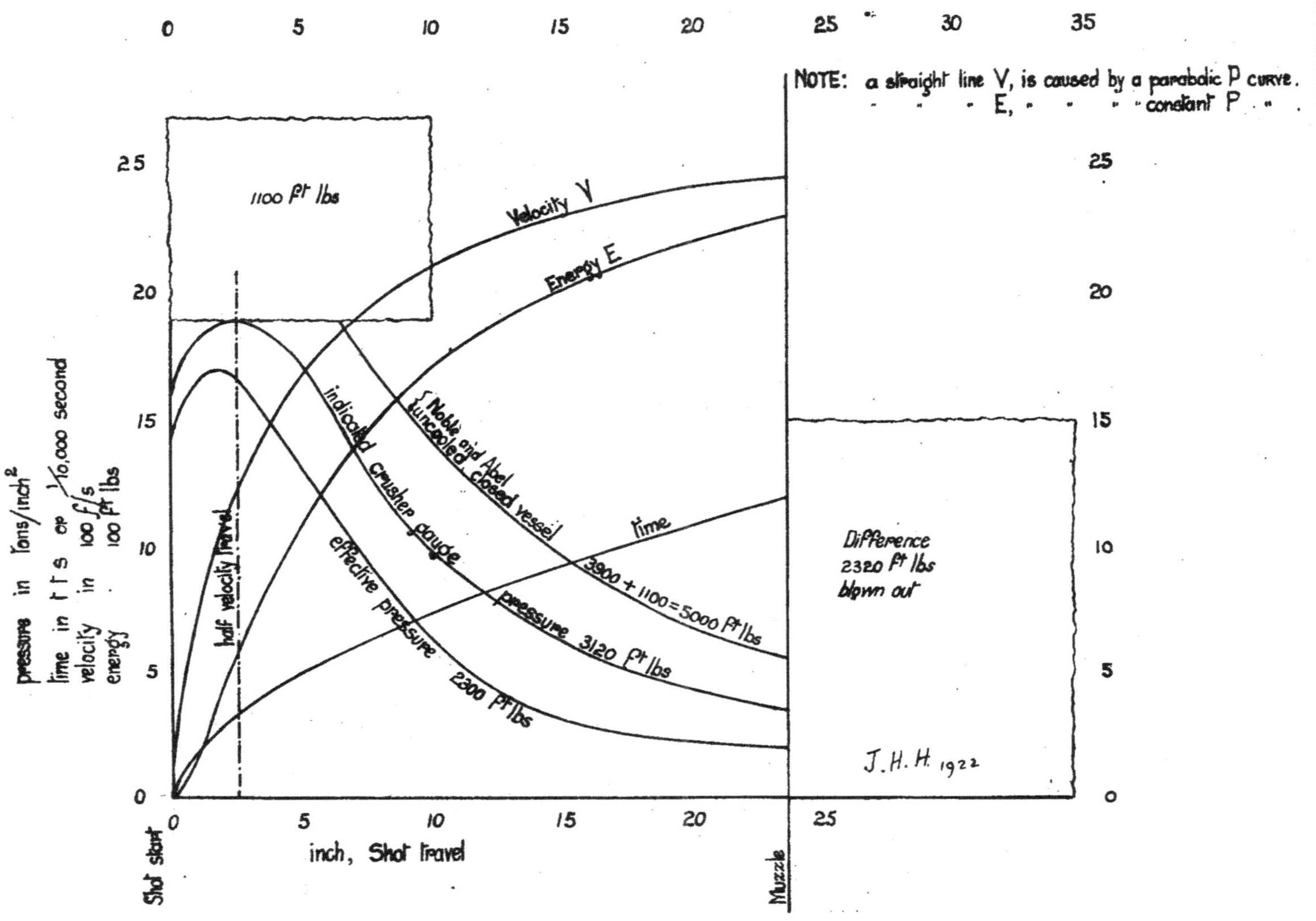

Malby & Sons Photo-Litho

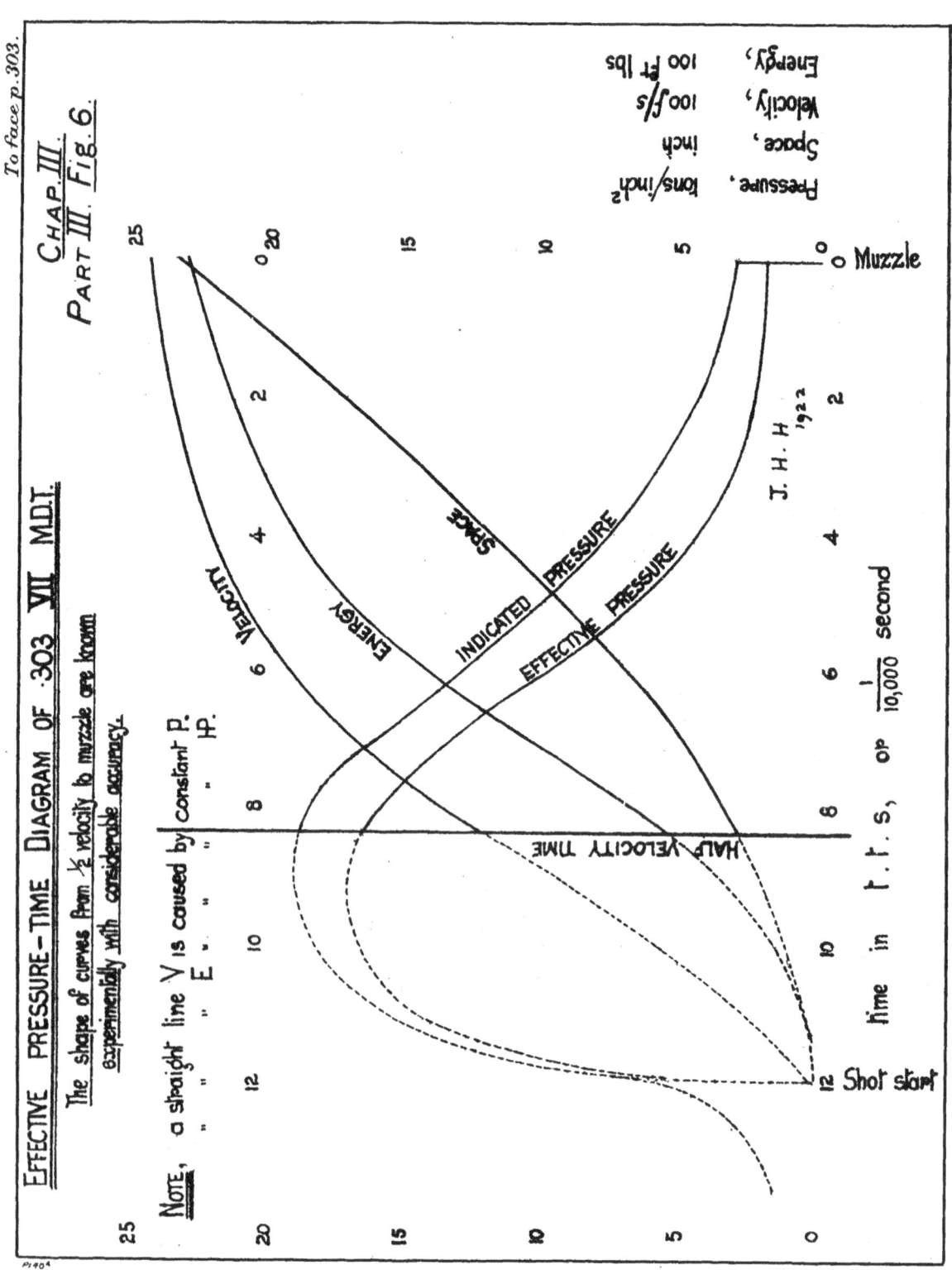

shown dotted. Using the same 1·4 per cent. law the pressure-time curve can be drawn for the effective pressure.

Owing to the uncertainty of the shape of the breech end of the curve below half velocity, the origin of time has been taken at the muzzle. From the muzzle backwards to about half momentum the curves are very near to the truth. After that they are largely guesswork, as no truly satisfactory machine has yet been devised for registering pressures while they rise at the rate of some 100,000 tons per second, more or less.

## Section 5
### Mr. F. W. Jones' Articles in *Arms and Explosives*

An English book dealing with the ballistics of rifles would be incomplete without a notice of Mr. Jones' writings during the last 20 years. No practical man can afford to ignore them, and every student is under an obligation to him as the only publicist in the trade since Housman's death.

He deals with full power rifles with the following characteristics :—

The loading density ($\Delta$) is between 0·6 and 0·85.
$W/d^2$ between 0·25 and 0·37.
The weight of the bullet is from 5 to 8 times that of the powder charge.
$c/d^2$ lies between 1·7 and 2·4 when $c$ is expressed in cubic inches.
The barrel length is about 30 inches.
The bullets are lead with cupro-nickel envelopes and flat based.

In 1912 he published a series of articles on the design of a new rifle, in which he showed the effect of variations of the loading of the cartridge on the height of the trajectory, recoil and striking energy. The idea of connecting the two subjects of the interior and exterior ballistics numerically was novel. Its direct result was to draw attention to the possibilities of such work, and to cause the production of the combined table at the end of this chapter, although by different methods. His Interior Ballistic articles are to be found in *Arms and Explosives* for February, 1904, and from November, 1915, to February, 1916. They treat the subject from the severely practical side, and result in empirical formulæ for velocity and pressure which, in practice, are sufficiently accurate for every-day use. They deal professedly with Mark I Cordite, size $3\frac{3}{4}$, but contain numerical constants which can be very simply adjusted for other sizes or other powders.

His fundamental assumption in 1904 was that the energy per grain of powder transmitted to the bullet is constant if the maximum pressure is constant, however the powder and shot are varied in weight, provided that the bullet is given a fictitious weight to account for frictional losses. He added (400 $d$) grains to the weight of the bullet, or 200 grains to a ·500 bore and 102·4 grains to a ·256 bore. Experience shows that this is a fair enough average.

As E/gr. increases with P we have

$$P^x \frac{(W + 400d) V^2}{2gw} = \text{a constant},$$

or $\qquad V = k^* P^x \sqrt{\dfrac{w}{W + 400d}}$ with W in grains.

Since V is also directly proportional to $w$ in rifles,

$P^x$ must contain the term $\sqrt{w}$.

In his pressure formula he had $P = k\, w^{1 \cdot 35}$ so that $x = \dfrac{0 \cdot 5}{1 \cdot 35} = 0 \cdot 37$.

His velocity formula then became $V = 2{,}380\, P^{0 \cdot 37} \dfrac{w}{W + 400d}$.

* $k$ is a conventional symbol for a numerical constant to be determined by experiment.

The great advantage of this formula is that it makes the velocity depend upon the actual realized pressure. In fact, it does so depend on pressure, and a worn barrel gives a lower velocity because it gives a lower pressure.

In 1915 he began by pointing out that no monomial formula could give good results over a very wide range of loads, because the indices of the prime variables were really changing with the load. But given a weight of powder that would produce 20 tons pressure in a given cartridge, it was possible from experience to write down the proportional charge which would give any other pressure, and the proportion was the same in all rifles. He gave a table and a curve which, for all practical purposes, can be expressed as

$$P = kw^{2.25}.$$

His 1916 pressure formula can be written with great precision as

$$P = k \ (w/d^2)^{2.25} \ (W/d^2)^{0.56} \ (d^2/c)^{1.69} \ (1/d)^{0.20}$$

where the last term includes his latest method of allowing for friction in the bore by adding, for the purposes of calculation, a certain amount to the weight of the bullet.

Putting P equal to 20, and including it in the constant $k$, and dividing all indices by 2·25, we have with $w$ as 20-ton charge

$$w_0/d^2 = k \ (d^2/W)^{0.25} \ (c/d^2)^{0.75} \ (d)^{0.10}$$

or

$$w_0 = k \ (1/W)^{0.25} \ (c)^{0.75} \ d^{1.10},$$

which is identical with his formula.

$$w_0 = 26 \cdot 2d \left\{ \frac{c^3}{W \ (1 + 1/5d)} \right\}^{1/4},$$

since $1 + 1/5d$ is indistinguishable from $\left(\frac{1 \cdot 36}{d}\right)^{0.34}$, 26·2 is his constant for size $3\frac{3}{4}$ cordite

when $w$ and W are in grains, $d$ is adjusted to take account of the depth of rifling and $c$ is the capacity of the chamber measured in grains weight of water.

The velocity for this 20-ton charge is obtained empirically from the 1904 velocity formula by putting P = 20 in the expression $2{,}380 \ (P)^{0.37}$, making it 7,210, so that

$$V = 7{,}210 \ \sqrt{\frac{w_0}{W \ (1 + 1/5d)}},$$

or if the barrel length, $u$, is not 30 inches

$$V = 3{,}650 \ u^{0.2} \ \sqrt{\frac{w_0}{W \ (1 + 1/5d)}}.$$

This constant 3,650 is suitable for size $3\frac{3}{4}$. If for another powder $w_0$ has to be increased to $1 \cdot 07 w_0$ to produce 20 tons, and $V_0$ is increased thereby to $1 \cdot 02 \ V_0$, then

$$3{,}650 \text{ becomes } \frac{1 \cdot 02}{\sqrt{1 \cdot 07}} \ 3{,}650 \text{ for that powder.}$$

The simplicity of the method is enough to recommend it even if put forward by a person of less experience than Mr. Jones.

Section 6

The Monomial Method

A monomial formula is one which has but one term only on each side of the equation, thus :—

$y = ax$ is monomial, and

$y = ax + b$ is binomial.

The particular form in which it is now proposed to cast the formulæ is the simple percentage formula for variations in the chief elements of loading when the result of one load is known. It has the advantage of extreme simplicity, substantial accuracy, and also it gives a direct method of linking the results of change of load to the effect on the performance of the bullet in its trajectory. It unites exterior with interior ballistics.

The simple statement that 1 per cent increase in the powder charge gives 2·3 per cent. increase of pressure is given by the formula

$$\log P = \text{a constant} + 2\cdot 3 \log w$$

or again by

$$P = kw^{2\cdot 3},$$

where 2·3 is called the index or power of $w$.

A formula of this shape can be shown as a straight line on squared paper ruled both ways in logarithms.

A straight line on ordinary squared paper represents the law of simple interest, just as a straight line on squared paper ruled one way in inches and the other way in logarithms represents the law of compound interest. Probably this last way is more likely to represent the true shape of the formulæ, but it is not very convenient to use for the present purpose.

As a result of very many experiments with all sorts of rifles and powders, a good average value of the indices or percentage effects of the change of each element of loading has been determined, both for velocity and pressure.

Table of effect of a 1 per cent increase for Military Rifles.

|  | $w$ | $W$ | $c$ | $L$ | $u$ | $d$ |
|---|---|---|---|---|---|---|
| Velocity ... ... ... | +1·00 | −0·25 | −0·65 | −0·40 | +0·16 | −0·06 |
| Pressure ... ... ... | +2·30 | +0·55 | −1·90 | −1·50 | — | −1·35 |

These indices give complete mechanical similitude by the general principle. If, however, the indices of $w$, $W$ and $c$ are determined by firing with one calibre only, using a constant value of $L$ and $d$, the implicit assumption is that the friction and cooling affect the index of $d$ for velocity by $-0\cdot 14$ and for pressure by $-0\cdot 55$, indicating that the change of friction, etc., consequent on an increase of calibre of 10 per cent reduces the maximum pressure by 5·5 per cent and the muzzle velocity by 1·4 per cent, both of which changes are nearly in accordance with known facts, but not necessarily so. One example will suffice to make this table intelligible. If the chamber is enlarged 10 per cent and all else is left constant, the velocity falls off 6·5 per cent and the pressure falls off 19·0 per cent.

The table above can be written (by using logarithms) as two formulas, where $k$ is the symbol for a numerical constant to be found by experiment.

 I. $\log V = \log k + 1\cdot 00 \; w/d^2 - 0\cdot 25 \log W/d^2 - 0\cdot 65 \log c/d^2 - 0\cdot 40 \log L + 0\cdot 16 \log u.$

 II. $\log P = \log k + 2\cdot 30 \log w/d^2 + 0\cdot 55 \log W/d^2 - 1\cdot 90 \log c/d^2 - 1\cdot 50 \log L.$

By firing one single cartridge of which all the weights and dimensions are known, and by observing its muzzle velocity and maximum pressure, the value of $k$ in the V and P formula can be settled. In practice, however, a large number of rifles and loads are used and an average value of $k$ is obtained.

As we are chiefly concerned with the cartridge of maximum power for its size, the cartridge case should always be filled completely with powder to avoid using an unnecessarily large case. In practice $\Delta = 0\cdot85$ is found to be about the greatest density possible when filling by machinery, so that a case which will hold 100 grains of water is filled by 85 grains of powder. In the above formula it is possible to cut out the term $w/d^2$ because the term $c/d^2$ and $\Delta = 0\cdot85$ actually specify the powder charge.

For convenience in calculation the powder size is better stated as $1/L = £$, so that a powder of $L = 0\cdot011$ has $£ = 91\cdot0$.

Making these changes and working out a good average value of $k$ for a powder of the same chemical nature as Dupont Z but of any value of £, the formulæ become for a barrel with a shot travel of 24 inches

$$\log V = 2\cdot351 - 0\cdot25 \log W/d^2 + 0\cdot35 \log c/d^2 + 0\cdot40 \log £$$

$$\log P = \overline{2}\cdot515 + 0\cdot55 \log W/d^2 + 0\cdot40 \log c/d^2 + 1\cdot50 \log £$$

To save calculation Col. Hezlet has made a nomogram for each formula (figures 7 and 8).

By actually plotting or working out these formulæ or more simply by reading results off the nomograms, Table A is prepared, and with its help Table B is filled in, showing exactly what maximum velocities can be obtained at any given pressure for any given bullet weight and cartridge case. It also gives the required granulation of the powder to produce these results.

TABLE A

*By plotting on V and £ by $W/d^2$ disregarding $c/d^2$*
*Table of V for £*

|  | $\dfrac{W/d^2}{£}$ | 0·2 | 0·3 | 0·4 |
|---|---|---|---|---|
| 18 tons | £<br>65<br>70<br>75<br>80<br>85<br>90<br>95<br>100 | —<br>—<br>—<br>3,350<br>3,200<br>3,020<br>2,870<br>2,740 | —<br>2,820<br>2,650<br>2,500<br>2,370<br>2,250<br>2,150<br>— | 2,440<br>2,290<br>2,150<br>2,050<br>1,930<br>—<br>—<br>— |
| 20 tons | 75<br>80<br>85<br>90<br>95<br>100<br>105 | —<br>—<br>—<br>3,310<br>3,150<br>3,010<br>2,880 | 2,890<br>2,730<br>2,590<br>2,460<br>2,340<br>—<br>— | 2,350<br>2,220<br>2,100<br>2,000<br>1,950<br>—<br>— |
| 22 tons | 80<br>85<br>90<br>95<br>100<br>105<br>110 | —<br>—<br>—<br>3,430<br>3,270<br>3,130<br>3,000 | 2,970<br>2,815<br>2,670<br>2,550<br>2,450<br>2,350<br>— | 2,410<br>2,280<br>2,180<br>2,090<br>2,010<br>—<br>— |

PART III.                                                          CHAP. III.
                                                                    Fig. 7.

$\Delta = 0.85$, SHOT TRAVEL 24 INCH
DUPONT N/C POWDER

$$\log V = 2.351 - 0.25 \log \frac{W}{d^2} + 0.35 \log \frac{C}{d^2} + 0.40 \log \pounds$$

Join $\frac{W}{d^2}$ to $\frac{C}{d^2}$. Join the point where this straight line cuts the central reference line to $\pounds$, and produce backwards to cut the V scale.

The value of V so obtained is the answer.

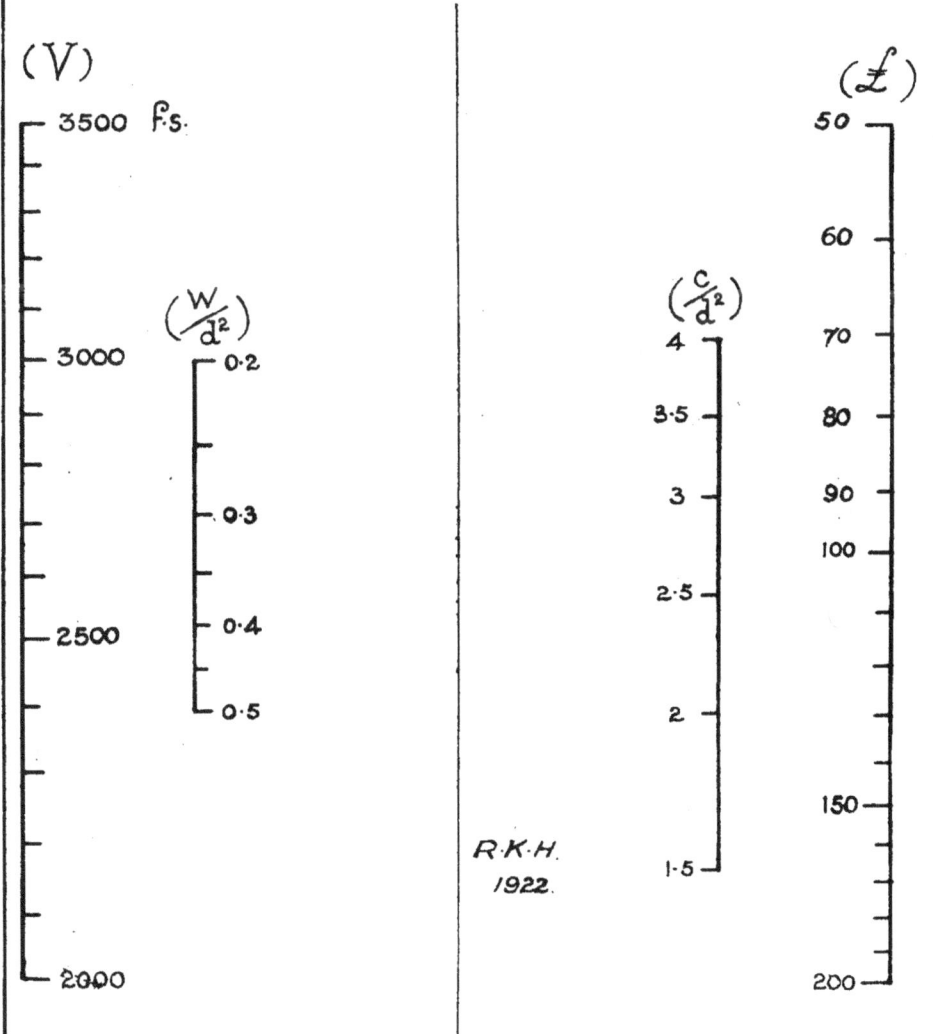

R.K.H.
1922.

*To face p. 307.*

**P.T III.**  **CHAP. III.**
$$\log P = \bar{2}\cdot 515 + 0\cdot 55 \log \frac{W}{d^2} + 0\cdot 40 \log \frac{c}{d^2} + 1\cdot 50 \log \pounds$$
**Fig. 8.**

Join $\frac{W}{d^2}$ to $\frac{c}{d^2}$. Join the point where this straight line cuts the central reference line to the given value of P and produce backwards to cut the $\pounds$ scale.

This gives the value of $\pounds$ required.

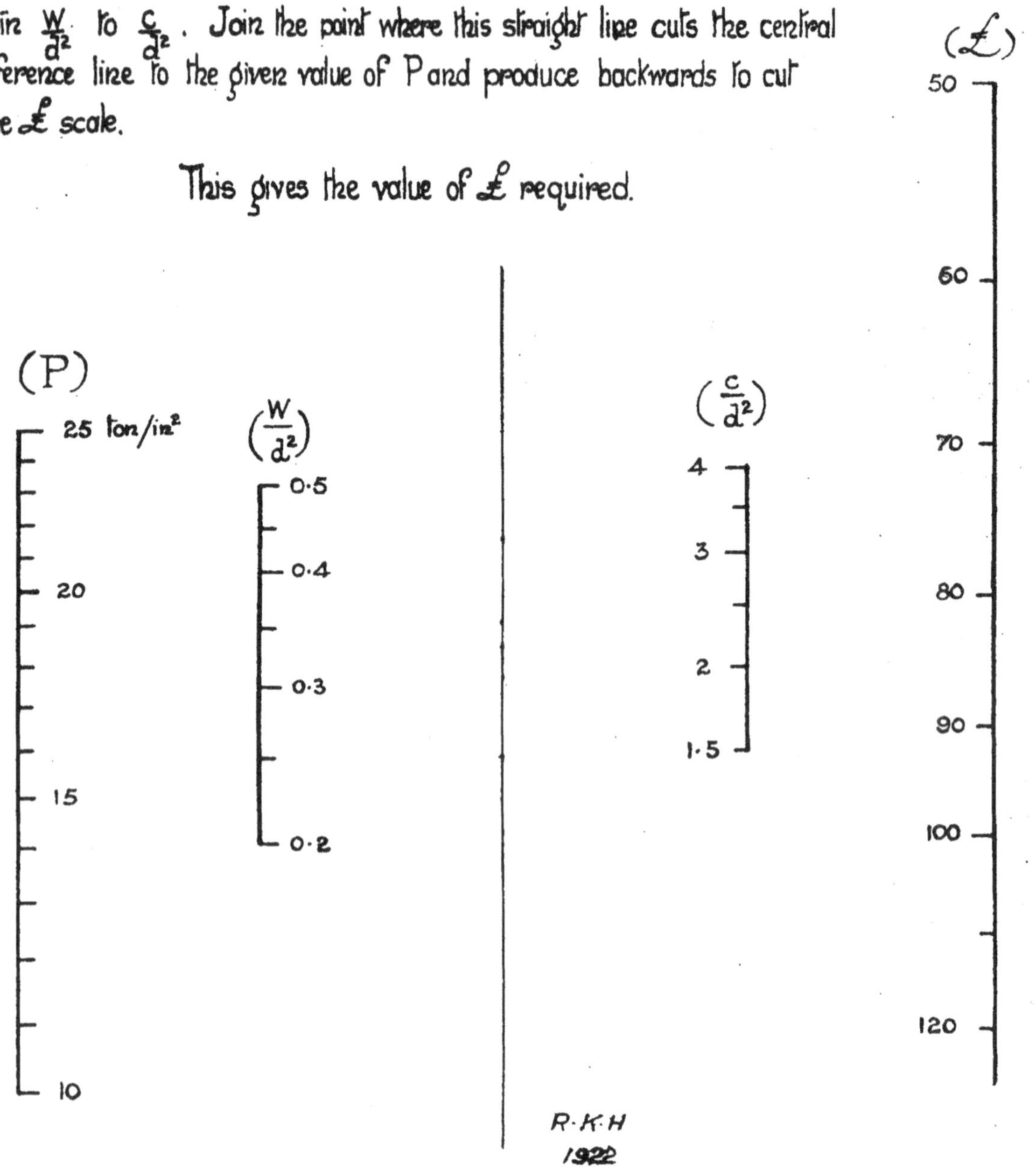

($\pounds$) scale: 50, 60, 70, 80, 90, 100, 120

(P) scale: 25 ton/in², 20, 15, 10

$\left(\frac{W}{d^2}\right)$ scale: 0.5, 0.4, 0.3, 0.2

$\left(\frac{c}{d^2}\right)$ scale: 4, 3, 2, 1.5

R. K. H
1922

## BALLISTIC CHART FOR RIFLE OF ANY CALIBRE (d inch) SHOWING GREATE[ST]
### CHAMBER FILLED WITH POWDER OF BEST SIZE

### PART III.

Powder size is given by number to one inch
Dupont N°16 is 0·011 inch or 91 to inch
Powder charge follows from $\Delta = 0.85$
·303 VII has $c = 0.190 = 48$ grains water
$48 \times 0.85 = 41.7$ grains

N.C.T. powder (Dupont quo[ted])
$\Delta$ (loading density) = 0[·85]

Bullet weight is given [by]
Mark VII is 174 grains

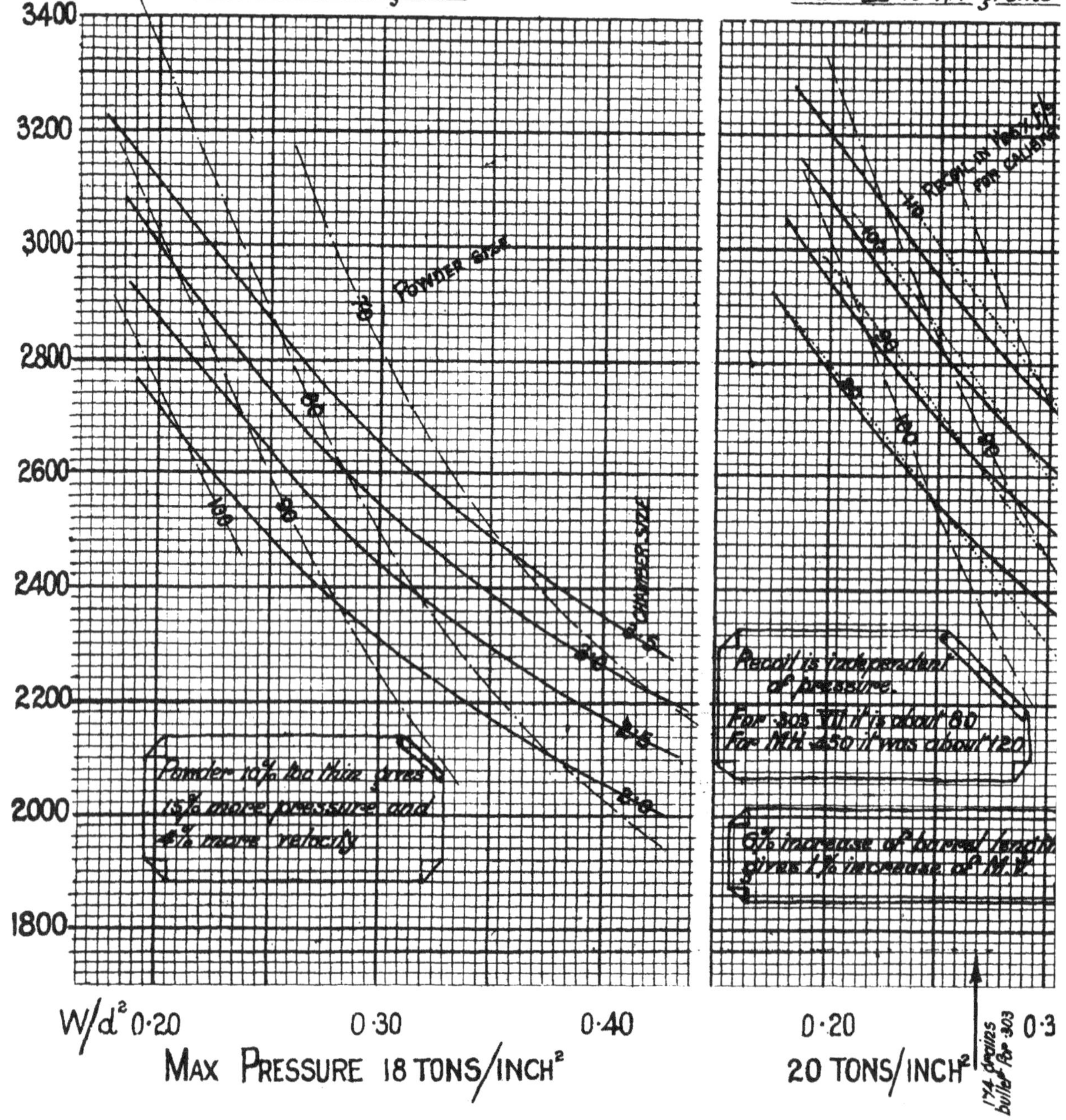

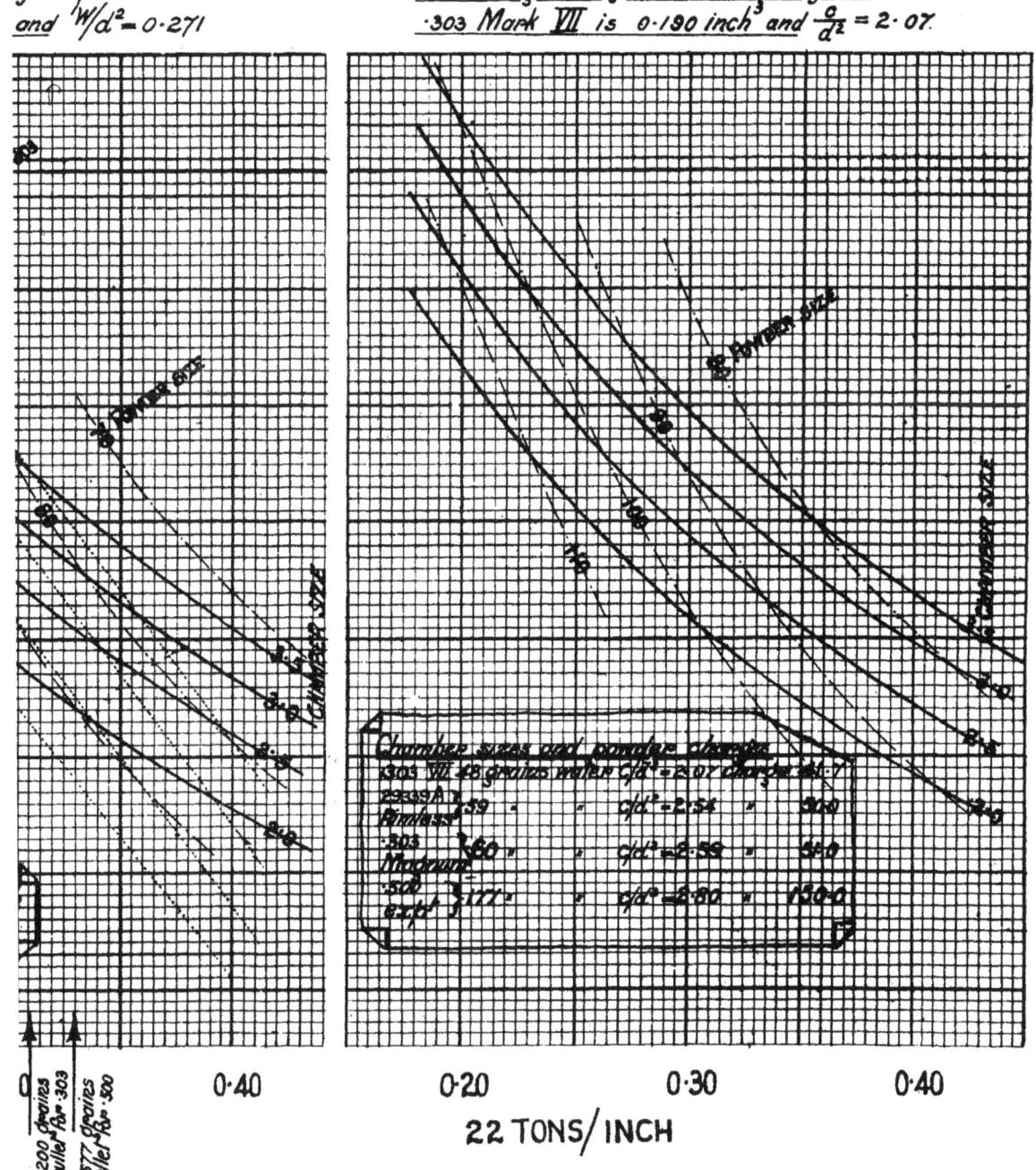

*To face page 307.*

**ST M.V. TO BE EXPECTED FOR ANY SIZE CHAMBER AND MAXIMUM PRESSURE GIVEN.**

CHAP. III.
FIG. 9.

TABLE B

| Pressure | $c/d^2$ | $W/d^2$ 0.2 | | | | 0.3 | | | | 0.4 | | | |
|---|---|---|---|---|---|---|---|---|---|---|---|---|---|
| | | 2.0 | 2.5 | 3.0 | 3.5 | 2.0 | 2.5 | 3.0 | 3.5 | 2.0 | 2.5 | 3.0 | 3.5 |
| 18 | £ | 101 | 95 | 91 | 87 | 87 | 82 | 78 | 75 | 78 | 74 | 70 | 67 |
| | V | 2,710 | 2,865 | 2,995 | 3,105 | 2,305 | 2,435 | 2,545 | 2,645 | 2,055 | 2,175 | 2,270 | 2,355 |
| 20 | £ | 108 | 102 | 97 | 93 | 93 | 88 | 84 | 81 | 84 | 79 | 75 | 72 |
| | V | 2,785 | 2,935 | 3,080 | 3,195 | 2,370 | 2,505 | 2,620 | 2,720 | 2,110 | 2,235 | 2,335 | 2,420 |
| 22 | £ | 115 | 108 | 103 | 99 | 99 | 94 | 89 | 86 | 89 | 84 | 80 | 77 |
| | V | 2,860 | 3,020 | 3,155 | 3,275 | 2,430 | 2,570 | 2,685 | 2,785 | 2,170 | 2,290 | 2,395 | 2,480 |

For convenience the whole of Table B has been drawn on Fig. 9 and on the 20-ton portion the recoil of a cartridge of ·303 calibre is also shown. From this diagram almost any question can be answered at once as to the possibility of making a rifle comply with certain conditions of velocity and pressure. The answer given will be as accurate as the information on which the diagram is based.

Besides velocity and pressure there are other points to consider, such as height of trajectory and striking energy, to name but two of them. In an article by Major Hardcastle in the *R.A. Journal*, November, 1915, entitled " Monomial formulas . . . " following on Mr. F. W. Jones' articles in *Arms and Explosives*, April and June, 1912, the monomial method of attack is given in considerable detail. The object sought by both writers was to bring into strong relief the effect of pushing up the power of a rifle, and to show the price to be paid, for instance, for every inch of flattening of the trajectory. The price has always to be paid, and at times it may be excessive. The simplest way to visualize the price is to reduce it to terms of maximum pressure, size of case and bullet weight and so forth.

The principal result of exterior ballistic calculations for modern military rifles can be expressed as a first approximation as monomials, and they can then be combined with the monomials of interior ballistics.

Calling R the range in yards
H the greatest height of trajectory in feet
$n$ the coefficient of reduction
T the time of flight in seconds
$v$ the remaining velocity in f/s.
E the angle of elevation in minutes
D the deflection in minutes due to a crosswind
M the speed of the cross wind in m.p.h.

and the other symbols as usual, the monomials of exterior ballistics (calling the numerical constant $k$) are

$$R = kV^{0.83}\left(\frac{W}{nd^2}\right)^{0.25} H^{0.36};$$

$$v = kV^{1.0}\left(\frac{W}{nd^2}\right)^{0.6}\left(\frac{1}{R}\right)^{0.5};$$

$$E = k\,(1/V)^{2.0}\left(\frac{nd^2}{W}\right)^{0.7} R^{1.33};$$

$$D = k\,(1/V)^{1.2}\left(\frac{nd^2}{W}\right)^{1.2} R^{1.2}\, M^{1.0}.$$

The limit of substantial accuracy of these formulæ is about the range at which accurate target practice can be expected, say, 1200 yards.

It will be at once noticed that H varies as $n^{0.695}$ or, say, $n^{0.7}$, pointing to the value of a good shape for the bullet if a flat trajectory is required.

From the table of average index, $+$ 10 per cent of P is given by the following per cent changes, $+ 100/230$ of $w$, $+ 100/5 \cdot 5$ of W, $- 100/19 \cdot 0$ of $c$, $- 100/15 \cdot 0$ of L, and $- 100/13 \cdot 5$ of $d$, and these changes produce changes of V as follows :—for $w+(100/23 \cdot 0)$ 1·00, for W $- (100/5 \cdot 5)$ 0·25, and so on.

Combining the table of effect of a 1 per cent increase with the monomials of exterior ballistics, the following table C results. A study of this table and the method seems likely to repay the student for the time spent upon it.

TABLE C

TABLE OF PERCENTAGE EFFECTS DUE TO + 10 PER CENT INCREASE OF PRESSURE

|  | Charge Weight | Bullet Weight | Chamber Capacity | Thickness powder grain | Calibre | Remarks |
|---|---|---|---|---|---|---|
| Per cent. | + 4·35 | + 18·2 | − 5·25 | − 6·67 | − 7·40 | Per cent changes to alter pressure by + 10 per cent. |
| ,, | + 4·35 | − 4·55 | + 3·42 | + 2·36 | + 0·45 | Per cent change in M.V. |
| ,, | + 3·16 | + 0·78 | + 2·84 | + 1·96 | + 4·08 | Per cent change in range of trajectory H feet high. |
| ,, | − 10·00 | − 2·27 | − 7·86 | − 5·43 | − 11·40 | Per cent change of height of trajectory of R yards. |
| ,, | + 4·35 | + 6·37 | + 3·42 | + 2·36 | + 9·35 | Per cent change in striking velocity at about 1,000 yards. |
| ,, | − 8·70 | − 3·64 | − 6·84 | − 4·72 | − 11·26 | Per cent change in angle of elevation. |
| ,, | − 5·22 | − 16·34 | − 4·10 | − 2·83 | − 18·34 | Per cent change in lateral wind allowance. |

EXAMPLE.—Reducing the chamber by 5·25 per cent increases the pressure by 10 per cent and the velocity by 3·42 per cent. It also reduces H by 7·86 per cent, E by 6·84 per cent, and D by 4·10 per cent, and increases R by 2·84 per cent.

## CHAPTER IV

### EXTERIOR BALLISTICS (NUMERICAL)

Within the limits of a single chapter the outlines only of the subject can be indicated if any space is to be reserved for actual numerical examples of the more important rifle problems.

The chapter has accordingly been divided into seven sections, as follows :—

(1) Methods not requiring ballistic tables.
(2) The ballistic tables and their use.
(3) Given the chief particulars, to construct a full range table.
(4) Detailed numerical example of all the trajectories of one rifle for elevations from 0° to 90°.
(5) Practical methods of determining the value of the ballistic coefficient.
(6) Long-range fire analysis in two arcs.
(7) The ballistics of low-power weapons and revolvers.

Besides the actual ballistic tables the appliances or tools required are: a slide-rule; a table of four-figure logarithms and natural sines, etc.; squared paper ruled in inches and tenths; two straight celluloid battens or splines about a yard long and ¾ inch wide, one about 0·08 inch thick and the other about half as thick for bending into curves; and, finally, a good supply of glass-headed pins about 1½ inches long for setting out the points on the curve to which the batten is to be bent. No other appliances were used in the writing of this chapter.

## Section 1

Methods not requiring ballistic tables:—
  (a) Parabolic formulæ.
  (b) The Froude-Metford and allied formulæ.
  (c) Three Notanda.

### (a) Parabolic Formulæ

(1) The formula for range, elevation, and M.V. is

$$R = \frac{V^2 \sin 2E}{3g} \text{ yards,}$$

and as $\sin 90° = 1·0$, and $\sin 30° = \frac{1}{2}$, and $3g$ is nearly 100,

$$R = (V/10)^2.$$

When $E = 45°$ the range is the greatest possible, and when $E = 15°$ the range is half the maximum range. This formula is useful for bombs and grenades. It gives the working rule, "The muzzle velocity in f/s. is equal to or greater than ten times the square root of the range in yards." Thus, if $R = 144$ yards, $V$ is at least 120 f/s.

(2) The formula for "drop" of the shot from the line of departure is

$$h = \tfrac{1}{2} gt^2 \text{ feet,}$$

and as $t = 3R/V$ and there are 12 inches in a foot, $d$, the drop in inches $= 193·2 \, (3R/V)^2$, or practically $d = 200 \, (S/V)^2$.

This formula is useful for short ranges up to about 100 yards. Owing to air resistance, the apparent value of $g$ in this formula falls off from 32·2 at the muzzle to a value of about 28 at 500 yards and 24 at 1,000 yards.

(3) The formula for elevation is

$$\sin 2E = 3Rg/V^2.$$

As the sine of 1 minute is 1/3,438, this formula when expressed in minutes becomes

$$E = 166{,}000 \, R/V^2 \text{ minutes,}$$

or, putting $t = 3 \, R/V$,

$$E = \frac{573}{R} gt^2 = \frac{18{,}450}{R} t^2.$$

Mr. A. Mallock, F.R.S., in *Proceedings of the Royal Society*, September, 1907, suggested that this formula might be used for a shot in air by sophisticating the value of "$g$." Calling the adjusted value of $g$ by the letter "$f$," he proposed

$$f = g - \frac{8·4}{V-850}(V-v).$$

This value is not suitable for rifles. Mr. F. W. Jones, examining the results of the Hodsock experiments with the Fremantle ballistic pendulum in 1922 (referred to in another chapter), concluded that a better value of "$f$" was

$$f = g\left\{1 - b\left(\frac{V-v}{V}\right)\right\},$$

and suggested
$$b = 3/7.$$

On analysing Table II and other tables it is found that $b = 3/7$ gives good results down to values of $v = V/3$, but that $b = 2/5$ is nearer the average. The necessary conditions to cause "$b$" to be really constant are not likely to occur in nature.

Since the drop of the bullet at the target when divided by the range gives the tangent of the angle of elevation, we have

$$\tan E = \tfrac{1}{2} ft^2/3R = ft^2/6R,$$

so that
$$f = 6R \tan E/t^2.$$

But as for all small angles $\tan E = E/3{,}438$,

$$f = RE/573\, t^2,$$

which is the same result as given above. As RE and $t$ are known from firing results for any rifle, $f$ can be worked out easily and then used instead of $g = 32 \cdot 2$ in parabolic formulæ to plot the actual trajectory by the drop.

Up to about 400 yards the approximate formula $E = 200{,}000\ R/V^2$ gives results for modern rifles which are correct to one minute.

(4) Sladen's formula for plotting a trajectory when the time curve is given is interesting, in that it states a property of a parabola which remained apparently undiscovered by anyone till Col. Sladen announced it about 50 years ago.

If T is the whole time of flight and $t$ is the time of flight to some intermediate point where $y$ is the height of the trajectory in feet, then $y = \tfrac{1}{2} gt\,(\mathrm{T} - t)$ feet. The time to the vertex is T/2, so that the height of the vertex is given by $y = 16 \cdot 1\ \mathrm{T}^2/4$, which is practically

$$y = 4\mathrm{T}^2 \quad \text{or} \quad (2\mathrm{T})^2.$$

This gives the rule for height of vertex "Double the time of flight in seconds and square the result for vertex height in feet."

With modern rifles it is better to use $y = (4\tfrac{1}{4}\ \mathrm{T}^2)$ for height of vertex, because the apparent value of $g$ in this formula when used in air, falls off from about 40 or 50 near the muzzle to about 34 to 38 at the vertex and about 20 to 30 near the target, the error getting greater as the range lengthens.

(b) The Froude-Metford and allied formulæ.

Before squared paper became cheap and plentiful it was a matter of considerable difficulty to produce a smooth elevation table showing the angle for each hundred yards of range. Sir Henry Halford, writing in the 1894 edition of this text-book, stated that the following formula was given by the late Mr. W. Froude to Mr. Metford :—

Elevation in minutes

$$= \mathrm{N}a + \mathrm{N}\frac{(\mathrm{N}-1)}{2}b + \mathrm{N}\frac{\mathrm{N}-1}{2}\frac{\mathrm{N}-2}{3}c + \ldots,$$

where N is the range in yards.

$a$ is the angular value of the "drop" in vacuo for 1 yard.

$b$, $c$, etc., are air resistance factors to be found by experiment, selecting ranges for firing which are distributed over the whole range required, including the extreme range obtainable.

From the formula $E = 166,000 \, R/V^2$ given above, $a = \dfrac{0\cdot 166}{(V/1,000)^2}$ when $R = 1$ and tables of $a$ and $V$, and of $N$, $N\dfrac{N-1}{2}$, $N\dfrac{N-1}{2}\dfrac{N-2}{3}$, were appended; $b$, $c$, etc., were found by determining the angles actually required at several distances. These give as many equations from which the best values of $b$ and $c$ can be obtained.

Writing in the *Journal of the Royal Artillery*, Mr. C. E. Wolff, in April, 1908, and Sir G. Greenhill, in February, 1909, went very fully into this formula.

With M, the minutes elevation for $m$ hundreds of yards, the Froude formula can be written as

$$M = Am + Bm^2 + Cm^3 + \ldots.$$

Taking the Mark VII bullet with 2,440 f/s. = V, and supposing the angle for 1,200 yards to be 86 minutes and for 2,400 yards to be 396 minutes

A is shown to be

$$\cdots \left(\frac{4,075}{V}\right)^2 = \left(\frac{4,075}{2,440}\right)^2 = 2\cdot 78,$$

where $(4,075)^2 = 16,600,000$.

We then have two equations to find B and C.

$$86 = 12\,(2\cdot 78) + 12^2\,B + 12^3\,C$$
$$396 = 24\,(2\cdot 78) + 24^2\,B + 24^3\,C$$

which give

$$B = 0\cdot 159$$
$$C = 0\cdot 0172,$$

calling

$$\alpha = A + B + C, = 2\cdot 956 = 2\cdot 96,$$
$$\beta = 2B + 6C, = 0\cdot 421 = 0\cdot 42,$$
$$\gamma = 6C, = 0\cdot 103 = 0\cdot 10.$$

These are the top figures of the difference columns of the subjoined table.

| Range. | E. | 1st Diff. or Rise. | 2nd Diff. | 3rd Diff. |
|---|---|---|---|---|
| 0 | 0 | | | |
| | | $2\cdot 96 = \alpha$ | | |
| 100 | 2·96 | | $0\cdot 42 = \beta$ | |
| | | 3·38 | | $0\cdot 10 = \gamma$ |
| 200 | 6·34 | | 0·52 | |
| | | 3·90 | | |
| 300 | 10·24 | | 0·62 | |
| | | 4·52 | | |
| 400 | 14·76 | | | |
| | | 13·08 | | |
| 1,200 | 85·24 | | 1·52 | |
| | | 14·60 | | |
| | | 37·92 | | |
| 2,400 | 399·36 | | 2·72 | |

The errors between 85·24 and 86 and 399·36 and 396 are due to cutting off the last figure of decimals. The use of 0·10 instead of 0·103 affects the 2,400 yard elevation by its cumulative effect, which actually amounts to 5·313 minutes.

The values of A, B and C can be reconstructed from the different columns by the following relations:—

$$A = \alpha - \tfrac{1}{2}\beta + \tfrac{1}{3}\gamma,$$
$$B = \tfrac{1}{2}\beta - \tfrac{1}{2}\gamma,$$
$$C = \tfrac{1}{6}\gamma.$$

Given a smooth table as above, Greenhill goes on to show that a first differentiation of M gives the angle of descent; a second differentiation of M gives the velocity; a third differentiation of M gives the retardation and air resistance; a fourth differentiation of M gives the rate of variation of the resistance.

These are best shown by numerical examples.

(1) Angle of descent = W minutes.
The formula is

$$W = m(A + 2Bm + 3Cm^2 + \dots).$$

For $\quad R = 1,200, \quad m = 12$

$$W = 12(2\cdot78 + 3\cdot82 + 7\cdot43) = 12 \times 14\cdot03$$
$$= 168 \text{ minutes}.$$

As it is not always convenient to make out a range and elevation table with a constant third difference, an alternative method is given, using only the hundred yard rises before and after the 1,200-yard elevation. Then

$$W = 12\left(\frac{13\cdot08 + 14\cdot60}{2}\right) = 12 \times 13\cdot84 = 166,$$

or in general terms

$$W = m \times \text{(average rise at } m \text{ hundred yards)}.$$

The formula in this form is due to Sir Henry Halford, and is of the greatest use, being by far the simplest and most accurate way of determining the angle of descent at any range up to 10° elevation.

(2) The remaining velocity = $v$ f/s.
The formula is

$$v = \frac{4{,}075}{\sqrt{A + 3Bm + 6Cm^2}}$$

at 1,000 yards, $m = 10$

$$v = \frac{4{,}075}{\sqrt{2\cdot78 + 4\cdot77 + 10\cdot32}} = \frac{4{,}075}{\sqrt{17\cdot87}} = 964 \text{ f/s}.$$

Here, again, it is necessary to have a perfectly smooth curve, and, unfortunately, there is no means of turning the difficulty by using the rise. All that can be done is to rewrite the term under the square root, as

$$\alpha + \tfrac{1}{2}\beta(3m-1) + \tfrac{1}{6}\gamma(2 - 9m + 6m^2).$$

This, however, gives the M.V. simply enough as

$$MV = \frac{4{,}075}{\sqrt{A}}, \quad A = \alpha - \tfrac{1}{2}\beta + \tfrac{1}{3}\gamma.$$

(3) Retardation and air resistance.

If $p \cdot n \cdot d^2$ is the resistance in lbs. to a bullet weighing W lbs. moving at $v$ f/s., the formula is

$$\frac{p \cdot n \cdot d^2}{W} = 257{,}850 \, \frac{B + 4Cm}{(A + 3Bm + 6Cm^2)^2}.$$

(4) Rate of variation of resistance.

If the resistance of the air at or about the muzzle velocity varies as the $n$th power of the velocity

$$n = 4 - \frac{8CA}{3B^2},$$

and this should give a value of $n$ of about 2 or $2\frac{1}{2}$.

Although only the formulæ for angle of descent and remaining velocity are of practical use, all four have been given to show that the operation of smoothing an elevation curve should be undertaken with care, and also that there is no necessary reason for expecting such a curve to come out exactly smooth.

In practice nowadays the observed points are plotted on a large sheet of squared paper by glass-headed pins, and then a batten is bent along the line of the pins as a guide to the pencil used to draw the complete curve. From this pencil curve intermediate ranges are read off as required and smoothed by differencing.

There are two other useful formulæ which require no ballistic tables, one to plot a trajectory by ordinates, and the second to decide whether a trajectory will clear a given height when firing by quadrant.

To calculate the ordinates of any trajectory up to 10° elevation a range and elevation table only is required.

If the total range is $R_1$ yards for which $E_1$ minutes of elevation are wanted, and the ordinate range is $R_2$ and elevation $E_2$ the formula is

$$Y = R_2 \frac{E_1 - E_2}{3{,}438} \text{ yards.}$$

The reason is easily seen if the $R_2$ trajectory is imagined as swung up till the point $R_2$ comes on to the $R_1$ trajectory.

This is a very accurate way of plotting a trajectory up to 10° elevation.

To decide whether a trajectory will clear a given obstacle. This is best considered under two headings :—

(a) On the level.
(b) By altitudes.

The solution of (a) is the same as that just given for the ordinate.

For (b) let the altitude (as shown by contour on the map) of the lower of the two, gun or target, be Z yards, and the altitude of the other be $Z + H$ yards, and let the altitude of the obstacle be $Z + h$. Let the range from gun to target be $R_1$ yards, for which $E_1$ minutes is the elevation ; and let $R_2$ be the range from gun to obstacle and $E_2$ the elevation.

The greatest height of $h$ to enable the trajectory to clear it is given by the formula

$$h = H \frac{R_1 - R_2}{R_1} + R_2 \frac{E_1 - E_2}{3{,}438} \text{ yards.}$$

This solution is practically exact up to $E_1 = 600$, and up to an angle of sight of 20°.

At angles of sight of 40° or 60° the ranges in mountainous countries can hardly exceed 1,000 yards, and then the second term of the formula becomes very small compared with the first term, so that an error of a few inches in it cannot affect the result.

The above formula is also applicable to "safety clearance" over our own troops.

If our troops are on the contour Z + Q their clearance is $h - Q$. If Q is greater than $h$ we shall hit them or the ground below them.

(c) Notanda.

(1) The working formula for converting distances into minutes of angle, whether for deflection or on a vertical target for elevation, is—

"Reduce the distances to inches and divide by the number of hundreds of yards in the range, the answer gives the minutes of deflection."

This rule gives an answer 4·7 per cent. too great, as it says virtually that: 1 inch subtends 1 minute at 100 yards or 3,600 inches; whereas really 1 inch subtends 1 minute at 3,438 inches or 95 yards 1 foot 6 inches.

So that if M is the number of minutes.
   N  ,,  ,,  100 yards.
   D  ,,  ,,  inches.

$$M = \frac{D}{1 \cdot 047 \, N} = \frac{0 \cdot 955 \, D}{N},$$

or

$$D = 1 \cdot 047 \, MN = \frac{M \cdot N}{0 \cdot 955}.$$

Another way of expressing the same thing is used for angles of depression or elevation for a height of H feet and a range R yards—

$$M = 1{,}146 \frac{H}{R} \text{ minutes.}$$

(2) The simplest method of correcting a sight curve for loss of velocity due to some such cause as a worn barrel is shown on Fig. 1.

Here the normal sight curve is shown by the heavy solid line marked OPQ for a supposed M.V. of 2,700 f/s. If the gun wears so as to have an M.V. of only 2,400 f/s., the dotted line OAB gives the new sight curve. This new curve is constructed on the supposition that 1 per cent. loss of velocity increases the elevation by 2 per cent. for the same range. Thus at the range OE the normal elevation is EP, and at the range OF it is FQ.

The new elevations are EA and FB, such that

$$\frac{EP}{EA} = \frac{FQ}{FB} = \left(\frac{2{,}400}{2{,}700}\right)^2 = \left(\frac{\text{velocity in worn gun}}{\text{velocity in new gun}}\right)^2.$$

The rule is not exact, but it is practically true at ranges up to a mile.

For a change in the density of the air, such as that due to a high altitude or very hot weather, the new curve OCD in Fig. 1 can be drawn by the method first published by Professor G. Forbes, F.R.S., in *Proc. R.S.*, 1905.

Supposing the air to be 10 per cent. lighter than normal, as would occur at 4,000 feet or 105° F.

Draw OP and OQ and produce them to C and D so that $\frac{OC}{OP} = \frac{OD}{OQ} = \frac{110}{100}$. Then C and D are points on the sight curve for the lighter air.

Incidentally, the range OG is 10 per cent. longer than OE, and OH is 10 per cent. longer than OF.

This rule is exact up to 10° elevation and also a little beyond. As shown later, it is of the greatest practical use in calculations.

(3) Ommundsen's theorem was the subject of a lecture by Sir G. Greenhill to members of the Junior Institute of Engineers on January 19th, 1912. This theorem was also called "The Negative Angle System," and even if of no practical use on service it deserves mention as an exercise in exterior ballistics.

P.ᵀ III. Chap. IV.

Fig I

### Diagram showing how to correct sight curve for,
### (a) Loss of velocity due to worn gun.
### (b) Increased ranging power due to lighter air.

OPQ is normal sight curve for 2700 f/s and 60° Fahrenheit and ground level.

OAB is curve for worn gun giving 2400 f/s.

O.C.D is curve for 105° Fahᵗ or 4000 feet altitude, where air is 10% lighter.

$$\frac{EP}{EA} = \frac{FQ}{FB} = \left(\frac{2400}{2700}\right)^2$$

$$\frac{OP}{OC} = \frac{OQ}{OD} = \frac{100}{110}$$

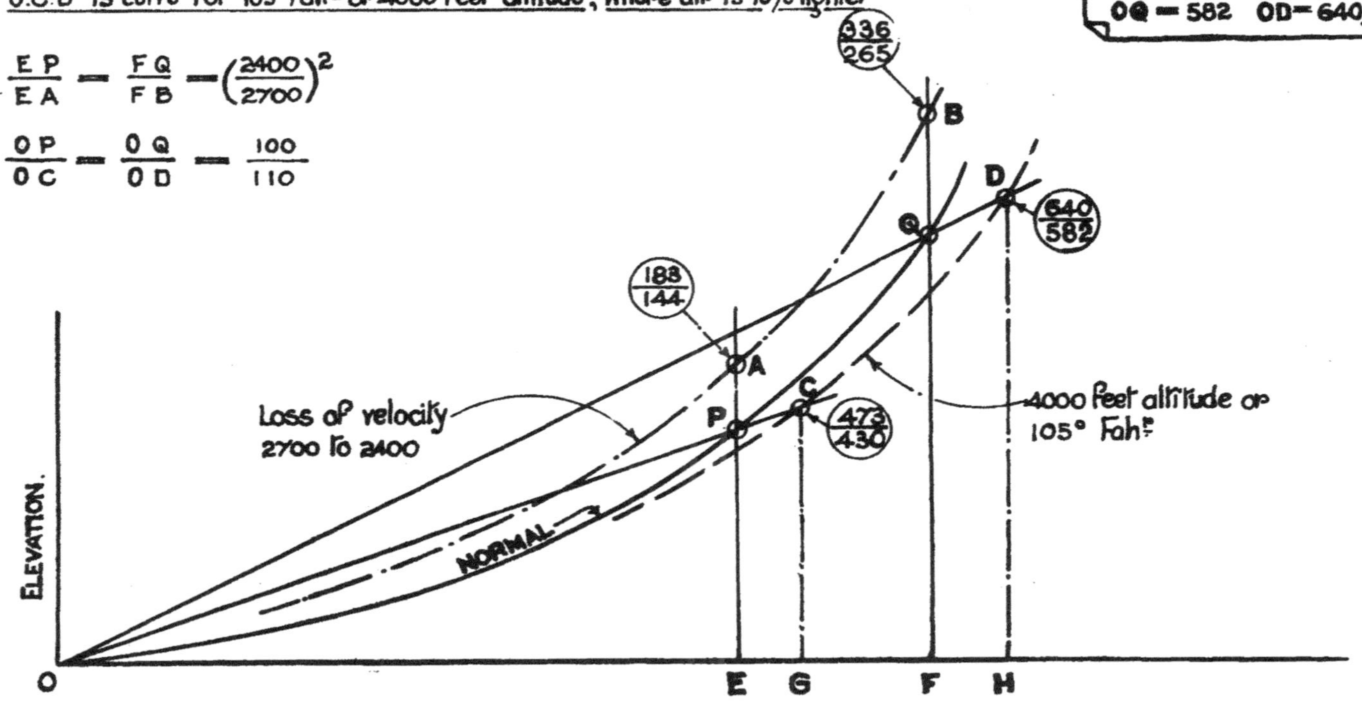

| | |
|---|---|
| EA = 183 | EP = 144 |
| FB = 336 | FQ = 265 |
| OP = 430 | OC = 473 |
| OQ = 582 | OD = 640 |

The problem he set himself to solve was how best to utilize a "fixed sight" or, as it is sometimes called, a "battle sight." Such a sight is used in several foreign armies; it gives a standard elevation when the back sight is in its normal position. This elevation may be suitable for any one range from, say, 200 to 600 yards. This range is arrived at by considering what is the greatest distance from the muzzle at which the whole height of the trajectory is not too great for the target (of some specified height) to be struck at every shorter distance without varying the aim. If a man standing upright is the target, the range is about 600 to 700 yards. If kneeling about 500 yards, and if the face only (say, 8 inches high) is the target, the range is only some 200 or 300 yards, according to the power of the rifle. With the sight in the fixed sight position a properly sighted rifle must hit the object up to the limit of the range, because the bullet never rises above that height in that range. All the soldier has to do is to use his battle sight at that target and get a certain hit for every well-aimed shot at any range up to the battle-sight range. To quote from *Rifles and Ammunition*, 1915, by Ommundsen and Robinson, pp. 268 *et seq.* :—

"Although the difficulties of judging distances are very formidable, yet these difficulties may be greatly reduced in military individual shooting and almost entirely eliminated in sporting rifle shooting. All that is required is a thorough knowledge of the trajectory of the particular rifle used."

Supposing that the heart of a stag is 6 inches in diameter and situated 6 inches clear above the junction of the belly line and foreleg, he shows that by aiming at the junction point with the sight set at 350 yards, a hit in the heart results, whenever the stag is between 50 and 300 yards away, whereas if aim is taken at the heart and the distance is judged a miss may very easily occur.

This apparent paradox may be explained in either of two ways :—

(a) Numerically by following every detail of the figures involved.
(b) By the geometry of a reflected image.

As (b) is the more actual presentment of the problem, it may be taken first, especially as it gives an absolutely correct solution.

Suppose that the greatest height of the trajectory of R yards is $(2h)$ inches and the height of the target is $(h)$ inches. Further suppose that the target is situated on a sheet of looking-glass or clear ice or still water, by which the image is reflected and visible upside down. If the sight is set to such a range R yards (for which the elevation is E minutes), that the maximum height is $(2h)$ inches and aim is taken at the lowest point of the reflected image, and also if the line of sight is coincident with the axis of the bore, *i.e.*, the height of the foresight above bore axis is "nil," it is obvious that the actual elevation on the gun is

$$E - \frac{95 \cdot 5h}{R} \text{ minutes,}$$

because the gun is depressed by the angle due to the reflected image at which aim is taken, and this angle of depression is

$$\frac{95 \cdot 5h}{R} \text{ minutes,}$$

varying inversely as R, so that the subtracted angle is bigger according as the range is smaller.

To follow out this effect it is convenient to take a case and a definite example.

Using the Mark VII 0·303 table of Hythe results given in Appendix I, the greatest height of any trajectory can be worked out by the formula for ordinates already given.

If the value of $h$ is given, the value of $\frac{95 \cdot 5h}{R}$ can be worked out and tabulated.

Since the value of $h$ is given, the value of $2h$ is known, and from the table of greatest heights the range for $2h$ can be got and also the elevation.

From these figures the value of $E - \frac{95 \cdot 5h}{R}$ can be tabulated and the results can be considered.

The whole work when laid out in an orderly way takes perhaps 10 minutes with a slide rule. The table is as follows :—

TABLE FOR NEGATIVE ANGLE FOR ·303 MK, VII.

$2h = 72$ inches, $h = 36$ inches.

| R yds. | E mins. | Heights (in inches) of Ordinates at Various Ranges (in yards). | | | | | Greatest Height, Inches. | $\frac{95 \cdot 5h}{R}$ | $32 \cdot 0 - \frac{95 \cdot 5h}{R}$ |
|---|---|---|---|---|---|---|---|---|---|
| | | 1,000 | 800 | 600 | 400 | 200 | | | |
| 0 | 0 | 0 | 0 | 0 | 0 | 0 | 0 | Infinite | Infinite neg. |
| 100 | 3 | — | — | — | 13 | *4·2* | 2 | 34·4 | − 2·4 |
| 200 | 7 | — | — | 38 | *17* | 0 | 5 | 17·2 | 14·8 |
| 300 | 11 | — | 91 | *44* | 13 | | 10 | 11·5 | 20·5 |
| 400 | 15 | 187 | 105 | 42 | 0 | | 17 | 8·6 | 23·4 |
| 500 | 19 | 214 | *110* | — | | | 28 | 6·9 | 25·1 |
| 600 | 25 | *220* | 95 | 0 | | | 44 | 5·7 | 26·3 |
| 700 | 32·0 | 205 | — | — | | | 72 | 4·8 | 27·2 |
| 800 | 40 | — | 0 | | | | 110· | — | — |
| 900 | 49 | — | | | | | 160 | — | — |
| 1,000 | 60 | 0 | | | | | 200 | — | — |

NOTE.—(1) Height of ordinates $= R_2 \left( \frac{E_1 - E_2}{95 \cdot 5} \right)$ inches.

(2) Greatest height obtained by plotting figures in italics against range thus, 220 against 1,000, 110 against 800, etc., and reading off intermediate figures from a smooth curve.

(3) $\frac{95 \cdot 5h}{R}$ is the angle automatically subtracted by aiming at the reflected image.

(4) 32·0 is the angle for 700 yards, at which range the greatest height of trajectory is 72 inches.

Then, if E is plotted against R and $32 \cdot 0 - \frac{95 \cdot 5h}{R}$ is also plotted against R on the same sheet, the curves will be seen to cross at 120 yards or 4 minutes, and again at 620 yards or $26\frac{1}{2}$ minutes. This signifies that by using the reflected aim the target will be hit at any range between 120 and 620 yards.

By the usual fixed sight method the target 36 inches high would only be hit up to 530 yards. The difficulties of judging distances are therefore postponed in this case from 530 yards to 620 yards.

The method has often been used for rabbit shooting with a 0·22 rim fire, taking the rabbit as $2\frac{1}{4}$ inches high. If the sight is adjusted to place the mean point of impact on the point aimed at, at a range of 100 yards, and aim is always taken $2\frac{1}{4}$ inches below the rabbit, the rabbit is hit at any range between 18 and 88 yards.

As a further exercise for the reader, it may be pointed out that if the sights are $x$ inches above the axis of the bore, the correct aim for the rabbit is $(2\frac{1}{4} - x)$ inches below or, in general, that the amount below to aim is affected in this exact ratio by the height of the foresight.

SECTION 2

THE BALLISTIC TABLES AND THEIR USE

Table I of $p$, T and S contains six columns, of which the first is the velocity of the shot at intervals of 10 f/s. from 100 to 3,500.

Column 2 is headed "$p$ lbs.," and gives the air resistance in lbs. to a shot of 1-inch calibre. The shape of one inch shot, which offers this resistance is usually called the standard shape. In making the experiments on which former tables of $p$ were based, the shape used had a flat base, a cylindrical body and an ogival head struck with a particular radius, $1\frac{1}{2}$ calibres for the original tables of Bashforth and 2 calibres for the 1909 tables. The resulting tables of $p$ represented the experimenter's own opinion of what the resistance probably was, so far as he could ascertain it. There is, however, no necessary connection in nature between his opinion and the actual facts. Really accurate instrumental determinations of the resistance of a shot moving at velocities between 200 f/s. and 1,000 f/s. have never yet been made, nor is there any immediate probability of their being arrived at. The resistance in that region can only be estimated. At velocities above 1,400 f/s. the experimental difficulties are easily surmounted.

In Table I the "standard" shape is that of the modern military bullet with a flat base and the usual pointed head. The value of $p$ has been arrived at by an inductive process. Probably no existing bullet shape fits this $p$ table at all velocities exactly, but all modern flat based bullets are quite close to it.

Column 3 is headed $\Delta$ T meaning "time-difference." It gives the time taken by a standard shaped 1-inch shot, weighing 1 lb., to change its velocity by 10 f/s. against a resistance of $p$ lbs. Using the first figure of the table $20 \cdot 60$ as an example, the resistance at 105 f/s. is $\frac{1}{2}$ $(0 \cdot 0137 + 0 \cdot 0165)$ lbs. $= 0 \cdot 0151$. It takes the shot $20 \cdot 60$ seconds to change its velocity from 110 f/s. to 100 f/s., against an average resistance of $0 \cdot 0151$ lbs.

Column 4, headed T, is the summation of the $\Delta$ T column. It happens to begin with 0, but any other number would have done as well, because it is not the actual value of T which is used, but the difference between its value at two different velocities.

Thus, $T_{2440} = 222 \cdot 885$
$T_{1605} = 220 \cdot 316$

T $= 2 \cdot 569$ seconds

which is the time taken by the 1 pounder of 1 inch to change its velocity from 2,440 to 1,605 f/s.

Column 5, headed $\Delta$ S gives the number of feet travel during which the same shot takes to change its velocity 10 f/s. Thus the first figure $2,163 \cdot 0$ means that the shot travels $2,163 \cdot 0$ feet, while its velocity is changing from 110 to 100 f/s, against an average resistance of $0 \cdot 0151$ lbs.

Column 6, headed S, has a similar meaning to that of the T column. It is the sum of all the values of $\Delta$ S preceding it. Here the first figure $1,489 \cdot 0$ is arbitrary. The second figure $3,652 \cdot 0$ is $1,489 \cdot 0 + 2,163 \cdot 0$.

Using the same velocities as in the T example

$S_{2440} = 69,010 \cdot 4$
$S_{1605} = 63,920 \cdot 5$

S $= 5,089 \cdot 9$ feet $= 1,696 \cdot 6$ yards,

which is the distance travelled in horizontal flight, while the velocity changes as above.

It is therefore evident that Table I gives the simple connection between the distance travelled, the time occupied and the change of velocity from one given value to another for the 1 pounder of 1 inch diameter of standard shape.

Incidentally, it shows that the expression standard shape is a misnomer. It would be accurate to say "of standard *resistance*," but to save confusion the old term "shape" is retained.

To use this table for bullets of dimensions other than standard, a multiplier termed the "ballistic coefficient" and denoted by C is employed. This takes account of the weight, diameter and shape or resistance factor of the bullet, so that

$$C = W/n\, d^2.$$

In this expression $W/d^2$ is the sectional density of the bullet, or the weight in pounds divided by the square of the nominal calibre in inches. For the Mark VII bullet of 174 grains and ·303 inch diameter

$$W/d^2 = \frac{174}{7{,}000 \times (0\cdot303)^2} = 0\cdot271,$$

since there are 7,000 grains in 1 lb.

The value of $n$ for use with these tables at 60° F. and 30-inch barometer is about 1·00, as the shape is that of the so-called standard, viz., a modern flat-based pointed bullet. Hence $C = 0\cdot271$.

Round-nosed bullets offer greater resistance, and the value of $n$ to be used is greater than unity. For the Mark VI bullet of 215 grains at short ranges, say, to 1,000 yards, it is about $1\tfrac{1}{2}$, so that $W/d^2$ for Mark VI is $0\cdot335$, and C is about $0\cdot335/1\cdot5$ or $0\cdot223$. C is therefore a measure of the ballistic efficiency of the bullet or, in ordinary language, it is the "bullet value," as it measures the power of the bullet to overcome the resistance of the air.

The factor, $n$ is termed technically the "coefficient of reduction." A numerical example will make its meaning clear. The resistance $p$ in Table I at 2,440 f/s. is given as 13·01 lbs. If a particular bullet of calibre $d$ inches in a particular locality offers a resistance of 1·23 lbs., the value of $n$ is there and then

$$n = \frac{1\cdot 23}{13\cdot 01 d^2}.$$

If $d = 0\cdot303$ and $d^2 = 0\cdot0918$, then $n = 1\cdot0298$ or $1\cdot03$. This same bullet in the same locality, moving at 1,000 f/s. where $p = 1\cdot49$, might offer a resistance of 0·1545 lbs., giving

$$n = \frac{0\cdot1545}{1\cdot49 \times 0\cdot0918} = 1.13.$$

Such variations of the value of $n$ are probably of frequent occurrence, but the values of $p$ chosen for this table have been purposely selected, so that each modern military bullet has a value of $n$ which is nearly constant for the first 2,000 yards of flight. The subject will be referred to in greater detail in a succeeding section.

The value of $n$ is affected by the conditions of the locality of the firing because the barometer and thermometer are always changing. The effect of the barometer and thermometer is taken into account in the tenuity factor, generally designated by the Greek letter "tau," and the special effect of the altitude above sea-level which affects the tenuity is often called $f$. With small arms there appears no object in keeping these terms separate, so the whole effect of density of the air, whether caused by altitude, or by temperature or other meteorological changes, will be in future called F.

A table of F for the "tau" effect is given in the appendix, Table V. The special effect of altitude is given by the formula

$$F = 1 + \frac{Y}{1{,}000}(0\cdot022),$$

where Y is the average height in feet of the bullet above sea level for the period under consideration. This average height may be taken as the height at the beginning of the period added to two-thirds of the increase of height during the period.

This altitude effect is of small consequence with small arms, because of their comparatively short effective range. Even in air fighting at 10,000 feet altitude it can be almost neglected in practice, because even at a fighting range of ¼ mile the unavoidable dispersion of the cone of fire is so much greater than the effect of altitude. Also, as shown in an example in Section 4 of this chapter, the effect on vertical fire is practically compensated by the action of gravity.

The value of F is used as a multiplier to the value of $n$ found in normal weather. Thus at 77° F. with a 28-inch barometer, $F = 0 \cdot 901$. If this was the weather when $n$ was found to be $1 \cdot 03$ at 2,440 f/s,. then the value of $n$ at 60° F. and 30-inch was $1 \cdot 143$, since $1 \cdot 143 \times 0 \cdot 901 = 1 \cdot 03$.

The ballistic coefficient $C = W/nd^2$ is used as follows:—

In the preceding examples of T and S the values for the standard shot were found to be:—

$$2 \cdot 569 \text{ seconds and } 5089 \cdot 9 \text{ feet.}$$

For any other shot having a ballistic coefficient of value (say) $C = 0 \cdot 271$, the time and distance are multiplied by $0 \cdot 271$, and become $0 \cdot 696$ seconds and 1,379 feet respectively.

The formulæ for use are then:—

$$t/C = T_V - T_v,$$
$$s/C = S_V - S_v,$$

each involving four items, of which, when any three are known, the fourth is found by a simple sum.

In a succeeding section examples are given of the working.

Table IA gives two auxiliary tables calculated from Table I. They are chiefly required for the computation of Table II, as briefly explained a little later. In this connection they require no comment, as we are not concerned with the method of constructing tables.

The $I_V$ table has, however, another use in the formula

$$\tan A = C (I_V - I_v),$$

giving the tangent of the angle turned through by the trajectory as the velocity changes from V to $v$.

The method of using the table is due to W. D. Niven, formerly Professor at the Artillery College, Woolwich (*vide Proc. R.S.*, Vol. XXVI, No. 181, 1877). He assumed that the time to the vertex was exactly half the total time of horizontal flight. Using the S and T columns of Table I, and knowing the value of V, C, and R, the remaining velocity and total time are evaluated. Half this time is then the time to the vertex. Knowing this, the velocity at the vertex is obtained, and the values of Iv and Iv are read off Table IA, whence tan A. Since the trajectory at the vertex is horizontal, the angle A is therefore the angle of elevation required. As, however, the assumption of half-time to the vertex is always untrue in air, this method is liable to produce considerable errors. If, however, the time to the vertex is assumed to be $0 \cdot 45$ of the total horizontal time, this method can be used up to 5° elevation, with an error seldom greater than 3 minutes of angle.

It is then useful for calculating elevation tables for muzzle velocities between 2,000 and 1,000 f/s. up to ordinary sporting or target ranges. It should, however, not be relied upon for obtaining the value of C from the observed elevation, range and muzzle velocity.

Table II. This table consists of 13 separate complete tables for the standard shot, one for every 100 f/s. M.V. from 3,200 f/s. to 2,000 f/s., covering all the useful M.V. of modern military rifles. Each table is made out for every 500 yards up to 10,000 yards under the heading $Z = R/C$ yards and gives $A = E/C$, the minutes elevation required,

$u$ the remaining velocity in f/s. and $T = t/C$, the time of flight in seconds, together with difference columns $\Delta z$ and $\Delta v$.

$\Delta z$ is the difference from one value of $z$ to the next higher one. $\Delta v$ is the difference from one M.V. to the next higher M.V.

This table is handled for other ballistic coefficients in the same way as Table I.

The Z, A and T column are multiplied by C to give the yards, minutes elevation and seconds time of flight for a bullet with known C.

To make this clear, and also to show the use of the difference columns, an example from the Mark VII elevation, as observed at Bisley in 1922 is given. The data are known to be

M.V. = 2,450, elevation at 1,000 yards = 58·0 minutes,

which is fitted by a value of $C = 0·263$.

The first operation is to construct a table for M.V. 2,450 by interpolation from Table II. The work is as follows:—

| Z | A V=2,400 f/s. | ΔV V+100 f/s. | A V=2,450 f/s. | $u$ V=2,400. | ΔV V+100 f/s. | $u$ V=2,450. | T V=2,450. |
|---|---|---|---|---|---|---|---|
| 0 | 0 | — | 0 | 2,400 | 100 | 2,450 | 0 |
| 500 | 15 | 1 | 14½ | 2,147 | 99 | 2,196 | 0·65 |
| 3,500 | 201 | 20 | 191 | 1,023 | 31 | 1,038 | 6·88 |
| 3,664 | 218 | 15* | — | — | — | — | — |
| 3,762* | 229* | 25 | 217* | 1,000 | — | 1,000 | 7·66 |
| 4,000 | 254 | 24 | 242 | 953 | 26 | 966 | 8·38 |
| 4,500 | 315 | 28 | 301 | 894 | 22 | 905 | 9·98 |

An explanation of the method by which the figures marked with an asterisk are obtained will make the above working easier to follow.

I.—The figure 15 is the difference in A caused by a change in velocity of 100 f/s., when $Z = 3,664$, and is obtained as follows:

A (for $Z = 3,500$ at 2,500 f/s. V.) = 181 (Table II).

A (for $Z = 3,860$ „ ) = 216 (Table II).

∴ $\Delta A$ = 35.

Z (for $u = 1,000$ f/s.) at 2,400 f/s.V. = 3,664 (Table II).

Z (for $u = 1,000$ f/s.) at 2,500 f/s.V. = 3,860 (Table II).

3,664 − 3,500 = 164.

3,860 − 3,500 = 360.

∴ $\Delta A$ for 2,500 f/s. = $\dfrac{164}{360} \times 35 = 15.$*

It will be seen that for V 2,500 f/s.

A (for $Z = 3,664$) = 181 + 15 = 196,

but this figure is not required in the calculation, and is omitted in the above working table.

II.—The figure 3,762* is an interpolation between 3,664 and 3,860, and is the value of Z corresponding to $u = 1,000$ f/s. when $V = 2,450$ f/s.

III.—The figure 217* is an interpolation for A corresponding to Z = 3,762 at V = 2,450 f/s. The first step in the working is to find A for Z = 3,762 at 2,400 f/s.

A (for Z = 3,664 at 2,400 V.) = 218 (Table II).

A (for Z = 4,000 at 2,400 V.) = 254 ,,

$\therefore \Delta$ A (for Z = 3,762 at 2,400 V.) = $\frac{98}{336} \times 36 = 10 \cdot 8$.

$\therefore$ A (for Z = 3,762 at 2,400) = 229*.

A (for Z = 3,860 at 2,500 V.) = 216 (Table II).

A (for Z = 3,500 at 2,500 V.) = 181 ,,

$\therefore \Delta$ A (for Z = 3.762 at 2,500 V.) = $\frac{262}{360} \times 35 = 25$.

$\therefore$ A (for Z = 3,762 at 2,500 V.) = 206.

$\therefore$ A (for Z = 3,762 at 2,650 V.) = $\frac{229 + 206}{2} = 217*$.

The Z, $u$ and T columns in Table II are calculated from Table I. The A column requires a long and heavy set of computations requiring both Table I and Table IA.

The formulæ are

$$\tan E = \tfrac{1}{2}aC.$$

or

$$E = AC = 3{,}438 \times \tfrac{1}{2}aC$$

Where

$$\tfrac{1}{2}a = I_V - \frac{A_V - A_\bullet}{S_V - S_\bullet}.$$

The table for C = 0·263, V = 2,450, then reads

| R | E | $v$ | $t$ |
|---|---|---|---|
| 0 | 0·0 | 2,450 | 0·00 |
| 131·5 | 3·8 | 2,196 | 0·17 |
| — | — | — | — |
| 920 | 50·3 | 1,038 | 1·81 |
| 991 | 57·1 | 1,000 | 2·02 |
| 1,052 | 63·7 | 966 | 2·21 |
| 1,183 | 79·2 | 905 | 2·63 |

*Note.*—The first figure in the R column 131·5 is 500 × 0·263.

In this way a range table can be made out very rapidly indeed as soon as a value of C is settled upon.

Table III is in the same shape as Table II for V = 1,000 f/s., but it is used principally for quite different purposes, and contains an extra column, headed B, for use with the formula $\tan W = \dfrac{1{,}719\ B}{A} \tan E$, where W = angle of descent. It is made out for the quadratic law where resistance varies exactly as the square of the velocity. The formulæ from which it is calculated are given in "Ballistic tables," printed for H.M. Stationery Office, 1910, (2,500 copies for official use only), Introduction pp. ix. and x.

The law used is, in Bashforth's notation,

$$1 \cdot 49 \text{ lbs.} = p = k/g \left(\frac{V}{1,000}\right)^2, \text{ whence } k/g = 1.49,$$

or in 1910 notation, $Aq = 48 \cdot 0$; $V = 1,000$.

The advantage of using the quadratic law is two-fold. First, it is antecedently probable that a bullet flying at any velocity below about 1,000 f.s. should encounter a resistance of this nature having a constant value of $k/g$, provided the bullet flies point first (i.e., without yaw) or assumes a position in flight having a small constant angle of yaw. Second, a table made out on this law for one particular muzzle velocity can be used for any other muzzle velocity very easily. This table is made out for $V = 1,000$, Suppose it is wished to use it for a revolver with $V = 615$ f/s., the table is multiplied as follows :—

$$\text{old A} \times \left(\frac{1,000}{615}\right)^2 = \text{new A};$$

$$\text{old } u \times \frac{615}{1,000} = \text{new } u;$$

$$\text{old T} \times \frac{1,000}{615} = \text{new T};$$

$$\text{old B} \times \left(\frac{1,000}{615}\right)^2 = \text{new B}.$$

It is then used exactly as Table II is used. The main object of Table III is, however, to enable rifle trajectories and especially striking velocities to be calculated with accuracy at long ranges up to 20° or 30° elevation and to investigate the effect of yaw at long ranges.

Table IV is Bashforth's table (extended by the present Sir A. G. Hadcock) for spherical shot.

It is useful for such projectiles as shrapnel bullets and No. 6 and similar shot for game to find striking velocity and time of flight. It is used in the same way as Table I. The numerical results obtained with it differ slightly from those given by Table III.

Table V is Bashforth's table of tenuity due to changing temperature and barometric pressure. It will be noticed that a rise of 16° F. just counteracts a rise of 1 inch in the barometer, and that they each affect the value of F by $3 \cdot 3$ per cent. For ordinary purposes the table is not required, as the simple rule suffices that a rise of 16° F. in thermometer, or a fall of 1 inch in barometer, or a rise of 1,000 feet above sea level, each and all reduce the value of F or of $n$ by $3 \cdot 3$ per cent and so tend to increase the range.

For very great altitudes ($h$ feet) such as aircraft use, say 10,000 feet or more, some such formula as

$$1/F = 1 \cdot 0 + 0 \cdot 000026h$$

may be used. The fighting ranges in the air are, however, as yet comparatively small, perhaps ¼-mile or so. At such short ranges an error in estimating the value of C has to be very large indeed to produce a vertical change of point of impact of a yard.

## Section 3

### Given the Values of C, R and V, to Construct a Full Range Table

A complete range table would contain full information for every 100 yards from 0 to extreme ranges on the following nine particulars for an angle of sight of zero :—

(1) Angle of departure required in minutes, E minutes.
(2) Angle of descent, W minutes, and also its tangent as a gradient.
(3) Dimensions of 90 per cent. cones, vertical and horizontal, called beaten zones
(4) Time of flight.
(5) Remaining velocity and striking energy.
(6) Effect in yards of five minutes' alteration of elevation or deflection.
(7) Dangerous space.
(8) Effect of wind on range and direction.
(9) Effect of tenuity.

The calculations are as follows :—

(A) Using Table II, and the given values of C and V, draw up a table of R, E, $v$ $t$, as explained in Section II.
(B) Plot each of the columns of E, $v$, and $t$ on squared paper against R and read off values of E, $v$, and $t$ at each 100 yards of R.
(C) Difference these columns of E, $v$ and $t$, and smooth the ragged places of the difference column by altering the main column slightly, thus :—

| R | E off Curve | Difference | New Difference | Smooth E | Smoothing |
|---|---|---|---|---|---|
| 0 | 0·0 |  |  |  | 0 |
|  |  | 3·0 | 3·0 |  |  |
| 100 | 3·0 |  |  | 3·0 | 0 |
|  |  | 4·0 | 3·7 |  |  |
| 200 | 7·0 |  |  | 6·7 | −0·3 |
|  |  | 4·0 | 4·3 |  |  |
| 300 | 11·0 |  |  | 11·0 | 0 |
|  |  | 5·0 | 5·0 |  |  |
| 400 | 16·0 |  |  | 16·0 | 0 |
|  |  | 6·0 | 6·0 |  |  |
| 500 | 22·0 |  |  | 22·0 | 0 |

It was not really necessary to smooth this curve, but the example shows the principle that $-0·3$ on one item of the main column gives $-0·3$ on the preceding first difference and $+0·3$ on the following first difference ; so that consecutive changes of the same amount but of opposite signs in the first difference column are equivalent to a change of the same amount in one, and only one, item of the main column.

This simple example contains the whole theory of smoothing curves of evenly spaced items, and is worth the most careful attention. It can be applied to the second difference column also by changing three consecutive second differences, No. 1 being $-1$, No. 2 $+2$, and No. 3 $-1$. Further applications will readily occur to the reader once the principle is grasped of operating on the difference column instead of on the main column.

A, B, and C above supply the information for the columns for Nos. (1), (4) and (5), since the remaining energy is $2 Wv^2/2g$ ft. lbs.

No. (2), the angle of descent W, is found by Halford's formula in Section I, or at long ranges it may be taken as double the angle of elevation as a first approximation. The slope of descent is found as a gradient by looking out tan W in the tables of natural tangents and changing the decimal to a fraction of one in  . . .   Thus, if W = 6° 31′, tan W = tan 6° 31′ = 0·1142 = 1/8·76, giving a gradient of one in 8·76.

No. (3).—The dimensions of the 90 per cent zone are found either by actual firing at several ranges with the actual gun mounting and ammunition, or else they are calculated from the known angular dimensions of the trumpet-shaped cone for the general run of modern military rifles, when these angular dimensions are plotted against the angle of elevation.

It should be noticed that the area comprised in the rectangle whose length is the 90 per cent length zone and whose breadth is the 90 per cent breadth zone, contains 81 per cent of the shots, and not 90 per cent of them.

The width and length of the 90 per cent zones at all ranges are experimental figures. When they are given, the height in yards of the 90 per cent length zone is obtained at any range by dividing it by the gradient of the slope of descent at that range. Thus, if at 2,800 yards the 90 per cent length zone is 210 yards and the gradient is 1 in 2·3, the height is 210/2·3 = 91½ yards very nearly.

No. (7).—The dangerous space or distance between the first catch and the first graze for any one bullet of a cone is equal to the height of the object in yards (say 2 for a man and 3 for a horseman) multiplied by the number in the gradient. Thus, if the gradient at 1,100 yards is 1/31, the dangerous space for a man is 62 yards inwards from the first graze, i.e., 1,100 to 1,038 yards, and 93 yards for a horseman, i.e., 1,100 to 1,007 yards. The length of the dangerous space decreases rapidly as the range increases, because the slope of descent gets steeper.

The dangerous zone is also a length. It equals the length of the 90 per cent zone together with the dangerous space. Thus, at 1,100 yards as above, if the 90 per cent zone is 270 yards long, the near edge of the 90 per cent zone is at 965 yards, where the gradient is 1/46 and the dangerous space for a man is 92 yards. The dangerous zone is then 362 yards long from 873 yards to 1,235 yards.

No. (6), the effect of 5 minutes alteration of deflection is as shown in Section I

$$\frac{5 \times R}{3,438} \text{ yards,}$$

or R/688 yards, or, say, 1 yard at 700 yards.

The effect of 5 minutes elevation depends upon the rise in minutes for an extra 100 yards at that range. If the rise is 17 minutes for the next 100 yards, then 5 minutes gives 5/17ths of 100 yards extra range.

No. (8), the formula for the effect of a lateral wind is in feet

$$W \sin A \left( t - \frac{3R}{V} \right),$$

where W is the wind in f/s.,

> A is the angle of the cross wind to the range. Reckoning a III or IX o'clock wind as 90°
> $t$ is the time of flight over the range R yards, and
> V is the muzzle velocity in f/s.

This is not a very convenient form for daily use. By taking any one of the muzzle velocities of Table II and using the Z and T columns with C = 1·0 and R = Z with W sin A = 1·0 and T = $t$, the value of $t - \frac{3R}{V}$ can be worked out for each value of Z in a few minutes. The wind deflection in feet is proportional to the result. If the result is divided by the range, this second result is proportional to the wind allowance in minutes of angle.

Thus, for V = 2,450, the answers are as below :—

| Z yards | T seconds | $\dfrac{3Z}{2,450}$ | $T - \dfrac{3Z}{2,450}$ | $\div Z$ | $\div Z^2$ |
|---|---|---|---|---|---|
| 0 | 0 | 0 | 0 | 0 | 0 |
| 1,000 | 1·37 | 1·22 | 0·15 | 15 | 14 |
| 2,000 | 3·16 | 2·45 | 0·71 | 35 | 17 |
| 3,000 | 5·48 | 3·68 | 1·80 | 60 | 20 |
| 4,000 | 8·35 | 4·90 | 3·45 | 86 | 21 |
| 5,000 | 11·68 | 6·12 | 5·56 | 111 | 22 |
| 6,000 | 15·42 | 7·35 | 8·07 | 135 | 22 |
| 7,000 | 19·60 | 8·57 | 11·03 | 158 | 23 |
| 8,000 | 24·30 | 9·81 | 14·49 | 181 | 22 |
| 9,000 | 29·57 | 11·03 | 18·54 | 206 | 23 |
| 10,000 | 35·50 | 12·25 | 23·25 | 233 | 22 |

Here the column headed $T - \dfrac{3Z}{2,450}$ is proportional to the wind deflection in feet.

The column headed ÷ Z (divided by Z) is proportional to the wind allowance in minutes.

If this column is again divided by Z the result is seen in the column headed ( ÷ $Z^2$ ), and after Z = 3,000 the answer is nearly constant and equal to about 22. This signifies that the wind allowance *in minutes* for a constant wind is practically proportional to the range up to the limit of the table. As C is generally in the region of one-quarter to one-third for modern bullets, Z = 2,000 means that the range is about 500 or 600 yards, and Z = 10,000, that it is 2,500 to 3,000 yards.

Hence, if the wind allowance for 600 yards, say, is known or judged correctly, all the other wind allowance can be got by proportion of the actual range to 600 yards, or the 1,200 yards allowance in minutes is double the 600 yards allowance, and the 2,400 yards allowance is fourfold. The 300 yards allowance is rather less than half the 600 yards allowance.

A simple and accurate way of comparing bullets for comparative lateral stiffness in a wind is to observe or calculate the allowance required on the same day and range, and then reduce the result to a figure showing the range at which a wind of 1 m.p.h. requires 1 minute deflection. This is easily done by dividing the range by the deflection required for each m.p.h. of wind. Some results are below :—

| Bullet | Range Employed | Deflection Obtained | Range Figure |
|---|---|---|---|
| ·303-inch Mark VI | 1,000 | 1·88 | 560 |
| ·303-inch ,, VII | 1,000 | 1·45 | 730 |
| ·303-inch 1921 Magnum flat base | 1,100 | 1·04 | 1,060 |
| ·303-inch 1922 ,, H long point | 1,100 | 0·82 | 1,340 |
| ·303-inch 1922 ,, streamline | 1,100 | 0·82 | 1,340 |
| ·296-inch Swiss | 1,100 | 0·88 | 1,250 |
| 375/303 1909 | 1,100 | 0·94 | 1,170 |
| ·280 Ross match | 1,100 | 0·79 | 1,390 |
| Martini Henry 1880 | 1,000 | 2·03 | 490 |
| ·461 Metford Match rifle | 1,000 | 1·18 | 850 |

Here the column headed "range figure" is the range at which one minute deflection is wanted for each m.p.h. of wind. The figures given are believed to be reliable. They are useful when choosing a rifle for very accurate work, such as sniping.

The calculation for wind effect on range is not so simple, and is rather long. For all ordinary purposes up to a mile or so the formula can be written as

$$\Delta R = \frac{3W}{CV} N^2 \text{ yards,}$$

where N is the number of hundreds of yards in the range and $\Delta R$ is the change of range due to the wind, added for a following wind and subtracted for a head wind.

With $C = 0.25$ and $V = 2,400$, this becomes $\Delta R = \frac{W}{200} N^2$, and with $W = 14.7$ f/s., or 10 m.p.h. $\Delta R = N^2/13\frac{1}{2}$.

The total effect then of a 10 m.p.h. wind on a bullet like Mark VII is as follows:—

| Range. | N | $N^2$ | Effect in yards. |
|---|---|---|---|
| 0 | 0 | 0 | 0 |
| 500 | 5 | 25 | 1·85 |
| 1,000 | 10 | 100 | 7·4 |
| 1,500 | 15 | 225 | 16·6 |
| 2,000 | 20 | 400 | 29·6 |
| 2,500 | 25 | 625 | 46·3 |

As the time of flight for 1,500 yards is about 4 seconds, and for 2,500 yards about 10 seconds, and 10 m.p.h. is 5 yards a second, the wind effect beyond a mile range is practically equal to the distance travelled by the wind in the time of flight. At distances less than a mile the effect is so small that it may be neglected. Hence a special formula is not really required. To bring these effects to minutes it is only necessary to multiply by the "rise" and divide by 100.

SECTION 4

DETAILED NUMERICAL EXAMPLE OF ALL THE TRAJECTORIES OF ONE RIFLE, FOR ELEVATIONS FROM 0° TO 90°

The example selected is taken from the dimensions of the Swiss rifle, calibre 0·296 inch, bullet 174 grains, $W/d^2 = 0.284$, M.V. 2,600 f/s. value of "$n$" taken as unity, giving $C = 0.284$, as for a flat-based bullet. The Swiss bullet is actually stream lined.

Using Tables I and II, and also the auxiliary tables referred to in Section II, the computing staff of the Ordnance Committee worked out in full detail the trajectories at 5°, 10°, 20°, 30°, 35°, and 40°, with the results abstracted in Appendices II and III. The elaborate nature and precision of this calculation may be inferred from the following facts:—

(1) The bullet was located at every 1/16th second of its flight in each trajectory.

(2) 66 sheets of paper, each having 8 columns of 30 lines of figures, were used, giving 528 columns, or 15,000 lines, or 100,000 digits.

(3) In addition, the auxiliary tables took nearly as much work to evaluate.

(4) No such calculation for a rifle was extant before.

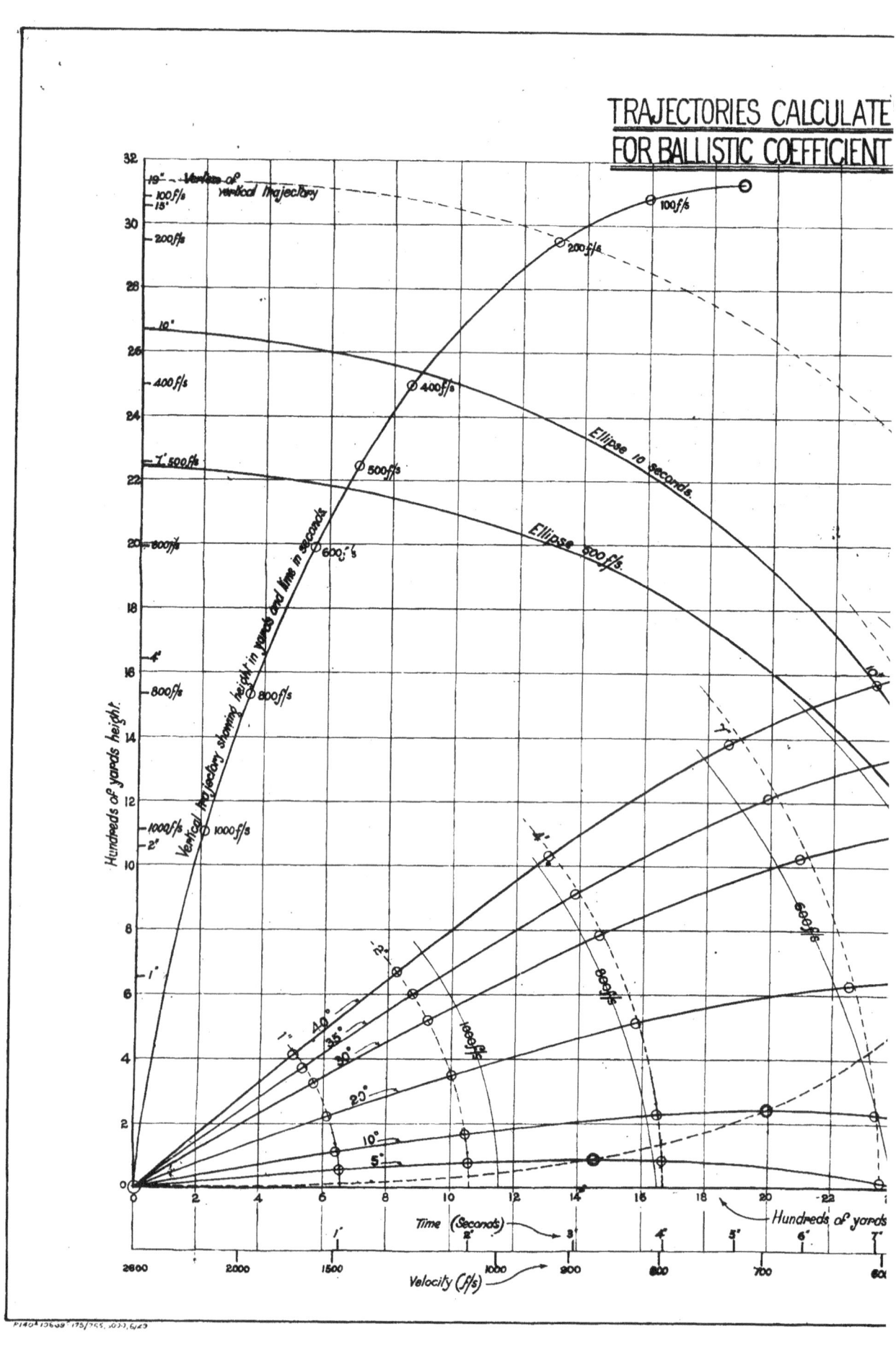

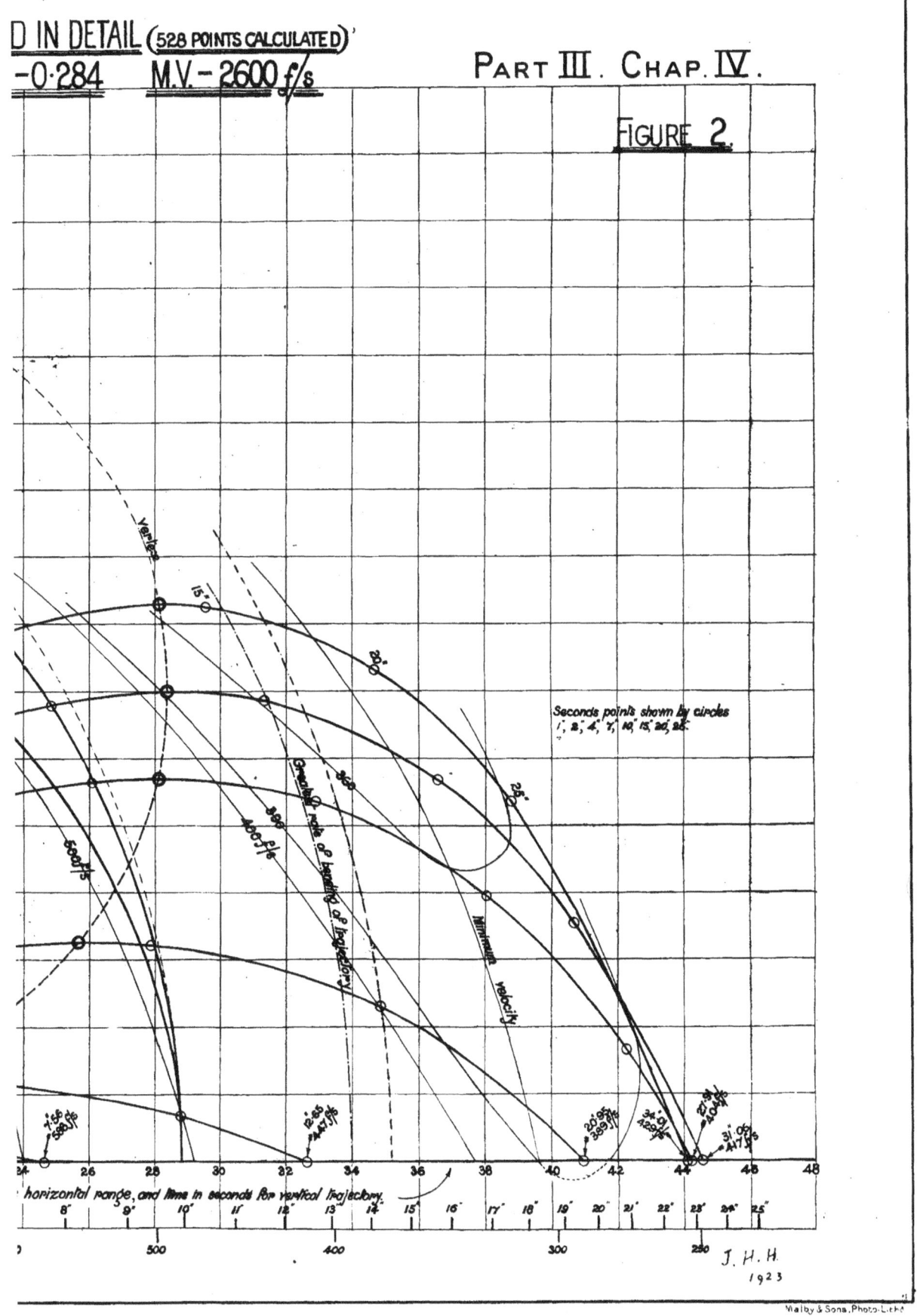

(5) It proves conclusively that the greatest range of small-arm projectiles must occur at about 35° to 30° elevation, and not at 45°. This was proved experimentally at Lydd in the S.A. Penetration Trials with Mark VI bullet in 1893, but it was not then thought to be necessarily true, as it is, for all small arms.

The complete results are also shown plotted in considerable detail in Fig. 2, where the trajectories are plotted to scale, including the trajectory when the bullet is fired vertically upwards.

The position of the bullet after 1, 2, 4, 7, 10, etc., seconds is given, as also its position when its velocity has fallen to 1,000, 800, . . . . , 360 f/s. In each case, the curve is not quite a quadrant of a circle or of an ellipse. The position of the bullet when it is changing its direction most rapidly is indicated by a curve marked "greatest bending of trajectory." The rate of change varies from about 1° per second near the muzzle to as much as $5\frac{1}{2}°$ per second in the 40° trajectory after 15 seconds' flight.

Running along the bottom of the diagram is a scale of time in seconds and velocity in f/s showing what would be the travel of the bullet along the horizontal range if gravity did not exist, and it only had to overcome air resistance. The working out of this scale is given in Appendix V. Similarly running up the left-hand side is a scale showing the same items when the bullet is ascending vertically against both gravity and air resistance, as, in fact, occurs when firing straight up. In the calculation of the vertical path an allowance is made for the decreased density of the air as the bullet ascends. By comparison with the horizontal scale, it will be seen that for the first few thousand feet vertically the tenuity effect practically wipes out the gravity effect, both being small compared with the air resistance effect.

Appendix IV gives the calculations required for range, elevation, time of flight, remaining velocity, angle of descent (W), and for the value of "$f$" the sophisticated value of gravity mentioned in Section I. Except for the column W, every figure is obtained by slide rule work directly from Table II, up to the limit of this table, viz., Z = 10,000 or R = 2,840 yards. Beyond this range special methods have to be used which are indicated in the following section. They are too complicated for ordinary use.

The value of W is obtained by Halford's method described in Section I. It requires the E and R columns to be plotted and a range table read off for each hundred yards of range to obtain the mean "rise" at each hundred yards exactly. An example is given later. It will be noticed from Appendices III and IV that W is about $2\frac{1}{4}$ to $2\frac{1}{2}$ times E for moderate elevations. As a rule, it is quite accurate enough to call it $2\frac{1}{4}$ times E, or W = $2\frac{1}{4}$ E minutes.

To work out the vertical trajectory it is best to divide it into two parts, the first from earth till $v = 1,000$ f/s. and the second from $v = 1,000$ f/s. to $v = 0$, which is the vertex.

The method is as follows:—

In the first part the trajectory is broken up into arcs or pieces, such that an equal velocity, namely 200 f/s., is lost in each arc. The 200 f/s. arcs are worked out by the "S" Table to find the distance in feet on a horizontal trajectory in which the velocity falls 200 f/s.

The assumption is then made that gravity and tenuity just neutralize one another. On this assumption the value of F for each arc is evaluated from the formula

$$F = 1 + \frac{Y}{1,000}(0 \cdot 022).$$

as shown in Table A.

## TABLE A

| V | Sv | $\Delta S = \frac{s}{c}$ | S feet ($\Delta S \times 0.284$) | $\Sigma S$ | 2/3 S | $\Sigma S + 2/3 S$ = Mean S or $\bar{y}$ | F |
|---|---|---|---|---|---|---|---|
| 2,600 | 69,947 | | | 0 | | | |
| | | 1,172 | 333 | | 222 | 222 | 1·0048 |
| 2,400 | 6,8775 | | | 333 | | | |
| | | 1,178 | 334 | | 222 | 555 | 1·0122 |
| 2,200 | 67,597 | | | 667 | | | |
| | | 1,184 | 336 | | 223 | 890 | 1·0196 |
| 2,000 | 66,413 | | | 1,003 | | | |
| | | 1,224 | 347 | | 231 | 1,234 | 1·0272 |
| 1,800 | 65,189 | | | 1,350 | | | |
| | | 1,302 | 370 | | 247 | 1,597 | 1·0351 |
| 1,600 | 63,887 | | | 1,720 | | | |
| | | 1,379 | 392 | | 262 | 1,982 | 1·0435 |
| 1,400 | 62,508 | | | 2,112 | | | |
| | | 1,701 | 483 | | 323 | 2,435 | 1·0536 |
| 1,200 | 60,807 | | | 2,595 | | | |
| | | 3,002 | 852 | | 569 | 3,164 | 1·0697 |
| 1,000 | 57,805 | | | 3,447 | | | |

In Table B, in which is continued the working out of the first part of the trajectory, the first column gives the arcs, the second gives the mean velocity in each arc, the third gives the value of "$p$" from Table I for the mean velocity, the fourth gives the value of "F" from Table A of the first part of the calculation. Then, since C = 0·284, for V = 2,500

$$\bar{p}g/FC \text{ is } 1511, \quad 1511 + g \text{ is } 1543,$$

and $\Delta t$ is 200/1,543, which is summed in the next column.

$\Delta H$ is 2,500 × 0·1297 = 324, and this is summed in the column headed $\varepsilon H$.

The total time for the velocity to fall to 1,000 f/s. is thus found to be 2·1683 seconds, during which time the bullet ascends 3,465 ft.

## TABLE B

$g = 32·2$; C = 0·284; $g/c = 1,134$.

| V | $\bar{v}$ | $\bar{p}$ | F | $pg/FC$ | $pg/FC + 32$ | $\Delta t$ | $\Sigma t$ | $\Delta H$ | $\Sigma H$ | 2/3 $\Delta H$ | $\Sigma H + 2/3 \Delta H$ | F |
|---|---|---|---|---|---|---|---|---|---|---|---|---|
| 2,600 | | | | | | | | | | | | |
| | 2,500 | 13·39 | 1·005 | 1,511 | 1,543 | ·1297 | | 324 | | 216 | 216 | 1·005 |
| 2,400 | | | | | | | ·1297 | | 324 | | | |
| | 2,300 | 12·14 | 1·012 | 1,361 | 1,393 | ·1435 | | 330 | | 220 | 544 | 1·012 |
| 2,200 | | | | | | | ·2732 | | 654 | | | |
| | 2,100 | 10·94 | 1·020 | 1,216 | 1,248 | ·1602 | | 336 | | 224 | 876 | 1·019 |
| 2,000 | | | | | | | ·4334 | | 990 | | | |
| | 1,900 | 9·57 | 1·027 | 1,056 | 1,088 | ·1835 | | 349 | | 231 | 1,221 | 1·027 |
| 1,800 | | | | | | | ·6169 | | 1,339 | | | |
| | 1,700 | 8·13 | 1·035 | 890 | 922 | ·2170 | | 369 | | 246 | 1,582 | 1·035 |
| 1,600 | | | | | | | ·8339 | | 1,708 | | | |
| | 1,500 | 6·76 | 1·044 | 734 | 766 | ·2612 | | 392 | | 261 | 1,966 | 1·043 |
| 1,400 | | | | | | | 1·0951 | | 2,100 | | | |
| | 1,300 | 4·78 | 1·054 | 514 | 546 | ·3662 | | 476 | | 318 | 2,415 | 1·053 |
| 1,200 | | | | | | | 1·4513 | | 2,576 | | | |
| | 1,100 | 2·33 | 1·070 | 247 | 279 | ·7170 | | 789 | | 526 | 3,099 | 1·068 |
| 1,000 | | | | | | | 2·1683 | | 3,465 | | | |

For the second part, from 1,000 f/s. to the vertex, the formula given in Greenhill's *Notes on dynamics*, 1908, page 83, is best used, taking the resistance to vary as the square of the velocity.

Calling the terminal velocity Q f/s., Q is found by looking out the value of "$p$," which is equal to $\overline{F}C$ in Table I.

For $\overline{F}$ a maximum height has to be estimated. Assuming the added height, during the second part of the trajectory, to be 6,000 feet, the mean height will be

$$3,465 + \tfrac{2}{3} \times 6,000 = 7,465 \text{ feet.}$$

$$\therefore F = 1 + (0\cdot 022 \times 7\cdot 465) = 1\cdot 164$$

$$p = 1\cdot 164 \times 0\cdot 284 \quad = 0\cdot 331$$

$$Q = 520 \text{ f/s.}$$

The formula then used for the evaluation of "$t$" is :—

$$t = \frac{Q}{g} \tan^{-1} \frac{V}{Q}.$$

Then

$$t = \frac{520}{32\cdot 2} \tan^{-1} \frac{1,000}{520} = \frac{520}{32\cdot 2} \tan^{-1} 1\cdot 924$$

$$= \frac{520}{32\cdot 2} \times 62\cdot 53°.$$

Since

$$62\cdot 53° = \frac{62\cdot 53}{57\cdot 3} \text{ radians}$$

$$\therefore t = \frac{520}{g} \times \frac{62\cdot 53}{57\cdot 3} = 18\cdot 38.$$

$\therefore$ total time to vertex $= 18\cdot 38 + 2\cdot 17 = 20\cdot 55$ seconds.

The height ascended in 20·55 seconds is given by

$$H = \frac{Q^2}{g} 2\cdot 30 (\log \sec. 62\cdot 53°).$$

Whence

$$H = \frac{520^2}{32\cdot 2} 2\cdot 30 (\log \sec. 62\cdot 53°)$$

$$= 6,490.$$

$\therefore$ Total height $= 6,490 + 3,465 = 9,955$ feet, or 3,315 yards. Thus the maximum height is approximately $\tfrac{3}{4}$ of the greatest range, instead of $\tfrac{1}{2}$, as in a vacuum.

The vertical height in Fig. 2 was obtained by an abbreviated method, giving 3,130 yards instead of 3,315, and 19 seconds.

## Section 5

### Practical Methods of Determining the Value of the Ballistic Coefficient

Since the weight (W lbs.) and the calibre ($d$ inches) of the bullet (taking $d$ always as the nominal calibre of the rifle for the sake of simplicity and uniformity of method) are both known, the operation of finding C would appear to resolve itself into finding $n$ only. This may be effected by merely estimating a likely value of $n$ by looking at the bullet and comparing its shape with that of other bullets of known $n$. In practice, however, guessing is not resorted to. The bullet is shot from the gun at known ranges, and with a known M.V. in known weather, that is on a calm day with a known value of the tenuity correction F of Table V. The angle of projection required for the range is noted, and also, if possible, the striking velocity. The elevation results are then plotted on a large sheet of squared paper on a scale of 5 inches to 1,000 yards and 50 minutes elevation to the inch.

The first thousand yards or so is generally plotted separately at double scale, or even larger, particular pains being taken to plot the 200 or 300 yard elevation correctly. Through the plotted points a curve is then drawn with a batten and pins, and from this curve the elevation required at each hundred yards is read off, as in Appendix I, which is the actual mean result of very many firings at all ranges at Hythe. These elevations are then differenced and slightly smoothed by alterations of one minute at a time here and there, using great care in the first 300 yards so as not to be more than a small amount greater than the vacuum figure, or

$$E \text{ (mins.)} = 2,000 \text{ N/V for N hundreds of yards.}$$

This elevation curve is then the best evidence obtainable of the actual average shooting of the rifle. It remains to reconcile calculations with it and not to alter it to suit calculated results. For this reconciliation Fig. 3 may be drawn up to the limits of Table II, or about 2,500 yards, approximately the limits of barrage fire. On this diagram the range and elevation curves of Appendix I are carefully drawn, as also the curve resulting from Table II with $V = 2,450$ and $C = 0 \cdot 250$ in a way similar to Appendix IV. The value of C used is a matter of no consequence as any value of C will end by giving the same information. V, however, must be very near the true M.V. of the rifle.

A pin is then inserted at the origin O of the curve, $R = O$, $E = O$, and a pin is stuck in at each 100 yards of the Hythe curve. Using a ruler against the two pins a line is ruled and produced to cut the calculated curve. Taking 2,400 yards as an example, the Hythe pin is at P and the calculated pin is at Q. O P measured on the original drawing is $364 \cdot 5$ mm. and O Q is $386 \cdot 5$ mm. Hence by Forbes's method the actual value of Hythe C for 2,400 is $\frac{364 \cdot 5}{386 \cdot 5} \times 0 \cdot 25 = 0 \cdot 236$. The value of C at other ranges is obtained in the same way with the results written on Fig. 3.

By plotting the time of flight column, results for C can be got in a similar way.

By a third plotting of columns Z and U from Table II for 2,450 f/s., the value of Z for any value of $u = v$ in the velocity column of Appendix I can be read off. Then as $Z = R/C$ and R is known, C is known at the range.

The values of C determined by these three methods do not as a rule agree. The difference may easily amount to 10 per cent or so. In consequence some convention has to be made to define the meaning of C. The usual convention is to adopt the value of C which fits the elevation table and to disregard the others except for some very special purpose. The values of $t$ and $v$ worked out from the elevation value of C differ so little from the observed values that the error is of no practical consequence. It should be remembered also that there is room for considerable error in observing the time of flight and still greater room in determining the striking velocity.

The values of C obtained as above from the Hythe firing are tabulated below :—

| R, Range | Values of C | | |
|---|---|---|---|
| | C by E | C by $t$ | C by $v$ |
| 1,000 | 0·255 | 0·247 | 0·264 |
| 1,500 | 0·244 | 0·241 | 0·267 |
| 2,000 | 0·244 | 0·232 | 0·246 |
| 2,500 | 0·231 | 0·210 | 0·231 |

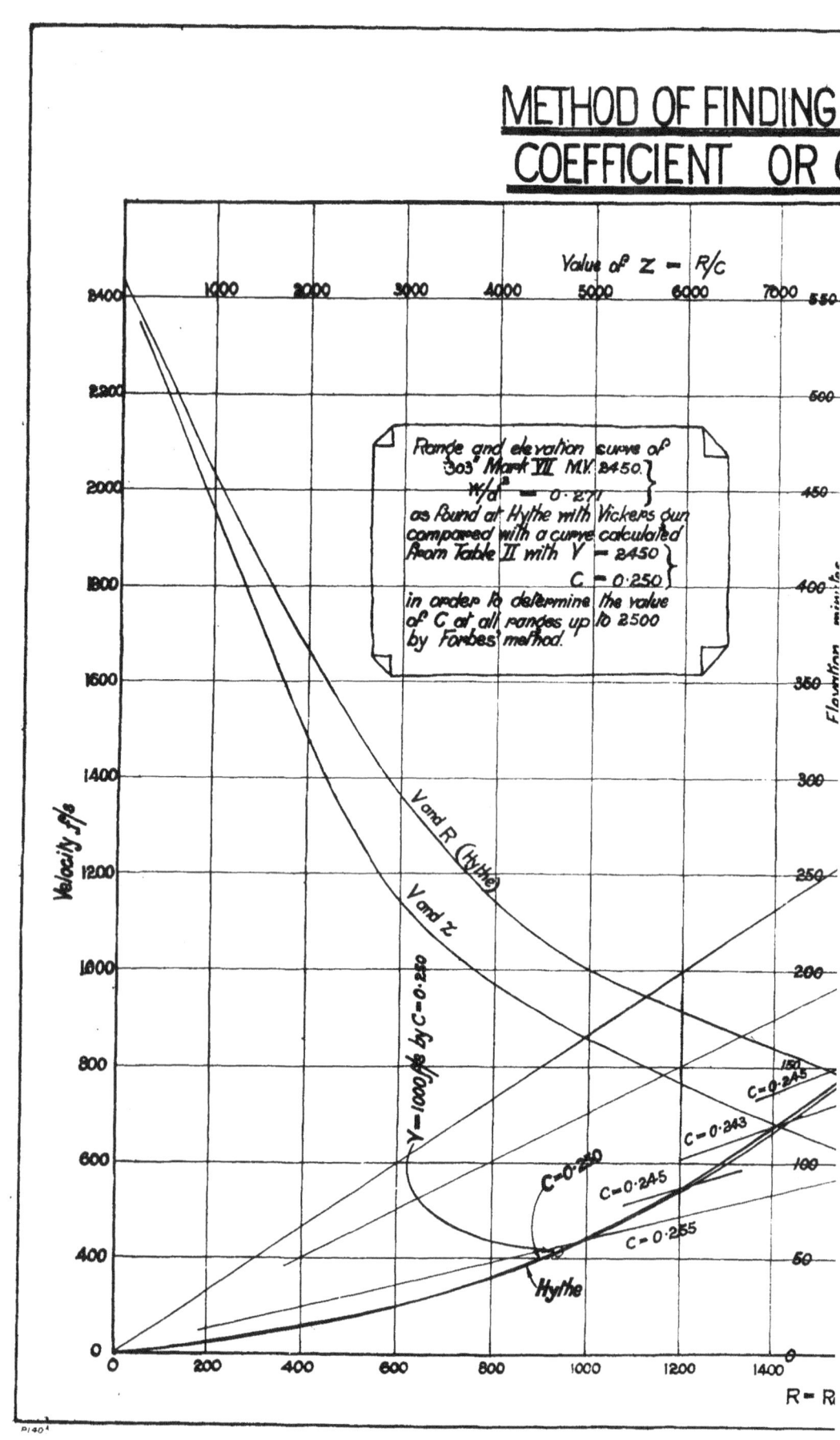

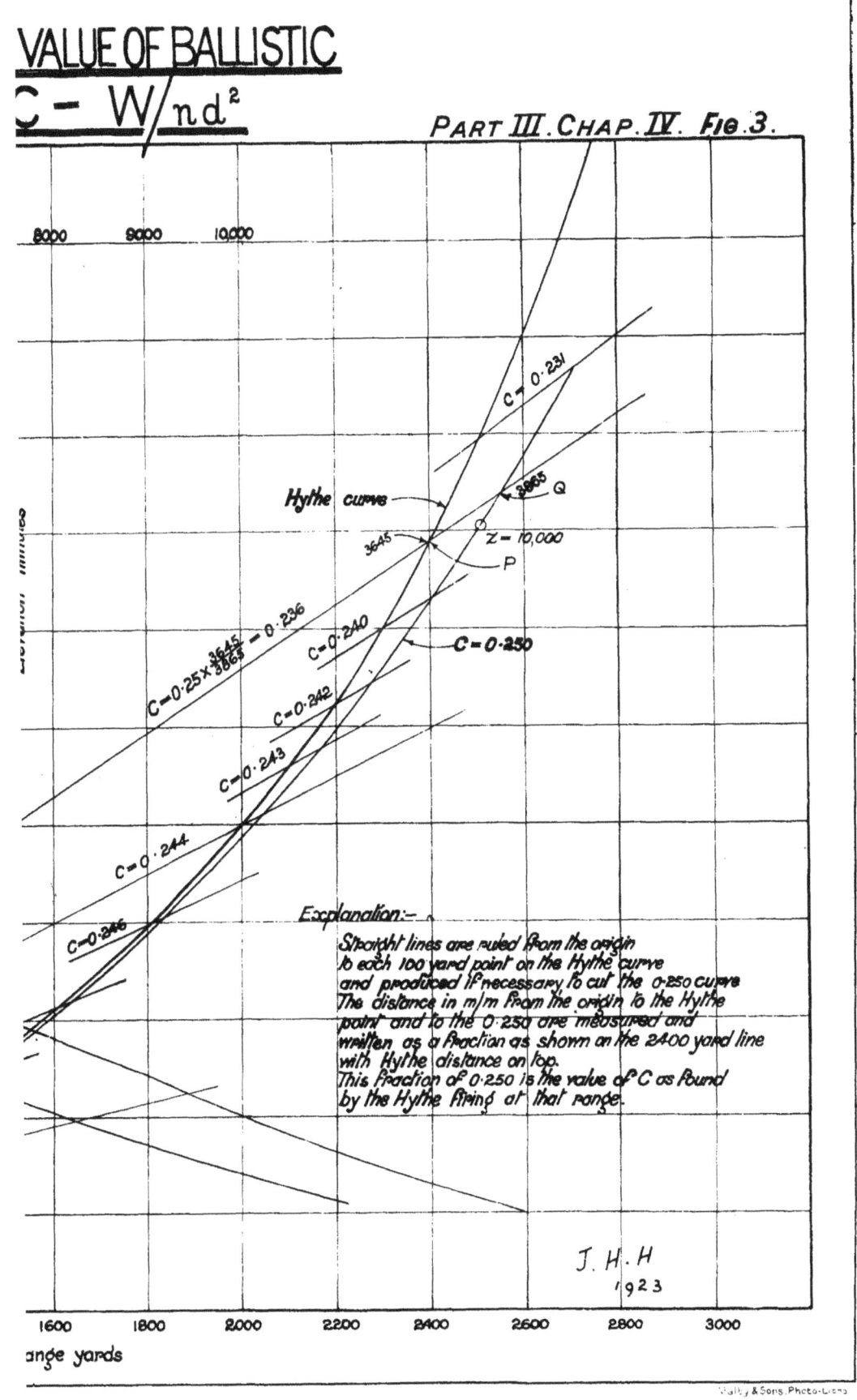

To face page 330.

PART III. CHAP. IV. FIG. 3.

Using the value of C determined from R and E and calculating the times and velocities the work is as follows :—

|   |   |   | $S_{2450} = 69,069$ $\quad T_{2450} = 222 \cdot 91$ | | | |
|---|---|---|---|---|---|---|
| R | C | 3R/C | $S_v \quad\quad v$ | $T_v$ | $t/c$ | $t$ |
| 1,000 | 0·255 | 11,770 | 57,299 $\quad$ 977 $\}$ <br> error $-$ 23 | 214·75 | 8·16 | 2·08 $\}$ <br> $-$ 0·02 |
| 1,500 | 0·244 | 18,420 | 50,649 $\quad$ 746 $\}$ <br> error $-$ 54 | 206·80 | 16·11 | 3·92 $\}$ <br> $+$ 0·02 |
| 2,000 | 0·244 | 24,600 | 44,469 $\quad$ 588 $\}$ <br> error $-$ 12 | 197·50 | 25·40 | 6·20 $\}$ <br> $-$ 0·20 |
| 2,500 | 0·231 | 32,480 | 36,589 $\quad$ 432 $\}$ <br> error $+$ 2 | 181·90 | 41·00 | 9·48 $\}$ <br> $-$ 0·42 |

The discrepancy of 54 f/s. at 1,500 yards is due to the figure 800 in Appendix I having been found by calculation using other tables and not by experiment. Either of the calculated results 800 or 746 may be right.

The discrepancy of 0·42 seconds at 2,500 yards may easily be due to the timekeeper hearing the strike of the bullet some hundreds of feet away from him, as sound only travels 1,100 feet in a second.

A further example is taken from the Hodsock (1922) experiments. Here the striking velocities were observed with great accuracy, and are given as reported. The elevations and time have been calculated from the value of C obtained using the ballistic tables, and for the purpose of this example have been recorded as if actually reported as observed figures. (*See* page 332.)

From other firing results the values of E given appear to be rather too small. This would cause C as found from E to appear too great. The figures are worth studying as showing the degree of accuracy to be expected in good experimental work, and the need for a certain measure of contempt for small discrepancies in the numerical work.

The comparative table which follows the Hodsock figures will further emphasize the need for this contempt when founded on accurate information. It was assumed that the Mark VI bullet had M.V. 2,050, that its $W/d^2$ was 0·335, and that it required an elevation of 1° 27·0' exactly at 1,000 yards in normal weather. It was required to find the calculated striking velocity and time of flight of this bullet under these precise conditions. For this calculation four different sets of ballistic tables were used, and from each of them the value of C was determined from V, R and E. Using this value and the formulæ

$$Sv - Sv = S/C$$
$$Tv - Tv = t/c$$

the value of $v$ was obtained from the S table and thence the value of $t$ from the T tables.

*Hodsock Observed Figures*

| Bullet | 1921 Flat Base | | | 1922 Streamline | | |
|---|---|---|---|---|---|---|
| | M.V. of both = 2,850 f/s., $W/d^2 = 0.271$ | | | | | |
| Range | $v$ | E | $t$ | $v$ | E | $t$ |
| 0 | 2,850 | 0 | 0 | 2,850 | 0 | 0 |
| 200 | 2,460 | — | — | 2,539 | — | — |
| 400 | 2,101 | — | — | 2,244 | — | — |
| 600 | 1,749 | 17 | 0·82 | 1,886 | 16 | 0·78 |
| 800 | 1,443 | 25 | 1·21 | 1,630 | 24 | 1·10 |
| 1,000 | 1,184 | 36½ | 1·66 | 1,349 | 33 | 1·55 |

| | Values of C | | | | | |
|---|---|---|---|---|---|---|
| 200 | 0·260 | — | — | 0·331 | — | — |
| 400 | 0·270 | — | — | 0·322 | — | — |
| 600 | 0·273 | 0·269 | 0·278 | 0·322 | 0·325 | 0·310 |
| 800 | 0·269 | 0·297 | 0·271 | 0·322 | 0·335 | 0·325 |
| 1,000 | 0·261 | 0·291 | 0·277 | 0·315 | 0·339 | 0·339 |

Finally, assuming $v = 895$ f/s., the S table was used again to find the value of C. The results are as follows :—

| Ballistic Table | C found | $v$ | $t$ | C found from $v = 895$ |
|---|---|---|---|---|
| Bashforth | 0·402 | 874 | 2·467 | 0·4220 |
| Ingalls | 0·360 | 895 | 2·442 | 0·3600 |
| Official, 1909 | 0·367 | 904 | 2·438 | 0·3582 |
| T.B. of S.A., 1929 | 0·254 | 880 | 2·459 | 0·2632 |

It is known from experimental firing at Hodsock that for this bullet $v$ is about 885 f/s. at 1,000 yards when M.V. is 2,060 f/s.

The 1929 tables which have been made primarily for bullets of higher velocity and sharper points seem, therefore, as reliable for the blunt bullet as any other set of tables.

## Section 6
### Long Range Fire Analysis in Two Arcs

There are two special reasons for wishing to apply calculation to the known results for range and elevation at angles greater than 10°. One is to be able to design a bullet of the best shape, and the other is to ascertain the striking velocity of a bullet whose ranging is known. The method adopted must be such that it is quick to work and reasonably accurate in its results. It has been found that it is enough to break each trajectory into two arcs, dividing it at the point where the velocity is exactly 1,000 f/s. Above 1,000 f/s. the value of C can be got with considerable precision by short range firing.

Figure 4 shows clearly the method of breaking up the trajectory and obtaining the new angle of elevation and the new range to complete it. The meaning of the terms employed is also given. For clearness the value of C for use with Table III is called in the first arc $C_1$ and in the second arc $C_2$. A sketch of the method used will now be given, but a complete discussion would be out of place. The problem is to find the best values of $C_2$.

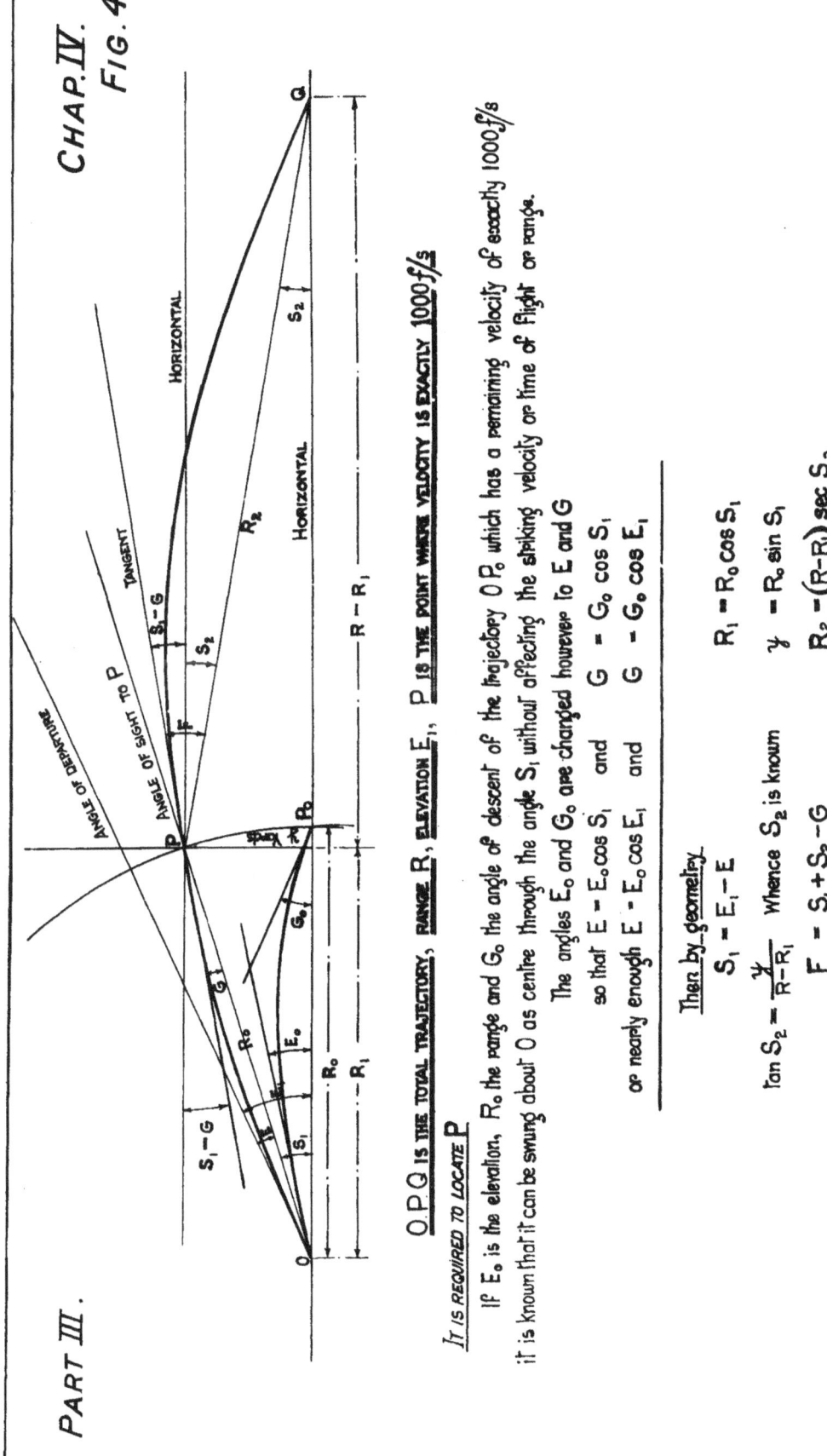

PART III.

CHAP. IV.
FIG. 4.

O.P.Q IS THE TOTAL TRAJECTORY, RANGE R, ELEVATION E₁, P IS THE POINT WHERE VELOCITY IS EXACTLY 1000 f/s

IT IS REQUIRED TO LOCATE P

If $E_o$ is the elevation, $R_o$ the range and $G_o$ the angle of descent of the trajectory $OP_o$ which has a remaining velocity of exactly 1000 f/s it is known that it can be swung about O as centre through the angle $S_1$ without affecting the striking velocity or time of flight or range.

The angles $E_o$ and $G_o$ are changed however to E and G

so that $E = E_o \cos S_1$ and $G = G_o \cos S_1$

or nearly enough $E = E_o \cos E_1$ and $G = G_o \cos E_1$

Then by geometry—

$S_1 = E_1 - E$

$\tan S_2 = \dfrac{y}{R - R_1}$ Whence $S_2$ is known

$F = S_1 + S_2 - G$

$R_1 = R_o \cos S_1$

$y = R_o \sin S_1$

$R_2 = (R - R_1) \sec S_2$

If $R_2$ is now revolved about P or Q till it is horizontal $F \sec S_2$ becomes $E_2$ the new angle of elevation for the new range $R_2$ and the new MV is 1000 f/s. Whence the new ballistic coefficient can be found.

The foundation of the method is the known calculated results tabulated in Appendices II and III. Using these results an approximate method is employed, and the correcting factors are determined by comparison.

In the following scheme for analysis of the known results the range on the flat, time of flight, angle of descent and elevation to produce a striking velocity of 1,000 f/s. are first tabulated with the suffix 0 to them. The paper is ruled for six whole trajectories from 5° to 40°, and the total range, angle of descent, time and striking velocity are filled in from Appendix III, as also the time from 1,000 f/s. to full range.

In practice it is found that just as good results are obtained by dropping the term sec $S_2$ in the formula for $R_2$ and $E_2$, which therefore become

$$R_2 = R - R_1$$
$$E_2 = S_2 + S_1 - G.$$

The values of $R_2$ and $E_2$ are shown worked out by straightforward but rather long arithmetical sums.

It is known from the value of $p$ given in Table I that below 1,000 f/s. $p$ varies almost as the square of the velocity so that

$$p = 1 \cdot 2 \, (V/1,000)^2.$$

Table III is made out exactly for

$$p = 1 \cdot 49 \, (V/1,000)^2,$$

so that using Table III with a value of C

$$C_2 = 1 \cdot 49 \, C_1/1 \cdot 20 = 1 \cdot 24 \, C_1 = 0 \cdot 353$$

and employing exact but very laborious methods $R_2$ and $E_2$ should be obtained with $C_2 = 0 \cdot 353$.

They are not so obtained by the direct method of handling Table III previously explained, but if E is divided by a factor β such that β is nearly sec $E_2$, or, what is the same thing with small arms, β = the ratio between the range along the curved path and the actual horizontal range (vide "Nature," August 2nd, 1924, p. 159), and is exactly as given in the table on the scheme, coincidence is secured.

SCHEME FOR V = 2,600 AND RANGE REPORT AS IN APPENDIX III.

$R_0$ = 1,152  $W_0$ = 1° 55'.
$E_0$ = 61·1  $t_0$ = 2·26.

| $E_1$ | 5 | 10 | 20 | 30 | 35 | 40 |
|---|---|---|---|---|---|---|
| R... | 2,464 | 3,273 | 4,097 | 4,423 | 4,456 | 4,413 |
| $W_1$ | 11° 14' | 24° 20' | 47° 34' | 62° 24' | 67° 25' | 71° 21' |
| t | 7·56 | 12·65 | 20·95 | 27·91 | 31·06 | 34·01 |
| v | 588 | 477 | 390 | 405 | 417 | 429 |
| t − 2·26 | 5·30 | 10·39 | 18·69 | 25·65 | 28·80 | 31·75 |
| Sec $E_1$ | 1·004 | 1·015 | 1·064 | 1·155 | 1·221 | 1·305 |
| $E_0 \cos E_1 = E$ | 1° 1' | 1° 0' | 0° 57' | 0° 53' | 0° 50' | 0° 47' |
| $E_1 - E = S_1$ | 3° 59' | 9° 0' | 19° 3' | 29° 7' | 34° 10' | 39° 13' |
| sin $S_1$ | 0·0695 | 0·1564 | 0·326 | 0·487 | 0·562 | 0·632 |
| cos $S_1$ | 0·998 | 0·988 | 0·945 | 0·874 | 0·827 | 0·775 |
| $R_0 \sin S_1 = y$ | 80 | 180 | 375 | 561 | 647 | 728 |
| $R_0 \cos S_1 = R_1$ | 1,150 | 1,140 | 1,088 | 1,006 | 952 | 892 |
| $R - R_1 = R_2$ | 1,314 | 2,133 | 3,009 | 3,417 | 3,504 | 3,521 |
| $y/R_2 = \tan S_2$ | 0·0609 | 0·0844 | 0·1246 | 0·1643 | 0·1847 | 0·2068 |
| $S_2$ | 3° 29' | 4° 50' | 7° 6' | 9° 20' | 10° 28' | 11° 41' |
| $W_0 \cos S_1 = G$ | 1° 55' | 1° 54' | 1° 49' | 1° 41' | 1° 35' | 1° 29' |
| $S_1 - G$ | 2° 4' | 7° 6' | 17° 14' | 27° 26' | 32° 35' | 37° 44' |
| $S_2 + S_1 - G = E_2$ | 5° 3' | 11° 56' | 24° 20' | 36° 46' | 43° 3' | 49° 25' |

Values of β.

| E | 0 | 5 | 10 | 15 | 20 | 25 | 30 | 35 | 40 | 45 | 50 |
|---|---|---|---|---|---|---|---|---|---|---|---|
| β | 1·00 | 1·01 | 1·02 | 1·03 | 1·04 | 1·065 | 1·11 | 1·17 | 1·25 | 1·37 | 1·50 |

The factor β is therefore the correcting factor, and has been found by trial and error from one particular case, where precision can be secured. It is found that it suits other bullets very well indeed, but at some future time a better set of values may be obtained.

If now we enter Table III with $C = 0·35$ and $E_2/\beta C$ and $R_2/C = Z$, it will be found that the value of Z against $E_2/\beta C = A$ in the table is indistinguishable from $R_2/C$, thus proving the correctness of the value of β.

The work is as follows:—

| $E_1$ | 5 | 10 | 20 | 30 | 35 | 40 |
|---|---|---|---|---|---|---|
| $E_2$ | 333 | 715 | 1,455 | 2,194 | 2,565 | 2,965 |
| $R_2$ | 1,314 | 2,133 | 3,009 | 3,417 | 3,504 | 3,521 |
| β for $E_2$ | 1·01 | 1·02 | 1·06 | 1·19 | 1·32 | 1·47 |
| $E_2/\beta$ | 330 | 701 | 1,370 | 1,850 | 1,940 | 2,000 |
| $E_2/0·35\beta$ | 943 | 2,000 | 3,918 | 5,290 | 5,550 | 5,710 |
| $R_2/0·35$ | 3,760 | 6,100 | 8,600 | 9,750 | 10,000 | 10,100 |
| Z for $E_2/0·35\beta$ | 3,780 | 6,120 | 8,650 | 9,780 | 9,990 | 10,100 |

*Note.*—Plot the curve of Z and A and read off value of Z for $E_2/0·35\beta$ to obtain the last line.

For the angle of descent the procedure is as follows:—
Call the whole angle of descent $W_1$, and the angle of descent of the tilted trajectory $W_2$. Then $W_1 = W_2 + S_2$ to use in the formula

$$\tan W_2 = B\frac{\beta C_2}{2},$$

where B is tabulated in the last column of Table III. This function B was first tabulated by Ingalls for use in the formula

$$\tan W = B^1 \tan E,$$

where $B^1 = 1,719 \, B/A$. This formula is not applicable to rifle trajectories beyond about 10° elevation, and up to that elevation it is not needed.

The working is given below with $C = 0·35$.

| E | 5 | 10 | 20 | 30 | 35 | 40 |
|---|---|---|---|---|---|---|
| $Z = R_2/0·35$ | 3,760 | 6,100 | 8,600 | 9,750 | 10,000 | 10,100 |
| B for Z | 0·775 | 2·070 | 5·070 | 7·420 | 8·030 | 8·307 |
| tan $W_2$ | 0·137 | 0·370 | 0·939 | 1·542 | 1·852 | 2·13 |
| $W_2$ | 7° 50′ | 20° 20′ | 43° 12′ | 57° 3′ | 61° 38′ | 64° 55′ |
| $S_2$ | 3° 29′ | 4° 50′ | 7° 6′ | 9° 20′ | 10° 28′ | 11° 41′ |
| $W_2 + S_2 = W_1$ | 11° 19′ | 25° 10′ | 50° 18′ | 66° 23′ | 72° 6′ | 76° 36′ |
| actual $W_1$ | 11° 14′ | 24° 20′ | 47° 34′ | 62° 24′ | 67° 25′ | 71° 21′ |

The coincidence is not very good, but the amount of error is of no practical consequence. For striking velocity, which is the important item, the coincidence is good. The formula is $v = u \cos E_1/\cos W_2$ where $u$ is taken from Table III against $Z = R_2/0.35$. The work is

| $R_2/0.35$ | 3,760 | 6,100 | 8,600 | 9,750 | 10,000 | 10,100 |
|---|---|---|---|---|---|---|
| $u$ | 583 | 416 | 291 | 246 | 237 | 232 |
| $\cos E_1$ | 0.996 | 0.985 | 0.940 | 0.866 | 0.819 | 0.766 |
| $\cos W_2$ | 0.991 | 0.938 | 0.729 | 0.544 | 0.475 | 0.424 |
| $v$ | 588 | 436 | 375 | 392 | 408 | 417 |
| actual $v$ | 588 | 447 | 390 | 405 | 417 | 429 |

For time of flight the formula is

$$t = T^1 \beta C + t_0$$

where $T^1$ is taken from Table III against

$$Z = R_2/C_2 = R_2/0.35$$

| $R_2/0.35$ | 3,760 | 6,100 | 8,600 | 9,750 | 10,000 | 10,100 |
|---|---|---|---|---|---|---|
| $T^1$ | 14.9 | 29.34 | 50.8 | 63.45 | 66.45 | 67.56 |
| $T^1 \beta C$ | 5.28 | 10.48 | 18.8 | 26.4 | 30.6 | 34.7 |
| $T^1 \beta C + 2.26$ | 7.54 | 12.74 | 21.06 | 28.66 | 32.86 | 36.96 |
| Actual time | 7.56 | 12.65 | 20.95 | 27.91 | 31.06 | 34.01 |

With a knowledge of the actual results of firing for range and elevation at long ranges, the procedure of analysis for C is as follows :—

(1) Obtain the value of $C_1$ to 1000 f/s. from short range work and calculate $R_0$, $W_0$, $E_0$ and $t_0$.

(2) By the method shown above work out $y$, R, E and G, whence $(S_1 - G)$ and $E_2$, $R_2$ and $\beta$.

(3) Then with a set of values of $E_1$, $R_2$ and $E_2/\beta$ try various values of $C_2$ until coincidence is got between the value of $Z = R_2/C_2$ and $A = E_2/\beta C_2$ in Table III.

Such coincidence indicates that the bullet is conforming to the quadratic law of resistance and gives its resistance. Out of some dozen actual examples not one has failed to give coincidences from 1,000 f/s. up to 40° elevation.

Four actual examples are given from the Hythe firings with enough details to enable the advanced student to frame his own exercises.

| Bullet | 0.303 VI | 0.303 VII | H 0.303 | Balle D |
|---|---|---|---|---|
| $W/d^2$ | 0.335 | 0.271 | 0.271 | 0.285 |
| M.V. | 2,050 | 2,450 | 2,650 | 2,400 |
| $C_1$ | 0.254 | 0.255 | 0.365 | 0.292 |
| $R_0$ | 761 | 957 | 1,520 | 1,070 |
| $E_0$ | 54½ | 55 | 79 | 64 |
| E for 1,000 yards | 87 | 58 | 37 | 58½ |

$E_1 = 10°$

| | 0.303 VI | 0.303 VII | H 0.303 | Balle D |
|---|---|---|---|---|
| $R_2$ | 1,965 | 1,795 | 1,419 | 2,342 |
| $E_2$ | 11° 3′ | 12° 8′ | 15° 25′ | 11° 10′ |

$$E_1 = 20°$$

| Bullet. | 0·303 VI | 0·303 VII | H 0·303 | Balle D |
|---|---|---|---|---|
| $R_2$ | 2,638 | 2,337 | 2,025 | 3,338 |
| $E_2$ | 22° 55′ | 25° 7′ | 29° 56′ | 22° 15′ |

$$E_1 = 30°$$

| | 0·303 VI | 0·303 VII | H 0·303 | Balle D |
|---|---|---|---|---|
| $R_2$ | 2,958 | 2,565 | 2,196 | 3,764 |
| $E_2$ | 24° 42′ | 27° 56′ | 45° 8′ | 35° 23′ |
| Mean $C_2$ | 0·31 | 0·24 | 0·113 | 0·45 |
| Value of $n$ for $C_1$ | 1·32 | 1·06 | 0·74 | 0·98 |
| Value of $n$ for $C_2$ | 1·08 | 1·13 | 2·5 | 0·63 |

The effect of lengthening the point of the bullet is obviously to diminish resistance at high velocities and increase it greatly, perhaps by yaw, at low velocities as shown by bullet H 0·303, which has a point 1 inch long. The effect of streamlining is shown by balle D, which has a very low value for $n$.

## Section 7

### The Ballistics of Low Power Weapons and Revolvers

Five examples are given of the shooting of obsolete or low power weapons. All the figures have been obtained by actual shooting.

| Weapon | Snider | Martini Henry | 0·220 Rim-fire | Webley Pistol (4½-inch) barrel | No. 6 shot 12-bore Shot Gun |
|---|---|---|---|---|---|
| $W/d^2$ | 0·20 | 0·339 | 0·12 | 0·20 | 0·020 |
| M.V. | 1,200 | 1,315 | 1,000 | 600 | 1,315 |
| E degrees | Ranges obtained at Various Elevations. | | | | |
| 0 | 0 | 0 | 0 | 0 | 0 |
| ½ | 206 | 250 | 155 | 65 | 26 |
| 1 | 360 | 450 | 275 | 120 | 47 |
| 1½ | 480 | 630 | 370 | 175 | 62 |
| 2 | 593 | 800 | 460 | 230 | 74 |
| 3 | 775 | 1,050 | 580 | 330 | 92 |
| 4 | 917 | 1,280 | 675 | 420 | 108 |
| 5 | 1,040 | 1,460 | 750 | 500 | 122 |
| Greatest range at about 30° | (2,500?) | (3,000?) | 1,300 | 1,300 | 167 at 10° / 230 at 30° |
| Value of C | 0·125 | 0·20 | 0·10 | 0·18 | 0·015 |
| Value of $n$ | 1·6 | 1·7 | 1·2 | 1·1 | 1·3 |

[*To face page* 336.

PART III, CHAP. V, FIG. 1.

PART III, CHAP. V, FIG. 2.

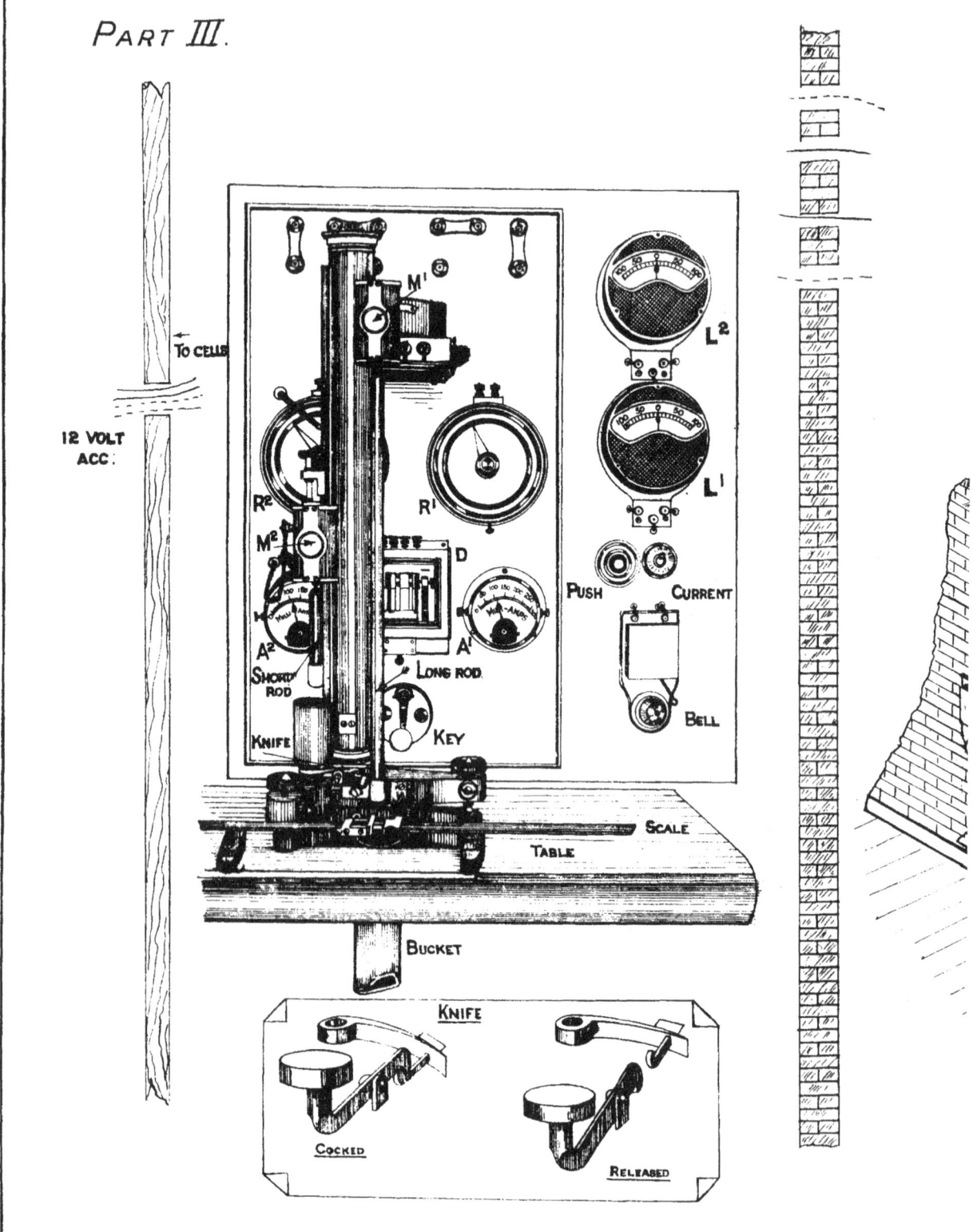

To face p. 337.

CHAP. V. FIG. 3.

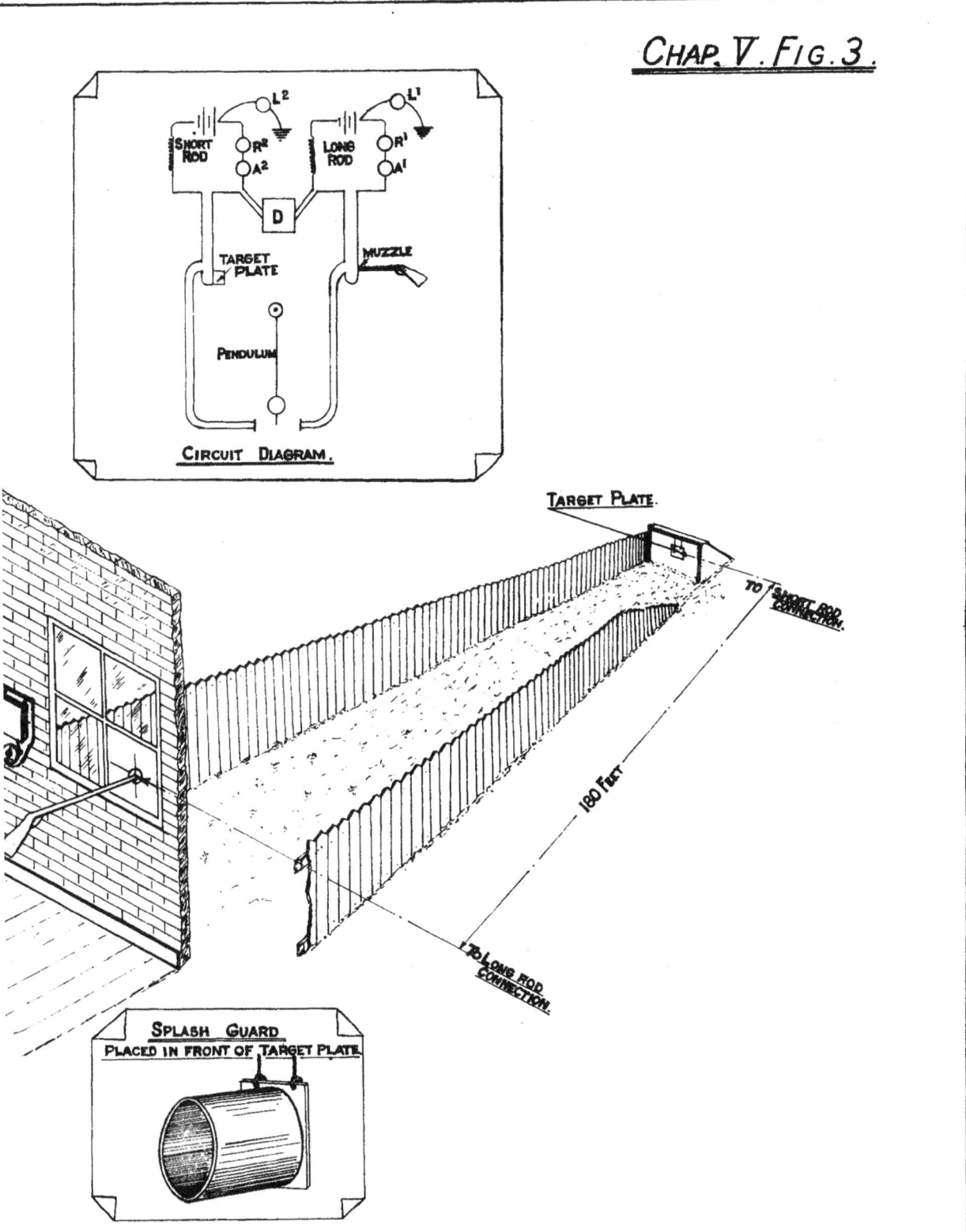

In these examples Table III is used after correcting A for the new muzzle velocity by the formula

$$\text{old A} \times \left(\frac{1,000}{\text{M.V.}}\right)^2 = \text{new A}.$$

The fact that one value of C will fit all the ranges of each bullet shows that it is fairly safe to use the quadratic law of resistance up to M.V. of about 1,300 f/s. The shot spends so little of its time in reducing its velocity to 1000 f/s. or lower, where the quadratic law really begins, that the error introduced by extrapolation above 1,000 f/s. is negligible.

With the values of C given above, it is a simple matter to work out the times of flight and striking velocities. They can then be compared with the results given by Bashforth's tables for spherical projectiles by supposing that $C = 0 \cdot 02$ fits Bashforth's tables for No. 6 spherical shot in the 12 bore for which $C = 0 \cdot 015$ for Table III. Bashforth's tables will not, however, give angles of elevation.

## CHAPTER V

### The "Le Boulengé" Chronograph

This is the instrument in general use for finding the velocity of a bullet. The handbook supplied with the instrument gives a general description of it. As stated in Chapter III, it was invented by a Belgian officer about 70 years ago. Retaining the general principles of action, it was modernized successively in the nineteenth century by Captain Bréger (France) and Major Holden (England). During the Great War further modifications were added for small arms purposes by the Inspection Department at Woolwich. When used for rifles only one instrument can be used, whereas with ordnance two (or more) can be used to record independently the velocity of each round fired, each instrument thus furnishing a check on the accuracy of the other. To obtain a check on the accuracy of a single instrument leak detectors have been included in the line circuits, and a pendulum adapted from one illustrated in the catalogue of Hahn (Germany) has been introduced to throw a constant time of about one-tenth of a second.

Fig. 1 gives a general view of the instrument room fitted for four chronographs with the pendulum on the left background and the switch cupboard in the right foreground. Fig. 2 is a close-up of one chronograph with its own adjuncts. Fig. 3 shows on the right a diagrammatic view of the shooting house and range with (inset above) the conventional diagram of the electric circuit and (inset below) the enlarged detail of the splash guard to catch the very dangerous flying fragments of the bullet from the target plate.

On the left side of Fig. 3 is a lettered diagram of Fig. 2 with (inset below) an enlargement of the recording knife. The distance between the shooting house and the instrument may be a few feet or many miles, according to convenience. The general principle of the chronograph is sufficiently well known to be dismissed in a few words. When the rifle is fired the bullet on leaving the muzzle breaks an electric circuit. On reaching the far end of the measured range it breaks another electric circuit. The chronograph measures the exact time between these two breaks by means of two falling rods and a knife actuated by the second, or short, or far-screen or registrar rod, to give it all its names. From the recorded time and the known range the velocity of the bullet at half range is inferred. To save calculation this time is translated into velocity by a scale specially engraved on a brass measuring bar.

Considered as an electrical instrument the chronograph is very simple and easy to keep in good order, provided that ordinary care is taken to look over its essential parts.

The instrument consists of a stout brass pillar about a yard high mounted on a solid work-bench, and adjusted to the vertical by three levelling screws. This pillar carries two electro-magnets $M^1$ and $M^2$ of which $M^1$ is fixed in height and $M^2$, is adjustable in height over a few inches. Each magnet has a vertical pole piece capable of carrying (when the proper current is on) the falling rod marked " Long rod " and " Short rod " respectively in Fig. 3. These rods weigh just under 1 lb. each, and are provided with soft iron tips for the pole pieces to attract.

For convenience, the working current is always 100 milliamperes, and the rod and magnet tips are ground in a template till they are of such a shape that the rods will drop off as soon as the current falls to 90 milliamperes. Ammeters A1 and A2 are provided to observe the current and rheostats R 1 and R 2 to regulate the current. The battery and circuit for each magnet is quite separate, and consists of six 2-volt accumulators giving 12 volts. The total resistance of the circuit is about 120 ohms, of which the magnet provides about half, the remainder being in the line and rheostat.

The long rod when dropped by the magnet $M_1$ falls into a bucket. The short rod dropped by $M_2$ falls on to the trigger plate of the knife and is prevented from bouncing off on to the floor by a smaller bucket. The knife, when released by the trigger, flies forward by its spring and cuts a small mark on the falling long rod. The distance fallen by the long rod is thus ascertained and translated into time by the formula $h = \frac{1}{2} g t^2$ with $g = 32 \cdot 19$.

The office of the disjunctor D is to break both circuits simultaneously so as to adjust the instrument to the scale. On the key being depressed an arm in D flies forward and is suddenly stopped. This arm carries two fly-over contacts which break momentarily at the shock of stopping and at once remake themselves. One contact is in each line, so that with ordinary care in manufacture to make the fly-overs symmetrical both circuits are broken exactly together. In order to ascertain whether one fly-over is stronger than the other a commutator is provided (shown as a box just below the letter $M_1$ in Fig. 1). This is a crossover switch to pass either circuit through either fly over. It is a necessary adjunct, although such an accidental error of strength is very rare indeed, and, in fact, almost impossible.

The muzzle break is arranged for in several ways. A common method is to stretch a wire or strip of tin foil across the muzzle, the wire being either of lead or very thin copper. This wire has to be made up after each round. A better method is shown in Fig. 4, where a very thin narrow steel spring is blown off its contact by the muzzle blast and remakes its contact automatically. With a little care in design and shape of spring it will last for several thousand rounds, and requires no attention. By actual experimental firing it is shown to be at least as quick and regular in breaking as any muzzle wire. In addition it can only be broken by the gas blast. Wires can be broken by the tip or by the side of the bullet or by the blast in advance of the bullet.

The far screen has been made in several ways; wire screens, tin discs and plates of several sorts have been used. The quickest break has been shown to occur with lead wire or with the target plate shown in Fig. 5. All wire screens are dangerous, as a man has to go out and mend them after each shot. An accident of some sort, even if not fatal, occurs about once in 100,000 rounds. The target plate re-connects itself automatically, and is therefore cheaper, safer and quicker. The plate illustrated is in constant use in Woolwich Arsenal, and was made by a small adaptation from Kynoch's design. The weight in the spring contact was added. The action is evident from the drawing. The shock of the bullet striking the hard armour plate is propagated through the metal and the weighted contact flies off its point for an instant. It is at least as rapid in its break as lead wire, and far more regular and rapid than any other device.

*Action on firing.*—The rods are hung up and the knife cocked. The screens are made up. The gun is fired, breaking both screens. The muzzle screen drops the long rod, and the far screen drops the short rod, which actuates the knife and so marks the long rod. The distance the long rod has fallen is then measured on the brass scale, and from it the time of falling is read off. From this time of falling, the time of falling of short rod and of knife acting is subtracted. The result is the time taken by the bullet to fly the measured range. It is translated into velocity by the scale on the brass bar.

To face p. 338.

PART III.                                   CHAP. V. FIG. 4.

To face p. 339.

CHAP. V. FIG. 5.

Pᴛ III.

## TARGET PLATE.

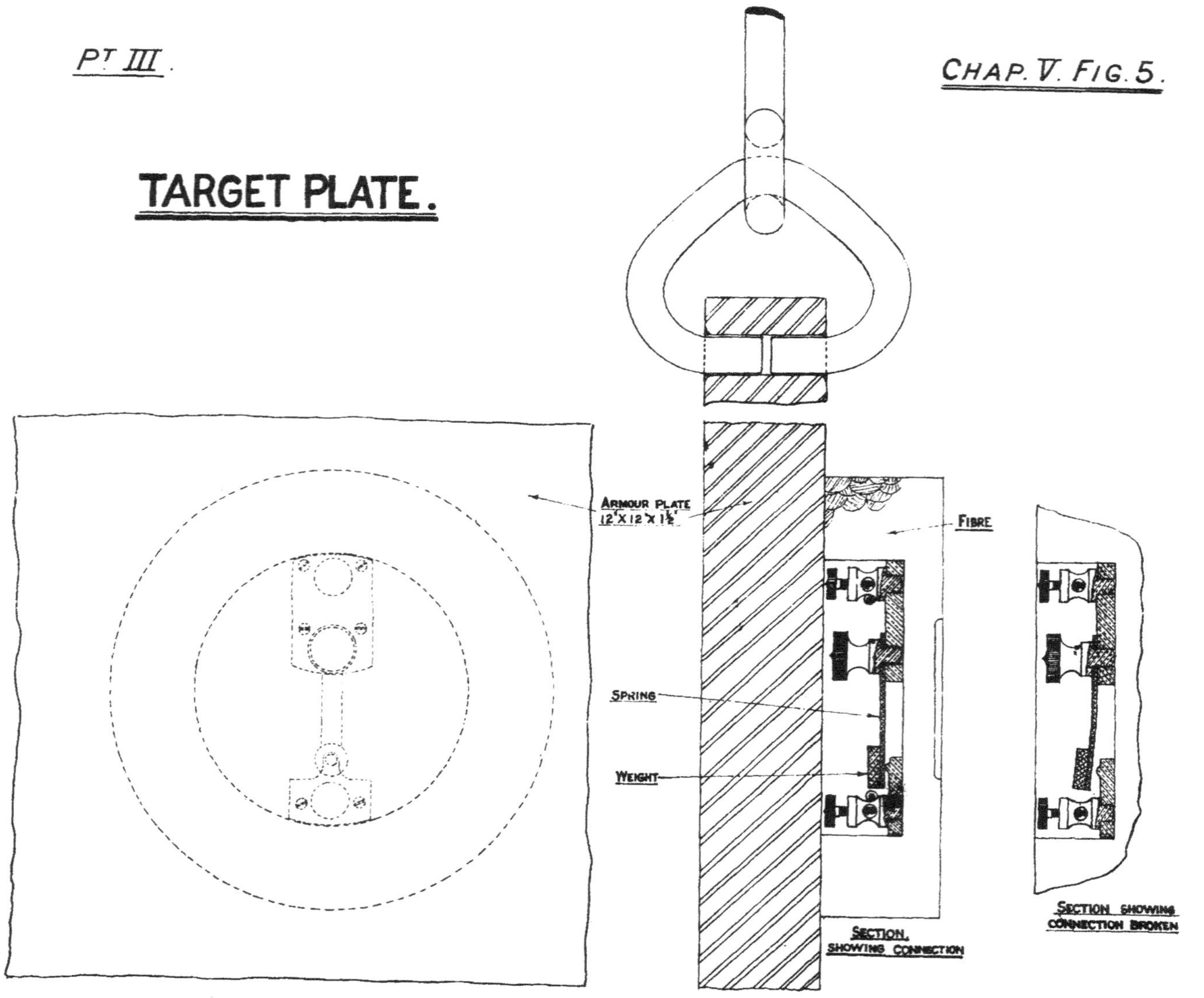

ARMOUR PLATE 12"×12"×1½"

FIBRE

SPRING

WEIGHT

SECTION SHOWING CONNECTION

SECTION SHOWING CONNECTION BROKEN

Maltby & Sons Photo Litho

Fig. 6 shows the brass scale with the long rod placed in position for measuring. A special fitment to ensure parallelism of rod and scale is shown. It is worth adding to the service pattern scale. Inset below is shown an enlargement of the measuring slide. The

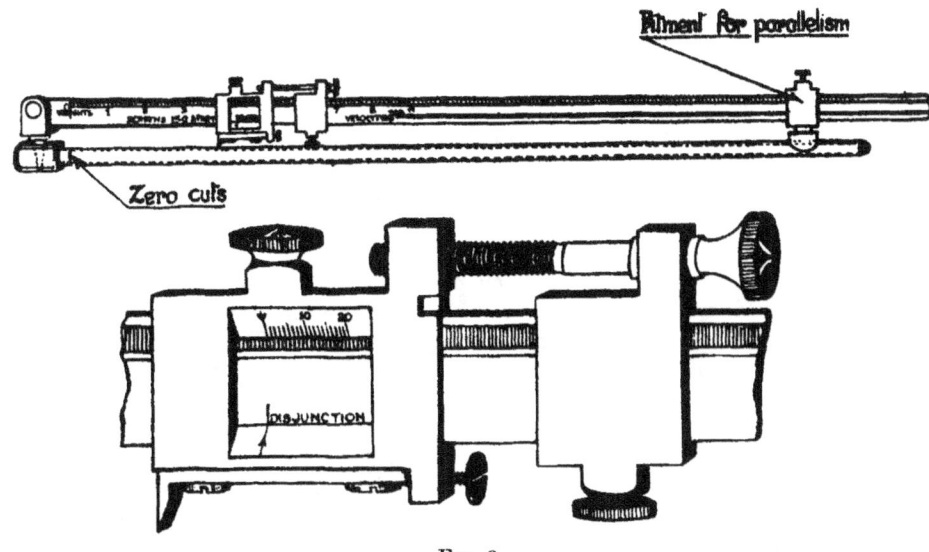

FIG. 6.

vernier and its arrow are not used except for checking, as they measure in inches. The lower arrow shown in line with the disjunction mark is used in combination with the special scale for reading velocity direct.

*Small Arms Modifications.*—The leak detectors L1 and L2 are of the usual trade pattern. The smallest movement of the index hand shows that the insulation of the line circuit is defective and that an "earth" exists. With ordnance, when two chronographs are used in pairs, a difference of reading as small as 5 f/s. in 2,500 f/s. is a sign of some defect. The effect of even a slight "earth" is very marked. If a 40-candle power 250-volt incandescent lamp is put in parallel with a screen so that the current is never entirely broken, the error of velocity has been observed to be as much as 200 f/s. in 2,500 f/s. The ordinary tests do not show up this source of error. An "earth" of the same conductivity would cause a large movement in the leak detector.

The pendulum shown on plate A and in the conventional diagram of Fig. 1 is held back to stops normally. In its swinging arc two light-rubbing or knock-over contacts are placed, spaced far enough apart for the pendulum to take about one-sixth of a second to break them. These circuits are put into series with the gun circuit by means of a plug. The gun screens being made up, the pendulum is released and drops the rods.

If all is in perfect order, the chronograph records some one standard average velocity within $\pm 1$ f./s. from one year's end to the next. The standard average is found by experiment with as many chronographs as possible, all in perfect electrical order. If there is only one chronograph in the place, the standard can be established, although by a lengthy process, and always by trial and error. A portable instrument to throw a constant time with unfailing regularity has not yet been devised. During the Great War it was often urgently needed.

The following practical notes were compiled during the Great War by the Small Arms Inspection Department in the Arsenal :—

The most important six points :—

(1) Measure the range exactly to nearest inch.

(2) Check zero of long rod and of scale.

(3) Disjunct accurately at 4·345 inch and see that the disjunction mark on scale is at this distance.

(4) Check regularity of disjunction.

(5) Work by equal currents on ammeter.

(6) Take extreme precautions to avoid the smallest earth.

Notes to supplement the usual sources of information.

When the instrument is taken over, it is advisable to look to the following processes :—

(1) Go over all the instrument and lines and batteries to see that all appears to be correct and electrically clean.

(2) Drop a plumb line from the long rod magnet down the bucket to see that there is a clear fall for this rod. Also put a pound or two of small shot in the bucket to prevent the rod from bouncing.

(3) Hang the long rod with the tube well pushed home to the square at its end and actuate the knife a dozen times to cut a line all round the tube, turning the tube at each cut a little bit so as to get a fresh surface for the knife. This proves that the knife has no vertical play and determines the zero of the long rod. If necessary tilt the pillar till the knife cuts deep enough but not too deep.

(4) Set the scale to zero on the inches scale and adjust the scale knife by loosening the two clamping screws till the scale knife cuts the ring just cut by the main knife. Take particular care to see that the carrier of the scale knife has no play on the scale. It should read zero on the velocity scale at the same time as it reads zero on the inch scale, and it should read 4·345 inch when the velocity scale reads "disjunct." If it does not, it should be made to do so.

(5) If ammeters are not provided for putting in series permanently in each line wire, indent for them.

(6) It leak detectors are not provided in series in each line wire, get good ones and put them in, or else obtain some means for ascertaining the existence of even the smallest "earth" in any line.

(7) Having ascertained the zero of the rod for the scale in use, proceed to disjunct, and do not be content with one disjunction, but take ten disjunctions turning the tube for each one. The whole ten should then make one continuous ring on the long rod.

(8) Make sure that the long rod is really straight, and that the detent on the knife is only just strong enough to hold the knife ; also that the long rod hangs truly enough to just clear the agates near the knife.

(9) When disjuncting, have both line wires in series and the screens made up, and note the readings of the ammeter of each line. Work with the same ammeter reading exactly. 100 milliamperes is generally a convenient reading. The tips of the rod should be ground to such a shape that they hang with (say) 90 milliamperes.

The provision of loose small tubular weights was intended to take the place of the ammeters and the grinding of the points. They are not, however, so good, because, unless used after each round, they give no information as to the constancy of the voltage of the batteries or the ohms in the line.

(10) Measure the 180 feet of the range with a steel tape and paint up permanently on the range the exact points used in the measurement. Stones sunk in the ground and engraved with bench marks are preferable.

(11) The silvered tube very seldom needs changing if it is used smoked by a taper. The worst of the dents can be knocked out by the careful use of a riveting hammer. A smoked tube is much more easy to read than an unsmoked one.

(12) If possible rig up a pendulum to throw a constant time corresponding to about 2,000 f/s.

## CHAPTER VI

### Instruments for Measuring the Pressure in the Rifle and how to Use Them

The earliest attempts to measure the pressure of the powder gas were made by Count Rumford. They are described in Chapter II.

The method was virtually that employed in the safety valve of a steam boiler. A weight was placed on the valve of known area of an explosive vessel, the pressure being inferred from the amount of weight necessary to prevent the valve from lifting or gas from blowing out on firing. This is also the principle employed in the deadweight apparatus still in use as the definitive standard in ballistic laboratories when using the closed-vessel method. Provided that the maximum pressure lasts long enough to lift the safety valve an appreciable amount (say, a few thousandths of an inch) the results are accurate if a given charge lifts a given weight on a valve of known area about one shot in every two fired. A balance is then obviously struck within very small limits. A maximum pressure lasting a few hundredths of a second, as is usually the case in a closed vessel, is ample, but the method fails when the pressure only lasts a few ten-thousandths of a second, because the lift of the valve is then too small to be detected with certainty.

The next apparatus was devised by Rodnam, of the United States, and, with certain modifications, is in daily use everywhere for ordnance under the name of the crusher gauge.

Fig. 1 shows a section of this gauge. The description is taken from *The Text Book of Gunnery*.

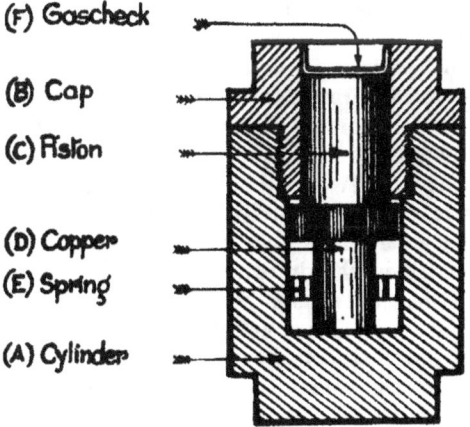

Fig. 1.

It consists of a short steel cylinder A, into which is screwed a cap B, through the centre of which is accurately bored a hole one-sixth of a square inch in area, and in this fits the piston C. By unscrewing the cap, the copper cylinder D can be inserted and held in place by a small piece of watch spring E. The copper is then central with one end in contact with the body of the gauge and the other with the head of the piston. A small gas check F is fitted to prevent the penetration of gas into the inside of the gauge.

The gauge is placed in the chamber of the gun before firing. On firing the pressure of the gas, acting on the end of the piston, compresses and shortens the copper. The gauge is then taken out of the gun, the cap unscrewed, and the copper removed and measured. The pressure corresponding to the shortening of the copper is read off from a table compiled originally for the compression of similar coppers in a pressing machine. An abbreviation of the table is as follows :—

| Length | Pressure | Length | Pressure |
|---|---|---|---|
| Inches | Tons/inch$^2$ | Inches | Tons/inch$^2$ |
| 0·500 | 0·0 | 0·300 | 19·0 |
| 0·450 | 7·0 | 0·250 | 24·0 |
| 0·400 | 11·0 | 0·200 | 32·0 |
| 0·350 | 15·0 | 0·150 | 46·0 |

The underlying principle of this gauge is much the same as for the deadweight apparatus after allowing for the fact that the resistance to motion of the piston when urged on by the powder gas is provided by the compression of the copper instead of by the lifting of an actual weight. The moving parts are therefore much lighter, and the gauge is in that sense more sensitive, as the lift or motion of the piston is larger and the maximum pressure need not be maintained for more than a few thousandths of a second.

For rifles this gauge is, however, much too large to be loaded into the cartridge case, so that it has been modified in detail in an ingenious way. This was first known as the "Appareil Maissin" described in *Mem. des Poudres et Salp.*, Vol. V, 1892, Part II, pp. 19-24.

It was independently invented by a workman in the Arsenal in 1895, and is now employed in England under the name of the "Ordnance Factory oiled-case method."

Fig. 2 shows the details.

The piston itself is abolished as a separate item.

The brass cartridge case itself performs the duty of the piston by being well oiled so that on firing it slides back freely against the face of the breeching. This breeching contains the copper crusher, which, however, has to have a central hole to permit of the striker passing through to fire the cap. A steel pad is placed between the cartridge case and the copper, and this is also perforated for the striker. The perforated copper therefore really hangs horizontally on the striker pin. The arrangements for opening and closing the breech are sufficiently indicated in the drawing. Formerly a specially heavy barrel was provided weighing about 11 lbs., but this is now found to be unnecessary as there is no danger of a barrel itself bursting if it has stood the proper proof rounds.

The area of the virtual piston in the oiled-case method is the cross section of the breech end of the gun chamber, that is, the area of the cartridge case "under head."

The whole cartridge case when oiled behaves like the head of a hydraulic ram, the gas pressure acting not only on the inside of the base, but also on the oil film and cross area of the walls of the case.

The action of this oiled-case method is precisely like that of the crusher gauge just described.

The table for reading off pressure is made out for a piston ·471-inch diameter or ·1743-square-inch area. If the diameter of case under head is ·461 inch, as in the service ·303 case of all marks, the pressure read off the table requires to be multiplied by the factor $\left(\frac{471}{461}\right)^2$, that is, by 1·043, to give the proper answer. This is invariably done for all cartridges other than the service ·303. The original Ordnance Factory table was compiled more than 30 years ago, before the effect of diameter of case was thoroughly understood.

The original O.F. gauge before the oiled-case method was thought of, required a special cartridge case with a piston (or Morse base) in the base of the case, which pressed directly

PT. III.   CHAP. VI. FIG. 2.   *To face p. 342.*

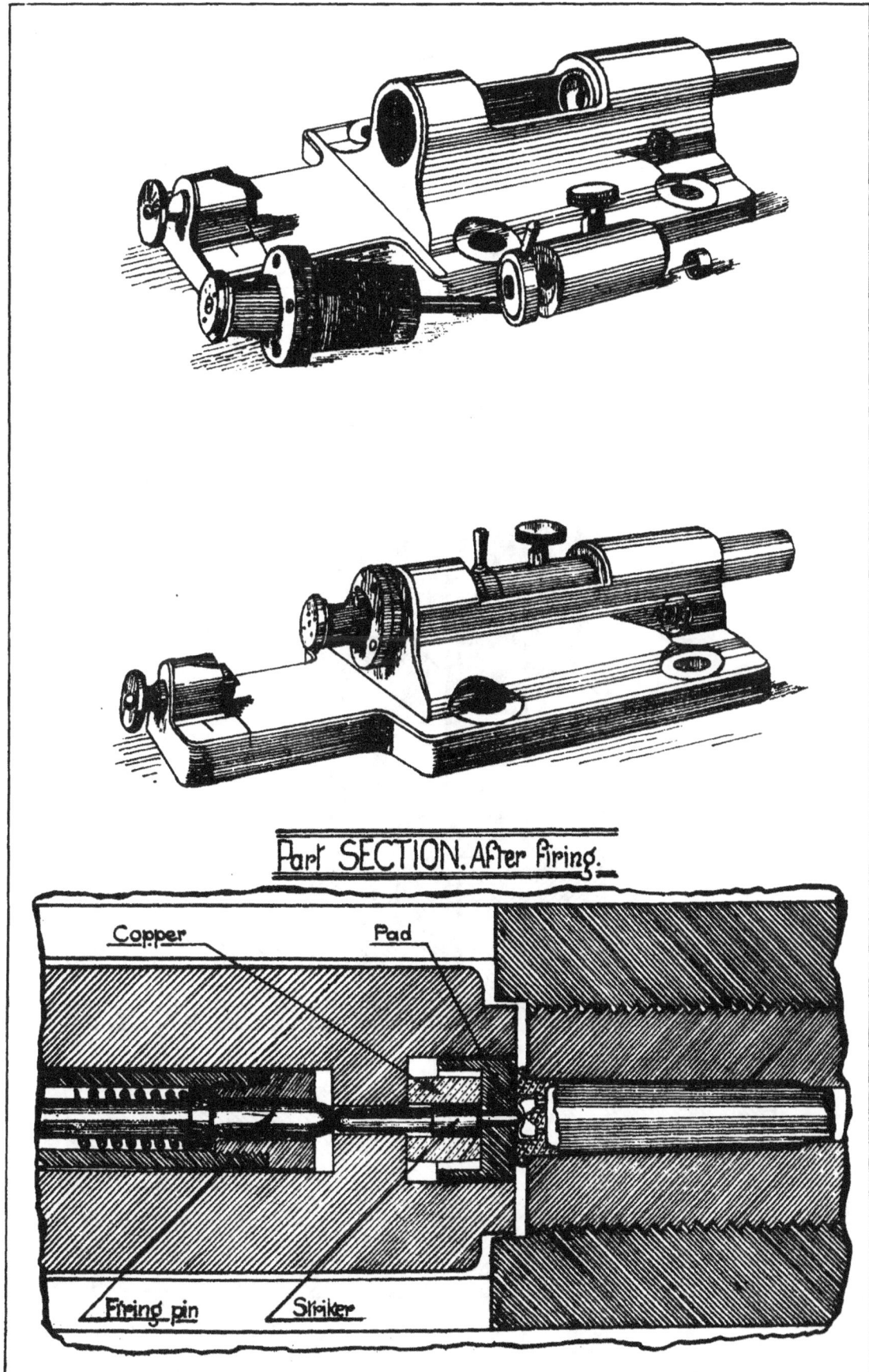

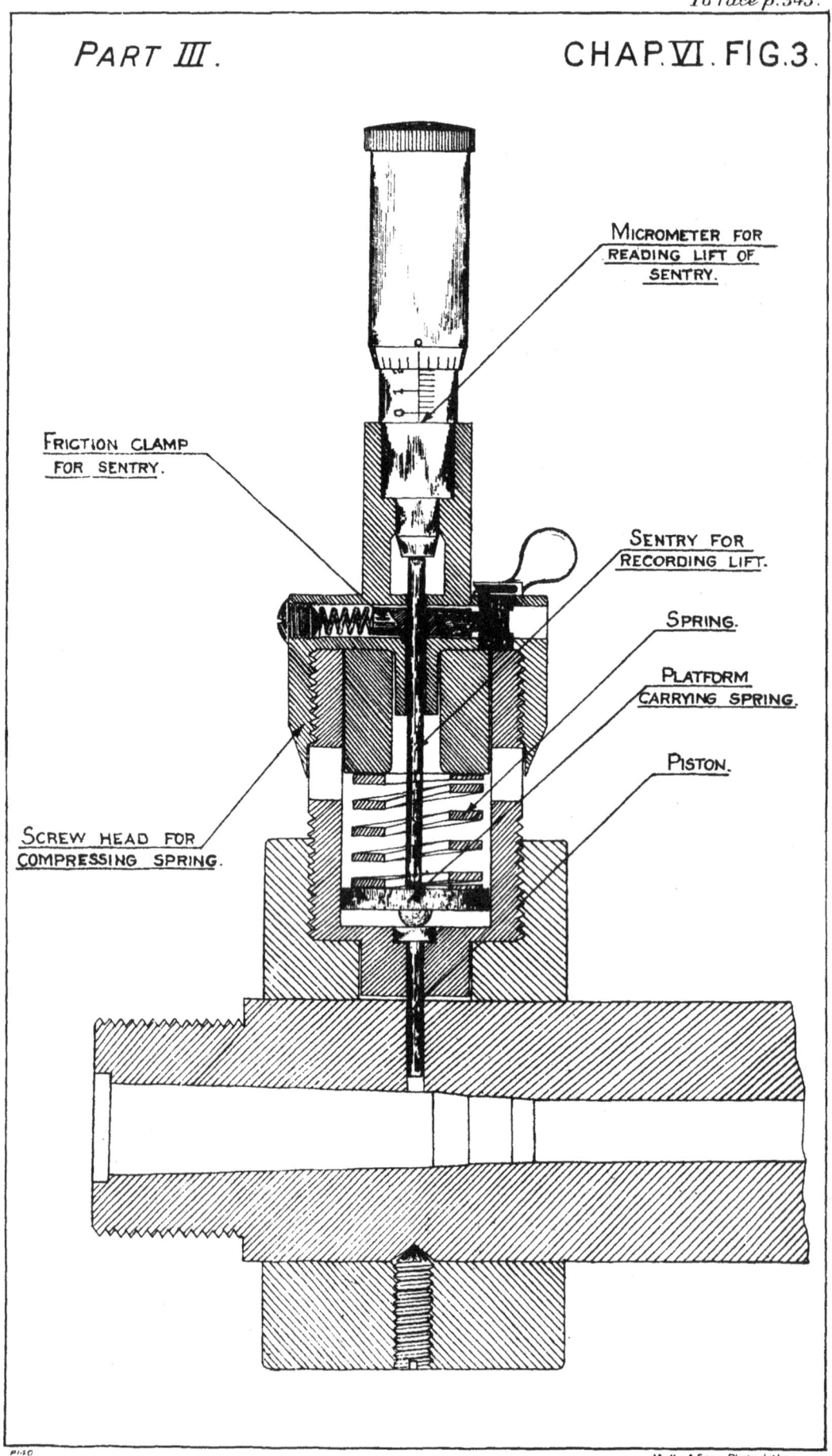

on to the copper besides carrying the cap. A table was made out for this size of piston. When the change over to oiled cases was made, an arbitrary multiplier was applied to the old table to allow for the change of piston, and was adopted as standard for ·303 cases. Although it is now known to be 4·3 per cent. wrong, it is found less inconvenient to leave the standard alone than to correct it.

The original pressing machine is still in use as well as the rules for preparation of the actual coppers. They are very reliable in the sense that they give the same results now as they did originally, but here, again, small errors have been located. In consequence, all the pressures should be regarded as conventional figures, and not as scientifically correct. The conventions apply only to this particular system when used in all its details.

The advantage of the O.F. oiled-case system is that it is extremely simple to work, and gives results which are as accurate as those of any of the alternative methods which have been proposed, with the exception of a high-frequency spring gauge giving a continuous record of the change of pressure from start to finish of the explosion. The disadvantage of the O.F. system is that it is too heavy and slow in its motion to be able to record a maximum pressure which lasts less than about one-thousandth of a second, i.e., ten times as long as it lasts in a rifle. It also appears that metallurgical difficulties occur, as it is very probable that the resistance of copper compressed in a pressing machine bears no relation at all to its resistance when forced to shorten itself at a rate of 10 to 100 feet per second. In other words, the internal friction of copper itself when compressed produces a large "damping" effect on the motion, which cannot easily be determined even by experiments, but which seriously vitiates the results. A further minor disadvantage is that the copper must be compressed before firing to such a length that on firing the added compression shall be less than $\frac{1}{20}$ inch or so. Otherwise the cartridge case is liable to blow through and cause danger, owing to its sliding backwards and out of the barrel far enough to leave its walls unsupported by the barrel. Such a blow-through releases a quantity of high-pressure gas, and is liable to cause serious damage. A similar accident involving the wreckage of the breech mechanism may occur also in a fully stocked rifle if the bolt head is short or breech space is unduly large.

Another form of the crusher gauge called the "radial" or "side-pressure" gauge is largely used for rifles. It is illustrated in Fig. 3. A small hole is bored through the metal of the barrel into the chamber, and the cartridge case is also bored in such a position that when loaded the two holes register together exactly. The hole in the barrel is fitted with a piston, on top of which is the rest of the apparatus, the whole working just like a crusher gauge. The gas pressure acts on one end of the piston, and the resistance at the other end is provided by a cylinder of either copper or lead, or even by a powerful spring, as in the instance illustrated. From the resulting compression of the crusher or spring the maximum pressure is inferred either by consulting a table or by using the apparatus as if it were a "deadweight" apparatus. In the latter case a quantity of rounds has to be fired, and the weight of the spring or the initial compression of the crusher employed has to be continually altered until a balance is struck allowing the piston to lift about one shot in every two fired. The faults of this gauge are the same as those of the dead-load or of the ordinary crusher, differing only in degree. The weights of the moving parts in relation to the strength of the spring or equivalent metal crusher can with difficulty be cut down sufficiently to be sensitive to a pressure lasting only a few ten-thousandths of a second, and even then the sluggishness of the apparatus is such that the record is completed and the piston is at the top of its stroke when the bullet is half-way down the barrel and the maximum pressure is long past and gone.

A very much better form of gauge is one in which the spring is so stiff and light that it can accurately follow (with small sinuosities) the actual rise and fall of the pressure. Owing to the extreme stiffness of the spring the total movement of the piston is so minute that it has to be enormously magnified by optical methods. This involves the use of a mirror and a long beam of light photographing itself on a revolving drum of sensitive paper.

Such an instrument has been devised by Mr. G. P. Thring, of Cambridge. It is fully described in his patent No. 163742/1921, and in a communication to the Society of Mechanical Engineers. With certain modifications, it has been used by the Research Department in Woolwich Arsenal. From the description it appears to be a practical instrument for laboratory use to take the pressure of small arms accurately. It is diagrammatically illustrated in Fig. 4.

In this instrument the piston is carried upwards in the form of a thin hollow steel cylinder about 2 inches long and ¼-inch diameter. This is closely fitted into another outer cylinder rigidly attached to the barrel and connected to the inner one by a very short strong screw thread at the top. The two together form the spring, the inner tube being compressed and the outer one extended by the thrust of the gas on the piston head. Inside the inner hollow tube is a leg or pin having at its top a tiny hinge. To this hinge is attached another light leg at right angles to it, passing through a hole in both tubes and ending in a tiny mirror. This mirror is attached to a small piece of watch spring attached flat and rigidly to the outside of the outer tube. The motion of the piston is imparted to the leg or pin and to the hinge and to the other light leg, and they revolve very slightly round the watch spring as fulcrum by means of the flexibility of the thin flat spring. This gives an angular motion to the mirror, on which a beam of light falls. The beam is reflected on to the revolving drum of sensitive paper, and its motion is magnified to a convenient size on the drum being placed at a distance from the mirror. The drum revolves on a pivot, and is provided in any convenient way with a time scale. The resulting trace is shown on Fig. 4, as an indicator diagram giving pressure one way and time the other way, each on a known scale. It will be noticed that the trace is not a smooth curve, but shows small jerks. These jerks or sinuosities are a necessary feature in any properly-designed gauge. Their frequency ought to be such that at least one or two of these appear in the steepest part of the rising pressure, otherwise the gauge is too sluggish to give a true record.

A brief account will now be given of the theory of the pressure gauge; it has been compiled from the articles by (1) Vieille and Sarrau (*Mem. des Poudres et Salp*, 1883); (2) Charbonnier (*Revue d'Artillerie*, Nov., 1908); and (3) from the paper written by the Inspection Department during the war, containing Mr. M. Segal's purely mathematical analysis of the motions of a piston (excluding damping) when subjected to such rapid changes of pressure as occur in a high-power rifle or in any small-arms weapon, such as a revolver or shot gun.

The ordinary spring balance used in any kitchen or shop can be made to behave in such a way as to illustrate the general principle. Suppose a 6-lb. weight makes the scale pan descend ¼-inch and the hand registers 6 lbs. when at rest. If the weight is released as soon as it touches the pan the hand will be observed to fly forward to some figure much greater than 6 and less than 12, then to oscillate rapidly on each side of the 6, each oscillation being smaller than the previous one till, finally, the hand steadies at 6 lbs.

Repeating with a 24-lb weight the hand will oscillate across the figure 24, but decidedly more slowly, in fact, at half the rate. The motion of the spring pan is affected by four causes:—

(1) The resistance of the spring, causing the hand to register the exact weight when at rest.

(2) The inertia of the whole system, causing the hand to overshoot and to vibrate.

(3) The internal friction or viscosity of the spring, which causes the oscillations to be "damped" out.

(4) The time constant of the system depending upon the relation between stiffness of spring and weight of moving parts causing the time of each oscillation to be identical.

If, instead of letting go the weight as soon as it touches the pan, the weight is lowered slowly enough, the hand will creep quietly forward to the figure 6 or 24, as the case may

PT. III. CHAP. VI. FIG. 4.    *To face p.344.*

Connecting screw thread.
Outer cylinder in extension
Beam of light.
Recording drum carrying sensitive paper.
Small hinge
Flat spring
Mirror
First leg or pin
Other leg
Source of light
Piston.

## THRING'S PRESSURE RECORDER.
## CONVENTIONAL SKETCH BASED ON THAT GIVEN IN PATENT N° 163742. 23/5/21.

be, and come to rest there without visible oscillation. This shows that the rate of application of the weight affects the movement of the index hand. Again, if the weight is placed on the pan and is partially lowered and then snatched off, the hand will move forward with or without visible oscillation, and then fly back to zero. If a loose non-return hand is fitted as an extra, it will remain at the greatest figure reached by the hand, and will be the only visible record of the attempt to weigh. Its reading will be quite unreliable, as it will measure a mixture of weight and time of application. If, however, several such attempts are made following some definite rule as to varying the time of snatching off the weight. it is conceivable that a skilful mathematician, having all exact details before him except the actual weight employed, might be able to arrive at the real value of the weight used. It is also clear that there is some relation between the rate of lowering the weight, or of snatching it off, and the time constant of the system. The greater the time of oscillation of the loaded pan, *i.e.*, the more sluggish the spring, the more slowly must the weight be lowered to avoid visible oscillations or to prevent the loose non-return hand over-registering. A copper crusher may be considered as a spring with a non-return ratchet, so that the only record it can give is equivalent to that given by the previously-mentioned loose non-return hand. In Fig. 3 the sentry and micrometer to measure its lift play the part of the loose hand. The internal friction or viscosity producing the damping effect exists in both copper and spring, and probably also in every substance. It is peculiarly objectionable in a crusher, because it cannot be measured by noting the successive decrements of the oscillations, since there can be none in crushed metal. According to Charbonnier, this damping effect varies with the speed, and this seems a very probable hypothesis, although it produces complete confusion in all our instrumental observations. It is, however, the simplest hypothesis, and the confusion of record must be accepted and encountered, however serious it may seem to be. This confusion is not, however, as serious as at first sight it might seem to be, because, after all is said and done, the compression of a copper placed behind a real rifle cartridge is a real physical phenomenon. The metal is crushed more or less, and a steel bolt head would have to sustain the action of the same forces for the same time, so that the compression of the copper crusher, whether designated in tons per square inch or in thousandths of an inch compression, does in the end represent a definite distortion of metal caused by the explosion. It is therefore not surprising to find that in the end the practical men who have to deal with the design and proof of cartridges and of barrels, and have to decide upon a fair proof charge for a given gun, agree that a certain crushing effect on a copper represents a certain definite effect on a gun, although they may and do differ very distinctly on the scale of tonnage to be used in describing the effect. The mathematician, being constrained to accept certain definititions as to tonnage and time and distance, is unable to accept any solution which ignores one or more of his terms. Since tonnage only considers weight per square inch, and since time is also a real factor in the distortion of metal, no terminological agreement is possible, although there is reasonable agreement in practice.

The apparent discrepancy is best resolved by considering the real forces and times involved. It can be proved that unless the time of application of the gas pressure until maximum is reached is two or three times as long as the time constant of the gauge, the gauge can only give a reasonably accurate record by a series of coincidences. In mathematical text books the time constant of a gauge is defined as follows (in ten-thousandths of a second) :—

$$T (t.t.s.) = \frac{1}{1\cdot 67} \sqrt{\frac{w}{k}}$$

where $w$ is the weight of the moving parts in grains and $k$ is the number of pounds required to crush the spring $0\cdot 001$ inch.

The values of these figures for various gauges are approximately as follows :—

| w | k | T | Gauges |
|---|---|---|---|
| 200 | 27 | 1·7 | Crusher gauge. |
| 500 | 35 | 2·4 | O.F. oiled case. |
| 500 | 5 | 6·0 | Commercial lead crusher. |
| 14 | 9 | 0·8 | Housman radial. |
| 120 | 80 | 0·8 | Thring gauge. |
| 42,000 | 0·024 | 800·0 | Kitchen scales weight 6 lbs. |
| 168,000 | 0·024 | 1600·0 | ,,   ,,   ,, 24 ,, |
| 350 | 1 | 12·0 | Spring gauge, Fig. 3. |

Fig. 6, Chapter III, shows a reasonable diagram of pressure and time, and here some three t.t.s. are occupied as the pressure rises from a small figure to maximum. Accordingly, the Housman radial gauge and the Thring gauge may be expected to give fairly good results provided no metallurgical difficulties occur in calibrating the spring or copper. No great difficulties occur with the Thring gauge, because the steel is always working within its elastic limits, and its speed of compression is low.

The problem which the French physicists considered was whether a crusher or similar gauge could definitely register the maximum pressure of ordnance. Except for the damping or metallurgical difficulties there appears to be no reason why it should not do so, but they did not contemplate such rapid time rises as occur with small arms. Here the problem is entirely different. The gauges cannot possibly follow the pressure rises and yet they give a record. The question is exactly what it is they record. A full reply to this question would occupy many pages, and would require the use of higher mathematics. But a very short answer, which is true enough for all practical purposes, can be given as a formula for a pressure gauge of the usual sluggish pattern.

The formula is :—

$$C = \frac{12g}{10^5 W} FT^2.$$

In which C is total compression in thousandths of an inch.

W is the weight of the moving parts in lbs.

F is the mean thrust in lbs. on the area of the piston in square inches.

T is the total time of application of this thrust in t.t.s.

In this formula there is only one term to be argued about, and that is the best value of the mean thrust F. The thrust is the difference between the gas pressure and the resistance of the crusher or spring. The instant the gas pressure becomes greater than the resistance the thrust begins to act till the resistance is greater than the gas pressure.

By working out a series of actual examples it appears that the best value for F is given by the formula

$$F = \tfrac{2}{3} P (1 - m)$$

where P is the real maximum pressure and $m$ is the ratio between the initial setting of the gauge and P. Then if P = 16 tons and the initial compression is 12 tons

$$m = 0·75, 1 - m = 0·25$$

and
$$F = 32/12 = 2\tfrac{2}{3}.$$

## CHAPTER VII

### BALLISTIC PENDULUM AND THEORY OF RECOIL

The ballistic pendulum was the earliest instrument of precision to be applied to gunnery. It was invented by Benjamin Robins in the reign of George II, a few years before the battle of Fontenoy. Fig. I is a photographic copy of Robins's own illustration of it.

Fig. 1.

The body of the machine is "the same with what is vulgarly used in the weighing and lifting of very heavy bodies, and is called by workmen the 'triangles.'" The block GHIK of wood fastened on to the iron T-shaped rocker, EFK, weighed about ½ cwt., and was some 6 feet long, making the whole "triangles" about twice as high as a man.

The 12-bore musket bullet "impinged on" the wood block GHIK, making it revolve about the axis EF, and pulling out the narrow ribbon LNW "with some minute reluctance" through the steel contrivance MNU "made something in the manner of a drawing pen." The amount of ribbon drawn through by the "stroke" of the ball is easily measured, and by computation the striking velocity can then be obtained.

The method of computation used by Robins was, of course, correct, but it was cumbersome owing to the rather crude design of the pendulum. He took the following measurements :—

(1) Weight of the whole pendulum, wood and all, 56·19 lbs.

(2) Centre of gravity of pendulum from axis EF, 52·00 inches.

(3) Centre of the wood GKIH to axis, 66·00 inches.

(4) Time of swings, 100 complete swings to and from in 253 seconds, or 23·7 in one minute, whence $n = 23·7$.

(5) From the value of $n$ he computed the distance L inches of the centre of oscillation from the axis EF by the formula

$$\frac{2\pi n}{60} = \sqrt{\frac{12g}{L}},$$

or $Ln^2 = 35,240·0$

whence $L = 62·66$ inches

(6) By the formula, "as geometers will know,"

$$\frac{62·66}{66·00} \times \frac{52·00}{66·00} \times 56·19$$

he obtained the virtual weight of the pendulum as 42·03 lbs.

(7) The weight of the bullet used being 1/12 lb., the virtual pendulum is 504 times as heavy as the bullet.

(8) "The velocity of the point of oscillation after the stroke will, by the laws observed in the congress of such bodies as rebound not from each other, be the 1/504 of the velocity the bullet moved with before the stroke."

(9) The velocity after the stroke is easily deduced from the chord of the arc through which it ascends by the blow. The chord is measured by the amount of ribbon pulled out, and after allowing for the distance of the ribbon from the centre of the wood this is nearly 16 inches on a radius of 66 inches, or a chord of 0·2425 or an angle of recoil 13° 56′.

The cosine of 13° 56′ is 0·9706 and the versed sine is 0·0294. Hence the height ascended is $66 \times 0·0294 = 1·939$ inches, called the versed sine of the arch.

(10) The velocity is that due to a body falling 1·939 inch, and by the formula

$$V^2 = 2gS. \quad \text{When } S = 1·939/12 \text{ feet}$$

the velocity is nearly that of $3\frac{1}{4}$ f/s.

(11) Hence the velocity of the bullet is

$$3·25 \times 505 = 1,641 \text{ f/s}.$$

Robins goes on to say, "But that those who may be disposed to try those experiments may not have unforeseen difficulties to struggle with, I shall subjoin a few observations which it will be necessary for them to attend to, both to ensure success to their trials and safety to their persons." His notes are as follows :—

(a) The wood-block is essential both for safety by catching splinters and for accuracy of result by preventing rebound.

(b) The weight and proportion of the wood block should be proportioned to the bullet in use.

(c) Beech is the best wood for the block.

(d) Avoid standing at the side of the block when firing is proceeding, because splinters of lead sometimes fly out and are dangerous.

(e) At velocities below 500 f/s. the bullet rarely lodges in the block, but rebounds, and the record is useless.

(f) Use a stout barrel with metal about 1 calibre thick all along the bore.

(g) The muzzle blast extends 16 to 18 feet, and occasionally to 25 feet. Ten yards is a good range at which to fire.

(h) Prevent the wind from blowing the pendulum and giving it an initial swing.

This form of pendulum was in use till about the date of the Crimean War, when it was superseded by electric instruments. It was resuscitated and re-designed by Metford for rifles about 1860, and was used by Mallock, Fremantle, and Mellish at the present Lord Cottesloe's private range at Wistow in 1895 and 1904 for the determination of the resistance of the air to rifle bullets up to 4,500 f/s. ("*Proc. Roy. Soc.*," Vol. 74, 1904, and Vol. 79, 1907). The Wistow modifications are most important from a practical point of view, especially when high-velocity jacketed bullets are used.

The most obvious modification is the substitution of a heavy iron pendulum bob about 1 foot in diameter faced with wood supported by about 5 feet of piano wiring with knife-edge hinges, for the T-shaped iron rocker with wood face. This at once gets rid of the necessity for working out the virtual weight of the pendulum, and again, " as geometers will know," the bob-weight plus one-third of the weight of the wires gives the weight to use in the calculations. As the bob-weight is 1 to 2 cwt. and the wires weigh about 1 lb., it is a matter of small consequence whether the wire weight is or is not correct. Again, when all the weight is concentrated in the bob, it hardly matters whether the bullet hits the centre of the bob or some inches away. In Robins's design it was very necessary to hit quite close to the centre of oscillation, not only to obtain correct results, but also to avoid cross-breaking of the axis EF.

By this one modification (due to Metford) the formula for velocity can be simplified and made tractable by the avoidance of several corrections. The main formula is unaltered, viz., weight of bullet multiplied by bullet velocity equals weight of recoiling parts multiplied by their velocity, but the items required are fewer and more easily measured.

The velocity of the recoiling parts when the angle of swing is small and when

$$v = \text{f/s. of recoil,}$$
$$a = \text{inches of recoil,}$$
$$n = \text{swings to and fro in one minute}$$

is as follows, $$v = \frac{2\pi n}{12 \times 60} \times a = \frac{na}{114 \cdot 7} \text{ f/s.}$$

The weight of the recoiling parts is the weight of the bob (W lbs.) plus the weight of the bullet ($w$ lbs.) embedded in it, so that with V as the striking velocity

$$w\text{V} = \frac{(\text{W} + w)\, na}{114 \cdot 7},$$

which is the accurate working formula for the ballistic pendulum. The value

$$\frac{2\pi n}{12 \times 60} = \frac{n}{114 \cdot 7}$$

is a mere numerical constant which can be evaluated once and for all by observing the number of complete swings (to and fro) made in one minute. As W is several thousand times as large as $w$, it suffices to weigh the bob about once a day, and then $V =$ a constant multiplied by inches recoil.

Fig. 2 gives a general view of the Wistow pattern pendulum bob swinging close above the timber baulk on to which it can be lowered when detached from the suspension, and shows (a) the face of the bob with bullet marks, (b) a yoke carrying the bob by wires, (c) the weigh beam with hook, (d) the paper screen let into the door, which door is closed before firing to stop draughts.

Figs. 3 and 4 show rear views of the same bob with measuring rod in position. This rod is a light aluminium sliding rod with vernier attached to read recoil to 0·001-inch. It should never be actually in contact with the bob when the bob is at rest.

The bob is about 1 ft. in diameter, making it possible to hit it at 1,000 yards or at any shorter distance. It has a flanged steel ring behind the wood face to stop lateral escape of pieces of bullet. The wood face of the bob is separated by an inch or two from the hard steel armour face behind it. This effectually prevents rebounce of bullet fragments through the wood face, which is a serious cause of error. In Fig. 4, a knock-over damper is shown to keep the bob at rest at zero.

A precisely similar instrument was used with great success in the Hodsock firings remarked on in Chapter V.

A modification of this pendulum has been in use at Woolwich since 1916. The chief changes have been in the nature of the " bob " and in the suspension.

It was found that the method of using armour plate to take the full shock of a powerful bullet was too expensive and dangerous, particularly in the case of a bullet with an armour-piercing core. The bob was therefore made of a piece of iron water piping, about 15 inches in diameter and 3 feet long, closed at the far end by boiler plate of $\frac{1}{2}$-inch thickness. The whole pipe was filled with sawdust held in place by a thin wooden face which could be patched up with cardboard. The result was that the bullet had lost a great deal of power before striking the boiler plate and was easily manageable.

The suspension was by brass chains instead of wires, so as to get rid of as much elasticity as possible. Theoretically 5 wires are enough to ensure the absence of wobble when the bob is not hit in the centre, but it is essential that they should be incapable of stretching. The simplest practical solution is to suspend the bob by the equivalent of two five-barred gates, each having two hinges at the top to connect to the scaffolding and two at the bottom to connect to the bob. A sixth bar is of some advantage in practice. The weight of the total bob used was about 200 lbs., and with a vernier measuring recoil to a thousandth of an inch, and a suspension giving about 25 complete oscillations a minute, almost any bullet could be used and accurately timed. It is very little trouble to keep a record of the weight of bullets fired into the bob, and to allow for their additional weight in the calculation.

Such instruments as those described above can also be used for determining the recoil of a given cartridge. The rifle is rigidly attached to the bob (pointing out of the house), and is fired by any convenient means such as tying forward the trigger by string whilst an elastic loop exerts a pull on the trigger. Burning the string fires the gun without imparting any motion to it. The recoil is measured in inches, and the velocity is obtained by the preceding formula

$$v = \frac{na}{114 \cdot 7} \text{ f/s.}$$

The momentum of recoil is therefore $W v$, and it is clear that this momentum is due to the cartridge and not to the gun.

If the cartridge is kept unaltered and the rifle barrel is shortened inch by inch, the value of $(W v)$ will be observed to decrease slightly. The muzzle velocity falls off steadily as each inch is cut off in the proportion of about 6 per cent. barrel length to 1 per cent. velocity. The muzzle blast gets more and more severe as each inch is cut, but the increase is rather less than the fall of muzzle velocity.

With a service S.M.L.E. rifle firing Mark VII ·303 cartridge the product of W and $v$ is about 81. As the rifle weighs about 9 lbs. the recoil velocity is about 9 f/s. also. If

[*To face page* 350.

Part III, Chap. VII, Fig. 2.

Part III, Chap. VII, Fig. 3.

[*To face page* 351.

PART III, CHAP. VII, FIG. 4.

the barrel is shortened by 20 per cent., or say 5 inches, the muzzle velocity (as can easily be measured) is reduced by about 3 per cent., from 2,450 f/s. to (say) 2,375 f/s. The momentum of recoil is, however, reduced from 81 to only about 80. Charbonnier explains the reason very clearly by pointing out that the "ejecta" of the rifle comprise not only the weight of the bullet and of the powder gas, but also the air forcibly blown forward by the muzzle blast.

In consequence we can assert that the momentum of recoil of a rifle is due to its cartridge and to its cartridge only, and that it can be expressed by the formula

$$M = (W_1 + k\, w_1)\, V$$

where M is the momentum in second poundals (i.e., lbs. × ft.-secs.)

$W_1$ is bullet weight in lbs.

$w_1$ is charge weight in lbs.

V is muzzle velocity in f/s.

and $k$ is a numerical constant to be found by experiment.

Experiments of an extensive nature with ordinary guns show that the value of $k$ lies between 1 and 2, with an average value of $1\frac{1}{2}$.

Some authorities give recoil in energy, but serious exception to this may be taken unless the actual weight of the gun is specified. Taking M = 81 and W as 9, the velocity of recoil is 9·0 f/s., and the energy of recoil is

$$\frac{WV^2}{2g} = \frac{9 \times 9^2}{64\cdot 4} = 11\cdot 33 \text{ ft.-lbs.}$$

If the gun weighs 90 lbs. or W = 90, then $v = 0\cdot 9$ and the energy is $\frac{90 \times 0\cdot 81}{64\cdot 4} = 1\cdot 133$ ft.-lbs. Such an extreme example shows clearly the danger of quoting the energy of recoil without gun weight.

As regards the sensation of recoil, it seems well established that the actual velocity of recoil of the gun is a very great factor. In shot guns weighing 6 to 7 lbs. 15 f/s. has been long established as a maximum above which gun-headache is sure to ensue. But with an elephant rifle weighing perhaps 15 lbs. such a velocity is unbearable for more than one or two shots.

The following table gives some useful measured figures from which to form an opinion :—

| Gun | Measured Recoil lbs. ×f/s. Service Load | Weight of Gun lbs. | Recoil Energy ft.-lbs. | H ft. | Recoil Velocity f/s. |
|---|---|---|---|---|---|
| S.M.L.E. | 81. | 9·0 | 11·0 | 1·22 | 9·0 |
| Webley pistol | 50·0 | 3·2 | 12·0 | 3·8 | 15·0 |
| 0·500 bore, exptl. | 285·0 | 18·0 | 70·0 | 3·9 | 16·0 |
| ,, with recoil reducer or silencer | 212·0 | 20·0 | 35·0 | 1·75 | 11·0 |
| 12 bore | 97·5 | 6·5 | 22·8 | 3·5 | 15·0 |

The value of H is what Robins calls the versed sine of the arc described by the pendulum. It is independent of the length of the pendulum.

It should be noted that since the energy of a blow equals the weight multiplied by the distance fallen, the gun punches the firer as if let fall from a height of H feet, as tabulated above. The proviso is, of course, that the butt hits the shoulder in exactly the place where the firer would have placed it for himself. The Webley pistol gives a considerable kick, but it is not so serious a proposition to handle as the ·500 bore weighing six times as much. Only a few exceptionally powerful and skilled men can fire more than one shot from the ·500 bore without recoil reducers, even with a thick rubber heel plate. With them any determined man can fire a dozen rounds. Many men can fire a thousand rounds in one day from the S.M.L.E. or a 12 bore.

## CHAPTER VIII

### PROBABILITY OF FIRE

In Part II, Chapter V, the point of mean impact (M.P.I.) and the figure of merit (F.O.M.) of a group of shots on a vertical target are defined. The method of obtaining these is also explained clearly. In brief, M.P.I. is the centre of gravity of the group and F.O.M. is the average radial distance from M.P.I. In *Small Arms Training*, Vol. I, the "cone of fire" is defined, with a diagram showing its similarity to the jet of a fire-engine, round at first and then elliptical, marking the ground over a space much longer than its width. The nucleus is defined as the breadth of ground with M.P.I. in the centre containing 50 per cent. or half the shots. The effective zone is the breadth for 75 per cent., and the beaten zone is for 90 per cent. of the shots.

Thus, if the line of fire is North to South and a river runs across it from East to West and a Vickers gun has its M.P.I. on the centre of the river, then if the river just catches half the shots, its breadth is equal to the nucleus. If it catches 75 per cent. of the shots its width equals the effective zone and 90 per cent. gives the beaten zone. The remaining 10 per cent. of the tailings are not considered.

In *Machine Gun Training*, 1925, Appendix I, the actual dimensions of the 90 per cent. or beaten zone for ·303 Vickers gun for Mark VII are given at every hundred yards up to 10° elevation or 2,800 yards.

The table may be abbreviated as follows :—

Dimensions in yards and in minutes of 90 per cent. cones.

| Range | Elevation degrees | Width | | Length Yards | Height Yards | Length or height Minutes |
|---|---|---|---|---|---|---|
| | | Yards | Minutes | | | |
| 0 | 0 | 0·0 | 0 | 0 | 0·0 | 0 |
| 400 | ¼ | 1·7 | 18 | 800 | 2·6 | 30 |
| 800 | ¾ | 3·8 | 17 | 450 | 5·3 | 30 |
| 1,200 | 1½ | 7·0 | 21 | 240 | 9·6 | 33 |
| 1,600 | 2¼ | 11·3 | 26 | 150 | 16·3 | 35 |
| 2,000 | 4¼ | 16·7 | 30 | 130 | 27·3 | 44 |
| 2,400 | 6¼ | 22·0 | 33 | 170 | 50·0 | 75 |
| 2,800 | 10¼ | 28·3 | 36 | 210 | 96·0 | 125 |

[*To face page* 353.

PART III, CHAP. VIII, FIG. 1.

For all practical purposes this table is fitted well enough by the following rules using the columns of dimensions in minutes :—

(1) The width in minutes is 20 minutes plus twice the elevation in degrees.

(2) The length or height in minutes is 20 minutes plus eight times the elevation in degrees.

In these rules the 20 minutes added may be considered as due to the average rattle or play of the Vickers Gun when firing rapid.

Cartridges other than ·303-inch, Mark VII, appear to follow these rules also, thus if a streamline bullet gave 5,000 yards range for 10° elevation, the 90 per cent width zone at 5,000 yards would probably be

$$20 + \text{twice } 10 = 40 \text{ minutes},$$

giving $\dfrac{40 \times 50}{36}$ yards $= 56$ yards.

The length would be $20 + 80 = 100$ minutes of elevation, which would be turned into yards by consulting the table of elevation for the streamline cartridge between 4,900 and 5,100 yards.

The principal object of this chapter is to show the relations between the dimensions of the nucleus, the effective zone and the beaten zone ; also to show the effect of displacing the M.P.I. from the centre of the target, and to touch lightly on the laws of chance or the theory of probabilities.

This theory was founded a few years before the great fire of London by two French mathematicians, Blaise Pascal and Pierre de Fermat. According to the story it arose out of the settlement of the fair division of the stakes in a game of cards which had to be left unfinished. In general, if an event may happen in " a " ways and may fail in " b " ways, and all the ways are equally likely, the odds are "a to b, on," or, in mathematical terms, the probability in favour is $\dfrac{a}{a+b}$ and the probability against is $\dfrac{b}{a+b}$, the sum of which two is $\dfrac{a+b}{a+b}$ or unity. When $a = b$ the probability is $0 \cdot 50$, or a half, and the betting expression is " an even money chance."

In order to visualize the laws of chance or the law of error, Sir Francis Galton made an instrument shown in Fig. 1, which he called the " Quincunx," from the Latin word used to describe the arrangement in the planting of trees in an orchard which is imitated by the pins in this instrument. A quantity of millet or turnip seed is poured through the funnel at the top. As they fall, the seeds knock against the pins and are scattered in an arbitrary way, but it is found that they group themselves in the stalls at the bottom in a manner which closely imitates the profile of the Probability Curve

From the shape taken on by the seeds in their stalls this curve is familiarly known as the " cocked hat," being highest in the middle and falling away in hollow curves to the sides like a hat of this sort. It has this peculiar property that if the total number of seeds used is known and the position of the line cutting the hat into two halves is determined by fixing the position of the exact centre of the crown of the hat, the shape of the hat can be drawn by counting the seeds in any two stalls. These counts give the percentage of seeds in two definite vertical slices of the hat.

Chauvenet's table of factors gives the rest of the information. He takes the width of the central band of stalls which contain exactly half the seeds, viz., 50 per cent., as his measure and calls it $1 \cdot 00$. He then gives the factor by which the width of this band has to be multiplied in order to contain any other percentage of seeds.

CHAUVENET'S TABLE OF PROBABILITY

| Per cent. | Factor | Per cent. | Factor | Per cent. | Factor | Per cent. | Factor | Per cent. | Factor |
|---|---|---|---|---|---|---|---|---|---|
| 1 | 0·02 | 21 | 0·40 | 41 | 0·80 | 61 | 1·27 | 81 | 1·94 |
| 2 | 0·04 | 22 | 0·41 | 42 | 0·82 | 62 | 1·30 | 82 | 1·98 |
| 3 | 0·06 | 23 | 0·43 | 43 | 0·84 | 63 | 1·33 | 83 | 2·03 |
| 4 | 0·07 | 24 | 0·45 | 44 | 0·86 | 64 | 1·36 | 84 | 2·08 |
| 5 | 0·09 | 25 | 0·47 | 45 | 0·89 | 65 | 1·39 | 85 | 2·13 |
| 6 | 0·11 | 26 | 0·49 | 46 | 0·91 | 66 | 1·42 | 86 | 2·18 |
| 7 | 0·13 | 27 | 0·51 | 47 | 0·93 | 67 | 1·45 | 87 | 2·24 |
| 8 | 0·15 | 28 | 0·53 | 48 | 0·95 | 68 | 1·48 | 88 | 2·30 |
| 9 | 0·17 | 29 | 0·55 | 49 | 0·98 | 69 | 1·51 | 89 | 2·37 |
| 10 | 0·18 | 30 | 0·57 | 50 | 1·00 | 70 | 1·54 | 90 | 2·44 |
| 11 | 0·20 | 31 | 0·59 | 51 | 1·02 | 71 | 1·57 | 91 | 2·52 |
| 12 | 0·22 | 32 | 0·61 | 52 | 1·04 | 72 | 1·60 | 92 | 2·60 |
| 13 | 0·24 | 33 | 0·63 | 53 | 1·07 | 73 | 1·64 | 93 | 2·69 |
| 14 | 0·26 | 34 | 0·65 | 54 | 1·09 | 74 | 1·67 | 94 | 2·78 |
| 15 | 0·28 | 35 | 0·67 | 55 | 1·12 | 75 | 1·71 | 95 | 2·91 |
| 16 | 0·30 | 36 | 0·70 | 56 | 1·14 | 76 | 1·74 | 96 | 3·04 |
| 17 | 0·32 | 37 | 0·72 | 57 | 1·17 | 77 | 1·78 | 97 | 3·22 |
| 18 | 0·34 | 38 | 0·74 | 58 | 1·19 | 78 | 1·82 | 98 | 3·45 |
| 19 | 0·36 | 39 | 0·76 | 59 | 1·22 | 79 | 1·86 | 99 | 3·82 |
| 20 | 0·38 | 40 | 0·78 | 60 | 1·25 | 80 | 1·90 | 100 | say 4 |

By an intelligent use of this table the curve of the "cocked hat" can be worked out numerically and plotted on squared paper.

Four columns are required :—

Column 1 is Chauvenet's factor at even intervals.
Column 2 is the percentage against each factor.
Column 3 is the difference of column 2.
Column 4 is half column 3.

| Column 1 | Column 2 | Column 3 | Column 4 |
|---|---|---|---|
| Factor | Per cent. | Difference | Half difference. |
| 0·0 | 0·0 | | |
| | | 26·5 | 13·2 |
| 0·5 | 26·5 | | |
| | | 23·5 | 11·7 |
| 1·0 | 50·0 | | |
| | | 18·5 | 9·4 |
| 1·5 | 68·7 | | |
| | | 13·7 | 6·8 |
| 2·0 | 82·4 | | |
| | | 8·4 | 4·2 |
| 2·5 | 90·8 | | |
| | | 4·9 | 2·5 |
| 3·0 | 95·7 | | |
| | | 2·4 | 1·2 |
| 3·5 | 98·1 | | |
| | | 1·8 | 0·9 |
| 4·0 | 99·9 | | |
| | Total. | 99·9 | 49·9 |

If the centre partition of the Quincunx is numbered 0·0 and the others 0·5, 1·0, and so on, outwards from the centre, and there are 13·2 per cent. of the total seeds in each of the two centre stalls, then the other stalls will have 11·7 per cent., 9·4 per cent., and so on.

This curve of probability, or mountain, gives the real profile of the mountain of mud

Part III, Chap. VIII, Fig. 2.

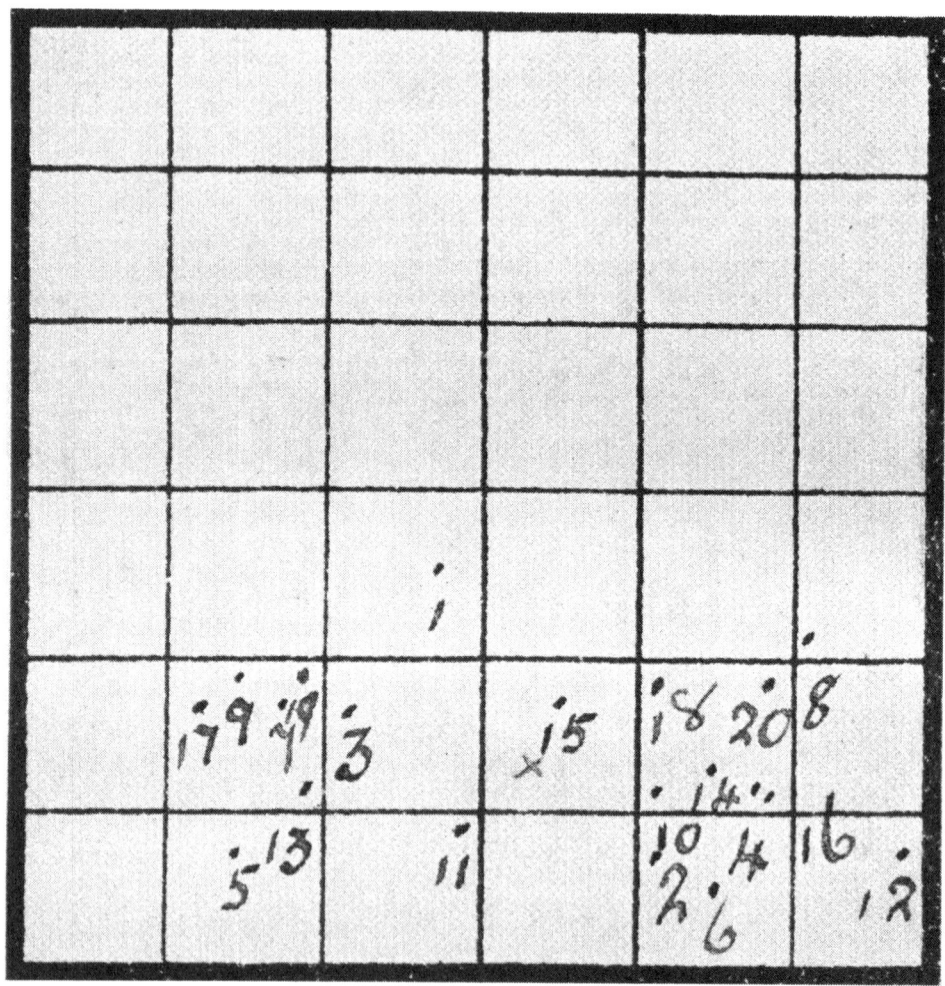

PART III, CHAP. VIII, FIG. 3.

which would be produced by a howitzer firing clay shot continually on a fixed elevation, provided, of course, that the gun did not wear or the wind change, etc., and that only range errors were considered. This is the main fact on which the whole theory of dispersion and intensity of fire and fire effect is based. Chauvenet's table is based on high mathematics, and the Quincunx proves its truth experimentally.

A striking exemplification of the theory is given by the photograph Fig. 2, of a 3 by 6-foot iron plate used during the war as a stop butt at 500 yards for ·303 tracer bullets in Woolwich Arsenal. The white square was the aiming mark used, the rest of the plate being painted black to show up the tracing effect. The centre of this white square has been cut clean out, although the plate is 2 inches thick and the dispersion of the bullets so large that 20 shots rarely kept on a square yard. The cumulative effect of bullets striking has a hardening effect on the plate, which in the long run tends to make it more and more brittle at the centre, but this only accentuates the position of the mean point of impact by letting daylight through earlier and more definitely.

The surprising thing about this iron plate is that no particular pains were taken to aim at the exact centre of the white square. The white square was chiefly used as a reference mark, but its very distinctness acted as a lure to the firer's eye, so that in the long run its actual centre became the M.P.I., and was pierced in consequence.

As a complete numerical example, a very interesting diagram, fired at Woolwich for the Royal Laboratory on November 1st, 1912, from the Whitworth fixed rest at 600 yards has been selected. The rifle and cartridges were of the ordinary Mark VII ·303 description, except that the bullets were specially selected as uneven in thickness of envelope. The cupro-nickel envelope was eccentric, being thicker by some ·005 inch on one side than the other. The thick side was marked, and in the loading of the cartridges into the rifle care was taken to load the odd numbers so that the thick side would be uppermost when the bullet left the muzzle. The even numbers were loaded so that the thick side was then lowest. Twenty rounds were fired, and the resulting target in facsimile is given as Fig. 3, the squares being 6 inches. The target looks as if there had been a change of wind, as there are two distinct groups 18 inches apart, except that the shot numbering is odd to the left and even to the right.

The measurement of the target is as shown on the following table :—

DEVIATION OF EACH SHOT

| No. of shot | Horizontal measurement | | Vertical measurement | | Absolute deviation from M.P.I. | | Difference from mean vertical |
|---|---|---|---|---|---|---|---|
| | feet. | ins. | feet. | ins. | feet. | ins. | ins. |
| 1 | — | 10 | 1 | 3 | — | 8 | 7 |
| 2 | 1 | 7 | — | 4 | — | 6 | 4 |
| 3 | — | 7 | — | 10 | — | 7 | 2 |
| 4 | 1 | 10 | — | 7 | — | 8 | 1 |
| 5 | — | 3 | — | 4 | 1 | 0 | 4 |
| 6 | 1 | 9 | — | 3 | — | 8 | 5 |
| 7 | — | 4 | — | 10 | — | 10 | 2 |
| 8 | 2 | 1 | 1 | 1 | 1 | 0 | 5 |
| 9 | — | 3 | — | 11 | 1 | 0 | 3 |
| 10 | 1 | 7 | — | 7 | — | 5 | 1 |
| 11 | — | 11 | — | 5 | — | 4 | 3 |
| 12 | 2 | 5 | — | 4 | 1 | 3 | 4 |
| 13 | — | 5 | — | 7 | — | 9 | 1 |
| 14 | 1 | 9 | — | 8 | — | 7 | 0 |
| 15 | 1 | 3 | — | 10 | — | 2 | 2 |
| 16 | 1 | 11 | — | 7 | — | 9 | 1 |
| 17 | — | 1 | — | 10 | 1 | 1 | 2 |
| 18 | 1 | 7 | — | 11 | — | 6 | 3 |
| 19 | — | 5 | — | 11 | — | 10 | 3 |
| 20 | 1 | 11 | — | 11 | — | 10 | 3 |
| Total | 23 | 9 | 14 | 0 | 14 | 5 | 56 |
| Mean | 1 | 2 | — | 8 | — | 8·65 | 2·80 |

The working out of the F.O.M. follows the usual method and results in F.O.M. for the complete group of 8·65 inches. The F.O.M. of the odd numbers by themselves is 4·20 inches, and of the even numbers 3·60, for which the working is not shown. The tightness of these two groups and the distance separating their M.P.I. incidentally demonstrate the value of the work done by Dr. F. W. Mann in America and published in his book, *The Bullet's Flight from Powder to Target*. He pointed out that wide groups were due largely to imperfect bullets, provided only that the imperfections were not identical and identically situated in each bullet. These bullets were very imperfect, but the imperfections were similar and identically situated for even or odd numbers as the bullet left the barrel.

In addition to working out the F.O.M., the average distance vertically of the shots from the average height of strike has been computed. The result is 2·80 inches. This average figure is the real criterion of the vertical accuracy. It is called the mean vertical error, or M.V.E. It has several remarkable properties, as it bears a definite numerical relation to the size of the nucleus, and effective and beaten zones. These properties are conveniently tabulated as follows :—

(1) M.V.E. multiplied by the abstract number 0·8453 gives the V.E. on which the chances are even. A zone twice as wide (that is, M.V.E. × 1·69) cut out of the centre of the pattern contains half the shots in the long run. Thus eleven shots have a V.E. greater than 2·80 × 0·8453, or 2·38 inches, and this is nearly half of twenty, and a horizontal band across the target 4·76 inches wide having M.P.I. in centre contains eleven shots. This 50 per cent zone is often called the probable zone.

(2) If a gun has a lateral error (L.E.) exactly equal to its V.E. it will throw a circular pattern whose F.O.M. will be half of 3·1416, or 1·57 times the M.V. (or L.) E. Thus, considering the 10-shot group of odd numbers with F.O.M. of 4·20 inches, and dividing 4·20 by 1·57, the result is 2·68 inches. For the even number group with F.O.M. 3·60 inches, the result is 2·29, and these figures are almost exactly the V.E. of each group if worked out separately. They are smaller than the V.E. 2·80 of the composite group, because the true M.P.I. of each group differs in position from that of the composite group.

(3) The circle drawn with the F.O.M. as radius and M.P.I. as centre contains 54·4 per cent of the shots if the V.E. equals the L.E., that is to say, if the pattern is circular and not elliptical. This can be proved either by mathematics or by measuring up a large number of targets. The circle containing the central eleven shots out of twenty will have a diameter equal to twice the F.O.M.

(4) Chauvenet's table gives the relation between the size of the probable zone and zones containing other percentages. Thus the beaten zone containing 90 per cent. of the shots has a factor of 2·44, meaning that it is 2·44 times as large as the nucleus or 50 per cent zone.

The effective zone of 75 per cent has a factor of 1·71, and is therefore 1·71 times as large as the nucleus.

A further numerical illustration may be given to make sure that the numbers are understood and to distinguish between diagrams on ordinary vertical targets and dispersion on ordinary flat ground.

Suppose that Fig. 3 is a map of the diagram of a machine gun situated due east of the M.P.I. and firing due west, and that the scale is such that the original measures of inches are really yards. The figures entered in the column headed horizontal measurements are now range errors, and 1 foot 2 inches means 14 yards. The M.E. (range) is then worked as the difference between this and the observations, 10, 19 (= 1' 7"), 7, 22 (= 1' 10"), etc.

The M.E. is then 7·9 yards of range.

The nucleus 50 per cent is 7·9 × 1·69 × 1·00 = 13·4 yards.

The effective zone 75 per cent is 13·4 × 1·71 = 22·9.

Pt. III. Fig. 4. Chap. VIII.

The beaten zone 90 per cent is $13\cdot4 \times 2\cdot44 = 32\cdot7$, and the 99 per cent zone may be taken as four times the nucleus or $53\cdot6$ yards. Or, again, a zone 20 yards wide has a factor of 20 divided by $13\cdot4$ or $1\cdot49$, and consequently contains $68\cdot3$ per cent of shots.

The width and height zones are handled in exactly the same way, and all three zones may be measured in yards or in minutes, as may be found convenient.

On a vertical target with circular patterns the following table is useful.

If the F.O.M. is R inches and the radius of the target or bull's-eye is $r$ inches, so that the target is $r/R$ times as large as the F.O.M., the percentage of hits follows the size of $r/R$ thus :—

| Per cent. | r/R | Per cent. | r/R |
| --- | --- | --- | --- |
| 0 | 0·000 | 70 | 1·238 |
| 10 | 0·365 | 80 | 1·431 |
| 20 | 0·533 | 90 | 1·708 |
| 30 | 0·674 | 95 | 1·950 |
| 40 | 0·806 | 96 | 2·100 |
| 50 | 0·940 | 97 | 2·200 |
| 54·4 | 1·000 | 98 | 2·300 |
| 60 | 1·079 | 99 | 2·421 |
| 70 | 1·238 | 100 | Infinity |

Hence a circle of diameter four times the F.O.M. holds more than 19 out of 20 shots, and one of diameter $3\cdot6$ times holds more than 14 out of 15 shots

$3\cdot4$ ,, ,, 9 ,, 10 ,,
$3\cdot0$ ,, ,, 6 ,, 7 ,,
$2\cdot0$ ,, ,, 1 ,, 2 ,,

showing what is obviously true, that the higher the " possible " the more difficult it is to get it, but also giving some idea of the numerical value of the extra difficulty.

The next point which requires consideration is the change of fire effect due to displacing the M.P.I. from the centre of the target. A few actual targets were shot for this purpose. Five good marksmen each fired five 7-shot groups at 300 yards at the same mark, and the size of the composite targets was recorded in minutes of angle. When the first man had fired his 35 shots, the group measured $3\cdot8$ Vertical and $2\cdot1$ Horizontal. Without painting out, the second man shot, and, of course, enlarged the group, and so on for all five men.

The tabulated results were :—

|  | Vertical. | Horizontal. |
| --- | --- | --- |
| First man finished | 3·8 minutes. | 2·1 minutes. |
| Second ,, ,, | 4·7 ,, | 2·6 ,, |
| Third ,, ,, | 5·3 ,, | 3·6 ,, |
| Fourth ,, ,, | 5·8 ,, | 4·1 ,, |
| Fifth ,, ,, | 6·5 ,, | 4·5 ,, |

The men and guns were of a quality to make consistent groups of $3\cdot0$ by $2\cdot0$, yet they ended by more than doubling this in 25 tries. Again, a Vickers gun was shot 50 rounds deliberately, single shot and re-aiming. The result in inches was $10\cdot8$ Vertical and $11\cdot6$ Horizontal, with F.O.M. of $3\cdot15$. The same gun was fired immediately afterwards 50 rounds rapid and gave $22\cdot0$ Vertical and $23\cdot0$ Horizontal with F.O.M. of $8\cdot34$. These figures, although real figures, should not be taken as in any way representative, but merely as illustrating the effect of displacing the M.P.I. or adding a fresh error.

The formula used for calculating the actual numerical effect of displacing the M.P.I. is to be found in the *Text Book of Gunnery*. The formula is too complicated to be given here, but it has been reduced to a graph, which is reproduced as Fig. 4. The unit of measurement is the size of the nucleus or 50 per cent. zone, which is called P. The size of the target is T and their ratio or fraction is T/P.

The horizontal scale is therefore T/P.

The vertical scale is the percentage of hits to be expected, and the curved defining lines are for various values of D/P where D is the size of the displacement. Thus, if P = 31 yards, T = 72 yards and D = 62 yards, T/P = 2·32 and the target affords a zone by Chauvenet of 88·3 per cent. hits if M.P.I. is central. D/P = 2·0, or the M.P.I. is displaced twice the length of the nucleus.

The graph gives the percentage of hits to be expected under such conditions as 13 per cent. instead of 88·3 per cent., as shown by the star mark on the graph.

A study of this graph provides an answer to the apparent paradox that under certain clearly defined conditions better fire effect is produced by unskilled shots than by marksmen. It is also obvious that these conditions are not infrequent on active service. It should, however, be noted that marksmen can easily scatter their fire on purpose, but unskilled shots cannot make close groups even if ordered to do so.

A simple way to study this graph is to extract from it and tabulate the figures given in the following table. Targets of four different sizes—1, 2, 5, and 10 (yards, say)—are considered, and the M.P.I. is displaced successively from zero to 10 yards. The percentage of hits then obtained by two firers of differing skill are looked out. The skill is given numerically by the size of the 50 per cent. zone made by each firer, and these are called P = 2 and P = 10 yards. It will be at once noticed that for all conditions below the heavy line across the following table the unskilled shot has the advantage.

It is also instructive to plot such a table on squared paper with the horizontal scale for D and the vertical scale for percentage. The curves to be plotted are for T/P, and if those for P = 2 are drawn solid and those for P = 10 are dotted, and the intersections of similar T/P curves are connected up, the line of demarcation between the utility of skilled and unskilled men is clearly evident. It is a graph of the heavy line across the table.

TABLE OF EFFECT OF DISPLACING M.P.I.

| | P = 2·0 | | | | | P = 10·0 | | | |
|---|---|---|---|---|---|---|---|---|---|
| T | — | 1 | 2 | 5 | 10 | — | 1 | 2 | 5 | 10 |
| T/P | — | 0·5 | 1·0 | 2·5 | 5·0 | — | 0·1 | 0·2 | 0·5 | 1·0 |

TABLE OF PER CENT. HITS FOR VARIOUS VALUES OF D

| D | D/P | | | | | D/P | | | | |
|---|---|---|---|---|---|---|---|---|---|---|
| 0 | 0 | 26½ | 50 | 91 | 100 | 0 | 5½ | 11 | 28 | 50 |
| 0·5 | 0·25 | 24 | 45 | 87 | 100 | 0·05 | 5 | 10 | 26 | 49 |
| 1·0 | 0·5 | 21 | 41 | 84 | 100 | 0·1 | 5 | 10 | 25 | 48 |
| 1·5 | 0·75 | 16 | 33 | 74 | 99 | 0·15 | 5 | 9 | 24 | 47 |
| 2 | 1·0 | 11 | 23 | 64 | 98 | 0·2 | 5 | 9 | 24 | 46 |
| 3 | 1·5 | 3½ | 9 | 36 | 91 | 0·3 | 5 | 9 | 23 | 44 |
| 4 | 2·0 | 1 | 2 | 16 | 75 | 0·4 | 5 | 9 | 22 | 42 |
| 5 | 2·5 | 1 | 1 | 6 | 50 | 0·5 | 5 | 9 | 21 | 41 |
| 6 | 3·0 | 1 | 1 | 2 | 25 | 0·6 | 4 | 8 | 19 | 36 |
| 7 | 3·5 | 1 | 1 | 1 | 9 | 0·7 | 4 | 8 | 17 | 32 |
| 8 | 4·0 | 1 | 1 | 1 | 3 | 0·8 | 3 | 7 | 15 | 29 |
| 9 | 4·5 | 1 | 1 | 1 | 1 | 0·9 | 3 | 6 | 13 | 26 |
| 10 | 5·0 | 1 | 1 | 1 | 1 | 1·0 | 2 | 5 | 11 | 23 |

Two minor points connected with the theory of probability require brief notice :—

(1) Combination of errors.
(2) Rejection of abnormal results.

The simplest instance of combined errors is seen daily at target practice.

Suppose a man is using a gun which is capable of F.O.M. = 2 inches at 500 yards when perfect ammunition is used, under ideal conditions. He uses ammunition which gives F.O.M. = 3 in a perfect gun and ideal conditions. He himself holds well enough to make F.O.M. = 2·5 with ideal conditions. The question is what F.O.M. to expect from all three errors combined. The simplest and an accurate answer is got by squaring each figure, adding them and taking the square root of the sum. Thus :—

$$(2)^2 + (3)^2 + (2 \cdot 5)^2 = 4 + 9 + 6\tfrac{1}{4} = 19\tfrac{1}{4}.$$

The square root of $19\tfrac{1}{4}$ is 4·6, and this is the F.O.M. to be expected. It is worth noticing that the largest error is by far the most important. Thus, if the figures had been 2, 1, 1, the sum of the squares is $4 + 1 + 1 = 6$ and the square root is 2·45. The answer to 2, 1, $\tfrac{1}{2}$ is 2·3, so that the effect of halving one of the smaller errors produces barely any amelioration.

As regards rejecting abnormal results, the case often occurs in a range report. One or perhaps two rounds in a series are very wide indeed. It is suspected that they are due to some abnormal reason, such as sheer carelessness, and not to the error of the gun. It is required to select some rule to decide whether such a round should be entirely disregarded. A procedure favoured by actuaries is to work to M.E. and then to apply the factor from the following table, which table is also due to Chauvenet.

The series consists of, say, 13 rounds; the M.E. is 8·4, and the actual error of the doubtful round is 29·0. The factor against 13 in the table is 3·07, and 3·07 multiplied by 8·4 is 25·788. The actual error is 29, which is larger than 25·788, so this round may be crossed out and treated as if it had never been fired.

TABLE FOR ABNORMAL ROUNDS

| Rounds | Factor | Rounds | Factor |
|---|---|---|---|
| 3 | 2·05 | 12 | 3·02 |
| 4 | 2·27 | 13 | 3·07 |
| 5 | 2·44 | 14 | 3·12 |
| 6 | 2·57 | 15 | 3·16 |
| 7 | 2·67 | 16 | 3·19 |
| 8 | 2·76 | 17 | 3·22 |
| 9 | 2·84 | 18 | 3·26 |
| 10 | 2·91 | 19 | 3·29 |
| 11 | 2·96 | 20 | 3·32 |

It should be noted that this table applies to only one round at a time.

## CHAPTER IX

### THE STRENGTH OF GUNS

Modern military small arms are now used for so many purposes for which they were not originally designed that unsuspected sources of trouble are apt to arise. In general their strength may be considered under three headings :—

(A) Strength for general purposes, such as shooting, bayonet fighting, grenade throwing, and so forth.

(B) Strength of bolt or breech action against the explosion of the charge.

(C) Strength of barrel against rupture by the explosion.

(A) *General purposes.*

This may be divided into three heads :—

(1) Continuous fair wear and tear of firing, whether rapid or deliberate.
(2) Bayonet fighting.
(3) Grenade throwing and similar rocket work.

A (1). The most usual casualty is that the wooden fore-end splits up from the front trigger guard screw and opens out at the rear as the result of continual use. The effective cause appears to be that the barrel and action in their recoil pull back the fore-end and the nosecap which is attached to the fore-end, at each shot, by means of a jerk on the two lugs which carry the magazine catch. In time this is too much for the wood and the wood splits.

The only other important effect of continual firing is to wear out the inside of the barrel, especially at the breech end. This process is generally termed erosion of the leed. It is caused by a combination of the great heat of the gas and by the velocity of the gas whilst it it leaking over and round the bullet. Pressure by itself unaccompanied by gas-velocity has no appreciable effect, but naturally the leakage has a higher velocity when the pressure is greatest. The enlargement of the bore at the muzzle end is due almost entirely to cleaning, especially cleaning with any abradant. It is interesting to gauge barrels which have fired ten or twenty thousand rounds of proof and have never been cleaned with anything but ammonia. The ·305 plug will not run through them, whereas barrels which have been in barracks for five years and have only fired perhaps 1,000 rounds will often permit the ·308 plug to run right through. Such barrack room barrels are, however, not worn out at the leed, and may, and often do, shoot well.

A (2). Bayonet fighting puts a very unfair strain on a rifle by the leverage of the bayonet on the thin fore-end and barrel. The fore-end usually suffers by breakage at the rear end of the nosecap. The mere bending of the barrel does not necessarily spoil its grouping power, but it entirely upsets the sighting.

A (3). Grenade firing is liable to produce all the troubles under A (1) and in an exaggerated way, and in addition the action may be severely strained and the butt be broken at the stock bolt. The normal recoil of Mark VII cartridge measured by momentum as explained in Chapters I and VI is about 9 lbs. multiplied by 9 f/s. or 81. When firing a grenade this may be increased threefold ; it is equivalent to dropping the rifle from some 12 or 14 feet on to its butt. The rifle was not designed for it, and yet stands it fairly well.

Rodded grenades occasionally produce a ring-bulge in the bore at the seat of the rod. The bulge occurs quite accidentally, it may be on the first round or it may be not till a thousand have been fired. The cause appears to be a wave of gas breaking like a sea-wave exactly on the face of the end of the rod. Such bulges do very little harm to the grouping power of the rifle. They probably do not improve the size of the group, but from actual *ad hoc* experiments with a dozen rifles during the war it was impossible to prove that the ring bulge did any positive harm.

(B) *Strength of breech action.*

Practically all military rifles are at least twice as strong as they need be, and many of them are ten times as strong. One great reason for this is that the cartridge case itself, when free from grease grips the wall of the chamber very tightly on firing and so relieves the dead weight thrust on the bolt head. This is very clearly shown when using the oiled-case method of pressure taking. The Mark VII cartridge is specified to give about 19 tons per square inch. If, however, the case is not dipped in oil it registers only about 10 or 12 tons per square inch. The "proof" cartridge for the ·303 barrel is specified as about 24 tons, yet when fired without oiling the case it registers much less pressure on the bolt head than the service cartridge properly oiled. It is therefore most necessary to remove all grease from the chamber before beginning to shoot if the action is not to be strained.

As is well known, the S.M.L.E. has a rear-locking bolt, and this is often supposed to be a source of weakness and possible danger. But it is often overlooked that the cartridge is a rimmed and not a rimless cartridge. As a necessary result there is much more metal in and near the head of a Mark VII cartridge than there is in a rimless case.

The rimless case has a deep groove cut in it for the extractor, so that the floor of the inside of the case is really only just inside the walls of the chamber. A stretched action or a badly fitting bolt head with a rimless cartridge may easily allow the floor of the case to be outside the chamber. If so, the cartridge case fails badly and allows a large rush of high-pressure gas to escape and to wreck the rifle. Such a casualty is almost unknown with the S.M.L.E., except when the cartridge case itself has a bad flaw in it. With good cases the S.M.L.E. can safely fire a few rounds at as much as 30 tons per square inch pressure whereas hardly any rifle using a rimless cartridge can stand one round at such an excessive pressure.

(C) *The strength of barrels.*

In the ordinary gunnery manuals the expression "strength of guns" refers almost entirely to rupture by excessive internal gas pressure. It is, however, a fact well known in the trade that if a rifle barrel has stood the usual "proof" rounds it is almost impossible to "open" it by any load, however unfair, unless an obstruction is placed in the barrel several inches in front of the bullet. Such an obstruction in the way of a fast-moving bullet creates a ring bulge of a most exaggerated type and blows the barrel into two, or at any rate opens it widely. Attempts have been made as serious experiments at Woolwich and elsewhere to overload a barrel enough to open it, but they have always failed. It is comparatively easy to wreck the bolt or action. Sixty-eight grains of sporting ballistite loaded as a blank cartridge into a ·303 case by excessive tapping and squeezing have wrecked the action. So has a steel rod two feet long when loaded from the muzzle to touch the bullet, but in each case the barrel itself has only been stretched at the screw thread and cracked open along the extractor way. A bullet purposely lodged half-way down the bore when firing Mark VII has cut the barrel in half by ring bulging, whereas the Springfield rifle has been wrecked by the failure of its rimless case merely by the use of a wrong bolt.

Such experiments make it unnecessary to go into the mathematical treatment of barrel strength such as is necessary with ordnance for which the treatment was devised. Robins's formula that the barrel thickness should be equal to the calibre is safe, *i.e.*, the minimum diameter of solid metal in a ·303 barrel should be ·909 inch. It may be as well, however, to touch on this treatment. It may be summarized as follows for a barrel consisting of one single homogeneous tube such as a rifle barrel.

If P is the maximum gas pressure in the tube and T is the maximum tension permissible in the steel without causing permanent stretch in the metal, also if R is the radius of the outside of the tube and $r$ the internal radius, then P must not exceed

$$\frac{R^2 - r^2}{R^2 + r^2} \times T.$$

Supposing $r = 0$, then P must be less than T; that is to say, the greatest permissible gas pressure must be less than the elastic limit of the steel.

As the elastic limit of barrel steel is in the region of 15 to 20 tons/inch$^2$, and as pressures of 20 and 25 tons/inch$^2$ are perfectly safe, and are daily exceeded by the "proof" charges, it is clear that this treatment is inapplicable to a rifle barrel. The reason is clear enough really from the wording "without causing permanent stretch." The proof rounds do produce permanent stretch and they set up an entirely new condition of affairs. The internal layers of the steel at the breech end of the barrel are deformed and remain deformed, causing the inside of the barrel to be squeezed by the outside. This produces an effect similar to that of wire-winding a piece of ordnance or even of binding the handle of a cricket bat. The superfluous strength of the outside layers is passed on to the inside layers and strengthens them.

The whole matter is now treated of under the name "auto-frettage" or automatic shrinkage, and it has recently been applied to cannon, especially by the French.

Col. L. Jacob has written a volume for the *Encyclopédie Scientifique* (O. Doin et fils, Paris), "Résistance et construction des Bouches à Feu, Auto Frettage," to which the student is referred.

It is often said that the earliest authorities who have written on the subject are Professor Perry, Monsieur Malaval and Professor Bridgeman, and that the earliest date of publication is 1905. It is, however, certain that a full exposition of the subject is to be found in the *Kynoch Journal*, No. 19, Vol. IV, October-November, 1902, over the signature of that remarkable man, R. H. Housman.

## CHAPTER X

## WOUNDING EFFECTS OF BULLETS

Theoretically the chief factor in the wounding effect of any missile is its energy of movement, expressed in ft.-lbs. by the formula $WV^2/2g$, where W is the weight and V the velocity of movement of the missile and $g$ the acceleration of gravity. In other words, the heavier the missile and the faster it moves the greater is its wounding power. Practically, however, as far as missiles from small arms are concerned, the problem is more complicated. Of three chief factors which go to make wounding power, viz., weight, cross-section area and velocity, ballistic and military considerations have reduced the first and second between 1851 and 1914 from 680 grains (English Minié) to 174 grains (Mark VII pointed bullet), and from ·69 inch to ·311 inch, while the muzzle velocity has increased up to nearly 3,000 feet per second. Under these circumstances the wounding effect of small-arm bullets has come to depend less upon the effect of the blow struck, and more upon the destruction which the bullet causes in the tissues. Penetration is assured, and the questions which arise are how far the reduction of cross-section area and weight have influenced the destructive action of the bullet, *i.e.*, its stopping power, and how far any loss is made up for by the increased velocity.

The wounding power of a missile may be estimated experimentally or by observation of actual wounds in war. Neither method gives entirely reliable results.

Experimentally much work has been done upon live and dead animals, and upon human cadavera with small-bore bullets, and some valuable information obtained. But it is evident that living animals present great differences in the size and hardness of their bones and the constitution of their soft parts. In particular, there is no living animal which could give any information of the effects of small-bore missiles on the human brain. In the case of dead tissues, *rigor mortis* alters the density entirely, and any measures used to preserve the tissues have a similar effect.

Indifferent tissues, such as wood, soap, clay and gelatine have also been used for experimentation. They have given a certain amount of information, wood as regards penetrative power, and the softer substances as regards the behaviour of the bullet on impact, and the effect of the spin on soft parts. But these results have no applicable value in estimating effects on the human body.

The results copiously recorded by big game hunters deserve mention, but are somewhat vitiated by the fact that the object and ideal of the hunter is quite different from those aimed at by the soldier. The aim of the big-game shot is to kill, neatly and effectively and without undue destruction of meat or trophies. The aim of the soldier is to "put out of action the largest possible number of men," * but not " by the use of

---

\* Quoted from text of Declaration of St. Petersburg of 1868.

arms which would uselessly aggravate the sufferings of men put out of action, or would render their death inevitable."

Sir Samuel Baker states the view of the big-game shot. He maintains that the greatest wounding power is possessed by a missile which can penetrate, but not traverse, the bodies of big game, can maintain its integrity when in the body, and be found lying flattened to a mushroom under the skin of the side of the body opposite to the wound of entry. He deprecates the use of lighter missiles, even if moving with a higher velocity.

Later experience has taught that the increased accuracy and flatter trajectory of the small calibre rifle was worth a small sacrifice of killing power. Major Whelen, in *The American Rifle*, summarizes the general experience when he says "A ·300 inch calibre expanding bullet of 170 grains driven with a muzzle velocity of 2,700 feet per second is certainly a most killing and satisfactory charge for all game with the exception of elephants, buffalos, rhinoceros and hippopotamus." But he thinks these large animals require a ·450, or ·465 rifle, shooting a ·450-grain jacketed bullet at 2,000 to 2,200 feet per second. The big-game hunter, in fact, foreshadows the difficulty of the small-bore weapon, that it tends to become more "humane" as the calibre lessens and the weight of the bullet is diminished.

Observation of actual wounds in battle would seem to be the best method of estimating wounding power, but it is beset with numerous difficulties. To be of any value it should include an examination of the dead, with accurate observations of the cause of death, a knowledge of the ranges at which all wounds are inflicted, standardized weapons, charges and missiles, and wounds inflicted on identical parts of the body in men of similar size and build. This is as a rule quite impracticable, but certain general principles can be laid down. Comparing, for example, the Great War with the South African campaign, it is quite clear that the wounding power of the pointed bullet is greater than that of the round-nose Mark II bullet. The through and through wound was rare, while " explosive " exits were common, and at greater ranges. The effects on bone also were more severe, the shattering being foretold by Fessler in 1905 as the result of his experiments with the pointed bullet.

The wounding power of the military small arm was considered satisfactory until the introduction of the small bore composite bullet. The older spherical or conoidal lead bullet had a large cross-section area, which increased on striking, and especially if it met bone. Hence probably the actual striking effect of the blow was an appreciable "stopping" factor. Even against fanatical tribesmen the Snider and the Martini-Henry bullets were satisfactory. But the ·303 bullet with its diminished cross-section area and greater velocity had a power of penetration which made its impact hard to feel in many cases, and its track through the tissues was a clean puncture without circumferential damage of any kind. Unless bone were hit, or unless the range was short enough to produce an " explosive " wound, " stopping power " was conspicuously lacking. What was lacking in the matter of deformation (increased impact on striking) could be made up for by specially made " expanding " bullets, and these were used against big game and also against savages and fanatics. But in civilized warfare these bullets are expressly ruled out, and it was not till the introduction of the pointed bullet that a more satisfactory " stopping power " was attained.

This leads to a consideration of what the chief factors are which influence the wounding power of a bullet.

ANATOMICAL AND SURGICAL CONSIDERATIONS

The severity of a wound varies much with the part which is hit. Soft tissues present little resistance to the passage of a bullet, but when bone is hit all the energy of the bullet is used up in smashing the bone. The fragments of bone form secondary missiles, and are carried onwards in the path of the bullet and assist in the laceration of the soft tissues.

Their size is roughly in inverse proportion to the velocity of the bullet, high velocity producing fine fragmentation of the bone.

Again there are certain parts, usually belonging to the vascular system, wounds of which end speedily in death. It is estimated that over 90 per cent. of deaths on the battle field are due to bleeding. When these so-called " vital " parts are wounded, the make or velocity of the bullet is a matter of indifference.

Gunshot wounds are of the punctured variety of wounds. That is to say, they are inflicted by missiles which pass deeply and into and through the body, carrying with them germs of infection which are thus deeply implanted in the body. The infection which arises from these germs is the chief cause of mortality and invalidity after gunshot injury. The pointed bullet, having a large area which is untouched, and therefore not scraped clean by the rifling of the barrel, is a more septic bullet than the Mark II bullet.

It has already been pointed out that in the composite round-nosed bullet the soft lead core is covered with a cupro-nickel jacket, and therefore cannot expand on impact on soft substances as the old lead bullet, whether spherical or conoidal, almost invariably did. The only way of obtaining increase of area on impact was by the method of the "expanding" bullet, a device which uncovered the lead at the tip and allowed it unfettered action on impact. This, however, was forbidden by international agreements. Increase of area on impact could, however, be attained by making the bullet unsteady so that it tilted on impact and moved sideways through the tissues. A bullet of very small bore, like the Japanese ·256-calibre bullet, was reported on by Macpherson in the Russo-Japanese war as being very unstable on impact, so that mere diminution of calibre without loss of length seems to make for unsteadiness in a bullet. But according to some authorities the pointed bullet presents the same unsteady character because by its structure the centre of gravity is thrown so far back. Unless the point is accurately in the line of the trajectory it immediately on impact comes under the influence of a couple tending to make it turn upon its short axis. A further lightening of the tip by the substitution of a lighter metal such as aluminium for part of the lead core would enhance this turning effect. How far this theory is correct it is a little difficult to say Observation of wounds of exit confirms the fact that the pointed bullet does turn in the tissues. Moreover, the increased wounding power of this bullet on soft parts as compared with the neat cylindrical track of the Mark II bullet seems to show that some enhancing factor is present.

But of all the factors which influence wounding-power the velocity of the bullet is the most important. It is intimately connected with that variety of wound incorrectly termed "explosive." In these wounds there is commonly a small wound of entry in the skin and an enormous crateriform opening at the exit from which protrude masses of damaged muscle and other tissues with tendons and fibrous structures, all bound together by blood clot. Fragments of bone are often found among the lacerated parts and even outside the wound. The parts present, in fact, the appearance of having been subjected to the effects of a local explosion. The destructive effects, moreover, are observed at some distance from the actual track of the bullet. Small hæmorrhages, separation of aponeurotic planes, laceration of muscle fibres, and destruction of cellular elements of glands have been observed an inch or more from the bullet track. There seems to have been a "tissuequake" of more or less severity. These wounds have been observed in all campaigns, since cylindroconoidal bullets have been used. And the point which all military surgeons make in discussing their causation is that they are an effect of great velocity. They are produced at ranges which are short for rifles with a comparatively low muzzle velocity, and which increase as the muzzle velocity increases. They came therefore particularly under notice in the South African war, and have increased in numbers in every succeeding campaign in association with the rising muzzle velocities of military rifles. The Martini-Henry and the Gras rifle caused explosive wounds up to 150 or 200 yards, while the Lee-Enfield and the Lebel caused them up to 300 or 400 yards. With the pointed bullet probably another 200 yards may be added.

These wounds are commonly observed in association with bony injury. The explanation of their causation is in this case quite simple. The bullet impinging upon a hard substance like bone, breaks up the bone into fragments, large and small, and then imparts some of its energy to these fragments, so that they become secondary missiles and magnify the destructive thrust of the bullet through the tissues. It is not at all necessary that the bullet should break up. It only seems necessary that it should possess sufficient velocity.

But a far more interesting class of explosive wounds is seen in association with soft parts only. Here the " secondary missile " explanation hardly seems sufficient. With sporting rifles using ordinary solid-nosed bullets explosive effects were first observed on animals when the Roumanian Mannlicher was introduced firing a bullet with a muzzle velocity of 2,300 f/s and falling 100 f/s per 100 yards. Now many sportsmen had used the British service rifle and Mk. VI ammunition, with a muzzle velocity of 2,000 f/s. without bursting their game. So it might seem that there is a critical velocity somewhere between 2,000 and 2,300 f/s. where bursting effects begin. It has been suggested that the rush of air following the bullet may be likened to the cavitation velocity with which air rushes in to fill a vacuum, and when this velocity reaches a limit somewhere beyond 2,100 f/s it becomes competent to produce disruptive effects on the tissues. More probably, however, there may be in the tissues themselves a commotion in the wake of a high velocity bullet comparable to that producing cavitation velocity in the air, and tending to produce a blowing-out effect at the exit, and some disruption results in the near vicinity of the track.

To sum up, the wounding effect of a small arm missile is due partly to the nature of the part struck, partly to the shape and structure of the missile, and chiefly to the velocity which it possesses on impact. The marked loss of " stopping power " of the non-deformable blunt-nosed ·303 bullet, as compared with leaden deformable bullets of larger calibre, has been quite regained by the pointed bullet. Experience of wounds in the Great War has shown that, as an effect of increased velocity, explosive wounds are seen up to ranges of 600 yards, and, as an effect of increased instability of the bullet after impact, all wounds show a higher degree of laceration of soft parts with finer comminution of bone, whether compact or cancellous.

## CHAPTER XI

## DEFINITIONS AND UNITS

*Exterior Ballistics.*—The theory of the motion of the projectile from the gun to the target.

*Interior Ballistics.*—The theory of the motion of the projectile in the gun.

*Gunnery.*—The practical application of the theory of ballistics.

*Axis of the Gun.*—The straight line passing down the centre of the bore; the longitudinal axis.

*Axis of the Trunnions.*—The straight line passing through the centre of the trunnions, at right angles to the axis of the gun; the transverse axis about which the gun is rotated when elevating. In machine guns the trunnions are generally part of the mounting.

*Calibre.*—The nominal diameter of the bore in inches, measured across the lands.

*Trajectory.*—The curve described by the centre of gravity of the projectile in flight.

*Muzzle Velocity.*—The velocity of the projectile at the muzzle of the gun; it is sometimes called *initial velocity*.

*Remaining Velocity.*—The velocity of the projectile at any point of the trajectory.

*Striking Velocity.*—The velocity of the projectile at the point of impact.

*Terminal Velocity.*—When the retardation due to air resistance is exactly equal to the acceleration of gravity there is no force acting on the projectile. The velocity is then constant, and is called "the terminal velocity." If C is known, it can be found from the ballistic tables by noting the velocity where the value of $p$ is equal to the ballistic coefficient $C = W/nd^2$.

*Line of Sight.*—The straight line passing through the sights and the target. (The line G.T., Fig. 1.)

*Vertical Plane of Sight.*—The vertical plane containing the line of sight.

*Lateral Plane of Sight.*—The plane passing through the line of sight at right angles to the vertical plane of sight.

*Line of Departure.*—The straight line representing the direction of motion of the projectile at the moment of leaving the gun. It is the tangent to the trajectory at the muzzle. It coincides with the axis of the gun when there is no *jump*, and when the gun recoils axially on a stationary mounting.

*Plane of Departure.*—The vertical plane containing the line of departure. It is sometimes called "vertical plane of fire."

*Jump.*—The vertical angle between the axis of the gun before firing and the line of departure. When the line of departure is above the axis of the gun the jump is said to be *positive;* when below it is *negative*. (The angle J, Fig. 1.)

*Elevation.*—An angle measured in a vertical plane. It is *positive* when measured upwards from the line of reference, and *negative* when measured downwards; it is then sometimes called "depression."

PT. III. CHAP. XI.   To face p. 366.

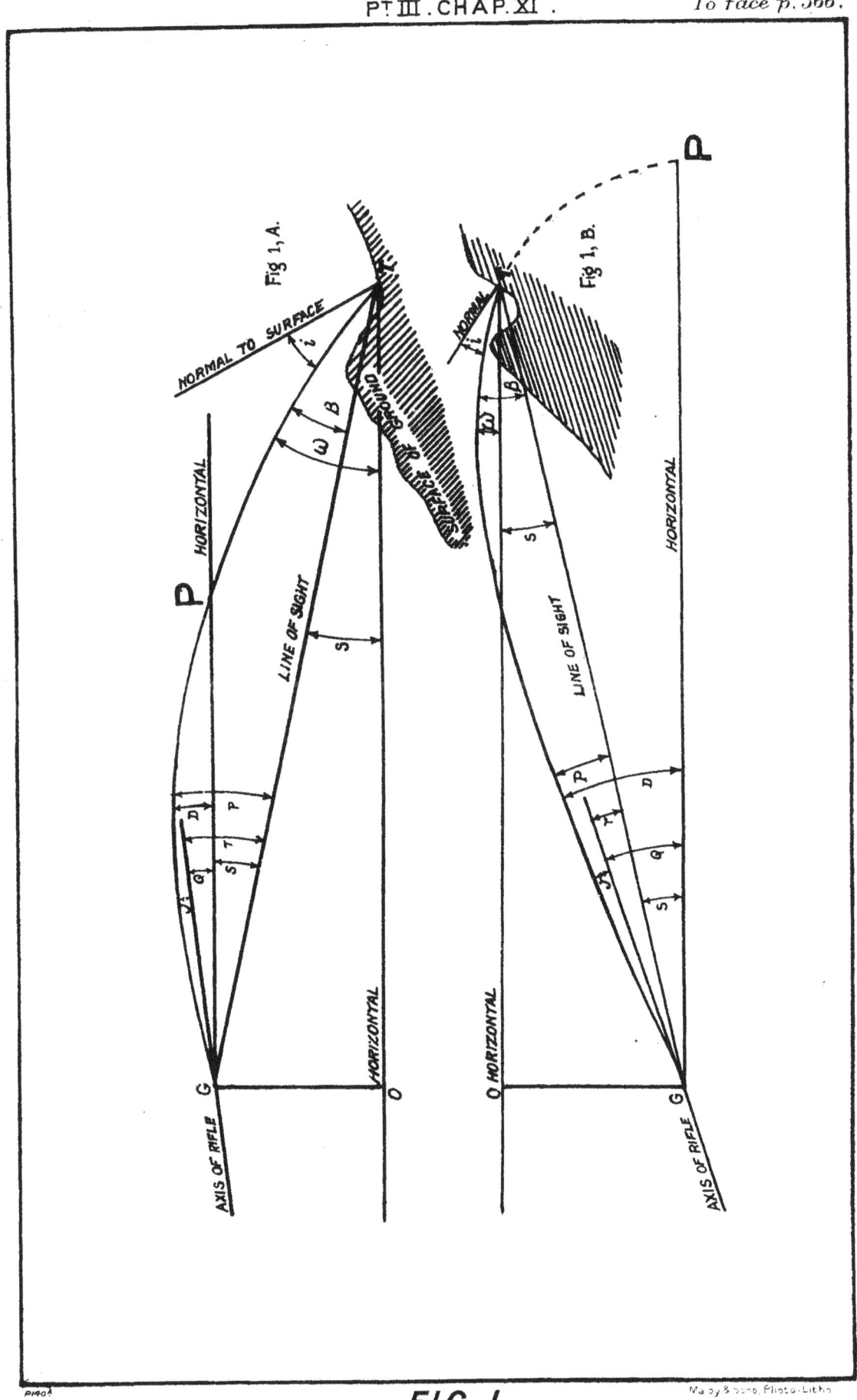

FIG. I.

*Angle of Sight.*—The " elevation " of the line of sight with reference to the horizontal plane. It is measured in the vertical plane of sight, and is positive when the line of sight is directed above the horizontal plane. (The angle S, Fig. 1.) When a *negative* angle of sight is given to a gun it is described as ". . . degrees . . . minutes depression."

*Angle of Departure (of the Projectile).*—The " elevation " of the line of departure, with reference to the horizontal plane. (The angle D, Fig. 1 ; *see* note 1).

*Angle of Projection (of the Projectile).*—The " elevation " of the line of departure with reference to the lateral plane of sight. When the angle of sight is zero, it is the same as the angle of departure. (The angle P, Fig. 1 ; *see* note 1.)

*Quadrant Elevation (of the Gun).*—The " elevation " of the axis of the gun immediately before firing with reference to the horizontal plane. It differs from the angle of departure by the angle of jump only. (The angle Q, Fig. 1 ; *see* note 1.)

*Tangent Elevation (of the Gun).*—The " elevation " of the axis of the gun immediately before firing, with reference to the lateral plane of sight. It differs from the angle of projection by the angle of jump only. (The angle T, Fig. 1 ; *see* note 1.)

*Rise.*—The mean alteration (in minutes) of elevation to give 100 yards alteration in range at any given distance.

*Deflection.*—A correction to direction. It is consequently an angle measured at right angles to the vertical plane of sight, either in the lateral plane of sight or in the horizontal plane (*see* note 2). It is applied in compensating for :—

(1) Lateral travel of gun or target during the time of flight.
(2) Deviation of the projectile due to cross wind.
(3) Drift, unless automatically allowed for by the sighting.
(4) Any other cause affecting direction.

*Point of Departure.*—The axis of the rifle barrel at the muzzle at the moment when the projectile leaves it. (The point G, Fig. 1.)

*Point of Arrival.*—The second point of intersection of the trajectory with the lateral plane of sight. (The point T, Fig. 1.)

*Point of Graze.*—The point of intersection of the trajectory with the horizontal plane through the gun. (The point P, Fig. 1).

*Drift.*—The deviation of the projectile from the plane of departure due to rotation.

*Vertex.*—The highest point of the trajectory ; it is also called " the culminating point."

*Angle of Descent.*—The angle measured in a vertical plane, which a tangent to the trajectory at the point of arrival makes with the lateral plane of sight. (The angle $\beta$, Fig. 1 ; *see* note 3.)

*Angle of Arrival.*—The angle measured in a vertical plane, which a tangent to the trajectory at the point of arrival makes with the horizontal plane. (The angle $\bar{\omega}$, Fig. 1 ; *see* note 3).

*Angle of Incidence.*—The angle which the tangent to the trajectory at the point of arrival makes with the normal to the surface struck. (The angle i, Fig. 1).

*Angle of Impact.*—The angle which the tangent to the trajectory at the point of arrival makes with the surface struck ; it is the complement of the angle of incidence.

*Vertical Fire.*—Fire directed vertically upwards or downwards, that is, at an angle of departure of 90°.

*Mean Point of Impact.*—A point which represents the mean position of the points of impact of a large number of rounds fired under the same conditions.

*Danger Space.*—The horizontal distance, measured towards the gun, in which a target would be hit by a given trajectory.

*Time of Flight.*—The time a projectile takes to reach the point of arrival, reckoned from the moment it leaves the muzzle.

*Range.*—A distance measured along the line of sight.

*Target Range.*—The "range" to the target; the distance between gun and target measured along the line of sight.

*Gun Range.*—The "range" to the point of arrival; the distance between the gun and the point of arrival. (The length G T, Fig. 1.) It is one of the functions of gunnery to make the gun range coincide with the target range.

*Notes.* (1) The angles of *departure* and *projection* refer to the motion of the projectile, not to the gun; they may therefore be called "ballistic" angles, and they depend upon ballistic theory. The angles of *quadrant elevation* and *tangent elevation* refer to the gun, not to the projectile; they may therefore be called "mechanical" angles. They are fundamental to the method of laying the gun, the object being to ensure that the projectile leaves the gun at the correct ballistic angle.

(2) Lateral deflection measured in a horizontal plane is sometimes called "lateral deflection in azimuth." The azimuth of a point is the bearing measured from true north. A difference in azimuth is therefore merely a difference in bearing, and represents a lateral deflection measured in the horizontal plane.

(3) At the point of graze the angle of descent is equal to the angle of arrival.

The definitions above are from *The Text Book of Ballistics and Gunnery*, prepared in the Military College of Science.

### Physical Definitions

*Fundamental Units.*—The three indefinables in nature are space, time and matter.

The British unit of length (space) generally employed in scientific work is the foot, the third part of the Standard yard defined by Act of Parliament, 18 and 19 Vict., 1855, the length between the lines on two gold plugs at 62° F., on a bronze bar kept by the Warden of the Standards.

The unit of weight (matter) is similarly defined as the pound, the weight of a piece of platinum marked " P.S. 1844 1 lb."

The unit of time is by universal consent the mean solar sexagesimal second. All the astronomical observatories in the world agree as to its length.

Velocity (V) if constant is feet described in one second written as f/s.; if variable it is measured at any instant by the feet which would have been described in one second if the velocity had remained constant.

Acceleration ($a$) if constant is the number of f/s. added in one second written as $f/s.^2$; retardation ($-a$) is the number of f/s. lost in one second. It is negative acceleration.

Acceleration or retardation if variable is measured in the same way as variable velocity.

The acceleration of gravity is the constant acceleration with which we are most familiar. Denoting it as usual by $g$ ($= 32\cdot2$ $f/s.^2$), then the velocity of a body falling freely will grow $g$ units of velocity (f/s.) per second. In metric units $g = 9\cdot81$ metres per second per second.

Force is that which produces change of motion either in amount or direction in a body. As the field of force in which we live is that due to the attraction of the earth, it is convenient to take the attraction of the earth on a lb. weight as the unit of force, and to call this the force of a pound.

Then a force of P lbs., acting on a weight of W lb., produces an acceleration of "$a$" f/s.² where

$$\frac{a}{g} = \frac{P}{W}$$

*Work* done by a force of P lb., acting through S feet, is given as *energy* by

$$PS = \frac{WV^2}{2g} \text{ ft.-lbs.}$$

The *impulse* of P lb., acting for $t$ seconds, is given as *momentum* by

$$Pt = \frac{WV}{g} \text{ second-lbs.}$$

The *thrust* of $p$ lb. per square inch on an area of A square inch is given by
$$pA = P \text{ lb.}$$

---

### CONVERSION TABLE OF METRIC AND BRITISH UNITS

#### WEIGHT

| Metric to British | | British to Metric | |
|---|---|---|---|
| 1 gram | = 15·432 grains. | 1 grain | = 0·064799 grams. |
| 1 gram | = 0·035274 ozs. | 1 oz. | = 28·349 grams. |
| 1 kg. | = 2·2046 lbs. | 1 lb. | = 0·45359 kg. |
| 1 kg. | = 0·019684 cwts. | 1 cwt. | = 50·802 kg. |
| 1 m. tonne | = 0·98421 tons. | 1 ton | = 1·0160 m. tonne. |

#### LENGTH

| | | | |
|---|---|---|---|
| 1 metre | = 1·0936 yards. | 1 yard | = 0·91438 metres. |
| 1 metre | = 3·2809 feet. | 1 foot | = 0·30479 metres. |
| 1 cm. | = 0·39371 in. | 1 inch | = 2·5400 cm. |
| 1 knot | = 1 nautical mile (6,080 ft.) an hour = 1·689 f/s. | | |

#### AREA

| | | | |
|---|---|---|---|
| 1 sq. cm. | = 0·15501 sq. in. | 1 sq. in. | = 6·4514 sq. cm. |
| 1 sq. m. | = 10·764 sq. ft. | 1 sq. ft. | = 0·092900 sq. m. |

#### VOLUME

| | | | |
|---|---|---|---|
| 1 c.c. | = 0·061027 c. in. | 1 c. in. | = 16·386 c.c. |
| 1 c. m. | = 35·317 c. ft. | 1 c. ft. | = 0·028315 c. m. |
| 1 litre | = 1·7608 pints. | 1 pint | = 0·56793 litres. |

## Pressure

| | |
|---|---|
| 1 atmosphere = 0·00656 tons per sq. in. | 1 ton per sq. in. = 152 atmospheres. |
| 1 atmosphere = 14·7 lbs. per sq. in. | 1 lb. per sq. in. = 0·0680 atmosphere. |
| 1 kg. per sq. cm. = 0·968 atmosphere. | 1 atmosphere = 1·03 kg. per sq. cm. |
| 1 kg. per sq. cm. = 14·223 lbs. per sq. in. | 1 lb. per sq. in. = 0·070309 kg. per sq. cm. |
| 1 kg. per sq. cm. = 0·0063493 tons per sq. in. | 1 ton per sq. in. = 157·49 kg. per sq. cm. |

## Energy

| | |
|---|---|
| 1 m. tonne = 3·2291 foot-tons. | 1 foot-ton = 0·30969 m. tonne. |
| 1 m. kg. = 7·2331 foot-lbs. | 1 foot-lb. = 0·13825 m. kg. |

*Note on Energy.*

1 British thermal unit raises 1 lb. water 1° F.
1 ,, ,, ,, = 777·2 ft.-lbs.
100,000 ,, ,, ,, = 1 therm (gas).
1 therm = 34,690 ft.-tons.
       = 39·24 horse-power hours.
       = 29·25 B.T.U. (electrical).
1 B.T.U. (electrical) = 1,185 foot-tons.
1 ,, ,, = 0·03419 therms.
1 ,, ,, = 1·345 horse-power hours.
1 small calorie (gram-calorie) = 3·085 ft.-lbs. and raises 1 gram of water 1° C.

# PART IV

## APPENDIX I

### Range Table of ·303 Mk. VII from Hythe Firings*

| Range, yds. | Elevation, mins. | Diff. | Descent, mins. | Diff. | Time, secs. | Diff. | Velocity, f/s. | Diff. | $f$. | Diff. |
|---|---|---|---|---|---|---|---|---|---|---|
| 0 | 0 |  | 0 |  | 0·0 |  | 2440 |  | 32·2 |  |
|  |  | 3 |  | 4 |  | 0·2 |  | 210 |  | 1·6 |
| 100 | 3 |  | 4 |  | 0·2 |  | 2230 |  | 30·6 |  |
|  |  | 4 |  | 4 |  | 0·1 |  | 200 |  | 1·2 |
| 200 | 7 |  | 8 |  | 0·3 |  | 2030 |  | 29·4 |  |
|  |  | 4 |  | 5 |  | 0·1 |  | 190 |  | 1·0 |
| 300 | 11 |  | 13 |  | 0·4 |  | 1840 |  | 28·4 |  |
|  |  | 4 |  | 7 |  | 0·2 |  | 180 |  | 0·9 |
| 400 | 15 |  | 20 |  | 0·6 |  | 1660 |  | 27·5 |  |
|  |  | 4 |  | 9 |  | 0·2 |  | 160 |  | 0·9 |
| 500 | 19 |  | 29 |  | 0·8 |  | 1500 |  | 26·6 |  |
|  |  | 6 |  | 11 |  | 0·2 |  | 140 |  | 0·6 |
| 600 | 25 |  | 40 |  | 1·0 |  | 1360 |  | 26·0 |  |
|  |  | 7 |  | 13 |  | 0·2 |  | 120 |  | 0·6 |
| 700 | 32 |  | 53 |  | 1·2 |  | 1240 |  | 25·4 |  |
|  |  | 8 |  | 15 |  | 0·3 |  | 100 |  | 0·6 |
| 800 | 40 |  | 68 |  | 1·5 |  | 1140 |  | 24·8 |  |
|  |  | 9 |  | 18 |  | 0·3 |  | 80 |  | 0·2 |
| 900 | 49 |  | 86 |  | 1·8 |  | 1060 |  | 24·6 |  |
|  |  | 11 |  | 21 |  | 0·3 |  | 60 |  | 0·1 |
| 1000 | 60 |  | 107 |  | 2·1 |  | 1000 |  | 24·5 |  |
|  |  | 12 |  | 24 |  | 0·3 |  | 40 |  | 0·1 |
| 1100 | 72 |  | 131 |  | 2·4 |  | 960 |  | 24·4 |  |
|  |  | 14 |  | 27 |  | 0·3 |  | 40 |  | 0·2 |
| 1200 | 86 |  | 158 |  | 2·7 |  | 920 |  | 24·2 |  |
|  |  | 16 |  | 31 |  | 0·4 |  | 40 |  | 0·2 |
| 1300 | 102 |  | 189 |  | 3·1 |  | 880 |  | 24·0 |  |
|  |  | 17 |  | 35 |  | 0·4 |  | 40 |  | 0·3 |
| 1400 | 119 |  | 224 |  | 3·5 |  | 840 |  | 23·7 |  |
|  |  | 18 |  | 40 |  | 0·4 |  | 40 |  | 0·3 |
| 1500 | 137 |  | 264 |  | 3·9 |  | 800 |  | 23·4 |  |
|  |  | 19 |  | 45 |  | 0·4 |  | 40 |  | 0·3 |
| 1600 | 156 |  | 309 |  | 4·3 |  | 760 |  | 23·1 |  |
|  |  | 20 |  | 51 |  | 0·5 |  | 40 |  | 0·4 |
| 1700 | 176 |  | 360 |  | 4·8 |  | 720 |  | 22·7 |  |
|  |  | 22 |  | 57 |  | 0·5 |  | 40 |  | 0·4 |
| 1800 | 198 |  | 417 |  | 5·3 |  | 680 |  | 22·3 |  |
|  |  | 24 |  | 63 |  | 0·5 |  | 40 |  | 0·4 |
| 1900 | 222 |  | 480 |  | 5·8 |  | 640 |  | 21·9 |  |
|  |  | 27 |  | 69 |  | 0·6 |  | 40 |  | 0·4 |
| 2000 | 249 |  | 549 |  | 6·4 |  | 600 |  | 21·5 |  |
|  |  | 30 |  | 75 |  | 0·6 |  | 40 |  | 0·4 |
| 2100 | 279 |  | 624 |  | 7·0 |  | 560 |  | 21·1 |  |
|  |  | 33 |  | 84 |  | 0·6 |  | 40 |  | 0·3 |

* In the various Training Manuals, a range table is given which differs slightly from the above table; it is one which has long been found satisfactory in use, particularly with the Vickers gun. The differences between the two tables lie well within the tolerations for the variation of ammunition.

APPENDIX I—Range Table of ·303 Mk. VII—*continued*

| Range, yds. | Elevation, mins. | Diff. | Descent, mins. | Diff. | Time, secs. | Diff. | Velocity, f/s. | Diff. | f. | Diff. |
|---|---|---|---|---|---|---|---|---|---|---|
| 2200 | 312 | | 708 | | 7·6 | | 520 | | 20·8 | |
| | | 39 | | 96 | | 0·7 | | 30 | | 0·3 |
| 2300 | 351 | | 804 | | 8·3 | | 490 | | 20·5 | |
| | | 45 | | 110 | | 0·8 | | 30 | | 0·3 |
| 2400 | 396 | | 914 | | 9·1 | | 460 | | 20·2 | |
| | | 51 | | 126 | | 0·8 | | 30 | | 0·2 |
| 2500 | 447 | | 1040 | | 9·9 | | 430 | | 20·0 | |
| | | 57 | | 142 | | 0·9 | | 30 | | 0·2 |
| 2600 | 504 | | 1182 | | 10·8 | | 400 | | 19·8 | |
| | | 63 | | 161 | | 1·0 | | 30 | | 0·1 |
| 2700 | 567 | | 1343 | | 11·8 | | 370 | | 19·7 | |
| | | 72 | | 182 | | 0·10 | | 30 | | 0·1 |
| 2800 | 639 | | 1525 | | 12·8 | | 340 | | 19·6 | |
| | | 86 | | 212 | | 0·11 | | 20 | | 0·1 |
| 2900 | 725 | | 1737 | | 13·9 | | 320 | | 19·5 | |
| | | 109 | | 272 | | 0·12 | | 20 | | 0·1 |
| 3000 | 834 | | 2009 | | 15·1 | | 300 | | 19·4 | |

Values of $f$ are for plotting trajectories by means of formula
H yards $= R \tan E - \frac{1}{4} f.t.^2$

## APPENDIX II

Abstract of Results of Ordnance Committee Calculations

$W/d^2 = 0·284$ $\qquad$ $V = 2600$ f/s.

| Elevation. | 5°. | 10°. | 20°. | 30°. | 35°. | 40°. |
|---|---|---|---|---|---|---|
| *To vertex.* | | | | | | |
| Range ... yds. | 1464 | 1992 | 2572 | 2813 | 2840·7 | 2812·9 |
| Height ... ft. | 258 | 717 | 1946·43 | 3407·93 | 4181·04 | 4962·2 |
| Time ... secs. | 3·2858 | 5·3914 | 8·7134 | 11·4167 | 12·61959 | 13·74161 |
| Velocity ... f/s | 865·89 | 694·81 | 523·64 | 426·29 | 388·40 | 354·17 |
| *To full range.* | | | | | | |
| *Maximum at 34° 42′ and 4457 yards.* | | | | | | |
| Range ... yds. | 2464 | 3273 | 4097 | 4423 | 4456·6 | 4412·9 |
| Descent, angle ... | 11° 14·44′ | 24° 20·03′ | 47° 34·9′ | 62° 24·50′ | 67° 24·84′ | 71° 21·11′ |
| Time ... secs. | 7·5559 | 12·6534 | 20·9479 | 27·9129 | 31·05973 | 34·0085 |
| Velocity ... f/s | 587·91 | 447·17 | 389·70 | 404·63 | 416·805 | 428·80 |
| Range with C increased 50 per cent. ... | — | 4168 | 5330 | 5821 | 5876 | — |

Resistance law taken from Table I.

## APPENDIX III

O.C. Calculations. $\dfrac{W}{d^2} = 0.284.$ $\quad V = 2600$ f/s.

Plotted and Pricked off

| E. | R yds. | t secs. | v f/s. | B°. | f. |
|---|---|---|---|---|---|
| 0 | 0 | 0.00 | 2600 | 0 | 32.2 |
| 1 | 1140 | 2.24 | 1005 | 1.9 | 23.8 |
| 2 | 1600 | 3.80 | 806 | 4.1 | 23.2 |
| 3 | 1955 | 5.17 | 709 | 6.4 | 23.0 |
| 4 | 2240 | 6.42 | 643 | 8.7 | 22.8 |
| 5 | 2464 | 7.56 | 588 | 11.25 | 22.6 |
| 6 | 2680 | 8.75 | 547 | 13.85 | 22.4 |
| 8 | 3010 | 10.8 | 487 | 19.05 | 22.0 |
| 10 | 3273 | 12.65 | 447 | 24.3 | 21.6 |
| 12 | 3500 | 14.5 | 422 | 29.3 | 21.3 |
| 14 | 3690 | 16.2 | 405 | 34.2 | 21.1 |
| 16 | 3845 | 17.8 | 395 | 39.3 | 20.8 |
| 18 | 3980 | 19.5 | 391 | (43.2) | 20.6 |
| 20 | 4097 | 20.95 | 390 | 47.68 | 20.4 |
| 22 | 4190 | 22.4 | 392 | (50.12) | 20.2 |
| 24 | 4275 | 23.9 | 394 | 54.1 | 20.1 |
| 26 | 4340 | 25.2 | 397 | 57.2 | 19.9 |
| 28 | 4390 | 26.55 | 400 | 59.8 | 19.8 |
| 30 | 4423 | 27.91 | 405 | 62.40 | 19.7 |
| 32 | 4440 | 29.2 | 411 | 64.5 | 19.6 |
| 34 | 4455 | 30.4 | 416 | 66.5 | 19.4 |
| 34° 42′ | 4457 | — | — | — | — |
| 35 | 4456 | 31.06 | 417 | 67.4 | 19.4 |
| 36 | 4454 | 31.6 | 420 | 68.5 | 19.3 |
| 38 | 4440 | 32.8 | 425 | 70.0 | 19.2 |
| 40 | 4413 | 34.01 | 429 | 71.35 | 19.2 |
| 50 | (4160) | — | — | — | — |
| 60 | (3600) | — | — | — | — |
| 80 | (1480) | — | — | — | — |
| 90 | 0 | — | — | — | — |

Height of vertex 9389 feet after 19.00 secs.

Note: $f = \dfrac{6R}{t^2} \tan E.$

## APPENDIX IV

### Table II used for

$$V = 2600 \qquad C = 0\cdot 284$$

| Z = R/C. | E = AC. | R = ZC. | t = TC. | v = u. | W (angle of descent). | $f = \dfrac{ZA}{573\, T^2}$ or $C\dfrac{ZE}{573\, t^2}$ |
|---|---|---|---|---|---|---|
| 0 | 0 | 0 | 0 | 2600 | 0 | 32·2 |
| 500 | 2·69 | 142 | 0·170 | 2345 | 3·2 | 31·6 |
| 1000 | 7·95 | 284 | 0·364 | 2090 | 9½ | 29·9 |
| 1500 | 13·08 | 426 | 0·582 | 1843 | 17 | 28·6 |
| 2000 | 19·3 | 568 | 0·83 | 1608 | 28 | 27·9 |
| 2500 | 27·0 | 710 | 1·115 | 1393 | 42 | 26·8 |
| 3000 | 36 | 852 | 1·44 | 1215 | 61 | 25·7 |
| 3500 | 46·8 | 994 | 1·82 | 1089 | 82 | 24·7 |
| 4000 | 59·3 | 1136 | 2·22 | 1007 | 109 | 23·9 |
| 4500 | 74·4 | 1278 | 2·65 | 940 | 146 | 23·6 |
| 5000 | 91·8 | 1420 | 3·13 | 882 | 187 | 23·2 |
| 5500 | 112 | 1562 | 3·45 | 832 | 234 | 23·2 |
| 6000 | 135 | 1704 | 4·15 | 785 | 288 | 23·2 |
| 6500 | 160 | 1846 | 4·71 | 741 | 340 | 23·3 |
| 7000 | 187 | 1988 | 5·30 | 700 | 401 | 23·1 |
| 7500 | 217 | 2130 | 5·93 | 661 | 487 | 22·9 |
| 8000 | 250 | 2272 | 6·59 | 624 | 550 | 22·7 |
| 8500 | 286 | 2414 | 7·30 | 589 | 642 | 22·6 |
| 9000 | 326 | 2556 | 8·04 | 555 | 740 | 22·5 |
| 9500 | 369 | 2698 | 8·83 | 524 | 863 | 22·2 |
| 10000 | 417 | 2840 | 9·66 | 494 | 1025 | 22·1 |

Since $f = \dfrac{RE}{573\, t^2}$ and $R = ZC$, $E = AC$, $t = TC$, the above formula for $f$ may be written

$$f = \frac{ZA}{573\, T^2} \quad \text{or} \quad C\frac{ZE}{573\, t^2}.$$

## APPENDIX V

Use of S and T Tables of Table I to determine R and $t$ when V, C and $v$ are given by Formulæ

$$C(S_V - S_v) = S = 3R$$
$$C(T_V - T_v) = t$$

given—

$$V = 2600 \qquad C = 0\cdot 284$$

| V. | $S_V$. | $S/C = S_V - S_v$. | $\Delta$ R. | R. | $T_V$. | $t/C = T_V - T_v$. | $\Sigma\, t/C$ | t. |
|---|---|---|---|---|---|---|---|---|
| 2600 | 69947 | — | — | 0 | 223·25 | — | 0 | 0 |
| 2000 | 66413 | 3534 | 334 | 334 | 221·71 | 1·54 | 1·54 | 0·438 |
| 1500 | 63217 | 3196 | 303 | 637 | 219·86 | 1·85 | 3·39 | 0·963 |
| 1000 | 57805 | 5412 | 514 | 1151 | 215·28 | 4·58 | 7·97 | 2·26 |
| 900 | 55444 | 2361 | 224 | 1375 | 212·79 | 2·49 | 10·46 | 2·97 |
| 800 | 52421 | 3023 | 286 | 1661 | 209·22 | 3·57 | 14·03 | 3·98 |
| 700 | 48932 | 3489 | 331 | 1992 | 204·56 | 4·66 | 18·69 | 5·31 |
| 600 | 44937 | 3995 | 378 | 2370 | 198·39 | 6·17 | 24·86 | 7·07 |
| 500 | 40267 | 4670 | 442 | 2812 | 189·85 | 8·54 | 33·40 | 9·50 |
| 400 | 34645 | 5622 | 533 | 3345 | 177·26 | 12·59 | 45·99 | 13·07 |
| 300 | 27503 | 7142 | 676 | 4021 | 156·58 | 20·68 | 66·67 | 18·94 |
| 250 | 23046 | 4457 | 422 | 4443 | 140·29 | 16·29 | 82·96 | 23·60 |
| 200 | 17658 | 5388 | 510 | 4953 | 116·16 | 24·13 | 107·09 | 30·65 |

Note.—By plotting the $t$ column against R, values of R can be read off for every whole second.

# APPENDIX VI

### TABLE I.—ABRIDGED TABLES OF $p$. T.S. FOR POINTED BULLETS

Computed by Mr. M. Segal, M.A., from resistance table supplied by Major J. H. Hardcastle

| V f/s. | $p$ lb. | $\Delta$ T. | T. | $\Delta$ S. | S. |
|---|---|---|---|---|---|
| 100 | 0·0137 | — | 0 | — | 1489·0 |
| 110 | 0·0165 | 20·60 | 20·60 | 2163·0 | 3652·0 |
| 120 | 0·0193 | 17·40 | 38·00 | 2001·0 | 5653·0 |
| 130 | 0·0227 | 14·80 | 52·80 | 1850·0 | 7503·0 |
| 140 | 0·0261 | 12·80 | 65·60 | 1728·0 | 9231·0 |
| 150 | 0·0297 | 11·15 | 76·75 | 1616·8 | 10847·8 |
| 160 | 0·0339 | 9·78 | 86·53 | 1515·9 | 12363·7 |
| 170 | 0·0381 | 8·65 | 95·18 | 1427·2 | 13790·2 |
| 180 | 0·0425 | 7·72 | 102·90 | 1351·0 | 15141·9 |
| 190 | 0·0469 | 6·96 | 109·86 | 1287·6 | 16429·5 |
| 200 | 0·0519 | 6·30 | 116·16 | 1228·5 | 17658·0 |
| 210 | 0·057 | 5·71 | 121·87 | 1170·6 | 18828·6 |
| 220 | 0·062 | 5·21 | 127·08 | 1120·1 | 19948·7 |
| 230 | 0·068 | 4·78 | 131·86 | 1075·5 | 21024·2 |
| 240 | 0·074 | 4·39 | 136·25 | 1031·6 | 22055·6 |
| 250 | 0·080 | 4·04 | 140·29 | 989·8 | 23045·6 |
| 260 | 0·086 | 3·73 | 144·02 | 951·2 | 23996·8 |
| 270 | 0·093 | 3·46 | 147·48 | 916·9 | 24913·7 |
| 280 | 0·099 | 3·23 | 150·71 | 888·2 | 25801·9 |
| 290 | 0·106 | 3·03 | 153·74 | 863·6 | 26665·5 |
| 300 | 0·113 | 2·84 | 156·58 | 837·8 | 27503·3 |
| 310 | 0·121 | 2·66 | 159·24 | 811·3 | 28314·6 |
| 320 | 0·129 | 2·49 | 161·73 | 784·3 | 29098·9 |
| 330 | 0·137 | 2·34 | 164·07 | 760·5 | 29859·4 |
| 340 | 0·145 | 2·21 | 166·28 | 740·3 | 30599·7 |
| 350 | 0·153 | 2·09 | 168·37 | 721·1 | 31320·8 |
| 360 | 0·162 | 1·97 | 170·34 | 699·4 | 32020·2 |
| 370 | 0·171 | 1·87 | 172·21 | 682·4 | 32702·6 |
| 380 | 0·180 | 1·77 | 173·98 | 663·8 | 33366·4 |
| 390 | 0·189 | 1·68 | 175·66 | 646·8 | 34013·2 |
| 400 | 0·199 | 1·60 | 177·26 | 632·0 | 34645·2 |
| 410 | 0·209 | 1·52 | 178·78 | 615·6 | 35260·8 |
| 420 | 0·219 | 1·45 | 180·23 | 601·8 | 35862·6 |
| 430 | 0·229 | 1·39 | 181·62 | 590·7 | 36453·3 |
| 440 | 0·239 | 1·33 | 182·95 | 578·5 | 37031·8 |
| 450 | 0·250 | 1·27 | 184·22 | 565·2 | 37597·0 |
| 460 | 0·261 | 1·22 | 185·44 | 555·1 | 38152·1 |
| 470 | 0·272 | 1·17 | 186·61 | 544·0 | 38696·1 |
| 480 | 0·283 | 1·12 | 187·73 | 532·0 | 39228·1 |
| 490 | 0·294 | 1·08 | 188·81 | 523·8 | 39751·9 |
| 500 | 0·306 | 1·04 | 189·85 | 514·8 | 40266·7 |
| 510 | 0·318 | 0·997 | 190·847 | 503·5 | 40770·2 |
| 520 | 0·330 | 0·959 | 191·806 | 493·9 | 41264·1 |
| 530 | 0·342 | 0·925 | 192·731 | 485·6 | 41749·7 |
| 540 | 0·354 | 0·894 | 193·625 | 478·3 | 42228·0 |
| 550 | 0·367 | 0·864 | 194·489 | 470·9 | 42698·9 |
| 560 | 0·380 | 0·834 | 195·323 | 462·9 | 43161·8 |
| 570 | 0·393 | 0·804 | 196·127 | 454·3 | 43616·1 |
| 580 | 0·406 | 0·778 | 196·905 | 447·4 | 44063·5 |
| 590 | 0·420 | 0·753 | 197·658 | 440·5 | 44504·0 |
| 600 | 0·434 | 0·728 | 198·386 | 433·2 | 44937·2 |
| 610 | 0·448 | 0·705 | 199·091 | 426·5 | 45363·7 |
| 620 | 0·462 | 0·683 | 199·774 | 420·0 | 45783·7 |
| 630 | 0·477 | 0·663 | 200·437 | 414·4 | 46198·1 |
| 640 | 0·492 | 0·643 | 201·080 | 408·3 | 46606·4 |

APPENDIX VI—TABLE I—continued

| V f/s. | p lb. | Δ T. | T. | Δ S. | S. |
|---|---|---|---|---|---|
| 650 | 0·507 | 0·623 | 201·703 | 401·8 | 47008·2 |
| 660 | 0·522 | 0·605 | 202·308 | 396·3 | 47404·5 |
| 670 | 0·538 | 0·587 | 202·895 | 390·3 | 47794·8 |
| 680 | 0·554 | 0·570 | 203·465 | 384·8 | 48179·6 |
| 690 | 0·570 | 0·553 | 204·018 | 378·8 | 48558·4 |
| 700 | 0·586 | 0·538 | 204·556 | 373·9 | 48932·3 |
| 710 | 0·602 | 0·524 | 205·080 | 369·4 | 49301·7 |
| 720 | 0·618 | 0·510 | 205·590 | 364·7 | 49666·4 |
| 730 | 0·635 | 0·496 | 206·086 | 359·6 | 50026·0 |
| 740 | 0·652 | 0·483 | 206·569 | 355·0 | 50381·0 |
| 750 | 0·669 | 0·471 | 207·040 | 350·9 | 50731·9 |
| 760 | 0·686 | 0·459 | 207·499 | 346·5 | 51078·4 |
| 770 | 0·704 | 0·447 | 207·946 | 342·0 | 51420·4 |
| 780 | 0·722 | 0·436 | 208·382 | 337·9 | 51758·3 |
| 790 | 0·740 | 0·425 | 208·807 | 333·6 | 52091·9 |
| 800 | 0·758 | 0·415 | 209·222 | 329·9 | 52421·8 |
| 810 | 0·776 | 0·405 | 209·627 | 326·0 | 52747·8 |
| 820 | 0·794 | 0·396 | 210·023 | 322·7 | 53070·5 |
| 830 | 0·812 | 0·387 | 210·410 | 319·3 | 53389·8 |
| 840 | 0·830 | 0·378 | 210·788 | 315·6 | 53705·4 |
| 850 | 0·864 | 0·367 | 211·155 | 310·1 | 54015·5 |
| 860 | 0·899 | 0·353 | 211·508 | 301·8 | 54317·3 |
| 870 | 0·935 | 0·339 | 211·847 | 293·2 | 54610·5 |
| 880 | 0·972 | 0·326 | 212·173 | 285·3 | 54895·8 |
| 890 | 1·01 | 0·314 | 212·487 | 277·9 | 55173·7 |
| 900 | 1·05 | 0·302 | 212·789 | 270·3 | 55444·0 |
| 910 | 1·09 | 0·290 | 213·079 | 262·5 | 55706·5 |
| 920 | 1·13 | 0·280 | 213·359 | 256·2 | 55962·7 |
| 930 | 1·17 | 0·270 | 213·629 | 249·8 | 56212·5 |
| 940 | 1·21 | 0·261 | 213·890 | 244·0 | 56456·5 |
| 950 | 1·25 | 0·252 | 214·142 | 238·1 | 56694·6 |
| 960 | 1·30 | 0·244 | 214·386 | 233·0 | 56927·6 |
| 970 | 1·34 | 0·236 | 214·622 | 227·7 | 57155·3 |
| 980 | 1·39 | 0·228 | 214·850 | 222·3 | 57377·6 |
| 990 | 1·44 | 0·220 | 215·070 | 216·7 | 57594·3 |
| 1000 | 1·49 | 0·212 | 215·282 | 210·9 | 57805·2 |
| 1010 | 1·54 | 0·205 | 215·487 | 206·0 | 58011·2 |
| 1020 | 1·58 | 0·199 | 215·686 | 202·0 | 58213·2 |
| 1030 | 1·62 | 0·194 | 215·880 | 198·9 | 58412·1 |
| 1040 | 1·65 | 0·190 | 216·070 | 196·6 | 58608·7 |
| 1050 | 1·75 | 0·182 | 216·252 | 190·2 | 58798·9 |
| 1060 | 1·86 | 0·172 | 216·424 | 181·5 | 58980·4 |
| 1070 | 1·97 | 0·162 | 216·586 | 172·5 | 59152·9 |
| 1080 | 2·08 | 0·153 | 216·739 | 164·5 | 59317·4 |
| 1090 | 2·20 | 0·145 | 216·884 | 157·3 | 59474·7 |
| 1100 | 2·33 | 0·137 | 217·021 | 150·0 | 59624·7 |
| 1110 | 2·46 | 0·130 | 217·151 | 143·6 | 59768·3 |
| 1120 | 2·60 | 0·123 | 217·274 | 137·1 | 59905·4 |
| 1130 | 2·75 | 0·116 | 217·390 | 130·5 | 60035·9 |
| 1140 | 2·90 | 0·110 | 217·500 | 124·9 | 60160·8 |
| 1150 | 3·06 | 0·104 | 217·604 | 119·1 | 60279·9 |
| 1160 | 3·23 | 0·099 | 217·703 | 114·3 | 60394·2 |
| 1170 | 3·40 | 0·094 | 217·797 | 109·5 | 60503·7 |
| 1180 | 3·58 | 0·089 | 217·886 | 104·7 | 60608·4 |
| 1190 | 3·77 | 0·085 | 217·971 | 100·7 | 60709·1 |
| 1200 | 3·85 | 0·082 | 218·053 | 98·0 | 60807·1 |
| 1210 | 3·94 | 0·080 | 218·133 | 96·4 | 60903·5 |
| 1220 | 4·03 | 0·078 | 218·211 | 94·8 | 60998·3 |
| 1230 | 4·12 | 0·076 | 218·287 | 93·1 | 61091·4 |
| 1240 | 4·21 | 0·075 | 218·362 | 92·6 | 61184·0 |

APPENDIX VI—Table I—*continued*

| V f/s. | p lb. | Δ T. | T. | Δ S. | S. |
|---|---|---|---|---|---|
| 1250 | 4·30 | 0·073 | 218·435 | 90·9 | 61274·9 |
| 1260 | 4·39 | 0·072 | 218·507 | 90·4 | 61365·3 |
| 1270 | 4·48 | 0·070 | 218·577 | 88·6 | 61453·9 |
| 1280 | 4·58 | 0·069 | 218·646 | 88·0 | 61541·9 |
| 1290 | 4·68 | 0·067 | 218·713 | 86·1 | 61628·0 |
| 1300 | 4·78 | 0·066 | 218·779 | 85·4 | 61713·4 |
| 1310 | 4·88 | 0·064 | 218·843 | 83·5 | 61796·9 |
| 1320 | 4·98 | 0·063 | 218·906 | 82·8 | 61879·7 |
| 1330 | 5·08 | 0·062 | 218·968 | 82·1 | 61961·8 |
| 1340 | 5·18 | 0·061 | 219·029 | 81·4 | 62043·2 |
| 1350 | 5·28 | 0·059 | 219·088 | 79·3 | 62122·5 |
| 1360 | 5·39 | 0·058 | 219·146 | 78·6 | 62201·1 |
| 1370 | 5·49 | 0·057 | 219·203 | 77·8 | 62278·9 |
| 1380 | 5·60 | 0·056 | 219·259 | 77·0 | 62355·9 |
| 1390 | 5·71 | 0·055 | 219·314 | 76·2 | 62432·1 |
| 1400 | 5·82 | 0·054 | 219·368 | 75·4 | 62507·5 |
| 1410 | 5·93 | 0·053 | 219·421 | 74·5 | 62582·0 |
| 1420 | 6·04 | 0·052 | 219·473 | 73·6 | 62655·6 |
| 1430 | 6·15 | 0·051 | 219·524 | 72·6 | 62728·2 |
| 1440 | 6·27 | 0·050 | 219·574 | 71·7 | 62799·9 |
| 1450 | 6·38 | 0·049 | 219·623 | 70·8 | 62870·7 |
| 1460 | 6·50 | 0·048 | 219·671 | 69·8 | 62940·5 |
| 1470 | 6·56 | 0·048 | 219·719 | 69·5 | 63010·0 |
| 1480 | 6·63 | 0·047 | 219·766 | 69·2 | 63079·2 |
| 1490 | 6·69 | 0·047 | 219·813 | 69·0 | 63148·2 |
| 1500 | 6·76 | 0·046 | 219·859 | 68·8 | 63217·0 |
| 1510 | 6·83 | 0·046 | 219·905 | 68·5 | 63285·5 |
| 1520 | 6·90 | 0·045 | 219·950 | 68·1 | 63353·6 |
| 1530 | 6·96 | 0·045 | 219·995 | 67·8 | 63421·4 |
| 1540 | 7·03 | 0·044 | 220·039 | 67·5 | 63488·9 |
| 1550 | 7·10 | 0·044 | 220·083 | 67·2 | 63555·1 |
| 1560 | 7·17 | 0·043 | 220·126 | 66·9 | 63622·0 |
| 1570 | 7·24 | 0·043 | 220·169 | 66·5 | 63688·5 |
| 1580 | 7·31 | 0·042 | 220·211 | 66·4 | 63754·9 |
| 1590 | 7·38 | 0·042 | 220·253 | 66·3 | 63821·2 |
| 1600 | 7·45 | 0·042 | 220·295 | 66·2 | 63887·4 |
| 1610 | 7·51 | 0·041 | 220·336 | 66·1 | 63953·5 |
| 1620 | 7·58 | 0·041 | 220·377 | 66·0 | 64019·5 |
| 1630 | 7·64 | 0·041 | 220·418 | 65·8 | 64085·3 |
| 1640 | 7·71 | 0·040 | 220·458 | 65·7 | 64151·0 |
| 1650 | 7·78 | 0·040 | 220·498 | 65·6 | 64216·6 |
| 1660 | 7·85 | 0·040 | 220·538 | 65·5 | 64282·1 |
| 1670 | 7·92 | 0·039 | 220·577 | 65·3 | 64347·4 |
| 1680 | 7·99 | 0·039 | 220·616 | 65·1 | 64412·5 |
| 1690 | 8·06 | 0·039 | 220·655 | 65·0 | 64477·5 |
| 1700 | 8·13 | 0·038 | 220·693 | 64·8 | 64542·3 |
| 1710 | 8·20 | 0·038 | 220·731 | 64·5 | 64606·9 |
| 1720 | 8·28 | 0·038 | 220·769 | 64·4 | 64671·3 |
| 1730 | 8·35 | 0·037 | 220·806 | 64·2 | 64735·5 |
| 1740 | 8·42 | 0·037 | 220·843 | 64·0 | 64799·5 |
| 1750 | 8·49 | 0·037 | 220·880 | 63·8 | 64863·3 |
| 1760 | 8·56 | 0·036 | 220·916 | 63·6 | 64926·9 |
| 1770 | 8·63 | 0·036 | 220·952 | 63·4 | 64990·3 |
| 1780 | 8·70 | 0·036 | 220·988 | 63·2 | 65053·5 |
| 1790 | 8·78 | 0·035 | 221·023 | 63·1 | 65116·5 |
| 1800 | 8·85 | 0·035 | 221·058 | 63·0 | 65179·6 |
| 1810 | 8·92 | 0·035 | 221·093 | 62·9 | 65242·5 |
| 1820 | 9·00 | 0·035 | 221·128 | 62·7 | 65305·2 |
| 1830 | 9·07 | 0·034 | 221·162 | 62·6 | 65367·8 |
| 1840 | 9·14 | 0·034 | 221·196 | 62·6 | 65430·4 |

APPENDIX VI—TABLE I—*continued*

| V f/s. | p lb. | ΔT. | T. | ΔS. | S. |
|---|---|---|---|---|---|
| 1850 | 9·21 | 0·034 | 221·230 | 62·5 | 65492·9 |
| 1860 | 9·28 | 0·034 | 221·264 | 62·5 | 65555·4 |
| 1870 | 9·35 | 0·033 | 221·297 | 62·4 | 65617·8 |
| 1880 | 9·43 | 0·033 | 221·330 | 62·2 | 65680·0 |
| 1890 | 9·50 | 0·033 | 221·363 | 62·1 | 65742·1 |
| 1900 | 9·57 | 0·033 | 221·396 | 62·0 | 65804·1 |
| 1910 | 9·64 | 0·032 | 221·428 | 61·9 | 65866·0 |
| 1920 | 9·71 | 0·032 | 221·460 | 61·7 | 65927·7 |
| 1930 | 9·78 | 0·032 | 221·492 | 61·5 | 65989·2 |
| 1940 | 9·86 | 0·032 | 221·524 | 61·3 | 66050·5 |
| 1950 | 9·93 | 0·031 | 221·555 | 61·0 | 66111·5 |
| 1960 | 10·01 | 0·031 | 221·586 | 60·8 | 66172·3 |
| 1970 | 10·09 | 0·031 | 221·617 | 60·6 | 66232·9 |
| 1980 | 10·17 | 0·031 | 221·648 | 60·3 | 66293·2 |
| 1990 | 10·24 | 0·030 | 221·678 | 59·9 | 66353·1 |
| 2000 | 10·32 | 0·030 | 221·708 | 59·6 | 66412·7 |
| 2010 | 10·38 | 0·030 | 221·738 | 59·3 | 66472·0 |
| 2020 | 10·44 | 0·030 | 221·768 | 59·3 | 66531·3 |
| 2030 | 10·50 | 0·030 | 221·798 | 59·3 | 66590·6 |
| 2040 | 10·56 | 0·029 | 221·827 | 59·3 | 66649·9 |
| 2050 | 10·62 | 0·029 | 221·856 | 59·3 | 66709·2 |
| 2060 | 10·69 | 0·029 | 221·885 | 59·3 | 66768·5 |
| 2070 | 10·75 | 0·029 | 221·914 | 59·3 | 66827·8 |
| 2080 | 10·82 | 0·029 | 221·943 | 59·3 | 66887·1 |
| 2090 | 10·88 | 0·029 | 221·972 | 59·3 | 66946·4 |
| 2100 | 10·94 | 0·028 | 222·000 | 59·2 | 67005·6 |
| 2110 | 11·00 | 0·028 | 222·028 | 59·2 | 67064·8 |
| 2120 | 11·06 | 0·028 | 222·056 | 59·2 | 67124·0 |
| 2130 | 11·12 | 0·028 | 222·084 | 59·2 | 67183·2 |
| 2140 | 11·18 | 0·028 | 222·112 | 59·2 | 67242·4 |
| 2150 | 11·24 | 0·028 | 222·140 | 59·2 | 67301·6 |
| 2160 | 11·30 | 0·028 | 222·168 | 59·2 | 67360·8 |
| 2170 | 11·36 | 0·027 | 222·195 | 59·2 | 67420·0 |
| 2180 | 11·42 | 0·027 | 222·222 | 59·2 | 67479·2 |
| 2190 | 11·48 | 0·027 | 222·249 | 59·2 | 67538·4 |
| 2200 | 11·54 | 0·027 | 222·276 | 59·0 | 67597·4 |
| 2210 | 11·60 | 0·027 | 222·303 | 59·0 | 67656·4 |
| 2220 | 11·66 | 0·027 | 222·330 | 59·0 | 67715·4 |
| 2230 | 11·72 | 0·027 | 222·357 | 59·0 | 67774·4 |
| 2240 | 11·78 | 0·026 | 222·383 | 59·0 | 67833·4 |
| 2250 | 11·84 | 0·026 | 222·409 | 59·0 | 67892·4 |
| 2260 | 11·90 | 0·026 | 222·435 | 59·0 | 67951·4 |
| 2270 | 11·96 | 0·026 | 222·461 | 59·0 | 68010·4 |
| 2280 | 12·02 | 0·026 | 222·487 | 59·0 | 68069·4 |
| 2290 | 12·08 | 0·026 | 222·513 | 59·0 | 68128·4 |
| 2300 | 12·14 | 0·026 | 222·539 | 58·8 | 68187·2 |
| 2310 | 12·20 | 0·026 | 222·565 | 58·8 | 68246·0 |
| 2320 | 12·27 | 0·025 | 222·590 | 58·8 | 68304·8 |
| 2330 | 12·33 | 0·025 | 222·615 | 58·8 | 68363·6 |
| 2340 | 12·40 | 0·025 | 222·640 | 58·8 | 68422·4 |
| 2350 | 12·46 | 0·025 | 222·665 | 58·8 | 68481·2 |
| 2360 | 12·52 | 0·025 | 222·690 | 58·8 | 68540·0 |
| 2370 | 12·58 | 0·025 | 222·715 | 58·8 | 68598·8 |
| 2380 | 12·64 | 0·025 | 222·740 | 58·8 | 68657·6 |
| 2390 | 12·70 | 0·025 | 222·765 | 58·8 | 68716·4 |
| 2400 | 12·76 | 0·024 | 222·789 | 58·8 | 68775·2 |
| 2410 | 12·82 | 0·024 | 222·813 | 58·8 | 68834·0 |
| 2420 | 12·88 | 0·024 | 222·837 | 58·8 | 68892·8 |
| 2430 | 12·94 | 0·024 | 222·861 | 58·8 | 68951·6 |
| 2440 | 13·01 | 0·024 | 222·885 | 58·8 | 69010·4 |

## APPENDIX VI—TABLE I—*continued*

| V f/s. | p lb. | ΔT. | T. | ΔS. | S. |
|---|---|---|---|---|---|
| 2450 | 13·07 | 0·024 | 222·909 | 58·7 | 69069·1 |
| 2460 | 13·14 | 0·024 | 222·933 | 58·7 | 69127·8 |
| 2470 | 13·20 | 0·024 | 222·957 | 58·7 | 69186·5 |
| 2480 | 13·26 | 0·023 | 222·980 | 58·7 | 69245·2 |
| 2490 | 13·32 | 0·023 | 223·003 | 58·7 | 69303·9 |
| 2500 | 13·39 | 0·023 | 223·026 | 58·7 | 69362·6 |
| 2510 | 13·45 | 0·023 | 223·049 | 58·7 | 69421·3 |
| 2520 | 13·52 | 0·023 | 223·072 | 58·7 | 69479·0 |
| 2530 | 13·58 | 0·023 | 223·095 | 58·7 | 69537·7 |
| 2540 | 13·64 | 0·023 | 223·118 | 58·7 | 69596·4 |
| 2550 | 13·70 | 0·023 | 223·141 | 58·4 | 69654·8 |
| 2560 | 13·77 | 0·023 | 223·164 | 58·4 | 69713·2 |
| 2570 | 13·83 | 0·022 | 223·186 | 58·4 | 69771·6 |
| 2580 | 13·89 | 0·022 | 223·208 | 58·4 | 69830·0 |
| 2590 | 13·95 | 0·022 | 223·230 | 58·4 | 69888·4 |
| 2600 | 14·02 | 0·022 | 223·252 | 58·4 | 69946·8 |
| 2610 | 14·09 | 0·022 | 223·274 | 58·4 | 70005·2 |
| 2620 | 14·16 | 0·022 | 223·296 | 58·4 | 70063·6 |
| 2630 | 14·23 | 0·022 | 223·318 | 58·4 | 70122·0 |
| 2640 | 14·31 | 0·022 | 223·340 | 58·4 | 70180·4 |
| 2650 | 14·38 | 0·022 | 223·362 | 58·2 | 70238·6 |
| 2660 | 14·45 | 0·021 | 223·383 | 58·2 | 70296·8 |
| 2670 | 14·52 | 0·021 | 223·404 | 58·2 | 70355·0 |
| 2680 | 14·60 | 0·021 | 223·425 | 58·2 | 70413·2 |
| 2690 | 14·67 | 0·021 | 223·446 | 58·2 | 70471·4 |
| 2700 | 14·74 | 0·021 | 223·467 | 58·2 | 70529·6 |
| 2710 | 14·81 | 0·021 | 223·488 | 58·2 | 70587·8 |
| 2720 | 14·89 | 0·021 | 223·509 | 58·2 | 70645·0 |
| 2730 | 14·96 | 0·021 | 223·530 | 58·2 | 70703·2 |
| 2740 | 15·04 | 0·021 | 223·551 | 58·2 | 70761·4 |
| 2750 | 15·11 | 0·021 | 223·572 | 58·0 | 70819·4 |
| 2760 | 15·18 | 0·020 | 223·592 | 58·0 | 70877·4 |
| 2770 | 15·25 | 0·020 | 223·612 | 58·0 | 70935·4 |
| 2780 | 15·33 | 0·020 | 223·632 | 58·0 | 70993·4 |
| 2790 | 15·40 | 0·020 | 223·652 | 58·0 | 71051·4 |
| 2800 | 15·48 | 0·020 | 223·672 | 58·0 | 71109·4 |
| 2810 | 15·55 | 0·020 | 223·692 | 58·0 | 71167·4 |
| 2820 | 15·63 | 0·020 | 223·712 | 58·0 | 71225·4 |
| 2830 | 15·70 | 0·020 | 223·732 | 58·0 | 71283·4 |
| 2840 | 15·78 | 0·020 | 223·752 | 58·0 | 71341·4 |
| 2850 | 15·85 | 0·020 | 223·772 | 57·5 | 71398·9 |
| 2860 | 15·92 | 0·020 | 223·792 | 57·5 | 71456·4 |
| 2870 | 15·99 | 0·019 | 223·811 | 57·5 | 71513·9 |
| 2880 | 16·07 | 0·019 | 223·830 | 57·5 | 71571·4 |
| 2890 | 16·14 | 0·019 | 223·849 | 57·5 | 71628·9 |
| 2900 | 16·22 | 0·019 | 223·868 | 57·5 | 71686·4 |
| 2910 | 16·29 | 0·019 | 223·887 | 57·5 | 71743·9 |
| 2920 | 16·37 | 0·019 | 223·906 | 57·5 | 71801·4 |
| 2930 | 16·44 | 0·019 | 223·925 | 57·5 | 71858·9 |
| 2940 | 16·52 | 0·019 | 223·944 | 57·5 | 71916·4 |
| 2950 | 16·59 | 0·019 | 223·963 | 57·0 | 71973·4 |
| 2960 | 16·67 | 0·019 | 223·982 | 57·0 | 72030·4 |
| 2970 | 16·74 | 0·019 | 224·001 | 57·0 | 72087·4 |
| 2980 | 16·82 | 0·019 | 224·020 | 57·0 | 72144·4 |
| 2990 | 16·89 | 0·018 | 224·038 | 57·0 | 72201·4 |
| 3000 | 16·97 | 0·018 | 224·056 | 57·0 | 72258·4 |
| 3010 | 17·04 | 0·018 | 224·074 | 56·2 | 72314·6 |
| 3020 | 17·12 | 0·018 | 224·092 | 56·2 | 72370·8 |
| 3030 | 17·20 | 0·018 | 224·110 | 56·2 | 72427·0 |
| 3040 | 17·28 | 0·018 | 224·128 | 56·2 | 72483·2 |

APPENDIX VI—TABLE I—*continued*

| V f/s. | p lb. | Δ T. | T. | Δ S. | S. |
|---|---|---|---|---|---|
| 3050 | 17·35 | 0·018 | 224·146 | 56·2 | 72539·4 |
| 3060 | 17·43 | 0·018 | 224·164 | 56·2 | 72595·6 |
| 3070 | 17·50 | 0·018 | 224·182 | 56·2 | 72651·8 |
| 3080 | 17·58 | 0·018 | 224·200 | 56·2 | 72708·0 |
| 3090 | 17·65 | 0·018 | 224·218 | 56·2 | 72764·2 |
| 3100 | 17·73 | 0·018 | 224·236 | 56·2 | 72820·4 |
| 3110 | 17·80 | 0·018 | 224·254 | 55·5 | 72875·9 |
| 3120 | 17·88 | 0·017 | 224·271 | 55·5 | 72931·4 |
| 3130 | 17·95 | 0·017 | 224·288 | 55·5 | 72986·9 |
| 3140 | 18·03 | 0·017 | 224·305 | 55·5 | 73042·4 |
| 3150 | 18·11 | 0·017 | 224·322 | 55·5 | 73097·9 |
| 3160 | 18·19 | 0·017 | 224·339 | 55·5 | 73153·4 |
| 3170 | 18·26 | 0·017 | 224·356 | 55·5 | 73208·9 |
| 3180 | 18·34 | 0·017 | 224·373 | 55·5 | 73264·4 |
| 3190 | 18·41 | 0·017 | 224·390 | 55·5 | 73319·9 |
| 3200 | 18·49 | 0·017 | 224·407 | 55·5 | 73375·4 |
| 3210 | 18·57 | 0·017 | 224·424 | 54·8 | 73430·2 |
| 3220 | 18·65 | 0·017 | 224·441 | 54·8 | 73485·0 |
| 3230 | 18·72 | 0·017 | 224·458 | 54·8 | 73539·8 |
| 3240 | 18·80 | 0·017 | 224·475 | 54·8 | 73594·6 |
| 3250 | 18·88 | 0·017 | 224·492 | 54·8 | 73649·4 |
| 3260 | 18·96 | 0·016 | 224·508 | 54·8 | 73704·2 |
| 3270 | 19·04 | 0·016 | 224·524 | 54·8 | 73759·0 |
| 3280 | 19·12 | 0·016 | 224·540 | 54·8 | 73813·8 |
| 3290 | 19·20 | 0·016 | 224·556 | 54·8 | 73868·6 |
| 3300 | 19·28 | 0·016 | 224·572 | 54·8 | 73923·4 |
| 3310 | 19·36 | 0·016 | 224·588 | 54·1 | 73977·5 |
| 3320 | 19·44 | 0·016 | 224·604 | 54·1 | 74031·6 |
| 3330 | 19·51 | 0·016 | 224·620 | 54·1 | 74085·7 |
| 3340 | 19·59 | 0·016 | 224·636 | 54·1 | 74139·8 |
| 3350 | 19·67 | 0·016 | 224·652 | 54·1 | 74193·9 |
| 3360 | 19·75 | 0·016 | 224·668 | 54·1 | 74248·0 |
| 3370 | 19·83 | 0·016 | 224·684 | 54·1 | 74302·1 |
| 3380 | 19·91 | 0·016 | 224·700 | 54·1 | 74356·2 |
| 3390 | 19·98 | 0·016 | 224·716 | 54·1 | 74410·3 |
| 3400 | 20·06 | 0·016 | 224·732 | 54·1 | 74464·4 |
| 3410 | 20·14 | 0·015 | 224·747 | 53·4 | 74517·8 |
| 3420 | 20·22 | 0·015 | 224·762 | 53·4 | 74571·2 |
| 3430 | 20·29 | 0·015 | 224·777 | 53·4 | 74624·6 |
| 3440 | 20·37 | 0·015 | 224·792 | 53·4 | 74678·0 |
| 3450 | 20·45 | 0·015 | 224·807 | 53·4 | 74731·4 |
| 3460 | 20·53 | 0·015 | 224·822 | 53·4 | 74784·8 |
| 3470 | 20·61 | 0·015 | 224·837 | 53·4 | 74838·2 |
| 3480 | 20·69 | 0·015 | 224·852 | 53·4 | 74891·6 |
| 3490 | 20·77 | 0·015 | 224·867 | 53·4 | 74945·0 |
| 3500 | 20·85 | 0·015 | 224·882 | 53·4 | 74998·4 |

# APPENDIX VI

TABLE IA.—COMPUTED BY BALLISTIC DEPARTMENT OF ORDNANCE COMMITTEE FROM TABLE I

*Primary Functions.*—Tables of "I" and "A" for velocities 100 f.s. to 3500 f.s.

| V. | $\Delta$ I. | I. | $\Delta$ A. | A. |
|---|---|---|---|---|
| 100 | — | 0 | — | 0 |
| 110 | 6·3155 | 6·3155 | 6830·2 | 6830 |
| 120 | 4·8706 | 11·1861 | 17510·4 | 24340 |
| 130 | 3·8114 | 14·9975 | 24220 | 48560 |
| 140 | 3·0522 | 18·0497 | 28553 | 77113 |
| 150 | 2·4754 | 20·5251 | 31184 | 108297 |
| 160 | 2·0311 | 22·5662 | 32661 | 140958 |
| 170 | 1·6876 | 24·2438 | 33404 | 174362 |
| 180 | 1·4201 | 25·6639 | 33713 | 208075 |
| 190 | 1·2111 | 26·8780 | 33826 | 241901 |
| 200 | 1·0400 | 27·9150 | 33657 | 275558 |
| 210 | 0·8966 | 28·8116 | 33202 | 308760 |
| 220 | 0·7801 | 29·5917 | 32709 | 341469 |
| 230 | 0·6839 | 30·2756 | 32194 | 373663 |
| 240 | 0·6014 | 30·8770 | 31543 | 405206 |
| 250 | 0·5308 | 31·4078 | 30825 | 436031 |
| 260 | 0·4709 | 31·8787 | 30099 | 466130 |
| 270 | 0·4203 | 32·2990 | 29422 | 495552 |
| 280 | 0·3781 | 32·6771 | 28856 | 524408 |
| 290 | 0·3422 | 33·0193 | 28368 | 552776 |
| 300 | 0·3099 | 33·3292 | 27793 | 580569 |
| 310 | 0·2807 | 33·6099 | 27154 | 607723 |
| 320 | 0·2545 | 33·8644 | 26460 | 634183 |
| 330 | 0·2318 | 34·0962 | 25842 | 660025 |
| 340 | 0·2124 | 34·3086 | 25320 | 685345 |
| 350 | 0·1950 | 34·5036 | 24810 | 710155 |
| 360 | 0·1786 | 34·6822 | 24194 | 734349 |
| 370 | 0·1649 | 34·8471 | 23723 | 758072 |
| 380 | 0·1519 | 34·9990 | 23182 | 781254 |
| 390 | 0·1405 | 35·1395 | 22683 | 803937 |
| 400 | 0·1304 | 35·2699 | 22249 | 826186 |
| 410 | 0·1208 | 35·3907 | 21749 | 847935 |
| 420 | 0·1125 | 35·5032 | 21332 | 869267 |
| 430 | 0·1053 | 35·6085 | 21003 | 890270 |
| 440 | 0·0984 | 35·7069 | 20628 | 910898 |
| 450 | 0·0919 | 35·7988 | 20208 | 931106 |
| 460 | 0·0863 | 35·8851 | 19896 | 951002 |
| 470 | 0·0810 | 35·9661 | 19544 | 970546 |
| 480 | 0·0759 | 36·0420 | 19154 | 989700 |
| 490 | 0·0717 | 36·1137 | 18898 | 1008595 |
| 500 | 0·0676 | 36·1813 | 18609 | 1027207 |
| 510 | 0·0635 | 36·2448 | 18233 | 1045440 |
| 520 | 0·0599 | 36·3047 | 17916 | 1063356 |
| 530 | 0·0567 | 36·3614 | 17643 | 1080999 |
| 540 | 0·0538 | 36·4152 | 17405 | 1098404 |
| 550 | 0·0510 | 36·4662 | 17160 | 1115564 |
| 560 | 0·0484 | 36·5146 | 16891 | 1132455 |
| 570 | 0·0458 | 36·5604 | 16609 | 1149064 |
| 580 | 0·0436 | 36·6040 | 16367 | 1165431 |
| 590 | 0·0414 | 36·6454 | 16133 | 1181564 |
| 600 | 0·0393 | 36·6847 | 15883 | 1197447 |
| 610 | 0·0375 | 36·7222 | 15654 | 1213101 |
| 620 | 0·0358 | 36·7580 | 15431 | 1228532 |
| 630 | 0·0342 | 36·7922 | 15240 | 1243772 |
| 640 | 0·0326 | 36·8248 | 15029 | 1258801 |

APPENDIX VI—TABLE IA—continued

| V. | Δ I. | I. | Δ A. | A. |
|---|---|---|---|---|
| 650 | 0·0311 | 36·8559 | 14802 | 1273603 |
| 660 | 0·0298 | 36·8857 | 14612 | 1288215 |
| 670 | 0·0289 | 36·9146 | 14402 | 1302617 |
| 680 | 0·0272 | 36·9418 | 14210 | 1316827 |
| 690 | 0·0260 | 36·9678 | 13998 | 1330825 |
| 700 | 0·0249 | 36·9927 | 13827 | 1344652 |
| 710 | 0·0240 | 37·0167 | 13670 | 1358322 |
| 720 | 0·0229 | 37·0396 | 13504 | 1371826 |
| 730 | 0·0220 | 37·0616 | 13323 | 1385149 |
| 740 | 0·0211 | 37·0827 | 13160 | 1398309 |
| 750 | 0·0203 | 37·1030 | 13016 | 1411325 |
| 760 | 0·0195 | 37·1225 | 12860 | 1424185 |
| 770 | 0·0188 | 37·1413 | 12699 | 1436884 |
| 780 | 0·0181 | 37·1594 | 12553 | 1449437 |
| 790 | 0·0174 | 37·1768 | 12399 | 1461836 |
| 800 | 0·0168 | 37·1936 | 12267 | 1474103 |
| 810 | 0·0162 | 37·2098 | 12128 | 1486231 |
| 820 | 0·0156 | 37·2254 | 12010 | 1498241 |
| 830 | 0·0151 | 37·2405 | 11888 | 1510129 |
| 840 | 0·0146 | 37·2551 | 11755 | 1521884 |
| 850 | 0·0140 | 37·2691 | 11555 | 1533439 |
| 860 | 0·0133 | 37·2824 | 11250 | 1544689 |
| 870 | 0·0126 | 37·2950 | 10933 | 1555622 |
| 880 | 0·0120 | 37·3070 | 10642 | 1566264 |
| 890 | 0·0114 | 37·3184 | 10369 | 1576633 |
| 900 | 0·0109 | 37·3293 | 10089 | 1586722 |
| 910 | 0·0103 | 37·3396 | 9800 | 1596522 |
| 920 | 0·0098 | 37·3494 | 9568 | 1606090 |
| 930 | 0·0094 | 37·3588 | 9333 | 1615423 |
| 940 | 0·0090 | 37·3678 | 9119 | 1624542 |
| 950 | 0·0086 | 37·3764 | 8898 | 1633440 |
| 960 | 0·0082 | 37·3846 | 8710 | 1642150 |
| 970 | 0·00785 | 37·39245 | 8513 | 1650663 |
| 980 | 0·00755 | 37·4000 | 8313 | 1658976 |
| 990 | 0·00720 | 37·40720 | 8105 | 1667081 |
| 1000 | 0·00685 | 37·41405 | 7890 | 1674971 |
| 1010 | 0·00655 | 37·42060 | 7708 | 1682679 |
| 1020 | 0·00630 | 37·42690 | 7560 | 1690239 |
| 1030 | 0·00610 | 37·4330 | 7445 | 1697684 |
| 1040 | 0·00590 | 37·4389 | 7360 | 1705044 |
| 1050 | 0·0056 | 37·4445 | 7121 | 1712165 |
| 1060 | 0·00525 | 37·44975 | 6797 | 1718962 |
| 1070 | 0·0049 | 37·45465 | 6461 | 1725423 |
| 1080 | 0·0046 | 37·45925 | 6162 | 1731585 |
| 1090 | 0·0043 | 37·46355 | 5893 | 1737478 |
| 1100 | 0·00405 | 37·46760 | 5620 | 1743098 |
| 1110 | 0·0038 | 37·47140 | 5381 | 1748479 |
| 1120 | 0·00355 | 37·47495 | 5138 | 1753617 |
| 1130 | 0·0033 | 37·47825 | 4891 | 1758508 |
| 1140 | 0·0031 | 37·48135 | 4681 | 1763189 |
| 1150 | 0·0029 | 37·48425 | 4464 | 1767653 |
| 1160 | 0·00275 | 37·48700 | 4285 | 1771938 |
| 1170 | 0·00260 | 37·48960 | 4105 | 1776043 |
| 1180 | 0·00245 | 37·49205 | 3925 | 1779968 |
| 1190 | 0·0023 | 37·49435 | 3776 | 1783744 |
| 1200 | 0·0022 | 37·49655 | 3675 | 1787419 |
| 1210 | 0·00215 | 37·49870 | 3615 | 1791034 |
| 1220 | 0·00205 | 37·50075 | 3555 | 1794589 |
| 1230 | 0·0020 | 37·50275 | 3491 | 1798080 |
| 1240 | 0·00195 | 37·50470 | 3473 | 1801553 |

## APPENDIX VI—TABLE Ia—continued

| V. | $\Delta$ I. | I. | $\Delta$ A. | A. |
|---|---|---|---|---|
| 1250 | 0·0019 | 37·50660 | 3409 | 1804962 |
| 1260 | 0·00185 | 37·50845 | 3391 | 1808353 |
| 1270 | 0·0018 | 37·51025 | 3323 | 1811676 |
| 1280 | 0·00175 | 37·51200 | 3301 | 1814977 |
| 1290 | 0·0017 | 37·51370 | 3230 | 1818207 |
| 1300 | 0·00165 | 37·51535 | 3204 | 1821411 |
| 1310 | 0·0016 | 37·51695 | 3133 | 1824544 |
| 1320 | 0·00155 | 37·51850 | 3106 | 1827650 |
| 1330 | 0·0015 | 37·52000 | 3080 | 1830730 |
| 1340 | 0·00145 | 37·52145 | 3054 | 1833784 |
| 1350 | 0·00140 | 37·52285 | 2976 | 1836760 |
| 1360 | 0·00140 | 37·52425 | 2949 | 1839709 |
| 1370 | 0·00135 | 37·52560 | 2919 | 1842628 |
| 1380 | 0·0013 | 37·52690 | 2890 | 1845518 |
| 1390 | 0·0013 | 37·52820 | 2860 | 1848378 |
| 1400 | 0·00125 | 37·52945 | 2830 | 1851208 |
| 1410 | 0·00125 | 37·53070 | 2796 | 1854004 |
| 1420 | 0·00120 | 37·53190 | 2762 | 1856766 |
| 1430 | 0·00115 | 37·53305 | 2725 | 1859491 |
| 1440 | 0·0011 | 37·53415 | 2691 | 1862182 |
| 1450 | 0·0011 | 37·53525 | 2657 | 1864839 |
| 1460 | 0·00105 | 37·53630 | 2620 | 1867459 |
| 1470 | 0·00105 | 37·53735 | 2609 | 1870068 |
| 1480 | 0·00105 | 37·53840 | 2598 | 1872666 |
| 1490 | 0·001 | 37·53940 | 2590 | 1875256 |
| 1500 | 0·001 | 37·54040 | 2583 | 1877839 |
| 1510 | 0·001 | 37·54140 | 2572 | 1880411 |
| 1520 | 0·00095 | 37·54235 | 2557 | 1882968 |
| 1530 | 0·00095 | 37·54330 | 2545 | 1885513 |
| 1540 | 0·0009 | 37·54420 | 2534 | 1888047 |
| 1550 | 0·0009 | 37·54510 | 2523 | 1890570 |
| 1560 | 0·0009 | 37·54600 | 2512 | 1893082 |
| 1570 | 0·0009 | 37·54690 | 2497 | 1895579 |
| 1580 | 0·00085 | 37·54775 | 2493 | 1898072 |
| 1590 | 0·00085 | 37·54860 | 2489 | 1900561 |
| 1600 | 0·00085 | 37·54945 | 2485 | 1903046 |
| 1610 | 0·0008 | 37·55025 | 2482 | 1905528 |
| 1620 | 0·0008 | 37·55105 | 2478 | 1908006 |
| 1630 | 0·0008 | 37·55185 | 2471 | 1910477 |
| 1640 | 0·0008 | 37·55265 | 2467 | 1912944 |
| 1650 | 0·0008 | 37·55345 | 2463 | 1915407 |
| 1660 | 0·0008 | 37·55425 | 2460 | 1917867 |
| 1670 | 0·00075 | 37·55500 | 2452 | 1920319 |
| 1680 | 0·00075 | 37·55575 | 2445 | 1922764 |
| 1690 | 0·00075 | 37·55650 | 2441 | 1925205 |
| 1700 | 0·0007 | 37·55720 | 2434 | 1927639 |
| 1710 | 0·0007 | 37·55790 | 2422 | 1930061 |
| 1720 | 0·0007 | 37·55860 | 2419 | 1932480 |
| 1730 | 0·0007 | 37·55930 | 2411 | 1934891 |
| 1740 | 0·0007 | 37·56000 | 2404 | 1937295 |
| 1750 | 0·0007 | 37·5607 | 2396 | 1939691 |
| 1760 | 0·00065 | 37·56135 | 2389 | 1942080 |
| 1770 | 0·00065 | 37·5620 | 2381 | 1944461 |
| 1780 | 0·00065 | 37·56265 | 2374 | 1946835 |
| 1790 | 0·00065 | 37·5633 | 2370 | 1949205 |
| 1800 | 0·00065 | 37·56395 | 2367 | 1951572 |
| 1810 | 0·00065 | 37·56460 | 2363 | 1953935 |
| 1820 | 0·0006 | 37·5652 | 2355 | 1956290 |
| 1830 | 0·0006 | 37·5658 | 2352 | 1958642 |
| 1840 | 0·0006 | 37·5664 | 2352 | 1960994 |

APPENDIX VI—TABLE IA—continued

| V. | Δ I. | I. | Δ A. | A. |
|---|---|---|---|---|
| 1850 | 0·0006 | 37·5670 | 2348 | 1963342 |
| 1860 | 0·0006 | 37·5676 | 2348 | 1965690 |
| 1870 | 0·00055 | 37·56815 | 2344 | 1968034 |
| 1880 | 0·00055 | 37·5687 | 2337 | 1970371 |
| 1890 | 0·00055 | 37·56925 | 2333 | 1972704 |
| 1900 | 0·00055 | 37·5698 | 2329 | 1975033 |
| 1910 | 0·00055 | 37·57035 | 2326 | 1977359 |
| 1920 | 0·00055 | 37·5709 | 2318 | 1979677 |
| 1930 | 0·00055 | 37·57145 | 2311 | 1981988 |
| 1940 | 0·00055 | 37·5720 | 2303 | 1984291 |
| 1950 | 0·0005 | 37·5725 | 2292 | 1986583 |
| 1960 | 0·0005 | 37·5730 | 2284 | 1988867 |
| 1970 | 0·0005 | 37·5735 | 2277 | 1991144 |
| 1980 | 0·0005 | 37·5740 | 2266 | 1993410 |
| 1990 | 0·0005 | 37·5745 | 2251 | 1995661 |
| 2000 | 0·0005 | 37·5750 | 2239 | 1997900 |
| 2010 | 0·0005 | 37·5755 | 2228 | 2000128 |
| 2020 | 0·0005 | 37·5760 | 2228 | 2002356 |
| 2030 | 0·0005 | 37·5765 | 2228 | 2004584 |
| 2040 | 0·00045 | 37·57695 | 2228 | 2006812 |
| 2050 | 0·00045 | 37·5774 | 2228 | 2009040 |
| 2060 | 0·00045 | 37·57785 | 2228 | 2011268 |
| 2070 | 0·00045 | 37·5783 | 2228 | 2013496 |
| 2080 | 0·00045 | 37·57875 | 2228 | 2015724 |
| 2090 | 0·00045 | 37·57920 | 2228 | 2017952 |
| 2100 | 0·00045 | 37·57965 | 2225 | 2020177 |
| 2110 | 0·00045 | 37·5801 | 2225 | 2022402 |
| 2120 | 0·00045 | 37·58055 | 2225 | 2024627 |
| 2130 | 0·00045 | 37·5810 | 2225 | 2026852 |
| 2140 | 0·00045 | 37·58145 | 2225 | 2029077 |
| 2150 | 0·00045 | 37·58190 | 2225 | 2031302 |
| 2160 | 0·00045 | 37·58235 | 2225 | 2033527 |
| 2170 | 0·0004 | 37·58275 | 2225 | 2035752 |
| 2180 | 0·0004 | 37·58315 | 2225 | 2037977 |
| 2190 | 0·0004 | 37·58355 | 2225 | 2040202 |
| 2200 | 0·0004 | 37·58395 | 2217 | 2042419 |
| 2210 | 0·0004 | 37·58435 | 2217 | 2044636 |
| 2220 | 0·0004 | 37·58475 | 2217 | 2046853 |
| 2230 | 0·00040 | 37·58515 | 2217 | 2049070 |
| 2240 | 0·00035 | 37·58550 | 2217 | 2051287 |
| 2250 | 0·00035 | 37·58585 | 2217 | 2053504 |
| 2260 | 0·00035 | 37·58620 | 2217 | 2055721 |
| 2270 | 0·00035 | 37·58655 | 2217 | 2057938 |
| 2280 | 0·00035 | 37·58690 | 2217 | 2060155 |
| 2290 | 0·00035 | 37·58725 | 2217 | 2062372 |
| 2300 | 0·00035 | 37·58760 | 2210 | 2064582 |
| 2310 | 0·00035 | 37·58795 | 2210 | 2066792 |
| 2320 | 0·00035 | 37·58830 | 2210 | 2069002 |
| 2330 | 0·00035 | 37·58865 | 2210 | 2071212 |
| 2340 | 0·00035 | 37·58900 | 2210 | 2073422 |
| 2350 | 0·00035 | 37·58935 | 2210 | 2075632 |
| 2360 | 0·00035 | 37·58970 | 2210 | 2077842 |
| 2370 | 0·00035 | 37·59005 | 2210 | 2080052 |
| 2380 | 0·00035 | 37·59040 | 2210 | 2082262 |
| 2390 | 0·00035 | 37·59075 | 2210 | 2084472 |
| 2400 | 0·0003 | 37·59105 | 2210 | 2086682 |
| 2410 | 0·0003 | 37·59135 | 2210 | 2088892 |
| 2420 | 0·0003 | 37·59165 | 2210 | 2091102 |
| 2430 | 0·0003 | 37·59195 | 2210 | 2093312 |
| 2400 | 0·0003 | 37·59225 | 2210 | 2095522 |

APPENDIX VI—TABLE IA—*continued*

| V. | Δ I. | I. | Δ A. | A. |
|---|---|---|---|---|
| 2450 | 0·0003 | 37·59255 | 2207 | 2097729 |
| 2460 | 0·0003 | 37·59285 | 2207 | 2099936 |
| 2470 | 0·0003 | 37·59315 | 2207 | 2102143 |
| 2480 | 0·0003 | 37·59345 | 2207 | 2104350 |
| 2490 | 0·0003 | 37·59375 | 2207 | 2106557 |
| 2500 | 0·0003 | 37·59405 | 2207 | 2108764 |
| 2510 | 0·0003 | 37·59435 | 2207 | 2110971 |
| 2520 | 0·0003 | 37·59465 | 2207 | 2113178 |
| 2530 | 0·0003 | 37·59495 | 2207 | 2115385 |
| 2540 | 0·0003 | 37·59525 | 2207 | 2117592 |
| 2550 | 0·0003 | 37·59555 | 2196 | 2119788 |
| 2560 | 0·0003 | 37·59585 | 2196 | 2121984 |
| 2570 | 0·0003 | 37·59615 | 2196 | 2124180 |
| 2580 | 0·0003 | 37·59645 | 2196 | 2126376 |
| 2590 | 0·00025 | 37·59670 | 2196 | 2128572 |
| 2600 | 0·00025 | 37·59695 | 2196 | 2130768 |
| 2610 | 0·00025 | 37·59720 | 2196 | 2132964 |
| 2620 | 0·00025 | 37·59745 | 2196 | 2135160 |
| 2630 | 0·00025 | 37·59770 | 2196 | 2137356 |
| 2640 | 0·00025 | 37·59795 | 2196 | 2139552 |
| 2650 | 0·00025 | 37·59820 | 2188 | 2141740 |
| 2660 | 0·00025 | 37·59845 | 2188 | 2143928 |
| 2670 | 0·00025 | 37·59870 | 2188 | 2146116 |
| 2680 | 0·00025 | 37·59895 | 2188 | 2148304 |
| 2690 | 0·00025 | 37·59920 | 2188 | 2150492 |
| 2700 | 0·00025 | 37·59945 | 2188 | 2152680 |
| 2710 | 0·00025 | 37·59970 | 2188 | 2154868 |
| 2720 | 0·00025 | 37·59995 | 2188 | 2157056 |
| 2730 | 0·00025 | 37·60020 | 2188 | 2159244 |
| 2740 | 0·00025 | 37·60045 | 2188 | 2161432 |
| 2750 | 0·00025 | 37·60070 | 2181 | 2163613 |
| 2760 | 0·00025 | 37·60095 | 2181 | 2165794 |
| 2770 | 0·00025 | 37·60120 | 2181 | 2167975 |
| 2780 | 0·00025 | 37·60145 | 2181 | 2170156 |
| 2790 | 0·00025 | 37·60170 | 2181 | 2172337 |
| 2800 | 0·00025 | 37·60195 | 2181 | 2174518 |
| 2810 | 0·00025 | 37·60220 | 2181 | 2176699 |
| 2820 | 0·00025 | 37·60245 | 2181 | 2178880 |
| 2830 | 0·00025 | 37·60270 | 2181 | 2181061 |
| 2840 | 0·00025 | 37·60295 | 2181 | 2183242 |
| 2850 | 0·00025 | 37·60320 | 2162 | 2185404 |
| 2860 | 0·00025 | 37·60345 | 2162 | 2187566 |
| 2870 | 0·0002 | 37·60365 | 2162 | 2189728 |
| 2880 | 0·0002 | 37·60385 | 2162 | 2191890 |
| 2890 | 0·0002 | 37·60405 | 2162 | 2194052 |
| 2900 | 0·0002 | 37·60425 | 2162 | 2196214 |
| 2910 | 0·0002 | 37·60445 | 2162 | 2198376 |
| 2920 | 0·0002 | 37·60465 | 2162 | 2200538 |
| 2930 | 0·0002 | 37·60485 | 2162 | 2202700 |
| 2940 | 0·0002 | 37·60505 | 2162 | 2204862 |
| 2950 | 0·0002 | 37·60525 | 2143 | 2207005 |
| 2960 | 0·0002 | 37·60545 | 2143 | 2209148 |
| 2970 | 0·0002 | 37·60565 | 2143 | 2211291 |
| 2980 | 0·0002 | 37·60585 | 2143 | 2213434 |
| 2990 | 0·0002 | 37·60605 | 2144 | 2215578 |
| 3000 | 0·0002 | 37·60625 | 2144 | 2217722 |
| 3010 | 0·0002 | 37·60645 | 2113 | 2219835 |
| 3020 | 0·0002 | 37·60665 | 2113 | 2221948 |
| 3030 | 0·0002 | 37·60685 | 2113 | 2224061 |
| 3040 | 0·0002 | 37·60705 | 2113 | 2226174 |

APPENDIX VI—TABLE IA—*continued*

| V. | Δ I. | I. | Δ A. | A. |
|---|---|---|---|---|
| 3050 | 0·0002 | 37·60725 | 2113 | 2228287 |
| 3060 | 0·0002 | 37·60745 | 2113 | 2230400 |
| 3070 | 0·0002 | 37·60765 | 2113 | 2232513 |
| 3080 | 0·0002 | 37·60785 | 2113 | 2234626 |
| 3090 | 0·0002 | 37·60805 | 2114 | 2236740 |
| 3100 | 0·0002 | 37·60825 | 2114 | 2238854 |
| 3110 | 0·0002 | 37·60845 | 2087 | 2240941 |
| 3120 | 0·0002 | 37·60865 | 2087 | 2243028 |
| 3130 | 0·0002 | 37·60885 | 2087 | 2245115 |
| 3140 | 0·0002 | 37·60905 | 2087 | 2247202 |
| 3150 | 0·00015 | 37·60920 | 2087 | 2249289 |
| 3160 | 0·00015 | 37·60935 | 2087 | 2251376 |
| 3170 | 0·00015 | 37·60950 | 2087 | 2253463 |
| 3180 | 0·00015 | 37·60965 | 2087 | 2255550 |
| 3190 | 0·00015 | 37·60980 | 2087 | 2257637 |
| 3200 | 0·00015 | 37·60995 | 2087 | 2259724 |
| 3210 | 0·00015 | 37·61010 | 2061 | 2261785 |
| 3220 | 0·00015 | 37·61025 | 2061 | 2263846 |
| 3230 | 0·00015 | 37·61040 | 2061 | 2265907 |
| 3240 | 0·00015 | 37·61055 | 2061 | 2267968 |
| 3250 | 0·00015 | 37·61070 | 2061 | 2270029 |
| 3260 | 0·00015 | 37·61085 | 2061 | 2272090 |
| 3270 | 0·00015 | 37·61100 | 2061 | 2274151 |
| 3280 | 0·00015 | 37·61115 | 2061 | 2276212 |
| 3290 | 0·00015 | 37·61130 | 2061 | 2278273 |
| 3300 | 0·00015 | 37·61145 | 2061 | 2280334 |
| 3310 | 0·00015 | 37·61160 | 2035 | 2282369 |
| 3320 | 0·00015 | 37·61175 | 2035 | 2284404 |
| 3330 | 0·00015 | 37·61190 | 2035 | 2286439 |
| 3340 | 0·00015 | 37·61205 | 2035 | 2288474 |
| 3350 | 0·00015 | 37·61220 | 2035 | 2290509 |
| 3360 | 0·00015 | 37·61235 | 2035 | 2292544 |
| 3370 | 0·00015 | 37·61250 | 2035 | 2294579 |
| 3380 | 0·00015 | 37·61265 | 2035 | 2296614 |
| 3390 | 0·00015 | 37·61280 | 2035 | 2298649 |
| 3400 | 0·00015 | 37·61295 | 2035 | 2300684 |
| 3410 | 0·00015 | 37·61310 | 2009 | 2302693 |
| 3420 | 0·00015 | 37·61325 | 2009 | 2304702 |
| 3430 | 0·00015 | 37·61340 | 2009 | 2306711 |
| 3440 | 0·00015 | 37·61355 | 2009 | 2308720 |
| 3450 | 0·00015 | 37·61370 | 2009 | 2310729 |
| 3460 | 0·00015 | 37·61385 | 2009 | 2312738 |
| 3470 | 0·00015 | 37·61400 | 2009 | 2314747 |
| 3480 | 0·00015 | 37·61415 | 2009 | 2316756 |
| 3490 | 0·00015 | 37·61430 | 2009 | 2318765 |
| 3500 | 0·00015 | 37·61445 | 2009 | 2320774 |

## APPENDIX VI

TABLE II—BALLISTIC TABLES FOR RIFLE BULLET TRAJECTORIES

Computed by Ballistic Department of Ordnance Committee from Table I

$V = 2000$ f.s.

$\Delta_v$ for use between $V = 2000$ f.s. and $V = 2100$ f.s.

| $Z = R/C$ yds. | $A = E/C$ | $\Delta z$ | $\Delta v$ | $u$ | $\Delta z$ | $\Delta v$ | $T = t/C$ | $\Delta_s$ | $\Delta v$ |
|---|---|---|---|---|---|---|---|---|---|
|  |  | + | − |  | − | + |  | + | − |
| 500 | 24 | 27 | 3 | 1757 | 228 | 95 | 0·80 | 0·93 | 0·04 |
| 1000 | 51 | 33 | 6 | 1529 | 205 | 89 | 1·73 | 1·06 | 0·09 |
| 1500 | 84 | 40 | 9 | 1324 | 162 | 76 | 2·79 | 1·20 | 0·15 |
| 2000 | 124 | 48 | 13 | 1162 | 106 | 59 | 3·99 | 1·35 | 0·21 |
| 2500 | 172 | 57 | 17 | 1056 | 73 | 36 | 5·34 | 1·47 | 0·26 |
| 2884 | 216 | (W = 343) |  | 1000 | — | — | 6·47 | — | — |
| 3000 | 229 | 65 | 22 | 983 | 65 | 27 | 6·81 | 1·59 | 0·30 |
| 3500 | 294 | 73 | 27 | 918 | 55 | 24 | 8·40 | 1·69 | 0·34 |
| 4000 | 367 | 81 | 32 | 863 | 48 | 21 | 10·09 | 1·79 | 0·39 |
| 4500 | 448 | 90 | 37 | 815 | 45 | 19 | 11·88 | 1·90 | 0·44 |
| 5000 | 538 | 99 | 42 | 770 | 43 | 17 | 13·78 | 2·00 | 0·48 |
| 5500 | 637 | 109 | 46 | 727 | 41 | 16 | 15·78 | 2·13 | 0·52 |
| 6000 | 746 | 120 | 51 | 686 | 38 | 16 | 17·91 | 2·25 | 0·57 |
| 6500 | 866 | 131 | 57 | 648 | 37 | 15 | 20·16 | 2·38 | 0·62 |
| 7000 | 997 | 145 | 64 | 611 | 34 | 14 | 22·54 | 2·52 | 0·67 |
| 7500 | 1142 | 158 | 72 | 577 | 33 | 13 | 25·06 | 2·68 | 0·72 |
| 8000 | 1300 | 170 | 80 | 544 | 31 | 13 | 27·74 | 2·84 | 0·78 |
| 8500 | 1470 | 186 | 87 | 513 | 29 | 12 | 30·58 | 3·01 | 0·85 |
| 9000 | 1656 | 205 | 94 | 484 | 28 | 11 | 33·59 | 3·20 | 0·91 |
| 9500 | 1861 | 223 | 104 | 456 | 27 | 11 | 36·79 | 3·39 | 0·99 |
| 10000 | 2084 | — | 113 | 429 | — | 11 | 40·18 | — | 1·07 |

$V = 2100$ f.s.

$\Delta_v$ for use between $V = 2100$ f.s. and $V = 2200$ f.s.

| $Z = R/C$ yds. | $A = E/C$ | $\Delta z$ | $\Delta v$ | $u$ | $\Delta z$ | $\Delta v$ | $T = t/C$ | $\Delta_s$ | $\Delta v$ |
|---|---|---|---|---|---|---|---|---|---|
|  |  | + | − |  | − | + |  | + | − |
| 500 | 21 | 24 | 2 | 1852 | 234 | 97 | 0·76 | 0·88 | 0·03 |
| 1000 | 45 | 30 | 4 | 1618 | 218 | 91 | 1·64 | 1·00 | 0·08 |
| 1500 | 75 | 36 | 8 | 1400 | 179 | 83 | 2·64 | 1·14 | 0·14 |
| 2000 | 111 | 44 | 11 | 1221 | 129 | 66 | 3·78 | 1·30 | 0·20 |
| 2500 | 155 | 52 | 15 | 1092 | 82 | 44 | 5·08 | 1·43 | 0·25 |
| 3000 | 207 | 60 | 19 | 1010 | 68 | 29 | 6·51 | 1·55 | 0·30 |
| 3074 | 216 | (W = 352) |  | 1000 | — | — | 6·74 | — | — |
| 3500 | 267 | 68 | 24 | 942 | 58 | 25 | 8·06 | 1·64 | 0·35 |
| 4000 | 335 | 76 | 30 | 884 | 50 | 22 | 9·70 | 1·74 | 0·39 |
| 4500 | 411 | 85 | 34 | 834 | 47 | 19 | 11·44 | 1·86 | 0·43 |
| 5000 | 496 | 95 | 39 | 787 | 44 | 18 | 13·30 | 1·96 | 0·47 |
| 5500 | 591 | 104 | 44 | 743 | 41 | 17 | 15·26 | 2·08 | 0·51 |
| 6000 | 695 | 114 | 49 | 702 | 39 | 16 | 17·34 | 2·20 | 0·56 |
| 6500 | 809 | 124 | 55 | 663 | 38 | 15 | 19·54 | 2·33 | 0·61 |
| 7000 | 933 | 137 | 61 | 625 | 35 | 15 | 21·87 | 2·47 | 0·66 |
| 7500 | 1070 | 150 | 67 | 590 | 33 | 14 | 24·34 | 2·62 | 0·71 |
| 8000 | 1220 | 163 | 74 | 557 | 32 | 13 | 26·96 | 2·77 | 0·77 |
| 8500 | 1383 | 179 | 81 | 525 | 30 | 12 | 29·73 | 2·95 | 0·83 |
| 9000 | 1562 | 195 | 90 | 495 | 28 | 12 | 32·68 | 3·12 | 0·91 |
| 9500 | 1757 | 214 | 100 | 467 | 27 | 11 | 35·80 | 3·31 | 0·98 |
| 10000 | 1971 | — | 110 | 440 | — | 10 | 39·11 | — | 1·05 |

388

## APPENDIX VI—TABLE II—continued

$V = 2200$ f.s.

$\Delta_v$ for use between $V = 2200$ f.s. and $V = 2300$ f.s.

| Z = R/C. yds. | A = E/C. | Δz. | Δv. | u. | Δz. | Δv. | T = t/C. | Δz. | Δv. |
|---|---|---|---|---|---|---|---|---|---|
| | | + | − | | | + | | + | − |
| 500 | 19 | 22 | 2 | 1949 | 240 | 99 | 0·73 | 0·83 | 0·04 |
| 1000 | 41 | 26 | 4 | 1709 | 226 | 92 | 1·56 | 0·94 | 0·08 |
| 1500 | 67 | 33 | 6 | 1483 | 196 | 87 | 2·50 | 1·08 | 0·13 |
| 2000 | 100 | 40 | 9 | 1287 | 151 | 72 | 3·58 | 1·25 | 0·18 |
| 2500 | 140 | 48 | 13 | 1136 | 97 | 53 | 4·83 | 1·38 | 0·24 |
| 3000 | 188 | 55 | 17 | 1039 | 72 | 33 | 6·21 | 1·50 | 0·29 |
| 3271 | 218 | (W = 361) | | 1000 | — | — | 7·02 | — | — |
| 3500 | 243 | 62 | 22 | 967 | 61 | 27 | 7·71 | 1·60 | 0·34 |
| 4000 | 305 | 72 | 26 | 906 | 53 | 23 | 9·31 | 1·70 | 0·38 |
| 4500 | 377 | 80 | 32 | 853 | 48 | 20 | 11·01 | 1·82 | 0·42 |
| 5000 | 457 | 90 | 37 | 805 | 45 | 19 | 12·83 | 1·92 | 0·46 |
| 5500 | 547 | 99 | 42 | 760 | 42 | 18 | 14·75 | 2·03 | 0·50 |
| 6000 | 646 | 108 | 48 | 718 | 40 | 17 | 16·78 | 2·15 | 0·55 |
| 6500 | 754 | 118 | 52 | 678 | 38 | 16 | 18·93 | 2·28 | 0·60 |
| 7000 | 872 | 131 | 58 | 640 | 36 | 15 | 21·21 | 2·42 | 0·65 |
| 7500 | 1003 | 143 | 64 | 604 | 34 | 14 | 23·63 | 2·56 | 0·71 |
| 8000 | 1146 | 156 | 71 | 570 | 33 | 13 | 26·19 | 2·71 | 0·77 |
| 8500 | 1302 | 170 | 78 | 537 | 30 | 13 | 28·90 | 2·87 | 0·83 |
| 9000 | 1472 | 185 | 87 | 507 | 29 | 11 | 31·77 | 3·05 | 0·89 |
| 9500 | 1657 | 204 | 95 | 478 | 28 | 11 | 34·82 | 3·24 | 0·96 |
| 10000 | 1861 | — | 105 | 450 | — | 11 | 38·06 | — | 1·04 |

$V = 2300$ f.s.

$\Delta_v$ for use between $V = 2300$ f.s. and $V = 2400$ f.s.

| Z = R/C. yds. | A = E/C. | Δz. | Δv. | u. | Δz. | Δv. | T = t/C. | Δz. | Δv. |
|---|---|---|---|---|---|---|---|---|---|
| | | + | − | | | + | | + | − |
| 500 | 17 | 20 | 2 | 2048 | 247 | 99 | 0·69 | 0·79 | 0·03 |
| 1000 | 37 | 24 | 4 | 1801 | 231 | 94 | 1·48 | 0·89 | 0·07 |
| 1500 | 61 | 30 | 6 | 1570 | 211 | 89 | 2·37 | 1·03 | 0·11 |
| 2000 | 91 | 36 | 8 | 1359 | 170 | 78 | 3·40 | 1·19 | 0·17 |
| 2500 | 127 | 44 | 11 | 1189 | 117 | 61 | 4·59 | 1·33 | 0·23 |
| 3000 | 171 | 50 | 16 | 1072 | 78 | 39 | 5·92 | 1·45 | 0·29 |
| 3462 | 217 | (W = 371) | | 1000 | — | — | 7·26 | — | — |
| 3500 | 221 | 58 | 20 | 994 | 65 | 29 | 7·37 | 1·56 | 0·33 |
| 4000 | 279 | 66 | 25 | 929 | 56 | 24 | 8·93 | 1·66 | 0·37 |
| 4500 | 345 | 75 | 30 | 873 | 49 | 21 | 10·59 | 1·78 | 0·41 |
| 5000 | 420 | 85 | 35 | 824 | 46 | 19 | 12·37 | 1·88 | 0·45 |
| 5500 | 505 | 93 | 40 | 778 | 43 | 18 | 14·25 | 1·98 | 0·50 |
| 6000 | 598 | 104 | 45 | 735 | 41 | 17 | 16·23 | 2·10 | 0·54 |
| 6500 | 702 | 112 | 49 | 694 | 39 | 16 | 18·33 | 2·23 | 0·59 |
| 7000 | 814 | 125 | 55 | 655 | 37 | 15 | 20·56 | 2·36 | 0·64 |
| 7500 | 939 | 136 | 61 | 618 | 35 | 14 | 22·92 | 2·50 | 0·69 |
| 8000 | 1075 | 149 | 68 | 583 | 33 | 13 | 25·42 | 2·65 | 0·75 |
| 8500 | 1224 | 161 | 76 | 550 | 32 | 12 | 28·07 | 2·81 | 0·81 |
| 9000 | 1385 | 177 | 83 | 518 | 29 | 12 | 30·88 | 2·98 | 0·87 |
| 9500 | 1562 | 194 | 91 | 489 | 28 | 11 | 33·86 | 3·16 | 0·94 |
| 10000 | 1756 | — | 100 | 461 | — | 11 | 37·02 | — | 1·02 |

APPENDIX VI—TABLE II—*continued*

V = 2400 f.s.

$\Delta_v$ for use between V = 2400 f.s. and V = 2500 f.s.

| Z = R/C. yds. | A = E/C. | Δz. | Δv. | u. | Δz. | Δv. | T = t/C. | $\Delta_s$. | Δv. |
|---|---|---|---|---|---|---|---|---|---|
| | | + | − | | − | + | | + | − |
| 500 | 15 | 18 | 1 | 2147 | 252 | 99 | 0·66 | 0·75 | 0·03 |
| 1000 | 33 | 22 | 3 | 1895 | 236 | 97 | 1·41 | 0·85 | 0·07 |
| 1500 | 55 | 28 | 5 | 1659 | 222 | 91 | 2·26 | 0·97 | 0·11 |
| 2000 | 83 | 33 | 8 | 1437 | 187 | 84 | 3·23 | 1·13 | 0·16 |
| 2500 | 116 | 39 | 11 | 1250 | 139 | 68 | 4·36 | 1·27 | 0·22 |
| 3000 | 155 | 46 | 15 | 1111 | 88 | 47 | 5·63 | 1·41 | 0·28 |
| 3500 | 201 | 53 | 20 | 1023 | 70 | 31 | 7·04 | 1·52 | 0·33 |
| 3664 | 218 | (W = 381) | | 1000 | — | — | 7·54 | — | — |
| 4000 | 254 | 61 | 24 | 953 | 59 | 26 | 8·56 | 1·62 | 0·37 |
| 4500 | 315 | 70 | 28 | 894 | 51 | 22 | 10·18 | 1·74 | 0·41 |
| 5000 | 385 | 80 | 32 | 843 | 47 | 19 | 11·92 | 1·83 | 0·45 |
| 5500 | 465 | 88 | 37 | 796 | 44 | 18 | 13·75 | 1·94 | 0·49 |
| 6000 | 553 | 100 | 41 | 752 | 42 | 16 | 15·69 | 2·05 | 0·53 |
| 6500 | 653 | 106 | 46 | 710 | 40 | 15 | 17·74 | 2·18 | 0·58 |
| 7000 | 759 | 119 | 52 | 670 | 38 | 15 | 19·92 | 2·31 | 0·63 |
| 7500 | 878 | 129 | 58 | 632 | 36 | 14 | 22·23 | 2·44 | 0·68 |
| 8000 | 1007 | 141 | 65 | 596 | 34 | 14 | 24·67 | 2·59 | 0·73 |
| 8500 | 1148 | 154 | 72 | 562 | 32 | 14 | 27·26 | 2·75 | 0·79 |
| 9000 | 1302 | 169 | 79 | 530 | 30 | 13 | 30·01 | 2·91 | 0·86 |
| 9500 | 1471 | 185 | 88 | 500 | 28 | 12 | 32·92 | 3·08 | 0·93 |
| 10000 | 1656 | — | 96 | 472 | — | 11 | 36·00 | — | 0·99 |

V = 2500 f.s.

$\Delta_v$ for use between V = 2500 f.s. and V = 2600 f.s.

| Z = R/C. yds. | A = E/C. | Δz. | Δv. | u. | Δz. | Δv. | T = t/C. | $\Delta_s$. | Δv. |
|---|---|---|---|---|---|---|---|---|---|
| | | + | − | | − | + | | + | − |
| 500 | 14 | 16 | 1 | 2246 | 254 | 99 | 0·63 | 0·71 | 0·03 |
| 1000 | 30 | 20 | 2 | 1992 | 242 | 98 | 1·34 | 0·81 | 0·06 |
| 1500 | 50 | 25 | 4 | 1750 | 229 | 93 | 2·15 | 0·92 | 0·10 |
| 2000 | 75 | 30 | 7 | 1521 | 203 | 87 | 3·07 | 1·07 | 0·15 |
| 2500 | 105 | 35 | 10 | 1318 | 160 | 75 | 4·14 | 1·21 | 0·21 |
| 3000 | 140 | 41 · | 13 | 1158 | 104 | 57 | 5·35 | 1·36 | 0·27 |
| 3500 | 181 | 49 | 16 | 1054 | 75 | 35 | 6·71 | 1·48 | 0·32 |
| 3860 | 216 | (W = 393) | | 1000 | — | — | 7·78 | — | — |
| 4000 | 230 | 57 | 21 | 979 | 63 | 28 | 8·19 | 1·58 | 0·36 |
| 4500 | 287 | 66 | 25 | 916 | 54 | 24 | 9·77 | 1·70 | 0·42 |
| 5000 | 353 | 75 | 30 | 862 | 48 | 20 | 11·47 | 1·79 | 0·45 |
| 5500 | 428 | 84 | 34 | 814 | 46 | 18 | 13·26 | 1·90 | 0·49 |
| 6000 | 512 | 95 | 38 | 768 | 43 | 17 | 15·16 | 2·00 | 0·53 |
| 6500 | 607 | 100 | 44 | 725 | 40 | 16 | 17·16 | 2·13 | 0·57 |
| 7000 | 707 | 113 | 49 | 685 | 39 | 15 | 19·29 | 2·26 | 0·62 |
| 7500 | 820 | 122 | 55 | 646 | 36 | 15 | 21·55 | 2·39 | 0·67 |
| 8000 | 942 | 134 | 61 | 610 | 34 | 14 | 23·94 | 2·53 | 0·72 |
| 8500 | 1076 | 147 | 69 | 576 | 33 | 13 | 26·47 | 2·68 | 0·78 |
| 9000 | 1223 | 160 | 76 | 543 | 31 | 12 | 29·15 | 2·84 | 0·84 |
| 9500 | 1383 | 177 | 85 | 512 | 29 | 12 | 31·99 | 3·02 | 0·90 |
| 10000 | 1560 | — | 92 | 483 | — | 11 | 35·01 | — | 0·97 |

## APPENDIX VI—TABLE II—continued

### V = 2600 f.s.

$\Delta_v$ for use between V = 2600 f.s. and V = 2700 f.s.

| Z = R/C. yds. | A = E/C. | Δz. | Δv. | u. | Δz. | Δv. | T = t/C. | $\Delta_z$. | Δv. |
|---|---|---|---|---|---|---|---|---|---|
| | | + | − | | − | + | | + | − |
| 500 | 13 | 15 | 1 | 2345 | 255 | 99 | 0·60 | 0·68 | 0·03 |
| 1000 | 28 | 18 | 2 | 2090 | 247 | 99 | 1·28 | 0·77 | 0·06 |
| 1500 | 46 | 22 | 3 | 1843 | 235 | 95 | 2·05 | 0·87 | 0·10 |
| 2000 | 68 | 27 | 6 | 1608 | 215 | 89 | 2·92 | 1·01 | 0·14 |
| 2500 | 95 | 32 | 8 | 1393 | 178 | 80 | 3·93 | 1·15 | 0·20 |
| 3000 | 127 | 38 | 11 | 1215 | 126 | 64 | 5·08 | 1·31 | 0·25 |
| 3500 | 165 | 44 | 15 | 1089 | 82 | 42 | 6·39 | 1·44 | 0·31 |
| 4000 | 209 | 53 | 19 | 1007 | 67 | 29 | 7·83 | 1·52 | 0·36 |
| 4052 | 215 | (W = 405) | | 1000 | — | — | 7·99 | — | — |
| 4500 | 262 | 61 | 23 | 940 | 58 | 25 | 9·35 | 1·67 | 0·39 |
| 5000 | 323 | 71 | 26 | 882 | 50 | 21 | 11·02 | 1·75 | 0·44 |
| 5500 | 394 | 80 | 31 | 832 | 47 | 19 | 12·77 | 1·86 | 0·48 |
| 6000 | 474 | 89 | 36 | 785 | 44 | 18 | 14·63 | 1·96 | 0·52 |
| 6500 | 563 | 95 | 42 | 741 | 41 | 18 | 16·59 | 2·08 | 0·56 |
| 7000 | 658 | 107 | 47 | 700 | 39 | 16 | 18·67 | 2·21 | 0·61 |
| 7500 | 765 | 116 | 52 | 661 | 37 | 15 | 20·88 | 2·34 | 0·66 |
| 8000 | 881 | 126 | 58 | 624 | 35 | 14 | 23·22 | 2·47 | 0·71 |
| 8500 | 1007 | 140 | 65 | 589 | 34 | 13 | 25·69 | 2·62 | 0·76 |
| 9000 | 1147 | 151 | 72 | 555 | 31 | 13 | 28·31 | 2·78 | 0·82 |
| 9500 | 1298 | 170 | 80 | 524 | 30 | 12 | 31·09 | 2·95 | 0·88 |
| 10000 | 1468 | — | 87 | 494 | — | 11 | 34·04 | — | 0·95 |

### V = 2700 f.s.

$\Delta_v$ for use between V = 2700 f.s. and V = 2800 f.s.

| Z = R/C. yds. | A = E/C. | Δz. | Δv. | u. | Δz. | Δv. | T = t/C. | $\Delta_z$. | Δv. |
|---|---|---|---|---|---|---|---|---|---|
| | | + | − | | − | + | | + | − |
| 500 | 12 | 14 | 1 | 2444 | 255 | 99 | 0·57 | 0·65 | 0·02 |
| 1000 | 26 | 17 | 2 | 2189 | 251 | 98 | 1·22 | 0·73 | 0·05 |
| 1500 | 43 | 19 | 4 | 1938 | 241 | 96 | 1·95 | 0·83 | 0·08 |
| 2000 | 62 | 25 | 5 | 1697 | 224 | 91 | 2·78 | 0·95 | 0·13 |
| 2500 | 87 | 29 | 8 | 1473 | 194 | 85 | 3·73 | 1·10 | 0·18 |
| 3000 | 116 | 34 | 11 | 1279 | 148 | 70 | 4·83 | 1·25 | 0·24 |
| 3500 | 150 | 40 | 14 | 1131 | 95 | 50 | 6·08 | 1·39 | 0·30 |
| 4000 | 190 | 49 | 17 | 1036 | 71 | 32 | 7·47 | 1·49 | 0·35 |
| 4254 | 215 | (W = 417) | | 1000 | — | — | 8·23 | — | — |
| 4500 | 239 | 58 | 22 | 965 | 62 | 26 | 8·96 | 1·62 | 0·38 |
| 5000 | 297 | 66 | 27 | 903 | 52 | 23 | 10·58 | 1·71 | 0·43 |
| 5500 | 363 | 75 | 30 | 851 | 48 | 19 | 12·29 | 1·82 | 0·47 |
| 6000 | 438 | 83 | 34 | 803 | 44 | 18 | 14·11 | 1·92 | 0·51 |
| 6500 | 521 | 90 | 40 | 759 | 43 | 17 | 16·03 | 2·03 | 0·55 |
| 7000 | 611 | 102 | 43 | 716 | 40 | 17 | 18·06 | 2·16 | 0·60 |
| 7500 | 713 | 110 | 49 | 676 | 38 | 16 | 20·22 | 2·29 | 0·65 |
| 8000 | 823 | 119 | 54 | 638 | 36 | 15 | 22·51 | 2·42 | 0·70 |
| 8500 | 942 | 133 | 60 | 602 | 34 | 14 | 24·93 | 2·56 | 0·75 |
| 9000 | 1075 | 143 | 67 | 568 | 32 | 13 | 27·49 | 2·72 | 0·80 |
| 9500 | 1218 | 163 | 74 | 536 | 31 | 12 | 30·21 | 2·88 | 0·86 |
| 10000 | 1381 | — | 82 | 505 | — | 12 | 33·09 | — | 0·93 |

## APPENDIX VI—TABLE II—continued

### V = 2800 f.s.

$\Delta_v$ for use between V = 2800 f.s. and V = 2900 f.s.

| Z = R/C. yds. | A = E/C. | $\Delta z$. | $\Delta v$. | u. | $\Delta z$. | $\Delta v$. | T = t/C. | $\Delta_x$. | $\Delta v$. |
|---|---|---|---|---|---|---|---|---|---|
| | | + | — | | — | + | | + | — |
| 500 | 11 | 13 | 1 | 2543 | 256 | 99 | 0.55 | 0.62 | 0.02 |
| 1000 | 24 | 15 | 2 | 2287 | 253 | 98 | 1.17 | 0.70 | 0.05 |
| 1500 | 39 | 18 | 3 | 2034 | 246 | 97 | 1.87 | 0.78 | 0.08 |
| 2000 | 57 | 22 | 4 | 1788 | 230 | 93 | 2.65 | 0.90 | 0.12 |
| 2500 | 79 | 26 | 6 | 1558 | 209 | 87 | 3.55 | 1.04 | 0.16 |
| 3000 | 105 | 31 | 9 | 1349 | 168 | 76 | 4.59 | 1.19 | 0.22 |
| 3500 | 136 | 37 | 13 | 1181 | 113 | 60 | 5.78 | 1.34 | 0.28 |
| 4000 | 173 | 44 | 16 | 1068 | 77 | 37 | 7.12 | 1.46 | 0.33 |
| 4442 | 212 | (W = 430) | | 1000 | — | — | 8.41 | — | — |
| 4500 | 217 | 53 | 18 | 991 | 65 | 28 | 8.58 | 1.57 | 0.37 |
| 5000 | 270 | 63 | 21 | 926 | 56 | 24 | 10.15 | 1.67 | 0.42 |
| 5500 | 333 | 71 | 26 | 870 | 49 | 21 | 11.82 | 1.78 | 0.46 |
| 6000 | 404 | 77 | 32 | 821 | 45 | 19 | 13.60 | 1.88 | 0.50 |
| 6500 | 481 | 87 | 37 | 776 | 43 | 17 | 15.48 | 1.98 | 0.54 |
| 7000 | 568 | 96 | 41 | 733 | 41 | 16 | 17.46 | 2.11 | 0.58 |
| 7500 | 664 | 105 | 46 | 692 | 39 | 15 | 19.57 | 2.24 | 0.63 |
| 8000 | 769 | 113 | 51 | 653 | 37 | 14 | 21.81 | 2.37 | 0.68 |
| 8500 | 882 | 126 | 56 | 616 | 35 | 13 | 24.18 | 2.51 | 0.73 |
| 9000 | 1008 | 136 | 62 | 581 | 33 | 13 | 26.69 | 2.66 | 0.79 |
| 9500 | 1144 | 155 | 69 | 548 | 31 | 13 | 29.35 | 2.81 | 0.85 |
| 10000 | 1299 | — | 77 | 517 | — | 12 | 32.16 | — | 0.91 |

### V = 2900 f.s.

$\Delta_v$ for use between V = 2900 f.s. and V = 3000 f.s.

| R = R/C. yds. | A = E/C. | $\Delta z$. | $\Delta v$. | u. | $\Delta z$. | $\Delta v$. | T = t/C. | $\Delta_x$. | $\Delta v$. |
|---|---|---|---|---|---|---|---|---|---|
| | | + | — | | — | + | | + | — |
| 500 | 10 | 12 | 1 | 2642 | 257 | 98 | 0.53 | 0.59 | 0.02 |
| 1000 | 22 | 14 | 2 | 2385 | 254 | 98 | 1.12 | 0.67 | 0.05 |
| 1500 | 36 | 17 | 2 | 2131 | 250 | 97 | 1.79 | 0.74 | 0.08 |
| 2000 | 53 | 20 | 4 | 1881 | 236 | 94 | 2.53 | 0.86 | 0.11 |
| 2500 | 73 | 23 | 6 | 1645 | 220 | 89 | 3.39 | 0.98 | 0.15 |
| 3000 | 96 | 27 | 8 | 1425 | 184 | 81 | 4.37 | 1.13 | 0.20 |
| 3500 | 123 | 34 | 9 | 1241 | 136 | 65 | 5.50 | 1.29 | 0.26 |
| 4000 | 157 | 42 | 13 | 1105 | 86 | 44 | 6.79 | 1.42 | 0.32 |
| 4500 | 199 | 50 | 17 | 1019 | 69 | 29 | 8.21 | 1.52 | 0.36 |
| 4638 | 213 | (W = 443) | | 1000 | — | — | 8.63 | — | — |
| 5000 | 249 | 58 | 20 | 950 | 59 | 25 | 9.73 | 1.63 | 0.41 |
| 5500 | 307 | 65 | 24 | 891 | 51 | 21 | 11.36 | 1.74 | 0.45 |
| 6000 | 372 | 72 | 28 | 840 | 47 | 19 | 13.10 | 1.84 | 0.49 |
| 6500 | 444 | 83 | 34 | 793 | 44 | 17 | 14.94 | 1.94 | 0.53 |
| 7000 | 527 | 91 | 39 | 749 | 42 | 16 | 16.88 | 2.06 | 0.57 |
| 7500 | 618 | 100 | 43 | 707 | 40 | 15 | 18.94 | 2.19 | 0.61 |
| 8000 | 718 | 108 | 48 | 667 | 38 | 15 | 21.13 | 2.32 | 0.66 |
| 8500 | 826 | 120 | 52 | 629 | 35 | 15 | 23.45 | 2.45 | 0.71 |
| 9000 | 946 | 129 | 58 | 594 | 33 | 14 | 25.90 | 2.60 | 0.76 |
| 9500 | 1075 | 147 | 63 | 561 | 32 | 12 | 28.50 | 2.75 | 0.82 |
| 10000 | 1222 | — | 71 | 529 | — | 12 | 31.25 | — | 0.88 |

## APPENDIX VI—TABLE II—continued

### V = 3000 f.s.

$\Delta_v$ for use between V = 3000 f.s. and V = 3100 f.s.

| Z = R/C. yds. | A = E/C. | Δz. | Δv. | u. | Δz. | Δv. | T = t/C. | $\Delta_g$. | Δv. |
|---|---|---|---|---|---|---|---|---|---|
|  |  | + | — |  | — | + |  | + | — |
| 500 | 9 | 11 | 0 | 2740 | 257 | 97 | 0·51 | 0·56 | 0·02 |
| 1000 | 20 | 14 | 1 | 2483 | 255 | 96 | 1·07 | 0·64 | 0·04 |
| 1500 | 34 | 15 | 3 | 2228 | 253 | 95 | 1·71 | 0·71 | 0·07 |
| 2000 | 49 | 18 | 3 | 1975 | 241 | 95 | 2·42 | 0·82 | 0·10 |
| 2500 | 67 | 21 | 5 | 1734 | 228 | 89 | 3·24 | 0·93 | 0·14 |
| 3000 | 88 | 26 | 7 | 1506 | 200 | 84 | 4·17 | 1·07 | 0·19 |
| 3500 | 114 | 30 | 9 | 1306 | 157 | 70 | 5·24 | 1·23 | 0·24 |
| 4000 | 144 | 38 | 11 | 1149 | 101 | 53 | 6·47 | 1·38 | 0·30 |
| 4500 | 182 | 47 | 14 | 1048 | 73 | 32 | 7·85 | 1·47 | 0·35 |
| 4829 | 213 | (W = 456) |  | 1000 | — | — | 8·82 | — | — |
| 5000 | 229 | 54 | 18 | 975 | 63 | 26 | 9·32 | 1·59 | 0·39 |
| 5500 | 283 | 61 | 22 | 912 | 53 | 23 | 10·91 | 1·70 | 0·43 |
| 6000 | 344 | 66 | 27 | 859 | 49 | 19 | 12·61 | 1·80 | 0·47 |
| 6500 | 410 | 78 | 31 | 810 | 45 | 18 | 14·41 | 1·90 | 0·51 |
| 7000 | 488 | 87 | 36 | 765 | 43 | 17 | 16·31 | 2·02 | 0·55 |
| 7500 | 575 | 95 | 41 | 722 | 40 | 16 | 18·33 | 2·14 | 0·59 |
| 8000 | 670 | 104 | 45 | 682 | 38 | 15 | 20·47 | 2·27 | 0·64 |
| 8500 | 774 | 114 | 50 | 644 | 36 | 14 | 22·74 | 2·40 | 0·69 |
| 9000 | 888 | 124 | 55 | 608 | 35 | 13 | 25·14 | 2·54 | 0·74 |
| 9500 | 1012 | 139 | 58 | 573 | 32 | 13 | 27·68 | 2·69 | 0·79 |
| 10000 | 1151 | — | 65 | 541 | — | 12 | 30·37 | — | 0·85 |

### V = 3100 f.s.

$\Delta_v$ for use between V = 3100 f.s. and V = 3200 f.s.

| Z = R/C. yds. | A = E/C. | Δz. | Δv. | u. | Δz. | Δv. | T = t/C. | $\Delta_g$. | Δv. |
|---|---|---|---|---|---|---|---|---|---|
|  |  | + | — |  | — | + |  | + | — |
| 500 | 9 | 10 | 0 | 2837 | 258 | 96 | 0·49 | 0·54 | 0·01 |
| 1000 | 19 | 12 | 0 | 2579 | 256 | 95 | 1·03 | 0·61 | 0·04 |
| 1500 | 31 | 15 | 1 | 2323 | 253 | 94 | 1·64 | 0·68 | 0·07 |
| 2000 | 46 | 16 | 3 | 2070 | 247 | 93 | 2·32 | 0·78 | 0·09 |
| 2500 | 62 | 19 | 4 | 1823 | 233 | 89 | 3·10 | 0·88 | 0·13 |
| 3000 | 81 | 24 | 6 | 1590 | 214 | 85 | 3·98 | 1·02 | 0·17 |
| 3500 | 105 | 28 | 9 | 1376 | 174 | 75 | 5·00 | 1·17 | 0·22 |
| 4000 | 133 | 35 | 11 | 1202 | 122 | 59 | 6·17 | 1·33 | 0·28 |
| 4500 | 168 | 43 | 13 | 1080 | 79 | 38 | 7·50 | 1·43 | 0·33 |
| 5000 | 211 | 50 | 16 | 1001 | 66 | 27 | 8·93 | 1·55 | 0·37 |
| 5008 | 212 | (W = 468) |  | 1000 | — | — | 8·95 | — | — |
| 5500 | 261 | 56 | 20 | 935 | 57 | 23 | 10·48 | 1·66 | 0·41 |
| 6000 | 317 | 62 | 24 | 878 | 50 | 20 | 12·14 | 1·76 | 0·45 |
| 6500 | 379 | 73 | 27 | 828 | 46 | 18 | 13·90 | 1·86 | 0·49 |
| 7000 | 452 | 82 | 33 | 782 | 44 | 17 | 15·76 | 1·98 | 0·53 |
| 7500 | 534 | 91 | 38 | 738 | 41 | 16 | 17·74 | 2·09 | 0·57 |
| 8000 | 625 | 99 | 43 | 697 | 39 | 15 | 19·83 | 2·22 | 0·61 |
| 8500 | 724 | 109 | 47 | 658 | 37 | 14 | 22·05 | 2·35 | 0·66 |
| 9000 | 833 | 121 | 51 | 621 | 35 | 13 | 24·40 | 2·49 | 0·71 |
| 9500 | 954 | 132 | 53 | 586 | 33 | 12 | 26·89 | 2·63 | 0·77 |
| 10000 | 1086 | — | 59 | 553 | — | 11 | 29·52 | — | 0·82 |

393

APPENDIX VI—TABLE II—continued

V = 3200 f.s.

| Z = R/C yds. | A = E/C. | Δz. | Δv. | u. | Δz. | Δv. | T = t/C. | Δ₂. | Δv. |
|---|---|---|---|---|---|---|---|---|---|
| | | + | − | | − | | | + | |
| 500 | 9 | 10 | — | 2933 | 259 | — | 0·48 | 0·51 | — |
| 1000 | 19 | 11 | — | 2674 | 257 | — | 0·99 | 0·58 | — |
| 1500 | 30 | 13 | — | 2417 | 254 | — | 1·57 | 0·66 | — |
| 2000 | 43 | 15 | — | 2163 | 251 | — | 2·23 | 0·74 | — |
| 2500 | 58 | 17 | — | 1912 | 237 | — | 2·97 | 0·84 | — |
| 3000 | 75 | 21 | — | 1675 | 224 | — | 3·81 | 0·97 | — |
| 3500 | 96 | 26 | — | 1451 | 190 | — | 4·78 | 1·11 | — |
| 4000 | 122 | 33 | — | 1261 | 143 | — | 5·89 | 1·28 | — |
| 4500 | 155 | 40 | — | 1118 | 90 | — | 7·17 | 1·39 | — |
| 5000 | 195 | 46 | — | 1028 | 70 | — | 8·56 | 1·51 | — |
| 5200 | 213 | (W=480) | | 1000 | — | — | 9·16 | — | — |
| 5500 | 241 | 52 | — | 958 | 60 | — | 10·07 | 1·62 | — |
| 6000 | 293 | 59 | — | 898 | 52 | — | 11·69 | 1·72 | — |
| 6500 | 352 | 67 | — | 846 | 47 | — | 13·41 | 1·82 | — |
| 7000 | 419 | 77 | — | 799 | 45 | — | 15·23 | 1·94 | — |
| 7500 | 496 | 86 | — | 754 | 42 | — | 17·17 | 2·05 | — |
| 8000 | 582 | 95 | — | 712 | 40 | — | 19·22 | 2·17 | — |
| 8500 | 677 | 105 | — | 672 | 38 | — | 21·39 | 2·30 | — |
| 9000 | 782 | 119 | — | 634 | 36 | — | 23·69 | 2·43 | — |
| 9500 | 901 | 126 | — | 598 | 34 | — | 26·12 | 2·58 | — |
| 10000 | 1027 | — | — | 564 | — | — | 28·70 | — | — |

APPENDIX VI

TABLE III.—SMALL ARM BALLISTIC TABLE. M.V. 1000 f/s.

Computed from Quadratic law p. 322 for 1000 f/s = 1·49 lb. by formulæ given in Ballistic Tables 1909

| Z = R/C yds. | A = E/C min. | Δ. | U f/s. | Δ. | T = t/C sec. | Δ. | B. | Δ. |
|---|---|---|---|---|---|---|---|---|
| 0 | 0 | | 1000 | | 0·00 | | 0·000 | |
| | | 87 | | 70 | | 1·55 | | 54 |
| 500 | 87 | | 930 | | 1·55 | | 0·054 | |
| | | 97 | | 64 | | 1·67 | | 64 |
| 1000 | 184 | | 866 | | 3·22 | | 0·118 | |
| | | 107 | | 60 | | 1·80 | | 77 |
| 1500 | 291 | | 806 | | 5·02 | | 0·195 | |
| | | 118 | | 55 | | 1·93 | | 92 |
| 2000 | 409 | | 751 | | 6·95 | | 0·287 | |
| | | 130 | | 52 | | 2·07 | | 109 |
| 2500 | 539 | | 699 | | 9·02 | | 0·396 | |
| | | 143 | | 50 | | 2·21 | | 129 |
| 3000 | 682 | | 649 | | 11·23 | | 0·525 | |
| | | 158 | | 46 | | 2·37 | | 155 |
| 3500 | 840 | | 603 | | 13·60 | | 0·680 | |
| | | 174 | | 41 | | 2·55 | | 183 |
| 4000 | 1014 | | 562 | | 16·15 | | 0·863 | |
| | | 193 | | 38 | | 2·77 | | 217 |
| 4500 | 1207 | | 524 | | 18·92 | | 1·080 | |
| | | 215 | | 36 | | 3·00 | | 258 |

## APPENDIX VI—TABLE III—continued

| Z = R/C yds. | A = E/C min. | Δ. | U f/s. | Δ. | T = t/C sec. | Δ. | B. | Δ. |
|---|---|---|---|---|---|---|---|---|
| 5000 | 1422 |     | 488 |    | 21·92 |      | 1·338 |      |
|      |      | 241 |     | 34 |       | 3·24 |       | 300  |
| 5500 | 1663 |     | 454 |    | 25·16 |      | 1·638 |      |
|      |      | 269 |     | 32 |       | 3·45 |       | 350  |
| 6000 | 1932 |     | 422 |    | 28·61 |      | 1·988 |      |
|      |      | 300 |     | 30 |       | 3·67 |       | 405  |
| 6500 | 2232 |     | 392 |    | 32·28 |      | 2·393 |      |
|      |      | 334 |     | 27 |       | 3·90 |       | 475  |
| 7000 | 2566 |     | 365 |    | 36·18 |      | 2·868 |      |
|      |      | 372 |     | 25 |       | 4·18 |       | 565  |
| 7500 | 2938 |     | 340 |    | 40·36 |      | 3·433 |      |
|      |      | 414 |     | 23 |       | 4·53 |       | 665  |
| 8000 | 3352 |     | 317 |    | 44·89 |      | 4·098 |      |
|      |      | 462 |     | 22 |       | 4·88 |       | 775  |
| 8500 | 3814 |     | 295 |    | 49·77 |      | 4·873 |      |
|      |      | 517 |     | 21 |       | 5·18 |       | 900  |
| 9000 | 4331 |     | 274 |    | 54·95 |      | 5·773 |      |
|      |      | 581 |     | 19 |       | 5·53 |       | 1045 |
| 9500 | 4912 |     | 255 |    | 60·48 |      | 6·818 |      |
|      |      | 658 |     | 18 |       | 5·98 |       | 1210 |
| 10000 | 5570 |    | 237 |    | 66·46 |      | 8·028 |      |
|      |      | 745 |     | 17 |       | 6·50 |       | 1395 |
| 10500 | 6315 |    | 220 |    | 72·96 |      | 9·423 |      |

## APPENDIX VI

### Table IV.—Ballistic Table for Spherical Shot

(Recalculated by Mr. Hadcock, R.A., from Bashforth's data, and extended to low velocities.)

For lower velocities this table is provisional, pending the results of further experiments

| v. f/s. | ΔT. | T. | ΔS. | S. |
|---|---|---|---|---|
| 300 | 1·2232 | 0·0000 | 366·91 | 0·00 |
| 310 | 1·1505 | 1·2232 | 356·67 | 366·91 |
| 320 | 1·0824 | 2·3737 | 346·37 | 723·58 |
| 330 | 1·0217 | 3·4561 | 337·22 | 1069·95 |
| 340 | 0·9647 | 4·4778 | 328·01 | 1407·17 |
| 350 | 0·9137 | 5·4425 | 319·78 | 1735·18 |
| 360 | 0·8653 | 6·3562 | 311·51 | 2054·96 |
| 370 | 0·8218 | 7·2215 | 304·07 | 2366·47 |
| 380 | 0·7805 | 8·0433 | 296·60 | 2670·54 |
| 390 | 0·7432 | 8·8238 | 289·84 | 2967·14 |
| 400 | 0·7076 | 9·5670 | 283·05 | 3256·98 |
| 410 | 0·6753 | 10·2746 | 276·88 | 3540·03 |

## APPENDIX VI—TABLE IV—continued

| v. f/s. | Δ T. | T. | Δ S. | S. |
|---|---|---|---|---|
| 420 | 0·6445 | 10·9499 | 270·69 | 3816·91 |
| 430 | 0·6151 | 11·5944 | 264·51 | 4087·60 |
| 440 | 0·5763 | 12·2095 | 253·59 | 4352·11 |
| 450 | 0·5508 | 12·7858 | 247·86 | 4605·70 |
| 460 | 0·5265 | 13·3366 | 242·20 | 4853·56 |
| 470 | 0·5035 | 13·8631 | 236·64 | 5095·76 |
| 480 | 0·4816 | 14·3666 | 231·18 | 5332·40 |
| 490 | 0·4609 | 14·8482 | 225·84 | 5563·58 |
| 500 | 0·4413 | 15·3091 | 220·63 | 5789·42 |
| 510 | 0·4227 | 15·7504 | 215·55 | 6010·05 |
| 520 | 0·4050 | 16·1731 | 210·61 | 6225·60 |
| 530 | 0·3883 | 16·5781 | 205·80 | 6436·21 |
| 540 | 0·3725 | 16·9664 | 201·14 | 6642·01 |
| 550 | 0·3575 | 17·3389 | 196·61 | 6843·15 |
| 560 | 0·3429 | 17·6964 | 192·01 | 7039·76 |
| 570 | 0·3291 | 18·0393 | 187·57 | 7231·77 |
| 580 | 0·3157 | 18·3684 | 183·11 | 7419·34 |
| 590 | 0·3028 | 18·6841 | 178·64 | 7602·45 |
| 600 | 0·2903 | 18·9869 | 174·19 | 7781·09 |
| 610 | 0·2786 | 19·2772 | 169·95 | 7955·28 |
| 620 | 0·2673 | 19·5558 | 165·75 | 8125·23 |
| 630 | 0·2567 | 19·8231 | 161·74 | 8290·98 |
| 640 | 0·2467 | 20·0798 | 157·92 | 8452·72 |
| 650 | 0·2371 | 20·3265 | 154·14 | 8610·64 |
| 660 | 0·2281 | 20·5636 | 150·53 | 8764·78 |
| 670 | 0·2195 | 20·7917 | 147·09 | 8915·31 |
| 680 | 0·2115 | 21·0112 | 143·80 | 9062·40 |
| 690 | 0·2038 | 21·2227 | 140·65 | 9206·29 |
| 700 | 0·1966 | 21·4265 | 137·63 | 9346·85 |
| 710 | 0·1898 | 21·6231 | 134·73 | 9484·48 |
| 720 | 0·1832 | 21·8129 | 131·88 | 9619·21 |
| 730 | 0·1770 | 21·9961 | 129·22 | 9751·09 |
| 740 | 0·1711 | 22·1731 | 126·59 | 9880·31 |
| 750 | 0·1653 | 22·3442 | 123·99 | 10006·90 |
| 760 | 0·1600 | 22·5095 | 121·57 | 10130·89 |
| 770 | 0·1547 | 22·6695 | 119·12 | 10252·46 |
| 780 | 0·1496 | 22·8242 | 116·72 | 10371·58 |
| 790 | 0·1447 | 22·9738 | 114·30 | 10488·30 |
| 800 | 0·1399 | 23·1185 | 111·89 | 10602·60 |
| 810 | 0·1352 | 23·2584 | 109·50 | 10714·49 |
| 820 | 0·1306 | 23·3936 | 107·07 | 10823·99 |
| 830 | 0·1201 | 23·5242 | 104·68 | 10931·06 |
| 840 | 0·1218 | 23·6503 | 102·33 | 11035·74 |
| 850 | 0·1177 | 23·7721 | 100·01 | 11138·07 |
| 860 | 0·1137 | 23·8898 | 97·76 | 11238·08 |

## APPENDIX VI—Table IV—continued

| v. f/s. | Δ T. | T. | Δ S. | S. |
|---|---|---|---|---|
| 870 | 0·1098 | 24·0035 | 95·53 | 11335·84 |
| 880 | 0·1062 | 24·1133 | 93·44 | 11431·37 |
| 890 | 0·1026 | 24·2195 | 91·35 | 11524·81 |
| 900 | 0·0993 | 24·3221 | 89·33 | 11616·16 |
| 910 | 0·0959 | 24·4214 | 87·32 | 11705·49 |
| 920 | 0·0928 | 24·5173 | 85·37 | 11792·81 |
| 930 | 0·0898 | 24·6101 | 83·48 | 11878·18 |
| 940 | 0·0869 | 24·6999 | 81·65 | 11961·66 |
| 950 | 0·0840 | 24·7868 | 79·83 | 12043·31 |
| 960 | 0·0813 | 24·8708 | 78·01 | 12123·14 |
| 970 | 0·0785 | 24·9521 | 76·19 | 12201·15 |
| 980 | 0·0759 | 25·0306 | 74·43 | 12277·34 |
| 990 | 0·0734 | 25·1065 | 72·67 | 12351·77 |
| 1000 | 0·0709 | 25·1799 | 70·87 | 12424·44 |
| 1010 | 0·0684 | 25·2508 | 69·08 | 12495·31 |
| 1020 | 0·0660 | 25·3192 | 67·31 | 12564·39 |
| 1030 | 0·0636 | 25·3852 | 65·55 | 12631·70 |
| 1040 | 0·0614 | 25·4488 | 63·81 | 12697·25 |
| 1050 | 0·0591 | 25·5102 | 62·08 | 12761·06 |
| 1060 | 0·0570 | 25·5693 | 60·42 | 12823·14 |
| 1070 | 0·0550 | 25·6263 | 58·82 | 12883·56 |
| 1080 | 0·0531 | 25·6813 | 57·31 | 12942·38 |
| 1090 | 0·0513 | 25·7344 | 55·89 | 12999·69 |
| 1100 | 0·0496 | 25·7857 | 54·59 | 13055·58 |
| 1110 | 0·0481 | 25·8353 | 53·36 | 13110·17 |
| 1120 | 0·0466 | 25·8834 | 52·21 | 13163·53 |
| 1130 | 0·0453 | 25·9300 | 51·15 | 13215·74 |
| 1140 | 0·0440 | 25·9753 | 50·16 | 13266·89 |
| 1150 | 0·0428 | 26·0193 | 49·23 | 13317·05 |
| 1160 | 0·0417 | 26·0621 | 48·35 | 13366·28 |
| 1170 | 0·0406 | 26·1038 | 47·53 | 13414·63 |
| 1180 | 0·0396 | 26·1444 | 46·73 | 13462·16 |
| 1190 | 0·0386 | 26·1840 | 45·97 | 13508·89 |
| 1200 | 0·0377 | 26·2226 | 45·27 | 13554·86 |
| 1210 | 0·0369 | 26·2603 | 44·61 | 13600·13 |
| 1220 | 0·0361 | 26·2972 | 44·00 | 13644·74 |
| 1230 | 0·0353 | 26·3333 | 43·43 | 13688·74 |
| 1240 | 0·0346 | 26·3686 | 42·87 | 13732·17 |
| 1250 | 0·0339 | 26·4032 | 42·36 | 13775·04 |
| 1260 | 0·0332 | 26·4371 | 41·85 | 13817·40 |
| 1270 | 0·0326 | 26·4703 | 41·39 | 13859·25 |
| 1280 | 0·0320 | 26·5029 | 40·94 | 13900·64 |
| 1290 | 0·0314 | 26·5349 | 40·49 | 13941·58 |
| 1300 | 0·0308 | 26·5663 | 40·04 | 13982·07 |
| 1310 | 0·0302 | 26·5971 | 39·59 | 14022·11 |

## APPENDIX VI—Table IV—continued

| v. f/s. | Δ T. | T. | Δ S. | S. |
|---|---|---|---|---|
| 1320 | 0·0297 | 26·6273 | 39·18 | 14061·70 |
| 1330 | 0·0291 | 26·6570 | 38·75 | 14100·88 |
| 1340 | 0·0286 | 26·6861 | 38·33 | 14139·63 |
| 1350 | 0·0281 | 26·7147 | 37·92 | 14177·96 |
| 1360 | 0·0276 | 26·7428 | 37·52 | 14215·88 |
| 1370 | 0·0271 | 26·7704 | 37·15 | 14253·40 |
| 1380 | 0·0267 | 26·7975 | 36·80 | 14290·55 |
| 1390 | 0·0262 | 26·8242 | 36·45 | 14327·35 |
| 1400 | 0·0258 | 26·8504 | 36·11 | 14363·80 |
| 1410 | 0·0264 | 26·8762 | 35·77 | 14399·91 |
| 1420 | 0·0250 | 26·9016 | 35·48 | 14435·68 |
| 1430 | 0·0246 | 26·9266 | 35·16 | 14471·16 |
| 1440 | 0·0242 | 26·9512 | 34·85 | 14506·32 |
| 1450 | 0·0238 | 26·9754 | 34·54 | 14541·17 |
| 1460 | 0·0235 | 26·9992 | 34·24 | 14575·71 |
| 1470 | 0·0231 | 27·0227 | 33·98 | 14609·95 |
| 1480 | 0·0228 | 27·0458 | 33·69 | 14643·93 |
| 1490 | 0·0224 | 27·0686 | 33·41 | 14677·62 |
| 1500 | 0·0221 | 27·0910 | 33·14 | 14711·03 |
| 1510 | 0·0218 | 27·1131 | 32·85 | 14744·17 |
| 1520 | 0·0214 | 27·1349 | 32·59 | 14777·02 |
| 1530 | 0·0211 | 27·1563 | 32·34 | 14809·61 |
| 1540 | 0·0208 | 27·1774 | 32·06 | 14841·95 |
| 1550 | 0·0205 | 27·1982 | 31·82 | 14874·01 |
| 1560 | 0·0202 | 27·2187 | 31·58 | 14905·83 |
| 1570 | 0·0200 | 27·2389 | 31·33 | 14937·41 |
| 1580 | 0·0197 | 27·2589 | 31·10 | 14968·74 |
| 1590 | 0·0194 | 27·2786 | 30·86 | 14999·84 |
| 1600 | 0·0191 | 27·2980 | 30·64 | 15030·70 |
| 1610 | 0·0189 | 27·3171 | 30·42 | 15061·34 |
| 1620 | 0·0186 | 27·3360 | 30·19 | 15091·76 |
| 1630 | 0·0184 | 27·3546 | 29·99 | 15121·95 |
| 1640 | 0·0182 | 27·3730 | 29·79 | 15151·94 |
| 1650 | 0·0179 | 27·3912 | 29·60 | 15181·73 |
| 1660 | 0·0177 | 27·4091 | 29·38 | 15211·33 |
| 1670 | 0·0175 | 27·4268 | 29·20 | 15240·71 |
| 1680 | 0·0173 | 27·4443 | 29·02 | 15269·91 |
| 1690 | 0·0171 | 27·4616 | 28·84 | 15298·93 |
| 1700 | 0·0168 | 27·4787 | 28·64 | 15327·77 |
| 1710 | 0·0167 | 27·4955 | 28·47 | 15356·41 |
| 1720 | 0·0165 | 27·5122 | 28·31 | 15384·88 |
| 1730 | 0·0163 | 27·5287 | 28·13 | 15413·19 |
| 1740 | 0·0161 | 27·5450 | 27·97 | 15441·32 |
| 1750 | 0·0159 | 27·5611 | 27·81 | 15469·29 |
| 1760 | 0·0157 | 27·5770 | 27·64 | 15497·10 |

## APPENDIX VI—Table IV—*continued*

| v f/s. | Δ T. | T. | Δ S. | S. |
|---|---|---|---|---|
| 1770 | 0·0155 | 27·5927 | 27·49 | 15524·74 |
| 1780 | 0·0154 | 27·6082 | 27·33 | 15552·23 |
| 1790 | 0·0152 | 27·6236 | 27·16 | 15579·56 |
| 1800 | 0·0150 | 27·6388 | 27·03 | 15606·72 |
| 1810 | 0·0148 | 27·6538 | 26·87 | 15633·75 |
| 1820 | 0·0147 | 27·6686 | 26·72 | 15660·62 |
| 1830 | 0·0145 | 27·6833 | 26·54 | 15687·34 |
| 1840 | 0·0143 | 27·6978 | 26·40 | 15713·88 |
| 1850 | 0·0142 | 27·7121 | 26·25 | 15740·28 |
| 1860 | 0·0140 | 27·7263 | 26·09 | 15766·53 |
| 1870 | 0·0139 | 27·7403 | 25·93 | 15792·62 |
| 1880 | 0·0137 | 27·7542 | 25·79 | 15818·55 |
| 1890 | 0·0136 | 27·7679 | 25·64 | 15844·34 |
| 1900 | 0·0134 | 27·7815 | 25·48 | 15869·98 |

## APPENDIX VI

### Table V.—Tenuity Correction F for Temperature and Pressure of Atmosphere Two-thirds Saturated with Moisture

From the Rev. F. Bashforth's Paper, "Proc. R.A.I.," Vol. XIII, No. 10

| Fahr. ° | 26 in. | 27 in. | 28 in. | 29 in. | 30 in. | 31 in. | Δ. |
|---|---|---|---|---|---|---|---|
| 0 | 0·983 | 1·021 | 1·059 | 1·097 | 1·134 | 1·172 | 38 |
| 1 | 0·981 | 1·019 | 1·056 | 1·084 | 1·132 | 1·170 | 38 |
| 2 | 0·979 | 1·017 | 1·054 | 1·092 | 1·130 | 1·167 | 38 |
| 3 | 0·977 | 1·015 | 1·052 | 1·090 | 1·127 | 1·165 | 38 |
| 4 | 0·975 | 1·012 | 1·050 | 1·087 | 1·125 | 1·162 | 38 |
| 5 | 0·973 | 1·010 | 1·047 | 1·085 | 1·122 | 1·160 | 37 |
| 6 | 0·971 | 1·008 | 1·045 | 1·083 | 1·120 | 1·157 | 37 |
| 7 | 0·969 | 1·006 | 1·043 | 1·080 | 1·118 | 1·155 | 37 |
| 8 | 0·966 | 1·004 | 1·041 | 1·078 | 1·115 | 1·152 | 37 |
| 9 | 0·964 | 1·001 | 1·039 | 1·076 | 1·113 | 1·150 | 37 |
| 10 | 0·962 | 0·999 | 1·036 | 1·073 | 1·110 | 1·147 | 37 |
| 11 | 0·960 | 0·997 | 1·034 | 1·071 | 1·108 | 1·145 | 37 |
| 12 | 0·958 | 0·995 | 1·032 | 1·069 | 1·105 | 1·142 | 37 |
| 13 | 0·956 | 0·993 | 1·029 | 1·066 | 1·103 | 1·140 | 37 |
| 14 | 0·954 | 0·991 | 1·027 | 1·064 | 1·101 | 1·137 | 37 |
| 15 | 0·952 | 0·989 | 1·025 | 1·062 | 1·098 | 1·135 | 37 |
| 16 | 0·950 | 0·986 | 1·023 | 1·060 | 1·096 | 1·133 | 37 |
| 17 | 0·948 | 0·984 | 1·021 | 1·057 | 1·094 | 1·130 | 37 |

## APPENDIX VI—TABLE V—continued

| Fahr. ° | 26 in. | 27 in. | 28 in. | 29 in. | 30 in. | 31 in. | Δ. |
|---|---|---|---|---|---|---|---|
| 18 | 0·946 | 0·982 | 1·019 | 1·055 | 1·091 | 1·128 | 36 |
| 19 | 0·944 | 0·980 | 1·017 | 1·053 | 1·089 | 1·125 | 36 |
| 20 | 0·942 | 0·978 | 1·014 | 1·051 | 1·087 | 1·123 | 36 |
| 21 | 0·940 | 0·976 | 1·012 | 1·048 | 1·084 | 1·121 | 36 |
| 22 | 0·938 | 0·974 | 1·010 | 1·046 | 1·082 | 1·118 | 36 |
| 23 | 0·936 | 0·972 | 1·008 | 1·044 | 1·080 | 1·116 | 36 |
| 24 | 0·934 | 0·970 | 1·006 | 1·042 | 1·078 | 1·114 | 36 |
| 25 | 0·932 | 0·968 | 1·004 | 1·039 | 1·075 | 1·111 | 36 |
| 26 | 0·930 | 0·966 | 1·001 | 1·037 | 1·073 | 1·109 | 36 |
| 27 | 0·928 | 0·964 | 0·999 | 1·035 | 1·071 | 1·106 | 36 |
| 28 | 0·926 | 0·962 | 0·997 | 1·033 | 1·069 | 1·104 | 36 |
| 29 | 0·924 | 0·960 | 0·995 | 1·031 | 1·066 | 1·102 | 36 |
| 30 | 0·922 | 0·958 | 0·993 | 1·028 | 1·064 | 1·099 | 36 |
| 31 | 0·920 | 0·956 | 0·991 | 1·026 | 1·062 | 1·097 | 35 |
| 32 | 0·918 | 0·954 | 0·989 | 1·024 | 1·059 | 1·095 | 35 |
| 33 | 0·916 | 0·952 | 0·987 | 1·022 | 1·057 | 1·093 | 35 |
| 34 | 0·914 | 0·950 | 0·985 | 1·020 | 1·055 | 1·090 | 35 |
| 35 | 0·913 | 0·948 | 0·983 | 1·018 | 1·053 | 1·088 | 35 |
| 36 | 0·911 | 0·946 | 0·981 | 1·016 | 1·051 | 1·086 | 35 |
| 37 | 0·909 | 0·944 | 0·979 | 1·013 | 1·048 | 1·083 | 35 |
| 38 | 0·907 | 0·942 | 0·977 | 1·011 | 1·046 | 1·081 | 35 |
| 39 | 0·905 | 0·940 | 0·974 | 1·009 | 1·044 | 1·079 | 35 |
| 40 | 0·903 | 0·938 | 0·973 | 1·007 | 1·042 | 1·077 | 35 |
| 41 | 0·901 | 0·936 | 0·971 | 1·005 | 1·040 | 1·075 | 35 |
| 42 | 0·899 | 0·934 | 0·968 | 1·003 | 1·038 | 1·072 | 35 |
| 43 | 0·898 | 0·932 | 0·967 | 1·001 | 1·036 | 1·070 | 35 |
| 44 | 0·896 | 0·930 | 0·964 | 0·999 | 1·033 | 1·068 | 34 |
| 45 | 0·894 | 0·928 | 0·963 | 0·997 | 1·031 | 1·066 | 34 |
| 46 | 0·892 | 0·926 | 0·960 | 0·995 | 1·029 | 1·063 | 34 |
| 47 | 0·890 | 0·924 | 0·958 | 0·993 | 1·027 | 1·061 | 34 |
| 48 | 0·888 | 0·923 | 0·957 | 0·991 | 1·025 | 1·059 | 34 |
| 49 | 0·886 | 0·920 | 0·955 | 0·989 | 1·023 | 1·057 | 34 |
| 50 | 0·884 | 0·919 | 0·953 | 0·987 | 1·021 | 1·055 | 34 |
| 51 | 0·883 | 0·917 | 0·951 | 0·985 | 1·019 | 1·053 | 34 |
| 52 | 0·881 | 0·915 | 0·949 | 0·983 | 1·017 | 1·051 | 34 |
| 53 | 0·879 | 0·913 | 0·947 | 0·981 | 1·015 | 1·048 | 34 |
| 54 | 0·877 | 0·911 | 0·945 | 0·978 | 1·012 | 1·046 | 34 |
| 55 | 0·875 | 0·909 | 0·943 | 0·976 | 1·010 | 1·044 | 34 |
| 56 | 0·874 | 0·907 | 0·941 | 0·974 | 1·008 | 1·042 | 34 |
| 57 | 0·872 | 0·905 | 0·939 | 0·972 | 1·006 | 1·039 | 34 |
| 58 | 0·870 | 0·904 | 0·937 | 0·970 | 1·004 | 1·037 | 34 |
| 59 | 0·868 | 0·902 | 0·935 | 0·968 | 1·002 | 1·035 | 33 |
| 60 | 0·866 | 0·900 | 0·933 | 0·966 | 1·000 | 1·033 | 33 |
| 61 | 0·865 | 0·898 | 0·931 | 0·964 | 0·998 | 1·031 | 33 |

APPENDIX VI—TABLE V—*continued*

| Fahr.° | 26 in. | 27 in. | 28 in. | 29 in. | 30 in. | 31 in. | Δ. |
|---|---|---|---|---|---|---|---|
| 62 | 0·863 | 0·896 | 0·929 | 0·962 | 0·996 | 1·029 | 33 |
| 63 | 0·861 | 0·894 | 0·927 | 0·960 | 0·993 | 1·027 | 33 |
| 64 | 0·859 | 0·892 | 0·925 | 0·958 | 0·991 | 1·024 | 33 |
| 65 | 0·857 | 0·890 | 0·923 | 0·956 | 0·989 | 1·022 | 33 |
| 66 | 0·855 | 0·889 | 0·921 | 0·954 | 0·987 | 1·020 | 33 |
| 67 | 0·854 | 0·887 | 0·919 | 0·952 | 0·985 | 1·018 | 33 |
| 68 | 0·852 | 0·885 | 0·918 | 0·950 | 0·983 | 1·016 | 33 |
| 69 | 0·850 | 0·883 | 0·916 | 0·949 | 0·981 | 1·014 | 33 |
| 70 | 0·849 | 0·881 | 0·914 | 0·946 | 0·979 | 1·012 | 33 |
| 71 | 0·847 | 0·879 | 0·912 | 0·944 | 0·977 | 1·010 | 33 |
| 72 | 0·845 | 0·878 | 0·910 | 0·943 | 0·975 | 1·008 | 33 |
| 73 | 0·843 | 0·876 | 0·908 | 0·941 | 0·973 | 1·006 | 32 |
| 74 | 0·842 | 0·874 | 0·906 | 0·939 | 0·971 | 1·004 | 32 |
| 75 | 0·840 | 0·872 | 0·904 | 0·937 | 0·969 | 1·001 | 32 |
| 76 | 0·838 | 0·870 | 0·902 | 0·935 | 0·967 | 0·999 | 32 |
| 77 | 0·836 | 0·868 | 0·901 | 0·933 | 0·965 | 0·997 | 32 |
| 78 | 0·834 | 0·867 | 0·899 | 0·931 | 0·963 | 0·995 | 32 |
| 79 | 0·833 | 0·865 | 0·897 | 0·929 | 0·961 | 0·993 | 32 |
| 80 | 0·831 | 0·863 | 0·895 | 0·927 | 0·959 | 0·991 | 32 |
| 81 | 0·829 | 0·861 | 0·893 | 0·925 | 0·957 | 0·989 | 32 |
| 82 | 0·827 | 0·859 | 0·891 | 0·923 | 0·955 | 0·987 | 32 |
| 83 | 0·826 | 0·858 | 0·889 | 0·921 | 0·953 | 0·985 | 32 |
| 84 | 0·824 | 0·856 | 0·887 | 0·919 | 0·951 | 0·983 | 32 |
| 85 | 0·822 | 0·854 | 0·885 | 0·917 | 0·949 | 0·980 | 32 |
| 86 | 0·821 | 0·852 | 0·884 | 0·915 | 0·947 | 0·978 | 32 |
| 87 | 0·819 | 0·850 | 0·882 | 0·913 | 0·945 | 0·976 | 32 |
| 88 | 0·817 | 0·848 | 0·880 | 0·911 | 0·943 | 0·974 | 31 |
| 89 | 0·815 | 0·847 | 0·878 | 0·909 | 0·941 | 0·972 | 31 |
| 90 | 0·814 | 0·845 | 0·876 | 0·908 | 0·939 | 0·970 | 31 |
| 91 | 0·812 | 0·843 | 0·874 | 0·905 | 0·937 | 0·968 | 31 |
| 92 | 0·810 | 0·841 | 0·872 | 0·903 | 0·935 | 0·966 | 31 |
| 93 | 0·808 | 0·839 | 0·870 | 0·902 | 0·933 | 0·964 | 31 |
| 94 | 0·806 | 0·837 | 0·868 | 0·900 | 0·931 | 0·962 | 31 |
| 95 | 0·805 | 0·836 | 0·867 | 0·898 | 0·929 | 0·960 | 31 |
| 96 | 0·803 | 0·834 | 0·865 | 0·896 | 0·926 | 0·957 | 31 |
| 97 | 0·801 | 0·832 | 0·863 | 0·893 | 0·924 | 0·955 | 31 |
| 98 | 0·799 | 0·830 | 0·861 | 0·891 | 0·922 | 0·953 | 31 |
| 99 | 0·797 | 0·828 | 0·859 | 0·889 | 0·920 | 0·951 | 31 |
| 100 | 0·796 | 0·826 | 0·857 | 0·888 | 0·918 | 0·949 | 31 |

# APPENDIX VII

## Index to Formulæ, Chap. IV, Part III

| Section. | Page. | Subject or Formula. |
|---|---|---|
| I | 309 | $R = \dfrac{V^2 \sin 2E}{3g}$ yards. |
|  | 309 | Max. $R = (V/10)^2$ yards. |
|  | 309 | Drop (inch) $= 193 \cdot 2 \, (3R/V)^2$. |
|  | 309 | $E = 166{,}000 \, R/V^2$ minutes $= \dfrac{573}{R} gt^2 = \dfrac{18450 \cdot 0}{R} t^2$. |
|  | 310 | $f = g\left\{1 - \dfrac{2}{5}\left(\dfrac{V-v}{V}\right)\right\}$ $= 6R \tan E/t^2$ $= RE/573 \, t^2$. |
|  | 310 | $y = (2T)^2$ feet in vacuo $= 4\frac{1}{4} T^2$ in air $= 50 T^2$ inches. |
|  | 311–12 | Froude-Metford formulæ. $M = Am + Bm^2 + Cm^3$ $A = a - \frac{1}{2}\beta + \frac{1}{3}\gamma.$ $B = \phantom{a-}\frac{1}{2}\beta - \frac{1}{2}\gamma.$ $C = \phantom{a-\frac{1}{2}\beta-}\frac{1}{3}\gamma.$ $W' = M \times$ average rise at M hundreds of yards. |
|  | 312 | $v = \dfrac{4075}{\sqrt{A + 3Bm + 6Cm^2}}$ |
|  | 312 | $V = \dfrac{4075}{\sqrt{a - \frac{1}{2}\beta + \frac{1}{3}\gamma}}$. |
|  | 313 | $\begin{cases} Y = R_2 \left(\dfrac{E_1 - E_2}{3438}\right) \\ \text{Inches} = \dfrac{R_2 \, \Delta E}{95 \cdot 5} = \dfrac{r \, \Delta E}{95 \cdot 5} \end{cases}$ |
|  | 313 | Contour formula— $h = H \dfrac{R_1 - R_2}{R_1} + R_2 \dfrac{E_1 - E_2}{3438}.$ |
|  | 314 | The Gunner's Rule— $\theta' = \dfrac{0 \cdot 955 \times \text{inches}}{N} = \dfrac{\text{inches}}{1 \cdot 047 \, N}$ $= 1146 \dfrac{H \text{ feet}}{R \text{ yards}}$ |
|  | 314 | $\dfrac{\phi_1}{\phi_2} = \left(\dfrac{V_2}{V_1}\right)^2$ |

## APPENDIX VII—continued

### INDEX TO FORMULÆ—continued

| Section. | Page. | Subject or Formula. |
|---|---|---|
| I | 314 | Forbes' method. |
|  | 315 | Negative angle. |
| II | 317 | Ballistic Tables and how to use them. |
|  | 319 | $t/c = T_V - T_v$ |
|  |  | $s/c = S_V - S_v$ |
|  | 318 | $\dfrac{W}{nd^2}$, $n = 1\cdot 0$ for modern bullets. $= 1\tfrac{1}{2}$ for blunt bullets. |
|  | 318 | Tenuity factor F is Bashforth's $\tau$. |
|  | 319 | $I_V$ table and vertex at $0\cdot 45 t$. |
|  | 321 | $\tan E = \tfrac{1}{2} a\, C$. |
|  | 321 | Table III. |
|  | 322 | Spherical Table. |
| III | 323 | Given C.R.V. to construct a full-range table. |
|  | 324 | Wind deflection in feet $= W \sin A \left(t - \dfrac{3R}{V}\right)$. |
|  | 325 | Wind tables for various rifles. |
|  | 326 | Effect of wind on range $\varDelta R = \dfrac{3W}{CV} N^2$ yards. |
| IV | 326 | Detail of $0°$–$90°$ traj. given $C + V$. |
|  | 327 | Angle of descent $W = 2\tfrac{1}{4}$ to $2\tfrac{1}{2}$ E. |
|  | 327 | Calculation of vertical trajectory. |
| V | 329 | Practical methods of determining $C = W/nd^2$. |
|  | 332 | Hodsock figures for stream-line bullets. |
|  | 332 | Disagreement of all tables. |
| VI | 332 | Long-range fire in two arcs. |
|  | 333 | $\beta \simeq \sec E_2$. |
|  | 335–6 | Examples: ·303 Mk. VI, VII, H, and French Balle D. |
| VII | 336 | Revolvers, etc., by quadratic law. Snider, M.H, 0·220, Webley, 12-bore. |

## APPENDIX VIII

### Details of the Rifles of Various Powers

| | Great Britain. | Austria. | Belgium. | Bulgaria. | Czechoslovakia. |
|---|---|---|---|---|---|
| 1. Designation and date of pattern | S.M.L.E., Mk. III | Mannlicher, 1895 | Mauser, 1889 | Mannlicher, 1895 | Mauser, 1924 |
| 2. Magazine system and capacity | Detachable vertical box, 10 rounds | Fixed vertical box, 5 rounds | Fixed vertical box, 5 rounds | *See under* "*Austria*" | Fixed vertical box, 5 rounds |
| 3. Charger or clip | Charger | Clip | Charger | | Charger |
| 4. Cut-off | Yes | No | No | | No |
| 5. Safety bolt | Yes | Yes | Yes | | Yes |
| 6. Weight without bayonet | 8 lbs. 10½ oz. | 8 lbs. 5¼ oz. | 8 lbs. 13 oz. | | 9 lbs. 2¾ oz. |
| 7.     ,, with bayonet | 9 lbs. 11½ oz. | 8 lbs. 15¾ oz. | 9 lbs. 12 oz. | | 10 lbs. ¾ oz. |
| 8. Length without bayonet | 3 ft. 8⅜ in. | 4 ft. 2 in. | 4 ft. 2¼ in. | | 3 ft. 7 in. |
| 9.     ,, with bayonet | 5 ft. 1·7 in. | 4 ft. 11½ in. | 5 ft. 7½ in. | | 4 ft. 6¼ in. |
| 10. Barrel, length (inches) | 25·19 | 30·12 | 30·67 | | 23·2 |
| 11.     ,, calibre, mm./inches | 7·7/0·303 | 8·0/0·315 | 7·65/0·301 | | 7·9/0·311 |
| 12.     ,, rifling, No. of grooves | 5 | 4 | 4 | | 4 |
| 13.     ,, twist, one turn in inches/calibres | 10/33 | 9·842/31 | 9·842/32·5 | | 9·45/30·38 |
| 14.     ,, direction of twist | To left | To right | To right | | To right |
| 15. Sights, type | "U" notch and blade | "V" notch and barleycorn | "V" notch and barleycorn | | "V" notch and barleycorn |
| 16.     ,, range of adjustment | 200–2000 yds. | 300–2600 m. | 500–2000 m. | | 300–2000 m. |
| 17.     ,, wind gauge | No | No | No | | No |

## APPENDIX VIII—continued

### Details of the Rifles of the Various Powers—continued

| | Denmark. | France. | Germany. | Greece. | Holland. |
|---|---|---|---|---|---|
| 1. Designation and date of pattern | Krag-Jorgensen, 1889 | Lebel, 1907-15, modified 1916 | Mauser, 1898 | Mannlicher-Schoenauer, 1903 | Mannlicher, 1895. |
| 2. Magazine system and capacity | Fixed horizontal box, 5 rounds | Fixed vertical box, 5 rounds | Fixed vertical box, 5 rounds | Fixed vertical box, rotary platform, 5 rounds | Fixed vertical box, 5 rounds |
| 3. Charger or clip | Neither | Clip | Charger | Charger | Clip |
| 4. Cut-off | No | No | No | No | No |
| 5. Safety bolt | No | No | Yes | Yes | Yes |
| 6. Weight without bayonet | 9 lbs. 12 oz. | 9 lbs. 3 oz. | 9 lbs. 14 oz. | 8 lbs. 5¼ oz. | 9 lbs. 4 oz. |
| 7. „ with bayonet | 10 lbs. 4¼ oz. | 10 lbs. 1¼ oz. | 9 lbs. | 9 lbs. | 9 lbs. 15 oz. |
| 8. Length without bayonet | 4 ft. 4¼ in. | 4 ft. 3·5 in. | 4 ft. 1·4 in. | 4 ft. ¼ in. | 4 ft. 3 in. |
| 9. „ with bayonet | 5 ft. 3 in. | 5 ft. 11·8 in. | 5 ft. 9¼ in. | 4 ft. 10¼ in. | 5 ft. 1 in. |
| 10. Barrel, length (inches) | 32·9 | 31·50 | 29·15 | 28·5 | 31·125 |
| 11. „ calibre, mm./inches | 8·0/0·315 | 8·0/0·315 | 7·9/0·311 | 6·5/0·256 | 6·5/0·256 |
| 12. „ rifling, No. of grooves | 6 | 4 | 4 | 4 | 4 |
| 13. „ twist, one turn in inches/calibres | 11·811/37·5 | 9·45/30 | 9·5/30·2 | 7·874/30·75 | 7·874/30·8 |
| 14. „ direction of twist | To right | To left | To right | To right | To right |
| 15. Sights, type | "V" notch and barleycorn | Square notch and grooved broad block foresight | "V" notch and barleycorn | "V" and barleycorn | "V" and barleycorn |
| 16. „ range of adjustment | 300–1900 m. | 400–2400 m. | 400–2000 m. | 400–2000 m. | 400–2000 m. |
| 17. „ wind gauge | No | No | No | No | No |

APPENDIX VIII—continued

DETAILS OF THE RIFLES OF VARIOUS POWERS—continued

|  | Hungary. | Italy. | Japan. | Norway. |
|---|---|---|---|---|
| 1. Designation and date of pattern | Mannlicher, 1895 | Mannlicher-Carcano, 1891 | Mauser, 1905, or "Year '38" | Krag-Jorgensen, 1910 |
| 2. Magazine system and capacity | (As under Austria) | Fixed vertical box, 6 rounds | Fixed vertical box, 5 rounds | Fixed horizontal box, 5 rounds |
| 3. Charger or clip | | Clip | Charger | Neither |
| 4. Cut-off | | No | No | No |
| 5. Safety bolt | | Yes | Yes | Yes |
| 6. Weight without bayonet | | 9 lbs. | 8 lbs. 12 oz. | 8 lbs. 15 oz. |
| 7. ,, with bayonet | | 9 lbs. 13 oz. | 9 lbs. 12 oz. | 9 lbs. 7¼ oz. |
| 8. Length without bayonet | | 4 ft. 2¾ in. | 4 ft. 3 in. | 4 ft. 2 in. |
| 9. ,, with bayonet | | 5 ft. 2½ in. | 5 ft. 5¼ in. | 5 ft. ¼ in. |
| 10. Barrel, length (inches) | | 30·75 | 31·30 | 30·07 |
| 11. ,, calibre, mm./inches | | 6·5/0·256 | 6·5/0·256 | 6·5/0·256 |
| 12. ,, rifling, No. of grooves | | 4 | 6 | 4 |
| 13. ,, twist, one turn in inches/calibres | | Increasing, breech to muzzle: 1 in 19¾–8¼ in., –32·2 calibres | 9/35·2 | 7·87/30·7 |
| 14. ,, direction of twist | | To right | To right | To left |
| 15. Sights, type | | "V" and barleycorn | "V" and barleycorn | "U" and barleycorn |
| 16. ,, range of adjustment | | 600–2000 m. | 400–2400 m. | 100–2200 m. |
| 17. ,, wind gauge | | No | No | No |

APPENDIX VIII—continued

DETAILS OF THE RIFLES OF VARIOUS POWERS—continued

| | Portugal. | Roumania. | Russia. | Spain. |
|---|---|---|---|---|
| 1. Designation and date of pattern | Mauser-Verguiero,* 1904 | Mannlicher,† 1893 | Mosin-Nagant, 1900 | Mauser, 1896 |
| 2. Magazine system and capacity | Fixed vertical box, 5 rounds | Fixed vertical box, 5 rounds | Fixed vertical box, with interrupter, 5 rounds | Fixed vertical box, 5 rounds |
| 3. Charger or clip | Charger | Clip | Charger | Charger |
| 4. Cut-off | No | No | No | No |
| 5. Safety bolt | Yes | Yes | Yes | Yes |
| 6. Weight without bayonet | 8 lbs. 13 oz. | 8 lbs. 12½ oz. | 8 lbs. 15 oz. | 9 lbs. 6¼ oz. |
| 7. ,, with bayonet | 9 lbs. 9½ oz. | 9 lbs. 9½ oz. | 9 lbs. 11 oz. | 10 lbs. 5½ oz. |
| 8. Length without bayonet | 4 ft. | 4 ft. ½ in. | 4 ft. 3·8 in. | 4 ft. ¾ in. |
| 9. ,, with bayonet | 4 ft. 11¼ in. | 4 ft. 10¼ in. | 5 ft. 9 in. | 4 ft. 10½ in. |
| 10. Barrel, length (inches) | 29·08 | 28·56 | 31·50 | 29·1 |
| 11. ,, calibre, mm./inches | 6·5/0·256 | 6·5/0·256 | 7·62/0·30 | 7·0/0·276 |
| 12. ,, rifling, No. of grooves | 4 | 4 | 4 | 4 |
| 13. ,, twist, one turn in inches/calibres | 7·78/30·76 | 8/31·3 | 9·5/31·6 | 8·6/31·4 |
| 14. ,, direction of twist | To right | To right | To right | To right |
| 15. Sights, type | "V" and barleycorn | "V" and barleycorn | "V" and barleycorn | "V" and barleycorn |
| 16. ,, range of adjustment | 200–2000 m. | 600–2000 m. | 400–2700 paces | 400–2000 m. |
| 17. ,, wind gauge | No | No | No | No |

\* Also various other types in use.
† Also Lebel, 1907–16; Mosin-Nagant, 1891–1916; Austrian Mannlicher, 1895.

APPENDIX VIII—continued

DETAILS OF THE RIFLES OF VARIOUS POWERS—continued

| | Sweden. | Switzerland. | Turkey. | U.S.A. |
|---|---|---|---|---|
| 1. Designation and date of pattern | Mauser, 1906 | Schmidt-Rubin,‡ 1909 | Mauser,§ 1905 | Springfield, 1903 |
| 2. Magazine system and capacity | Fixed vertical box, 5 rounds | Detachable vertical box, 6 rounds | Fixed vertical box, 5 rounds | Fixed vertical box, 5 rounds |
| 3. Charger or clip | Charger | Charger | Charger | Charger |
| 4. Cut-off | No | No | No | Yes |
| 5. Safety bolt | Yes | Yes | Yes | Yes |
| 6. Weight without bayonet | 8 lbs. 14¼ oz. | 10 lbs. | 9 lbs. 6 oz. | 8 lbs. 8 oz. |
| 7. „ with bayonet | 9 lbs. 8¼ oz. | 11 lbs. | 10 lbs. 12 oz. | 9 lbs. 8 oz. |
| 8. Length without bayonet | 4 ft. 1¼ in. | 4 ft. 3½ in. | 4 ft. 1 in. | 3 ft. 7¼ in. |
| 9. „ with bayonet | 4 ft. 10 in. | 5 ft. 2¼ in. | 5 ft. 9¼ in. | 4 ft. 11¼ in. |
| 10. Barrel, length (inches) | 29·1 | 30·8 | 29·1 | 23·79 |
| 11. „ calibre, mm./inches | 6·5/0·256 | 7·5/0·295 | 7·65/0·301 | 7·62/0·3 |
| 12. „ rifling, No. of grooves | 4 | 4 | 4 | 4 |
| 13. „ twist, one turn in inches/calibres | 7·87/30·7 | 10·5/35·6 | 9·84/32·7 | 10/33·3 |
| 14. „ direction of twist | To right | To right | To right | To right |
| 15. Sights, type | "V" and barleycorn | "U" and blade | "V" and barleycorn | "U" and blade |
| 16. „ range of adjustment | 300–2000 m. | 300–2000 m. | 400–2000 m. | 200–2850 yds. |
| 17. „ wind gauge | No | No | No | Yes |

‡ Also carbine, similar but shorter.
§ Also other types in use.

APPENDIX

DETAILS OF THE MACHINE GUNS AND LIGHT

| (1) Country and designation. | (2) Calibre. | (3) Length of barrel. | (4) Weight of Gun (without mounting). | (5) Type of mounting. | (6) Weight of mounting. | (7) Accessories on mounting. | (8) Method of operation. |
|---|---|---|---|---|---|---|---|
| **GREAT BRITAIN.** | | | | | | | |
| Vickers ·303-in. machine gun | ·303-in. | 28¼ in. | 30 lbs. | Tripod | 52 lbs. | Elevating gear, direction dial. | Recoil, aided by gas. |
| Lewis ·303-in. light machine gun | ·303-in. | 26¼ in. | 26 lbs. | Bipod rest | 2¾ lbs. | — | Gas. Piston working in cylinder. |
| Hotchkiss ·303-in. light machine gun. | ·303-in. | 23¼ in. | 27 lbs. | Tripod rest | 4 lbs. 2 ozs. | — | Gas. Cupped piston head fitting over gas nozzle. |
| **AMERICA, UNITED STATES OF.** | | | | | | | |
| Browning machine gun | ·300-in. | 23⅞ in. | 31¼ lbs. (air cooled for Air Service, 24¼ lbs.). | Tripod | 54 lbs. | Elevating and traversing gears. | Recoil. |
| *Colt machine gun | ·300-in. | 28 in. | 35 lbs. | Tripod | 61¼ lbs. | Elevating and traversing gear. | Gas, actuating radial lever. |
| Browning light machine gun | ·300-in. | Cavalry, 18 in. Infantry, 24 in. | Cavalry, 15 lbs. Infantry (1919), 15 lbs. 14 ozs. (1924), 18 lbs. | Bipod rest (1919 pattern, no mounting). | Cavalry, 12 ozs. Infantry (1924) 1 lb. 3 ozs. | — | Gas—piston in short cylinder. |
| *Certain States for Police Work.* | | | | | | | |
| Thompson " Sub-machine gun " or automatic rifle (three models) | ·45-in. and other smaller calibres | (a) 10·5 in. (b) and (c) 14¼ in. | (a) 10 lbs. ozs. (with magazine). (b) 11 lbs. (c) 11 lbs. 6 ozs. (approx.). | (a) and (b) No mounting or rest. (c) Bipod rest. | (c) About 1 lb. | — | Recoil of spent case acting on bolt face. |

\* Used by some British Forces (with calibre ·303 in.) during the War, 1914–1918, but in U.S.A. for training purposes only. Now obsolete.

## IX

### MACHINE GUNS OF VARIOUS POWERS

| (9) Type of ammunition conveyor. | (10) ? Cartridge supported by definitely locked action. | (11) Cooling system. | (12) Means of storing energy. | (13) Sights and range of adjustment. | (14) ? Whether gun can be adjusted to fire single shots. | (15) ? Whether special means of regulating rate of fire provided. | (16) Rate of fire (rounds per minute). | (17) Details of ammunition if different from that for the rifle. |
|---|---|---|---|---|---|---|---|---|
| Fabric belt, 250 rounds. Disintegrating belt for aircraft. | Yes | Ground Service— Water (in jacket). Air Service— Plain barrel in perforated casing. | Spiral spring in extension. | Aperture and blade. Up to 2,900 yards. | Yes (by method of loading). | No | 500 | — |
| Flat circular magazine on top. 47 rounds (96 rounds for aircraft). | Yes | Induced draught, aluminium radiator. (Plain barrel for Air Service). | Clock spring. | Aperture and blade. Up to 1,900 yards. | No | No | 550 | — |
| Strips, 9 and 30 rounds. Folding belt (50 rounds) for Tanks. | Yes | Heavy exhangeable barrel, with radiator rings. | Spiral spring in compression. | "V" notch and barleycorn. Up to 2,000 yards. | Yes | No | 550 | — |
| Canvas belt (Disintegrating link belt for aircraft). | Yes | Ground Service— water jacket. Air Service— Plain barrel in perforated casing. | Spiral spring in compression. | "U" and aperture backsight, blade foresight. Up to 2,800 yards. | No | No | 500 to 550. (Air Service, 1100 to 1200). | — |
| Canvas belt. | Yes | Air, radiating rings on barrel. | Two spiral springs in compression. | "U" and blade. Up to 1,500 yards. | No | No | 400 | — |
| Box magazine underneath. 20 rounds. | Yes | Air, barrel ringed only in Cavalry and 1924 patterns. | Spiral spring in compression. | Aperture and blade. Up to 1,300 yards. | Yes | No | 550 | — |
| Box magazine underneath. 5 and 20 rounds. Drum magazine underneath. 50 rounds. | No. Support given by main spring and through medium of slipping inclined surfaces in breech action. | Air. Radiating rings on barrel at breech end in model (c) only. | Spiral spring under compression. | Lyman. Up to 750 yards. | Yes | No | 1,500 | Rimless case. Cartridges similar to those of self-loading pistols. M.V. varies according to cartridge and calibre from 918 f.s. to 1860 f.s. Very high pressures are impracticable. |

APPENDIX

DETAILS OF THE MACHINE GUNS AND LIGHT

| (1) Country and designation. | (2) Calibre. | (3) Length of barrel. | (4) Weight of Gun (without mounting). | (5) Type of mounting. | (6) Weight of mounting. | (7) Accessories on mounting. | (8) Method of operation. |
|---|---|---|---|---|---|---|---|
| AUSTRIA. | | | | | | | |
| Schwarzlose, M. 7/12 | ·315-in. | 20¾-in. | 44 lbs. | Tripod. | 43¾ lbs. | Elevating and traversing gear, each with "quick release." | Recoil of spent case acting on the bolt face. |
| BELGIUM. | | | | | | | |
| Fusil Mitrailleur (C.S.R.G.), 1915, French pattern. | ·301-in. | — | 18 lbs. | Bipod. | — | Nil | Gas. Run up operated by two springs. |
| CZECHO-SLOVAKIA. | | | | | | | |
| Schwarzlose M. 7/12. Converted to take 7·9 mm. ammunition. | (For details, see AUSTRIA.) | | | | | | |
| Praga light machine gun, Type 24 | 7·9 mm., or ·31-in. | 23¼ in. | 17·6 lbs. | Tripod. | — | — | — |
| DENMARK. | | | | | | | |
| Eriksen light machine gun | 6·5 mm. = ·256-in. | 26 7/16 in. | 24 lbs. 1 oz. | Bipod rest, 10½ in. from muzzle. | 2¼ lbs. | — | Gas. Piston surrounded by main spring located below barrel, offset connecting rod, and sliding breech block connected to a rotary crank. |
| Madsen light machine gun | 8 mm. (·315-in.) | Cavalry, 18¼ in. Infantry, 23¼ in. | Cavalry, 16 lbs. 14 ozs. Infantry, 18 lbs. | Bipod rest. | Cavalry, 1 lb. 10 ozs. Infantry, 2 lbs. | — | Recoil. |
| FRANCE. | | | | | | | |
| Chatellerault 7·5 mm. light machine gun. | 7·5 mm. or ·295-in. (nominal). | 50 cm., or 19·68 in. | 9 kilogs., or 19 lbs. 13½ ozs. (with mounting) (?) | Bipod rest of light Hotchkiss type attached to the barrel. The legs of bipod close rearwardly. | — | — | Gas. Piston in cylinder. |

## IX—continued.

### Machine Guns of Various Powers—continued

| (9) Type of ammunition conveyor. | (10) ? Cartridge supported by definitely locked action. | (11) Cooling system. | (12) Means of storing energy. | (13) Sights and range of adjustment. | (14) ? Whether gun can be adjusted to fire single shots. | (15) ? Whether special means of regulating rate of fire provided. | (16) Rate of fire (rounds per minute). | (17) Details of ammunition if different from that for the rifle. |
|---|---|---|---|---|---|---|---|---|
| Canvas belt. | No. Inertia of breech block assisted by spring and mechanical leverage. | Water jacket with portable tank and pump for forced circulation. | Spiral spring in compression. | "V" and barleycorn. Up to 2,400 m. | No | No | 400 | — |
| Charger. 20 rounds. | — | Air cooled. Ribbed aluminium radiator fixed to barrel. | — | 1,200 m. (inaccurate over 400). | Yes | — | — | — |
| Charger of 20–25 rounds. | — | — | — | Up to 1,645 metres. | — | — | — | — |
| Rectangular box magazine on left side of gun, holding 50 rounds in 10 steel strips, placed horizontally one above the other. The strips are automatically fed into gun, and when empty are ejected on the right. | Yes | Air | Spiral spring in compression. | Radial backsight on top of breech casing. Adjustable barleycorn foresight. Up to 2,200 m. | Yes, by a lever on top of spade handle grip. | No | — | — |
| Arc-shaped magazine on top. 25 and 40 rounds. | Yes | Air; ringed barrel in perforated casing. | Spiral spring in compression. | "V" and barleycorn. Up to 1,000 m. | Yes | No | 450 | — |
| Box magazine of straight type, holding 25 rounds, inserted vertically into top of gun | Yes | Air | Spiral spring in compression. | Radial backsight on receiver of gun. An aperture arm which is swung to left when required for use is pivoted to the head of the sight. Adjustable foresight on left side of barrel. | Yes—two triggers are arranged in the guard, the front for single shots and the rear for auto. | Yes—slowing device in the butt. | 650 without slowing device. 450 with slowing device. | Rimless, special 7·5 mm. Experimental. Weight 24 grammes, or 370·37 grains, Weight of bullet, 140 grains. Length overall, 81 mm., or 3·189 ins. |

412

APPENDIX

DETAILS OF THE MACHINE GUNS AND LIGHT

| (1) Country and designation. | (2) Calibre. | (3) Length of barrel. | (4) Weight of Gun (without mounting). | (5) Type of mounting. | (6) Weight of mounting. | (7) Accessories on mounting. | (8) Method of operation. |
|---|---|---|---|---|---|---|---|
| FRANCE—continued. | | | | | | | |
| Darne machine gun, ·303-in. model (for aircraft). | ·303-in. | About 26 in. | 15¼ lbs. | — | — | — | Gas. Piston in cylinder. |
| Fusil Mitrailleur (Chauchard), C.S.R.G., 1915. | See BELGIUM. | Obsolescent; being replaced by Chatellerault. | | | | | |
| Hotchkiss machine gun (1914 model). Remains the standard machine gun. The 1915 Hotchkiss and the St. Etienne machine guns are no longer issued. | ·315-in. | 30¼ in. | 52 lbs. | Tripod. Hotchkiss Model, 1916, or Universal model, 1915. | 70 lbs. | Elevating clamp and support for shield. | Gas; cupped headed piston fitting on gas nozzle. |
| GERMANY. | | | | | | | |
| Gast (originally used in aircraft) | ·311-in. | Two barrels each 28¼ in. long. | 42¼ lbs. | Used mainly in aircraft. | — | — | Recoil assisted by gas. The recoil of one barrel supplying the motive power for the forward action of the other. |
| Maxim machine gun, M.G. '08 | ·311-in. | 28¼ in. | 40¼ lbs. | Sleigh and tripod. | Tripod, 65¼ lbs. Sleigh, 83 lbs. | Elevating gear with quick action release and clamp, and traversing slide with clamp. | Recoil aided by gas. |
| Maxim light machine gun, 08/15 | ·311-in. | 28¼ in. | Adapted for tripod, 31 lbs. Adapted for bipod rest, 30 lbs. | Tripod or light bipod rest. | Tripod, 51 lbs. Bipod, 2¼ lbs. | (Tripod) Elevating gear with quick action release and clamp, and traversing slide with clamp. | Recoil. |
| Parabellum (originally used in aircraft). | ·311-in. | 28 in. | 22 lbs. | Used mainly in aircraft. | — | — | Recoil. |
| HOLLAND. | | | | | | | |
| Schwarzlose | ·256-in. | 27·55 in. | 43·54 lbs. | Tripod. | (For further details, see AUSTRIA.) | | |
| HUNGARY. | | | | | | | |
| Schwarzlose M. 7. Also M. 7/12, a later model with increased rate of fire. | (For details, see AUSTRIA.) | | | | | | |

IX—continued.

MACHINE GUNS OF VARIOUS POWERS—continued

| (9) Type of ammunition conveyor. | (10) ? Cartridge supported by definitely locked action. | (11) Cooling system. | (12) Means of storing energy. | (13) Sights and range of adjustment. | (14) ? Whether gun can be adjusted to fire single shots. | (15) ? Whether special means of regulating rate of fire provided. | (16) Rate of fire (rounds per minute). | (17) Details of ammunition if different from that for the rifle. |
|---|---|---|---|---|---|---|---|---|
| Belt feed (disintegrating links) from left of gun. | Yes | Air | Spiral spring in compression. | No sights on gun. Experimental model. | Yes | No | — | Detail taken from experimental model for British Service ammunition. |
| Strip. 30 rounds. | Yes | Air; barrel has 5 thick rings close to breech. | Spiral spring in compression. | "V" and barleycorn. | No | No | — | — |
| Two drum-shaped magazines, each holding 189 rounds. | Yes | Air; plain barrels in perforated casings. | (See column 8.) Balance maintained by spiral springs in compression. | "V" backsight, barleycorn foresight. No adjustment for range. | No | No | 1,200 (approx.). | — |
| Fabric belt. 100 or 250 rounds. | Yes | Water jacket. | Spiral spring in extension. | "V" and barleycorn. Up to 2,000 m. | By method of loading only. | No | 500 (approx.). | — |
| Belt feed from box on right of gun. | Yes | Water jacket. | Spiral spring in extension. | "V" and barleycorn. Up to 2,000 m. | By method of loading only. | No | 500 | — |
| Canvas belt. | Yes | Air—plain barrel in perforated casing. | Spiral spring under compression. | Aircraft patterns. | No | No | — | — |

414

APPENDIX

Details of the Machine Guns and Light

| (1) Country and designation. | (2) Calibre. | (3) Length of barrel. | (4) Weight of Gun (without mounting). | (5) Type of mounting. | (6) Weight of mounting. | (7) Accessories on mounting. | (8) Method of operation. |
|---|---|---|---|---|---|---|---|
| **Italy.** | | | | | | | |
| Fiat, 1914 model (Revelli) | 6·5 mm. (·256-in.) | 25¾ in. | 37·4 lbs. | Tripod. | 49·5 lbs. | Elevating and traversing gear. | Recoil. |
| S.I.A. | 6·5 mm. | Not known | 23¼ lbs. | Special, provided with straps, which allow it to act as a carrier. | 12¼ lbs. | See Type of Mounting (col. 5). | Gas. |
| Revelli (Villar Parosa) automatic rifle or pistol. | Various pistol calibres. | Two barrels, each 12¾ in. long. | 13¼ lbs. | Bipod barrel rest. | 3½ lbs. | — | Recoil of spent case acting on bolt face. |
| Fiat 25 mm. Self-loading gun (Canon Automatique de 25 mm. 4. F.I.A.T.). | 4·25 mm. (·984-in. not actually measured). | 20 calibres, approximately 19¾ in. | 40 kilogs., or 88·18 lbs. | Various, according to employment of weapon. Trunnions are provided on the gun. | — | — | Recoil (short). |
| **Japan.** | | | | | | | |
| 1914 model Hotchkiss pattern air-cooled gun. | ·256-in. | — | 67 lbs. | Tripod. | 60 lbs. | Elevating and traversing gear. | Gas. |
| Light machine gun | ·256-in. | 19·10-in. | 22 lbs. 12¼ ozs. (with mounting permanently attached to the gun). | Bipod rest at front of gun. | About 1¼ lbs. (not detachable). | — | Gas. Piston in cylinder. |

IX—*continued*

MACHINE GUNS OF VARIOUS POWERS—*continued*

| (9) Type of ammunition conveyor. | (10) ? Cartridge supported by definitely locked action. | (11) Cooling system. | (12) Means of storing energy. | (13) Sights and range of adjustment. | (14) ? Whether gun can be adjusted to fire single shots. | (15) ? Whether special means of regulating rate of fire provided. | (16) Rate of fire (rounds per minute). | (17) Details of ammunition if different from that for the rifle. |
|---|---|---|---|---|---|---|---|---|
| Box magazine on left. 50 rounds in columns of 5. | Breech opening definitely controlled, but action not locked in the accepted sense of the term. | Water. The water jacket can be connected with a receptacle containing 2·6 gallons of water, which, by means of a pump can be made to circulate in the water jacket. | Spiral spring in extension. | "V" and barleycorn. 200 to 2,000 m. | Yes | No | 500 | — |
| Clips of 25 rounds. | — | Air cooled | — | — | Yes | No | 700 | — |
| Box magazine, 25 rounds. One above each barrel. | No | Air. Plain barrels. | Spiral spring in compression. | Aperture backsight. Series of 5 different heights of barleycorn foresight. From 100 up to 500 m. | No | No | The two barrels being separate, units can be fired separately or together. Rate of fire therefore can be very high. | Ammunition similar to that used in self-loading pistols alone is suitable, as very high pressures are impracticable. |
| Box magazine containing 8 rounds, inserted into gun from above. | Yes | Air; barrel partly covered by a jacket. | Spiral spring in compression. | Backsight on breech casing. Foresight on water jacket. The sights are offset to left of gun. | Single shots only. | — | One per second at most | Weight of bullet, 200 gr. M.V., 440 m , Maximum range 4,000 m. |
| Brass strips. 30 rounds. | Yes | Air | Spiral spring in compression. | Up to 2,400 m. | Yes | No | 300 to 400 | — |
| Hopper, with hinged lid on left side of gun, to contain six clips of 5 rounds each. Conveyor mechanism is provided for feeding the clips through the gun from left to right, the empty clip being ejected on the right. | Yes | Air; barrel enclosed for major portion of length in sheath with cooling rings. | Spiral spring in compression. | Radial backsight sighted to 1,500 m. Adjustable barleycorn foresight. Offset approx. 1¼ in. to right of gun to clear oil pump device. | No | No | 600 to 800 | — |

APPENDIX

DETAILS OF THE MACHINE GUNS AND LIGHT

| (1) Country and designation. | (2) Calibre. | (3) Length of barrel. | (4) Weight of Gun (without mounting). | (5) Type of mounting. | (6) Weight of mounting. | (7) Accessories on mounting. | (8) Method of operation. |
|---|---|---|---|---|---|---|---|
| **NORWAY.** | | | | | | | |
| Heavy machine gun, type 6·5 mm. Hotchkiss 1905. | (For detail, see FRANCE.) | | | | | | |
| Light machine gun, 6·5 mm. Madsen, 1914. | (For detail, see DENMARK.) | | | | | | |
| **PORTUGAL.** | | | | | | | |
| Light automatic, Lewis ·303-in. | (For detail, see GREAT BRITAIN.) | | | | | | |
| Vickers ·303-in. machine gun | (For detail, see GREAT BRITAIN.) | | | | | | |
| **RUSSIA.** | | | | | | | |
| Maxim machine gun (also Colt, Lewis, Chauchat and Thompson). | ·300-in. | 28¼ in. | 58 lbs. | Tripod with detachable wheels. | 70 lbs. | Elevating and traversing gear. | Recoil. |
| **SPAIN.** | | | | | | | |
| Light machine gun. A few Hotchkiss 1922 under trial. | 7 mm. | — | — | — | — | — | — |
| Hotchkiss 1914 machine gun | 7 mm. | (See FRANCE.) | | | | | |
| **SWEDEN.** | | | | | | | |
| Heavy machine gun, Schwarzlose | 6·5 mm. | (For detail, see AUSTRIA.) | | | | | |
| Light machine gun, Browning | (For detail, see AMERICA.) | | | | | | |
| **SWITZERLAND.** | | | | | | | |
| Furrer 7·45 mm. machine gun (Swiss light machine gun, pattern 1925). | 7·45 mm. (·2935-in.) | 23 in. (with extension = 33 4/16 in.). | 18 lbs. 6 ozs. | (a) Bipod rest at front of barrel casing. (b) Elevating rest on butt. | (a) Bipod rest, 2 lbs. 5¼ ozs. (b) Elevating rest, 2 lbs. 4 ozs. | — | Recoil (2·68 in.). Toggle action on the lines of the Borchardt Luger pistol. |
| EXPERIMENTAL | 1919—Not proceeded with. | | | | | | |
| Lewis light machine gun | Rifle (·303-in.) | 26¼ in. | 19 lbs. 12 ozs. | Monopod rest. | 1 lb. | — | Gas; piston in cylinder. |
| Hotchkiss light machine gun | Rifle (·303-in.) | 19¾ in. | Strip feed, 16 lbs. 8 ozs. Magazine feed, 16 lbs. | Bipod rest | 1 lb. 1 oz. | — | Gas; cupped headed piston fitting over gas nozzle. |

## IX—continued

### MACHINE GUNS OF VARIOUS POWERS—continued

| (9) Type of ammunition conveyor. | (10) ? Cartridge supported by definitely locked action. | (11) Cooling system. | (12) Means of storing energy. | (13) Sights and range of adjustment. | (14) ? Whether gun can be adjusted to fire single shots. | (15) ? Whether special means of regulating rate of fire provided. | (16) Rate of fire (rounds per minute). | (17) Details of ammunition if different from that for the rifle. |
|---|---|---|---|---|---|---|---|---|
| Fabric belt. | Yes | Water jacket. | Spiral spring in extension. | "V" and barleycorn. 400 to 2,000 paces. | By method of loading only. | No | 500 (approx.). | — |
| — | — | — | — | — | — | — | — | — |
| Box type magazine, 30 rounds, inserted into right side of gun. | Yes | Air | Spiral spring in compression. | Radial backsight and blade foresight, 100–2,000 m. Sight radius = 23·10 in. at 100 m. | Yes | No | 480 (300, allowing change of magazine, and 200 allowing change of magazine with extension). | Rimless. Weight, 412 grains. Weight of bullet, 174·70 grains. Weight of charge, 50·30 grains. Length overall, 3·05 in. Length of bullet, 1·271 in. (pointed with boat tail and waxed). Cap metal, brass. Propellent: nitro-cellulose (?) graphite flake. |
| Circular drum magazine, 47 rounds (as for British Service model). | Yes | Air; plain barrel. | Clock spring | Aperture and blade. Up to 1,900 yds. | Yes | No | 525 to 550 | — |
| Two types: (a) Box magazine, 20 rounds; (b) Strip 15 rounds. | Yes | Air; plain barrel. | Spiral spring in compression. | Rectangular notch and broad blade. | No, but single shots can be fired owing to slow rate of fire. | Rate of fire permanently controlled by pendulum escapement acting on sear. | 268 | — |

## APPENDIX X

### DETAILS OF RIFLE AMMUNITION OF VARIOUS POWERS

| Country. | Great Britain. | Austria. | Belgium. | Czecho-Slovakia. | Denmark. | France. | Germany. |
|---|---|---|---|---|---|---|---|
| Cartridge— | Mk. VII. | | | | | Balle "D." | |
| Length (inches) | 3·0375 (mean) | 3·0 | 3·055 | — | 3·2 | 2·948 | 3·17 |
| Weight (grains) | 384 (mean) | 437·5 | 441 | 379·6 | 450·6 | 426·2 | 369 |
| Rim or rimless | Rim | Rim | Rimless | — | Rim | Rim | Rimless |
| Bullet— | | | | | | Bi-ogival | |
| Shape of nose | Pointed | Round | Round | — | Pointed | Pointed | Pointed |
| ,, base | Flat | Flat | Conical cavity | — | Concave | Stream line | Concave |
| Material of envelope | Cupro-nickel | Steel | Cupro-nickel | — | Steel, nickel and copper alloy | No envelope | Steel |
| Material of core | Lead and antimony alloy. Alluminium tip. | Lead | Lead | — | Lead | Bullet of copper alloy | Lead |
| Length (inches) | 1·28 (mean) | 1·251 | 1·205 | — | 1·25 | 1·542 | 1·1 |
| Diameter (max.) | ·312 | ·323 | ·31 | — | ·325 | ·327 | ·323 |
| Weight (grains) | 174 (mean) | 244 | 215 | 154·3 | 196 | 197·6 | 154·2 |
| Charge— | | | | | | | |
| Weight (grains) | 37·5 | 42·3 | 42·5 | 49·3 | 50 | 46·3 | 49·5 |
| Propellent | Cordite | Nitro-cellulose | Nitro-glycerine & Nitro-cellulose | — | Nitro-cellulose | Nitro-cellulose | Nitro-cellulose |
| Value of $w/d^2$ | 0·255 | 0·334 | 0·319 | — | 0·265 | 0·267 | 0·211 |
| Muzzle velocity (f/s) | 2440 | 2034 | 2034 | 2624 | 2530 | 2380 | 2882 |
| Chamber press. (Tons/sq. inch) | 19·5 | 19·7 | 19·7 | — | — | 17·75 | 17·5 |

Appendix X.

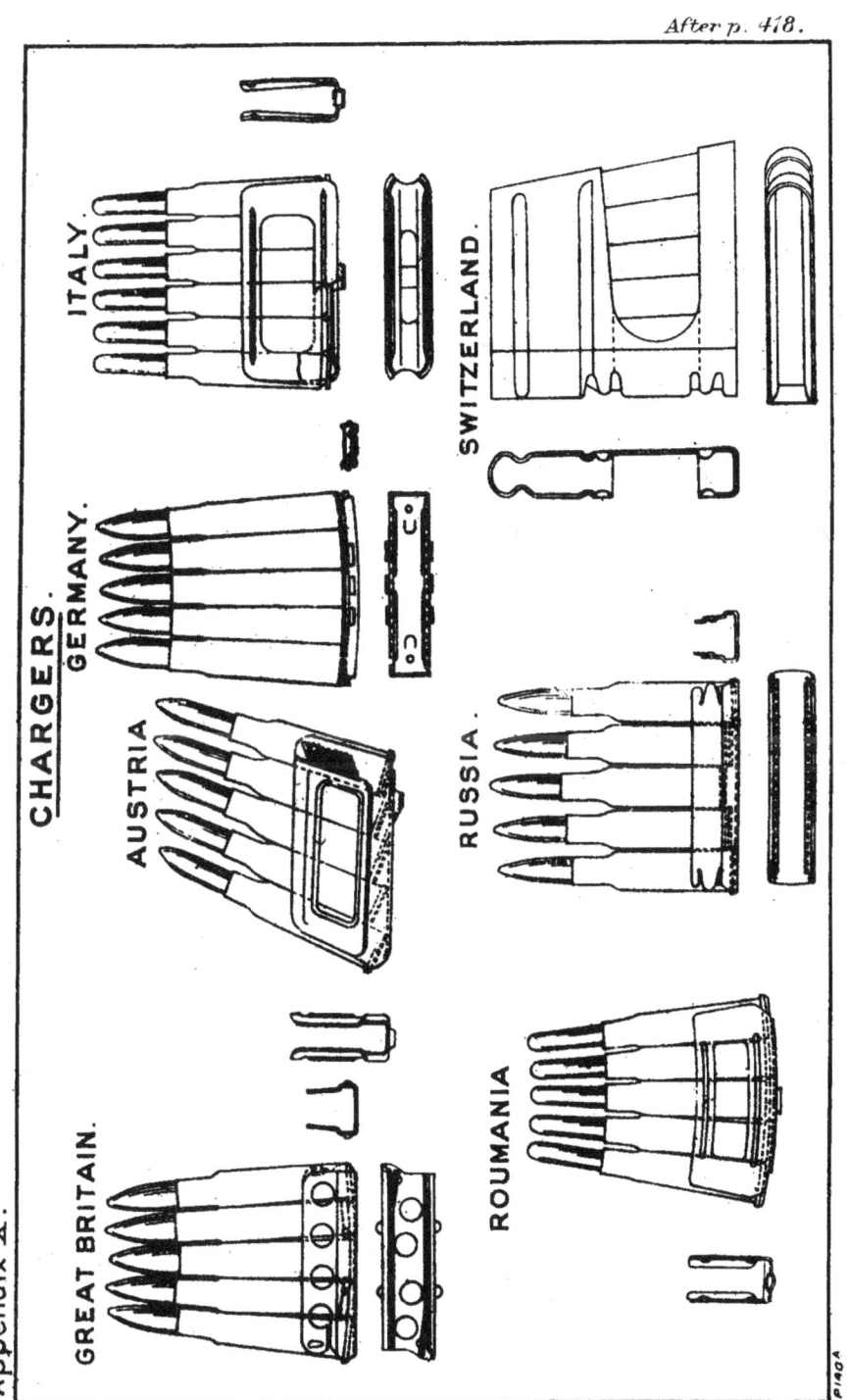

After p. 418.

Appendix X.

## CARTRIDGES.

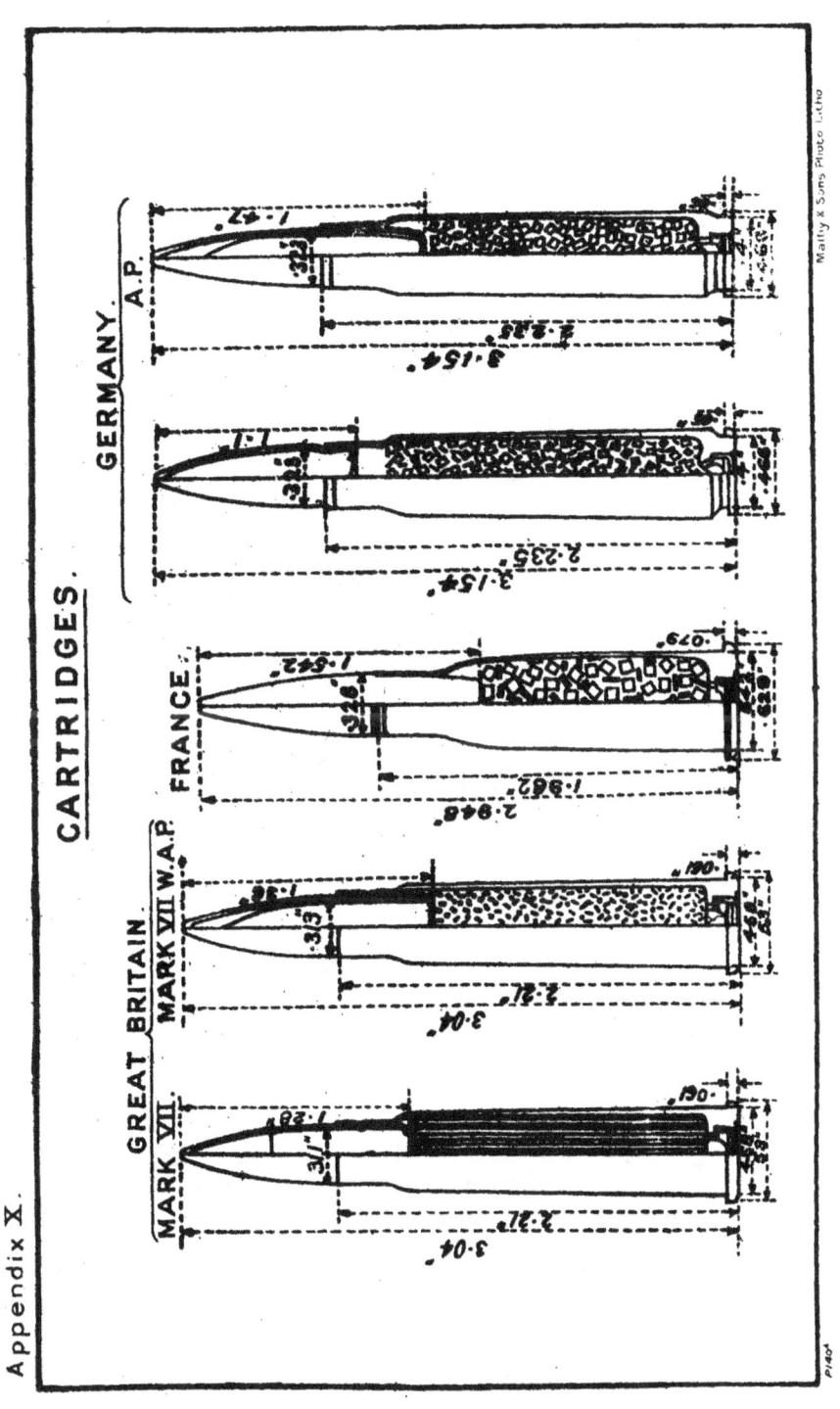

Appendix X.

## CARTRIDGES

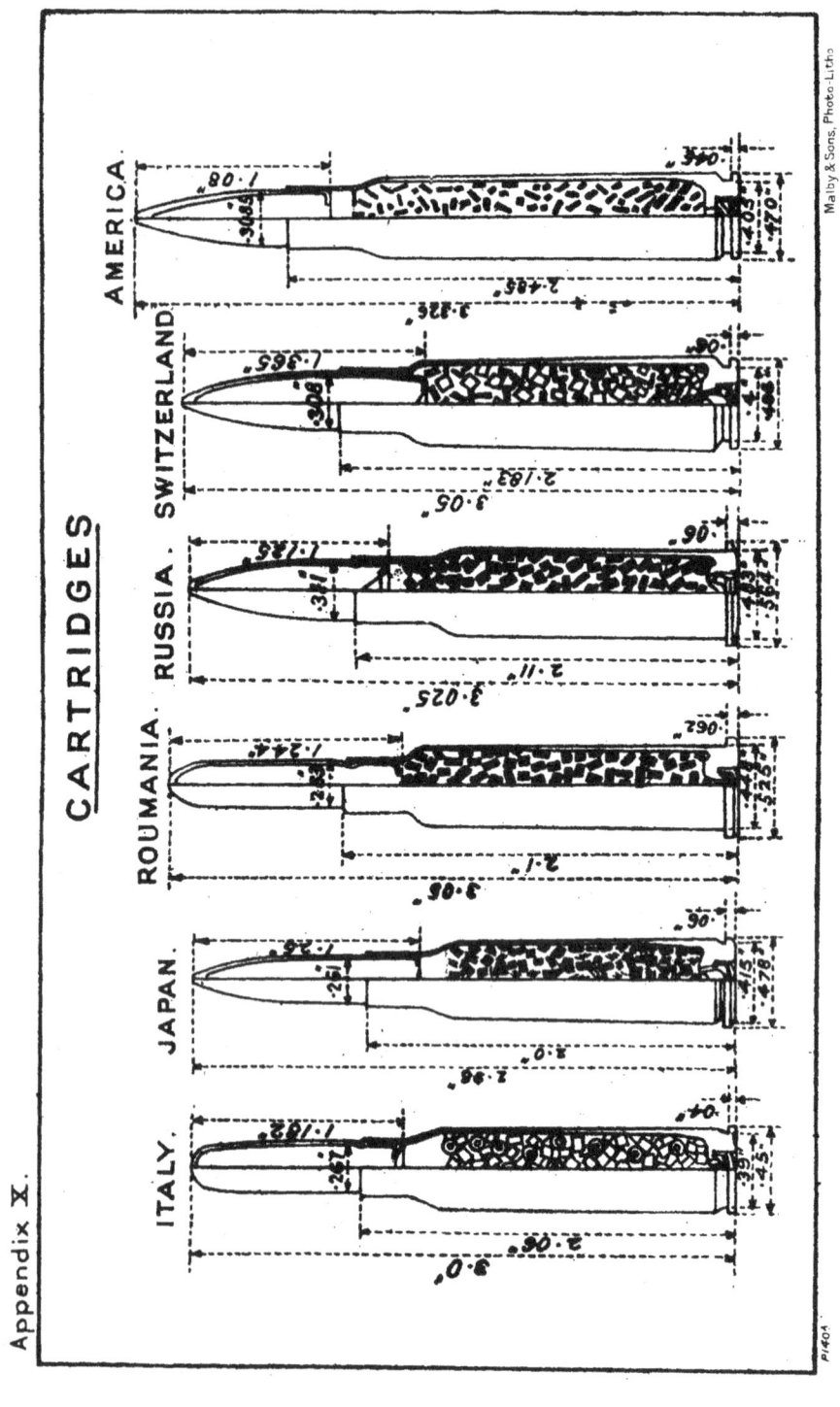

419

APPENDIX X—continued

DETAILS OF RIFLE AMMUNITION OF VARIOUS POWERS—continued

| Country. | Greece. | Holland. | Hungary. | Italy. | Japan. | Norway. | Portugal. |
|---|---|---|---|---|---|---|---|
| Cartridge— | | | | | | | Pattern 1912. |
| Length (inches) | 3·05 | 3·05 | 2·92 | 3·0 | 2·98 | 3·13 | 3·26 |
| Weight (grains) | 348 | 349 | 455 | 350 | 326 | 372·1 | 371·8 |
| Rim or rimless | Rimless | Rim | Rim | Rimless | Semi-rimless | Rimless | Rimless |
| Bullet— | | | | | | | |
| Shape of nose | Round | Round | Blunt | Round | Pointed | Pointed | Round |
| ,,    base | Flat | Flat | — | Concave | Flat | Flat | Flat |
| Material of envelope | Steel | Steel | Steel | Cupro-nickel | Cupro-nickel | Cupro-nickel | Cupro-nickel |
| Material of core | Lead | Lead | Lead | Lead | Lead antimony | Lead | Lead |
| Length (inches) | 1·24 | 1·23 | 1·24 | 1·182 | 1·26 | 1·3 | 1·255 |
| Diameter (max.) | ·263 | ·263 | — | ·267 | ·262 | ·263 | ·263 |
| Weight (grains) | 159·3 | 159 | 244 | 161·8 | 139 | 156·4 | 155·3 |
| Charge— | | | | | | | |
| Weight (grains) | 36·0 | 38 | — | 34·6 | 33·0 | 36·0 | 37·7 |
| Propellent | Nitro-cellulose | Nitro-cellulose | — | Ballistite | Nitro-cellulose | Nitro-cellulose | Nitro-glycerine & Nitro-cellulose (Tubular) |
| Value of $w/d^2$ | 0·329 | 0·329 | — | 0·324 | 0·289 | 0·323 | 0·321 |
| Muzzle velocity (f/s) | 2223 | 2433 | 2032 | — | 2500 | — | 2347 |
| Chamber press. (Tons/sq. inch). | 20·18 | — | — | — | — | — | — |

APPENDIX X—continued

DETAILS OF RIFLE AMMUNITION OF VARIOUS POWERS—continued

| Country. | Roumania. | Russia. | Spain. | Switzerland. | Turkey. | U.S.A. |
|---|---|---|---|---|---|---|
| Cartridge— | | | | | | |
| Length (inches) | 3.05 | 3.025 | 3.08 | 3.05 | 2.96 | 3.34 |
| Weight (grains) | 360 | 348 | 377.4 | 404.0 | 370 | 397 |
| Rim or rimless | Rim | Rim | Rimless | Rimless | Rimless | Rimless |
| Bullet— | | | | | | |
| Shape of nose | Round | Pointed | Round | Pointed | Pointed | Pointed |
| ,, base | Flat | Conical cavity | Conical cavity | Streamline concave (boat-tailed) | Flat | Flat |
| Material of envelope | Cupro-nickel | Cupro-nickel | Steel | Steel | Steel | Cupro-nickel |
| Material of core | Lead | Lead | Lead | Lead | Lead | Lead |
| Length (inches) | 1.22 | 1.10 | 1.21 | 1.37 | 1.06 | 1.08 |
| Diameter (max.) | .263 | .310 | .284 | .307 | .311 | .3085 |
| Weight (grains) | 159 | 148 | 172.8 | 174 | 154.3 | 150 |
| Charge— | | | | | | |
| Weight (grains) | 37.4 | 50.0 | 38.3 | 49.3 | 46.0 | 50.5 |
| Propellent | Nitro-cellulose | Nitro-cellulose (flake) | Nitro-cellulose | Nitro-cellulose (fulminaton) | Nitro-cellulose | Nitro-cellulose |
| Value of $w/d^2$ | 0.321 | 0.221 | 0.305 | 0.262 | 0.228 | 0.238 |
| Muzzle velocity (f/s) | 2400 | — | 2296 | 2720 | — | 2600 |
| Chamber press. (Tons/sq. inch) | — | — | 22.3 | — | — | 19.78 |

# INDEX

## A

| | PAGE |
|---|---|
| Aasen grenade | 106 |
| Abel heat test for propellents | 249 |
| Abel, Sir Frederick | 217 |
| Acceleration, definition of | 368 |
| Action, rifle, definition of | 68 |
| Actions, straight pull | 24 |
| " turning bolt | 20 |
| Accuracy tests, machine guns | 196 |
| " trials, rifles | 53 |
| " " of Mark VII ammunition | 238 |
| " of the rifle, considerations affecting | 53 |
| Adams revolver | 87 |
| " striker mechanism (grenades) | 119 |
| Adjustments, and tests, machine guns | 193 |
| Air resistance and retardation, formulæ | 313 |
| " " and velocity, relation between | 279, 284, 313 |
| Allways fuzes | 120 |
| Aluminium tips in bullets | 232 |
| Ammunition, accuracy trials of Mk. VII | 238 |
| " armour piercing | 257 |
| " blank | 262 |
| " casualty proof | 246 |
| " detection of defects in new | 242, 253 |
| " drill or dummy | 263 |
| " hangfire proof | 246 |
| " hard to extract proof | 246 |
| " history of small arm | 205 |
| " inspection of | 247 |
| " machine guns and defective | 243, 244, 245, 246, 250 |
| " manufacture of | 223 |
| " Mark VI | 257 |
| " miniature rifle | 265 |
| " miscellaneous military | 257 |
| " packing and labelling of | 253 |
| " pistol and revolver | 263 |
| " pressure trials | 239 |
| " proof | 237 |
| " rifle grenade | 261 |
| " sawdust proof | 246 |
| " special for R.A.F. | 259 |
| " standard | 239 |
| " storage | 256 |
| " tracer | 258 |
| " velocity trials | 239 |
| Angle of arrival, definition of | 367 |
| " departure, definition of | 367 |
| " descent, definition of | 367 |
| " descent, formulæ for | 312, 327 |
| " elevation, for extreme range | 277, 327 |
| " impact, definition of | 367 |
| " incidence, definition of | 367 |
| " projection, definition of | 367 |
| " sight, definition of | 367 |

| | PAGE |
|---|---|
| Angular measurement, conversion of to linear | 314 |
| Annealing, effect of, on cartridge brass | 227 |
| Atmospheric pressure, effect on sight curve | 314 |
| Automatic pistols | 95 |
| Auto-frettage | 362 |
| Axis of the gun, definition of | 366 |
| " trunnions, definition of | 366 |
| Azimuth, definition of | 368 |

## B

| | PAGE |
|---|---|
| Baker rifle | 3, 5 |
| Ballistic coefficient, determination of value of | 329 |
| Ballistic coefficient, notes on | 285 |
| " " use of | 318 |
| " definitions and units | 366 |
| " pendulum | 273, 347 |
| " tables and their use | 317 |
| Ballistics, exterior, definition of | 366 |
| " " descriptive | 275 |
| " " numerical | 308 |
| " interior, definition of | 366 |
| " " descriptive | 267 |
| " " numerical | 292 |
| " of low power weapons | 336 |
| " revolver | 93, 336 |
| Ballistite, invention of | 217 |
| " manufacture of | 222 |
| Barrel, definition of | 67 |
| " examination of | 58 |
| " injuries to | 58, 360 |
| " length, effect on velocity of variation in | 274, 302 |
| Barrel, obstructions, removal of from | 58 |
| " vibrations | 54 |
| " wear and erosion | 57, 157, 220, 360 |
| " worn, effect on bullet | 245, 270 |
| " " correction to sight curve for | 314 |
| Barrels, rifle, comparison of various | 29 |
| " strength of | 361 |
| Bashforth chronograph | 281 |
| Battle sights, ballistic considerations | 315 |
| Bayonet, history of the | 78 |
| " manufacture and inspection of the service | 84 |
| Berthier light machine gun, breech support | 166 |
| Blank cartridge Mk. V | 262 |
| " firing attachments for machine guns | 202 |
| Blowbacks, causes of | 243 |
| Body, rifle, definition of | 68 |
| Bolt, rifle, definition of | 67 |
| " actions, rifle | 10 |
| " heads, definition of | 16, 17 |

|   | PAGE |
|---|---|
| Borchardt-Luger self-loading pistol | 101, 102, 103 |
| Bore, definition of | 68 |
| ,, examination of | 58 |
| ,, removal of obstructions from | 58 |
| Boring and drilling | 50, 51 |
| Boulengé chronograph, Le | 337 |
| Boxer cartridge, introduction of | 6, 152, 206 |
| Brass, composition of, for cartridge case | 227 |
| Breech actions, strength of | 360 |
| ,, definition of | 67 |
| ,, loading rifles, early | 5, 9, 205 |
| Breech support, in machine guns | 166 |
| Brinnell hardness test | 226 |
| Browning and blueing | 51 |
| Browning light machine gun— | |
| breech support | 166, 168 |
| buffer system | 178 |
| cooling | 175 |
| details | 408 |
| feed | 164 |
| firing mechanism | 178 |
| safety | 172 |
| Browning machine gun— | |
| breech support | 166, 169 |
| buffer system | 177 |
| cooling | 176 |
| cartridge head space | 195 |
| details | 408 |
| feed | 164 |
| firing mechanism | 178 |
| regulation of rate of fire | 182 |
| safety | 173 |
| sights | 181 |
| Brunswick rifle | 3, 5 |
| Bullet, armour piercing | 257 |
| ,, British Mk. VII | 210 |
| ,, core, manufacture of Mk. VII | 232 |
| ,, defects in | 245, 253 |
| ,, envelope, manufacture of Mk. VII | 230 |
| ,, expanding | 211 |
| ,, flight of the, notes on | 276, 279 |
| ,, French Balle D | 209 |
| ,, lubrication | 214 |
| ,, manufacture of Mk. VII | 230 |
| ,, "set up" on firing | 269 |
| ,, sound of, in flight | 280 |
| ,, spin, effect on flight | 288 |
| ,, streamlined | 210 |
| ,, symmetry, importance of | 230, 238, 355 |
| ,, tips, manufacture of Mk. VII | 232 |
| ,, unsymmetrical, effect of | 230, 355 |
| ,, wounding, effects of | 362 |
| ,, burst cases, causes of | 244 |
| ,, bursters, grenade | 118 |

## C

|   | |
|---|---|
| Calibre, definition | 68 |
| ,, reduction of | 6, 206, 207 |
| Cant, effect of | 64 |
| Cap, percussion, development of | 6, 214 |
| ,, Mark VII cartridge | 214 |
| ,, ,, ,, defects | 242, 243 |
| ,, ,, ,, manufacture and filling | 233 |

|   | PAGE |
|---|---|
| Cap, Mark VII cartridge, priming composition | 233 |
| Carbines | 32 |
| Cartridge, blank, Mark V | 262 |
| ,, case, defects in | 244 |
| ,, ,, effect of firing on | 225, 270 |
| ,, ,, manufacture | 226 |
| ,, ,, modern | 211 |
| ,, ,, solid drawn, introduction of | 152, 206 |
| Cartridge, case, various types | 212, 420 |
| ,, central fire | 6, 206 |
| ,, design | 208, 292 |
| ,, drill or dummy | 263 |
| ,, early patterns of | 6, 205 |
| ,, ·303 Mark VI | 257 |
| ,, ·303 Mark VII, accuracy trials | 238 |
| ,, ,, ,, armour piercing | 257 |
| ,, ,, ,, casualty proof | 246 |
| ,, ,, ,, defects in | 242, 253 |
| ,, ,, ,, functioning trials | 242 |
| Cartridge, ·303 Mark VII, gauging | 250 |
| ,, ,, ,, hangfire proof | 246 |
| ,, ,, ,, hard to extract proof | 246 |
| Cartridge, ·303 Mark VII, inspection | 247 |
| ,, ,, ,, loading | 235 |
| ,, ,, ,, manufacture | 223 |
| ,, ,, ,, packing | 253 |
| ,, ,, ,, proof selection | 237 |
| ,, ,, ,, sawdust proof | 246 |
| ,, ,, ,, storage | 256 |
| ,, ,, ,, visual examination | 253 |
| Cartridge, ·303 Mark VII, weighing tests | 249 |
| ,, ,, ,, weights of components | 248 |
| Cartridge, Martini Henry | 206 |
| ,, miniature rifle | 265 |
| ,, pinfire | 208 |
| ,, pistol and revolver | 263 |
| ,, belted or Accles | 212 |
| ,, rifle grenade | 261 |
| ,, rim | 212 |
| ,, rimfire, notes on | 208 |
| ,, rimless and semi-rimless | 212 |
| ,, sporting rifle | 208 |
| Charge, combustion of | 213, 219, 268 |
| ,, ignition of | 268 |
| ,, shape of | 213, 220, 272 |
| ,, velocity and pressure, relationship between | 295, 301, 303, 308 |
| Charger, definition of | 68 |
| Chargers and clips | 215 |
| Chronograph, Bashforth | 281 |
| ,, Le Boulengé | 337 |
| Clip, definition of | 68 |
| Colt, machine gun— | |
| breech support | 167, 169 |
| cartridge head space | 195 |
| cooling | 175 |
| details | 408 |
| feed | 167 |
| safety | 172 |
| sights | 181 |

423

| | PAGE |
|---|---|
| Colt, revolver | 89, 91 |
| ,, ,, early pattern | 87 |
| ,, self-loading pistol | 100 |
| Co-efficient, ballistic | 285, 318, 329 |
| ,, of reduction | 318 |
| Combustion of charge | 213, 219, 268 |
| ,, ,, rate of | 219, 272 |
| ,, ,, temperature of, various propellents | 300 |
| Compensation | 55, 56 |
| Conversion table, British and metric units | 369 |
| Cordite, invention of and development | 5, 217 |
| ,, manufacture | 222 |
| ,, method of loading into case | 235 |
| ,, shape of | 213, 219, 272 |
| Cord wear | 57 |

### D

| | |
|---|---|
| Danger space, definition of | 368 |
| Definitions and units, ballistics | 366 |
| ,, physical | 368 |
| ,, rifle and rifle shooting | 67 |
| Deflection, definition of | 367 |
| Density of the atmosphere, effect on sight curve | 314 |
| Detonation | 213, 216 |
| Discharger, 2-inch grenade | 125 |
| Distribution of energy from explosion of charge | 270, 301, 302 |
| Distribution of heat from explosion of charge | 270, 301 |
| Dreyse needle gun | 6, 205 |
| Drift, causes and effect of | 288 |
| ,, definition of | 367 |
| Drilling and boring | 50, 51 |
| Dynamite | 217 |

### E

| | |
|---|---|
| Ejection, tests for, in machine guns | 197 |
| Elevation, and velocity, relationship between | 309, 314 |
| Elevation, angle of, formulæ | 309, 310 |
| ,, definition of | 366 |
| ,, for extreme range | 277, 327 |
| ,, quadrant, definition of | 367 |
| ,, tangent, definition of | 367 |
| Enfield rest | 66 |
| ,, rifle | 4, 5 |
| ,, ,, conversion to breech loading | 6 |
| Erosion of the bore | 57, 157, 220, 360 |
| Explosion and detonation | 213, 216 |
| ,, distribution of energy of | 270, 301, 302 |
| ,, ,, heat from | 270, 301 |
| ,, temperature and pressure | 274, 300 |
| Explosives, general remarks and classification | 215 |
| Explosives, historical outline of | 215 |
| Exterior ballistics, definition of | 366 |
| ,, ,, descriptive | 275 |
| ,, ,, numerical | 308 |

| | PAGE |
|---|---|
| Extraction, primary | 24, 68 |
| ,, tests for, in machine guns | 197 |

### F

| | |
|---|---|
| Feed, tests for, in machine guns | 197 |
| Ferguson, Lt.-Col. Patrick | 3, 5 |
| Figure of merit | 239 |
| Fire power, development of | 9, 102 |
| Flintlock musket | 2, 8, 9 |
| Force, definition of | 368 |
| ,, of explosives | 215 |
| Forsyth, Rev. Alexander | 205 |
| Fouling, metallic | 57, 210, 245 |
| Franco-Prussian War, the machine gun in the | 151 |
| French revolver | 92 |
| Free travel of bullet | 68 |
| Friction, tests for, in machine guns | 197 |
| Fuzes of grenades | 118 |

### G

| | |
|---|---|
| Galand revolver | 87 |
| Gardner machine gun | 152, 164 |
| Gast light machine gun— | |
| breech support | 169 |
| feed arrangements | 164 |
| means of storing energy | 177 |
| Gatling machine gun | 151 |
| Granatenwerfer (German) | 125 |
| Gravity, effect on trajectory of | 277, 278 |
| Grenades, British | 131, 134 |
| ,, ,, hand, Mark I | 106 |
| ,, ,, ,, Mark II | 107 |
| ,, ,, ,, or rifle No. 36 | 131 |
| ,, ,, No. 37. W.P. | 131 |
| ,, ,, No. 28. Gas | 132 |
| ,, ,, No. 42 signal day | 132 |
| ,, ,, No. 55 smoke, 2-inch | 133 |
| ,, ,, signal, 2-inch | 133 |
| ,, ,, No. 54, H.E. 2-inch | 133 |
| ,, ,, in the Great War | 107 |
| ,, considerations affecting design | 116 |
| ,, discharger, 2-inch (British) | 125 |
| ,, French | 136 |
| ,, fuzes, igniters, bursters | 118 |
| ,, German | 134 |
| ,, history of | 105 |
| ,, means of projection | 124 |
| ,, modern requirements | 117 |
| Grenade firing, effect on rifle of | 125, 360 |
| Guncotton, discovery of | 216 |
| Gunnery, definition of | 366 |
| Gunpowder, discovery of | 1, 216, 272 |
| Gun range, definition of | 368 |

### H

| | |
|---|---|
| Hale grenade | 106 |
| ,, rifle grenade | 124 |
| Handguards, stocks, &c., rifles | 30, 52 |
| Hangfires, causes and detection of | 243, 246, 260 |

|   |   |
|---|---|
| Hardening and tempering | 51 |
| Hard extraction | 245, 246 |
| Head space (cartridge), definition | 68, 194 |
| Henry rifle | 7 |
| Hotchkiss light machine gun (British)— | |
| breech support | 166, 167 |
| cartridge head space | 195 |
| cooling system | 175 |
| details | 408 |
| extraction and ejection | 179 |
| feed | 164, 165 |
| introduction of | 154 |
| safety devices | 172 |
| Hotchkiss light machine gun (Experimental)— | |
| cartridge head space | 195 |
| cooling | 175 |
| details | 416 |
| feed | 164 |
| regulation of rate of fire | 182 |
| safety devices | 172, 174 |
| Hotchkiss machine gun, 1914 (French)— | |
| breech support | 168 |
| cartridge head space | 195 |
| cooling | 175 |
| details | 412 |
| feed | 164 |
| Housman's experiments, charge, velocity and pressure | 295 |

### I

|   |   |
|---|---|
| Igniters, grenade | 118 |
| Ignition of charge, process of | 268 |
| Inspection of ammunition, ·303 Mark VII | 247 |
| Inspection of rifle, short. magazine, Lee-Enfield, Mark III | 52 |
| Interior ballistics, definition | 366 |
| ,, ,, descriptive | 267 |
| ,, ,, numerical | 292 |
| Italian revolver | 92 |

### J

|   |   |
|---|---|
| Japanese revolver | 92 |
| ,, rifle | 15 |
| ,, rifle details | 405 |
| Jones, F. W. (interior ballistics) | 303 |
| Jump, definition of | 366 |
| ,, determination of | 55 |

### K

|   |   |
|---|---|
| Kieselguhr | 217 |
| Knox form | 45 |
| Krag Jorgensen rifle | 22 |
| Kropatshek rifle | 7 |

### L

|   |   |
|---|---|
| Lancaster rifle | 3 |
| Lance, history of the | 70 |
| Lands, definition | 68 |

|   |   |
|---|---|
| Lapping | 52 |
| Lateral deflection in azimuth, definition | 368 |
| ,, plane of sight, definition | 366 |
| Lebel rifle | 20 |
| Leed (or lead), definition | 68 |
| ,, ,, effect of worn | 270 |
| Lee-Enfield rifle | 7 |
| Lee magazine and action, adoption of | 7 |
| Lee-Metford rifle | 4, 5, 7 |
| Length, unit of | 368 |
| Lewis light machine gun— | |
| breech support | 166, 167 |
| cartridge head space | 195 |
| cooling | 175 |
| details | 408 |
| extraction and ejection | 179 |
| feed | 164 |
| firing mechanism | 178 |
| introduction of | 154 |
| safety devices | 172 |
| tests and adjustments | 193 |
| Line of sight | 63 |
| ,, ,, definition | 366 |
| ,, departure | 366 |
| Long range fire analysis | 332 |
| Lubrication of bullet | 214 |
| Luger (parabellum) self-loading pistol | 101, 102 |

### M

|   |   |
|---|---|
| Machine guns— | |
| adoption by Britain | 152, 153 |
| barrel wear | 157 |
| blank firing attachments | 202 |
| breech support, various methods | 166 |
| British and Foreign, details (appendix) | 408 |
| characteristics | 155 |
| classification of different types | 159 |
| conditions affecting design | 155 |
| cooling systems | 156, 175 |
| defective ammunition in | 243, 246, 250 |
| development in the Great War | 154 |
| extraction and ejection | 179 |
| feed systems | 164 |
| history | 150 |
| Franco-Prussian War | 151, 152 |
| Russo-Japanese War | 153 |
| South African War | 153 |
| Zulu War, 1879 | 152 |
| means of storing energy | 177 |
| methods of control | 182 |
| mountings | 198 |
| operation by gas | 160 |
| ,, recoil | 162 |
| regulation of rate of fire | 182 |
| safety arrangements | 158, 172 |
| sighting | 180 |
| tests and adjustments | 193 |
| trigger and firing mechanism | 178 |
| Machining | 50 |
| Madsen light machine gun— | |
| breech support | 167, 169 |
| cartridge head space | 195 |
| cooling | 175 |
| details | 410 |
| extraction and ejection | 179 |

| | PAGE | | PAGE |
|---|---|---|---|
| Striker, tests for protrusion of, in machine guns | 193 | Velocity, relation between air resistance and | 279, 284, 313 |
| Striking velocity, definition | 366 | Velocity, remaining, definition | 366 |
| "Sweating" of barrels after firing | 220 | " " formulæ for | 312 |
| Sword, history | 69 | " relationship between elevation and | 314 |
| " the parts of the | 69 | " striking, definition | 366 |
| | | " terminal, definition | 366 |
| | | " " of projectiles | 277, 278 |
| | | " trials of new S.A.A. | 239 |

### T

| | | | |
|---|---|---|---|
| | | Vertex, definition | 367 |
| | | Vertical fire, definition | 367 |
| Tangent elevation, definition | 367 | " plane of sight, definition | 366 |
| Target range, definition | 368 | Vetterli rifle | 7 |
| Telescopic sights | 60 | Vibrations of barrel on firing | 54 |
| Tenuity factor | 318 | Vickers ·303-inch machine gun— | |
| Terminal velocity, definition | 366 | breech support | 167 |
| " of projectiles | 277, 278 | cooling | 176 |
| Thomas revolver | 87 | details | 408 |
| Thompson sub-machine gun— | | extraction and ejection | 179 |
| breech support | 171 | feed | 164 |
| details | 408 | firing mechanism | 178 |
| feed | 164 | safety arrangements | 173, 174 |
| Thring pressure gauge | 344 | tests and adjustments | 193 |
| Time of flight, calculation of | 317 | Villa Perosa | 172 |
| " definition | 368 | | |
| " unit of | 368 | | |
| Toggle joint locking system in machine guns | 167 | ### W | |
| Tracer ammunition | 258 | Wads, cartridge | 214 |
| Trajectory, calculation | 326 | Webley-Fosbery revolver | 96 |
| " clearance of obstacles by | 313 | Webley revolver, Mark VI ·455 | 90 |
| " definition | 366 | " " ·38 | 92 |
| " effect of air resistance on | 278 | Webley-Scott self-loading pistol | 101 |
| " " earth's curvature on | 278 | " " ammunition | 264 |
| " " gravity on | 278 | Webley self-loading pistol | 99 |
| " height, and interior ballistics | 307 | Weight, unit of | 368 |
| " Sladen's formulæ for plotting | 310 | Wheel lock musket | 1 |
| " plotting by formulæ | 317 | Whitworth rifle | 4 |
| " rigidity of the | 287 | Winchester rifle | 7 |
| " subdivision | 276 | Wind, effect on bullet's flight | 285 |
| " unresisted or parabolic | 277 | Work, calculation of | 369 |
| Tranter revolver | 87 | Wounding effects of bullets | 362 |

### V

| | | ### Y | |
|---|---|---|---|
| | | Yaw, causes and effects | 276, 288 |
| Velocity and barrel length | 274, 302 | | |
| " measurement of | 280, 337, 368 | ### Z | |
| " muzzle, definition | 366 | | |
| " of sound in air | 280 | Zones, calculation of percentage | 352 |
| " pressure and charge, relationship between | 295, 301, 303, 308 | Zulu War, 1879, the machine gun in the | 152 |